**Get The Fast**

D0087126

Software for Serious Engineering

Have you ever wanted to use your personal computer to finish homework, perform research, or design subassemblies for a prize-winning vehicle?

You can with the Student Edition of I-deas™. IBM, Ford, and Lockheed-Martin already use I-deas software. And thanks to a special agreement between McGraw-Hill and SDRC, you can use the equally powerful student edition.

I-deas Student Edition™ is designed to help you with engineering design, manufacturing, and concurrent engineering skills. Online tutorials and instructional projects link the capabilities of I-deas to your engineering design course project work.

Features:

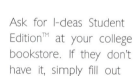

- 3-D component, assembly, and sculptured-surface modeling
- High productivity 3-D drafting software
- VGX variational geometry methods for teaching design and assembly
- Generated tool parts for milling and turning
- Comprehensive online tutorials

Ask for I-deas Student Edition™ at your college bookstore. If they don't have it, simply fill out this form and send it in.

You may also phone in your order by dialing **1-800-262-4729** or fax your order to **(614) 759-3644**.

☐ **Please send me I-deas Student Edition**™   ISBN 0072373903

Name _____

Address _____

School _____

City _____ **State** _____ **Zip** _____

E-mail _____

Payment Method: Price $110⁰⁰ (plus shipping, handling, & tax)

Amount Enclosed $_____   Delivery: ☐ Standard UPS ☐ Rush/UPS 2-day
                                                    (Additional Charge)

☐ Check   ☐ Visa   ☐ Mastercard   ☐ American Express   ☐ Discover

Account Number _____ Exp. Date _____

Signature _____

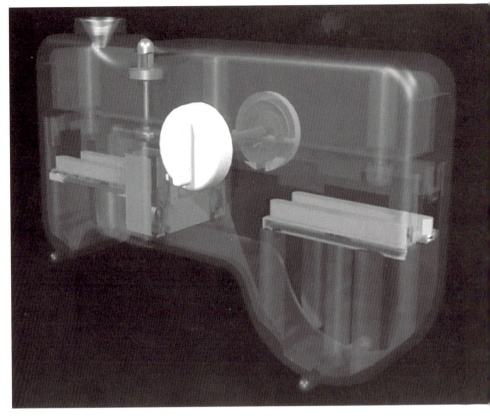

For a mechanical engineering class project, Dan Hassler of the University of Illinois at Urbana-Champaign used I-deas to redesign a bicycle chain degreaser to improve its rugged environment operation and to lower its manufacturing costs.

* I-deas uses four-color graphics that are not illustrated in this one-color promotion.

Place
Postage
Here

McGraw-Hill
Order Services
Post Office Box 182604
Columbus, OH 43272-3031

SDRC

ENGINEERING DESIGN

A Materials and Processing Approach

McGraw-Hill Series in Mechanical Engineering

Anderson
Computational Fluid Dynamics: The Basics with Applications

Anderson
Modern Compressible Flow: With Historical Perspective

Arora
Introduction to Optimum Design

Borman and Ragland
Combustion Engineering

Çengel
Thermodynamics: An Engineering Approach

Çengel
Heat Transfer: A Practical Approach

Çengel
Introduction to Thermodynamics & Heat Transfer

Dieter
Engineering Design: A Materials & Processing Approach

Doebelin
Engineering Experimentation: Planning, Execution, Reporting

Driels
Linear Control Systems Engineering

Gibson
Principles of Composite Material Mechanics

Hamrock
Fundamentals of Machine Elements

Hamrock
Fundamentals of Fluid Film Fabrication

Heywood
Internal Combustion Engine Fundamentals

Histand and Alciatore
Introduction to Mechatronics and Measurement Systems

Holman
Experimental Methods for Engineers

Jaluria
Design and Optimization of Thermal Systems

Kays and Crawford
Convective Heat and Mass Transfer

Kelly
Fundamentals of Mechanical Vibrations

Krieder and Rabl
Heating and Cooling of Buildings

Mattingly
Elements of Gas Turbine Propulsion

Modest
Radiative Heat Transfer

Norton
Design of Machinery

Oosthuizen and Carscallen
Compressible Fluid Flow

Oosthuizen and Naylor
Introduction to Convective Heat Transfer Analysis

Reddy
An Introduction to Finite Element Method

Schey
Introduction to Manufacturing Processes

Schlichting
Boundary Layer Theory

Shames
Mechanics of Fluids

Shigley and Uicker
Theory of Machines and Mechanisms

Stoecker
Design of Thermal Systems

Stoecker and Jones
Refrigeration and Air Conditioning

Turns
An Introduction to Combustion: Concepts and Applications

Ullman
The Mechanical Design Process

Wark
Advanced Thermodynamics for Engineers

Wark and Richards
Thermodynamics

White
Fluid Mechanics

White
Viscous Fluid Flow

ENGINEERING DESIGN

A Materials and Processing Approach

THIRD EDITION

George E. Dieter
University of Maryland

Boston   Burr Ridge, IL   Dubuque, IA   Madison, WI   New York   San Francisco   St. Louis
Bangkok   Bogotá   Caracas   Lisbon   London   Madrid   Mexico City   Milan   New Delhi   Seoul
Singapore   Sydney

McGraw-Hill Higher Education ⚛
*A Division of The **McGraw-Hill** Companies*

ENGINEERING DESIGN: A Materials and Processing Approach

Copyright © 2000, 1991, 1983 by The McGraw-Hill Companies, Inc. All rights reserved. Printed in the United States of America. Except as permitted under the United States Copyright Act of 1976, no part of this publication may be reproduced or distributed in any form or by any means, or stored in a data base or retrieval system, without the prior written permission of the publisher.

This book is printed on acid-free paper.

4 5 6 7 8 9 0 DOC/DOC 9 0 9 8 7 6 5 4

ISBN 0-07-366136-8

Vice president/Editor-in-Chief: Kevin T. Kane
Publisher: Thomas Casson
Senior sponsoring editor: Jonathan Plant
Developmental editor: Maja Lorkovic
Senior marketing manager: John T. Wannemacher
Project manager: Kimberly D. Hooker
Senior production supervisor: Heather D. Burbridge
Freelance design coordination: Pam Verros
Cover illustration: Paul D. Turnbaugh
Photo research coordinator: Sharon Miller
Compositor: Lachina Publishing Services
Typeface: 10.5/12 Times Roman
Printer: R. R. Donnelley & Sons Company

Library of Congress Cataloging-in-Publication Data

Dieter, George Ellwood.
    Engineering design : a materials and processing approach / George
E. Dieter. — 3rd ed.
      p.   cm.
    ISBN 0-07-366136-8
    1. Engineering design.   I. Title.
TA174.D495 2000
620′.0042—dc21   99-24646

http://www.mhhe.com

ABOUT THE AUTHOR

GEORGE E. DIETER is Glenn L. Martin Institute Professor of Engineering at the University of Maryland. The author received his B.S. Met.E. degree from Drexel University and his D.Sc. degree from Carnegie Mellon University. After a stint in industry with the DuPont Engineering Research Laboratory, he became Head of the Metallurgical Engineering Department at Drexel University, where he later became Dean of Engineering. Professor Dieter later joined the faculty of Carnegie Mellon University as Professor of Engineering and Director of the Processing Research Institute. He moved to the University of Maryland in 1977 as Professor of Mechanical Engineering and Dean of Engineering, serving as dean until 1994.

Professor Dieter is a fellow of ASM International, TMS, AAAS, and ASEE. He has received the education award from ASM, TMS, and SME, as well as the Lamme Medal, the highest award of ASEE. He has been chair of the Engineering Deans Council, and president of ASEE. He is a member of the National Academy of Engineering. He also is the author of *Mechanical Metallurgy,* published by McGraw-Hill, now in its third edition.

CONTENTS

PREFACE TO THIRD EDITION

THE THIRD EDITION of *Engineering Design* represents a major reorganization and expansion. The revision has resulted from the realization that engineering students need more structure to guide them through the design process. Thus, we have provided separate chapters on problem definition, concept generation and evaluation, the embodiment design process, and detail design. Chapters have been reordered to be more in the natural progression of the design process. For example, the chapter on gathering information has been moved from Chap. 14 to Chap. 4. A chapter on team behavior has been added as Chap. 3 to help the students in their design project work. A separate chapter on legal and ethical issues in engineering has been added.

As before, *Engineering Design* is intended to provide the senior engineering student with a realistic understanding of the design process. It is broader in content than most design texts, but it now contains much more prescriptive guidance in how to carry out design.

Many new topics have been added for the first time or have been expanded. These include marketing and customer surveys, quality function deployment, problem solving using TQM tools, the use of the Internet in information gathering, the basic concepts of TRIZ, the AHP method for concept evaluation, determination of product architecture, industrial design, human factors in design, design for the environment, design for assembly and manufacturability, failure modes and effects analysis, robust design and the Taguchi methodology, activity-based costing, and the product profit model.

The author hopes that students will consider this book to be a valuable part of their professional library. In order to enhance its usefulness for that purpose, many references to the literature have been included, as well as tips on useful design software, and references to web sites. In a book that covers such a wide sweep of material, it has not always been possible to go into the necessary depth for every topic. In these instances, reference is made to at least one authoritative source for further study.

Special thanks goes to Linda Schmidt of the Mechanical Engineering faculty, University of Maryland, for carefully reviewing and commenting on the manuscript.

Thanks also go to Shapour Azarm, Vincent Brannigan, S. K. Gupta, Bharat Kaku, Neal Kaske, Richard McCuen, and Guangming Zhang, my colleagues at the University of Maryland, for advice and consultation. I must also thank the following reviewers for their many helpful comments and suggestions: A. Sherif El-Gizawy, University of Missouri–Columbia; James Moller, Miami University; Keith Rouch, University of Kentucky; and Siva Thangam, Stevens Institute of Technology.

George E. Dieter
College Park, MD
1999

NOTE TO THE INSTRUCTOR

I-deas Student Edition is a software program designed to help your students with engineering design, manufacturing, and concurrent engineering skills. I-deas Student Edition allows your students to complete homework problems, perform research, and develop design team projects all on their personal computer. Currently, I-deas Professional Edition is widely used in the professional arena by companies such as IBM, Ford, Lockheed-Martin, and others.

Dieter's *Engineering Design,* **3/e,** is available packaged with I-deas software at a reduced price. If you would like to examine the software in consideration for adoption for your classes, please contact your McGraw-Hill sales representative or call **1-800-338-3987** and refer to **ISBN 007-237390-3.** If you would like to order this textbook packaged with I-deas Student Edition, please provide your bookstore with **ISBN 007-402718-2.**

PREFACE TO THE SECOND EDITION

THE THEME OF the first edition of *Engineering Design* was the importance of the connection between design and manufacturing. In the eight years since the first edition was published, this connection has become one of the driving forces of engineering in the World. Also in the intervening years, the role of computers in design has accelerated and become all-pervasive. These two aspects of design have been reinforced and strengthened in this new edition.

As before, *Engineering Design* is intended to provide the senior engineering student with a broad realistic understanding of the design process. It draws on a diverse set of topics: decision making, optimization, engineering economy, planning, applied statistics, reliability, and quality engineering, and focuses them on the design process.

A number of topics have been added or greatly expanded. These include: ethics in engineering, societal considerations in engineering, technological innovation, market identification, competitive benchmarking, protection of intellectual property, human factors in design, industrial design, expert systems, and Taguchi methods. A major revision has been given to Chapter 3, which has been retitled "Design Methods." Emphasis is given to conceptual design to introduce the ideas of French and Pugh's ideas of the product design specification and the concept selection technique. Also, the student is introduced to Nam Suh's principles of design. The section in this chapter on assessing alternatives has been completely rewritten.

Chapter 13 has been greatly expanded and focused on the quality issues of engineering design. It opens with Deming's philosophy of total quality and includes an in-depth discussion of Taguchi's methods, including an example on robust design. The coverage of control charts and statistical process control is increased.

Special thanks goes to Shapour Azarm, Herbert Foerstel, Richard McCuen, Ioannis Pandelidis, and Marvin Roush, my colleagues at the University of Maryland, for advice and consultation. I must also thank the following reviewers for their many helpful comments and suggestions: Louis Bucciarelli, Massachusetts Institute of

Technology; Kevin Craig, Rochester Polytechnic Institute; Darrell Gibson, Rose-Hulman Institute of Technology; Gaza Kardos, Carleton University–Canada; and George Schade, University of Nebraska.

George E. Dieter

PREFACE TO THE FIRST EDITION

ENGINEERING DESIGN IS intended to provide the senior engineering student with a realistic understanding of the engineering design process. It is written from the viewpoint that design is the central activity of the engineering profession, and it is more concerned with developing attitudes and approaches than in presenting design techniques and tools. Like other texts on this subject, it develops design as an interdisciplinary activity that draws on such diverse subjects as decision making, optimization, engineering economy, planning, and applied statistics. Chapters are presented on each of those subjects, all of which are needed to some degree by the designer.

However, this text goes beyond other design books in giving special emphasis to materials selection and materials processing and manufacturing. These are critical aspects of the design process that have been given little attention in other design texts. Moreover, because they have been deemphasized in most engineering curricula, it is felt that special attention to them is warranted and that they can be learned successfully within a design context. Although the text should be applicable to students in all fields of engineering, special emphasis has been given to the materials, mechanical, and metallurgical fields in the selection of examples and illustrations.

The only real way to learn design is to do design. The best way to use this book is as part of a project design course in which the students are engaged in a major design problem. When possible, the actual experience in doing design should be supplemented by lectures on such subjects as engineering economy and reliability, depending on the particular mix of other courses the students have taken. When the class schedule will not permit both lectures and design experience, selected chapters should be assigned as outside reading as the design progresses. Clearly, in a subject as broad as engineering design, everything that is needed cannot be included in a book of modest size. This is not a handbook or cookbook. Rather, it aims more at developing good design attitudes and habits. One of those habits is self-reliance—the ability of the student to learn independently. Therefore, special emphasis has been given to selected further readings and far more references have been included than is usual in a basic textbook so as to provide a convenient launching point for the student's independent study.

The contents of this text and their use in a project design approach have been shaped by over fifteen years of experience in teaching design to metallurgical and mechanical engineering students at three institutions. Special recognition is due to Howard A. Kuhn, John C. Purcupile, Dwight A. Baughman, Clifford L. Sayre, Jr., and Richard W. McCuen for their valuable ideas, suggestions, and interactions. The writer also acknowledges the special association of nearly twenty years with the many fine engineers at the Materials Technology Laboratory of TRW, Inc., Cleveland, who have done much to add realism to this book. Finally, special thanks goes to Jean Beckmann for her painstaking efforts to create a perfect manuscript.

George E. Dieter

1

THE PRODUCT DESIGN PROCESS

1.1
INTRODUCTION

What is design? If you search the literature for an answer to that question, you will find about as many definitions as there are designs. Perhaps the reason is that the process of design is such a common human experience. Webster's dictionary says that to design is "to fashion after a plan," but that leaves out the essential fact that to design is to create something that has never been. Certainly an engineering designer practices design by that definition, but so does an artist, a sculptor, a composer, a playwright, or many another creative member of our society.

Thus, although engineers are not the only people who design things, it is true that the professional practice of engineering is largely concerned with design; it is frequently said that design is the essence of engineering. To design is to pull together something new or arrange existing things in a new way to satisfy a recognized need of society. An elegant word for "pulling together" is *synthesis*. We shall adopt the following formal definition of design:[1] "Design establishes and defines solutions to and pertinent structures for problems not solved before, or new solutions to problems which have previously been solved in a different way." The ability to design is both a science and an art. The science can be learned through techniques and methods to be covered in this course, but the art is best learned by doing design. It is for this reason that your design experience must involve some realistic project experience.

The emphasis that we have given to the creation of new things in our discussion of design should not unduly alarm you. To become proficient in design is a perfectly attainable goal for an engineering student, but its attainment requires the guided experience that we intend this course to provide. Design should not be confused with discovery. *Discovery* is getting the first sight of, or the first knowledge of something, as

---

1. J. F. Blumrich, *Science,* vol. 168, pp. 1551–1554, 1970.

A number of factors serve to protect a product from competition. A product which requires high capital investment to manufacture or which requires complex manufacturing processes tends to be resistant to competition. At the other end of the product chain, the need for an extensive distribution system may be a barrier to entry. A strong patent position may keep out competition, as may strong brand identification and loyalty on the part of the customer.

1.3
THE DESIGN PROCESS—A SIMPLIFIED APPROACH

We frequently talk about "designing a system." By a system we mean the entire combination of hardware, information, and people necessary to accomplish some specified mission. A system may be an electric power distribution network for a region of the nation, a complex piece of machinery like a newspaper printing press, or a combination of production steps to produce automobile parts. A large system usually is divided into *subsystems,* which in turn are made up of *components.*

There is no single universally acclaimed sequence of steps that leads to a workable design. Different writers or designers have outlined the design process in as few as 5 steps or as many as 25. One of the first to write introspectively about design was Morris Asimow.[1] He viewed the heart of the design process as consisting of the elements shown in Fig. 1.2. As portrayed there, design is a sequential process consisting of many design operations. Examples of the operations might be (1) exploring the alternative systems that could satisfy the specified need, (2) formulating a mathematical model of the best system concept, (3) specifying specific parts to construct a component of a subsystem, and (4) selecting a material from which to manufacture a part. Each operation requires information, some of it general technical and business information that is expected of the trained professional and some of it very specific

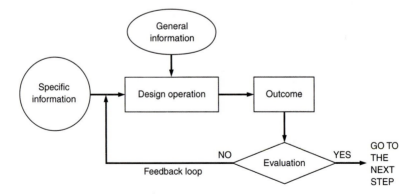

FIGURE 1.2
Basic module in the design process. (*After Asimow.*)

---

1. M. Asimow, "Introduction to Design," Prentice-Hall, Englewood Cliffs, NJ, 1962.

information that is needed to produce a successful outcome. An example of the last kind of information might be (1) a manufacturer's catalog on miniature bearings, (2) handbook data on high-temperature alloys, or (3) personal experience gained from a trip to observe a new manufacturing process. Acquisition of information is a vital and often very difficult step in the design process, but fortunately it is a step that usually becomes easier with time. (We call this process experience.)[1] The importance of developing sources of information is considered more fully in Chap. 4.

Once armed with the necessary information, the design engineer (or design team) carries out the design operation by using the appropriate technical knowledge and computational and/or experimental tools. At this stage it may be necessary to construct a mathematical model and conduct a simulation of the component's performance on a digital computer. Or it may be necessary to construct a full-size prototype model and test it to destruction at a proving ground. Whatever it is, the operation produces a design outcome that, again, may take many forms. It can be a ream of computer printout, a rough sketch with critical dimensions established, or a complete set of engineering drawings. At this stage the design outcome must be evaluated, often by a team of impartial experts, to decide whether it is adequate to meet the need. If so, the designer may go on to the next step. If the evaluation uncovers deficiencies, then the design operation must be repeated. The information from the first design is fed back as input, together with new information that has been developed as a result of questions raised at the evaluation step. We call this *iteration*.

The final result of the chain of design modules, each like Fig. 1.2, is a new working object (often referred to as *hardware*) or a collection of objects that is a new system. However, many design projects do not have as an objective the creation of new hardware or systems. Instead, the objective may be the development of new information that can be used elsewhere in the organization. It should be realized that few system designs are carried through to completion; they are stopped because it has become clear that the objectives of the project are not technically and/or economically feasible. However, they create new information, which, if stored in retrievable form, has future value, since it represents experience.

The simple model shown in Fig. 1.2 illustrates a number of important aspects of the design process. First, even the most complex system can be broken down into a sequence of design objectives. Each objective requires evaluation, and it is common for the decision-making phase to involve repeated trials or iterations. The need to go back and try again should not be considered a personal failure or weakness. Design is a creative process, and all new creations of the mind are the result of trial and error. Of course, the more knowledge we have and can apply to the problem the faster we can arrive at an acceptable solution. This iterative aspect of design may take some getting used to. You will have to acquire a high tolerance for failure and the tenacity and determination to persevere and work the problem out one way or the other.

The iterative nature of design provides an opportunity to improve the design on the basis of a preceding outcome. That, in turn, leads to the search for the best possible technical condition, e.g., maximum performance at minimum weight (or cost).

---

1. Experience has been defined, perhaps a bit lightheartedly, as just a sequence of nonfatal events.

Many techniques for *optimizing* a design have been developed, and some of them are covered in Chap. 12. And although optimization methods are intellectually pleasing and technically interesting, they often have limited application in a complex design situation. In the usual situation the actual design parameters chosen by the engineer are a compromise among several alternatives. There may be too many variables to include all of them in the optimization, or nontechnical considerations like available time or legal constraints may have to be considered, so that trade-offs must be made. The parameters chosen for the design are then close to but not at optimum values. We usually refer to them as *optimal values,* the best that can be achieved within the total constraints of the system.

In your scientific and engineering education you may have heard reference to the scientific method, a logical progression of events that leads to the solution of scientific problems. Percy Hill[1] has diagramed the comparison between the scientific method and the design method (Fig. 1.3).

The scientific method starts with a body of existing knowledge. Scientists have curiosity that causes them to question these laws of science; and as a result of their questioning, they eventually formulate a hypothesis. The hypothesis is subjected to logical analysis that either confirms or denies it. Often the analysis reveals flaws or inconsistencies, so that the hypothesis must be changed in an iterative process.

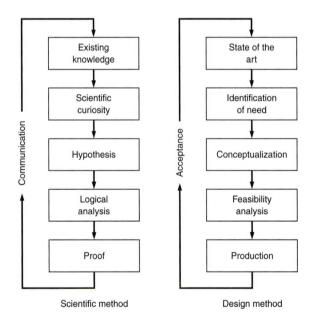

FIGURE 1.3
Comparison between the scientific method and the design method. (*After Percy Hill.*)

1. P. H. Hill, "The Science of Engineering Design," Holt, Rinehart and Winston, New York, 1970.

Finally, when the new idea is confirmed to the satisfaction of its originator, it must be accepted as proof by fellow scientists. Once accepted, it is communicated to the community of scientists and it enlarges the body of existing knowledge. The knowledge loop is completed.

The design method is very similar to the scientific method if we allow for differences in viewpoint and philosophy. The design method starts with knowledge of the state of the art. That includes scientific knowledge, but it also includes devices, components, materials, manufacturing methods, and market and economic conditions. Rather than scientific curiosity, it is really the needs of society (usually expressed through economic factors) that provide the impetus. When a need is identified, it must be conceptualized as some kind of model. The design concept must be subjected to a feasibility analysis, almost always with iteration, until an acceptable product is produced or the project is abandoned. When the design enters the production phase, it begins to compete in the world of technology. The design loop is closed when the product is accepted as part of the current technology and thereby advances the state of the art of the particular field.

In the brief outline of the design method shown above, the identification of a need requires further elaboration. Needs are identified at many points in a business or agency. Most organizations have research or development components whose job it is to create ideas that are relevant to the goals of the organization. A very important avenue for learning about needs is the customers for the product or services that the company sells. Managing this input is usually the job of the marketing organization of the company (see Sec. 1.8). Other needs are generated by government agencies, trade associations, or the attitudes or decisions of the general public. Needs usually arise from dissatisfaction with the existing situation. The need drivers may be to reduce cost, increase reliability or performance, or just change because the public has become bored with the product.

1.3.1 A Problem-Solving Methodology

A problem-solving methodology that is useful in design consists of the following steps:[1]

- Definition of the problem
- Gathering of information
- Generation of alternative solutions
- Evaluation of alternatives
- Communication of the results

This problem-solving method can be used at any point in the design process, whether at the conception of a product or the detailed design of a component.

---

1. A similar process called the guided iteration methodology has been proposed by J. R. Dixon; see J. R. Dixon and C. Poli, "Engineering Design and Design for Manufacturing," Field Stone Publishers, Conway, MA, 1995. A different but very similar problem-solving approach using TQM tools is given in Sec. 3.7.

Definition of the problem

The most critical step in the solution of a problem is the problem definition or formulation. The true problem is not always what it seems at first glance. Because this step seemingly requires such a small part of the total time to reach a solution, its importance is often overlooked. Figure 1.4 illustrates how the final design can differ greatly depending upon how the problem is defined.

The formulation of the problem should start by writing down a *problem statement.* This document should express as specifically as possible what the problem is. It should include objectives and goals, the current state of affairs and the desired state, any constraints placed on solution of the problem, and the definition of any special technical terms. The problem-definition step in a design project is covered in detail in Chap. 2.

Gathering information

Perhaps the greatest frustration you will encounter when you embark on your first design problem will be due to the dearth or plethora of information. No longer will your responsibility stop with the knowledge contained in a few chapters of a text. Your assigned problem may be in a technical area in which you have no previous background, and you may not have even a single basic reference on the subject. At the other extreme you may be presented with a mountain of reports of previous work and

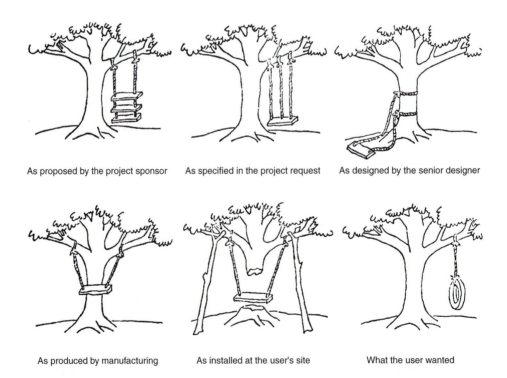

As proposed by the project sponsor   As specified in the project request   As designed by the senior designer

As produced by manufacturing   As installed at the user's site   What the user wanted

FIGURE 1.4
Note how the design depends on the viewpoint of the individual who defines the problem.

your task will be to keep from drowning in paper. Whatever the situation, the immediate task is to identify the needed pieces of information and find or develop that information.

An important point to realize is that the information needed in design is different from that usually associated with an academic course. Textbooks and articles published in the scholarly technical journals usually are of lesser importance. The need often is for more specific and current information than is provided by those sources. Technical reports published as a result of government-sponsored R&D, company reports, trade journals, patents, catalogs, and handbooks and literature published by vendors and suppliers of material and equipment are important sources of information. The Internet is becoming a very useful resource, too. Often the missing piece of information can be supplied by a telephone call or an e-mail to a key supplier. Discussions with in-house experts (often in the corporate R&D center) and outside consultants may prove helpful.

The following are some of the questions concerned with obtaining information:

What do I need to find out?
Where can I find it and how can I get it?
How credible and accurate is the information?
How should the information be interpreted for my specific need?
When do I have enough information?
What decisions result from the information?

The topic of information gathering is discussed in Chap. 4.

Generation of alternative solutions

Generating alternative solutions involves the use of creativity stimulation methods, the application of physical principles and qualitative reasoning, and the ability to find and use information. Of course, experience helps greatly in this task. The ability to generate high-quality alternative solutions is vital to a successful design. This important subject is covered in Chap. 5.

Evaluation of alternatives

The evaluation of alternatives involves systematic methods for selecting the best among several designs, often in the face of incomplete information. Engineering analysis procedures provide the basis for making decisions about service performance. Cost estimation (Chap. 14) and design for manufacturing analyses (Chap. 9) provide other important information. Various other types of engineering analysis also provide information. Simulation of performance with computer models is finding wide usage (Sec. 7.8). Simulated service testing of an experimental model and testing of full-sized prototypes often provide critical data. Without this quantitative information, it is not possible to make valid evaluations.

Several methods for evaluating design concepts, or any other problem solution, are given in Chap. 5.

An important consideration at every step in the design, but especially as the design nears completion, is checking. In general, there are two types of checks that can be made: mathematical checks and engineering-sense checks. Mathematical checks are concerned with checking the arithmetic and the equations used in the analytical model. Incidentally, the frequency of careless math errors is a good reason why you should adopt the practice of making all your design calculations in a bound note-book. In that way you won't be missing a vital calculation when you are forced by an error to go back and check things out. Just draw a line through the part in error and continue. It is of special importance to ensure that every equation is dimensionally consistent.

Engineering-sense checks have to do with whether the answers "feel right." Even though the reliability of your feeling of rightness increases with experience, you can now develop the habit of staring at your answer for a full minute, rather than rushing on to do the next calculation. If the calculated stress is 10^6 psi, you know something went wrong! Limit checks are a good form of engineering-sense check. Let a critical parameter in your design approach some limit (zero, infinity, etc.), and observe whether the equation behaves properly.

We have stressed the iterative nature of design. An optimization technique aimed at producing a *robust design* that is resistant to environmental influences (water vapor, temperature, vibration, etc.) most likely will be employed to select the best values of key design parameters (see Chap. 12). The management decision as to when to "freeze the design" will be dictated chiefly by considerations of time and money.

Communication of the results

It must always be kept in mind that the purpose of the design is to satisfy the needs of a customer or client. Therefore, the finalized design must be properly com-municated, or it may lose much of its impact or significance. The communication is usually by oral presentation to the sponsor as well as by a written design report. A recent survey showed that design engineers spend 60 percent of their time in dis-cussing designs and preparing written documentation of designs, while only 40 per-cent of the time is spent in analyzing designs and doing the designing. Detailed engi-neering drawings, computer programs, and working models are frequently part of the "deliverables" to the customer. It hardly needs to be emphasized that communication is not a one-time occurrence to be carried out at the end of the project. In a well-run design project there is continual oral and written dialog between the project manager and the customer. This extremely important subject is considered in greater depth in Chap. 17.

1.4
CONSIDERATIONS OF A GOOD DESIGN

Design is a multifaceted process. To gain a broader understanding of engineering design, we group various considerations in good design into three categories: (1) design requirements, (2) life-cycle issues, and (3) regulatory and social issues.

1.4.1 Design Requirements

It is obvious that to be satisfactory the design must demonstrate the required per-formance. Acceptable performance is the first, but far from only, design requirement.

Performance measures both function and behavior of the design, i.e., how well the device does what it is designed to do. Performance requirements can be divided into *functional* performance requirements and *complementary* performance requirements. Functional requirements address such capacity measures as forces, strength, energy or material flows, power, and deflection. They also are concerned with the efficiency of the design, its accuracy and sensitivity. Complementary performance requirements are concerned with the useful life of the design, its robustness to factors in the service environment (see Chap. 11), its reliability (see Chap. 12), and ease, economy, and safety of maintenance. Issues such as built-in safety features and the noise level in operation must be considered. Finally, the design must conform to all legal requirements and design codes (see Chap. 15).

A variety of analysis techniques must be employed in arriving at the *features* of a component in the design. By feature we mean specific physical attributes, such as shape, dimensions, or material properties. The digital computer has had a major impact in this area by providing powerful analytical tools based on finite-element analysis and finite difference. Calculations of stress, temperature, and other field-dependent variables can be made rather handily for complex geometry and loading conditions. When these analytical methods are coupled with interactive computer graphics, we have the exciting capability known as computer-aided engineering (CAE); see Sec. 1.9.

Next to performance requirements we have physical requirements. These pertain to such issues as size, weight, shape, and surface finish.

Environmental requirements deal with two separate aspects. The first concerns the service conditions under which the product must operate. The extremes of temperature, humidity, corrosive conditions, dirt, vibration, noise, etc., must be predicted and allowed for in the design. The second aspect of environmental requirements pertains to how the product will behave with regard to maintaining a safe and clean environment, i.e., green design. Among these issues is the disposal of the product when it reaches its useful life.

Aesthetic requirements refer to "the sense of the beautiful." They are concerned with how the product is perceived by a customer because of its shape, color, surface texture, and also such factors as balance, unity, and interest. This aspect of design usually is the responsibility of the industrial designer, as contrasted with the engineering designer. The industrial designer is an applied artist. Decisions concerning the appearance of the product should be an integral part of the initial design concept.

Manufacturing technology must be intimately connected with product design. There may be restrictions on the manufacturing processes that can be used, because of either selection of material or availability of equipment within the company.

The final major design requirement is cost. Every design has requirements of an economic nature. These include such issues as product development cost, initial product cost, life cycle product cost, tooling cost, and return on investment. In many cases cost is the most important design requirement, for if preliminary estimates of product cost look unfavorable, the design project may never be initiated. Cost enters into every aspect of the design process. Therefore, we have considered the subject of economic decision making (engineering economics) in some detail in Chap. 13. Procedures for estimating costs are considered in Chap. 14.

1.4.2 Total Life Cycle

The total life cycle of a part starts with the conception of a need and ends with the retirement and disposal of the product.

Material selection is a key element in the total life cycle (see Chap. 8). In selecting materials for a given application, the first step is evaluation of the service conditions. Next, the properties of materials that relate most directly to the service requirements must be determined. Except in almost trivial conditions, there is never a simple relation between service performance and material properties. The design may start with the consideration of static yield strength, but properties that are more difficult to evaluate, such as fatigue, creep, toughness, ductility, and corrosion resistance, may have to be considered. We need to know whether the material is stable under the environmental conditions. Does the microstructure change with temperature? Does the material corrode slowly or wear at an unacceptable rate?

Material selection cannot be separated from *producibility* (see Chap. 9). There is an intimate connection between design and material selection and the production processes. The objective in this area is a trade-off between the opposing factors of minimum cost and maximum durability. *Durability* is concerned with the number of cycles of possible operation, i.e., the useful life of the product.

Current societal issues of energy conservation, material conservation, and protection of the environment result in new pressures in selection of materials and manufacturing processes. Energy costs, once nearly totally ignored in design, are now among the most prominent design considerations. Design for materials recycling also is becoming an important consideration.

1.4.3 Regulatory and Social Issues

Specifications and standards have an important influence on design practice (see Chap. 4). The standards produced by such societies as ASTM and ASME represent voluntary agreement among many elements (users and producers) of industry. As such, they often represent minimum or least-common-denominator standards. When good design requires more than that, it may be necessary to develop your own company or agency standards. On the other hand, because of the general nature of most standards, a standard sometimes requires a producer to meet a requirement that is not essential to the particular function of the design.

The code of ethics of all professional engineering societies requires the engineer to protect public health and safety. Increasingly, legislation has been passed to require federal agencies to regulate many aspects of safety and health. The requirements of the Occupational Safety and Health Administration (OSHA), the Consumer Product Safety Commission (CPSC), and the Environmental Protection Agency (EPA) place direct constraints on the designer. Several aspects of the CPSC regulations have far-reaching influence on product design. Although the intended purpose of a particular product normally is quite clear, the unintended uses of that product are not always obvious. Under the CPSC regulations, the designer has the obligation to foresee as many unintended uses as possible, then develop the design in such a way as to prevent hazardous use of the product in an unintended but foreseeable manner. When unintended use cannot be prevented by functional design, clear, complete, unambiguous warnings must be permanently attached to the product. In addition, the designer must be cognizant of all advertising material, owner's manuals, and operating instructions that relate to the product to ensure that the contents of the material are consistent with safe operating procedures and do not promise performance characteristics that are beyond the capability of the design.

An important design consideration is adequate attention to human factors engineering, which uses the sciences of biomechanics, ergonomics, and engineering psychology to assure that the design can be operated efficiently by humans. It applies physiological and anthropometric data to such design features as visual and auditory display of instruments and control systems. It is also concerned with human muscle power and response times. For further information, see Sec. 6.7.

1.5
DETAILED DESCRIPTION OF DESIGN PROCESS

Morris Asimow[1] was among the first to give a detailed description of the complete design process in what he called the *morphology of design.* His seven phases of design are described below, with slight changes of terminology to conform to current practice.

1.5.1 Phase I. Conceptual Design

Conceptual design is the process by which the design is initiated, carried to the point of creating a number of possible solutions, and narrowed down to a single best concept. It is sometimes called the feasibility study. Conceptual design is the phase which requires the greatest creativity, involves the most uncertainty, and requires coordination among many functions in the business organization. The following are the discrete activities that we consider under conceptual design.

- *Identification of customer needs:* The goal of this activity is to completely understand the customers' needs and to communicate them to the design team.
- *Problem definition:* The goal of this activity is to create a statement that describes what has to be accomplished to satisfy the needs of the customer. This involves analysis of competitive products, the establishment of target specifications, and the listing of constraints and trade-offs. Quality function deployment (QFD) is a valuable tool for linking customer needs with design requirements. A detailed listing of the product requirements is called a product design specification (PDS). Problem definition, in its full scope, is treated in Chap. 2.
- *Gathering information:* Engineering design presents special requirements over engineering research in the need to acquire a broad spectrum of information. This subject is covered in Chap. 4.

---

1. M. Asimow, "Introduction to Design," Prentice-Hall, Englewood Cliffs, NJ, 1962.

- *Conceptualization:* Concept generation is involved with creating a broad set of concepts that potentially satisfy the problem statement. Team-based creativity methods, combined with efficient information gathering, are the key activities. This subject is covered in Chap. 5.
- *Concept selection:* Evaluation of the design concepts, modifying and evolving into a single preferred concept, are the activities in this step. The process usually requires several iterations. This is covered in Chap. 5.
- *Refinement of the PDS:* The product design specification is revisited after the concept has been selected. The design team must commit to achieving certain critical values of design parameters and to living with trade-offs between cost and performance.
- *Design review:* Before committing funds to move to the next design phase, a design review will be held. The design review will assure that the design is physically realizable and that it is economically worthwhile. It will also look at a detailed product-development schedule. This is needed so as to devise a strategy to minimize product cycle time and to identify the resources in people, equipment, and money needed to complete the project.

1.5.2 Phase II. Embodiment Design

In this phase a structured development of the design concept takes place. It is the place where flesh is placed on the skeleton of the design concept. An embodiment of all the main functions that must be performed by the product must be undertaken. It is in this design phase that decisions are made on strength, material selection, size, shape, and spatial compatibility. Beyond this design phase major changes become very expensive. This design phase is sometimes called preliminary design. Embodiment design is concerned with three major tasks—product architecture, configuration design, and parametric design.

- *Product architecture:* Product architecture is concerned with dividing the overall design system into subsystems or modules. In this step it is decided how the physical components of the design are to be arranged and combined to carry out the functional duties of the design.
- *Configuration design of parts and components:* Parts are made up of *features* like holes, ribs, splines, and curves. Configuring a part means to determine what features will be present and how those features are to be arranged in space relative to each other. While modeling and simulation may be performed in this stage to check out function and spatial constraints, only approximate sizes are determined to assure that the part satisfies the PDS. Also, more specificity about materials and manufacturing is given here. The generation of a physical model of the part with rapid prototyping processes may be appropriate.
- *Parametric design of parts and components:* Parametric design starts with information on the configuration of the part and aims to establish its exact dimensions and tolerances. Final decisions on the material and manufacturing processes are also established if this has not been done previously. An important aspect of parametric design is to examine the part, assembly, and system for *design robustness.* Robustness refers to how consistently a component performs under variable condi-

tions in its service environment. The methods developed by Dr. Genichi Taguchi for achieving robustness and establishing the optimum tolerance are discussed in Chap. 12. Parametric design also deals with determining the aspects of the design that could lead to failure (see Chap. 11). Another important consideration in parametric design is to design in such a way that manufacturability is enhanced (see Chap. 9).

1.5.3 Phase III. Detail Design

In this phase the design is brought to the stage of a complete engineering description of a tested and producible product. Missing information is added on the arrangement, form, dimensions, tolerances, surface properties, materials, and manufacturing processes of each part. This results in a specification for each special-purpose part and for each standard part to be purchased from suppliers. Detailed engineering drawings suitable for manufacturing are prepared. Frequently these are computer-generated drawings and may include three-dimensional solid models. Assembly drawings and assembly instructions will also be determined. Detail design often includes the building and testing of several preproduction versions of the product. Finally, detail design concludes with a design review before the information is passed on to manufacturing.

The activities described above are displayed in Fig. 1.5. This eight-step process is our representation of the basic design process. The purpose of this graphic is to remind you of the logical sequence of activities that leads from problem definition to the detail design. It constitutes the *primary design.* As we consider various aspects of

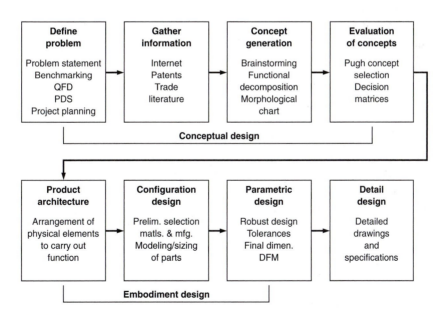

FIGURE 1.5
Discrete steps in engineering design process from problem definition to detail design. The chief tools or techniques applicable in each step are given.

this design process in much more detail as you progress through this text, we shall display this graphic to remind you of where you are in this process. However, remember that design does not normally proceed in a straight-line fashion. Many iterations or branching back (see Fig. 1.1) will be necessary and can be expected for final success.

Phases I, II, and III take the design from the realm of possibility to probability to the real world of practicality. However, the design process is not finished with the delivery of a set of detailed engineering drawings and specifications. Many other technical and business decisions must be made that are really part of the design process. A great deal of thought and planning must go into how the design will be manufactured, how it will be marketed, how it will be maintained during use, and finally, how it will be retired from service and replaced by a new, improved design. Generally these phases of design are carried out elsewhere in the organization than in the engineering department or product development department. As the project proceeds into the new phases, the expenditure of money and personnel time increases greatly.

1.5.4 Phase IV. Planning for Manufacture

A great deal of detailed planning must be done to provide for the production of the design. A method of manufacture must be established for each component in the system. As a usual first step, a *process sheet* is established; it contains a sequential list of manufacturing operations that must be performed on the component. Also, it specifies the form and condition of the material and the tooling and production machines that will be used. The information on the process sheet makes possible the estimation of the production cost of the component.[1] High costs may indicate the need for a change in material or a basic change in the design. Close interaction with manufacturing, industrial, materials, and mechanical engineers is important at this step. This topic is discussed more fully in Chap. 9.

The other important tasks performed in phase IV are the following:

1. Designing specialized tools and fixtures
2. Specifying the production plant that will be used (or designing a new plant) and laying out the production lines
3. Planning the work schedules and inventory controls (production control)
4. Planning the quality assurance system
5. Establishing the standard time and labor costs for each operation
6. Establishing the system of information flow necessary to control the manufacturing operation

All of these tasks are generally considered to fall within industrial or manufacturing engineering.

---

1. Precise calculation of manufacturing cost cannot be made until the process sheet is known. However, reasonable part cost estimates are made in conceptual and embodiment design. These are important elements for decision making at early stages of design. For more detail on costs, see Chap. 14.

1.5.5 Phase V. Planning for Distribution

Important technical and business decisions must be made to provide for the effective distribution to the consumer of the systems that have been produced. In the strict realm of design, the shipping package may be critical. Concepts such as shelf life may also be critical and may need to be addressed in the earlier stages of the design process. A system of warehouses for distributing the product may have to be designed if none exists.

The economic success of the design often depends on the skill exercised in marketing the product. If the product is of the consumer type, the marketing effort is concentrated on advertising and news media techniques, but highly technical products may require that the marketing step be a technical activity supported by specialized sales brochures and performance test data.

1.5.6 Phase VI. Planning for Use

The use of the design by the consumer is all-important, and considerations of how the consumer will react to the product pervade all steps of the design process. The following specific topics can be identified as being important user-oriented concerns in the design process: ease of maintenance, reliability, product safety, convenience in use (human factors engineering), aesthetic appeal, economy of operation, and duration of service.

Obviously, these consumer-oriented issues must be introduced into the design process at the very beginning. Phase VI of design is less well defined than the others, but it is becoming increasingly important with the advent of consumer protection and product safety legislation. More strict interpretation of product liability laws is having a major impact on design.

An important phase VI activity is the acquisition of reliable data on failures, service lives, and consumer complaints and attitudes to provide a basis for product improvement in the next design.

1.5.7 Phase VII. Planning for Retirement of the Product

The final step in the design process is the disposal of the product when it has reached the end of its useful life. *Useful life* may be determined by actual deterioration and wear to the point at which the design can no longer function, or it may be determined by technological obsolescence, in which a competing design performs the function either better or cheaper. In consumer products, it may come about through changes in fashion or taste.

In the past, little attention has been given in the design process to product retirement. This is rapidly changing, as people the world over are becoming concerned about environmental issues. There is concern with depletion of mineral and energy resources, and with pollution of the air, water, and land as a result of manufacturing and technology advancement. This has led to a formal area of study called *industrial ecology.* Design for the environment, also called *green design,* is becoming an important consideration in design (Sec. 6.8). As a result, the design of a product should include a plan for either its disposal in an environmentally safe way or, better, the recycling of its materials or the remanufacture or reuse of its components.

1.6
MARKETING

Marketing is a somewhat foreign concept to many engineers. However, it should whet your appetite to learn more about this mysterious thing called marketing when you learn that inadequate understanding of the market in which they will compete is the number one cause of new product failure. There are two different aspects of marketing. The first, to which we give major attention, deals with the identification of customer needs, product opportunities, and an understanding of market segments. Acquiring this information is often called market research. The second aspect of marketing deals with the introduction of the product into the marketplace and the development of an ongoing relationship with the customer.

The marketing department in a company creates and manages the company's relationship with its customers.[1] It is the company's window on the world with its customers. It translates customer needs into requirements for products and influences the creation of services that support the product and the customer. It is about understanding how people make buying decisions and using this information in the design, building, and selling of products. Marketing does not make sales; that is the responsibility of the sales department.

What is this thing we call the market? Simply stated, a market is a collection of people who refer to each other about their buying decisions. The amount of interaction that occurs will depend on the degree of risk that the buyer perceives in the product. For many purchases we depend on advertising or word of mouth recommendations from friends. For the purchase of major appliances, we may consult references like *Consumer Reports,* but if we are purchasing a major computer system, upon which storage of all company records will depend, we would make a major study and talk with existing users of the product. In this case the market (users group) would be people who use the product to run the same kinds of applications that we expect to run.

The marketing department can be expected to do a number of tasks. First is a preliminary marketing assessment, a quick scoping of the potential sales, competition, and market share at the very early stages of the product development. Then they will do a detailed market study. This involves face-to-face interviews with potential customers to determine their needs, wants, preferences, likes, and dislikes. This will be done before detailed product development is carried out. A common method for doing this is the *focus group.* In this method a group of people with a prescribed knowledge about a product or service is gathered around a table and asked their feelings and attitudes about the product under study. If the group is well selected and the leader is experienced, the sponsor can expect to receive a wealth of opinions and attitudes which can be used to determine important attributes of a potential product. In a vari-

---

1. R. D. Hisrich and M. P. Peters, "Marketing a New Product," Benjamin/Cummins Pub. Co., 1984.

Adam Osborne's Computer

Adam Osborne saw an unfilled need for a portable computer in the embryonic personal computer market. He created a basic computer in 4 months, and within the year had sales of several millions of dollars.

However, he failed to keep in touch with the market, which had been entered by the IBM PC. Failure to achieve IBM compatibility doomed his product, and within 2 years the Osborne computer was history.

An accurate perception of the market made Osborne, but failure to understand a quickly moving market killed his product.

---

ant of this, called *scenario analysis,* persons familiar with a product are asked to write down actual scenarios for the use of the product. Other methods include one-on-one interviews, observing a similar product in use, and detailed examination of competitor's products. Quantitative market research involves careful statistical analysis of customer surveys that are administered by mail, phone interviews, or shopping mall intercepts.[1]

In making a marketing study it is important to understand the various segments of the market. Markets can be segmented with respect to age, sex, race, education, geographic location, and customer income. With technical products there is also a segmentation regarding receptivity to new things. Thus, customers can be classified as:

- *Early adopters:* People who are intrigued by new things and find ways to adopt the product despite the risk.
- *Mainstream adopters:* People who carefully evaluate what their peers in the same market are doing.
- *Laggards:* People who want a product that is low-risk and easy to use. These people do not buy until the product is at the commodity stage at a low price.

Marketing experts also believe that the purchasers of a product are making a statement about their self-images by purchasing a specific brand of a product. Much of this is created by advertising. It is important to have a clear understanding of the market segments your product is targeted to before beginning the product-development activity.

Thus, market research should provide information of the following type:

- Define the market segments
- Identify the early adopters
- Identify competitive products
- Establish the market size ($)
- Determine the breadth of the product line and number of versions
- Determine the product price-volume relationships
- Establish the customer needs and wants

---

1. C. Gevirtz, "Developing New Products with TQM," Chap. 4, McGraw-Hill, New York, 1994.

The marketing department also plays a vital role in assisting with the introduction of the product into the marketplace. They perform such functions as undertaking customer tests or field trials (beta test) of the product, planning for test marketing (sales) in restricted regions, advising on product packaging and warning labels, preparing user instruction manuals and documentation, arranging for user instruction, and advising on advertising. Marketing may also be responsible for providing for a product support system of spare parts, service representatives, and a warranty system.

1.6.1 Classification of Products Based on Market

Implicit in the discussion to this point is that the product is being developed in response to an identified market need, i.e., a *market pull* situation. There are other situations that need to be recognized.[1] The opposite of market pull is *technology push.* This is the situation where the company starts with a new proprietary technology and looks for a market in which to apply this technology. Often successful technology push products involve basic materials or basic process technologies, because these can be deployed in thousands of applications, and the probability of finding successful applications is therefore high. The discovery of nylon by the DuPont Company and its successful incorporation into thousands of new products is a classic example. The development of a technology-push product begins with the assumption that the new technology will be employed. This can entail risk, because unless the new technology offers a clear competitive advantage to the customer the product is not likely to succeed.

A *platform product* is built around a preexisting technological subsystem. Examples of such a platform are the Apple Macintosh operating system or the Black & Decker doubly insulated universal motor.[2] A platform product is similar to a technology-push product in that there is an a priori assumption concerning the technology to be employed. However, it differs in that the technology has already been demonstrated in the marketplace to be useful to a customer, so that the risk for future products is less. Often when a company plans to utilize a new technology in their products they plan to do it as a series of platform products. Obviously, such a strategy helps justify the high cost of developing a new technology.

For certain products the manufacturing process places strict constraints on the properties of the product, so that product design cannot be separated from the design of the production process. Examples of *process-intensive products* are automotive sheet steel, food products, semiconductors, chemicals, and paper. Process-intensive products typically are made in high volume, often with continuous flow processes, as opposed to discrete goods. With such a product, it might be more typical to start with a given process and design the product within the constraints of the process.

Customized products are those in which variations in configuration and content are created in response to a specific order of a customer. Often the customization is with regard to color or choice of materials, but more frequently it is with respect to

---

1. K. T. Ulrich and S. D. Eppinger, "Product Design and Development," McGraw-Hill, New York, 1995, pp. 20–22.
2. M. H. Meyer and A. P. Lehnerd, "The Power of Product Platforms," The Free Press, New York, 1997.

content, as when a person orders a personal computer by phone, or the accessories with a new car. Customization requires consideration of modular design and depends heavily on information technology to convey the customer's wishes to the production line. In a highly competitive marketplace, customization is one of the major trends, so much so that futurists predict that we are moving to a world of *mass customization.*[1]

1.7
ORGANIZATION FOR DESIGN

The organization of a business enterprise can have a major influence on how effectively design and product development is carried out. There are two fundamental ways for arranging the organization: with regard to *function* or with respect to *projects*.

A simple grouping of engineering practice into its various functions is given in Fig. 1.6. We start with research, which is closest to the academic experience; and as we progress downward in the hierarchy, we find that more emphasis on the job is given to financial and administrative matters and less emphasis is given to strictly technical matters.

Research and development (R&D) often is considered the glamour end of the engineering spectrum. We tend to think of research as limited to a scientist isolated in a laboratory and motivated by curiosity to explore nature. But that is basic research of a very pure type. Actually, R&D itself covers quite a spectrum of effort. The Department of Defense (DOD), which sponsors a large amount of R&D, classifies its activities as follows:

6.1 Research

6.2 Exploratory development

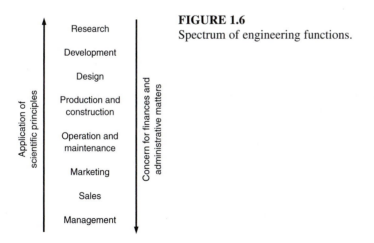

FIGURE 1.6
Spectrum of engineering functions.

---

1. E. B. Magrab, "Integrated Product and Process Design and Development," CRC Press, Boca Raton, FL, 1997, pp. 15–16.

6.3 Advanced development

6.4 Engineering development

6.5 Management and support

With downward progression in this classification, the work becomes more directed (applied) to a specific objective.

A project is a grouping of activities aimed at accomplishing a defined objective, like introducing a particular product into the marketplace. It requires certain activities: identifying customer needs, creating product concepts, building prototypes, designing for manufacture, etc. These tasks require people with different functional specialties. Thus, the two organizational arrangements, by function or project, are at opposite poles of possible organizational structures.

An important aspect of how an enterprise should be organized is concerned with the links between individuals. These links have to do with:

- *Reporting relationships:* A subordinate is concerned about who his or her supervisor is, since the supervisor influences evaluations, salary increases, promotions, work assignments, etc.
- *Financial arrangements:* Another type of link is budgetary. The source of funds to advance the project, and who controls these funds, is a vital consideration.
- *Physical arrangement:* Studies have shown that communication between individuals is enhanced if their offices are within 50 ft. of each other. Thus, physical layout, whether individuals share the same office, floor, or building, can have a major impact on the spontaneous encounters that occur and hence the quality of the communication.

Figure 1.7 shows the organizational chart for a manufacturing company organized along functional lines. All research and engineering reports to a single vice president; all manufacturing activity is the responsibility of another vice president; etc. Take the time to read the many functions under each vice president that are needed even in a manufacturing enterprise that is modest in size. A chief characteristic of a functional organization is that each individual has only one boss. By concentrating activities in units of common professional background, there are economies of scale, opportunities to develop deep expertise, and clear career paths for specialists. Generally, people gain satisfaction from working with colleagues who share similar professional interests. Since the organizational links are primarily among those who perform similar functions, formal interaction between different functional units, e.g., engineering and manufacturing, is forced to the level of the unit manager. This may be acceptable for a business with a narrow and slowly changing set of product lines, but the inevitable slow and bureaucratic decision making which this type of structure imposes can be a problem in a dynamic product situation.

The other extreme in organizational structure is the project organization, where people with the various functional expertise needed for the product development are grouped together to focus on the development of a specific product or product line (Fig. 1.8). Each development group reports to a project manager, who is responsible for overall success of the project, usually the ongoing development of a certain product line. The chief advantage of a project organization is that it focuses the needed

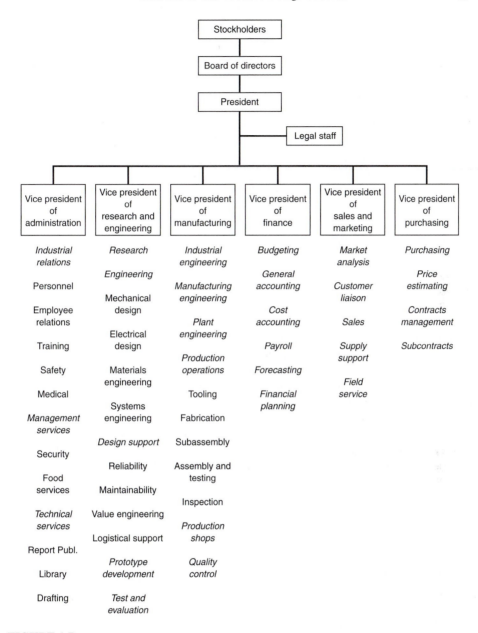

FIGURE 1.7
Example of a functional organization.

specialty talents on the attainment of the goal of the project. Often a project organization is time-limited; once the goal of the project is achieved, the people are reassigned back to functional units. Under this circumstance the organization is usually called a development *team.* This helps to address the chief disadvantage of this type of organization, that technical experts tend to lose their "cutting edge" functional capabilities with such intense focus on the project goal. Another disadvantage is that

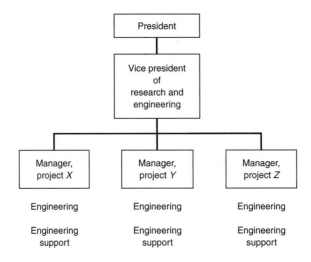

FIGURE 1.8
A simplified project organization.

the project organization is not as economical in utilization of scarce technical expertise as the functional organization. The project organization is very common in start-up companies, where indeed, the project and the company are synonymous. Often large corporations establish project organizations for large critical projects. The Lockheed "skunk works" that developed the U-2 spy plane is a well-known example.

Midway between these two types of organization is the *matrix organization,* which attempts to combine the advantages of each. In the matrix organization each person is linked to others according to both their function and the project they work on. As a consequence, each individual has two supervisors, one a functional manager and the other a project manager. While this may be true in theory, in practice either the functional manager or the project manager predominates.[1] In the *lightweight project organization* the functional links are stronger than the project links (Fig. 1.9*a*). The project manager is responsible for scheduling, coordination, and arranging meetings, but the functional managers are responsible for budgets, personnel matters, and performance evaluations. In the *heavyweight project organization* the project manager has complete budgetary authority, makes most of the resource allocation decisions, and plays a strong role in evaluating personnel (Fig. 1.9*b*). Although each participant belongs to a functional unit, the functional manager has little authority and control. The functional organization or the lightweight project organization works well in a stable business environment, especially one where the product predominates in its market because of technical excellence. A heavyweight project organization has advantages in introducing radically new products, especially where speed is important.

---

1. R. H. Hayes, S. C. Wheelwright, and K. B. Clark, "Dynamic Manufacturing: Creating the Learning Organization," The Free Press, New York, 1988, pp. 319–323.

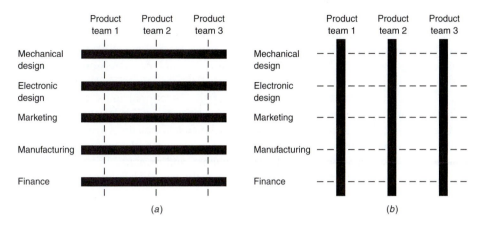

FIGURE 1.9
(*a*) A lightweight project organization; (*b*) a heavyweight project organization.

1.7.1 Concurrent Engineering

The conventional way of doing product design has been to carry out all of the steps serially. Thus, product concept, product design, and product testing have been done prior to process planning, manufacturing system design, and production. Commonly these serial functions have been carried out in distinct and separate organizations with little interaction between them. Thus, it is easy to see how the design team will make decisions, many of which can be changed only at great cost in time and money, without adequate knowledge of the manufacturing process. Refer to Fig. 1.1 to reinforce the concept that a large percentage of a product's cost is committed during the conceptual and embodiment stage of design. Very roughly, if the cost to make a change at the product concept stage is $1, the cost is $10 at the design stage and $100 at the production stage. The use of a serial design process means that as changes become necessary there is a doubling back to pick up the pieces, and the actual process is more in the nature of a spiral.

Starting in the 1980s, as companies met increasing competitive pressure, a new approach to integrated product design evolved, which is called *concurrent engineering.* The impetus came chiefly from the desire to shorten product-development time, but other drivers were the improvement of quality and the reduction of product life-cycle costs. Concurrent engineering is a systematic approach to the integrated concurrent design of products and their related processes, including manufacture and support. With this approach, product developers, from the outset, consider all aspects of the product life cycle, from concept to disposal, including quality, cost, schedule, and user requirements. A main objective is to bring as many viewpoints and talents to bear in the design phase so that these decisions will be valid for downstream parts of the product-development cycle like manufacturing and field service. Toward this end, computer-aided engineering (CAE) tools have been very useful (see Sec. 1.8). Concurrent engineering has three main elements: cross-functional teams, parallel design, and vendor partnering.

Of the various organizational structures for design that were discussed above, the heavyweight project organization, usually called just a *cross-functional design team,* is used most frequently with concurrent engineering. Having the skills from the functional areas embedded in the team provides for quick and easy decision making, and aids in communication with the functional units. For cross-functional teams to work, they must be empowered by the managers of the functional units with decision-making authority. It is important that the team leader engender loyalty of the team members toward the product and away from the functional units from which they came. Functional units and cross-functional teams must build mutual respect and understanding for each other's needs, requirements, and responsibilities. The importance of teams in current design practice is such that Chap. 3 is devoted to an in-depth look at team behavior.

Parallel design, sometimes called simultaneous engineering, refers to each functional area implementing their aspect of the design at the earliest possible time, roughly in parallel. For example, the manufacturing process development group starts its work as soon as the shape and materials for the product are established, and the tooling development group starts its work once the manufacturing process has been selected. These groups have had input to the development of the product design specification and to the early stages of design. Of course, nearly continuous communication between the functional units and the design team is necessary in order to know what the other functional units are doing. This is decidedly different from the old practice of completely finishing a design package of drawings and specifications before transmitting it to the manufacturing department.

Vendor partnering is a form of parallel engineering where the technical expertise of the vendor for certain components is employed as an integral member of the cross-functional design team. Traditionally, vendors have been selected by a bidding process after the design has been finalized. In the concurrent engineering approach, key vendors, known for proficient technology, reliable delivery, and reasonable cost, are selected early in the design process before the parts have been designed. A strategic partnership is developed in which the vendor becomes responsible for both the design and production of parts, in return for a major portion of the business. Vendor partnering has several advantages. It reduces the amount of part design that must be done in-house, it integrates the vendor's manufacturing expertise into the design, and it ensures a degree of allegiance and cooperation that should minimize the time for receipt of parts.

1.8
COMPUTER-AIDED ENGINEERING

The advent of plentiful computing is producing a major change in the way engineering design is practiced. While engineers were one of the first professional groups to adapt the computer to their needs, the early applications chiefly were computationally intensive ones, using a high-level language like FORTRAN. The first computer applications were conducted in batch mode, with the code prepared on punch cards. Overnight turnaround was the norm. Later, remote access to computer mainframes through terminals became common, and the engineer could engage in interactive (if still slow) computation. The development of the microprocessor and the proliferation of personal computers and engineering workstations with computational power equivalent to that of a mainframe ten years ago has created a revolution in the way an engineer approaches and carries out problem solving and design. This great change has not progressed uniformly in all dimensions, but the trend is clear and sure.

The greatest impact of computer-aided engineering to date has been in engineering drawing. The automation of drafting in two dimensions has become commonplace. The ready ability to make changes and to use parts of old designs in new drawings is a great saving in time. Currently, three-dimensional modeling is becoming more prevalent, as it has become available on desktop computers. Three-dimensional solid modeling provides a complete geometric and mathematical description of the part geometry. Solid models can be sectioned to reveal interior details, or they can be readily converted into conventional two-dimensional engineering drawings. Such a model is very rich in intrinsic information so that it can be used not only for physical design but for analysis, design optimization, simulation, rapid prototyping, and manufacturing. For example, geometric three-dimensional modeling ties in nicely with the extensive use of finite-element modeling (FEM) and makes possible interactive simulations in such problems as stress analysis, fluid flow, kinematics of mechanical linkages, and numerically controlled tool-path generation for machining operations. The ultimate computer simulation is *virtual reality,* where the viewer feels a part of the graphical simulation on the computer screen.

The computer extends the designer's capabilities in several ways. First, by organizing and handling time-consuming and repetitive operations, it frees the designer to concentrate on more complex design tasks. Second, it allows the designer to analyze complex problems faster and more completely. Both of these factors make it possible to carry out more iterations of design. Finally, through a computer-based information system the designer can share more information sooner with people in the company, like manufacturing engineers, process planners, tool and die designers, and purchasing agents. The link between computer-aided design (CAD) and computer-aided manufacturing (CAM) is particularly important, and often difficult to achieve.

Concurrent engineering is greatly facilitated by the use of computer-aided engineering (CAE). A computer database in the form of a solid model that can be accessed by all parties of the design team, as in the Boeing 777 example on page 30, is an important method for making sure that everyone is "on the same page." As seen in the example, even teams on different continents can participate. More and more the Internet, with appropriate security, is being used to transmit three-dimensional solid models to tool designers, part vendors, and numerical-control programmers for manufacturing development.

While commercially available database management systems (DBMS) were developed for business users, they can be helpful in many engineering design situations. For example, a designer could ask the DBMS to list all steel beams in a structure where the static load is greater than an allowable maximum load. Commercially DBMS can be used for such purposes as keeping lists of vendors, managing literature references, keeping track of an equipment inventory, or developing a database of material properties. However, engineering database management has several important differences from a business database. Interaction with a business database usually deals with a complete database, one that contains essentially all of the data and does

---

BOEING 777

The boldest example of the use of CAD is with the Boeing 777 long-range transport. Started in fall 1990 and completed in April 1994, this was the world's first completely paperless transport design. Employing the CATIA 3-D CAD system, it linked all of Boeing's design and manufacturing groups in Washington, as well as suppliers of systems and components worldwide. At its peak, the CAD system served some 7000 workstations spread over 17 time zones.

As many as 238 design teams worked on the project at a single time. Using conventional paper design, this would be expected to lead to numerous potential interferences among hardware systems, requiring costly design changes and revised drawings. This is a major cost factor in designing a complex system. The advantage of being able to see what everyone else was doing, through an integrated solid model and digital data system, saved in excess of 50 percent of the change orders and rework expected for a design of this magnitude.

The Boeing 777 has more than 130,000 unique engineered parts, and when rivets and other fasteners are counted, there are more than 3 million individual parts. The ability of the CAD system to identify interferences eliminated the need to build a physical model (mockup) of the airplane. Nevertheless, those experienced with transport design and construction reported that the parts of the 777 fit better the first time than those of any earlier commercial airliner.

---

not undergo significant change. Engineering design does not deal with complete databases. A design database starts out nearly empty and is filled up as application programs, specifications, and other constraints generate more data. Only when the design is finished is a complete database achieved.[1]

Spreadsheet programs are useful because of their ability to quickly make multiple calculations without requiring the user to reenter all of the data. Each combination of row and column in the spreadsheet matrix is called a cell. The quantity in each cell can represent either a number entered as input or a number that the spreadsheet program calculates according to a prescribed equation. The power of the spreadsheet is due to its ability to automatically recalculate results when new inputs have been entered in some cells. This can serve as a simple optimization tool as the values of one or two variables are changed and the impact on the output is readily observed. The usefulness of a spreadsheet in cost evaluations is self-evident. Most spreadsheets contain built-in mathematical functions that permit engineering and statistical calculations. It is also possible to use them to solve problems in numerical analysis.[2]

The solution of an equation with a spreadsheet requires that the equation be set up so that the unknown term is on one side of the equal sign. In working with equations it

---

1. W. J. Rasdorf, "Computers in Mechanical Engineering," March 1987, pp. 62–69; D. N. Chorafas and S. J. Legg, "The Engineering Database," Butterworths, Boston, 1988.
2. S. Jayamaran, "Computer-Aided Problem Solving for Scientists and Engineers," McGraw-Hill, New York, 1991; B.S. Gottfried, "Spreadsheet Tools for Engineers," McGraw-Hill, New York, 1996.

often is useful to be able to solve for any variable. Therefore, a class of equation-solving programs has been developed for small computations on the personal computer. The best-known examples are TK Solver, MathCAD, and Eureka.[1] Another important set of computational tools are the symbolic languages that manipulate the symbols representing the equation. Most common are Mathematica, Maple, and MatLab.

Specialized application programs to support engineering design are appearing at a rapid rate. These include software for finite-element modeling, QFD, creativity enhancement, decision making, and statistical modeling. Useful software packages of this type will be mentioned as these topics are introduced throughout the text.

1.9
DESIGNING TO CODES AND STANDARDS

While we have often talked about design being a creative process, the fact is that much of design is not very different from what has been done in the past. There are obvious benefits in cost and time saved if the best practices are captured and made available for all to use. Designing with codes and standards has two chief aspects: (1) it makes the best practice available to everyone, thereby ensuring efficiency and safety, and (2) it promotes interchangeability and compatibility. With respect to the last point, anyone who has traveled widely in other countries will understand the compatibility problems with connecting plugs and electrical voltage and frequency, when trying to use small appliances.

A code is a collection of laws and rules that assists a government agency in meeting its obligation to protect the general welfare by preventing damage to property or injury or loss of life to persons. A standard is a generally agreed-upon set of procedures, criteria, dimensions, materials, or parts. Engineering standards may describe the dimensions and sizes of small parts like screws and bearings, the minimum properties of materials, or an agreed-upon procedure to measure fracture toughness. Standards and specifications are sometimes used interchangeably. The distinction is that standards refer to generalized situations while specifications refer to specialized situations. Codes tell the engineer what to do and when and under what circumstances to do it. Codes usually are legal requirements, as in the building code or the fire code. Standards tell the engineer how to do it and are usually regarded as recommendations that do not have the force of law. Codes often incorporate national standards into them by reference, and in this way standards become legally enforceable.

There are two broad forms of codes: performance codes and prescriptive codes. Performance codes are stated in terms of the specific requirement that is expected to be achieved. The method to achieve the result is not specified. Prescriptive or specification codes state the requirements in terms of specific details and leave no discretion to the designer. A form of code is government regulations. These are issued by agencies (federal or state) to spell out the details for implementation of vaguely written

---

1. K. R. Foster, *Science,* June 3, 1988, pp. 1353–1358.

laws. An example is the OSHA Regulations developed by the U.S. Department of
Labor to implement the Occupational Safety and Health Act (OSHA).

Design standards fall into three categories: performance, test methods, and codes
of practice. There are published *performance standards*[1] for many products such as
seat belts, lumber, and auto crash safety. Test *method standards* set forth methods for
measuring properties such as yield strength, thermal conductivity, or resistivity. Most
of these are developed for and published by the American Society for Testing and
Materials (ASTM). Another important set of testing standards for products are developed by the Underwriters Laboratories (UL). *Codes of practice* give detailed design
methods for a repetitive technical problem, such as the design of piping, heat
exchangers, and pressure vessels. Many of these are developed by the American
Society of Mechanical Engineers (ASME Boiler and Pressure Vessel Code), the
American Nuclear Society, and the Society of Automotive Engineers.

Standards are often prepared by individual companies for their own proprietary
use. They address such things as dimensions, tolerances, forms, manufacturing
processes, and finishes. In-house standards are often used by the company purchasing
department when outsourcing. The next level of standard preparation involves groups
of companies in the same industry to arrive at industry consensus standards. Often
these are sponsored through an industry trade association, such as the American
Institute of Steel Construction (AISC) or the Door and Hardware Institute. Industry
standards of this type are usually submitted to the American National Standards
Institute (ANSI) for a formal review process, approval, and publication. A similar
function is played by the International Organization for Standardization (ISO) in
Geneva, Switzerland. Another important set of standards are government (federal,
state, and local) specification standards.[2] Because the government is such a large purchaser of goods and services, it is important for the engineer to have access to these
standards. Engineers working in high-tech defense areas must be conversant with
MIL standards and handbooks in their product line. A more detailed guide to sources
of codes and standards is given in Chap. 4.

In addition to protecting the public, standards play an important role in reducing
the cost of design and of products. The use of standard components and materials
leads to cost reduction in many ways. The use of design standards saves the designer,
involved in original design work, from spending time on finding solutions to a multi-
tude of recurring identical problems. Moreover, designs based on standards provide a
firm basis for negotiation and better understanding between the buyer and seller of a
product. Failure to incorporate up-to-date standards in a design may lead to difficulties with product liability (see Chap. 15). The price that is paid with standards is that
they can limit the freedom to incorporate new technology in the design (see box on
page 33).

The engineering design process is concerned with balancing four goals: proper
function, optimum performance, adequate reliability, and low cost. The greatest cost

1. V. L. Roberts, "Products Standards Index," Pergamon Press, New York, 1986.
2. "Index of Federal Specifications and Standards," Government Printing Office, Washington, D.C.

Standards as a Limit to Technology Advancement

On balance, standards are necessary to advancement of technology, but they can be an inhibiting factor as well. Consider the ASME Boiler and Pressure Vessel Code that has been adopted by all 50 states to regulate machinery using gases or liquids operating under pressure. Formulated during the early 1900s to prevent catastrophic failures and explosions, it spells out in detail the types of material that may be used and the performance specifications a new material must meet.

The materials specifications are nearly the same as they were 50 years ago, despite the fact that much stronger, more fracture-resistant materials are now available. This is because the performance criteria are so stringent that it would take tens of millions of dollars of testing to qualify a new material. No one company can afford to underwrite such costs. But the costs of failure are so high that no one wants to risk changing the code without these tests.

---

saving comes from reusing existing parts in design. The main savings come from eliminating the need for new tooling in production and from a significant reduction in the parts that must be stocked to provide service over the lifetime of the product. In much of new product design only 20 percent of the parts are new, about 40 percent are existing parts used with minor modification, while the other 40 percent are existing parts reused without modification.

Computer-aided design has much to offer in design standardization. A 3D model represents a complete mathematical representation of a part which can be readily modified with little design labor. It is a simple task to make drawings of families of parts which are closely related.

A formal way of recognizing and exploiting similarities in design is through the use of *group technology* (GT). GT is based on similarities in geometrical shape and/or similarities in their manufacturing process. Coding and classification systems[1] are used to identify and understand part similarities. A computerized GT database makes it possible to easily and quickly retrieve designs of existing parts that are similar to the part being designed. This helps combat the tendency toward part proliferation which is encouraged by the ease of use of a CAD system. The installation of a GT system aids in uncovering duplicative designs; it is a strong driver for part standardization. GT may also be used to create standardization in part features. For example, the GT database may reveal that certain hole diameters are used frequently in a certain range of parts while others are infrequently used. By standardizing on the more frequently used design features simplifications and cost savings in tooling can be achieved. Finally the information on manufacturing costs should be fed back to the designer so that high-cost design features are avoided.

---

1. W. F. Hyde, "Improving Productivity by Classification, Coding, and Data Base Standardization," Marcel Dekker, New York, 1981.

An important aspect of standardization in CAD–CAM is in interfacing and communicating information between various computer devices and manufacturing machines. The National Institute of Standards and Technology (NIST) has been instrumental in promulgating the Initial Graphics Exchange Specification (IGES). This is being replaced by the Product Data Exchange Specification (PDES). Both of these represent a neutral data format for transferring geometric data between equipment from different vendors of CAD systems.

1.10
DESIGN REVIEW

The design review is a vital aspect of the design process. It provides an opportunity for specialists from different disciplines to interact with generalists to ask critical questions and exchange vital information. A design review is a retrospective study of the design up to that point in time. It provides a systematic method for identifying problems with the design, aids in determining possible courses of action, and initiates action to correct the problem areas.

To accomplish these objectives the review team should consist of representatives from design, manufacturing, marketing, purchasing, quality control, reliability engineering, and field service. The chairman of the review team is normally a chief engineer or project manager with broad technical background and broad knowledge of the company's products. In order to ensure freedom from bias the chairman of the design review team should not have direct responsibility for the design under review.

Depending on the size and complexity of the product, design reviews should be held from three to six times in the life of the project. The minimum review schedule consists of conceptual, interim, and final reviews. The conceptual review occurs once the conceptual design (Chap. 5) has been established. This review has the greatest impact on the design, since many of the design details are still fluid and changes can be made at this stage with least cost. The interim review occurs when the embodiment design is finalized and the product architecture, subsystems, and performance characteristics are established. It looks critically at the interfaces between the subsystems. The final review takes place at completion of the detail design and establishes whether the design is ready for transfer to manufacturing.

A more sophisticated product would require reviews to assess safety and hazard elimination (Chap. 11), manufacturing issues (Chap. 9), and a review of the performance of the prototype tests.

Each review has two aspects. The first is concerned with elements of the design itself, while the second is concerned with the business aspects of the product. The essence of the technical review of the design is to compare the findings against the detailed *product design specification* (PDS) that is formulated at the problem-definition phase of the project. The PDS is a detailed document that describes what the design must be in terms of performance requirements, the environment in which it must operate, the product life, quality, reliability, cost, and a host of other design requirements (see Sec. 2.7). The PDS is the basic reference document for both the product design and the design review. The business aspect of the review is concerned with tracking the costs incurred in the project, projecting how the design will affect the expected marketing and sales of the product, and maintaining the time schedule. An important outcome of the review is to determine what changes in resources, people, and money are required to produce the appropriate business outcome. It must be realized that a possible outcome of any review is to withdraw the resources and terminate the project.

A formal design review process requires a commitment to good documentation of what has been done, and a willingness to communicate this to all parties involved in the project. The minutes of the review meeting should clearly state what decisions were made and a list of "action items" for future work. Since the PDS is the basic control document, care must be taken to keep it always updated.

1.10.1 Redesign

A common situation is redesign. As a result of decisions made at design reviews the details of the design are changed many times as prototypes are developed and tested. There are two categories of redesigns: *fixes* and *updates*. A fix is a design modification that is required due to less than acceptable performance once the product has been introduced into the marketplace. On the other hand, updates are usually planned as part of the product's life cycle before the product is introduced to the market. An update may add capacity and improve performance to the product or improve its appearance to keep it competitive.

The most common situation in redesign is the modification of an existing product to meet new requirements. For example, the banning of the use of fluorinated hydrocarbon refrigerants because of the "ozone-hole problem" required the extensive redesign of refrigeration systems. Often redesign results from failure of the product in service. A much simpler situation is the case where one or two dimensions of a component must be changed to match some change made by the customer for that part. Yet another situation is the continuous evolution of a design to improve performance. An extreme example of this is shown in Fig. 1.10. The steel railroad wheel has been in its present design for nearly 150 years. In spite of improvements in metallurgy and the understanding of stresses, the wheels still failed at about 200 per year, often causing disastrous derailments. The chief cause of failure is thermal buildup caused by failure of a railcar's braking system. Long-term research by the Association of American Railroads has resulted in the improved design. The chief design change is that the flat plate, the web between the bore and the rim, has been replaced by an S-shaped plate. The curved shape allows the plate to act like a spring, flexing when overheated, avoiding the buildup of stresses that are transmitted through the rigid flat plates. The wheel's tread has also been redesigned to extend the rolling life of the wheel. Car wheels last for about 200,000 miles. Traditionally, when a new wheel was placed in service it lost from 30 to 40 percent of its tread and flange while it wore away to a new shape during the first 25,000 miles of service. After that the accelerated wear stopped and normal wear ensued. In the new design the curve between the

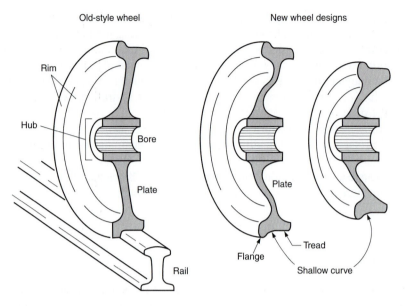

FIGURE 1.10
An example of a design improvement. Old design of railcar wheel vs. improved design.

flange and the tread has been made less concave, more like the profile of a "worn" wheel. The new wheels last for many thousand of miles longer, and the rolling resistance is lower, saving on fuel cost.

TECHNOLOGICAL INNOVATION AND THE DESIGN PROCESS

1.11
TECHNOLOGICAL INNOVATION

The advancement of technology has three phases:

Invention: The creative act whereby an idea is conceived
Innovation: The process by which an invention or idea is brought into successful practice and is utilized by the economy
Diffusion: The successive and widespread initiation of successful innovation

Without question, innovation is the most critical and most difficult of the three phases. Many studies have shown that the ability to introduce and manage technological change is a major factor in a country's leadership in world markets and also a major factor in raising the standard of living at home. Science-based innovation in the United States has spawned such key industries as aircraft, computers, plastics, and television. Relative to other nations, however, the importance of the United States' role in innovation appears to be decreasing. If the trend continues, it will affect our own well-being. Likewise, the nature of innovation has changed with time. Opportunities for the lone inventor and entrepreneur have become relatively more limited. As one indication, independent investigators obtained 82 percent of all U.S. patents in 1901, whereas the corresponding number in 1967 was 24 percent. Nevertheless, small companies do make a major contribution to innovation in this country.

The purpose of this section is to acquaint you with the innovation process and the steps in new product development. Traditionally, engineers play the major role in technological innovation, yet they often do not view themselves in that role. Engineering design and technological innovation are inseparable. It is hoped that you will view your study of design in that spirit so you can make your own strong contribution as an innovator.

The steps in a technological innovation activity can be considered to be:

This model differs from one that would have been drawn in the 1960s, which would have started with basic research in the innovation chain. The research results would have led to research ideas that in turn would have led to commercial development. Although strong basic research obviously is needed to maintain the storehouse of new knowledge and ideas, it has been well established that innovation in response to a market need has greater probability of success than innovation in response to a technological research opportunity. Market pull is far stronger than technology push when it comes to innovation.

The introduction of new products into the marketplace is like a horse race. The odds of picking a winner at the inception of an idea is about 5 or 10 to 1. The failure rate of new products that actually enter the marketplace is around 35 to 50 percent. Most of the products that fail stumble over market obstacles, such as not appreciating the time it takes for customers to accept a new product. The next generic cause is management problems, while technical problems comprise the smallest category for failure.

Studies of successful products delineate four factors that lead to success.[1]

1. *Product planning and research:* Products where adequate time was spent in problem definition (Chap. 2) and concept development (Chap. 5) achieved significantly higher success rate and profitability. These critical "up-front" activities include initial screening of the concept, preliminary market assessment, detailed market research, preliminary technical assessment, and a business and financial review—all before deciding to move to the development phase, i.e., embodiment and detail design.
2. *Product superiority:* Having a superior high-quality product that delivers real value to the customer makes all the difference between winning and losing. Such a product is superior to competing products in meeting customer needs, and these product attributes are easily perceived as being useful by the customer.

1. R. G. Cooper, "Winning at New Products," Addison-Wesley, Reading, MA, 1986; R. G. Cooper, "Research Technology Management," July-August, 1994, pp.40–50.

3. *Quality marketing:* High in importance is how well the marketing activities were executed from concept of the idea to the launch of the product in the marketplace (see Sec. 1.6).
4. *Proper organizational design:* Successful products are most often developed by a cross-functional team (Sec. 1.7), led by a strong product champion, supported by top management, and accountable for the entire project from beginning to end.

An approach to business strategy dealing with innovation and investment uses the colorful terminology advanced by the Boston Consulting Group in their portfolio management technique. Business projects are placed in one of four categories:

Star businesses: High growth potential, high market share
Wildcat businesses: High growth potential, low market share
Cash-cow businesses: Low growth potential, high market share
Dog businesses: Low growth potential, low market share

In this context, the break between high and low market share is the point at which a company's share is equal to that of its largest competitor. For a cash-cow business, cash flow should be maximized but investment in R&D and new plant should be kept to a minimum. The cash these businesses generate should be used in star and wildcat businesses. Heavy investment is required in star businesses so they can increase their market share. By pursuing this strategy, a star becomes a cash-cow business and eventually a dog business. Wildcat businesses require generous funding to move into the star category. That only a limited number of wildcats can be funded will bring about the survival of the fittest. Dog businesses receive no investment and are sold or abandoned as soon as possible. This whole approach is artificial and highly stylized, but it is a good characterization of corporate action concerning business investment. Obviously, the innovative engineer should avoid becoming associated with the dogs and cash cows; for there will be little incentive for creative work.

There are other business strategies that can have a major influence on the engineering design. A company that follows a *first in the field* strategy is usually a high-tech innovator. Some may prefer to let others pioneer and develop the market, with the strategy of being a *fast follow on* that is content to have a lower market share at the avoidance of the heavy R&D expense of the pioneer. Other companies may emphasize process development with the goal of becoming the *high-volume, low-cost producer.* Yet other companies adopt the strategy of being the supplier to a few major customers that market the product to the public.

A company with an active research program usually has more potential products than the resources to develop them into marketable products. To be considered for development a product should fill a need that is presently not adequately served, or serve a current market for which the demand exceeds the supply, or has a differential advantage over an existing product (such as better performance, improved features, or lower price). A screening matrix that can be used[1] to select the best prospects for product development is shown in Fig. 1.11 on page 40 and 41. Examine particularly the kind of business criteria that are used to make this decision. The range of expec-

---

1. R. J. Bronikowski, "Managing the Engineering Design Function," Van Nostrand Reinhold, New York, 1986.

tations for each criterion is given in the five columns, from excellent to poor. A weighting factor is applied to certain criteria. In this rating scheme a total of 16 criteria are considered, such that a perfect product would receive a rating of 100 and a poor product would score 20. Most potential products would range from 40 to 80 on this scale, with a rating of 70 typically being required for further consideration. A screening matrix like this should be completed by managers from marketing, product design, R&D, and manufacturing, each working independently.

Studies of the innovation process by Roberts[1] have identified five kinds of people who are needed for technological innovation.

Idea generator: The creative individual

Entrepreneur: The person who "carries the ball" and takes the risks

Gatekeepers: People who provide technical communication from outside to inside the organization

Program manager: The person who manages without inhibiting

Sponsor: The person who provides financial and moral support, often senior management

Roughly 70 to 80 percent of the people in a technical organization are routine problem solvers and are not involved in innovation. Therefore, it is important to be able to identify and nurture the small number who give promise of becoming technical innovators.

Innovators tend to be the people in a technical organization who are most current with technology and who have well-developed contacts with technical people outside the organization.[2] Thus, the innovators receive information directly and then diffuse it to other technical employees. Innovators tend to be predisposed to "do things differently" as contrasted with "doing things better." They are able to deal with unclear or ambiguous situations without feeling uncomfortable. That is because they tend to have a high degree of self-reliance and self-esteem. Age is not an important factor in innovation, nor is experience in an organization so long as it has been sufficient to establish credibility and social relationships. It is important for an organization to identify the true innovators and provide a management structure that helps them develop. Innovators respond well to the challenge of diverse projects and the opportunity to communicate with people of different backgrounds.

A successful innovator is a person who has a coherent picture of what needs to be done, not necessarily a detailed picture. Innovators emphasize goals, not methods of achieving the goal. They can move forward in the face of uncertainty because they do not fear failure. Many times the innovator is a person who has failed in a previous venture—and knows why. The innovator is a person who identifies what he or she needs in the way of information and resources—and gets them. The innovator aggressively overcomes obstacles—by breaking them down, or hurdling over them, or running around them. Frequently the innovator works the elements of the problem in parallel—not serially.

---

1. E. B. Roberts and H. A. Wainer, *IEEE Trans. Eng. Mgt.,* Vol. EM-18, no. 3, pp. 100–109, 1971; E. B. Roberts (ed.), "Generation of Technological Innovation," Oxford University Press, New York, 1987.
2. R. T. Keller, *Chem. Eng.,* Mar. 10, 1980, pp. 155–158.

Considerations	wt. Factor	Excellent 5
Est. gross profit	2	50% and over/yr.
Est. IROI	1	90% and over
Current annual available business	2	$10 million +
Market potential 5 yr.	1	In growth stage. Increasing sales & demand at an increasing rate
Est. market share 1 yr.	1	25% and over
Est. market share 5 yr.	2	50% and over
Stability	1	Product resistant to economic change
Degree of competition	1	No competitive products
Product leadership	1	Fills a need not currently satisfied. Is original
Customer acceptance	1	Readily accepted
Influence on other products	1	Complements and reinforces an otherwise incomplete line
Manufacturing content	1	Completely manufactured in-house
Patent position	1	Impregnable. Exclusive license or rights
Sales force qualification	1	Qualified sales force available
Time to introduction	2	Less than 6 mo.
Technical ability to develop and produce	1	Present technical know-how available and qualified

FIGURE 1.11

Screening matrix for selection of new products (*From R. J. Bronikowski, "Managing the Engineering Design Function," Van Nostrand Reinhold, New York, 1986; reprinted by permission of Carol Bronikowski.*)

Above average 4	Average 3	Below average 2	Poor 1
49–30%/yr	29–20%/yr	19–10%/yr	9% and less/yr
89–75%	74–60%	59–39%	39% and less
$10–$7.5 million	$7.5–$5.million	$5–$2.5 million	Less than $2.5 million
Reaching maturity. Increasing sales but at a decreasing rate	Turning from maturity to saturation. Leveling of sales	Declining sales & Profits	Demand, sales & and profits declining at an increasing rate
24–15%	14–10%	9–5%	4% and less
49–30%	29–20%	19–10%	9% and less
Some resistance to economic change and out of phase	Sensitive to economic change—but out of phase	Sensitive to economic change and in phase	Highly sensitive to economic change and in phase
Only slight competition from alternative	Several competitors to different extents	Many competitors	Firmly entrenched competition
Improvement over existing competition	Some individual appeal, but basically a copy	Barely distinguished from competitors	Copy with no advantages, possibly some disadvantages
Slight resistance	Moderate resistance	Appreciable customer education needed	Extensive customer education need
Easily fits current line, but not necessary	Fits current line, but may compete with it	Competes with, and may decrease sales of current line	Endangers or replaces an otherwise successful line
Partially mfg'd, assembled & packaged in-house	Assembled and packaged in-house	Packaged in-house	No manufacturing content
Some resistance to infringement, few firms with similar patents	Probably not patentable; however, product difficult to duplicate	Not patentable, can be copied	Product may infringe on other patents
Sales force has basic know-how. Minor product orientation required	Sales force has basic know-how, requires product and application education	Sales force requires extensive product and application education	Existing sales force inadequate to handle market and/or product
6-12 mo.	12-24 mo.	24-36 mo.	over 36 mo.
Most technical know-how available	Some know-how available	Extensive technical support required	Ability to develop and produce with present technology questionable

A successful technological innovation requires a good idea or concept that satisfies a societal need. It requires a business champion in the form of supportive top management willing to take the financial risk. The necessary human and technical resources must be assembled. Clear goals and objectives which include milestones for technical accomplishment, and financial and time expenditures, must be established. The final ingredient in a successful innovation is a little bit of luck.

Chief among the factors which lead to failure of a technological innovation are loss of the business champion and/or resources, or oscillation of the commitment to the project. Other factors could be the collapse of the window of opportunity and inadequate attention to technical weaknesses in the concept. Often failure to properly gauge the market is a major contributing factor in failure. Much of the success of the Japanese in world markets is attributed to their overwhelming emphasis in producing the product that is demanded by the market at the price the market is willing to pay, with constant emphasis on raising quality and lowering costs.

1.12
PRODUCT AND PROCESS CYCLES

Every product goes through a cycle from birth, into an initial growth stage, into a relatively stable period, and finally into a declining state that eventually ends in the death of the product (Fig. 1.12).

In the introductory stage the product is new and consumer acceptance is low, so sales are low. In this early stage of the product life cycle the rate of product change is rapid as management tries to maximize performance or product uniqueness in an attempt to enhance customer acceptance. When the product has entered the *growth stage,* knowledge of the product and its capabilities has reached a growing number of customers. There may be an emphasis on custom tailoring the product for slightly different customer needs. At the *maturity stage* the product is widely accepted and sales are stable and are growing at the same rate as the economy as a whole. When the product reaches this stage, attempts should be made to rejuvenate it by incremental innovation or the development of still new applications. Products in the maturity stage usually experience considerable competition. Thus, there is great emphasis on reduc-

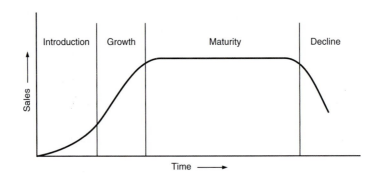

FIGURE 1.12
Product life cycle.

ing the cost of a mature product. At some point the product enters the *decline stage.* Sales decrease because a new and better product has entered the market to fulfill the same societal need.

In the product introduction phase, market uncertainty and the high cost of advanced productivity processes act as barriers to product innovation. Because product volume is low, expensive but flexible manufacturing processes are used and product cost is high. As we move into the period of product market growth, higher volume manufacturing processes reduce the unit cost. In the product maturity stage emphasis is on prolonging the life of the product by product improvement and significant reduction in unit cost. The high investment cost of advanced productivity processes becomes the barrier to further product innovation.

If we look more closely at the product life cycle, we will see that the cycle is made up of many individual processes (Fig. 1.13). In this case the cycle has been divided into the premarket and market phases. The former extends back to the idea concept and includes the research and development and marketing studies needed to bring the product to the market phase. The investment (negative profits) needed to create the product is shown along with the profit. The numbers along the profit vs. time curve correspond to the following processes in the product life cycle. This brief introduction should serve to emphasize that innovation leading to a new product is a complex, costly, and time-consuming process.

Premarket phase	Market phase
1. Idea generation	9. Product introduction
2. Idea evaluation	10. Market development
3. Feasibility analysis	11. Rapid growth
4. Technical R&D	12. Competitive market
5. Product (market) R&D	13. Maturity
6. Preliminary production	14. Decline
7. Market testing	15. Abandonment
8. Commercial production	

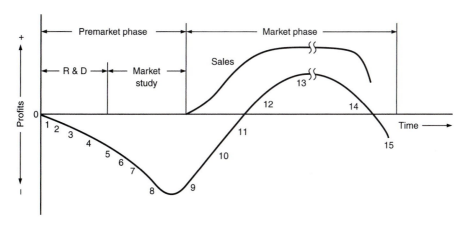

FIGURE 1.13
Expanded product life cycle.

1.12.1 Technology Development Cycle

The development of a new technology follows an S-shaped curve (Fig. 1.14*a*). In its early stage, progress is limited by the lack of ideas. A single good idea can make several other good ideas possible, and the rate of progress becomes exponential. During this period a single individual or a small group of individuals can have a pronounced effect on the direction of the technology. Gradually the growth becomes linear when the fundamental ideas are in place and progress is concerned with filling in the gaps between the key ideas. This is the period when commercial exploitation flourishes. Specific designs, market applications, and manufacturing develop rapidly in a field that has not yet settled down. Smaller entrepreneurial firms can have a large impact and capture a dominant share of the market. However, with time the technology begins to run dry and increased improvements come with greater difficulty. Now the market tends to become stabilized, manufacturing methods become fixed in place, and more capital is expended to reduce the cost of manufacturing. The business becomes capital-intensive; the emphasis is on production know-how and financial expertise rather than scientific and technological expertise. The maturing technology grows slowly, and it approaches a limit asymptotically. The limit may be set by a social consideration, such as the fact that the legal speed of automobiles is set by safety and fuel economy considerations, or it may be a true technological limit, such as the fact the speed of sound defines an upper limit for the speed of a propeller-driven aircraft.

The success of a technology-based company lies in recognizing when the core technology on which the company's products are based is beginning to mature and, through an active R&D program, transferring to another technology growth curve that offers greater possibilities (Fig. 1.14*b*). To do so, the company must manage across a *technological discontinuity*. Past examples of technological discontinuity are the change from vacuum tubes to transistors and from the three- to the two-piece metal can. Changing from one technology to another may be difficult because it requires different kinds of technical skill. Technology usually begins to mature before profits,

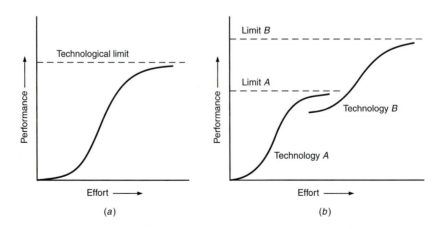

FIGURE 1.14
(*a*) Simple technology development cycle. (*b*) Transfer from one technology growth curve
(A) to another developing technology (B).

so that there is a management reluctance to switch to a new technology when business is going so well.

Occasionally a major scientific discovery will open up new opportunities for great advances in performance and reduction in cost. The microcomputer is such an example. However, more frequently small, almost imperceptible improvements will add up to equally great progress. These improvements occur through changes in operating procedures, materials, small variations in manufacturing processes, redesign of products for easier production, or substitution of less expensive components for those used in earlier design. Thus, a great deal of technological innovation is made by incremental advances in cost, performance, and quality improvements.

One can generalize that design and development can be broadly divided into idea dominated design and incremental dominated design. In recent years when the United States has not been competitive in an area of technology we have lost, usually not to radical new technology, but to better incremental improvements. Incremental development is also cyclical. When the current version of the product is in production a development team is working on the next product generation. And, when that next generation goes into production the following generation is started through the development and manufacturing cycle to build a significant product lead and achieve technological leadership.[1]

In highly competitive consumer industries, like personal computers, product cycle time is becoming recognized as a key factor for success.[2] Important in reducing cycle time is a close link between product development and manufacturing. Close ties between design and manufacturing result in early knowledge of technical problems, which when overcome lead to speedy market introduction and higher quality, because the product is easier to manufacture. Because of this cycle, there is a right and a wrong time to introduce new ideas. An idea must be produced at the beginning of the cycle; halfway through is too late because it would cause many changes and delay. Thus, in a technological development there is usually a *window of opportunity* that needs to be recognized. The window of opportunity can have an important pacing function on the development of technology. In areas like consumer electronics, where the development cycle is short, new ideas can be implemented at frequent intervals and the technology develops rapidly. However, in military aircraft, where the cycle time is of the order of 15 years the technology can only advance by large increments. Many of the problems associated with these systems are due to the difficulty of applying incremental developments.

1.12.2 Process Development Cycle

Three stages can be identified[3] in the development of a manufacturing process.

1. *Uncoordinated development:* The process is composed of general-purpose equipment with a high degree of flexibility. Since the product is new and is developing, the process must be kept fluid.

---

1. R. E. Gomory and R. W. Schmitt, *Science,* vol. 240, pp. 1131–1132, 1203–1204, May 27, 1988.
2. S. C. Wheelwright and K. B. Clark, "Revolutionizing Product Development," The Free Press, New York, 1992.
3. E. C. Etienne, *Research Management,* vol. 24, no. 1, pp. 22–27, 1981.

2. *Segmental:* The manufacturing system is designed to achieve higher levels of efficiency in order to take advantage of increasing product standardization. This results in a high level of automation and process control. Some elements of the process are highly integrated; others are still loose and flexible.
3. *Systemic:* The product has reached such a high level of standardization that every process step can be described precisely. Now that there is a high degree of predictability in the product, a very specialized and integrated process can be developed.

Process innovation is emphasized during the maturity stage of the product life cycle. In the earlier stages the major emphasis is on product development, and generally only enough process development is done to support the product. However, when the process development reaches the systemic stage, change is disruptive and costly. Thus, process innovations will be justified only if they offer large economic advantage.

We also need to recognize that process development often is an enabler of new products. Typically, the role of process development is to reduce cost so that a product becomes more competitive in the market. However, revolutionary processes can lead to remarkable products. An outstanding example is the creation of microelectromechanical systems (MEMS) by adapting the fabrication methods from integrated circuits.

1.12.3 Production and Consumption Cycle

The life cycle of production and consumption that is characteristic of all products is illustrated by the *materials cycle* shown in Fig. 1.15. This starts with the mining of a mineral or the drilling for oil or the harvesting of an agricultural fiber such as cotton. These raw materials must be processed to extract or refine bulk material (e.g., an aluminum ingot) that is further processed into a finished engineering material (e.g., an aluminum sheet). At this stage an engineer designs a product that is manufactured from the material, and the part is put into service. Eventually the part wears out or becomes obsolete because a better product comes on the market. At this stage, one option is to junk the part and dispose of it in some way that eventually returns the material to the earth. However, society is becoming increasingly concerned with the depletion of natural resources and the haphazard disposal of solid materials. Thus, we look for economical ways to recycle waste materials (e.g., aluminum beverage cans).

1.13
SOCIETAL CONSIDERATIONS IN ENGINEERING

The first fundamental canon of the ABET Code of Ethics states that "engineers shall hold paramount the safety, health, and welfare of the public in the performance of their profession." A similar statement has been in engineering codes of ethics since the early 1920s, yet there is no question that what society perceives to be proper treatment by the profession has changed greatly in the intervening time. Today's mass communications make the general public, in a matter of hours, aware of events taking place anywhere in the world. That, coupled with a generally much higher standard of

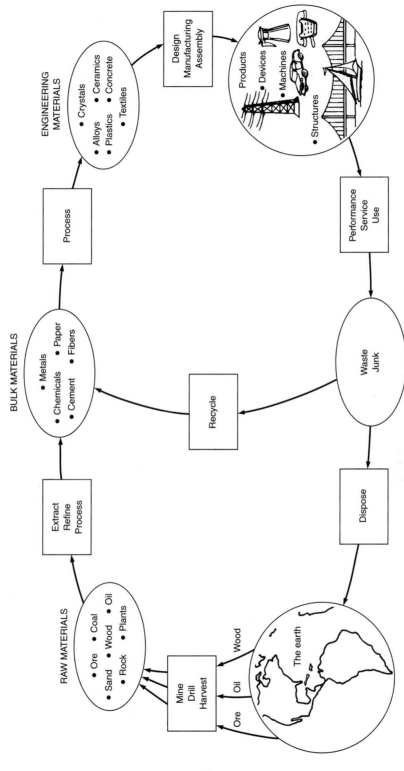

FIGURE 1.15
The total materials cycle. (*Reproduced from "Materials and Man's Needs," National Academy of Sciences, Washington, D.C., 1974.*)

47

education and standard of living, has led to the development of a society that has high expectations, reacts to achieve change, and organizes to protest perceived wrongs. At the same time, technology has had major effects on the everyday life of the average citizen. Whether we like it or not, all of us are intertwined in complex technological systems: an electric power grid, a national network of air traffic controllers, and a gasoline distribution network. Much of what we use to provide the creature comforts in everyday life has become too technologically complex or too physically large for the average citizen to comprehend.

Thus, in response to real or imagined ills, society has developed mechanisms for countering some of the ills and/or slowing down the rate of social change. The major social forces that have had an important impact on the practice of engineering are occupational safety and health, consumer rights, environmental protection, the anti-nuclear movement, and the freedom of information and public disclosure movement. The result of those social forces has been a great increase in federal regulations (in the interest of protecting the public) over many aspects of commerce and business and/or a drastic change in the economic payoff for new technologically oriented ventures. Those new factors have had a profound effect on the practice of engineering and the rate of innovation.

The following are some general ways in which increased societal awareness of technology, and subsequent regulation, has influenced the practice of engineering:

- Greater influence of lawyers on engineering decisions
- More time spent in planning and predicting the future effects of engineering projects
- Increased emphasis on "defensive research and development," which is designed to protect the corporation against possible litigation
- Increased effort expended in research, development, and engineering in environmental control and safety—areas that generally do not directly enhance corporate profit but can affect profits in a negative way because of government regulation

Clearly, the societal pressures described above have placed much greater constraints on how engineers can carry out their designs. Moreover, the increasing litigiousness of U.S. society requires a greater awareness of legal and ethical issues on the part of each engineer (see Chap. 15).

One of the most prevalent societal pressures at the present time is the environmental movement. Originally, governmental regulation was used to clean up rivers and streams, to ameliorate smog conditions, and to reduce the volume of solid waste that is sent to landfills. Today, there is a growing realization that placing environmental issues at a high priority (not doing them because the government demands it) represents smart business.[1] One major oil producer publicly takes seriously the link between carbon dioxide emissions and rising global temperatures caused by the burning of oil and has embarked on a major effort to become the leader in renewable energy sources like solar power. A major chemical company has given major emphasis to developing environmentally friendly products. Its biodegradable herbicides allow for a hundredfold reduction in the herbicide that must be applied per acre, greatly reducing toxic runoff into streams. This reorientation of business thinking

1. *Business Week,* Nov. 10, 1997, pp. 98–106.

toward environmental issues is often called *sustainable development,* businesses built on renewable materials and fuels.

The change in thinking, from fixing environmental problems at the discharge end of the pipe or smokestack to sustainable development, places engineering design at the heart of the issue. Environmental issues are given higher priority in design. Products must be designed to make them easier to reuse, recycle, or incinerate—a concept often called *green design.*[1] Green design also involves the detailed understanding of the environmental impact of products and processes over their entire life cycle. For example, life-cycle analysis would be used to determine whether paper or plastic grocery bags are more environmentally benign. Table 1.1 gives the chief aspects of an environmentally responsible design.

It seems clear that the future is likely to involve more technology, not less, so that engineers will face demands for innovation and design of technical systems of unprecedented complexity. While many of these challenges will arise from the requirement to translate new scientific knowledge into hardware, many of these challenges will stem from the need to solve problems in "socialware." By the term socialware is meant the patterns of organization and management instructions necessary to effective functioning of hardware.[2] Such designs will have to deal not only with the limits of hardware, but also with the vulnerability of any system to human ignorance, human error, avarice, and hubris. A good example of this point is the delivery system for civilian air transportation. While the engineer might think of the modern jet transport, with all of its complexity and high technology, as the main focus of concern, such a marvellous piece of hardware only satisfies the needs of society when embedded in an intricate system that includes airports, maintenance facilities, traffic controllers, navigation aids, baggage handling, fuel supply, meal service, bomb detection, air crew training, and weather monitoring. It is important to realize that almost all of these socialware functions are driven by federal or local rules and regulations. Thus, it should be clear that the engineering profession is required to deal with much more

TABLE 1.1
Characteristics of an Environmentally Responsible Design

- Easy to dissassemble
- Able to be recycled (see Sec. 8.14)
- Contains recycled materials
- Uses identifiable and recyclable plastics
- Reduces use of energy and natural materials in its manufacture
- Manufactured without producing hazardous waste
- Avoids use of hazardous materials
- Reduces product chemical emissions
- Reduces product energy consumption

1. Office of Technology Assessment, "Green Products by Design: Choices for a Cleaner Environment," OTA-E-541, Government Printing Office, Washington, DC, 1992.
2. E. Wenk, Jr., *Engineering Education,* November 1988, pp. 99–102.

than technology. Techniques for dealing with the complexity of large systems have been developed in the discipline of *systems engineering*.

Another area where the interaction between technical and human networks is becoming stronger is in consideration of risk, reliability, and safety (see Chap. 11). No longer can safety factors simply be looked up in codes or standards. Engineers must recognize that design requirements depend on public policy as much as industry performance requirements. This is an area of design where government influence has become much stronger.

There are five key roles of government in interacting with technology:[1]

- As a stimulus to free enterprise through manipulation of the tax system
- By influencing interest rates and supply of venture capital through changes in fiscal policy to control growth of the economy
- As a major customer for high technology
- As a funding source (patron) for research and development
- As a regulator of technology

Wenk[2] has expanded on the future interactions between engineering and society. The major conclusions of this study are summarized in Table 1.2.

Because of the growing importance of technology to society, a methodology for systematically determining the impact of technology on the social, political, economic, and physical environment is being evolved. It is called technology assessment (TA).[3] Technology assessment is an attempt to determine the benefits and risk inherent in the range of technological alternatives. Its practitioners try to provide an early-warning system for environmental mishaps, define the necessary monitoring and surveillance mechanisms, and provide the decision-making tools for setting technological priorities and allocating resources. Technology assessment, although still an evolving science, already has a number of characteristics that differentiate it from the more traditional methods of engineering analysis.

1. Technology assessment (TA) is mostly concerned with the second-, third-, and higher-order effects or impacts that are rarely considered in engineering analysis. Remote impacts often can be more important than the intended primary variable in social issues.
2. TA considers the needs of a wide range of constituencies.
3. TA is interdisciplinary. There is a need to be able to integrate different intellectual traditions and diverse methods of treating data.
4. TA probably is more closely related to policymaking than to technical problem solving.

Engineering is concerned with problems whose solution is needed and/or desired by society. The purpose of this section was to reinforce that point, and hopefully to show the engineering student how important a broad knowledge of economics and social science is to modern engineering practice.

---

1. E. Wenk, Jr., op. cit.
2. E. Wenk, Jr., "Tradeoffs: Imperatives of Choice in a High-Tech World," The Johns Hopkins University Press, Baltimore, 1986.
3. M. A. Borough, K. Chen, and A. N. Christakis, "Technology Assessment: Creative Futures," North Holland Publishing Co., New York, 1980.

TABLE 1.2
Future trends in interaction of engineering with society

- The future will entail more technology, not less.

- Because all technologies generate side effects, designers of technological delivery systems will be challenged to prevent, or at least mitigate, adverse consequences.

- The capacity to innovate, manage information, and nourish knowledge as a resource will dominate the economic domain as natural resources, capital, and labor once did. This places a high premium on the talent to design not simply hardware, but entire technological delivery systems.

- Cultural preferences and shifts will have more to do with technological choice than elegance, novelty, or virtuosity of the hardware.

- Acting as an organizing force, technology will promote concentration of power and wealth, and tendencies to large, monopolistic enterprises.

- The modern state will increasingly define the political space for technological choice, with trends becoming more pronounced toward the "corporate state." The political-military-industrial complex represents a small-scale model of such evolution.

- Distribution of benefits in society will not be uniform, so disparity will grow between the "haves" and the "have nots."

- Conflicts between winners and losers will become more strenuous as we enter an age of scarcity, global economic competition, higher energy costs, increasing populations, associated political instabilities, and larger-scale threats to human health and the environment.

- Because of technology, we may be moving to "one world," with people, capital, commodities, information, culture, and pollution freely crossing borders. But as economic, social, cultural, and environmental boundaries dissolve, political boundaries will be stubbornly defended. The United States will sense major economic and geopolitical challenges to its position of world leadership in technology.

- Complexity of technological delivery systems will increase, as will interdependencies, requiring management with a capacity for holistic and lateral conceptual thinking for both systems planning and trouble-free, safe operations.

- Decision-making will become more difficult because of increases in the number and diversity of interconnected organizations and their separate motivations, disruptions in historical behavior, and the unpredictability of human institutions.

- Mass media will play an ever more significant role in illuminating controversy and publicizing technological dilemmas, especially where loss of life may be involved. Since only the mass media can keep everyone in the system informed, a special responsibility falls on the "fourth estate" for both objective and courageous inquiry and reporting.

- Amidst this complexity and the apparent domination of decision-making by experts and the commercial or political elite, the general public is likely to feel more vulnerable and impotent. Public interest lobbies will demand to know what is being planned that may affect people's lives and environment, to have estimates of a wide range of impacts, to weigh alternatives, and to have the opportunity to intervene through legitimate processes.

- Given the critical choices ahead, greater emphasis will be placed on moral vision and the exercise of ethical standards in delivering technology to produce socially satisfactory results. Accountability will be demanded more zealously.

From E. Wenk, Jr., "Tradeoffs," Johns Hopkins University Press, 1986. Reprinted with permission from *Engineering Education,* November 1988, p. 101.

1.14
SUMMARY

Engineering design is a challenging activity because it deals with largely unstructured problems that are important to the needs of society. An engineering design creates something that did not exist before, requires choices between many variables and parameters, and often requires balancing multiple and sometimes conflicting requirements. Product design has been identified as the real key to world-competitive business.

The steps in the design process are:

Phase I: Conceptual design
- Recognition of a need
- Definition of the problem
- Gathering of information
- Developing a design concept
- Choosing between competing concepts (evaluation)

Phase II: Embodiment design
- Product architecture—arrangement of the physical functions
- Configuration design—preliminary selection of materials, modeling and sizing of parts
- Parametric design—creating a robust design, and selection of final dimensions and tolerances

Phase III: Detail design—creation of final drawings and specifications

While many consider that the engineering design process ends with detail design, there are many issues that must be resolved before a product can be shipped to the customer. These additional phases of design are often folded into what is called the product development process.

Phase IV: Planning for manufacture—design of tooling and fixtures, designing the process sheet and the production line, planning the work schedules, the quality assurance system, and the system of information flow.
Phase V: Planning for distribution—planning for packaging, shipping, warehousing, and distribution of the product to the customer.
Phase VI: Planning for use—the decisions made in phases I through III will determine such important factors as ease of use, ease of maintenance, reliability, product safety, aesthetic appeal, economy of operation, and product durability.
Phase VII: Planning for product retirement—again, decisions made in phases I through III must provide for safe disposal of the product when it reaches its useful life, or recycling of its materials or reuse or remanufacture.

Engineering design must consider many factors, which are documented in the product design specification (PDS). Among the most important of these factors are required functions with associated performance characteristics, environment in which it must operate, target product cost, service life, provisions for maintenance and logistics, aesthetics, expected market and quantity to be produced, man-machine interface requirements (ergonomics), quality and reliability, safety and environmental concerns, and provision for testing.

BIBLIOGRAPHY

Dixon, J. R., and C. Poli: "Engineering Design and Design for Manufacturing," Field Stone Publishers, Conway, MA, 1995.

Ertas, A., and J. C. Jones: "The Engineering Design Process," 2d ed., Wiley, New York, 1996.

Hales, C.: "Managing Engineering Design," Longman Scientific, Essex, England, 1993.

Magrab, E. B.: "Integrated Product and Process Design and Development," CRC Press, Boca Raton, FL, 1997.

Pahl, G., and W. Beitz: "Engineering Design," 2d ed., Springer-Verlag, New York, 1996.

Pugh, S.: "Total Design," Addison-Wesley, Reading, MA, 1991.

Ullman, D. G.: "The Mechanical Design Process," 2d ed., McGraw-Hill, New York, 1997.

Ulrich, K. T., and S. D. Eppinger: "Product Design and Development," McGraw-Hill, New York, 1995.

PROBLEMS AND EXERCISES

1.1. A major manufacturer of snowmobiles needed to find new products in order to keep the workforce employed all year round. Starting with what you know or can find out about snowmobiles, make reasonable assumptions about the capabilities of the company. Then develop a needs analysis that leads to some suggestions for new products that the company could make and sell. Give the strengths and weaknesses of your suggestions.

1.2. Take a problem from one of your engineering science classes, and add and subtract those things that would frame it more as an engineering design problem.

1.3. There is a need in underdeveloped countries for building materials. One approach is to make building blocks (4 by 6 by 12 in) from highly compacted soil. Your assignment is to design a block-making machine with the capacity for producing 600 blocks per day at a capital cost of less than $300. Develop a needs analysis, a definitive problem statement, and a plan for the information that will be needed to complete the design.

1.4. The need for material conservation and decreased cost has increased the desirability of corrosion-resistant coatings on steel. Develop several design concepts for producing 12-in-wide low-carbon-steel sheet that is coated on one side with a thin layer, e.g., 0.001 in, of nickel.

1.5. The support of thin steel strip on a cushion of air introduces exciting prospects for the processing and handling of coated steel strip. Develop a feasibility analysis for the concept.

1.6. The steel wheel for a freight car has three basic functions: (1) to act as a brake drum, (2) to support the weight of the car and its cargo, and (3) to guide the freight car on the rails. Freight car wheels are produced by either casting or rotary forging. They are subject to complex conditions of dynamic thermal and mechanical stresses. Safety is of great importance, since derailment can cause loss of life and property. Develop a broad systems approach to the design of an improved cast-steel car wheel.

1.7. Consider the design of aluminum bicycle frames. A prototype model failed in fatigue after 1600 km of riding, whereas most steel frames can be ridden for over 60,000 km. Describe a design program that will solve this problem.

1.8. Discuss the spectrum of engineering job functions with regard to such factors as (*a*) need for advanced education, (*b*) intellectual challenge and satisfaction, (*c*) financial reward, (*d*) opportunity for career advancement, and (*e*) people vs. "thing" orientation.

1.9. Strong performance in your engineering discipline ordinarily is one necessary condition for becoming a successful engineering manager. What other conditions are there?

1.10. Discuss the pros and cons of continuing your education for an MS in an engineering discipline or an MBA on your projected career progression.

1.11. Discuss in some detail the relative roles of the project manager and the functional manager in the matrix type of organization.

1.12. List the factors that are important in developing a new technologically oriented product.

1.13. List the key steps in the technology transfer (diffusion) process. What are some of the factors that make technology transfer difficult? What are the forms in which information can be transferred?

1.14. (*a*) Discuss the societal impact of a major national program to develop synthetic fuel (liquid and gaseous) from coal. (It has been estimated that to reach the level of supply equal to the imports from OPEC countries would require over 50 installations, each costing several billion dollars.)

(*b*) Do you feel there is a basic difference in the perception by society of the impact of a synthetic fuel program compared with the impact of nuclear energy? Why?

(*c*) The reason synthetic fuel from coal has not yet become a developed technology is that the cost still exceeds that of comparable natural fuel. What are some of the alternatives to synthetic fuel that may solve our nation's long-term energy problem?

2

NEED IDENTIFICATION AND
PROBLEM DEFINITION

2.1
INTRODUCTION

Of all the steps in the engineering design process, problem definition is the most important. Understanding the problem thoroughly at the beginning aids immeasurably in reaching an outstanding solution. Of course, this axiom holds for all kinds of problem solving, whether it be math problems, production problems, or design problems. However, in product design, where the ultimate test is whether the product sells well in the marketplace, it is vital to work hard to understand and provide what it is that the customer wants. This chapter gives a special emphasis to that aspect of problem definition, an approach not always taken in engineering design.

The information in this chapter draws heavily on developments within the total quality management (TQM) movement, where customer satisfaction is given major emphasis. The TQM tool of *quality function deployment* (QFD) will be presented in detail. The chapter ends with in-depth discussion of the *product design specification* (PDS), which serves as the governing documentation for the product design.

2.2
BEFORE THE PROBLEM-DEFINITION STEP

We start the product design process with problem definition. This is to emphasize the importance of this step, and because it is the logical first step in many situations. Suppose your boss says that your next assignment is to redesign the turboencabulator unit in the M35 system so as to improve thermal efficiency by 20 percent. The assignment would be clear to you, and you would set about defining the problem according to the methods described in this chapter. Or, your design team may have responded to a request for proposal (RFP) from a government agency, and having won the contract, the specifications of what is required are well spelled out. Again, the problem-definition step is the place where the design process begins. However, not all design tasks are as well defined. Often there is development work that precedes the point where problem definition starts. We might call this *planning for the design process* or *new product business development*.

To put this into context, review Sec. 1.11, Technological Innovation, Sec. 1.12, Product and Process Cycles, and Sec. 1.6, Marketing. The chief emphasis is to assemble enough information to decide whether the venture is a good investment for the company, and to decide what time to market and level of resources are required. The documentation might range from a one-page memorandum describing a simple product change to a business plan of several hundred pages giving details on such things as the business objectives, a product description and available technology base, the competition, expected volume of sales, marketing strategy, capital requirements, development cost and time, expected profit over time, and return to the shareholders.

Design projects commonly fall into one of five types:[1]

- *Variation of an existing product:* This involves the change of at most a few parameters, such as the power of a motor or the design of a fastening bracket. The required level of technical expertise is very modest.
- *Improvement of an existing product:* This, more major redesign, can be brought about by the need to improve performance or update features because of competition, the requirement to improve quality or cost in manufacturing, the failure of a vendor to be able to supply specified materials or components, or the development of new technology that allows for an improved product.
- *Development of a new product for a low-volume production run:* Many products are made only a few times in volumes less than 100,000 total units. The prospect of low-quantity production, as opposed to mass production, constrains the selection of manufacturing processes to those with cheaper tooling costs. There is more emphasis on buying off-the-shelf components than in designing special items. Often the first item produced is shipped to the customer; so it is the prototype.
- *Development of a new product for mass production:* This is the category of the automobile, major appliance, or top-of-the-line PC. These design projects allow an engineer flexibility in selecting materials and manufacturing processes, but they require careful planning for manufacture and assembly.
- *One-of-a-kind design:* These design projects can vary from a quick, simple design, using a minimum of analysis, as in the design of a welding fixture to hold parts in assembly, to a large expensive system like a 100-MW steam turbine. In the latter situation, because of the cost and complexity of the product, the testing of prototypes is not very affordable and we must learn as much as possible from analysis and from field experience. Design evolution in situations like this is likely to be incremental. We note that the design of most large buildings, process plants, and power plants is one-of-a-kind design.

Wheelwright and Clark[2] present a model for the product development process that would be applicable for a large technology-oriented company interested in bringing a steady flow of new products quickly to market. They visualize a widemouth

---

1. D. G. Ullman, "The Mechanical Design Process," 2d ed., McGraw-Hill, New York, 1997, pp. 78–79.
2. S. C. Wheelwright and K. B. Clark, "Revolutionizing Product Development," The Free Press, New York, 1992, pp. 124–127.

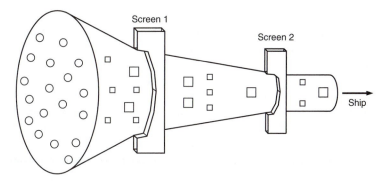

FIGURE 2.1
Wheelwright-Clark model of product development process. (*Reprinted with the permission of The Free Press a Division of Simon and Schuster, Inc., from "Revolutionizing Product Development" by Steven C. Wheelwright and Kim B. Clark, copyright 1992.*)

development funnel (Fig. 2.1). In the first part of the process there is a highly stimulated activity to gather new product ideas from a wide range of sources. These include the R&D division of the corporation but also include employees anywhere in the company, as well as customers, competitors, and suppliers. The first screen, where the funnel narrows down, is not a go-no-go decision point but instead is a review by midlevel managers to determine what additional information is needed before a go-no-go decision can be made at screen 2. When ideas are reviewed at screen 1, they should be checked for their fit with the technology and product market strategies of the company, and their likely commitment of resources. If an idea passes this screen, it passes into the area where project bounds are detailed and required knowledge is specified. If the idea is not ready to move on, then the specific tasks needed to pass screen 1 are agreed upon and assignments and a schedule are made for completing them. A second important function of screen 1 is that it begins to identify competing concepts and ideas that might be integrated into platform development projects. *Platform projects* are development projects that establish the basic product architecture that leads to a succession of follow-on derivative products.

Screen 2 is a go-no-go review in which senior management selects the product and process options that will become product development projects. Any project that passes screen 2 will be funded and staffed with every expectation that it will be carried through to introduction in the marketplace. The time between screens 1 and 2 is usually only 1 to 2 months and is spent taking the data and information developed for Screen 1 and putting it into a form that will enable senior management to choose among a set of competing projects. This process provides much important input to the project problem definition.

2.3
IDENTIFYING CUSTOMER NEEDS

Increasing worldwide competitiveness brings greater focus on the customer's wishes. Engineers and businesspeople are seeking answers to such questions as: Who are my customers? What does the customer want? and How can I provide it?

Webster defines a customer as "one that purchases a product or service." This is the definition of the customer that most people have in mind, the *external customer.* These are the people or organizations that buy what the company sells. Another way to state this is that these are the groups that can decide whether or not to pay their money. All others are constituencies. From a total quality management viewpoint, the definition of customer can be broadened to "anyone who receives or uses what an individual or organization provides." For example, the design engineer who receives information on the properties of three potential materials for his or her design is an *internal customer* of the materials specialist. From the viewpoint of developing the product design specification for an improved product, we focus on external customers, such as end users. From the viewpoint of completing the myriad of decisions and details that make up a successful product design, we must deal with a host of internal customers. Such internal customers as corporate management, manufacturing personnel, the sales staff, and field service personnel must be considered.

2.3.1 Gathering Information from Customers

It is the customer's desires that ordinarily drive the development of the product, not the engineer's vision of what the customer should want.[1] Information on the customer's needs is obtained through a variety of channels:[2]

- *Interviews with customers:* An active sales force should be continuously meeting with present and potential customers. Some corporations have account teams whose responsibility is to visit key customer accounts to probe for problem areas and to cultivate and maintain friendly contact. They should feed back information on current product strengths and weaknesses that will be helpful in product upgrades. An even better approach is for the design team to interview single customers in the service environment where the product will be used. Key questions to ask are: What do you like or dislike about this product? What factors do you consider when purchasing this product? What improvements would you make to this product?
- *Focus groups:* A focus group is an arranged discussion with 6 to 12 customers or potential customers of a product. Usually, the meeting is led by a facilitator who uses prepared questions to guide the discussion about the merits and disadvantages of the product. Often the focus group is held in a room with a one-way window that provides for videotaping of the discussion. In both the interviews and the focus groups it is important to record the customer's response in his or her own words. Any interpretation is held until the analysis of results. The facilitator should not hold slavishly to the prepared questions but should follow up on any surprise

---

1. Of course, in some instances new technology makes possible designs and products for which the customer has no conception. These are situations that the engineer delights in, but they are the exception, not the rule.
2. K. T. Ulrich and S. D. Eppinger, "Product Design and Development," McGraw-Hill, New York, 1995, pp. 33–51.

What Is Our Real Business?

Often it takes some time for a company to realize what business it should be in. Consider the case of the Black & Decker Corp. In 1984, Black & Decker purchased the General Electric Co.'s line of small appliances, including toasters, steam irons, and coffee makers. This was a move toward diversification, at a time when the company's sales of hand power tools appeared stagnated and severely threatened by overseas competition.

But small appliances is a low profit margin business, especially when compared with its DeWalt brand of high-end power tools, launched in 1992. As a result, Black & Decker announced in early 1998 that it would sell most of its household products division and concentrate on its high-end power tool business. What had happened?

This is a clear example of market stratification and brand identification. The identification of Black & Decker with small appliances created an unfavorable image in the minds of the high-end customers of power tools, carpenters and contractors. They reasoned, if B & D makes toasters and popcorn poppers then their tools must be "wimpy." It was only when Black & Decker created the line of high-quality DeWalt tools, with their distinctive bright yellow color, that major inroads began to be made in this high-end market segment, where profit margins average 30 percent compared with 5 percent in appliances. Today the DeWalt brand enjoys 45 percent of the U.S. market for professional power tools. This shift in business strategy goes hand in hand with other market shifts, where more homeowners and other amateur crafters are trading up to the professional level of power tools.

---

answers in an attempt to uncover latent needs of which the customer is not consciously aware.

- *Customer surveys:* A written questionnaire is best used for gaining opinions about the redesign of existing products or new products that are well understood by the public. Innovative new products are better explored with interviews or focus groups. Other common reasons for conducting a survey are to identify or prioritize problems and to assess whether an implemented solution to a problem was successful. A survey can be administered by mail, over the telephone, or in face-to-face interviews. One type of question is to list the features of the product and ask the customer to rank order them according to their preferences. If there are more than six or eight features, this is hard to do, and it is better to ask the customer to compare each feature, one by one, i.e., pairwise comparison. There should always be several open-ended questions to allow for unanticipated responses. It is important to give the survey a pilot run to check for ambiguities and misunderstanding in the questions before the survey is distributed.
- *Customer complaints:* A sure way to learn about needs for product improvement is from customer complaints that are received either through returned product, from dealer input or input from service centers, or with a customer telephone or e-mail hot line. Statistics on warranty claims can pinpoint design defects.

2.3.2 Constructing a Survey Instrument

Regardless of the method used to gain information from customers, considerable thought needs to go into developing the survey instrument.[1] The following steps should be followed.

1. Determine the survey purpose. Write a short paragraph stating the purpose of the survey and what will be done with the results. Be clear about who will use the results and whether they want hard statistics from surveys or would prefer more anecdotal results from focus groups.
2. Determine the type of data-collection method to be used.
3. Identify what specific information is needed. Each question should have a clear goal. Write this down so you are clear about what you are trying to learn. You should have no more questions than the absolute minimum you need to learn what you need to learn.
4. Design the questions. Each question should be unbiased, unambiguous, clear, and brief. There are three categories of questions: (1) attitude questions—how the customers feel or think about something; (2) knowledge questions—questions asked to determine whether the customer knows the specifics about a product or service; and (3) behavior questions—usually contain phrases like "how often," "how much," or "when." Some general rules to follow in writing questions are:

   Do not use jargon or sophisticated vocabulary.
   Focus very precisely. Every question should focus directly on one specific topic.
   Use simple sentences. Two or more simple sentences are preferable to one compound sentence.
   Do not lead the customer toward the answer you want.
   Avoid questions with double negatives because they may create misunderstanding.
   Always include the choice of Other _____. This ensures that the list of choices is inclusive.
   Always include one open-ended question. Open-ended questions can reveal insights and nuances, and tell you things you would never think to ask.

   Questions can have the following types of answers:

   yes—no—don't know
   strongly agree—mildly agree—neutral—mildly disagree—strongly disagree. (On a 1–5 scale such as this, always set up the numerical scale so that a high number means a good answer.)
   rank order—list in descending order of preference
   unordered choices—choose (b) over (d) or (b) from a, b, c, d, e.

   The number of questions should be such that they can be answered in 30 minutes.

---

1. P. Salant and D. A. Dillman, "How to Conduct Your Own Survey," Wiley, New York, 1994; R. B. Frary, "A Brief Guide to Questionnaire Development," **http://www.ericae.net//ft/tamu/vpiques3.htm**.

Design the printed survey form so that tabulating and analyzing data will be easy. Be sure to include instructions for completing and returning it.

5. Arrange the order of questions so that they provide context to what you are trying to learn from the customer. Group the questions by topic, and start with easy ones.
6. Pretest the survey. Before distributing the survey to the customer, always pretest it on a smaller sample. This will tell you whether any of the questions are poorly worded so that they may be misunderstood, whether the rating scales are adequate, and whether the questionnaire is too long.
7. Administer the survey. Key issues in administering the survey are whether the people surveyed constitute a representative sample for fulfilling the purpose of the survey, and what size sample must be used to achieve statistically significant results. Answering these questions requires special expertise and experience. Consultants in the area of marketing should be used for really important situations.

EXAMPLE. A student design team[1] selected the familiar "jewel case" that protects a compact disc in storage as a product needing improvement. As a first step the team brainstormed (see Sec. 3.6) to develop ideas for possible improvements to the CD case (Fig. 2.2). The following ideas were generated in response to the question: What functions or attributes of a CD case need improvement?

1. Case more resistant to cracking
2. Easier to open
3. Add color
4. Better waterproofing
5. Make it lighter
6. More scratch-resistant
7. Easier extraction of CD from the circular fastener
8. Streamlined look
9. Case should fit the hand better
10. Easier to take out leaflet describing the CD
11. Use recyclable plastic
12. Make interlocking cases so they stack on top of each other without slipping
13. Better locking case
14. Hinge that doesn't come apart

Next, the ideas for improvement were grouped into common areas by using an *affinity diagram*. A good way to achieve this is to write each of the ideas on a Post-it note and place them randomly on a wall. The team then examines the ideas and arranges them into columns of logical groups. Place a header card to denote the category of the group.

---

1. The original concept for this problem was developed by a team of business and engineering students at the University of Maryland, College Park. Team members were Barry Chen, Charles Goldman, Annie Kim, Vikas Mahajan, Kathy Naftalin, Max Rubin, and Adam Waxman. The results of their study have been modified significantly by the author.

Compact Disc Case
Product Improvement Survey

A group of students in ENES 190 is attempting to improve the design and usefulness of the standard storage case for compact discs. Please take 10 minutes to fill out this customer survey and return it to the student marketer.

Please indicate the level of importance you attach to the following aspects of a CD case.
1 = low importance   5 = high importance

1. A more crack-resistant case	1	2	3	4	5
2. A more scratch-resistant case	1	2	3	4	5
3. A hinge that doesn't come apart	1	2	3	4	5
4. A more colorful case	1	2	3	4	5
5. A lighter case	1	2	3	4	5
6. A streamlined look (aerodynamically sleek)	1	2	3	4	5
7. A case that fits your hand better	1	2	3	4	5
8. Easier opening CD case	1	2	3	4	5
9. Easier extraction of the CD from the circular fastener	1	2	3	4	5
10. Easier to take out leaflet describing contents of the CD	1	2	3	4	5
11. A more secure locking case	1	2	3	4	5
12. A waterproof case	1	2	3	4	5
13. Make the case from recyclable plastic	1	2	3	4	5
14. Make it so cases interlock so they stack on each other without slipping	1	2	3	4	5

Please list any other improvement features you would like to see in a CD case. _____

_____

_____

Would you be willing to pay more for a CD if the improvements you value with a 5 or 4 rating are available on the market?   yes   no

If you answered yes to the previous question, how much more would you be willing to pay? _____

How many CD's do you own (approximately)? _____

FIGURE 2.2
Customer survey for the compact disc case.

Stronger	Aesthetics	Opening and Extracting	Environment	Other
1	3	2	4	12
6	5	7	11	
14	8	10		
	9	13		

The affinity diagram was used to organize the questions on the survey that was distributed to potential customers.

2.3.3 Evaluating Customer Needs

To evaluate the customer responses, we could calculate the average score for each question, using a 1–5 scale. Those questions scoring highest would represent aspects of the product ranked highest in the mind of the customers. Alternatively, we can take the number of times a feature or attribute of a design is mentioned in the survey, and divide by the total number of customers surveyed. For the questionnaire shown above, we might use the number of responses to each question rating a feature as either a 4 or a 5. For the questions given above, this result is shown in Table 2.1.

It is worth noting that a response to a questionnaire of this type really measures the need obviousness as opposed to need importance. To get at true need importance, it is necessary to conduct face-to-face interviews or focus groups, and to record the actual words used by the persons interviewed. These responses need to be studied in depth, a tedious process. Also, it is important to realize that often respondents will omit talking about factors that are very important to them, because they seem so obvious. Safety or durability are good examples.

It is important to divide customer needs into two groups: hard constraints that absolutely must be satisfied (*musts*) and softer needs that can be traded off against other customer needs (*wants*). Customer needs can best be identified from focus group surveys or from the higher-ranking items in the written survey.

The relative frequency of responses from a survey can be displayed in a bar graph or a Pareto diagram (Fig. 2.3). In the bar graph the frequency of responses to each of the questions is plotted in order of the question number. In the Pareto diagram the frequency of responses is arranged in order of decreasing frequency, with the item of highest frequency at the left-hand side of the plot. Questions with less than 40 percent response rate have been omitted. This plot clearly identifies the most important customer requirements—the vital few. From these plots and Table 2.1 we conclude that the customer is most concerned with a more crack-resistant case (number 1) and that the convenience features of being able to stack the cases in a stable, interlocking way (number 14), and making it easier to extract the leaflet (number 10) and extract the CD (number 9) from the case appeal to the customer.

TABLE 2.1
Summary of responses from customer survey for CD case

Question number	Number of responses with 4 or 5 rating	Relative frequency
1	70	81.4
2	38	44.2
3	38	44.2
4	17	19.8
5	17	19.8
6	20	23.2
7	18	20.9
8	38	44.2
9	40	46.5
10	43	50.0
11	24	27.9
12	36	41.8
13	39	45.3
14	47	54.6

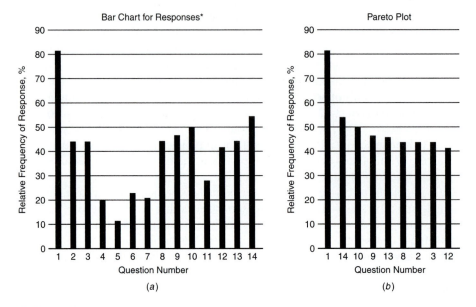

FIGURE 2.3

(*a*) Frequency of response plotted against question number in a conventional bar graph.
(*b*) Same data plotted as a Pareto diagram.
*Counts responses for each question that scored either 4 or 5.

2.4
BENCHMARKING

Benchmarking is a method for measuring a company's operations against the best companies both inside and outside of the industry.[1] It takes its name from the surveyor's benchmark or reference point from which elevations are measured. Benchmarking, as used in modern business context, is the search for industry best practices that lead to superior performance. Benchmarking can be used for product comparisons in the design and manufacturing of products, in service functions such as delivery and warranty issues, or in business areas like order entry, billing, financing, and repair.

Regardless of the focus of the benchmarking effort, it usually includes the following steps:

- Select the product, process, or functional area of the company that is to be benchmarked.

---

1. R. C. Camp, "Benchmarking," 2d ed., Quality Press, American Society for Quality, Milwaukee, 1995; M. J. Spendolini, "The Benchmarking Book," Amacom, New York, 1992; M. Zairi, "Effective Benchmarking: Learning from the Best," Chapman & Hall, New York, 1996 (many case studies).

- Identify the key performance *metrics* that will be measured and used for comparison. For example, for a product some metrics might be number of parts, estimated product cost, or material utilization for each component. For a manufacturing process the metrics might be yield of good parts, cycle time per part, or setup time. From a business viewpoint, metrics might be fraction of sales to repeat customers, percent of returns, or return on investment.
- Identify the *best-in-class companies* for each product or process to be benchmarked. A best-in-class company is one that performs the process at the lowest cost with the highest degree of customer satisfaction, or has the largest market share of a product type. This search should be broad, and in addition to including direct competitors, it can include companies in the same industry but who are not direct competitors, future or latent competitors, and companies in a totally different industry but which perform similar functions. For example, a consumer electronics company wishing to improve its direct sales to customers might benchmark a retail mail-order clothing company. If benchmarking is focused only on competitors, it may not lead to the best-in-class ranking. The number of companies chosen should be from four to six, should be as diverse a set as possible, and should exhibit good financial performance.
- Compare the best-in-class products or processes with the in-house equivalent using the performance metrics. The objective is to identify gaps in performance between the in-house product or process and the best-in-class companies.
- Specify programs and actions to meet and exceed the competition. The goal is for the in-house product or process to become the best in class. This requires a definitive action plan, with suitable resource commitment, and an agreed-upon schedule of milestones for achieving this goal.

A process similar to but more narrow than benchmarking is *reverse engineering.* This entails the dismantling of a product to determine its technology and how it is made, with the purpose of replication. The "tear-down" of a product is often a part of product benchmarking, but without the intent of copying the design. However, the collection of this type of benchmark information provides a better understanding of the solutions selected by the competition.

Collection of benchmark data is a critical issue. Much business data is available from published sources, government publications, and trade associations. Establishing an internal network of product experts within the company can often provide a pool of information about competing products and companies. Other sources of information are industry consultants and suppliers. Usually, detailed information will require the mailing of questionnaires and visits to companies. Sometimes trade or professional associations can facilitate such a visit, but arranging such visits usually comes down to having good contacts and offering information from your own company that may seem useful to the companies you wish to visit. Benchmarking operates most effectively on a quid pro quo basis.

Finally, it is important to realize that benchmarking is not a one-time effort. The competition will be working hard to improve, just as you have been doing. Benchmarking should be viewed as the first step in a process of continuous improvement.

2.5
CUSTOMER REQUIREMENTS

From a global viewpoint we should recognize that there is a hierarchy of human needs that motivate individuals in general.[1]

1. *Physiological needs* such as thirst, hunger, sex, sleep, shelter, and exercise. These constitute the basic needs of the body; and until they are satisfied, they remain the prime influence on the individual's behavior.
2. *Safety and security needs,* which include protection against danger, deprivation, and threat. When the bodily needs are satisfied, the safety and security needs become dominant.
3. *Social needs* for love and esteem by others. These needs include belonging to groups, group identity, and social acceptance.
4. *Psychological needs* for self-esteem and self-respect and for accomplishment and recognition.
5. *Self-fulfillment needs* for the realization of one's full potential through self-development, creativity, and self-expression.

As each need in this hierarchy is satisfied, the emphasis shifts to the next higher need.

Our design problem should be related to the basic human needs, some of which may be so obvious that in our modern technological society they are taken for granted. However, within each basic need there is a hierarchy of problem situations.[2] As the type I problem situations are solved, we move to the solution of higher-level problems within each category of basic need. It is characteristic of our advanced affluent society that, as we move toward the solution of type II and III problem situations, the perception of the need by society as a whole becomes less universal.

Basic need	Problem situation		
	I	**II**	**III**
Food	Hunger	Vitamin deficiency	Food additives
Shelter	Freezing	Cold	Comfort
Work	Availability	Right to work	Work fulfillment

Problem situation	Analysis of problem	Societal perception of need
I	None required	Complete agreement
II	Definition of problem	Some disagreement in priorities
	Calculation of cost	
	Setting of priorities	
III	Analysis of present and future costs	Strong disagreement on most issues
	Analysis of present and future risks	
	Environmental impact	

---

1. A. H. Maslow, *Psych. Rev.,* vol. 50, pp. 370–396, 1943.
2. Based on ideas of Prof. K. Almenas, University of Maryland.

Many current design problems deal with type III situations in which there is strong societal disagreement over needs and the accompanying goals. The result is protracted delays and increasing costs.

The customer requirements should be characterized as to performance, time, cost, and quality. *Performance* deals with what the design should do when it is completed and in operation. The *time* dimension includes all time aspects of the design. Currently, much effort is being given to reducing the cycle time to market for new products.[1] In many consumer products, the first to market with a great product captures the market. *Cost* pertains to all monetary aspects of the design. It is a paramount consideration, for everything else being roughly equal, cost determines most customers' buying decisions. *Quality* is a complex characteristic with many aspects and definitions. For now we will define quality as the totality of features and characteristics of a product or service that bear on its ability to satisfy stated or implied needs.

A more inclusive customer requirement than the four listed above is *value.* Value is the worth of a product or service. It can be expressed by the function provided divided by the cost, or the quality provided divided by the cost.

Studies of large successful companies have shown that the return on investment correlated with high market share and high quality. Garvin[2] identified the eight basic dimensions of quality for a manufactured product.

- *Performance:* The primary operating characteristics of a product. This dimension of quality can be expressed in measurable quantities, and therefore can be ranked objectively.
- *Features:* Those characteristics that supplement a product's basic functions. Features are frequently used to customize or personalize a product to the customer's taste.
- *Reliability:* The probability of a product failing or malfunctioning within a specified time period. See Chap. 11.
- *Durability:* A measure of the amount of use one gets from a product before it breaks down and replacement is preferable to continued repair. Durability is a measure of product life. Durability and reliability are closely related.
- *Serviceability:* Ease and time to repair after breakdown. Other issues are courtesy and competence of repair personnel and cost and ease of repair.
- *Conformance:* The degree to which a product's design and operating characteristics meet both customer expectations and established standards. These standards include industry standards and safety and environmental standards.
- *Aesthetics:* How a product looks, feels, sounds, tastes, and smells. The customer response in this dimension is a matter of personal judgment and individual preference. This area of design is chiefly the domain of the *industrial designer,* who is more artist than engineer. An important technical issue that affects aesthetics is *ergonomics,* how well the design fits the human user.
- *Perceived quality:* This dimension generally is associated with reputation. Advertising helps to develop this dimension of quality, but it is basically the quality of similar products previously produced by the manufacturer that influences reputation.

---

1. G. Stalk, Jr., and T. M. Hout, "Competing against Time," The Free Press, New York, 1990.
2. D. A. Garvin, *Harvard Business Review,* November–December, 1987, pp. 101–109.

The dimensions of performance, features, and conformance are interrelated. When competing products have essentially the same performance and many of the same features, customers will tend to expect that all producers of the product will have the same quality dimensions. In other words, customer expectations set the baseline for the product's conformance. Table 2.2 illustrates this by giving the performance criteria and features commonly found in three common household products.

We need to recognize that there are four levels of customer requirement:[1] (1) expecters, (2) spokens, (3) unspokens, and (4) exciters. These requirements must be satisfied at each level before addressing those at the next level.

- *Expecters:* These are the basic attributes that one would expect to see in the product, i.e., standard features. Expecters are frequently easy to measure and are used often in benchmarking.
- *Spokens:* These are the specific features that customers say they want in the product. Because the customer defines the product in terms of these attributes, the designer must be willing to provide them to satisfy the customer.
- *Unspokens:* These are product attributes the customer does not generally talk about, but are nevertheless important to him or her. They cannot be ignored. They may be attributes the customer simply forgot to mention or was unwilling to talk

TABLE 2.2
Performance criteria and features of three common products

Washing machines	Refrigerators	Self-propelled lawn mowers
Performance	*Performance*	*Performance*
Amount of water used	Efficiency	Motor horsepower
Cleanliness of clothes	Temperature	Handling
	Temperature distribution	Maneuverability
Features		Sharpness of turns
Automatic shutoff for		Ease of use
unbalanced load	*Features*	Starting the engine
Number of agitator speeds	Freezer light	Handling of grass catcher
Number of spin-dry speeds	Size (interior volume)	Shifting gears
Number of water fill levels	Automatic icemaker	Changing cutter height
Bleach dispenser	Location and size of freezer	Vacuum action to draw
Automatic control of water	Adjustable door shelves	cuttings into catcher
temperature	Solid or wire shelves	Cuts evenly
Porcelain lid	Humidity controlled crisper	Works well in tall grass
Discharge pump can lift water		Mulching action uniformly
to 6 ft (2 m) above washer	*Conformance*	disperses chips
	Compressor noise level	
	Chlorofluorocarbon (CFC) or	*Features*
	hydrofluorocarbon (HFC)	Grass catcher capacity

From E. B. Magrab, "Integrated Product and Process Design and Development," CRC Press, Boca Raton, 1997. With permission.

---

1. E. A. Magrab, "Integrated Product and Process Design," CRC Press, Boca Raton, 1997, pp. 91–92.

about or simply does not realize he or she wants. It takes great skill on the part of the design team to identify the unspoken requirements.

- *Exciters:* Often called *delighters,* these are product features that make the product unique and distinguish it from the competition. Note that the absence of an exciter will not make customers unhappy, since they do not know what is missing.

Customer satisfaction increases as the product fulfills requirements higher up in this hierarchy. Expecters must be satisfied first because they are the basic characteristics that a product is expected to possess. Spokens give greater satisfaction because they go beyond the basic level and respond to specific customer desires. Unspokens are an elusive category, while true exciters will serve to make a product unique. As we have seen in Sec. 2.3, not all customer requirements are equal. It is important to identify those requirements which are most important and ensure they are delivered in the product.

Customer complaints tend to be about expecter-type requirements. Therefore, a product development strategy aimed at solely eliminating complaints may not result in highly satisfied customers. To obtain this, you must adopt a strategy for actively seeking the "voice of the customer."

2.6
QUALITY FUNCTION DEPLOYMENT

Quality function deployment (QFD) is a planning and problem-solving tool that is finding growing acceptance for translating customer requirements into the engineering characteristics of a product.[1] It is a largely graphical method that systematically looks at all of the elements that go into the product definition as a group effort. As you will see below, the nature of the information that is required for the QFD diagram forces the design team to answer questions that might be glossed over in a less rigorous methodology and to learn what it does not know about the problem. Because it is a group decision-making activity, it creates a high level of buy-in and group understanding of the problem. Finally, QFD requires that the customers' requirements be expressed as measurable design targets in terms of engineering parameters. Thus, QFD is a natural precursor to establishing the product design specification (see Sec. 2.7).

QFD was developed in Japan in the early 1970s, with its first large-scale application in the Kobe Shipyard of Mitsubishi Heavy Industries. It was rapidly adopted by the Japanese automobile industry. By the mid-1980s many U.S. auto, defense, and electronic companies were using QFD. A recent survey of 150 U.S. companies showed that 71 percent of these have adopted QFD since 1990. These companies reported that 83 percent believed that using QFD had increased customer satisfaction with their products, and 76 percent felt it facilitated rational design decisions. It is important to remember these statistics because using QFD requires considerable

---

1. As will be discussed below, QFD can be rolled out all the way from the product planning stage, which is the focus here, through the detail part design to the production planning stage.

commitment of time and effort. Most, however, report that the time spent in QFD saves time later in design, especially in minimizing changes caused by poorly understanding the problem.

The layout of the QFD diagram is shown in Fig. 2.4. Because of this configuration, it is often called the *house of quality*.[1] The following is a description of what is found in each of the "rooms" of this house.

1. *Customer requirements (whats)* are gathered by the team as discussed in Sec. 2.3. To aid in understanding, group these requirements as identified by an affinity diagram.
2. *Competitive assessment* shows how the top two or three competitive products rank with respect to the customer requirements. This section starts with ranking each customer requirement on a scale of 1 to 5 and then, by considering the

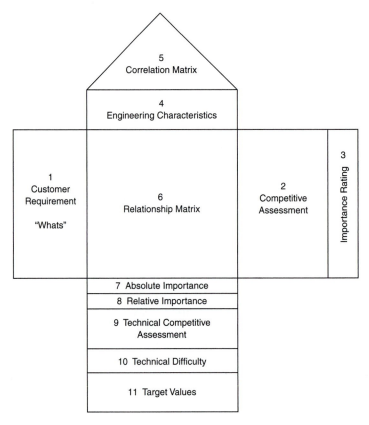

FIGURE 2.4

House of quality format for QFD. Numbers of "rooms" correlate with description in text.

1. J. R. Hauser and D. Clausing, "The House of Quality," *Harvard Business Review,* May–June 1988, pp. 63–73.

planned improvement and any requirements that are planned for special attention (sales points), it builds up to an importance rating.

3. *Importance rating:* For details on how this is arrived at, see the example that follows.

4. *Engineering characteristics (hows)* that enable satisfying the customer requirements are listed in columns. A way to arrive at the **ECs** is to ask the question: "What can I control that allows me to meet my customer's needs?" These must not be specific design details or solutions but must be characteristics that can be measured and given target values like weight, force, velocity, etc. It is desirable to label each **EC** to indicate the preferred direction of the magnitude of the characteristic. Thus a ⇑ or a + indicates that a higher value is better, and a ⇓ or a − indicates that a lower value is better.

5. *The correlation matrix* shows the degree of interdependence among the engineering characteristics in the "roof of the house." It is better to recognize these coupling relationships early in the design process so that appropriate trade-offs can be made. Usually a 9 or ● is used to signify a strong relationship and a 3 or ○ is for a medium relationship.

6. *The relationship matrix* determines the correlation between the engineering characteristics, **EC**, and the customer requirements, **CR**. To do this take each **CR**, and for each **EC**, in turn, ask whether it will significantly, moderately, slightly, or not at all impact the customer need. Generally, a nonlinear 9/3/1/0 (blank) scale[1] is used to weight disproportionately those **ECs** that strongly affect customer requirements.

7. To determine the *absolute importance,* first multiply the numerical value in each of the cells of the relationship matrix (6) by the importance rating (3). Then, sum the numbers in the cells of each column. These totals show the absolute importance of each engineering characteristic in meeting the customer requirements.

8. *Relative importance* is the absolute importance, normalized on a scale from 1 to 100. To arrive at this, total the values of absolute importance. Then, take each value of absolute importance, divide it by the total, and multiply by 100. Those **ECs** with the highest rating should be given special attention, for these are the ones that have the greatest effect upon customer satisfaction.

9. *The technical competitive assessment* benchmarks your company performance against two or three top competitors for each of the engineering characteristics. Generally a scale of 1–5 (best) is used. Often this information is obtained by getting examples of the competitor's product and testing them. Note that the data in this room compares each of the product performance characteristics with those of the closest competitors. This is different from the competitive assessment in room 2, where we compared the closest competitors with respect to each of the customer requirements.

10. *Technical difficulty* indicates the ease with which each of the engineering characteristics can be achieved. Basically, this comes down to an estimate by the design

---

1. The original Japanese workers in QFD used the symbols ●, ○, and Δ, taken from the racing form symbols for win, place, and show, for 9, 3, and 1.

team of the probability of doing well in attaining desired values for each **EC**. Again a 1 is a low probability and a 5 represents a high probability of success.

11. Setting *target values* is the final step in the QFD. By knowing which are the most important **ECs**, understanding the technical competition, and having a feel for the technical difficulty, the team is in a good position to set the targets for each engineering characteristic. Setting targets at the beginning of the design process provides a way for the design team to gauge the progress they are making toward satisfying the customer's requirements as the design proceeds.

You can see that the QFD summarizes a great deal of information in a single diagram. It will be one of the important reference documents during the progress of the design. Like most design documents, as more information is developed about the design the QFD should be updated. Not all design situations will call for a complete QFD as described here. However, as a minimum, rooms 1, 3, 4, 6, 8, and 11 should be used.

> **EXAMPLE.** The design of an improved case to protect and store compact discs, which was introduced in Sec. 2.3, will be continued. Based on the customer responses recorded in Fig. 2.3*a*, the "whats" are listed in room 1, grouped according to major category of customer requirements. Only those **CRs** that received a frequency of 40 percent or higher are listed here. One additional requirement, cost, is added because it is the major requirement of the recording and distribution companies, namely, that a CD case with improved features cost them no more than they are now paying. See Fig. 2.5.
>
> Next turn to room 2. The *customer importance* is established by taking the results from the customer surveys and allocating them along a 1–5 scale, where 5 is the highest. This region of the QFD is devoted chiefly to competitive assessment of the product, comparing the best of existing products with the proposed product. For the CD case little benchmarking data is available, since CD cases are a very generic product. However, we can rate the existing CD cases on the market against the new CD case that is planned to be offered for sale, again on a 1–5 scale. The ratio of planned to existing is called the *improvement ratio.* Since we plan to include some new features in the new CD case design, we expect to have some "talking points" or "sales features" that will aid in introducing the new product into the market. These *sales points* are rated, with a 1.5 given to the highest sales points and 1.3 to lower-rated features. For the CD case, we rate as high sales points the higher toughness case and easier opening case, and the ability to stack many CDs without tumbling.
>
> The *improvement ratio* (room 3) is given by the product of *customer importance* × *improvement ratio* × *sales points.* The *relative weight* is each value of importance weight divided by the sum of all values of importance weight. Note that the sum of the relative weights equals unity. Five **CRs** rank highest: 1, 3, 4, 5, and 10. The **CR** cost does not appear in this list, because we rated it low on improvement ratio and sales points; our goal is to design a CD case with improved features at the same cost as the current case. The way to look at this is that cost is a *must requirement,* while all the others are *want requirements.* Strictly speaking, the QFD should not mix needs that are musts and those that are wants. However, we included cost in the QFD so it would not be forgotten.
>
> Room 4 lists the engineering characteristics (the "hows") that enable the design team to meet the customer requirements. These are the technical requirements that the design must possess, without getting to a specific level of detail. In different words, they are the translation of the **CRs** into the internal or technical language of the organization, sometimes called the *substitute quality characteristics.*[1] They are arrived at by the team steep-

---

1. L. Cohen, "Quality Function Deployment," Addison-Wesley, Reading, MA, 1995.

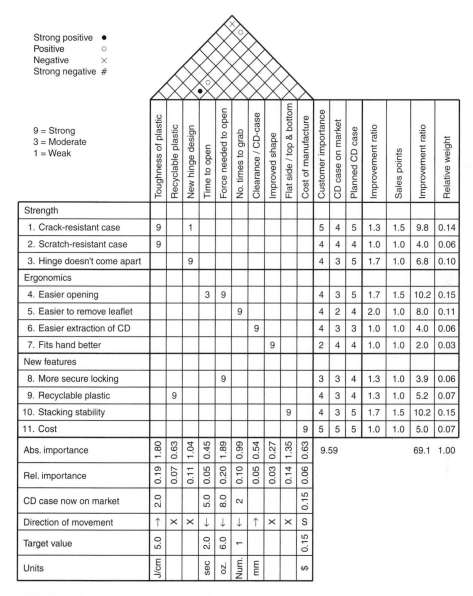

Strong positive ●
Positive ○
Negative ×
Strong negative #

9 = Strong
3 = Moderate
1 = Weak

	Toughness of plastic	Recyclable plastic	New hinge design	Time to open	Force needed to open	No. times to grab	Clearance / CD-case	Improved shape	Flat side / top & bottom	Cost of manufacture	Customer importance	CD case on market	Planned CD case	Improvement ratio	Sales points	Improvement ratio	Relative weight
Strength																	
1. Crack-resistant case	9		1								5	4	5	1.3	1.5	9.8	0.14
2. Scratch-resistant case	9										4	4	4	1.0	1.0	4.0	0.06
3. Hinge doesn't come apart			9								4	3	5	1.7	1.0	6.8	0.10
Ergonomics																	
4. Easier opening				3	9						4	3	5	1.7	1.5	10.2	0.15
5. Easier to remove leaflet						9					4	2	4	2.0	1.0	8.0	0.11
6. Easier extraction of CD							9				4	3	3	1.0	1.0	4.0	0.06
7. Fits hand better								9			2	4	4	1.0	1.0	2.0	0.03
New features																	
8. More secure locking					9						3	3	4	1.3	1.0	3.9	0.06
9. Recyclable plastic		9									4	3	4	1.3	1.0	5.2	0.07
10. Stacking stability									9		4	3	5	1.7	1.5	10.2	0.15
11. Cost										9	5	5	5	1.0	1.0	5.0	0.07
Abs. importance	1.80	0.63	1.04	0.45	1.89	0.99	0.54	0.27	1.35	0.63	9.59					69.1	1.00
Rel. importance	0.19	0.07	0.11	0.05	0.20	0.10	0.05	0.03	0.14	0.06							
CD case now on market	2.0			5.0	8.0	2				0.15							
Direction of movement	↑	×	×	↓	↓	↓	↑	×	×	S							
Target value	5.0			2.0	6.0	1				0.15							
Units	J/cm			sec	oz.	Num.	mm			$							

FIGURE 2.5
QFD table for the compact disc case example.

ing itself in the relevant technologies of the problem and then using brainstorming to generate ideas about the required product functions. Often more **ECs** than are shown in Fig. 2.5 are used, but a smaller number is used here because of the simplicity of the product, and for brevity in the example. Ideally, the **ECs** are characteristics that can be measured and given target values. A few of the **ECs** in Fig. 2.5 violate that precept because of the desire to embed an idea in the QFD which is important but not readily quantified.

The correlation matrix, room 5, records possible interactions between **ECs** for future trade-off decisions. For example, increasing the toughness of the plastic from which the case is made may have a negative impact on the cost of the manufactured CD case. A new hinge design will have a positive impact on the force to open the CD case.

The relationship matrix, room 6, correlates the engineering characteristics to the customer requirements (**ECs** vs. **CRs**). In the systems used in Fig. 2.5, a strong correlation is worth 9, a medium correlation 3, and a weak correlation 1. The objective is to make sure that every customer requirement is addressed by at least one engineering characteristic. If a row of **CRs** is blank, then the team should reexamine whether this is a real customer need, and if it is, then they should identify one or more **ECs** that address this need. Similarly, if any of the columns are blank, then it indicates that the design is imposing technical requirements that satisfy no customer requirements.

The importance of the **ECs** is determined by multiplying each of the cells in the matrix by its relative weight, and summing each column to give the *absolute importance* (room 7). For the **EC** "number of times to grab the leaflet" the absolute importance is $9 \times 0.11 = 0.99$. The sum of all the importance ratings is 9.59, so the *relative importance* for this **EC** is $(0.99/9.59) = 0.10$. We note that the most important technical requirements are the toughness of the plastic and the force to open the CD case.

We now need to establish target values for those **ECs** that can be quantified. We record the units in which **EC** is expressed, and the direction which indicates improvement. This can be increase ↑ or decrease ↓, the same S, or yes Y or no N, or simply whether the characteristic will be considered in the design, X. In this example we give the values for the **ECs** of the current CD cases on the market, and the target values for the new design.[1]

The QFD method is used most often in the planning of a product, as has just been illustrated above. However, QFD can be used throughout the product design process.[2] Figure 2.6 shows how the product planning "house of quality" feeds into the design of individual parts, and this into the process planning, and finally into production planning. We note that the **ECs** of the house of quality become the input for the part design QFD, with the target values of the house of quality becoming the constraints for that matrix. Thus, as shown in Fig. 2.6, each QFD feeds critical design information to the matrix that is downstream from it.

2.7
PRODUCT DESIGN SPECIFICATION

The product design specification (PDS) is the basic control and reference document for the design and manufacture of the product.[3] The PDS is a document which contains all of the facts related to the outcome of the product development. It should avoid forcing the design and predicting the outcome, but it should also contain the realistic constraints that are imposed on the design. Creating the PDS finalizes the process of establishing the customer needs and wants, prioritizing them, and beginning to cast them into a technical framework so that design concepts can be estab-

---

1. Special software usually is used to construct the QFD diagram. Three packages are QFD/Capture, International Techne Group, 5303 DuPont Circle, Milford, OH, 45150; QFD Scope, Integrated Quality Dynamics, and QFD Designer from American Supplier Institute.
2. D. Clausing, "Total Quality Development," ASME Press, New York, 1994; J. B. ReVelle, J. W. Moran, and C. A. Cox, "The QFD Handbook," Wiley, New York, 1998.
3. S. Pugh, "Total Design," Chap. 3, Addison-Wesley, Reading, MA, 1990.

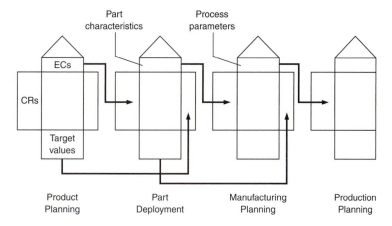

FIGURE 2.6
Example of how the results of product planning QFD feed into QFD for designing the individual parts, and this in turn feeds a QFD for manufacturing planning, which ultimately leads to a production planning sheet.

lished. The process of group thinking and prioritizing that developed the QFD diagram provides an excellent input to writing the PDS. However, it must be understood that the PDS is evolutionary and will change as the design process proceeds. But, at the end of the process the PDS will describe in writing the product that is intended to be manufactured and marketed.

Listed below are the elements that are to be found in a product design specification. Not every product will require consideration of every item in this list, but many will. The list demonstrates the complexity of product design.

In-Use Purposes and Market

- Product title
- Purpose or function the product is to perform
- Predictable unintended uses the product may be put to
- Special features of the product
- What types of products will the product compete against and who makes them?
- What is the intended market?
- Why is there a need for this product?
- Relationship of the product to other company products
- Anticipated market demand (units per year)
- Target company selling price and estimated retail price

Functional Requirements

- Functional performance—flow of energy, information, materials; operational steps; efficiency; accuracy
- Physical requirements—size, weight, shape, surface finish

- Service environment. The total environment that the product must operate in from the factory floor, through storage, transportation, and use must be considered. This includes range of temperature, pressure range, wind velocity, rain and salt spray, humidity, dirt and dust, corrosive environments, shock loading, vibration, noise level, insect and bird damage, degree of abuse by operators. Product must have acceptable shelf life.
- Life-cycle issues

  Useful life
  Reliability (mean time to failure)
  Robustness
  Maintainability
  Diagnosability
  Testability
  Repairability
  Installability
  Retirement from service and recyclability
  Cost of operation (energy costs, crew size, etc.)

- Human factors

  Aesthetics
  Man-machine interface, *ergonomics*
  User training

Corporate Constraints

- Time to market. Is there adequate time, given the resources, to design a quality product and its manufacturing process?
- Manufacturing requirements (in-house, in-country issues). Will it be required to use in-house manufacturing capabilities, and if so, will this limit available manufacturing processes and constrain cost reduction? Are there limitations on materials selection because of corporate policies?
- Suppliers. Do existing relationships with suppliers limit choices?
- Trademark, logo, brand name. What are the constraints, if any, in using these marketing tools?
- Financial performance. What are the corporate criteria on profitability and return on investment (ROI) that must be met?
- Corporate ethics. The product team should exhibit high professional ethics in dealing with suppliers, owners of intellectual property, corporate citizens, society at large, and each other.

Social, Political, and Legal Requirements

- Safety and environmental regulations. Not only must all U.S. regulations be met, but those of the European Community (EC) need to be considered for products that are expected to be exported.
- Standards. Search for, list, and use all pertinent product standards that may be applicable.

- Safety and product liability. Strong documentation of the design process, careful design of warning labels, in appropriate languages, and adherence to safety standards are the main defenses against product liability suits. Critical parts should be documented in the PDS so that the designer may give priority to ensuring these parts will be as reliable and safe as possible.
- Patents and intellectual property. All areas of useful information should be consulted prior to launching the design. Knowledge of the patent art is needed to avoid infringement. Arrange for licenses to use critical pieces of technology.

The PDS should be as complete as possible about what the design should do, but it should say as little as possible about how the requirements are to be met. Whenever possible the specification should be expressed in quantitative terms, and when appropriate it should give limits within which acceptable performance lies. For example: The power output of the engine should be 5 hp, plus or minus 0.25 hp. Remember that the PDS is a dynamic document. While it is important to make it as complete as possible at the outset of design, do not hesitate to change it as you learn more as the design evolves.

EXAMPLE. The CD jewel case given in Secs. 2.3 and 2.6 is continued here. From the information in Figs. 2.3 and 2.5 we can write the preliminary product design specification.

Product title
Compact disc jewel case.

Purpose
To provide an improved way to store and protect compact discs.

New or special features
Stronger; less susceptible to cracking on dropping.
Easier and quicker to open and extract the CD.
Easier to remove the descriptive leaflet.
More stable in stacking.

Competition
Will compete against standard hinged CD case produced by many plastics manufacturers.

Intended market
We will sell direct to largest producers of prerecorded music. Approximately 500 million CD jewel boxes are sold each year in the United States. Secondary market will be CD for computer games.

Need for product
User survey has shown customer interest in new features; 50 percent of people surveyed expressed willingness to pay a bit more for an improved product. Our business strategy is to produce a superior product at the existing cost to music producers.

Relationship to existing products line

This is a start-up venture. No other products currently exist.

Market demand

Current U.S. market is about 500 million units annually. We anticipate a 5 percent market share by year 2 (25 million units), growing to 20 percent share by year 5.

Price

We anticipate selling CD jewel cases at a unit price of $0.15 in bulk lots. The manufacturing cost should be no more than $0.12 per unit.

Functional performance

Protects the CD from dirt, scratches, spilled liquids.
Secures the CD firmly in the case.
Allows for easy opening of case.
Allows for stable stacking of many CDs.
Allows easy removal of descriptive leaflet.
Able to be dropped on floor from height of 3 ft without opening or cracking.

Physical requirements

Same size as regular CD case ($5.5 \times 4.87 \times 0.4$ in).
Approximately same weight as regular CD case.
Rectangular shape with rounded corners.
Smooth, but not slippery surface.
Transparent, so that identification material can be read on both sides.

Service environment

Case material should be stable from -20 to $120°F$, 20 to 100 percent relative humidity.

Life-cycle issues

Opening/closure mechanism must not fail for 1000 cycles.
Case made from recyclable material.

Human factors

Allows display of artwork or advertisements on two large flat surfaces.
No sharp corners or edges to cause cuts or snag clothing.
Rounded edges give good "feel."
Opening of case must be simple. Closure must be positive, and give an audible click.

Corporate constraints

Must be in market within 6 months.
Manufacturing will be contracted to suppliers.
Will use the trademark CD-EASE.
Must conform to corporate code of ethics.

Legal requirements

No toxic materials to be associated with manufacture.

Use of rosette to hold CD firmly in cradle will require license to U.S. patent. See U.S. patents 4613044, 5450951, and 5425451.

2.8
SUMMARY

Problem definition is the most crucial step in the engineering design process. If the problem is not properly defined, then all subsequent effort may prove to be for naught. This is especially true in product design, where considerable time and effort must be spent in listening to and analyzing the *voice of the customer.* The steps in problem definition for product design are:

- Position the product in the corporate product line as part of a product platform.
- Identify the customer needs: use interviews, focus groups, surveys, customer complaints.
- Evaluate the customer needs: separate into musts and wants, and prioritize the wants.
- Conduct benchmarking with best-in-class companies.
- Begin to identify the customer requirements. Recognize the four levels of requirements: expecters, spokens, unspokens, exciters.
- Build a QFD diagram as a team exercise to clearly establish the relationship between the customer requirements and the engineering characteristics of the design.
- With all of the above information, write a product design specification (PDS). The PDS becomes the controlling documentation for the design.

For a detailed description of a product development, starting with the voice of the customer and ending with a QFD, see Shiba et al.[1] The product in this case is a stripping basket, a device used by saltwater fly fishermen to collect their line before they cast it out.

BIBLIOGRAPHY

Customer Needs and Product Alignment

Meyer, M. H., and A. P. Lehnerd: "The Power of Product Platforms," The Free Press, New York, 1997.
Smith P. G., and D. G. Reinertsen: "Developing Products in Half the Time: New Rules, New Tools," 2d ed., Wiley, New York, 1996.

---

1. S. Shiba, A. Graham, and D. Walden, "A New American TQM," pp. 201–239, Productivity Press, Portland, OR, 1993.

Ulrich, K. T., and S. D. Eppinger: "Product Design and Development," McGraw-Hill, New
    York, 1995, Chaps 2, 3, 4.
Urban, G. L., and J. R. Hauser: "Design and Marketing of New Products," 2d ed., Prentice-
    Hall, Englewood Cliffs, NJ, 1993.

Quality Function Deployment

Bickell, B. A., and K. D. Bickell: "The Road Map to Repeatable Success: Using QFD to
    Implement Change," CRC Press, Boca Raton, FL, 1995.
Clausing, D.: "Total Quality Development," ASME Press, New York, 1995.
Cohen, L.: "Quality Function Deployment," Addison-Wesley, Reading, MA, 1995.
Day, R. G.: "Quality Function Deployment," ASQC Quality Press, Milwaukee, WI, 1993.
Guinta, L. R., and N. C. Praizler: "The QFD Book," Amacom, New York, 1993.
King, B.: "Better Designs in Half the Time," 3d ed., GOAL/QPC, Methuen, MA, 1989.

Customer Requirements and PDS

McGrab, E. A.: "Integrated Product and Process Design," CRC Press, Boca Raton, FL, 1997.
Pugh S.: "Total Design," Addison-Wesley, Reading, MA, 1990.
Ullman, D. G.: "The Mechanical Design Process," 2d ed., McGraw-Hill, New York, 1997.

PROBLEMS AND EXERCISES

2.1. Select 10 products from a catalog for a supplier of household items (not clothing) and
decide which needs in Maslow's hierarchy of human needs they satisfy. Then, identify
the particular product features that make the products attractive to you.

2.2. The demand for most edible fish exceeds the supply. While fish can be raised in ponds
on land or in ocean enclosures close to shore, there are limitations of scale. The next step
is mariculture—fish farming in the open sea. Develop a new product business development plan for such a venture.

2.3. The transistor, followed by the microprocessor, is one of the most far-reaching products
ever developed. Make a list of the major products and services that have been impacted
by these inventions.

2.4. Write a survey to find what customers want in a washing machine.

2.5. Take 10 min and individually write down small things in your life, or aspects of products that you use, that bother you. You can just name the product, or better yet, give an
attribute of the product that "bugs you." Be as specific as you can. You are really creating a needs list. Combine this with other lists prepared by members of your design team.
Perhaps you have created an idea for an invention.

2.6. Suppose you are the inventor of a new device called the helicopter. By describing the
functional characteristics of the machine, list some of the societal needs that it is
expected to satisfy. Which of these have come to fruition, and which have not?

2.7. A focus group of housewives was convened to show them an innovative clothespin and ask what characteristics they want in a clothespin. The comments were as follows:

> It needs to grip tightly.
> I have arthritis. They shouldn't be too hard to open or close.
> I don't like pins that get tangled in my clothes basket.
> It better not stain the clothes.
> If I'm going to buy new clothespins at that price, they better last a long time.
> I don't want them to look shabby after long use.

Translate these customer requirements into engineering characteristics of the product.

2.8. Complete the relationship matrix (room 6) and the correlation matrix (room 5) for a house of quality for a heating and air-conditioning design project. The customer requirements are lower operating costs; improved cash flow; managed energy use; increased occupant comfort; and easy to maintain. The engineering characteristics are energy efficiency ratio ≥ 10; zonal controls; programmable energy management system; payback ≤ 1 year; and 2-hr spare parts delivery.

2.9. A product design team is designing an improved flip-lid trash can such as would be found in a family kitchen. The problem statement is as follows:

> Design a user-friendly, durable flip-lid trash can that opens and closes reliably. The trash can must be lightweight yet tip-resistant. It must combat odor, fit standard kitchen trash bags, and be safe for all users in a family environment.

With this information, and a little research and imagination where needed, construct a QFD for this design project.

2.10. Write a problem statement for cross-country skis that allow skiing on dirt or grass. List the "musts" and "wants" separately.

2.11. Write a product design specification for the flip-lid trash can described in Prob. 2.9.

3

TEAM BEHAVIOR AND TOOLS

3.1
INTRODUCTION

A recent column in *The Wall Street Journal* was headed "Engineering Is Re-engineered into a Team Sport." The article went on to say, "These firms want people who are comfortable operating in teams and communicating with earthlings who know nothing about circuit-board design or quantum mechanics." This is to emphasize that when industry leaders are asked what they would like to see changed in engineering curricula they invariably respond, "Teach your students to work effectively in teams." A more near term reason for devoting this chapter to team behavior is that the engineering design courses for which this text is intended are mostly focused around team-based projects. All too often we instructors thrust you students into a team situation without providing proper understanding of what it takes to achieve a smooth functioning team. Most often things work out just fine, but at a cost of extra hours of trial and error to find the best way to function as a team. Indeed, the greatest complaint that students have about project design courses is *it takes too much time.* This chapter is designed to give you an understanding of the team building process and introduce you to some tools that people have found helpful in getting results through teams.

Why all the fuss about teams? Basically, it is because, properly handled, teams outperform individuals. Certainly, in most engineering design, because of the complexity of the problem, teams are a necessity. No one person could possess all of the knowledge and skill needed for a successful solution, and no one person working 20 hours per day could complete all of the tasks that need to be done.

A team is a small number of people with complementary skills who are committed to a common purpose, performance goals, and approach for which they hold themselves mutually accountable.[1] There are two general types of teams: teams that do real

---

1. J. R. Katzenbach and D. K. Smith, "The Wisdom of Teams," HarperCollins, New York, 1994.

TABLE 3.1
Differences between a working group and a team

Working group	Team
Strong, clearly focused leader	Individual and mutual accountability
The group's purpose is the same as the broader organizational mission	Specific team purpose that the team itself develops
Individual work products	Collective work products
Runs efficient meetings	Encourages open-ended discussion and active problem-solving meetings
Measures its effectiveness indirectly by its influence on others	Measures performance directly by assessing collective work products
Discusses, decides, and delegates	Discusses, decides, and does real work together

From J. R. Katzenbach and D. K. Smith, "The Wisdom of Teams," HarperCollins, New York, 1994.

work, like design teams, and teams that make recommendations. Both are important, but we focus here on the former. Most people have worked in groups, but a working group is not necessarily a team. Table 3.1 clearly defines the differences. We see from Table 3.1 that a team is a high order of group activity. Many groups do not reach this level, but it is a goal truly worth achieving.

3.2
WHAT IT MEANS TO BE AN EFFECTIVE TEAM MEMBER

There is a set of attitudes and work habits that you need to adopt to be a good team member. First and foremost, you need to *take responsibility for the success of the team.* Without this commitment, the team is weakened by your presence. Without this commitment, you shouldn't be on the team.

Next, you need to *be a person who delivers on commitments.* This means that you consider membership on the team as something worthwhile and that you are willing to rearrange your job and personal responsibilities to satisfy the needs of the team. On occasions when you cannot complete an assignment, always notify the team leader as soon as possible so other arrangements can be made.

Much of the team activities takes place in meetings where members share their ideas. Learn to *be a contributor to discussions.* Some of the ways that you can contribute are by asking for explanations to opinions, guiding the discussion back on track, and pulling together and summarizing ideas.

Listening is an art that not all of us have learned to practice. Learn to *give your full attention to whomever is speaking and demonstrate this by asking helpful questions.* To help focus on the speaker, take notes and never do distracting things like reading unrelated material, writing letters, walking around, or interrupting the speaker.

Develop techniques for getting your message across to the team. This means thinking things through briefly in your own mind before you speak. Always speak in a loud, clear voice. Have a positive message, and avoid "put-downs" and sarcasm. Keep focused on the point you are making. Avoid rambling discussion.

Learn to give and receive useful feedback. The point of a team meeting is to benefit from the collective knowledge and experience of the team to achieve an agreed-upon goal. Feedback is of two types. One is a natural part of the team discussion. The other involves corrective action for improper behavior by a member of the team[1] (see Sec. 3.6).

The following are characteristics of an effective team:

- Team goals are as important as individual goals.
- The team understands the goals and is committed to achieving them.
- Trust replaces fear and people feel comfortable taking risks.
- Respect, collaboration, and open-mindedness are prevalent.
- Team members communicate readily; diversity of opinions is encouraged.
- Decisions are made by consensus and have the acceptance and support of the members of the team.

I hope you will want to learn how to become an effective team member. Most of this chapter is devoted to helping you do that. Being a good team member is not a demeaning thing at all. Rather, it is a high form of group leadership. Being recognized as an effective team member is a highly marketable skill. Corporate recruiters say that the traits they are looking for in new engineers are communication skills, team skills, and problem-solving ability.

3.3
TEAM ROLES

We have just discussed the behavior that is expected of a good team member. Within a team members assume different roles in addition to being an active team member.

An important role that is external to the team but vital to its performance is the *team sponsor.* The team sponsor is the manager who has the need for the output of the team. He or she selects the team leader, negotiates the participation of team members, provides any special resources needed by the team, and formally commissions the team.

The *team leader* convenes and chairs the team meetings using effective meeting management practices (see Sec. 3.5). He or she guides and manages the day-to-day activity of the team by tracking the team's accomplishment toward stated goals, helping team members develop their skills, communicating with the sponsor about progress, trying to remove barriers toward progress, and helping to resolve conflict within the team. In general, there are three styles of team leadership: the traditional or autocratic leader, the passive leader, and the facilitative leader. Table 3.2 lists some major characteristics of these types of leaders. Clearly, the facilitative leader is the modern type of leader who we wish to be leading teams.

Many teams in industry include a *facilitator,* a person trained in group dynamics who assists the leader and the team in achieving its objectives by coaching them in

---

1. P. R. Scholtes et al., "The Team Handbook," Joiner Associates, Madison, WI, 1988; "The Team Memory Jogger," Joiner Associates, 1995.

TABLE 3.2
Characteristics of three leadership types

Traditional leader	Passive leader	Facilitative leader
Directive and controlling	Hands off	Creates open environment
No questions—just do it	Too much freedom	Encourages suggestions
Retains all decision-making authority	Lack of guidance and direction	Provides guidance
Nontrusting	Extreme empowerment	Embraces creativity
Ignores input	Uninvolved	Considers all ideas
Autocratic	A figurehead	Maintains focus; weighs goals vs. criteria

team skills and problem-solving tools, and assisting in data-collection activities. Sometimes the facilitator leads the meeting, especially if a controversial subject is being discussed. While the facilitator functions as a team member in most respects, she or he must remain neutral in team discussions and stand ready to provide interventions to attain high team productivity and improved participation by team members or, in extreme situations, to resolve team disputes. A key role of the facilitator is to keep the group focused on its task.

Sometimes teams have a *process observer.* The process observer is a member of the team appointed on a rotating basis to observe the process and progress of the meeting. He or she assists the facilitator in keeping the discussion on track, encouraging full participation of team members, and encouraging listening. Often, the facilitator also serves in the role of process observer. One task of the process observer is to look for hidden agendas that prevent an effective team process, like individuals who continually shirk work or who are overly protective of their organizational unit. When serving as process observer, the team member does not take part actively in the discussion.

3.4
TEAM DYNAMICS

Students of team behavior have observed that most teams go through five stages of development.[1]

1. *Orientation (forming):* The members are new to the team. They are probably both anxious and excited, yet unclear about what is expected of them and the task they are to accomplish. This is a period of tentative interactions and polite discourse, as the team members undergo orientation and acquire and exchange information.
2. *Dissatisfaction (storming):* Now the challenges of forming a cohesive team become real. Differences in personalities, working and learning styles, cultural

---

1. R. B. Lacoursiere, "The Life Cycle of Groups," Human Service Press, New York, 1980; B. Tuckman, Developmental Sequence in Small Groups, *Psychological Bulletin,* no. 63, pp. 384–399, 1965.

We Don't Want a General Patton

Many student design teams have difficulty with team leadership. Unless the instructor insists on each team selecting a leader, the natural egalitarian student spirit tends to work against selecting a team leader. Often students prefer to rotate the leadership assignment. While this procedure has the strong benefit of giving each student a leadership experience, it often leads to spotty results and is definitely a time-inefficient procedure.

One approach that works well for semester-long projects is to start out by rotating the leadership assignment for about 1 month. This gives everyone in the team a chance at leadership, and it also demonstrates which students have the strongest leadership talents. Often a natural leader emerges. The team should embrace such a person and make him or her their leader. Of course, in this enlightened era, we want nothing other than a facilitative leader.

---

backgrounds, and available resources (time to meet, access to and agreement on the meeting place, access to transportation, etc.) begin to make themselves known. Disagreement, even conflict, may break out in meetings. Meetings may be characterized by criticism, interruptions, poor attendance, or even hostility.

3. *Resolution (norming):* The dissatisfaction abates when team members establish group norms, either spoken or unspoken, to guide the process, resolve conflicts, and focus on common goals. The norms are given by rules of procedure and the establishment of comfortable roles and relationships among team members. The arrival of the resolution stage is characterized by greater consensus[1] seeking, and stronger commitment to help and support each other.

4. *Production (Performing):* This is the stage of team development we have worked for. The team is working cooperatively with few disruptions. People are excited and have pride in their accomplishments, and team activities are fun. There is high orientation toward the task, and demonstrable performance and productivity.

5. *Termination (Adjourning):* When the task is completed, the team prepares to disband. This is the time for joint reflection on how well the team accomplished its task, and reflection on the functioning of the team.

It is important for teams to realize that the dissatisfaction stage is perfectly normal and that they can look forward to its passing. Many teams experience only a brief stage 2 and pass through without any serious consequences. However, if there are serious problems with the behavior of team members, they should be addressed quickly. Also, some teams can be expected to lose a member or add a member after the team formation has begun. They must all recognize that changing even one team member makes it a new team and that they must again all go through the five stages of team development on an accelerated schedule.

One way or another, a team must address the following set of psychosociological conditions.

---

1. Consensus means general agreement or accord. Consensus does not require 100 percent agreement of the group. Neither is 51 percent agreement a consensus.

- *Safety:* Are the members of the team safe from destructive personal attacks? Can team members freely speak and act without feeling threatened?
- *Inclusion:* Team members need to be allowed equal opportunities to participate. Rank is not important inside the team. Make special efforts to include new, quiet members in the discussion.
- *Appropriate level of interdependence:* Is there an appropriate balance between the individuals' needs and the team needs? Is there a proper balance between individual self-esteem and team allegiance?
- *Cohesiveness:* Is there appropriate bonding between members of the team?
- *Trust:* Do team members trust each other and the leader?
- *Conflict resolution:* Does the team have a way to resolve conflict?
- *Influence:* Do team members or the team as a whole have influence over members? If not, there is no way to reward, punish, or work effectively.
- *Accomplishment:* Can the team perform tasks and achieve goals? If not, frustration will build up and lead to conflict.

It is important for the team to establish some guidelines for working together. Guidelines will serve to ameliorate the dissatisfaction stage and are a necessary condition for the resolution stage. The team should begin to develop these guidelines early in the orientation stage. Table 3.3 lists some suggested guidelines that the team could discuss and modify until there is consensus.

TABLE 3.3
Suggested guidelines for an effective team

- We will be as open as possible but will honor the right of privacy.
- Information discussed in the team will remain confidential.
- We will respect differences between individuals.
- We will respect the ideas of others.
- We will be supportive rather than judgmental.
- We will give feedback directly and openly, in a timely fashion. Feedback will be specific and focus on the task and process and not on personalities.
- We will all be contributors to the team.
- We will be diligent in attending team meetings. If an absence is unavoidable, we will promptly notify the team leader.
- When members miss a meeting we will share the responsibility for bringing them up to date.
- We will use our time wisely, starting on time, returning from breaks, and ending our meetings promptly.
- We will keep our focus on our goals, avoiding sidetracking, personality conflicts, and hidden agendas. We will acknowledge problems and deal with them.
- We will not make phone calls or interrupt the team during meetings.
- We will be conscientious in doing assignments between meetings and in adhering to all reasonable schedules.

TEAM SIGNATURES

_____   _____

_____   _____

_____   _____

_____   _____

TABLE 3.4
Different behavioral roles found in groups

Helping roles		Hindering roles
Task roles	Maintenance roles	
Initiating: proposing tasks; defining problem	Encouraging	Dominating: asserting authority or superiority
Information or opinion seeking	Harmonizing: attempting to reconcile disagreement	Withdrawing: not talking or contributing
Information or opinion giving	Expressing group feeling	Avoiding: changing the topic; frequently absent
Clarifying	Gate keeping: helping to keep communication channels open	Degrading: putting down others' ideas; joking in barbed way
Summarizing	Compromising	Uncooperative: Side conver-
Consensus testing	Standard setting and testing: checking whether group is satisfied with procedures	sations: whispering and private conversations across the table

People play various roles during a group activity like a team meeting. It should be helpful in your role as team leader or team member to recognize some of the behavior listed briefly in Table 3.4. It is the task of the team leader and facilitator to try to change the hindering behavior and to encourage team members in their various helping roles.

3.5
EFFECTIVE TEAM MEETINGS

Much of the work of teams is accomplished in team meetings. It is in these meetings that the collective talent of the team members is brought to bear on the problem, and in the process, all members of the team "buy in" to its solution. Students who complain about design projects taking too much time often are really expressing their inability to organize their meetings and manage their time effectively.

At the outset it is important to understand that an effective meeting requires planning. This is the responsibility of the person who will lead the meeting. Meetings should begin on time and last for about 90 min, the optimum time to retain all members' concentration. A meeting should have a written agenda, with the name of the designated person to present each topic and an allotted time for discussion of the topic. If the time allocated to a topic proves to be insufficient, it can be extended by the consent of the group, or the topic may be given to a small task group to study further and report back at the next meeting of the team. In setting the agenda, items of greatest urgency should be placed first on the agenda.

The team leader directs but does not control discussion. As each item comes up for discussion on the agenda, the person responsible for that item makes a clear statement of the issue or problem. Discussion begins only when it is clear that every participant understands what is intended to be accomplished regarding that item. One reason for keeping teams small is that every member has an opportunity to contribute to the discussion. Often it is useful to go around the table in a round robin fashion, asking each person for their ideas or solutions, while listing them on a flip chart or blackboard. No criticism or evaluation should be given here, only questions for clarification. Then the ideas are discussed by the group, and a decision is reached. It is important that this be a group process and that an idea become disassociated from the individual who first proposed it.

Decisions made by the team in this way should be consensus decisions. When there is a consensus, people don't just go along with the decision, they invest in it. Arriving at consensus requires that all participants feel that they have had their full say. Try to help team members to avoid the natural tendency to see new ideas in a negative light. However, if there is a sincere and persuasive negative objector, try to understand their real objections. Often they have important substance, but they are not expressed in a way that they can be easily understood. It is the responsibility of the leader to keep summing up for the group the areas of agreement. As discussion advances, the area of agreement should widen. Eventually you come to a point where problems and disagreement seem to melt away, and people begin to realize that they are approaching a decision that is acceptable to all.

3.5.1 Simple Rules for Meeting Success

1. Pick a regular meeting location and try not to change it.
2. Pick a meeting location that: *(a)* is agreeable and accessible to all (unless your team is trying to "get away"), *(b)* has breathing room when there is full attendance plus a guest or two, *(c)* has a pad or easel in the room, *(d)* isn't too hot, too cold, or too close to noisy distractions.
3. Regular meeting times are not as important as confirming the time of meetings. Once a meeting time has been selected, confirm it immediately in writing (e-mail or memo). Remain flexible on selecting meeting length and frequency. Shape the time that the team spends together around the needs of the work to be accomplished.
4. Send an e-mail reminder to team members just before the first of several meetings.
5. If you send materials out in advance of a meeting, bring extra copies just in case people forget to bring theirs, or it did not arrive. Do not send out agendas or reading materials in advance unless you give people at least four business days to look things over.
6. Start on time, or no later than 5 to 7 min from the stated starting time.
7. Pass out an agenda at the beginning of the meeting and get the team's concurrence to the agenda. Start every meeting with "what are we trying to accomplish today?"
8. Rotate the responsibility for writing meeting summaries of each meeting. The summaries should document: *(a)* when did the team meet, *(b)* what were the issues discussed (in outline form), *(c)* decisions, agreements, or apparent consensus on issues, *(d)* next meeting date and time, *(e)* "homework" for next meeting. In general, meeting summaries should not exceed one page, unless you are attaching results from group brainstorming, lists of issues, ideas, etc. Meeting summaries should be distributed by the assigned recorder within 48 h of the meeting.

9. Notice members who come late, leave early, or miss meetings. Ask if the meeting time is inconvenient or competing demands are keeping them from meetings. Ask if the team sponsor could help by talking with their supervisor.

10. Observe team members who are not speaking. Near the end of the discussion, ask them directly for their opinion on an issue. Consult them after the meeting to be sure that they are comfortable with the team and discussion.

11. Occasionally use meeting evaluations (perhaps every second or third meeting) to gather anonymous feedback on how the group is working together. Meeting evaluations should be turned in to the facilitator, who should summarize the results, distribute a copy of those results to everyone, and lead a brief discussion at the next meeting on reactions to the meeting evaluations and any proposed changes in the meeting format.

12. Do not bring guests or staff support or add team members without seeking the permission of the team.

13. Avoid canceling meetings. If the team leader cannot attend, an interim discussion leader should be designated.

14. End every meeting with an "action check": *(a)* what did we accomplish/agree upon today? *(b)* what will we do at the next meeting? *(c)* what is everyone's "homework," if any, before the next meeting?

15. Follow up with any person who does not attend, especially people who did not give advance notice. Call to update them about the meeting and send them any materials that were passed out at the meeting. Be sure they understand what will take place at the next meeting.

For smooth team operation, it is important to:

- Create a team roster. Ask team members to verify mailing addresses, e-mail addresses, names, and phone numbers of administrative support staff. Include information about the team sponsor. Use e-mail addresses to set up a distribution list for your team.
- Organize important material in team binders. Include the team roster, team charter, essential background information, data, critical articles, etc.

3.6
PROBLEMS WITH TEAMS

A well-functioning team achieves its objectives quickly and efficiently in an environment that induces energy and enthusiasm. However, it would be naive to think that everything will always go well with teams. Therefore, we spend a little time in discussing some of the common problems encountered with teams, and possible solutions. As a starting point, review Table 3.4, for the helping and hindering roles that people play in groups.

The characteristics of a good team member are:

- Respects other team members without question
- Listens carefully to the other team members
- Participates but does not dominate

- Self-confident but not dogmatic
- Knowledgeable in his or her discipline
- Communicates effectively
- Disagrees but with good reason and in good taste

The characteristics of a disruptive team member are:

- Shows lack of respect for others
- Tends to intimidate
- Stimulates confrontation
- Is a dominant personality type
- Talks all the time, but does not listen
- Does not communicate effectively
- Overly critical

Handling a disruptive member requires a skilled team leader or facilitator. What can we do about the team member who dominates the team discussion? Such people often are quick-thinking idea people who make important contributions. One way to deal with this is to acknowledge the important contributions from the person and then shift the discussion to another member by asking them a question. If the domination continues, talk to the member outside of the meeting.

Another disruptive type is the member who is overly critical and constantly objects to point after point. If this type of behavior is allowed to go on, it will destroy the spirit of openness and trust that is vital for a good team performance. This behavior is harder to control. The leader should continually insist that the comments be restated to be more positive, and if the offender can't or won't do this, then the leader should do it. Again, a strong talk outside of the meeting to point out the destructive nature of the behavior is called for, and if there is no improvement, then this member should be asked to leave the team.

A less disruptive type is the person who obstinately disagrees with some point. If this is based on information that the member is sharing with the team, then it is a good part of the process. However, if the disagreement becomes focused on personalities or an unwillingness to reach consensus, then it becomes disruptive behavior. To combat this, ask members to summarize the position they disagree with, to be sure they understand the group's position. Then, ask them to make positive recommendations to see whether there is an area of agreement. If these steps fail, then change the subject and move on, returning to the subject another time.

A common team problem occurs when the team strays too far from the topic. This happens when the leader is not paying strict attention and suddenly finds the team "out in left field." The team can be brought back by asking whether the current discussion is leading to the agreed-upon objective, as guided by the agenda. The leader should introduce new material into the discussion that is more closely related to the objective. The literature is replete with additional suggestions on how to handle problem situations in teams.[1]

---

1. R. Barra, "Tips and Techniques for Team Effectiveness," Barra International, New Oxford, PA, 1987, pp. 60–67; D. Harrington-Mackin, "The Team Building Tool Kit," American Management Association, New York, 1994.

3.7
PROBLEM SOLVING TOOLS

In this section we present some common problem-solving tools that are useful in any problem situation, whether as part of your overall design project or in any other business situation—as in trying to identify new sources of income for the student ASME chapter. These tools are especially well suited for problem solving by teams. They have a strong element of common sense and do not require sophisticated mathematics, so they can be learned and practiced by any group of educated people. They are easy to learn, but a bit tricky to learn to use with real expertise. These tools have been codified within the discipline called *total quality management.*[1]

Many strategies for problem solving have been proposed. The one that we have used and found effective is a simple three-phase process.[2]

- Problem definition
- Cause finding
- Solution finding and implementation

Table 3.5 lists the tools which are most applicable in each phase of the problem-solving process. Most are described below in a long exercise that illustrates their use. A few are found in other sections of this text.

Having read Chap. 2, it will come as no surprise that we view problem definition as the critical phase in any problem situation. A problem can be defined as the difference between a current state and a more desirable state. Often the problem is posed by management or the team sponsor, but until the team redefines it for itself, the problem has not been defined. The problem should be based on data, which may reside in the reports of previous studies, or in surveys that the team undertakes to define the problem. In working toward an acceptable problem definition, the team uses *brainstorming* and the *affinity diagram.* The process by which this is accomplished within a team is called the *nominal group technique.* The outcome of the problem-definition stage is a well-crafted problem statement.

The objective of the cause-finding stage is to identify all of the possible causes of the problem and to narrow them down to the most probable *root causes.* This phase starts with the gathering of data and analyzing the data with simple statistical tools. The first step in data analysis is the creation of a *check sheet* in which data is recorded by classifications. Numeric data may lend itself to the construction of a histogram, while a Pareto chart or simple bar chart may suffice for other situations. Run charts may show correlation with time, and scatter diagrams show correlation with critical parameters. Once the problem is understood with data the *cause-and-effect diagram*

---

1. J. W. Wesner, J. M. Hiatt, and D. C. Trimble, "Winning with Quality: Applying Quality Principles in Product Development," Addison-Wesley, Reading, MA, 1995; C. C. Pegels, "Total Quality Management," Boyd & Fraser, Danvers, MA, 1995; W. J. Kolarik, "Creating Quality," McGraw-Hill, New York, 1995; S. Shiba, A. Graham, and D. Walden, "A New American TQM," Productivity Press, Portland, OR, 1993.
2. Ralph Barra, "Tips and Techniques for Team Effectiveness," Barra International, PO Box 325, New Oxford, PA.

TABLE 3.5
Problem-solving tools

Problem definition	Cause finding	Solution planning and
Brainstorming	*Gathering data*	*implementation*
Affinity diagram	Interviews	Brainstorming
Nominal group technique	Focus groups (see Sec. 2.3)	How-how diagram
	Surveys	Concept selection
	Analyzing data	method (see Sec. 5.9)
	Check sheet	Force field analysis
	Histogram (see Sec. 10.4)	Implementation plan
	Search for root causes	
	Cause-and-effect diagram	
	Why-why diagram	
	Interrelationship digraph	

and the *why-why diagram* are effective tools for identifying possible causes of the problem. The *interrelationship digraph* is a useful tool for identifying root causes.

With the root causes identified, the objective of the solution-finding phase is to generate as many ideas as possible as to how to eliminate the root causes. Brainstorming clearly plays a role, but this is organized with a *how-how diagram.* With solutions identified, the pros and cons of a strategy for implementing them is identified with *force field analysis.* Finally, the specific steps required to implement the solution are identified and written into an *implementation plan.* Then, as a last step, the implementation plan is presented to the team sponsor.

We have outlined briefly a problem-solving strategy that utilizes a number of tools that are often associated with total quality management[1] (TQM). These tools are described below within the context of a single problem.

> **EXAMPLE.** The problem-solving methodology and tools listed in Table 3.5 are presented through a long example. First we describe the tool and then illustrate it with a continuing example.

Problem Definition

A group of engineering honors students[2] was concerned that more engineering seniors were not availing themselves of the opportunity to do a senior research project. All engineering departments listed this as a course option, but only about 5 percent of the students chose this option. To properly define the problem, the team brainstormed around the question "Why do so few senior engineering students choose to do a research project?"

---

1. M. Brassard and D. Ritter, "The Memory Jogger™II, A Pocket Guide of Tools for Continuous Improvement," GOAL/QCP, Methuen, MA, 1994; N. R. Tague, "The Quality Toolbox," ASQC Quality Press, Milwaukee, WI, 1995.
2. The team of students making this study in 1994 was Brian Gearing, Judy Goldman, Gebran Krikor, and Charnchai Pluempitiwiriyawej. The results of the team's study have been modified appreciably by the author.

Brainstorming. Brainstorming is a group technique for generating ideas in a non-threatening, uninhibiting atmosphere. It is a group activity in which the collective creativity of the group is tapped and enhanced. The objective of brainstorming is to generate the greatest number of alternative ideas from the uninhibited responses of the group. Brainstorming is most effective when it is applied to specific rather than general problems. It is frequently used in the problem-definition phase and solution-finding phase of problem solving.

There are four fundamental brainstorming principles.

1. *Criticism is not allowed.* Any attempt to analyze, reject, or evaluate ideas is postponed until after the brainstorming session. The idea is to create a supportive environment for free-flowing ideas.
2. *Ideas brought forth should be picked up by the other people present.* Individuals should focus only on the positive aspects of ideas presented by others. The group should attempt to create chains of mutual associations that result in a final idea that no one has generated alone. All output of a brainstorming session is to be considered a group result.
3. *Participants should divulge all ideas entering their minds without any constraint.* All members of the group should agree at the outset that a seemingly wild and unrealistic idea may contain an essential element of the ultimate solution.
4. *A key objective is to provide as many ideas as possible within a relatively short time.* It is not unusual for a group to generate 20 to 30 ideas in 1/2 hour of brainstorming. Obviously, to achieve that output the ideas are described only roughly and without details.

There are some generalized questions that have proved helpful. By posing them to yourself or to the group during a brainstorming session, you can stimulate the flow of ideas.

Combinations: What new ideas can arise from combining purposes or functions?
Substitution: What else? Who else? What other place? What other time?
Modification: What to add? What to subtract? Change color, material, motion, shape?
Elimination: Is it necessary?
Reverse: What would happen if we turn it backward? Turn it upside down? Inside out? Oppositely?
Other use: Is there a new way to use it?

A brainstorming session must have a facilitator to control the group and to record the ideas. Write down the ideas verbatim on a flip chart or blackboard. Large Post-it notes are good because they can be used in subsequent phases of problem solving without imposing a need for transcribing. Start with a clear, specific written statement of the problem. Allow a few minutes for members to collect their thoughts, and then begin. Go around the group, in turn, asking for ideas. Anyone may pass, but all should be encouraged to contribute. Build on (piggyback on) the ideas of others. Encourage creative, wild, or seemingly ridiculous notions. There is no questioning, discussion, or criticism of ideas. Generally the ideas build slowly, reach a point where they flow faster than they can be written down, and then fall off. When the group has exhausted all ideas, stop. After a pause for refreshment, review the list for comprehension by seeking clarification on each idea generated. Go through each idea asking, "Does this idea deserve further consideration?" If not, put a bracket around it but do not cross it off the list. This allows any member of the group to reinstate the idea later.

When the student group brainstormed, they obtained the following results.

> *Problem:* Why do so few engineering seniors do a research project?
> Students are too busy.
> Professors do not talk up research opportunity.
> They are thinking about getting a job.
> They are thinking about getting married.
> They are interviewing for jobs.
> They don't know how to select a research topic.
> I'm not interested in research. I want to work in manufacturing.
> I don't know what research the professors are interested in.
> The department does not encourage students to do research.
> I am not sure what research entails.
> It is hard to make contact with professors.
> I have to work part-time.
> Pay me and I'll do research.
> I think research is boring.
> Lab space is hard to find.
> Faculty just use undergraduates as a pair of hands.
> I don't know any students doing research.
> I haven't seen any notices about research opportunities.
> Will working in research help me get into grad school?
> I would do it if it was required.

An alternative form of brainstorming, called *brainwriting,* is sometimes used when the topic is so controversial or emotionally charged that people will not speak out freely in a group. In brainwriting the team members sit around a table and each person writes four ideas on a sheet of paper. Then she or he places the sheet in the center of the table and selects a sheet from another participant to add four additional ideas. That sheet goes back in the center, and another sheet is chosen. The process ends when no one is generating more ideas. Then the sheets are collected, and the ideas collated and discussed.

Affinity diagram. The affinity diagram identifies the inherent similarity between items. It is used to organize ideas, facts, and opinions into natural groupings. In Sec. 2.3 we used the affinity diagram to organize the questions in the customer requirement survey. There we pointed out that a way to do this was to record the ideas on Post-it notes or file cards. Next, each idea is "scrubbed," i.e., each person explains what they wrote on each card so that each team member understands it the same. This often identifies more than one card with the same thought, or reveals cards that have more than one idea on them. If this happens, additional cards are made up. Then the notes or cards are sorted into loosely related groupings. If an idea keeps being moved between two groups because of disagreement as to where it belongs, make a duplicate and put it in both groups. Also, create a group called "Other" for ideas that do not seem to fall in any of the other categories.

As the team becomes more comfortable with the organization, create a header card that broadly describes the content of the group. This often shows that an idea has been put in the incorrect group. This is a time when discussion is allowed, and people may be called upon to defend their idea or where it is placed. When we do this for the brainstorming exercise, we get:

Time constraints
   Students are too busy.
   They are interviewing for jobs.
   I have to work part-time.
Faculty issues
   Professors don't talk up research opportunities.
   The department does not encourage students to do research.
   It is hard to make contact with professors.
   Faculty just use undergraduates as a pair of hands.
   I would do it if it was required.
Lack of interest
   They are thinking about getting a job.
   [They are thinking about getting married.]
   I'm not interested in research. I want to work in manufacturing.
   [Pay me and I'll do research.]
   I think research is boring.
   I would do it if it was required.
Lack of information
   They don't know how to select a research topic.
   I don't know what research the professors are interested in.
   I'm not sure what research entails.
   I don't know any students doing research.
   I haven't seen any notices about research opportunities.
   Will working in research help me get into graduate school?
Other
   Lab space is hard to find.

Note that in the discussion a few of the ideas have been bracketed and removed from active consideration.

Nominal group technique (NGT). The nominal group technique (NGT) is a method of group idea generation and decision making. Often it starts with silent brainstorming (brainwriting) to generate the ideas. The use of the term "nominal" in this method comes from the fact that it often starts out with a nominal, i.e., silent and independent idea generation, group activity, and as we shall see, independent evaluation by each team member.

If the number of choices generated by brainstorming is large, it may be useful to employ some *list reduction* methods. Start with the entire list of ideas displayed so that everyone can see them. For each item ask the question, "Should this item continue to be considered?" A simple majority vote keeps the item on the list; otherwise it is marked with brackets. At the end of voting, any item marked with brackets can be put back on the list by a single team member. Next, each idea is compared with all others, in a pairwise fashion, to decide whether they are different ideas. If all team members feel they are essentially the same, then the ideas are combined with a new wording.

The last step of the NGT involves decision making, with the members of the team acting independently and anonymously. If the number of choices is relatively small, then each person can *rank order* the choices. For example, if there are five choices, A, B, C, D, E, each person would associate a value of 1 to 5 to each choice, where 5 is best. The ranking for all members of the team would be combined, and the choice with the highest score would be the team's first choice. When the number of choices is large, e.g., 20, it becomes difficult to rank order so many items. Here, the "one-half plus one" approach is often used. The team is asked to pick the top 11 items (20/2)+1 in rank order. Again, the ranking of each team member is combined to arrive at the overall team decision.

A variation on decision making by ranking is rating by *multivoting*. Each team member receives a number of votes, usually about one-third of the total number of choices. You can distribute these votes among as many or as few choices as you wish. Often the voting is done by giving each team member the appropriate number of colored sticky dots, and the voting is done by going to the flip chart and pasting them beside your choice(s). Multivoting usually proceeds in stages. In the first round those choices with only a few votes are eliminated. The number of votes per member is adjusted, and a second round of voting is held. The process is repeated until a clear favorite emerges. If the list is reduced to only a few choices with no clear favorite, then the multivoting process should stop, and the team should discuss their options and make a decision by ranking.

The advantage of the NGT is that team members with differing styles of providing input are treated equally because the process imposes the same format requirement on each member. Strong personalities do not unduly influence the outcome. The volume of the loudmouth is turned down while the soft-spoken voice is more clearly heard.

The student team clarified their understanding of the problem by using the NGT to eliminate some extraneous ideas about the low student participation in research projects. Using the affinity diagram as a guide, they carried out silent brainstorming to arrive at the ideas in the following table.

Ideas	Brian	Judy	Gebran	Charn	Total
A. Lack of readily available information about research topics	2	3	1	4	10
B. Lack of understanding of what it means to do research	4	4		1	9
C. Time constraints			4		4
D. No strong tradition of undergraduate research					0
E. Lack of mandatory research course	1				1
F. Lack of student interest		1	2	3	6
G. Lack of incentives	3	2	3	2	10

Using ranking by "half plus one" the students concluded that ideas A, B, and G were very close in terms of contributing to the low participation in research projects. (Note that straight ranking of seven ideas would have been perfectly feasible.) They carried out a second round of ranking with the results given in the table below.

Ideas	Brian	Judy	Gebran	Charn	Total
A. Lack of readily available information about research topics	2	1	1	2	6
B. Lack of understanding of what it means to do research	3	3	3	3	12
C. Lack of incentives	1	2	2	1	6

As a result of a second round of ranking, the team of four students formed the tentative impression that a lack of understanding on the part of undergraduates about what it means to do research is a strong contributor to the low participation by students in research projects. However, they realized that they were but four honors students, whose ideas might be at variance with a wider group of engineering students. They realized that a larger database was needed as they went into the cause-finding stage of problem solving.

Cause finding

Pareto chart. As a first step in data collection the students prepared a survey of what undergraduate students thought about research. They asked whether they were interested in doing research, whether they were currently doing research, and asked them to give an importance ranking for the seven possible causes shown in the above table, with idea D omitted. A very similar survey was given to faculty.

The results of the survey are best displayed by a *Pareto chart.* This is a bar chart used to prioritize causes or issues, in which the cause with the highest frequency of occurrence is placed at the left, followed by the cause with the next frequency of occurrence, and so on. It is based on the Pareto principle, which states that a few causes account for most of the problems, while many other causes are relatively unimportant. This is often stated as the *80/20 rule,* that roughly 80 percent of the problem is caused by only 20 percent of the causes—or 80 percent of the sales—come from 20 percent of the customers—or 80 percent of the tax income comes from 20 percent of the taxpayers, etc. A Pareto chart is a way of analyzing the data that *identifies the vital few in contrast to the trivial many.*

Of the 75 surveys received from undergraduate students, a surprising 93 percent said they were interested in doing a research project, while 79 percent felt there was a lack of undergraduate involvement in research. The Pareto chart for the student ranking of the causes why they do not do research is shown in Fig. 3.1. Lack of understanding of what it means to do research has moved to second place, to be replaced in first place by "lack of information about research topics." However, if one thinks about these results they would conclude that "no mandatory research course" is really a subset of "lack of understanding about research," so that this remains the number one cause of the problem. It is interesting that the Pareto chart for the faculty surveys showed lack of facilities and funding, and lack of incentives, in the one/two position.

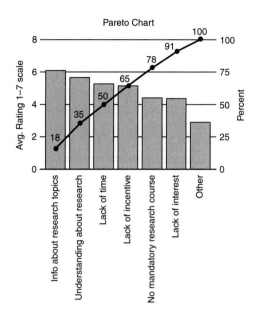

FIGURE 3.1
Pareto chart for average rating of reasons why undergraduate students do not do research projects. Based on survey of students with 75 responses.

Otherwise the order of causes of the problem was about the same. Referring again to Fig. 3.1, note that this contains another piece of information in addition to relative importance. Plotted along the right axis is the cumulative percent of responses. We note that the first five categories (first four when the above correction is made) contain 80 percent of the responses.

Cause-and-effect diagram. The cause-and-effect diagram, also called the fishbone diagram (after its appearance), or the Ishikawa diagram (after its originator), is a powerful graphical way of identifying the factors that cause a problem. It is used after the team has collected data about possible causes of the problem. It is often used in conjunction with brainstorming to collect and organize all possible causes and converge on the most probable root causes of the problem.

Constructing a cause-and-effect diagram starts with writing a clear statement of the problem (effect) and placing it in a box to the right of the diagram. Then the backbone of the "fish" is drawn horizontally out from this box. The main categories of causes, "ribs of the fish," are drawn at an angle to the backbone, and labeled at the ends. These may be categories specific to the problem, or more generic categories such as *methods, machines* (equipment), *materials,* and *people* for a problem dealing with a production process, and *policies, procedures, plant,* and *people* for a service-related process. Ask the team, "What causes this?" and record the cause, not the symptom, along one of the ribs. Dig deeper, and ask what causes the cause you just recorded, so the branches develop subbranches and the whole chart begins to look like the bones of a fish. In recording ideas from the brainstorming session, be succinct but use problem-oriented statements to convey the sense of the problem. As the diagram builds up, look for root causes. One way to identify root causes is to look for causes

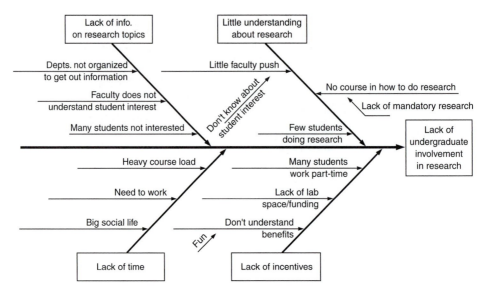

FIGURE 3.2
Cause-and-effect diagram for lack of undergraduate student involvement in research.

that appear frequently within or across categories. Possible root causes are circled on the chart, and the team discusses them and may vote on them. Every attempt is made to use data to verify root causes.

Figure 3.2 shows the cause-and-effect diagram generated by the students to understand the causes for the low student involvement in research. We note that time pressures caused by heavy course loads and necessity to work part-time are one possible root cause, while others center around the lack of understanding of students about what it means to do research and the lack of appreciation by faculty of student interest in doing research.

Why-why diagram. To delve deeper into root causes, we turn to the why-why diagram. This is a tree diagram, which starts with the basic problem and asks "Why does this problem exist?" in order to develop a tree with a few main branches and several smaller branches. The team continues to grow the tree by repeatedly asking "why" until patterns begin to show up. Root causes are identified by causes that begin to repeat themselves on several branches of the why-why tree.

The Pareto chart, when reinterpreted, shows that student lack of understanding about research was the most important cause of low student participation in research. The cause-and-effect diagram also shows this as a possible root cause. To dig deeper we build the why-why diagram shown in Figure 3.3. This begins with the clear statement of the problem. The lack of understanding about research on the part of the undergraduates is two-sided: the faculty don't communicate with the students about opportunities, and the students don't show initiative to find out about it. The team, in asking why, came up with three substantial reasons. Again, they asked why, about

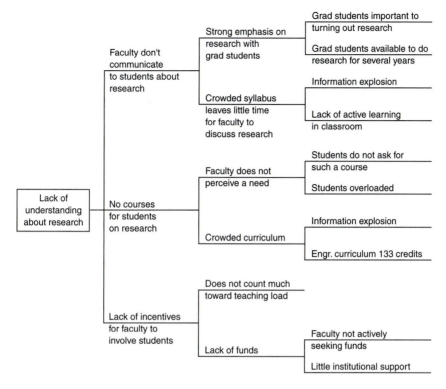

FIGURE 3.3
Why-why diagram for lack of student understanding about research.

each of these three causes, and asking why yet a third time builds up a tree of causes. At this stage we begin to see patterns of causes appearing in different branches of the tree—a sign that these are possible root causes. These are:

- Students and curriculum are overloaded.
- The information explosion is a major cause of the above.
- The faculty don't perceive a need to provide information about research.
- The faculty perceive a low student interest in doing research.
- A lack of resources funding and space limits faculty involvement in undergraduate research.

Narrowing down this set of causes to find the root cause is the job of the next tool.

Interrelationship digraph This is a tool that explores the cause-and-effect relationships among issues and identifies the root causes. The major causes (from 4 to 10) identified by the cause-and-effect diagram are laid out in a large circular pattern (Fig. 3.4). The cause and influence relationships are identified by the team between each cause or factor in turn. Starting with A (chosen at random) we ask whether a causal relationship exists between A and B, and if so, whether the direction is stronger from

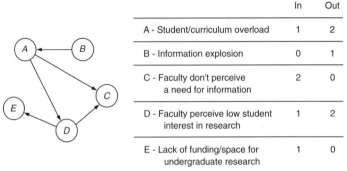

	In	Out
A - Student/curriculum overload	1	2
B - Information explosion	0	1
C - Faculty don't perceive a need for information	2	0
D - Faculty perceive low student interest in research	1	2
E - Lack of funding/space for undergraduate research	1	0

FIGURE 3.4
Interrelationship digraph to identify root causes from why-why diagram (Fig. 3.3).

A to B or B to A. If the causal relationship is stronger from B to A, then we draw an arrow in that direction. Next we explore the relationship between A and C, A and D, etc., in turn, until causal relationships have been explored between all of the factors. Note that there will not be a causal relationship between all factors. For each cause or factor, the number of arrows going in and coming out should be recorded. A high number of outgoing arrows indicates the cause or factor is a root cause or driver. A factor with a high number of incoming arrows indicates that it is a key indicator and should be monitored as a measure of improvement.

In the example in Fig. 3.4, the root causes are the overloaded students and curriculum, and the fact that the faculty perceive that there is a low undergraduate student interest in doing research. The key input is that the faculty do not perceive a need to supply information on research to the undergraduates. Solutions to the problem should then focus on ways of reducing student overload and developing a better understanding of the student interest in doing research.

Solution planning and implementation

While this is the third of three phases, it does not consume one-third of the time in the problem-solving process. This is because, having identified the true problem and the root causes, we are most of the way home to a solution. The objective of solution finding is to generate as many ideas as possible on "how" to eliminate the root causes and to converge on the best solution. To do this we first employ brainstorming and then use multivoting or other evaluation methods to arrive at the best solution. The concept-selection method and other evaluation methods are discussed in Sec. 5.9.

How-how diagram. A technique that is useful for exposing gaps in the causal chain of action is the how-how diagram.[1] Like the why-why diagram, the how-how diagram is a tree diagram, but it starts with a proposed solution and asks the question "How do we do that?" The how-how diagram is best used after brainstorming has generated a set of solutions and an evaluation method has narrowed them to a small set.

---

1. R. Barra, op. cit.

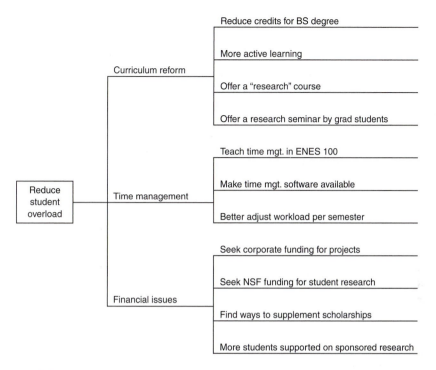

FIGURE 3.5
How-how diagram for problem of reducing student overload, so more students will be able to engage in research projects.

A how-how diagram is constructed for the question "How can we reduce the overload on students?" Brainstorming and multivoting had shown the main issues to be:

- Curriculum reform
- Student time management
- Student and faculty financial issues

Specific solutions that would lead to improvements in each of these areas are recorded in Fig. 3.5.

Force field analysis. Force field analysis is a technique that identifies those forces that both help (drive) and hinder (restrain) the implementation of the solution of a problem. In effect, it is a chart of the pros and cons of a solution, and as such, it helps in developing strategies for implementation of the solution. This forces team members to think together about all the aspects of making the desired change a permanent change, and it encourages honest reflection on the root causes of the problem and its solution. The first step in constructing the force field diagram (Fig. 3.6) is to draw a large T on a flip chart. At the top of the T, write a description of the problem that is being addressed. To the far right of the T, write a description of the ideal solution that we would like to achieve. Participants then list forces (internal and external)

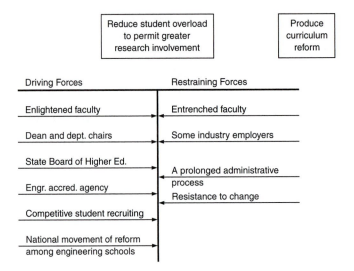

FIGURE 3.6
Force field diagram for implementing solutions to reducing student overload.

that are driving the organization toward the solution on the left side of the vertical line. The forces that are restraining movement toward the ideal solution are listed on the right side of the vertical line. Using a consensus voting method, prioritize the driving forces that should be strengthened to achieve the most movement toward the ideal solution state. Also, identify the restraining forces that would allow the most movement toward the goal if they were removed. This last step is important, because change is more often achieved by removing barriers than by simply pushing the positive factors for change.

Figure 3.6 shows that the key to achieving the needed curriculum reform is to bring aboard some recalcitrant faculty, with help from the dean and departmental chairs. The change process should be expected to be administratively protracted, but doable.

Implementation plan. The problem-solving process should end with the development of specific actions to implement the solution. In doing this, think hard about maximizing the driving forces and minimizing the restraining forces listed in Fig. 3.6. The implementation plan takes the specific actions listed on the how-how diagram and lists the specific steps, in order, that must be taken. It also assigns responsibility to each task, and gives a required completion date. The implementation plan also gives an estimate of the resources (money, people, facilities, material) required to carry out the solution. In addition, it prescribes what level of review and frequency of review of the problem implementation will be followed. A final, but a very important part of the plan, is to list the metrics that will measure a successful completion of the plan.

The team working to increase undergraduate student participation in research evaluated the three major issues for "reduction of student overload" in the how-how

IMPLEMENTATION PLAN

Date: 8/10/00

PROBLEM STATEMENT: Increase the undergraduate student participation in research.

PROPOSED SOLUTION: Create an action team of faculty and students within the college to produce major curriculum reform, to include reduction of credits for the BS degree from 133 to 123 credits, more teaching by active learning, and more opportunity for undergraduate students to do research.

SPECIFIC STEPS:

	Responsibility	Completion date
1. Create curriculum reform action team	Dean	9/30/00
2. Discuss issues with Faculty Council/Dept. Chairs	Dean	10/30/00
3. Hold discussion with dept. faculty	Team	11/15/00
4. Discuss with College Industrial Advisory Council	Dean/Team	11/26/00
5. Discuss with Student Council	Team	11/30/00
6. Day of learning about active learning	Team	1/15/01
7. Dept. curriculum committees begin work	Dept. Chairs	1/30/01
8. Teach "research course" as honors seminar	Team	5/15/01
9. Organize "research seminar," taught by grad students	Team	5/15/01
10. Preliminary reports by dept. curriculum committees	Dean/Team	6/2/01
11. Fine-tuning of curriculum changes	Curric. Com.	9/15/01
12. Faculty votes on curriculum	Dept. Chairs	10/15/01
13. Submittal of curriculum to Univ. Senate	Dean	11/15/01
14. Vote on curriculum by Univ. Senate		2/20/02
15. Implementation of new curriculum	Dean/Chairs	9/1/02

RESOURCES REQUIRED

Budget: $15,000. Speakers for Day of Learning
People: None additional; redirection of priorities is needed.
Facilities: Reserve Dean's Conference Room, each month, 1st and 3rd Wed, 3-5 pm.
Materials: covered in budget above.

REVIEWS REQUIRED

Monthly meeting between team leader and Dean.

MEASURES OF SUCCESSFUL PROJECT ACHIEVEMENT

Reduction in credits for BS degree from 133 to 123 credits.
Increase in number of undergraduates doing research project from 8% to 20%.
Increase in number of engineering students graduating in 4 years.
Increase in number of undergraduates going to graduate school.

FIGURE 3.7
Implementation plan for creating curriculum reform.

diagram (Fig. 3.5). They decided that curriculum reform and financial issues were of equal importance, with time management of lesser importance. The implementation plan for the issue of curriculum reform is shown in Fig. 3.7. A similar plan dealing with financial issues would also be developed. It is important to realize that close communication with the team sponsor, or the manager responsible for the problem solution, is needed to avoid overstepping the limits of authority.

Epilogue. This was not just an isolated student exercise. Over the next 3 years the number of credits for a BS degree was reduced from 133 to 122 credits in all engineering programs. Most of them adopted active learning modes of instruction. A major corporate grant was received to support undergraduate student projects and many faculty included undergraduates in their research proposals. The level of student participation in research projects doubled.

Plan-do-check-act. The plan-do-check-act (PDCA) cycle is a basic concept of TQM. Its origin goes back to Walter Shewhart, an early pioneer of statistical quality control. The idea behind PDCA is that once a solution is arrived at we try it out in a small way to see how it works. This is the *do* stage. Everything that preceded it in problem solving is the *plan* stage. In the *do* stage we collect data to compare with the preexisting situation. In the *check* stage we compare the results with the new solution with the old conditions (baseline data) to determine whether the change has produced the intended improvement. This is an important yet subtle point. Much problem solving neglects this check and assumes that because considerable thought and effort has gone into finding a solution, it will produce the intended improvement. If, indeed, the change is positive, then we *act* to standardize the change in appropriate policies and procedures. If the change is not positive or if it is not as great as we need, then we return to the *plan* stage. The PDCA cycle is a continuous process. In cases where the change "checked out" and we act to implement the change, we are never satisfied with the result. Depending upon priorities and time we revisit the problem topic to search for even better performance in the spirit of continuous improvement.

3.8
TIME MANAGEMENT

Time is an invaluable and irreplaceable commodity. You will never recover the hour you squandered last Tuesday. All surveys of young engineers making an adjustment to the world of work point to personal time management as an area that requires increased attention. The chief difference between time management in college and as a practicing engineer is that time management in the world of work is less repetitive and predictable than when you are in college. For instance, you are not always doing the same thing at the same time of the day, as you do when you are taking classes as a college student. If you have not done so, you need to develop a personal time management system that is compatible with the more erratic time dimension of professional practice. Remember, effectiveness is doing the right things, but efficiency is doing those things the right way, in the shortest possible time.

An effective time management system is vital to help you focus on your long-term and short-term goals. It helps you decipher urgent tasks from important tasks. It is the only means of gaining free time for yourself. Each of you will have to work out a time management system for yourself. The following are some time-tested points to achieve it:

- Start with written goals of what you want to accomplish for the year and for the month.
- Next find out where you spend your time. Start by keeping a log of how you spend your time for a period of at least 1 week, divided into 30-min increments. Classify each activity into one of four categories: (1) important and urgent; (2) important but

not urgent; (3) urgent but not important; and (4) not important and not urgent. From this study you should identify things you are spending time on that you should not be doing. Also, this study should get you in the habit of prioritizing the tasks you have to do, and focusing on the important and urgent items.

- Make a written plan for each day, with the tasks you want to accomplish in priority order. Do this on an 8 1/2 by 11 sheet of paper, not on a lot of little notes to yourself. You may decide to invest in a paper-based or computer-based personal planning system, but ordinary sheets of paper will get the job done. Be cognizant of the 80/20 rule, that 80 percent of your positive results will come from the vital 20 percent of your activities, the urgent and important.
- Set personal deadlines, in addition to business-imposed deadlines, to inspire action and avoid procrastination.
- Learn to act immediately and constructively. After reading a memo, hanging up from a phone conversation, or talking with a visitor in your office, take a specific action like responding to the memo, scheduling a meeting, or digging out a follow-up correspondence file. When doing paperwork, try to handle each paper no more than once.
- Avoid a cluttered desk and office. This requires a good filing system and perseverance.
- Schedule an entire block of time for a major project. Make sure there are no distractions at this time.
- Identify your best time of day, in terms of energy level and creative activity, and try to schedule your most challenging tasks for that time period.
- Group like tasks, e.g., returning phone calls or writing memos, into periods of common activity for more efficient performance.
- Occasionally make appointments with yourself to reflect on your work habits, and think creatively about the future.

3.9
PLANNING AND SCHEDULING

It is an old business axiom that time is money. Therefore, planning future events and scheduling them so they are accomplished with a minimum of time delay is an important part of the engineering design process. For large construction and production projects, detailed planning and scheduling is a must. Computer-based methods for handling the large volume of information have become commonplace. However, engineering design projects of all magnitudes of scale can profit greatly by applying the simple planning and scheduling techniques discussed in this chapter.

One of the most common criticisms leveled at the young graduate engineer is an overemphasis on technical perfection of the design and not enough concern for completing the design on time and below the estimated cost. Therefore, the planning and scheduling tools presented in this chapter can profitably be applied at the personal level as well as to the more complex engineering project.

In the context of engineering design, *planning* consists of identifying the key activities in a project and ordering them in the sequence in which they should be performed. *Scheduling* consists of putting the plan into the time frame of the calendar.

The major decisions that are made over the life cycle of a project fall into four areas: performance, time, cost, and risk.

Performance: The design must possess an acceptable level of operational capability or the resources expended on it will be wasted. The design process must generate satisfactory specifications to test the performance of prototypes and production units.

Time: In the early phases of a project the emphasis is on accurately estimating the length of time required to accomplish the various tasks and scheduling to ensure that sufficient time is available to complete those tasks. In the production phase the time parameter becomes focused on setting and meeting production rates, and in the operational phase it focuses on reliability, maintenance, and resupply.

Cost: The importance of cost in determining what is feasible in an engineering design has been emphasized in earlier chapters. Keeping costs and resources within approved limits is one of the chief functions of the project manager.

Risk: Risks are inherent in anything new. Acceptable levels of risk must be established for the parameters of performance, time, and cost, and they must be monitored throughout the project. The subject of risk is considered in detail in Chap. 11.

The first step in developing a plan is to identify the activities that need to be controlled. The usual way to do that is to start with the entire system and identify the 10 or 20 activities that are critical. Then the larger activities are broken down into subactivities, and these in turn are subdivided until you get to tasks performed by single persons. Generally the work breakdown proceeds in a hierarchical fashion from the system to the subassembly to the component to the individual part.

3.9.1 Bar chart

The simplest scheduling tool is the bar, or Gantt, chart (Fig. 3.8). The activities are listed in the vertical direction, and elapsed time is recorded horizontally. This shows clearly the date by which each activity should start and finish, but it does not make clear how the ability to start one activity depends upon the successful completion of other activities.

The dependence of one activity on another can be shown by a network logic diagram like Fig. 3.9. This diagram clearly shows the precedence relations, but it loses the strong relation with time that the bar chart displays.

Note that a *critical path* through the network can be determined. In this case it is the 20 weeks required to traverse the path *a–b–c–d–e–f–g,* and it is shown on the

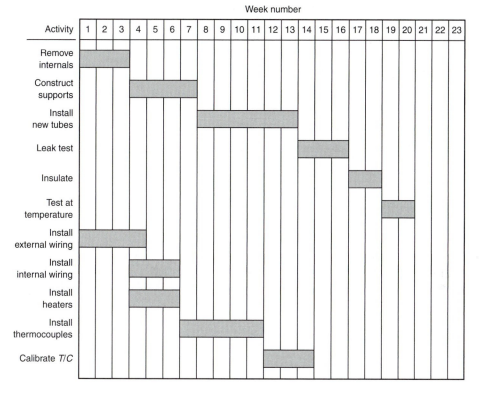

FIGURE 3.8
Bar chart for prototype testing a heat exchanger.

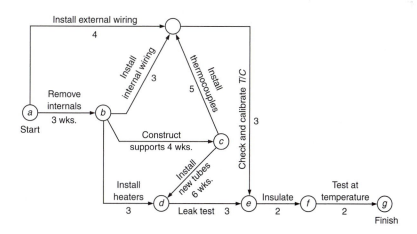

FIGURE 3.9
Network logic diagram for heat exchanger.

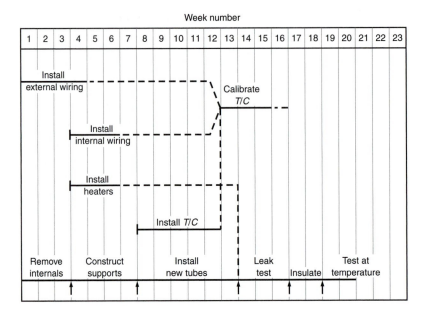

FIGURE 3.10
Modified bar chart for heat exchanger tests.

modified bar chart (Fig. 3.10). The parts of the schedule that have slack time are shown dashed. *Slack* is the time by which an activity can exceed its estimated duration before failure to complete the activity becomes critical. For example, for the activities of installing heaters, there is a 7-week slack before the activities must be completed to proceed with the leak testing. Thus, the identification of the longest path focuses attention on the activities that must be given special management attention, for any delay in those activities would critically lengthen the project. Conversely, identification of activities with slack indicates the activities in which some natural slippage can occur without serious consequences. This, of course, is not license to ignore the activities with slack.

3.9.2 Critical-Path Method

Two computer-based scheduling systems based on networks were introduced in the late 1950s to aid in scheduling large engineering projects. The critical-path method (CPM), developed by Du Pont and Remington Rand, is a deterministic system that uses the best estimate of the time to complete a task. The program evaluation and review technique (PERT), developed for the U.S. Navy, uses probabilistic time estimates. The techniques have much in common. We shall start by considering CPM.

The basic tool of CPM is an arrow network diagram similar to Fig. 3.9. The chief elements of this diagram are:

1. An *activity*—time-consuming effort that is required to perform part of a project. An activity is shown on an arrow diagram by a line with an arrowhead pointing in the direction of progress in completion of the project.
2. An *event*—the end of one activity and the beginning of another. An event is a point of accomplishment and/or decision. A circle is used to designate an event.

There are several logic restrictions to constructing the network diagram.

1. An activity cannot be started until its tail event is reached. Thus, if $\overset{A}{\longrightarrow}\!\!\!\!\bigcirc\!\overset{B}{\longrightarrow}$

   activity B cannot begin until activity A has been completed. Similarly, if

   $\overset{C}{\longrightarrow}\!\!\!\!\bigcirc\!\!<\!\!\overset{D}{\underset{E}{}}$ activities D and E cannot begin until activity C has been completed.

2. An event cannot be reached until all activities leading to it are complete. If

   $\underset{G}{\overset{F}{>\!\!\!\bigcirc\!\overset{H}{\longrightarrow}}}$ activities F and G must precede H.

3. Sometimes an event is dependent on another event preceding it, even though the two events are not linked together by an activity. In CPM we record that situation by introducing a dummy activity, denoted ----►. A *dummy activity* requires zero time and has zero cost. Consider two examples:

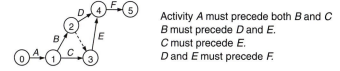

$A \longrightarrow \bigcirc \longrightarrow C$ Activities *A* and *B* must both be
$\qquad \vdots$ completed before Activity *D*, but
$B \longrightarrow \bigcirc \longrightarrow D$ Activity *C* depends only on *A* and
                   is independent of Activity *B*.

Activity *A* must precede both *B* and *C*
B must precede *D* and *E*.
C must precede *E*.
D and *E* must precede *F*.

The longest time through the network (the critical path) may be determined by inspection for a relatively simple network like the one in Fig. 3.10, but a methodology for the much more complex problems found in engineering project management must be established. To do so we establish the following parameters.

Earliest start time (ES): The earliest time an activity can begin when all preceding activities are completed as rapidly as possible.
Latest start time (LS): The latest time an activity can be initiated without delaying the minimum completion time for the project.
Earliest finish time (EF): EF = ES+D, where *D* is the duration of each activity.

Latest finish time (LF): LF $=$ LS$+D$

Total float (TF): The slack between the earliest and latest start times. TF $=$ LS$-$ES. An activity on the critical path has zero total float.

In CPM the estimate of each activity duration is based on the most likely estimate of time to complete the activity. All durations should be expressed in the same time units, such as days or weeks. The sources of time estimates are records of similar projects, calculations involving the manpower needs, legal restrictions, and technical considerations.

The network diagram in Fig. 3.10 has been redrawn as a CPM network in Fig. 3.11. To facilitate solution with computer methods, the events that occur at the nodes have been numbered serially. The node number at the tail of each activity must be less than that at the head. The ES times are determined by starting at the first node and making a forward pass through the network while adding each activity duration in turn to the ES of the preceding activity. The details are shown in Table 3.6

The LS times are calculated by a reverse procedure. Starting with the last event, a backward pass is made through the network while subtracting the activity duration from the limiting LS at each event. The calculations are detailed in Table 3.7. We note that, for calculating LS, each activity starting from a common event can have a different late start time, whereas all activities starting from the same event had the same early start time.

The chief work is in establishing ES and LS times. Once that is accomplished, the remaining boundary time parameters can be determined by routine operations; see Table 3.8. The critical path is identified by the activities with zero total float.

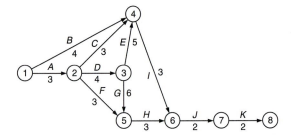

FIGURE 3.11
CPM diagram for heat exchanger project.

TABLE 3.6
Calculation of early start time based on Fig. 3.11

Event	Activity	ES	Comment
1	A, B	0	Conventional to use ES $=$ 0 for the initial event
2	C, D, F	3	$ES_2 = ES_1 + D = 0 + 3 = 3$
3	E, G	7	$ES_3 = ES_2 + D = 7$
4	I	12	At a merge like 4 the largest ES$+D$ of the merging activities is used
5	H	13	$ES_5 = ES_3 + 6 = 13$
6	J	16	$ES_6 = ES_5 + 3 = 16$
7	K	18	
8	—	20	

TABLE 3.7
Calculation of late start times based on Fig. 3.11

Event	Activity	LS	Event	Activity	LS
8	—	20	5-2	F	10
8-7	K	18	4-3	E	8
7-6	J	16	4-2	C	10
6-5	H	13	4-1	B	9
6-4	I	13	3-2	D	3
5-3	G	7	2-1	A	0

TABLE 3.8
Summary of boundary timetable

Activity	Description	D, weeks	ES	LS	EF	LF	TF
A	Remove internals	3	0	0	3	3	0
B	Install external wiring	4	0	9	4	13	9
C	Install internal wiring	3	3	10	6	13	7
D	Construct supports	4	3	3	7	7	0
E	Install thermocouples	5	7	8	12	13	1
F	Install heaters	3	3	10	6	13	7
G	Install new tubes	6	7	7	13	13	0
H	Leak test	3	13	13	16	16	0
I	Check thermocouples	3	12	13	15	16	1
J	Insulate	2	16	16	18	18	0
K	Test prototype at temperature	2	18	18	20	20	0

ES is determined by forward pass through network.
LS is determined by backward pass through network.
EF = ES+D.
LF = LS+D.
TF = LS−ES.

Generally in a CPM problem we are interested in two classes of solutions:

1. The least-cost solution using costs associated with the normal time to complete the activities.
2. The least-time solution in which crash costs are incurred to reduce the time, e.g., by employing overtime, extra workers, or bringing in extra production equipment.

3.9.3 PERT

The program evaluation and review technique (PERT) uses the same ideas as CPM; but instead of using the most likely time estimate, it uses a probabilistic estimate of time for completion of an activity. The designer is asked to make an optimistic time estimate o if everything goes smoothly and a pessimistic time estimate p if everything goes badly. The most likely time m is bracketed between those values. The time estimates are assumed to follow a beta frequency distribution that gives the expected time as

$$t_e = \frac{o + 4m + p}{6} \tag{3.1}$$

The expected time is a mean value that divides the area under the frequency distribution into two equal parts. In PERT the expected time is computed for each activity, and the expected times are used to determine the critical path and the boundary times as illustrated for the CPM technique.

The expected time for each activity also has a standard deviation (see Sec. 10.5), which describes its scatter, given by

$$\sigma \frac{p - o}{6} \tag{3.2}$$

The standard deviation along a path in the PERT network is the square root of the sum of the individual variances for the separate activities along that path.

$$\sigma_{\text{path}} = \sqrt{\Sigma \sigma^2} \tag{3.3}$$

Knowing the variance for each activity permits the calculation of the probability that a certain scheduled event will be completed on schedule. If SS is the scheduled start of a particular event, called a milestone, and ES is the earliest start time for the event, then

$$z = \frac{\text{SS} - \text{ES}}{\sigma_{\text{path}}} \tag{3.4}$$

where z is the standard normal deviate and represents the area under the standardized normal frequency distribution (see Sec. 10.5). If, for example, $z = 0$, there is a 50 percent probability of completing the event on the scheduled date. If $z = -0.5$, there is a 30 percent probability.

PERT/COST is an attempt to include cost data in the CPM-PERT type of network scheduling program. The original concept involved costs at a very high level of detail, but that has proved very cumbersome because of the need for continual updating, re-estimating, and cost changes due to design changes. In most cases, PERT/COST is operated with costs aggregated to a considerable degree.

Project management software is common for the personal computer and workstation. The three scheduling techniques discussed in this chapter can be found in many software versions in a range of complexity and price. They are often reviewed in computer magazines.

3.10
SUMMARY

This chapter considered methods for making you a more productive engineer. Some of the ideas, time management and scheduling, are aimed at the individual, but most of this chapter deals with helping you work more effectively in teams. Most of what is covered here falls into two categories: attitudes and techniques.

Under attitudes we stress:

• The importance of delivering on your commitments, and of being on time
• The importance of preparation, for a meeting, for a field test, etc.

- The importance of giving and learning from feedback
- The importance of using a structured problem-solving methodology
- The importance of managing your time

With regard to techniques, we have presented information on the following:

Team processes:
- Team guidelines (rules of the road for teams)
- Rules for successful meetings
  Problem-solving tools (TQM):
- Brainstorming
- Affinity diagram
- Nominal group technique
- Multivoting
- Pareto chart
- Cause-and-effect diagram
- Why-why diagram
- Interrelationship digraph
- How-how diagram
- Force field analysis
- Implementation plan
  Scheduling tools:
- Bar chart (Gantt chart)
- Critical path method (CPM)
- Program evaluation and review technique (PERT)

Further information on these tools can be found in the references listed in the Bibliography. Also given there are software packages for applying some of these tools.

BIBLIOGRAPHY

Team Methods

Cleland, D. I.:"Strategic Management of Teams," Wiley, New York, 1996.
Harrington-Mackin, D.: "The Team Building Tool Kit," American Management Association, New York, 1994.
Katzenbach, J. R., and D. K. Smith: "The Wisdom of Teams," HarperBusiness, New York, 1993.
Quick, T. L.: "Successful Team Building," American Management Association, New York, 1992.
Scholtes, P. R., et al.: "The Team Handbook," Joiner Associates, Madison, WI, 1988.

Problem-Solving Tools

Barra, R.: "Tips and Techniques for Team Effectiveness," Barra International, New Oxford, PA, 1987.
Brassard, M., and D. Ritter: "The Memory Jogger™ II," Goal/QPC, Methuen, MA, 1994.
Folger, H. S., and S. E. LeBlanc: "Strategies for Creative Problem Solving," Prentice-Hall, Englewood Cliffs, NJ, 1995.

Ozeki, K., and T. Asaka: "Handbook of Quality Tools: The Japanese Approach," Productivity Press, Inc., Cambridge, MA, 1990.
Tague, N. R.: "The Quality Toolbox," ASQC, Quality Press, Milwaukee, WI, 1995.

Planning and Scheduling

Cleland, D. I., and W. R. King: "Systems Analysis and Project Management," 2d ed., McGraw-Hill, New York, 1975.
Lewis, J. P.: "Mastering Project Management," McGraw-Hill, New York, 1988.
Martin, P., and K. Tate, "Project Management Memory Jogger™," Goal/QPC, Methuen, MA, 1997.
Meredith, D. D., K. W. Wong, P. W. Woodhead, and R. H. Wortman: "Design and Planning of Engineering Systems," 2d ed., Prentice-Hall, Englewood Cliffs, NJ, 1985.
Rosenau, M. D.: "Successful Project Management," 3d ed., Wiley, New York, 1998.

Project Management Software

LOWER-END SOFTWARE PACKAGES

Can schedule tasks in Gantt, update schedules over time. More like an advanced personal planner.

Milestones, Etc. 5.0, KIDAS Software, Austin, TX.
Schedule+, Microsoft Corp., Redmond, WA.

MIDRANGE PACKAGES

Can schedule tasks and manage resources within a fairly large project.

Microsoft Project 98, Microsoft Corp., Redmond, WA.
Project Scheduler 7, Scitor Corp., Menlo Park, CA.
SureTrak Project Manager 2.0, Primavera Systems, Bala Cynwyd, PA.
TurboProject Professional, IMSI, San Rafael, CA.

HIGH-END PACKAGES

Handle larger projects and include resource assignment and leveling, people scheduling and time sheets, interface with financial data.

Primavera Project Planner, Primavera Systems, Bala Cynwyd, PA.
SuperProject 4.0, Computer Associates International, Islandia, NY.

PROBLEMS AND EXERCISES

3.1. For your first meeting as a team do some team building activities to help you get acquainted.

    (a) Ask a series of questions, with each person giving an answer in turn. Start with the first question and go completely around the team, then the next, etc. Typical questions might be: (1) What is your name? (2) What is your major and class?

(3) Where did you grow up or go to school? (4) What do you like best about school? (5) What do you like least about school? (6) What is your hobby? (7) What special skills do you feel you bring to the team? (8) What do you want to get out of the course? (9) What do you want to do upon graduation?

(b) Do a brainstorming exercise to come up with a team name and a team logo.

3.2. Early in the process of forming a team, have a serious discussion to draw up team ground rules. These are rules of agreement about behavior at team meetings and agreement on how team members will give and receive feedback. These are distinct from the rules for an effective meeting discussed in Sec 3.5.

3.3. Teams often find it helpful to create a team charter between the team sponsor and the team. What topics should be covered in the team charter?

3.4. To learn to use the TQM tools described in Sec. 3.7, spend about 4 h total of team time to arrive at a solution for some small problem that is familiar to the students and they feel needs improvement. Look at some aspect of an administrative process in the department or campus. Be alert to use the TQM tools in your design project.

3.5. After about 2 weeks of team meetings, invite a disinterested and knowledgeable person to attend a team meeting as an observer. Ask them to give a critique of what they found. Then invite them back in 2 weeks to see if you have improved your team performance.

3.6. Develop a rating system for effectiveness of team meetings.

3.7. Keep a log of how you spend your time over the next week. Break it down by 30-min intervals.

3.8. The following restrictions exist in a scheduling network. Determine whether the network is correct; and if it is not, draw the correct network.

(a) A precedes C
B precedes E
C precedes D and E

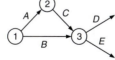

(b) A precedes D and E
B precedes E and F
C precedes F

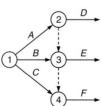

3.9. The development of an electronic widget is expected to follow the following steps.

Activity	Description	Time est., weeks	Preceded by
A	Define customer needs	4	
B	Evaluate competitor's product	3	
C	Define the market	3	
D	Prepare product specs	2	B
E	Produce sales forecast	2	B
F	Survey competitor's marketing methods	1	B
G	Evaluate product vs. customer needs	3	A, D
H	Design and test the product	5	A, B, D
I	Plan marketing activity	4	C, F
J	Gather information on competitor's pricing	2	B, E, G
K	Conduct advertising campaign	2	I
L	Send sales literature to distributors	4	E, G
M	Establish product pricing	3	H, J

Establish the arrow network diagram for this project and determine the critical path by using the CPM technique.

4

GATHERING INFORMATION

4.1
THE INFORMATION PROBLEM

We have already seen in Chap. 2 that the need for information about potential markets can be crucial in a design project. There are many, many other pieces of information that you will need to find and validate quickly. For example, we might need to find the suppliers and costs of fractional-horsepower motors with a certain torque and speed. At a lower level of detail, we would need to know the geometry of the mounting brackets for the motor we select for the design. At a totally different level, we might need to know whether the totally new trade name we created for a new product line infringes on any existing trade names, and further, whether it will cause any cultural problems when pronounced in Spanish, Japanese, and Mandarin Chinese. Clearly, the information needed for an engineering design is more diverse and less readily available than that needed for conducting a research project, for which the published technical literature is the main source of information. We choose to emphasize the importance of the information-gathering step in design by placing this chapter early in this text and in the sequence of design steps (Fig. 4.1).

This chapter gives some suggestions for coping with your information needs. It is not intended to be encyclopedic or contain all the information on how and where to look. The first step you should take is to become familiar with your local information sources. Visit your university or company library and make friends with the librarian. Find out what is available and what your organization is prepared to do to help you with your information needs.

The next thing you should do is develop a personal plan for coping with information. The world technical literature is doubling every 10 to 15 years. That amounts to about 2 million technical papers a year, or a daily output that would fill seven sets of the Encyclopaedia Britannica. This tremendous flood of information aids greatly in the development of new knowledge, but in the process it makes obsolete part of what you already know. To develop a personal plan for information processing is one of the

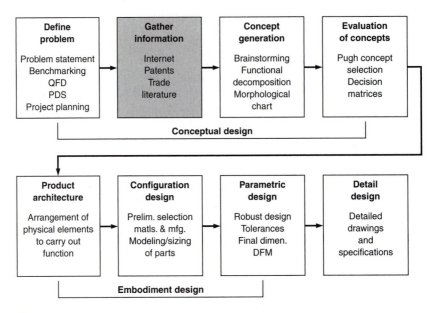

FIGURE 4.1

Steps in design process, showing early placement of the gathering information step.

most effective things you can do to combat your own technological obsolescence. Such a plan begins with the recognition that you cannot leave it entirely to your employer to finance your needs in this area. As a professional, you should be willing to allocate a small portion of your resources, e.g., 1 percent of your net annual salary, for adding to your technical library and your professional growth. This includes the purchase of new textbooks in fields of current or potential interest, specialized monographs, software, membership in professional societies, and subscriptions to technical journals and magazines. In one way or another, you should attend conference and technical meetings where new ideas on subjects related to your interest are discussed.

You should develop your own working files of technical and business information that is important to your work. A good way to do so is to put articles and information you want to have as ready reference into large three-ring binders with page dividers marking off different areas of interest. Material of less current interest may be stored in folders in filing cabinets. A common difficulty, once you start this activity, is compulsive saving. In order that your files will not grow without bound, be selective. Adopt the policy of discarding outdated material when you replace it with newer information. If you are concerned with losing track of possibly useful articles by discarding older material, you might compromise by keeping only the title page, which contains the abstract, and also possibly the last page, which contains the conclusions and references.

To have current awareness of your technical field, you should take a three-pronged approach:[1] (1) read the core journals in your chief area of interest, (2) utilize current awareness services, (3) participate in selective dissemination programs.

---

1. B. E. Holm, "How to Manage Your Information," Reinhold Book Corp., New York, 1968.

Every professional must read enough journals and technical magazines to keep up with the technology in the field and be able to apply the new concepts that have been developed. These journals, which should be read on a monthly basis, should come from three categories:

1. General scientific, technical, and economic (business) news. The monthly magazine of your main professional society would fit here.
2. Trade magazines in your area of interest or business responsibility.
3. Research-oriented journals in your area of interest.

Reading regularly in the above three categories is a major aspect of keeping current in your field. However, for many people this will not cover as wide a spectrum of the published literature as is required. Therefore, secondary current awareness services have been developed. Some of them provide abstracts of articles, others just the titles of articles. The abstract services available in each discipline, e.g., *Engineering Index* and *Metals Abstracts,* are scanned by some serious professionals each month to see what new information has been published. This usually requires spending several hours in the library. *Current Contents: Engineering and Technology* is a weekly publication of the Institute for Scientific Information, Philadelphia, which reproduces the title pages of a large number of engineering publications. A computerized index provides the addresses of authors if you want to write for reprints. The advantage of *Current Contents* is that it is less expensive than a full abstract service, so you can have an individual or shared subscription. Your technical library should have a copy.

Selective dissemination is concerned with sending specific information to the individual who has a need for and interest in it. Many company librarians provide such a service. Researchers in a common field will often develop a "community of interest" and keep each other informed by sharing their papers and ideas. As more and more technical information is put into computer database, it becomes easier to provide selective dissemination.

There is general recognition that engineers as professionals make much less use of the published technical literature than do scientists and other technical professionals. Some of this difference may be due to differences in education, but mostly it is due to a basic difference in how technical information is organized and stored.[1] Almost all major indexing systems index by subjects, i.e., the names of things. This is ideal for the expert concerned with research or analysis, who is very familiar with the vocabulary in a narrow technical field. However, indexing by name is a severe handicap for the design engineer, who is involved in synthesis. The designer, concerned with finding the best way to solve a problem, wants to start the search with the function of what is needed. There is a vast array of devices for attaining those functions, but there is no organized way to help the designer identify the functions by name so they can be uncovered in a literature search. Also, the synthesis-oriented design engineer is concerned with attributes, e.g., higher energy efficiency and lower material cost. Although they frequently are described in technical articles, the attributes very seldom are included in indexing terms.

---

1. E. J. Breton, *Mech. Eng.,* pp. 54–57, March 1981.

Thus, it is not surprising that a recent study revealed that engineers collect most of their advanced technical information from discussions with vendors, salespeople, consultants, and other engineers. Partly that is because information on new products, the buildings blocks of technology, is poorly handled in the technical literature. A 12-month study[1] of news releases on new products in trade magazines showed that over 65 percent were actually old products or were old products with cosmetic alterations.

4.2
COPYRIGHT AND COPYING

A copyright is the exclusive legal right to publish a tangible expression of literary or artistic work, and it is therefore the right to prevent the unauthorized copying by another of that work. In the United States a copyright is awarded for a period of the life of the copyright holder plus 50 years. It is not necessary to publish a copyright notice for a work to be copyrighted. A copyright comes into existence when one fixes the work in "any tangible medium of expression." Unlike for a patent, with a copyright there is no extensive search to ensure the degree of originality of the work.

A major revision of the copyright law of 1909 went into effect on January 1, 1978. The present copyright law covers original works of authorship that are literary works as well as pictorial, graphic, and sculptural works. Important for engineering design is the fact that the new law is broad enough to cover for the first time written engineering specifications, sketches, drawings, and models.[2] However, there are two important limitations to this coverage. Although plans, drawings, and models are covered under the copyright law, their mechanical or utilitarian aspects are expressly excluded. Thus, the graphic portrayal of a useful object may be copyrighted, but the copyright would not prevent the construction from the portrayal of the useful article that is illustrated.

The other limitation pertains to the fundamental concept of copyright law that one can copyright not an idea, but only its tangible expression. The protection offered the engineer under the new law lies in the ability to restrict the distribution of plans and specifications by restricting physical copying. An engineer who retains ownership of plans and specifications through copyrighting can prevent a client from using them for other than the original, intended use and can require that they be returned after the job is finished.

A major impetus for revising the copyright law was to make the law compatible with the technology of fast, inexpensive copying machines. The new law retains the principle of *fair use* in which an individual has the right to make a single copy of copyrighted material for personal use for the purpose of criticism, comment, news reporting, teaching, scholarship, or research. Copying which does not constitute fair use must pay a royalty fee to the Copyright Clearance Center. While the U.S. Copyright Act does not directly define fair use, it does base it on four factors:[3]

---

1. Ibid.
2. H. K. Schwentz and C. J. Hardy, *Professional Engineer,* pp. 32–33, July 1977.
3. D. V. Radack, *JOM,* February 1996, p. 74.

- The purpose and character of the use—is it of a commercial nature or for nonprofit educational purposes?
- The nature of the copyrighted work—is it a highly creative work or a more routine document?
- The amount of the work used in relation to the copyrighted work as a whole.
- The effect of the use on the potential market value of the copyrighted work. Usually this is the most important of the factors.

4.3
HOW AND WHERE TO FIND IT

The search for information can be performed more efficiently if a little thought and planning are used at the outset. First, be sure you understand the purposes for which the information is being sought. If you are looking for a specific piece of information, e.g., the yield strength of a new alloy or the cost of a miniature ball bearing, you should pursue one set of information sources. However, if your purpose is to become familiar with the state of the art in an area that is new to you, you should follow a different course. Table 4.1, based on a listing by Woodson,[1] shows the many sources of information that are open to you.

TABLE 4.1
Sources of information for engineering design

I. Public sources
   A. Federal departments and agencies (Defense, Commerce, Energy, NASA, etc.)
   B. State and local government (highway department, departments dealing with land use, consumer safety, building codes, etc.)
   C. Libraries—community, university, special
   D. Universities, research institutions, museums
   E. Foreign governments—embassies, commercial attaches
   F. Internet—Much information is free. Some requires fees.

II. Private sources
   A. Nonprofit organizations and services
      1. Professional societies
      2. Trade and labor associations
      3. Membership organizations (motorists, consumers, veterans, etc.)
   B. Profit-oriented organizations
      1. Vendors (include manufacturers, suppliers, financiers). Catalogs, samples, test data, cost data and information on operation, maintenance, servicing and delivery
      2. Other business contacts with manufacturers and competitors
      3. Consultants
   C. Individuals
      1. Direct conversation or correspondence
      2. Personal friends, associates, "friends of friends"
      3. Faculty

1. T. T. Woodson, "Introduction to Engineering Design," chap. 5, McGraw-Hill, New York, 1966.

In reviewing this list, you can divide the sources of information, into (1) people who are paid to assist you, e.g., the company librarian or consultant, (2) people who have a financial interest in helping you, e.g., a potential supplier of equipment for your project, and (3) people who help you out of professional responsibility or friendship.

All suppliers of materials and equipment provide sales brochures, catalogs, technical manuals, etc., that describe features and operation of their products. Usually this information can be obtained at no cost by checking the reader service card that is enclosed in most technical magazines. Much of this information is now available on the Internet. Practicing engineers commonly build up a file of such information. Generally a supplier who has reason to expect a significant order based on your design will most likely provide any technical information about the product that is needed for you to complete your design.

It is only natural to concentrate on searching the published technical literature for the information you need, but don't overlook the resources available among your colleagues. The professional files or notebooks of engineers more experienced than you can be a gold mine of information if you take the trouble to communicate your problem in a proper way. Remember, however, that the flow of information should be a two-way street. Be willing to share what you know, and above all, return the information promptly to the person who lent it to you. The surest way to get shut off is to gain a reputation as a moocher.

In seeking information from sources other than libraries (see Sec. 4.4), a direct approach is best. Whenever possible, use a phone call rather than a letter. A direct dialogue is vastly superior to the written word. However, you may want to follow up your conversation with a letter. Open your conversation by identifying yourself, your organization, the nature of your project, and what it is you need to know. Preplan your questions as much as possible, and stick to the subject of your inquiry. Don't worry about whether the information you seek is confidential information. If it really is confidential, you won't get an answer, but you may get peripheral information that is helpful. Above all, be courteous in your manner and be considerate of the time you are taking from the other person. Some companies employ an outside service that networks technical experts to supply pieces of information.[1]

It may take some detective work to find the person to call for the information. You may find the name of a source in the published literature or in the program from a recent conference you attended. The Yellow Pages in the telephone directory or an Internet search engine are good places to start. For product information, you can start with the general information number that is listed for almost every major corporation or check their homepage on the worldwide web. To locate federal officials, it is helpful to use one of the directory services that maintain up-to-date listings and phone numbers.

It is important to remember that information costs time and money. It is actually possible to acquire too much information in a particular area, far more than is needed to make an intelligent decision. One can consider that each decision in the design process is a balance between the risk of proceeding with what you have versus the cost of gaining more information to minimize the risk.

However, do not underestimate the importance of information gathering or the effort required in searching for information. Many engineers feel that this isn't real

---

1. B. Boardman, *Research Technology Management,* July-August 1995, pp. 12–13.

engineering, but surveys of how design engineers use their time show that they spend about 15 to 20 percent of their time searching for information, a like percent in meetings, and a similar percent writing reports and engaging in other forms of communication. The time spent actually designing is often less than 50 percent of the total effort.

4.4
LIBRARY SOURCES OF INFORMATION

In the preceding section we considered the broad spectrum of information sources and focused mostly on the information that can be obtained in the business world. In this section we shall deal with the type of information that can be obtained from library sources. The library is the most important resource for students and young engineers who wish to develop professional expertise quickly.

A library is a repository of information that is published in the open or unclassified literature. Although the scope of the collection will vary with the size and nature of the library,[1] all technical libraries will have the capability of borrowing books and journals for you or providing, for a fee, copies of needed pages from journals and books. Many technical libraries also carry selected government publications and patents, and company libraries will undoubtedly contain a collection of company technical reports (which ordinarily are not available outside the company).

When you are looking for information in the library you will find a hierarchy of information sources, as shown in Table 4.2. These sources are arranged in increasing order of specificity. Where you enter the hierarchy depends on your own state of knowledge about the subject and the nature of the information you want to obtain. If

TABLE 4.2
**Hierarchy of library
information sources**

Technical dictionaries

Encyclopedias

Handbooks

Textbooks and monographs

Bibliographies

Indexing and abstract services

Technical and professional journals

Translations

Technical reports

Patents

Catalogs and manufacturers' brochures

1. If you do not have a good technical library at your disposal, you can avail yourself via mail of the fine collection of the Engineering Societies Library, now located at the Linda Hall Library, 5109 Cherry Street, Kansas City, MO 64110-2498; (800)-662-1545; e-mail: requests@lhl.lib.mo.us; Internet web page: http://www.lhl.lib.mo.us/.

you are a complete neophyte, it may be necessary to use a technical dictionary and read an encyclopedia article to get a good overview of the subject. If you are quite familiar with the subject, then you may simply want to use an index or abstract service to find pertinent technical articles. Most sources of information will be found in the reference section of the library.

The search for information can be visualized along the paths shown in Fig. 4.2. Starting with a limited information base, you should consult technical encyclopedias and library's public access catalog, today automated in most libraries, to search out broad introductory texts. As you become expert in the subject, you should move to more detailed monographs and/or use abstracts and indexes to find pertinent articles in the technical literature. Reading these articles will suggest other articles (cross references) that should be consulted. Another route to important design information is the patent literature (Sec. 4.7).

The task of translating your own search needs into the terminology that appears in the library catalog is often difficult. As mentioned previously, library catalogs,

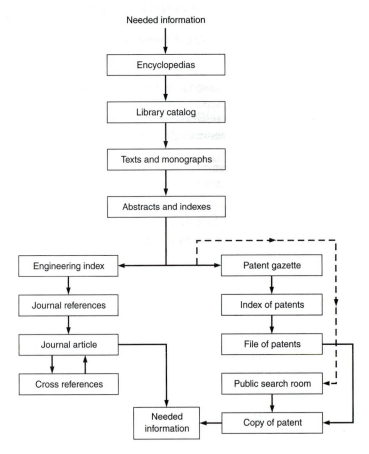

FIGURE 4.2
Flow diagram for an information search.

whether in card form or on-line, are developed for more traditional scholarly and research activities. The kinds of questions raised in the context of engineering design may cut through the card catalog at an "oblique section." When trying to convey the needs and objectives of your search to the librarian, the best tactic is to tell the librarian *what it is you do not know* rather than describe what you already know. Two parameters that describe the efficiency of your information search are:

$$\text{Precision} = \frac{\text{number of relevant documents retrieved}}{\text{total number retrieved}}$$

$$\text{Recall} = \frac{\text{number of relevant documents retrieved}}{\text{number of relevant documents in the collection}}$$

4.4.1 Dictionaries and Encyclopedias

At the outset of a project dealing with a new technical area there may be a need to acquire a broad overview of the subject. English language technical dictionaries usually give very detailed definitions. Also, they often are very well illustrated. Some useful references are:

Davis, J. R. (ed.): "ASM Materials Engineering Dictionary," ASM International, Materials Park, OH, 1992.

Nayler, G. H. F.: "Dictionary of Mechanical Engineering," 4th ed., Butterworth-Heinemann, Boston, 1996.

Parker, S. P. (ed.): "McGraw-Hill Dictionary of Engineering," McGraw-Hill, New York, 1997.

Parker, S. P. (ed.): "McGraw-Hill Dictionary of Scientific and Technical Terms," 5th ed., McGraw-Hill, New York, 1994.

Technical encyclopedias are written for the technically trained person who is just beginning to learn about a new subject. Thus, encyclopedias are a good place to start out if you are only slightly familiar with a subject because they give a broad overview rather quickly. In using an encyclopedia, spend some time checking the index for the entire set of volumes to discover subjects you would not have looked up by instinct. Some useful technical encyclopedias are:

Bever, M. B. (ed.): "Encyclopedia of Materials Sciences and Engineering," 8 vols., The MIT Press, Cambridge, MA, 1986.

"McGraw-Hill Encyclopedia of Environmental Science and Engineering," 3d ed., McGraw-Hill, New York, 1993.

"McGraw-Hill Encyclopedia of Physics," 2d ed., McGraw-Hill, New York, 1993.

"McGraw-Hill Encyclopedia of Science and Engineering," 8th ed., 20 vols., McGraw-Hill, New York, 1997. Also available on CD-ROM.

4.4.2 Handbooks

Undoubtedly, some place in your engineering education a professor has admonished you to reason out a problem from "first principles" and not be a "handbook engineer." That is sound advice, but it may put handbooks in a poor perspective that is undeserved.

Handbooks are compendia of useful technical data. Many handbooks also provide ample technical description of theory and its application, so they are good refreshers of material once studied in greater detail. You will find that an appropriately selected collection of handbooks will be a vital part of your professional library.

An extensive list of handbooks dealing with the material properties of metals, ceramics, and polymers is given in Sec. 8.4. A compendium of handbooks and reference books concerned with various manufacturing processes is given in Sec. 9.2. We list below some basic engineering handbooks that cover most of the other engineering disciplines. Many handbooks are becoming available on CD-ROM so that keyword searches and searches for specific concepts or terms can be made. This greatly increases the usefulness of a large handbook.

Baumeister, T. (ed.): "Marks' Standard Handbook for Mechanical Engineering," 10th ed., McGraw-Hill, New York, 1996.

Christiansen, D. (ed.): "Electronics Engineers' Handbook," 4th ed., McGraw-Hill, New York, 1996.

Fink, D. G., and H. Beaty: "Standard Handbook for Electrical Engineers," 13th ed., McGraw-Hill, New York, 1993.

Karassik, I. J., et al.: "Pump Handbook," 2d ed., McGraw-Hill, New York, 1986.

Kobayashi, A. S. (ed.): "Handbook of Experimental Mechanics," 2d ed., Wiley, New York, 1993.

Kreith, F. (ed.): "The Mechanical Engineering Handbook," CRC Press, Boca Raton, FL, 1997.

Kutz, M. (ed.): "Mechanical Engineers' Handbook," 2d ed., Wiley, New York, 1998.

Loftin, M. K. (ed.): "Standard Handbook for Civil Engineers," 4th ed., McGraw-Hill, New York, 1995.

Maynard, H. B. (ed.): "Industrial Engineering Handbook," 3d ed., McGraw-Hill, New York, 1971.

Parmley, R. O.: "Standard Handbook of Fastening and Joining," 3d ed., McGraw-Hill, New York, 1997.

Perry, R. H., (ed.): "Perry's Chemical Engineers' Handbook," 7th ed., McGraw-Hill, New York, 1997.

Salvendy, G. (ed.): "Handbook of Industrial Engineering," 2d ed., Wiley, New York, 1992.

Smith, E. H. (ed.): "Mechanical Engineer's Reference Book," 12th ed., Butterworth-Heinemann, London, 1994.

Tapley, B. D. "Eshbach's Handbook of Engineering Fundamentals," 4th ed., Wiley, New York, 1990.

Tucker, A. B. (ed.): "The Computer Science and Engineering Handbook," CRC Press, Boca Raton, FL, 1996.

Walsh, R. A.: "Electromechanical Design Handbook," 2d ed., McGraw-Hill, New York, 1994.

Woodson, W. E. (ed.): "Human Factors Design Handbook," 2d ed., McGraw-Hill, New York, 1991.

Handbooks are often highly specialized. For example, there is the "Metering Pump Handbook," "Dudley's Gear Handbook," and the long-running favorite "Roark's Formulas for Stress and Strain."

4.4.3 Textbooks and Monographs

New technical books are continually being published. A good way to keep up to date is to scan the books-in-print column of your professional society's monthly magazine, or to belong to a technical book club. If you want to find out what books are available in a particular field consult "Books in Print," available in nearly every library, or use an Internet book selling service such as amazon.com.

4.4.4 Indexing and Abstracting Services

Indexing and abstracting services provide current information on periodical literature, and they also provide a way to retrieve published literature. An *indexing service* cites the article by title, author, and bibliographic data. An *abstracting service* also provides a summary of the contents of the article. Although indexing and abstracting services primarily are concerned with articles from periodicals, many often include books and conference proceedings, and some list technical reports and patents. The following is a list of indexes and abstracts that cover most of the engineering disciplines.

> *Applied Mechanics Reviews*
> *Applied Science and Technology Index* (formerly *Industrial Arts Index*)
> *Ceramic Abstracts*
> *Chemical Abstracts*
> *Computing Reviews*
> *Corrosion Abstracts*
> *Engineered Materials Abstracts*
> *Engineering Index*
> *Environment Index*
> *Fuels and Energy Abstracts*
> *Highway Research Abstracts*
> *International Aerospace Abstracts*
> *Mechanical Engineering Abstracts*
> *Metals Abstracts* [combines ASM *Review of Metals Literature* (U.S.) and
>   *Metallurgical Abstracts* (British)]
> *Science Abstracts A: Physics Abstracts*
> *Science Abstracts B: Electrical and Electronics Abstracts*
> *Science Abstracts C: Computer and Control Abstracts*
> *Solid State and Superconductivity Abstracts*

A useful source to much detailed information is *Dissertation Abstracts,* which gives abstracts of most doctoral dissertations completed in the United States and Canada. A copy of the dissertation can be ordered at nominal cost.

The reader should know that the situation with respect to indexes and abstracts is in a state of flux at the present time. Many abstracting services have made their products available on CD-ROM, or more recently directly on the Internet (see Sec. 4.6). Others are going direct from print to the Internet, while others are dropping print copies altogether. The advantage of using CD-ROM or the Internet is speed of searching, but also it gives the ability to do "pinpoint searching" at the intersection of several databases.

Conducting a search of the published literature is like putting together a complex puzzle. One has to select a starting place, but some starts are better than others. A good strategy[1] is to start with the most recent subject indexes and abstracts and try to find a current review article or general technical paper. The references cited in it will be helpful in searching back along the "ancestor references" to find the research that led to the current state of knowledge. However, this search path will miss any references that were overlooked or ignored by the original researchers. Therefore, the next step should involve citation searching to find the "descendant references" using *Science Citation Index.*[2] Once you have a reference of interest, you can use *Citation Index* to find all other references published in a given year that cited the key reference. Because they are on CD-ROM or on-line, such searches can be done quickly and precisely. These two search strategies will uncover as many references as possible about the topic. The next step is to identify the key documents. One way to do this is to identify the references with the greatest number of citations, or those that other experts in the field cite as particularly important. You must remember that it takes 6 to 12 months for a reference to be included in an index or abstract service, so current research will not be picked up using this strategy. Current awareness can be achieved by searching *Current Contents* on a regular basis using keywords, subject headings, journal titles, and authors already identified from your literature search. One must also be aware that much information needed in engineering design cannot be accessed through this strategy because it is never listed in scientific and technical abstract services. For this information, the Internet is becoming a vital resource (see Sec. 4.6).

The scientific and technical literature is increasing exponentially. A natural response to this explosion of knowledge is the division into subfields, and this brings with it the development of new journals. The resulting cost from this proliferation means that not every library can afford to carry every journal. If your library does not have a journal or book that you need, ask the librarian to obtain it for you by the Interlibrary Loan (ILL) service. A detailed directory to periodicals and indexing services is *Ulrich's International Periodicals Directory,* published annually in five volumes.

4.4.5 Translations

Although English is the predominant world language for scientific literature, about one-third of the world's scientific and technical literature is produced in the Soviet Union, China, and Japan in languages that are unfamiliar to over 95 percent of scientists and engineers in the United States. Therefore, to avail yourself of information published in those languages, as well as in more common scientific languages like German and French, you should be familiar with sources of translations.

Unfortunately, machine translation by computer, while improving steadily, has not yet reached the point where a finished translation can be done routinely. Therefore, most technical translation must be performed by an experienced translator. Working unaided, a translator typically can complete 5 pages a day. Starting with a machine translation as a first draft, this can be increased to about 15 pages. Because technical

---

1. L. G. Ackerson, *RQ,* vol. 36, pp. 248–260, 1996.
2. Note that there are different version of *Citation Index* for biological, chemical, physical, and engineering and technology publications.

translation is expensive, it is important to make the existence of translations widely available. The best source of this information is the National Translations Center, Library of Congress, Washington, DC. Phone (202)-707-0100.

4.4.6 Catalogs, Brochures, and Business Information

An important kind of design information is catalogs, brochures, and manuals giving information on materials and components that can be purchased from outside suppliers. Most engineers build up quite a collection of this trade literature, often using the reply cards in trade magazines as a way of obtaining new information. Visits to trade shows are an excellent way to become acquainted quickly with the products offered by many vendors. When faced with the problem of where to turn to find information about an unfamiliar new component or material, start with the *Thomas Register of American Manufacturers*. This large multivolume annual publication gives the names and addresses of manufacturers by product and service, as well as company profiles and an extensive catalog file. Many trade publications also publish annual product locators. Rather than worry about keeping a large collection of supplier catalogs, many companies subscribe to a "package library" that provides this information on microfilm, CD-ROM, or via the Internet.

Most technical libraries also contain certain types of business or commercial information that is important in design. Information on the consumption or sales of commodities and manufactured goods by year and state is collected by the federal government and is available in the U.S. Department of Commerce *Census of Manufacturers* and the Bureau of the Census *Statistical Abstract of the United States*. This type of statistical information, important for marketing studies, is also sold by commercial vendors. In using this data it is arranged by industry according to the *Standard Industrial Classification* codes (SIC code). For example, products in the primary metals industries start with 33, those in transportation equipment 37, etc.

4.5
GOVERNMENT SOURCES OF INFORMATION

The federal government either conducts or pays for about 35 percent of the research and development in this country. That generates an enormous amount of information, mostly in the form of technical reports. This R&D enterprise is concentrated in defense, space, environmental, medical, and energy-related areas. It is an important source of information, but all surveys indicate that it is not utilized nearly as much as it ought to be.[1]

The Government Printing Office (GPO) is the federal agency with the responsibility for reproducing and distributing federal documents. Although it is not the sole source of government publications, it is the place to start. It publishes the *Monthly*

---

1. J. S. Robinson, "Tapping the Government Grapevine," 2d ed., The Oryx Press, Phoenix, AZ, 1993.

Catalog of United States Government Publications, the most comprehensive bibliography of unclassified federal publications. The *Publications Reference File,* published bimonthly on microfiche, lists the publications for sale by the GPO.

Reports prepared under contract by industrial and university R & D organizations ordinarily are not available from the GPO. These reports may be obtained from the National Technical Information Service (NTIS), a branch of the Department of Commerce. NTIS, a self-supporting agency through sale of information, is the nation's central clearinghouse for U.S. and foreign technical reports, federal databases, and software. Searches for technical reports should begin with *Government Reports Announcements and Index* (GRA & I) or its computerized counterpart, NTIS Bibliographic Database. This database is searchable by titles, authors, accession number, contract number, or subject.

About one-quarter of the NTIS collection is foreign reports and translations. The Japanese Technical Literature Act of 1986 requires NTIS to monitor Japanese technical activity and collect and translate Japanese technical reports. These and other translations are listed in the weekly *NTIS Foreign Abstract Newsletter.*

In searching for government sources of information, the GPO will give a broader spectrum of information, while NTIS will focus you on the technical report literature. However, even with the vast collection at NTIS it does not have all federally sponsored technical reports. Because agency submissions are voluntary rather than legally mandated, about one-third of the federal technical reports do not reach NTIS. Its collection is chiefly based on NASA and the Departments of Defense and Energy. Another large collection of technical reports, mostly on microfiche, is in the Science and Technology Division of the Library of Congress. This contains reports from 60 countries. Information on how to access these sources is given in Table 4.3.

TABLE 4.3
Sources of government technical information

Source	Addresses
Government Printing Office www.gpo.gov/su_docs/locate.html	Superintendent of Documents, Government Printing Office, Washington, DC 20402 (202)-783-3238
	Mail orders go to: Box 371594, Pittsburgh, PA 15250-7954
National Technical Information Service www.ntis.gov	5285 Port Royal Road Springfield, VA 22161 (703)-487-4650
Library of Congress Science Reference Section www.loc.gov	10 First Street, S.E. Washington, DC 20540 (202)-707-5580
National Institute for Standards and Technology National Center for Standards and Certification Information www.nist.gov	Administration Building, Room A629 Gaithersburg, MD 20899 (301)-975-4040

4.6
INFORMATION FROM THE INTERNET

The fastest-growing communication medium is the Internet. Not only is this becoming the preferred form of personal and business communication via e-mail, but it is rapidly becoming a major source for data retrieval and a channel of commerce. The Internet is a computer network interconnecting numerous computers or local computer networks. Worldwide, the number of host computers on the Internet has increased from less than 100,000 in 1992 to over 18 million 5 years later. These computer networks are linked by a set of common technical protocols so that users in a Macintosh network can communicate with or use the services located, for example, in a Unix network. These protocols are known as the *transmission control protocol/Internet protocol,* or the TCP/IP protocol suite. The Internet functions with a communications technology called *packet-switching,* which breaks the data into small fragments. Each fragment is packed, coded for its source and designation, and sent onto the transmission line. Thus, packets of digital data from various sources pour into the Internet and find each destination. Upon arriving at the destination, the packets are unpacked and reassembled to recover the original data. This data can be text, graphics, pictures, sound, video, or computer code.[1]

The Internet consists of many pathways. A frequently used service is e-mail, a worldwide electronic mail system. You can send and receive e-mail from almost any part of the Internet with almost any on-line software. Many of the public access files, databases, and software on the Internet are available in its FTP archives using the File Transfer Protocol. Another Internet service is Telnet, which allows your computer to enter the files of another computer. Remote access of your library's public access catalog system is most likely through Telnet. Usenet is the part of the Internet devoted to on-line discussion groups, or "newsgroups." The most recent and most rapidly growing component of the Internet is the World Wide Web, an enormous, far-flung collection of colorful onscreen documents that are linked to each other by highlighted words called hypertext.

The World Wide Web (WWW) was initially developed to build a distributed hypermedia system.[2] A hypermedia system is written in a *hypertext language* such as HTML (hypertext markup language), such that "point and click" connections allow the user to jump from one information source to another on the Internet. The system aims at giving global access to a universe of documents. As one hypertext link in a document leads to another and yet another, the links form a web of information, i.e., a worldwide web. The World Wide Web is a subset, although a very important subset, of the Internet. Its popularity comes from the fact that it makes distributing and accessing digital information simple and inexpensive. Its weakness is that it is just a

---

1. For a history of the Internet see http://www.isoc.org/internet-history. A comprehensive site at the Library of Congress gives information on Internet search tools, collections of Internet resources, U.S. government resources, and tutorials on the Internet. http://lcweb.loc.gov/global/.

2. For a history of the WWW see http://www.w3.org/pub/WWW/History.html. For an introduction to the web see http://www.robelle.com/www-paper/intro.html.

huge collection of documents arranged in no defined order. Using the web therefore requires a search engine.

When people say they are "surfing the web" they mean they are randomly seeking and reading Internet addresses to see what is there. While this can be exhilarating for a first-time user of the Web, it is akin to a person attempting to find a book in a 2 million volume library without first consulting the catalog. Locations on the Internet are identified by *universal resource locators* (URL). For example, a URL that gives a brief history of the Internet is http://www.isoc.org/internet-history. The prefix http://www indicates we are trying to access an HTTP server on the World Wide Web at a computer with the domain name "isoc.org" (the nonprofit organization known as The Internet Society). The document in question is stored in that computer in a file called "internet-history."

Before the advent of the World Wide Web there were tools that allowed users to search the Internet.[1] Archie is software designed to find computer files from Internet servers using FTP. Gopher is a menu-driven interface that allows users to access a huge electronic library of documents on the Internet, especially on FTP, Telnet, and Usenet servers. The search engine for Gopher is called Veronica. These search tools are still used, but they have been replaced in popularity by a number of user-friendly, highly graphical search tools—often called *web browsers.* The most commonly used web browsers today are the Netscape Communicator and the Microsoft Internet Explorer. Most browsers allow access to FTP, Telnet, and Usenet servers.

To search the World Wide Web requires a *search engine.* Most of these search engines work by preindexing and regularly updating pointers to a huge number of URLs on the Web. As a result, when you enter your search criteria you are searching a powerful database of the search engine, but not the Web itself, which would be impossibly slow. Each search engine works differently in combing through the Web. Some scan for information in the title or header of the document, while others look at the bold headings on the page. In addition the way the information is sorted, indexed, and categorized differs between search engines. Therefore, all search engines will not produce the same result for a specific inquiry.

Fortunately, most search engines are supported by advertisements, and so they are free to users. The most commonly used general-purpose search engines are listed below.

Web search engines

Name	Web address	Description
AltaVista	http://www.altavista.digital.com	Huge database. Updated daily. Most eclectic of search engines
Excite	http://www.excite.com	A search query on Excite searches the entire Web for documents containing related concepts, not just the keywords
HotBot	http://www.hotbot.com	A new addition among search engines. Offers many advanced search features

1. J. He, Search Engines on the Internet, *Experimental Techniques,* January/February 1998, pp. 34–38.

InfoSeek	http://infoseek.com	One of largest databases. Covers www, Gopher, and FTP. One of largest followings among science/engineering web sites
Lycos	http://lycos.com	One of largest Web indexes. Database updated frequently
Yahoo	http://www.yahoo.com	Attracts largest number of users. Strong topical index of Web sites. Provides jump points to other large search engines

It is important to understand how search engines obtain and maintain their databases. One method is registration, in which you must register your web site using an on-line form. In your submittal you suggest keywords by which the site is indexed. This method has the advantage of providing some kind of review or screen to what will be found by the search engine. This increases the relevance of what is found by the search engine, but it also limits the scope of the web that is covered. Yahoo is the chief user of this method. The other extreme is a search engine that acquires its database with a robot crawler (or spider). This is sophisticated software that crawls over all portions of the Web, far and wide, to collect and index its contents. AltaVista is the best example of this type of data acquisition. Most search engines are combinations of these two methods, actively soliciting site registrations but also sending out robot crawlers.

There are two general types of search methodologies.[1] One, exemplified by Yahoo, acts as a *search tree*. Its contents are manually updated lists of Web resources structured in a hierarchical fashion. Any resource on the Web can be found only if a reference to it has been entered actively into the search tree. Suppose we are interested in finding vendors of centrifugal pumps, and design characteristics of these pumps. The home page of Yahoo opens to a list of 14 major categories. We select the entry for Science (it is unfortunate that most search engines list Engineering as a subcategory under science). The sequence going lower into the classification system is: Science → Engineering → Mechanical Engineering → Indices → Mechanical Engineering-World Wide Web Virtual Library. At this site we have multiple branches. Selecting Mechanical Engineering Vendor Pages, we find a limited set of vendors, about a total of 60, but no vendors of centrifugal pumps. As an alternate choice we find a category University ME Departments, and check on MIT. Drilling down to Lectures, we click on Sketch Modeling and find a nicely illustrated tutorial on how to make product models from cardboard and foam board (http://me.mit.edu/lectures/sketch-modelling). Thus we have illustrated the joy of occasional net surfing but we have not satisfied our utilitarian objective of finding detailed information on centrifugal pumps.

Next we turn to automatic searching on Yahoo. Entering "centrifugal pump(s)" as the search topic returned 32 Web sites. Most of these were the names and hyperlink connections to pump manufacturers, but it did turn up a forum about ideas for pump sizing and total head calculation in centrifugal pumps.

---

1. L. Perrochon, *IEEE Communications Magazine,* June 1996, pp.142–145.

The second strategy for building a search engine is to build a Web *search index.* As the robot crawlers traverse the Web they build a huge index of the words in the pages covered. The search engine permits the search of this index. Obviously, the search engine can only find what has been indexed. However, since the indexing of pages is performed automatically, more pages are indexed by indexes than by trees. Also, because any page is added by the crawler without any human judgment, the number of documents returned for a query can be very large. For example, we continued our search for information on vendors of centrifugal pumps on the search engine HotBot. Entering Centrifugal pumps in the search box, and telling the engine to look for "all the words" in the "last two years" in "North America(.com)" returned 7154 matches. Selecting "all the words" tells HotBot to search for pages that contain "centrifugal" and "pumps" at least once on every page, but not necessarily in the order "centrifugal pumps." This is similar to a Boolean AND or + search. Searching for "any of the words" will return pages that contain one or more of the words. This is similar to a Boolean OR or − search. For this example it returned 202,134 pages. Selecting "the exact phrase" tells HotBot to find documents that contain Centrifugal pumps in the exact order shown here. This gave 2543 responses.

It is important to realize that much of the information retrieved from the Internet is "raw data" in the sense that it has not been reviewed for correctness by peers or an editor. There is a tendency to think that everything on the Web is current material, but that may not be so. Much material gets posted and is never updated. Another problem is the volatility of Web pages. Web pages disappear when their webmaster changes job or loses interest. With increasing use of advertisement on the Internet there is a growing concern about the objectivity of the information that is posted there. All of these are points that the intelligent reader must consider when enjoying and utilizing this fast-growing information resource.

Another important point is the degree to which a search engine covers all of the material out on the web. A recent study[1] estimated that even the best search engine (HotBot) covered about one-third of the 320 million Web pages. Therefore, the best strategy for maximizing the retrieval of information from the Internet is to pick two or three search engines. Take the time to learn the details of search strategies with each one, and then get practice at using them.

4.6.1 Engineering URLs

There are many URLs devoted to engineering topics.[2] The table on p. 137 lists some URLs that might be most useful for a student starting a design project.

---

1. Science, Apr. 1, 1998, *Wall Street Journal,* Apr. 3, 1998.
2. B. J. Thomas, "The World Wide Web for Scientists and Engineers," American Society of Mechanical Engineers, Fairfield, NJ, 1998.

Internet Starter Kit on Mechanical Design

Name and organization	URL	Description
Technical information:		
WWW Virtual Library: Mechanical Engineering at Stanford University	http://CDR.stanford.edu/ html/WWW-ME/home.html	A good place to start. Provides access to University ME departments, ME vendor pages, and a variety of on-line services
Mechanical Engr. Electronic Design Library at U. Mass	http://www.ecs.umass.edu/mie/ labs/mda/dlib/dlib.html	Everything to support many student projects. Materials data/design of std. machine components/design for manufacturing/fits and tolerances/standards and codes/vendor catalogs/ergonomics
National Technical Information Service	http://ntis.gov	3 million reports on file. Abstract of 370,000 reports received since 1990 can be searched
NASA Technical Reports Server	http://techreports.larc.nasa.gov/ cgi-bin/NTRS	Covers all NASA centers. Abstracts only. Reports must be ordered
Mechanical Engineering magazine	http://www.memagazin	Back issues of ME magazine to August 1996
Journal of Mechanical Design—ASME	http://www-jmd.engr.ucdavis .edu/jmd/	Can search back issues of JMD by subject
Machine Design magazine on-line	http://www.penton.com/md/	On-line version of *Machine Design*
Patents (see Sec 4.7)		
Commercial and marketing information:		
Switch Board	http://www.switchboard.com	Will find telephone number of persons without knowing city and state. Must know street and city for a company phone number
Thomas Register	http://thomasregister.com	Allows search of 155,000 companies, 60,000 product and service categories, and 124,000 brand names. Requires registration
Machine Design Product Locator	http://www.pdem.net	Product and manufacturers' directory. Web address directory
Industry Net	http://www.industry.net	Information on new products
Government Printing Office	http://access.gpo.gov	Information on products available from GPO
Commerce Business Daily	http://cbd.cos.com	Daily publication of U.S. Department of Commerce giving procurement opportunities, contract awards, and surplus property for sale
STAT-USA	http://stat-usa.gov	One-stop site for U.S. business, economic, and trade statistics and information. Cost is modest
Gale Business Resources	http://galenet.gale.com	Database containing information on 450,000 U.S. and international companies. Statistical data and info. on market share. May be cost for service, but available in many college libraries

Given the complexity and heterogeneity of the Web,[1] it is not surprising that various information-processing companies have created specialized subscription information packages from the Web and other on-line databases. The Engineering Information Village, created by Engineering Information Inc., is a good example. From its Web site (http://www.ei.org) one can connect to nearly 30 databases, including Ei Compendex, Ei Manufacturing, and Ei MechDisc. Access to financial and business statistics is also provided.

Information Handling Services Group (IHS Group) is a major international publisher of electronic information databases for the technical and business markets. The IHS Engineering Resource Center provides a single source via the World Wide Web of vendor catalogs, specifications, and standards (http://www.ihs.com/erc). The IHS Group's British affiliate ESDU provides electronically more than 200 volumes of validated engineering design data containing more than 1200 design guides with supporting software (http://www.esdu.com). These cover some 20 subject areas in structural, mechanical, aeronautical, and chemical engineering.

It also is clear that as Web-based commerce increases engineers will utilize the Web in design to a greater degree. Computer-based design catalogs available on the Internet are beginning to play a big role.[2] These will provide ready access to what is available, its technical characteristics and cost. Technical calculations for strength, capacity, etc., will be done in the Web page. The graphics will allow checking the component for its interface with other components in the system that is being designed. Finally, the graphics will be capable of being imported into the CAD model of the design. Parts of this component design system already exist on the Internet, but the integrated system is yet to come.

4.7
PATENT LITERATURE

The U.S. patent system is the largest body of information on technology in the world. At present there are over 5 million U.S. patents, and the number is increasing at a rate close to 100,000 each year. Old patents can be very useful for tracing the development of ideas in an engineering field, while new patents describe what is happening at the frontiers of the field. Only about 20 percent of the technology that is contained in U.S. patents can be found elsewhere in the published literature.[3] Therefore, the engineer who ignores the patent literature is aware of only the tip of the iceberg of information.

---

1. For a guide to finding business statistics on the web see P. Bernstein, "Finding Statistics Online," Information Today, Inc., Medford, NJ, 1998.
2. The DesignSuite from InPart (www.inpart.com) contains over 150,000 models of standard components and their performance data. The database can be searched by a hierarchical file structure (bearing/radial bearing/double-pillow block bearing) or with a graphical file that gives thumbnail pictures of the components. Once the component has been selected, its geometric model can be downloaded into the user's computer-aided design program. This database not only aids the component selection process but eliminates the need to draw the part in the design drawings.
3. P. J. Terrago, *IEEE Trans. Prof. Comm.*, vol. PC-22, no. 2, pp. 101–104, 1979.

4.7.1 Intellectual Property

The term *intellectual property* refers to the protection of ideas with patents, copyrights, trademarks, and trade secrets. These entities fall within the broad area of property law and, as such, can be sold or leased just like other forms of property. Also, just as property can be stolen or trespassed upon, so intellectual property can be infringed. We have already learned something about copyrights (Sec. 4.2). A *trademark* is any name, word, symbol, or device that is used by a company to identify their goods or services and distinguish them from those made or sold by others. The right to use trademarks is obtained by registration and extends indefinitely so long as the trademark continues to be used. A *trade secret* is any formula, pattern, device, or compilation of information which is used in a business to create an opportunity over competitors who do not have this information. Sometimes trade secrets are information which could be patented but for which the corporation chooses not to obtain a patent because it expects that defense against infringement will be difficult. Since a trade secret has no legal protection, it is essential to maintain the information in secret.

4.7.2 Patents

Article 1, Section 8 of the Constitution of the United States states that Congress shall have the power to promote progress in science and the useful arts by securing for limited times to inventors the exclusive right to their discoveries. A patent granted by the U.S. government gives the patentee the right to prevent others from making, using, or selling the patented invention. Any patent application filed since 1995 has a term of protection that begins on the date of the grant of the patent and ends on a date 20 years after the filing date of the application. The 20-year term from the date of filing brings the United States into harmony with most other countries in the world in this respect. The most common type of patent, the *utility patent,* may be issued for a new and useful machine, process, article of manufacture, or composition of matter. In addition *design patents* are issued for new ornamental designs and *plant patents* are granted on new varieties of plants. Computer software generally is protected by copyright, but there is a growing tendency for patenting where the software is embedded in the computer hardware or where the computational circuitry is identifiable from other portions of the circuitry.

In patent law a process is defined as an operation performed by rule to produce a certain result. In addition, patent law defines a patentable process to include *a new use* of a known process, machine, manufacture, or composition of matter. Thus, a new use for a known compound which is not analogous to a known use may be a patentable process. However, not all processes are patentable. Methods of doing business or natural laws or phenomena, as well as mathematical equations and methods of solving them are not patentable subject matter.

There are three general criteria for awarding a patent.

1. The invention must be *new* or *novel.*
2. The invention must be *useful.*
3. It must be nonobvious to a person skilled in the art covered by the patent.

A key requirement is novelty. Thus, if you are not the first person to propose the idea your cannot expect to obtain a patent. If the invention was made in another country but it was known or used in the United States before the date of the invention in the United States it would not meet the test of novelty. Finally, if the invention was published anywhere in the world before the date of invention but was not known to the inventor it would violate the requirement of novelty. The requirement for usefulness is rather straightforward. For example, the discovery of a new chemical compound (composition of matter) which has no useful application is not eligible for a patent. The final requirement, that the invention be unobvious, can be subject to considerable debate. A determination must be made as to whether the invention would have been the next logical step based on the state of the art at the time the discovery was made. If it was, then there is no patentable discovery.

The requirement for novelty places a major restriction on publication prior to filing a patent application. In the United States the printed publication of the description of the invention anywhere in the world more than one year before the filing of a patent application results in automatic rejection by the Patent Office. It should be noted that to be grounds for rejection the publication must give a description detailed enough so that a person with ordinary skill in the subject area could understand and make the invention. Also, public use of the invention or its sale in the United States one year or more before patent application results in automatic rejection. The patent law also requires diligence in *reduction to practice.* If development work is suspended for a significant period of time, even though the invention may have been complete at that time, the invention may be considered to be abandoned. Therefore, a patent application should be filled as soon as it is practical to do so.

In the case of competition for awarding a patent for a particular invention, the patent is awarded to the inventor who can prove the earliest date of conception of the idea and can demonstrate reasonable diligence in reducing the idea to practice.[1] The date of invention can best be proved in a court of law if the invention has been recorded in a bound laboratory notebook with prenumbered pages and if the invention has been witnessed by a person competent to understand the idea. For legal purposes, corroboration of an invention must be proved by people who can testify to what the inventor did and the date when it occurred. Therefore, having the invention disclosure notarized is of little value since a notary public usually is not in a position to understand a highly technical disclosure. Similarly, sending a registered letter to oneself is of little value. For details about how to apply, draw up, and pursue a patent application the reader is referred to the literature on this subject.[2]

---

1. A major difference between U.S. patent law and almost every other country's laws is that in the United States a patent is awarded to the first person to invent the subject matter, while in other countries the patent is awarded to the first inventor to file a patent application. Another difference is that in any country but the United States public disclosure of the invention before filing the applications results in loss of patent rights on grounds of lack of novelty.

2. W. G. Konold, "What Every Engineer Should Know about Patents," 2d ed., Marcel Dekker, New York, 1989; D. S. Goldstein, "Intellectual Property Protection," Professional Publishers, Belmont, CA, 1992; M. A. Lechter (ed.), "Successful Patents and Patenting for Engineers and Scientists," IEEE Press, New York, 1995.

4.7.3 Technology Licensing

The right to exclusive use of technology that is granted by a patent may be transferred to another party through a licensing agreement. A license may be either an exclusive license, in which it is agreed not to grant any further licenses, or a nonexclusive license. The licensing agreement may also contain details as to geographic scope, e.g., one party gets rights in Europe, another gets rights in South America. Sometimes the license will involve less than the full scope of the technology. Frequently consulting services are provided by the licensor for an agreed-upon period.

Several forms of financial payment are common. One form is a paid up license which involves a lump sum payment. Frequently the licensee will agree to pay the licensor a percentage of the sales of the products that utilize the new technology, or a fee based on the extent of use of the licensed process. Before entering into an agreement to license technology it is important to make sure that the arrangement is consistent with U.S. antitrust laws or that permission has been obtained from appropriate government agencies in the foreign country.

4.7.4 The Patent Literature

The *Gazette of the U.S. Patent Office* is issued each Tuesday with the weekly issuance of patents. It contains an abstract and selected figures from each patent that is issued that week. The *Gazette* is very helpful for keeping current on the patent literature and for getting a quick overview of a patent. It is widely disseminated in the United States.

The "Annual Index" of patents is published each year in two volumes, one an alphabetical index of patentees and the other an index of inventions by subject matter. These can be used to obtain the patent number, and then the *Gazette* can be used to learn about the patent. All U.S. patents may be examined in the public search room in Crystal City, VA (near Washington, DC). The main advantage of visiting the search room is that patents are grouped according to classes and subclasses. However, 58 libraries nationwide are designated as Patent Depository Libraries and contain most U.S. patents. In these libraries, the patents are arranged according to number.

Patents have been arranged into about 400 classes, and each class is subdivided into many subclasses.[1] The "Index to Classification" is a loose-leaf volume that lists the major subject headings into which patents have been divided. Once the Index identifies the appropriate class numbers, the searcher should go to the "Manual of Classification." This loose-leaf volume lists each class, with its subclasses, in numerical order. By searching the "Manual of Classification," you can identify the classes and subclasses that are likely to be of interest. It is important to realize that subjects will often be found in more than one patent class.

An on-line computerized classification system called CASSIS (Classification and Search Support Information System) is available at all of the Patent Depository

---

1. K. J. Dodd, *IEEE Trans. Prof. Comm.,* vol. PC-22, no. 2, pp. 95–100, 1971.

Libraries around the nation. For a given patent number CASSIS will display all its locations in the Patent Classification System (PCS). For a given technology as described by a PCS classification, CASSIS will display the numbers of all patents assigned to that classification. Also, CASSIS will identify all classifications whose full titles contain designated key words. Finally, CASSIS will search the alphabetical list of subject headings in the index to the PCS to identify the classifications containing subject matter of interest.

A *patentability search* is a search of the patent literature to determine whether an invention can be patented and what the scope of the patent protection would be. This search draws on the information assembled in the search, especially the prior art and background of the invention. By carefully studying the claims in the prior art, the patent attorney can construct claims that are neither too broad (so as to be precluded by prior patents) or too narrow (so as to limit the usefulness of the patent).

An *infringement search* is an exhaustive search of the patent literature to determine whether an idea is likely to infringe on patents held by others. Often it is undertaken when making a new product is contemplated.

If you find your patent or idea in possible infringement with an existing patent, you may fight back by challenging the validity of the patent in the courts. To do so requires a *validity search.* More than half of all patents challenged are ruled invalid by the courts. A patent is really not good until it is tried and held valid by the courts, although it is presumed valid when issued by the Patent Office.

The homepage for the U.S. Patent and Trademark Office is http://www.uspto.gov. The full text of the patents from the present to 1976 is available. Other sites are http://patents.ibm.com/ and http://sunsite.unc.edu/patents/. All of these sites provide patent information without charge. They also allow access to the "Manual of Classification" and will suggest strategies for patent searching.[1] Patent searching can be found on the Internet for a fee where there are special search requirements.

4.7.5 Reading a Patent

Because a patent is a legal document, it is organized and written in a style much different from the style of the usual technical paper. Patents must stand on their own and contain sufficient disclosure to permit the public to practice the invention after the patent expires. Therefore, each patent is a complete exposition on the problem, the solution to the problem, and the applications for the invention in practical use.

Figure 4.3 shows the first page of a patent for a compact disc case. This page carries bibliographic information, information about the examination process, an abstract, and a general drawing of the invention. At the very top we find the inventor, the patent number, and the date of issuance. Below the line on the left we find the title of the invention, the inventor(s) and address(es), the date the patent application was filed, and the application number. Next are listed the class and subclass for both the

---

1. A tutorial on the basics of patent searching is available from the University of Texas at Austin via the Internet at http://www.lib.utexas.edu/Libs/ENG/PTUT/ptut.html. This tutorial focuses on the use of the CASSIS system of classification.

United States Patent [19]

Blase

[11] **Patent Number:** 5,425,451

[45] **Date of Patent:** Jun. 20, 1995

[54] **COMPACT DISC CASE**

[76] Inventor: **William F. Blase,** 1409 Golden Leaf Way, Stockton, Calif. 95209

[21] Appl. No.: **238,695**

[22] Filed: **May 5, 1994**

[51] Int. Cl.[6] **B65D 85/57**

[52] U.S. Cl. **206/313;** 206/309

[58] Field of Search 206/307–313, 206/387, 444

[56] **References Cited**

U.S. PATENT DOCUMENTS

3,042,469	7/1962	Lowther	206/311
3,265,453	8/1966	Seide	206/311
4,613,044	9/1986	Saito et al.	
4,694,957	9/1987	Ackeret	206/309
4,736,840	4/1988	Deiglmeier	
4,875,743	10/1989	Gelardi et al.	206/309
4,998,618	3/1991	Borgions	206/307
5,099,995	3/1992	Karakane et al.	206/309
5,168,991	12/1992	Whitehead et al.	
5,176,250	1/1993	Cheng	206/313
5,205,405	4/1993	O'Brien et al.	
5,244,084	9/1993	Chan	206/309
5,332,086	7/1994	Chuang	206/444

FOREIGN PATENT DOCUMENTS

3440479	5/1986	Germany	206/309

Primary Examiner—Jimmy G. Foster

[57] **ABSTRACT**

A new and improved compact disc case apparatus includes a lower case assembly and an upper case assembly which are placed in registration with each other to form an enclosure assembly. The enclosure assembly includes a side which contains a slot. A pivot assembly is connected between the lower case assembly and the upper case assembly adjacent to a first lower corner and a first upper corner. A disc retention tray is positioned between the lower case assembly and the upper case assembly. The disc retention tray pivots on the pivot assembly such that the disc retention tray can be selectively moved to an open position or a closed position. In the closed position, the disc retention tray is housed completely in the enclosure assembly. In the open position, the disc retention tray is substantially outside the enclosure assembly such that a disc can be selectively taken off of and placed on the disc retention tray. The disc retention tray includes a handle portion. The enclosure assembly includes a truncated corner which is distal to the first lower corner and the first upper corner and which is adjacent to the slotted side. The handle portion of the disc retention tray projects from the truncated corner of the enclosure assembly when the disc retention tray is in a closed position. The disc retention tray includes a recessed edge portion. The recessed edge portion of the disc retention tray is located adjacent to the handle portion of the disc retention tray

3 Claims, 4 Drawing Sheets

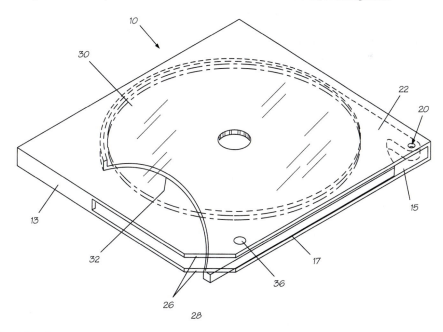

FIGURE 4.3

The first page of a United States patent for a compact disc case.

U.S. patent system and the international classification system and the U.S. classes in which the examiner searched for prior art. The references are the patents that the examiner cited as showing the most prior art at the time of the invention. The rest of the page is taken up with a detailed abstract and a key drawing of the invention. Additional pages of drawings follow, each keyed to the description of the invention.

The body of the patent starts with a section on the Background of the Invention, followed by the Summary of the Invention and a Brief Description of the Drawings. Most of the patent is taken up by the description of the Preferred Embodiment. This comprises a detailed description and explanation of the invention, often in legal terms and phrases that are strange-sounding to the engineer. The examples cited show as broadly as possible how to practice the invention, how to use the products, and how the invention is superior to prior art. Not all examples describe experiments that were actually run, but they do provide the inventor's teaching of how they should best be run. The last part of the patent comprises the *claims* of the invention. These are the legal description of the rights of invention. The broadest claims are usually placed first, with more specific claims toward the end of the list. The strategy in writing a patent is to aim at getting the broadest possible claims. The broadest claims are often disallowed first, so it is necessary to write narrower and narrower claims so that not all claims are disallowed.

There is a very important difference between a patent disclosure and a technical paper. In writing a patent, inventors and their attorneys purposely broaden the scope to include all materials, conditions, and procedures that are believed to be equally likely to be operative as the conditions that were actually tested and observed. The purpose is to develop the broadest possible claims. This is a perfectly legitimate legal practice, but it has the risk that some of the ways of practicing the invention that are described in the embodiments might not actually work. If that happens, then the way is left open to declare the patent to be invalid.

Another major difference between patents and technical papers is that patents usually avoid any detailed discussion of theory or why the invention works. Those subjects are avoided to minimize any limitations to the claims of the patent that could arise through the argument that the discovery would have been obvious from an understanding of the theory.

4.8
CODES AND STANDARDS

The importance of codes and standards in design was discussed in Sec. 1.12. The United States is the only industrialized country in which the national standards body is not a part of or supported by the national government. The American National Standards Institute (ANSI) is the coordinating organization for the voluntary standards system of the United States. It certifies the standards-making processes of other organizations,[1] initiates new standards-making projects, represents the United States on the International Standards Committees of the International Organization for Standardization (ISO), and examines the standards prepared by other organizations to

---

1. For an extensive list of standards sponsoring organizations see T. A. Hunter, Designing to Codes and Standards, "ASM Handbook," vol. 20, ASM International, Materials Park, OH.

determine whether they meet the requirements for consensus so as to be included as an ANSI standard.

The American Society for Testing and Materials (ASTM) is the major organization that prepares standards in the field of materials and product systems. It is the source of more than half of the existing ANSI standards.

The Standards Development Services (SDSS) of the National Institute of Standards and Technology (NIST) manages the voluntary product standards program established by Part 10, Title 15 of the Code of Federal Regulations.

Trade associations produce or review voluntary standards. Those that have produced a substantial number of standards include:

- American Petroleum Institute
- Association of American Railroads
- Electronics Industries Association
- Manufacturing Chemists Association
- National Electrical Manufacturers Association

A number of professional and technical societies have made important contributions through standards activities. The most active are:

- American Association of State Highway and Transportation Officials
- American Concrete Institute
- American Society of Agricultural Engineers
- American Society of Mechanical Engineers
- Institute of Electrical and Electronics Engineers
- Society of Automotive Engineers

The ASME prepares the well-known Boiler and Pressure Vessel Code that is incorporated into the laws of most states. The ASME Codes and Standards Division also published performance test codes for turbines, combustion engines, and other large mechanical equipment.

Several other important standards-making organizations are:

- National Fire Protection Association (NFPA)
- Underwriters Laboratories, Inc. (UL)
- Factory Mutual Engineering Corp. (FMEC)

The Department of Defense (DOD) is the most active federal agency in developing specifications and standards. The General Services Agency (GSA) is charged with preparing standards for common items such as light bulbs and hand tools.

The following are key reference sources for information about specifications and standards.

- "Annual Catalog"
  American National Standards Institute
  10 East 40th Street
  New York, NY 10016
- "Annual Book of ASTM Standards"
  ASTM
  1916 Race Street
  Philadelphia, PA 19107

The Book of Standards is in 48 volumes and contains over 8500 standards.

- "Index of Specifications and Standards"
  U.S. Department of Defense
  Washington, DC
- "Index of Federal Specifications and Standards"
  U.S. General Services Administration
  GPO, Washington, DC
- "Index and Directory of U.S. Industry Standards," 2d ed., 1984
  Information Handling Services
  Englewood, CO 80150

  References to over 400 U.S. standards organizations and more than 26,000 individual standards

- "World Industrial Standards Speedy Finder," 1983
  ASTM
  Philadelphia, PA 19107

  Can locate a specific standard by either number or product name for U.S., U.K., W. Germany, France, or Japan.

- R. B. Toth (ed.): "Standards Management," ANSI, New York, 1989

The following specialized volumes dealing with metals are available.

- "Unified Numbering System for Metals and Alloys," 5th ed., ASTM, 1989.
- "Worldwide Guide to Equivalent Irons and Steels," ASM, 3d ed., 1993.
- "Worldwide Guide to Equivalent Nonferrous Metals and Alloys," ASM, 3d ed., 1996.

The National Source for Global Standards is a Web-based information service sponsored by ANSI. It provides free bibliographic access to more than 250,000 standards at http://www.nssn.org. This includes reference to ASTM, SAE, and MIL standards, as well as British and German (V.D.I.) standards. Information on where to order the standard is provided.

4.9
EXPERT SYSTEMS

A very different source of information is the new field of *expert systems*[1] or *knowledge-based systems.* These computer-based systems simulate the role of an expert in solving some problem using an information database provided by the expert and decision rules for interpreting the data. Expert systems is an active area of the field of *artificial intelligence.*[2] Other areas of artificial intelligence are automated reasoning,

---

1. J. Liebowitz (ed.), "The Handbook of Applied Expert Systems," CRC Press, Boca Raton, FL, 1997; J. N. Siddall, "Expert Systems for Engineers," Marcel Dekker, New York, 1990; M. Green, "Knowledge Aided Design," Academic Press, New York, 1992.
2. P. H. Winston, "Artificial Intelligence," Addison-Wesley, Reading, MA, 1977.

intelligent databases, knowledge acquisition, knowledge bases and knowledge representation, machine learning, natural languages, and vision and sensing.

An expert system acquires knowledge through knowledge-acquisition software tools from a trained specialist called a knowledge engineer. The knowledge engineer obtains his or her knowledge from one or more experts in the technical area, called domain experts. Once the expert system has been constructed this relationship need no longer exist. A prime advantage of expert systems is that they capture the knowledge of experts that may otherwise be lost through death or retirement. Moreover, they can contain the cumulative knowledge of several experts, they are available any time of day or night, and they can be distributed widely throughout an organization. However, it should be quickly added that expert systems are not a substitute for a human expert. Unless a problem is fully understood, which can come only from humans, the expert system project will fail.

The user of an expert system works through a keyboard interface. The input consists of system facts and suppositions of varying degrees of validity. The user interface tends to be highly interactive, following the format of a question and answer session. The expert system returns answers, recommendations, or diagnoses. In a design expert system a graphics interface may be required in order to visualize the object being designed. At present most expert systems utilize a specialized database, but they are moving rapidly to be able to draw upon generalized databases.

The elements of an expert system are shown in Fig. 4.4. The two major divisions are the knowledge base and the inference engine. The knowledge base is unique to a particular domain but the inference engine may be common to many domains of knowledge. The *knowledge acquisition facility* is the component responsible for entering the knowledge into the database. At its simplest level this facility acts as an editor and knowledge is entered directly in a form acceptable by the language in which the expert system is written. On a more sophisticated level this facility can translate an input in natural language into the representation that the knowledge base can understand. This is an important feature in design, for unless the user is willing to learn the programming language in which the expert system is written it would be impossible to customize the knowledge base.

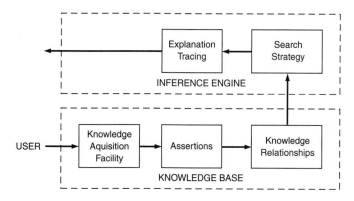

FIGURE 4.4
General structure of an expert system.

The *assertions* component, also called the working memory or temporary data store, contains the knowledge about the particular problem being solved. Data are represented by predicate logic, by frames, or by semantic networks. The *knowledge relationships* component contains formulas showing the relationship among several pieces of information. The "if-then" production rule is the most common relationship. This rule has the form:

IF condition THEN action

For example: IF stress level exceeds 85 ksi THEN part will fail. In any knowledge base there is a balance between the assertion of facts (declarative knowledge) and the rules for manipulating those facts (procedural knowledge). Generally, the less knowledge declared the greater the procedural knowledge required, and vice versa.

The inference engine contains the control mechanisms for the expert system. In a production rules expert system the AI reasoning is responsible for choosing which rule to perform next. This constitutes the *search strategy*. Since the probability of one rule following another is less than 100 percent in most cases, it has been necessary to incorporate uncertainty into the rules. With a small number of rules it is practical to search in random order, but if the number of rules is large it is necessary to partition into sublists on some logical basis. *Explanation tracing* is provided to retrace the chain of production rules that led to the development of the system. This greatly enhances the credibility of the expert system.

Note that an expert system goes well beyond the usual bibliographic source of information in that it contains information, but in addition, it contains decision rules that allow that information to be applied to some specific problem. They are finding use in applications like assisting manufacturing personnel in operating equipment and in preliminary designs of fairly standard problems where the expert system substitutes for the experience of the designer. The benefits to be expected from growth in the use of expert systems in design include:

- The ability to capture valuable expertise and then to put it comfortably into the hands of a novice.
- The ability to improve the consistency of designs within an organization.
- The ability to eliminate errors in problem solving.
- The ability to interface the expert system with advanced software for engineering analysis. By further increasing the analytical capacity the amount of detail the system can cope with is expanded.
- The ability to search large databases for optimal selection of concepts, components, and materials.
- The ability to search design libraries for similar designs, so that engineers can learn from past experience and avoid duplication.
- The ability to reduce the cost of design while at the same time improving quality.

4.10
SUMMARY

The gathering of design information and data is not a trivial task. It requires knowledge of a wide spectrum of information sources. These sources are, in increasing order of specificity:

- The World Wide Web, and its access to digital databases
- Business catalogs and other trade literature
- Government technical reports and business data
- Published technical literature, including trade magazines
- Network of professional friends, aided by e-mail
- Network of professional colleagues at work
- Corporate consultants

At the outset it is a smart move to make friends with a knowledgeable librarian or information specialist in your company or at a local library who will help you become familiar with the information sources and their availability. Also, devise a plan to develop your own information resources of handbooks, texts, tearsheets from magazines, computer software, Web sites, etc., that will help you grow as a true professional.

BIBLIOGRAPHY

Anthony, L. J.: "Information Sources in Engineering," Butterworth, Boston, 1985.
"Guide to Materials Engineering Data and Information," ASM International, Materials Park, OH, 1986.
Mildren, K. W., and P. J. Hicks: "Information Sources in Engineering," 3d ed., Bowker Saur, London, 1996.
Wall, R. A. (ed.): "Finding and Using Product Information," Gower, London, 1986.

PROBLEMS AND EXERCISES

4.1. Prepare in writing a personal plan for combating technological obsolescence. Be specific about the things you intend to do and read.

4.2. Select a technical topic of interest to you.
   (a) Compare the information that is available on this subject in a general encyclopedia and a technical encyclopedia.
   (b) Look for more specific information on the topic in a handbook.
   (c) Find five current texts or monographs on the subject.

4.3. Use the indexing and abstracting services to obtain at least 20 current references on a technical topic of interest to you. Use appropriate indexes to find 10 government reports related to your topic.

4.4. Where would you find the following information?
   (a) The services of a taxidermist.
   (b) A consultant on carbon-fiber-reinforced composite materials.
   (c) The price of an X3427 semiconductor chip.
   (d) The melting point of osmium.
   (e) The proper hardening treatment for AISI 4320 steel.

4.5. Discuss how priority is established in a patent litigation.

4.6. How would you obtain information about a U.S. patent given the following conditions:
  (*a*) you know the patent number?
  (*b*) you know only the patentee's name?
  (*c*) you know the patent number and want to know prior development in the field?
  (*d*) no patent numbers or names are given?

4.7. What is the distinction between copyright and patent protection for software?

5

CONCEPT GENERATION AND EVALUATION

5.1
INTRODUCTION

With a clear product design specification developed in Chap. 2 we have arrived at the point where we are ready to generate design concepts, evaluate them, and decide which one will be carried forward to a final product. Figure 5.1 shows where we are in the eight-step design process. The principle that guides this work is that put forth

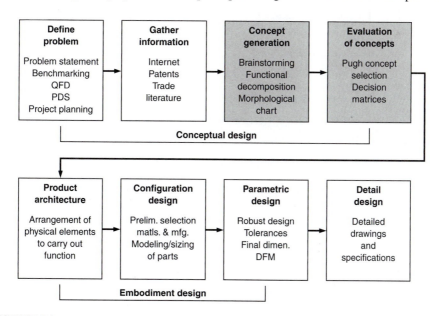

FIGURE 5.1
Steps in the design process, showing concept generation and evaluation as the essential and completing steps in conceptual design.

by the noted American architect-engineer Louis Henri Sullivan, "form follows function." By this we mean, if the functions of the design are clearly understood, then its appropriate form or structure will be easier to determine.

At the outset it is important to understand what is meant by concept generation. A design concept is an idea that is sufficiently developed that it can be evaluated in terms of physical realizability; i.e., the means of performing each major function has been determined. This may take the form of simple calculations, sketches, circuit diagrams, proof-of-concept models, or a detailed written description of the concept. Concept generation does not mean seizing upon the first feasible concept and beginning to refine it into a product design. This might be appropriate for a simple redesign, but in general this procedure will lead to a poor design. The process that we will employ in this chapter will result in the generation of multiple design concepts. Then, with a set of design concepts we will subject them to an evaluation scheme to determine the best concept or small subset of best concepts. Finally, a decision process will be used to decide on the best concept to develop into the final design.

Everyone would like to enhance their creativity and feel more secure in their ability to generate creative design concepts. For this reason Sec. 5.2 deals with some of the rapidly growing knowledge about creativity, and Sec. 5.3 discusses methods for enhancing creativity. The process for concept generation that we shall use follows closely that suggested by Ulrich and Eppinger[1] while the evaluation process is similar to that used by Ullman.[2] These are shown in Fig. 5.2.

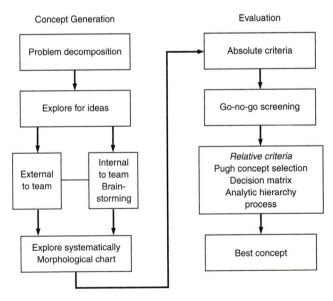

FIGURE 5.2
Steps that will be discussed in the concept generation and evaluation process.

1. K. T. Ulrich and S. D. Eppinger, "Product Design and Development," McGraw-Hill, New York, 1995.
2. D. G. Ullman, "The Mechanical Design Process," 2d ed., McGraw-Hill, New York, 1997.

5.2
CREATIVITY AND PROBLEM SOLVING

Creative thinkers are distinguished by their ability to synthesize new combinations of ideas and concepts into meaningful and useful forms. Engineering creativity is more akin to inventiveness than research. We would all like to be called "creative," yet most of us, in our ignorance of the subject, feel that creativity is reserved for only the chosen few. There is the popular myth that creative ideas arrive with flashlike spontaneity—the flash of lightning and clap of thunder routine. However, students of the creative process[1] assure us that most creative ideas occur by a slow, deliberate process that can be cultivated and enhanced with study and practice. We are all born with an inherent measure of creativity, but the process of maturation takes a toll on our native capacity.

A characteristic of the creative process is that initially the idea is only imperfectly understood. Usually the creative individual senses the total structure of the idea but initially perceives only a limited number of the details. There ensues a slow process of clarification and exploration as the entire idea takes shape. The creative process can be viewed as moving from an amorphous idea to a well-structured idea, from the chaotic to the organized, from the implicit to the explicit. Engineers, by nature and training, usually value order and explicit detail and abhor chaos and vague generality. Thus, we need to train ourselves to be sensitive and sympathetic to those aspects of the creative process. We need, also, to understand that the flow of creative ideas cannot be turned on upon command. Therefore, we need to recognize the conditions and situations that are most conducive to creative thought. We must also recognize that creative ideas are elusive, and we need to be alert to capture and record our creative thoughts.

Listed below are some positive steps you can take to enhance your creative thinking. A considerable literature has been written on creativity,[2] but the steps given here[3] encompass most of what has been suggested:

1. *Develop a creative attitude:* To be creative it is essential to develop confidence that you can provide a creative solution to a problem. Although you may not visualize the complete path through to the final solution at the time you first tackle a problem, *you must have self-confidence;* you must believe that a solution will develop before you are finished. Of course, confidence comes with success, so start small and build your confidence up with small successes.

2. *Unlock your imagination:* You must rekindle the vivid imagination you had as a child. One way to do so is to begin to question again. Ask "why" and "what if,"

---

1. E. Raudsepp, *Chem. Eng.,* pp. 101–104, Aug. 2, 1976; pp. 95–102, Aug. 16, 1976.

2. J. R. M. Alger and C. V. Hays, "Creative Synthesis in Design," Prentice-Hall, Englewood Cliffs, NJ, 1964; E. Van Frange, "Professional Creativity," Prentice-Hall, Englewood Cliffs, NJ, 1959; A. F. Osborne, "Applied Imagination," 3d ed., Charles Scribner's Sons, New York, 1965; E. Lumsdaine and M. Lumsdaine, "Creative Problem Solving," McGraw-Hill, New York, 1995; E. DeBono, "Serious Creativity," HarperCollins, New York, 1992.

3. R. J. Bronikowski, *Chem. Eng.,* July 31, 1978, pp. 103–108.

even at the risk of displaying a bit of naïveté. Scholars of the creative process have developed thought games that are designed to provide practice in unlocking your imagination and sharpening your power of observation.[1]

3. *Be persistent:* We already have dispelled the myth that creativity occurs with a lightning strike. On the contrary, it often requires hard work. Most problems will not succumb to the first attack. They must be pursued with persistence. After all, Edison tested over 6000 materials before he discovered the species of bamboo that acted as a successful filament for the incandescent light bulb. It was also Edison who made the famous comment, "Invention is 95 percent perspiration and 5 percent inspiration."

4. *Develop an open mind:* Having an open mind means being receptive to ideas from any and all sources. The solutions to problems are not the property of a particular discipline, nor is there any rule that solutions can come only from persons with college degrees. Ideally, problem solutions should not be concerned with company politics. Because of the NIH factor (not invented here) many creative ideas are not picked up and followed through.

5. *Suspend your judgment:* We have seen that creative ideas develop slowly, but nothing inhibits the creative process more than critical judgment of an emerging idea. Engineers, by nature, tend toward critical attitudes, so special forbearance is required to avoid judgment at an early stage.

6. *Set problem boundaries:* We place great emphasis on proper problem definition as a step toward problem solution. Establishing the boundaries of the problem is an essential part of problem definition. Experience shows that this does not limit creativity, but rather focuses it.

A creative experience often occurs when the individual is not expecting it and after a period when they have been thinking about something else. The secret to creativity is to fill the mind and imagination with the context of the problem and then relax and think of something else. As you read or play a game there is a release of mental energy which your preconscious can use to work on the problem. Frequently there will be a creative "Ah-ha" experience in which the preconscious will hand up into your conscious mind a picture of what the solution might be. Since the preconscious has no vocabulary, the communication between the conscious and preconscious will be by pictures or symbols. This is why it is important for engineers to be able to communicate effectively through three-dimensional sketches.

To achieve a truly creative solution to a problem a person must utilize two thinking styles: *vertical* or *convergent thinking* and *lateral* or *divergent thinking.* Vertical thinking is the type of analytical thought process reinforced by most engineering courses where one moves forward in sequential steps after a positive decision has been made about the idea. If a negative decision is made at any point in the process, you must retrace your steps along the analysis trail until the original concept statement is reached. In lateral thinking your mind moves in many different directions, combining different pieces of information into new patterns (synthesis) until several solution concepts appear. Table 5.1 compares the characteristics of vertical and lateral thinking.

---

1. E. Raudsepp, "Creative Growth Games," 2d ed., Jove Publications, New York, 1982.

TABLE 5.1
Characteristics of vertical and lateral thinking

Vertical or convergent thinking	Lateral or divergent thinking
Only one correct solution (selective)	Many possible solutions (generative)
Analytical process (judgmental)	Nonjudgmental
Movement is made in a sequential, rule-based manner	Movement is made in a more random pattern
If a positive decision cannot be made at a step, progress stops	If a positive decision cannot be made at a step, thinking jumps
Follows only most likely decision path	Follows all paths
Deals only with reality as science knows it today	Can create its own reality (fantasy)
Classifications and labels are rigid	Reclassifies objects to generate ideas

Source: B. Lee Tuttle, Creative Concept Development, "ASM Handbook," vol. 20, 1997.

Cognitive studies have shown that the left hemisphere of the brain concentrates in processing information that is analytical, logical, or sequential in nature. The right hemisphere functions by recognizing relationships, integrating and synthesizing information, and arriving at intuitive insights. Engineers often are characterized as being left-brained, but there is no reason that you cannot become a whole-brain individual.

5.2.1 Invention

An invention is something novel and useful. As such, we generally can consider it to be the result of creative thought. A study of a large number of inventions[1] showed that inventions can be classified into seven categories:[2]

1. *The simple or multiple combination:* The most elementary form of invention is a simple combination of two existing inventions to produce a new or improved result.
2. *Labor-saving concept:* This is a higher level of invention sophistication in which an existing process or mechanism is changed in order to save effort, produce more with the same effort, or dispense with a human operator.
3. *Direct solution to a problem:* This category of invention is more typical of what we can consider to be engineering problem solving. The inventor is confronted with a need and sets out deliberately to design a system that will satisfy the need.
4. *Adaptation of an old principle to an old problem to achieve a new result:* This is a variation of category 3. The problem (need) has been in existence for some time, and the principle of science or engineering that is key to its solution also has been known. The creative step consists in bringing the proper scientific principle to bear on the particular problem so as to achieve the useful result.

---

1. G. Kivenson, "The Art and Science of Inventing," 2d ed., pp. 14–20, Van Nostrand Reinhold, New York, 1982.
2. For a Web Site that provides information on evaluation of inventions, tips on finding financial support for inventions and patenting inventions, see html://www.inventorsdigest.com.

5. *Application of a new principle to an old problem:* A problem is rarely solved for all time; instead, its solution is based on the then current limitations of knowledge. As knowledge (new principles) becomes available, its application to old problems may achieve startling results. As an example, the miniaturization of electronic and computer components is creating a revolution in many areas of technology.

6. *Application of a new principle to a new use:* People who are broadly knowledgeable about new scientific and engineering discoveries often are able to apply new principles in completely different disciplinary areas or areas of technology.

7. *Serendipity:* The mythology of invention is full of stories about accidental discoveries that led to great inventions. Lucky breaks do occur, but they are rare. Also, they hardly ever happen to someone who is not already actively pursuing the solution of a problem. Strokes of good fortune seem to be of two types. The first occurs when the inventor is actively engaged in solving a problem but is stymied until a freak occurrence or chance observation provides the needed answer. The second occurs when inventors suddenly gain a valuable insight or discover a new principle that is not related to the problem they are pursuing. They then apply the discovery to a new problem, and the result is highly successful.

5.2.2 A Psychological View of Problem Solving

Some psychologists describe the creative problem-solving process in terms of a simple four-stage model.

Preparation (*stage* 1): The elements of the problem are examined and their interrelations are studied.
Incubation (*stage* 2): You "sleep on the problem."
Inspiration (*stage* 3): A solution or a path toward the solution suddenly emerges.
Verification (*stage* 4): The inspired solution is checked against the desired result.

This is obviously a simplified model, since we know that many problems are solved more by perspiration than by inspiration. Nevertheless, there is great value in letting a problem lie fallow so as to give the preconscious mind a chance to operate.

It is useful to understand something about how the mind stores and processes information. We can visualize the mind as a three-element computer.

1. The *preconscious mind* is a vast storehouse of information, ideas, and relations based on past education and experience.

2. The *conscious mind* compares the information and ideas stored in the preconscious mind with external reality.

3. The *unconscious mind* acts on the other two elements. It may distort the relation of the conscious and preconscious through its control of symbols and the generation of bias.

The exact details of how the human mind processes information are still the subject of much active research, but it is known that the mind is very inferior to modern computers in its data-processing capacity. It can picture or grasp only about seven or eight things at any instant. Thus, the mind can be characterized as a device with extremely low information-processing capacity combined with a vast subliminal store of information. Those characteristics of the mind have dominated the development of problem-solving methods.

Our attempts at problem solving often are stymied by the mind's low data-processing rate, so that it is impossible to connect with the information stored in the preconscious mind. Thus, an important step in problem solving is to study the problem from all angles and in as many ways as possible to understand it completely. Most problems studied in that way contain more than the seven or eight elements that the mind can visualize at one time. Thus, the elements of the problem must be "chunked together" until the chunks are small enough in number to be conceptualized simultaneously. Obviously, each chunk must be easily decomposed into its relevant parts.

Another important step in problem solving is the generation of divergent ideas and relations. The brainstorming technique described in Sec. 3.6 is one of the generation methods. The objective of this step is to stir up the facts in the preconscious mind so that unusual and creative relationships will be revealed.

5.3
CREATIVITY METHODS

Improving creativity is a popular endeavor. Over two dozen techniques for developing creative thinking have been proposed.[1] A check on the Internet at amazon.com revealed several hundred books on the subject. There are four factors that help improve a person's creativity, and creativity methods aim at strengthening these characteristics.

- *Sensitivity:* The ability to recognize that a problem exists
- *Fluency:* The ability to produce a large number of alternative solutions to a problem
- *Flexibility:* The ability to develop a wide range of approaches to a problem
- *Originality:* The ability to produce original solutions to a problem

5.3.1 Mental Blocks

Before looking at ways of enhancing creativity it is important to understand how mental blocks interfere with creative problem solving.[2] A *mental block* is a mental wall that prevents the problem solver from correctly perceiving a problem or conceiving its solution. There are many different types of mental blocks.

Perceptual blocks

The most frequently occurring perceptual blocks are:

- *Stereotyping:* Thinking conventionally or in a formulaic way about an event, person, or way of doing something. Not thinking "out of the box."

---

1. See the Web site http://www.ozemail.com.au/~caveman/Creative/Techniques/index.
2. J. L. Adams, "Conceptual Blockbusting," 3d ed., Addison-Wesley, Reading, MA, 1986.

Space Capsule Heat Shield

In the early days of the NASA space program a critical unsolved problem was how to protect the space capsule from burning up due to the frictional heating from reentry into the earth's atmosphere. Initially the problem was stated as "Find a metal capable of withstanding the heat of reentry."

After months of frustrating failure the problem was stated more broadly as "How can we keep the space capsule from burning up in the earth's atmosphere?" This broader problem formulation led to solutions based on
- High melting point, low thermal conductivity nonmetallic materials
- Ablative materials, that burn slowly and fall away
- Solutions that do not use a heat shield

- *Information overload:* You become so overloaded with minute details that you are unable to sort out the critical aspects of the problem.
- *Limiting the problem unnecessarily:* Broad statements of the problem help keep the mind open to a wider range of ideas.

Emotional blocks

These are obstacles that reduce the freedom with which you can explore and manipulate ideas. They also interfere with your ability to conceptualize readily.

- *Fear of risk taking:* This is inbred in us by the educational process.
- *Unease with chaos:* People, and many engineers in particular, are uncomfortable with highly unstructured situations.
- *Adopting a judgmental attitude:* We often approach problem solving with a negative attitude and jump too quickly to judgment.
- *Unable or unwilling to incubate:* In our busy lives, we don't take the time to let ideas lie dormant so they can incubate properly.

Cultural blocks

People acquire a set of thought patterns from living in a culture. For example, the U.S. culture currently is one which avoids problem solving around issues that are termed "politically incorrect."

Environmental blocks

The environment in which we work can have a strong influence on creativity. If the workplace is noisy, hot, or subject to frequent distractions, creativity will suffer. The attitude in your organization also contributes to this environment. Some organizations place a higher importance on being creative than others, and this affects the level of creativity.

Intellectual blocks

Lacking the necessary information base or the necessary intellectual skills can be a barrier to creativity.

5.3.2 Brainstorming

The most common method for creating ideas is brainstorming. The brainstorming process was described in Sec. 3.6. Two additional pointers for effective brainstorming are:

- Carefully define the problem at the start. Time spent here can avoid wasting time generating solutions to the wrong problem.
- Allow 5 min for each individual to think through the problem on their own before starting the group process. This avoids a follow-the-leader type of thought process.

One way to help brainstorming participants is to enlarge their search space by using a checklist to help them develop a budding idea. The originator of brainstorming proposed such a list,[1] which has been modified by Eberle[2] into the acrostic SCAMPER. See Table 5.2. The questions in the SCAMPER checklist are applied to the problem in the following way:[3]

- Read aloud the first SCAMPER question.
- Write down ideas or sketch ideas that are stimulated by the question.
- Rephrase the question and apply it to the other aspects of the problem.
- Continue applying the questions until the ideas cease to flow.

Because the SCAMPER questions are very generalized, they sometimes will not apply to a specific technical problem. Therefore, if a question fails to evoke ideas, move on quickly to the next question. A group that will be doing product development

TABLE 5.2
SCAMPER checklist to aid in brainstorming

Proposed change	Description
Substitute	What if used in a different material, process, person, power source, place, or approach?
Combine	Could I combine units, purposes, or ideas?
Adapt	What else is like this? What other idea does it suggest? Does the past offer a parallel? What can I copy?
Modify, magnify, minify	Could I add a new twist? Could I change the meaning, color, motion, form, or shape? Could I add something? Make stronger, higher, longer, thicker? Could I subtract something?
Put to other uses	Are there new ways to use this as is? If I modify it, does it have other uses?
Eliminate	Can I remove a part, function, person without affecting outcome?
Rearrange, reverse	Could I interchange components? Could I use a different layout or sequence? What if I transpose cause and effect? Could I transpose positive and negative? What if I turn it backward, upside down, or inside out?

1. A. Osborne, "Applied Imagination," Charles Scribner & Sons, New York, 1953.
2. R. Eberle, "SCAMPER: Games for Imagination Development," D.O.K. Press, Buffalo, NY, 1990.
3. B. L. Tuttle, Creative Concept Development, "ASM Handbook," vol. 20, pp. 39–48, ASM International, Materials Park, OH 1997.

over time in a particular area should attempt to develop their own questions tailored to the situation.

Another way to stimulate ideas is with a random input. The random input can come from opening the dictionary, or any book, and randomly selecting a word. The word is used to act as a trigger to change the pattern of thought that has resulted in a mental block. Creativity pioneer Edward de Bono breaks blocks by requiring the participants to find a connection between the problem and some object, like a clock or a vase.

5.3.3 Synectics

Synectics[1] is a technique for creative thinking which draws on analogical thinking, i.e., on the ability to see parallels or connections between apparently dissimilar topics. Four types of analogies are used.

1. *Direct analogies:* Most of these are found in biological systems.[2]
2. *Personal analogies:* The designer imagines what it would be like to use one's body to produce the effect that is being sought, e.g., what it would feel like to be a helicopter rotor?
3. *Symbolic analogies:* These are poetic metaphors and similes in which one thing is identified with aspects of another, e.g., the *mouth* of a river, a *tree* of decisions.
4. *Fantasy analogies:* Here we let our imagination run wild and wish for things that don't exist in the real world.

5.3.4 Force-Fitting Methods

There are a large number of techniques that encourage new ideas by forcing the mind to make creative leaps. They are most often used when a team becomes stuck and can't come forth with many ideas. The SCAMPER checklist shown in Table 5.2 is one of the most widely used methods.

Another trigger to break mind blocks and open the search space is use the analogy and fantasy concepts from synectics to ask participants to explore possibilities that otherwise seem impossible, by asking what would happen if they were possible. For example, "What if gravity didn't exist?" One approach is to first define the problem's constraints, and then ask "what if" as these are removed in turn. Once free from constraints, novel ideas often flow. Then the real world constraints are reintroduced and the solution is modified to work within the constraints.

---

1. W. J. J. Gordon, "Synectics," Harper & Row, New York, 1961.
2. T. W. D'Arcy, "Of Growth and Form," Cambridge Univ. Press, 1961; S. A. Wainwright *et al.,* "Mechanical Design in Organisms," Arnold, London, 1976; M. J. French, "Invention and Evolution: Design in Nature and Engineering," 2d ed., Cambridge University Press, New York, 1994; S. Vogel, "Cat's Paws and Catapults: Mechanical Worlds of Nature and People," W. W. Norton & Co., New York, 1998.

Note that the inventor of the Velcro fastener, George de Mestral, conceived the idea after wondering why cockleburrs stuck to his wool socks. Under the microscope he found that hook-shaped projections on the burrs grasped loops in the wool sock. After a long search he found that a nylon material could be formed into hooks that would retain their shape.

Attribute listing creates a checklist by listing the important attributes, parts, or functions of a problem. The group focuses on each attribute, in turn, and answers questions such as "Why does it have to be this way?"

5.3.5 Mind Map

A very useful technique for note taking and the generation of ideas by association is the mind map[1] and its close relation, the *concept map*.[2] A mind map is created on a large sheet of paper. The problem is drawn at the center. Think about what factors, ideas, or concepts are directly related to the problem. Write them down surrounding the central problem. Underline or circle them and connect them to the central problem. Figure 5.3 shows a mind map or concept map created for a project on the recycling of steel and aluminum scrap.[3] Note that this is a good method to use before you start to write a report or major written document to ensure that all necessary topics will be covered in the document.

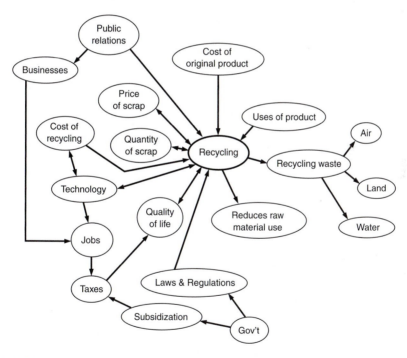

FIGURE 5.3
Concept map for the recycling of a metal like steel or aluminum.

1. Tony Buzan, "Use Both Sides of Your Brain," Dutton, New York, 1983.
2. J. D. Novak and D. B. Gowan, "Learning How to Learn," Cambridge Press, New York, 1984.
3. I. Nair, Decision Making in the Engineering Classroom, *J. Engr. Education,* vol. 86, no. 4, pp. 349–356, 1997.

5.4
CREATIVE IDEA EVALUATION

As we saw in Sec. 3.7, the objective in brainstorming is to generate as many ideas as possible. Quantity counts above quality, and wild ideas are encouraged. Now we need to look at these ideas more critically, with the objective of shaping them into a smaller subset of ideas that can be practically implemented as solutions.[1] The objective of creative idea evaluation is not to come up with a single or very small number of solutions. This is achieved by the evaluation methods considered in Sec. 5.9. The purpose of this step in concept generation is synthesis of creative yet practical ideas.

The first step is to sort the ideas into categories. See discussion on the affinity diagram in Sec. 3.7. Now a second round of brainstorming takes place. The team works with one category at a time, or if the team is large it breaks into subgroups and each works on a category. The objective is to reduce the category to fewer ideas that are more fully developed and of high quality. During this round of brainstorming discussion is encouraged as the team elaborates, adds detail, and hitchhikes on ideas. Some of the force-fitting methods discussed in the previous section are employed. This is a period of *synthesis* as ideas are combined, force-fitted, and integrated into new, better-quality ideas. It is not unusual for new ideas to be generated in this session, and for these to prove to be the best ideas. Creative idea evaluation should not be rushed. It can easily take two or three times as long as the first brainstorming.

Finally, with new and better-developed ideas in each category, the team should try to take a good idea from each category and force-fit them into yet another small set of ideas. These ideas are likely to be highly creative solutions to the problem, since the categories will have very little in common. It should be realized that this step is not possible with all problems. In this case you are left with lists of valuable ideas that when implemented will solve the problem.

5.5
THEORY OF INVENTIVE PROBLEM SOLVING (TRIZ)

The theory of inventive problem solving, better known by its Russian acronym TRIZ, is a creative problem-solving methodology especially tailored for scientific and engineering problems. This approach was developed by Genrich Altshuller and his coworkers in Russia,[2] starting around 1946. They studied over 1.5 million patents, worldwide, and discovered that often the same problems had been solved in various technical fields using only a few dozen inventive principles. They organized the problem solutions from the patent literature into five levels.

- Level 1: Routine design solutions arrived at by methods well known in the specialty area. This category constituted about 30 percent of the total.

---

1. E. Lumsdaine and M. Lumsdaine, "Creative Problem Solving," chap. 7, McGraw-Hill, New York, 1995.
2. G. S. Altshuller, "Creativity as an Exact Science," Gordon and Breach, New York, 1988.

- Level 2: Minor corrections to an existing system by methods known in the industry. About 45 percent of the total.
- Level 3: Fundamental improvements to an existing system which resolve *contradictions* within the industry. About 20 percent of the total. This is where creative design solutions would appear.
- Level 4: Solutions based on application of new scientific principle to perform the primary functions of the design. About 4 percent of total.
- Level 5: Pioneering inventions based on rare scientific discovery. Less than 1 percent of the total.

TRIZ is aimed at improving design concepts at levels 3 and 4, where straight application of good engineering practice doesn't yield acceptable results. Such problems typically contain a fundamental technical contradiction, where improving one attribute of a system degrades another. An example would be the contradiction between aircraft crashworthiness and light weight. Traditionally technical contradictions are resolved with trade-offs, but TRIZ aims at eliminating the need for such compromise. Because TRIZ is more structured than brainstorming and creativity techniques, and is supported by well-developed software packages,[1] it has begun to find acceptance and use in U.S. industry.

The full development of TRIZ[2] consists of a series of tools:

- A formalized problem-formulation system
- The solution in terms of either a physical or technical contradiction
- The concept of the ideal state of a design
- Substance-field (Su-Field) analysis
- The algorithm of inventive problem solving (ARIZ)

Space considerations allow us to introduce only the idea of contradictions and to give a brief introduction to ARIZ. While this is just a beginning introduction to TRIZ, it can serve as a significant stimulation to creativity in design and to further study of the subject.

At the start of a problem-solving process you are faced with an initial situation involving certain disadvantages that need to be eliminated. These disadvantages can be eliminated by changing the system or one of its subsystems, or by modifying some higher-level system. In the TRIZ method, all problems are divided into *mini-problems* and *maxi-problems*. A mini-problem exists when the shortcoming is improved or disappears and the system remains unchanged. A maxi-problem is one where a new system is conceived of based on a novel principle of functioning. A *system conflict* or *contradiction* occurs when attempts to improve some attribute of the system lead to deterioration in other system attributes. Typical conflicts are reliability vs. complexity, productivity vs. accuracy, strength vs. ductility, etc. Conventional engineering

---

1. TechOptimizer™, Invention Machine Corp., Boston, MA; Innovation Workbench™, Ideation International, Inc., Southfield, MI.
2. V. R. Fey and E. I. Rivin, "The Science of Innovation: A Managerial Overview," TRIZ Group, Southfield, MI, 1997; J. Terninko, A. Zusman, and B. Zlotin, "Systematic Innovation: An Introduction to TRIZ," St. Lucie Press, Boca Raton, FL, 1998; see the TRIZ Web site http://www.jps.net/triz/triz.html.

practice attempts to solve such a problem by finding an acceptable compromise through trade-offs. TRIZ attempts to use a creative solution to overcome a system conflict. Altshuller found, surprisingly, that given the great diversity of technology, there were only about 1250 typical system conflicts. Furthermore, these could be overcome by the application of only 40 inventive principles.

Early in his patent research Altshuller identified 39 engineering parameters or product attributes that engineers generally try to improve. These are listed in Table 5.3. Also, the 40 inventive principles, often called the techniques for overcoming system conflicts (TOSC), are given in Table 5.4. As you read these through, several, like combining and changing the color, could have been derived from the blockbusters in the SCAMPER listing (Table 5.2). However, most of the inventive principles listed above have special technical meaning, or special meaning introduced by Altshuller. For example:[1]

3. Local quality
   a. Provide transition from a homogeneous structure of an object or outside environment to a heterogeneous structure.
   b. Have different parts of the object carry out different functions.
   c. Place each part of the object under conditions most favorable for its operation.

TABLE 5.3
Engineering parameters commonly used in TRIZ

1. Weight of moving object	21. Power
2. Weight of nonmoving object	22. Waste of energy
3. Length of moving object	23. Waste of substance
4. Length of nonmoving object	24. Loss of information
5. Area of moving object	25. Waste of time
6. Area of nonmoving object	26. Amount of substance
7. Volume of moving object	27. Reliability
8. Volume of nonmoving object	28. Accuracy of measurement
9. Speed	29. Accuracy of manufacturing
10. Force	30. Harmful factors acting on object
11. Tension, pressure	31. Harmful side effects
12. Shape	32. Manufacturability
13. Stability of object	33. Convenience of use
14. Strength	34. Repairability
15. Durability of moving object	35. Adaptability
16. Durability of nonmoving object	36. Complexity of device
17. Temperature	37. Complexity of control
18. Brightness	38. Level of automation
19. Energy spent by moving object	39. Productivity
20. Energy spent by nonmoving object	

---

1. See for example, Terninko, Zusman, and Zlotin, op. cit., Appendix C. The TRIZ software provides these definitions, along with illustrated examples—and much more.

TABLE 5.4
The Inventive Principles of TRIZ

1. Segmentation	21. Rushing through
2. Extraction	22. Convert harm into benefit
3. Local quality	23. Feedback
4. Asymmetry	24. Mediator
5. Combining	25. Self-service
6. Universality	26. Copying
7. Nesting	27. An inexpensive short-lived object instead of an expensive durable one
8. Counterweight	28. Replacement of a mechanical system
9. Prior counteraction	29. Use of a pneumatic or hydraulic construction
10. Prior action	30. Flexible film or thin membranes
11. Cushion in advance	31. Use of porous material
12. Equipotentiality	32. Change the color
13. Inversion	33. Homogeneity
14. Spheroidality	34. Rejecting and regenerating parts
15. Dynamicity	35. Transformation of physical and chemical states of an object
16. Partial or overdone action	36. Phase transition
17. Moving to a new dimension	37. Thermal expansion
18. Mechanical vibration	38. Use strong oxidizers
19. Periodic action	39. Inert environment
20. Continuity of useful action	40. Composite materials

These 40 principles have a remarkably broad range of application. However, they do require considerable study to understand them fully.

A *contradiction table* can be built to guide the engineer in applying TRIZ principles to a design problem. There are two types of contradictions. A *physical contradiction* occurs when some element of the system is subject to two opposing requirements. A *technical contradiction* occurs when an improvement in a desired attribute or characteristic of the system results in a deterioration in another attribute. The first step in a contradiction solution is to phrase the problem statement to reveal the contradiction. Then the parameters to be improved are identified, and the parameters which are being degraded are identified. The row of the *contradiction table* is entered with the parameter that it is desired to improve, and this is intersected with the column of the parameter that is producing an undesired result. The cell in the intersection gives the numbers of the inventive principles that are suggested as being able to resolve the contradiction (Fig. 5.4).

> **EXAMPLE.**[1] A metal pipe was used to pneumatically transport plastic pellets. A change in the process required that metal powder now be used with the pipe instead of plastic. The harder metal powder causes erosion of the inside of the pipe at the elbow where the metal particles turn 90° (Fig. 5.5). Conventional solutions to this problem might include

---

1. Terninko, Zusman, and Zlotin, op. cit., pp. 76–79.

Feature to improve \ Undesired result		**1** Weight of moving object	**2** Weight of nonmoving object	* *	**9** Speed	**10** Force	* *	**38** Level of automation	**39** Productivity
	* *								
8	Volume of nonmoving object		35,10 19,14			2,18 37			35,37 10,2
9	Speed	2,28 13,38				13,15 19,28		10 18	
	* *								
27	Reliability	3,8 10,40	3,10 8,28			21,35 11,28		11,13 27	1,35 29,38
	* *								
39	Productivity	35,26 24,37	28,27 15,3			28,15 10,36		5,12 35,26	

FIGURE 5.4
A sample of a contradiction table. Row and column numbers refer to Table 5.3. The numbers in the cells refer to Table 5.4.

FIGURE 5.5
Erosion of pipe elbow due to impingement of metal particles.

reinforcing the inside of the elbow with abrasion-resistant hard-facing alloy, providing for an elbow that could be easily replaced after it has eroded, or redesigning the shape of the elbow. However, all of these solutions require significantly extra costs, so a more creative solution was sought.

First we think about the function that the elbow serves. Its primary function is to change the direction of the flow of metal particles. However, we want to increase the speed with which the particles are delivered, and at the same time reduce the energy requirements. The first requirement involves parameter 9 in Table 5.3, and the second involves parameter 19.

If we think about increasing the speed of the particles, we can envision that other parameters of the system will be degraded, or affected in a negative way. For example, increasing the speed increases the force with which the particles strike the inside wall of the elbow, and erosion increases. This and other degraded parameters are listed below. Also included in the table are the inventive principles taken from a contradiction table for each pair of parameters. For example, Fig. 5.4 shows that to improve speed (9) when acted upon by the undesirable effect of force (10), the suggested inventive principles to apply are 13, 15, 19, and 28. A full contradiction table is needed to develop the inventive principles for the other parameters. A similar analysis is made for improving energy, parameter 19.

Improving speed (parameter 9)

Degraded parameter	Parameter number	Inventive principles used
Force	10	13, 28, 15, 19
Durability	15	8, 3, 26, 14
Temperature	17	28, 30, 36, 2
Energy	19	8, 15, 35, 38
Loss of matter	23	10, 13, 28, 38
Quantity of substance	26	10, 19, 29, 38

Improving energy (parameter 19)

Degraded parameter	Parameter number	Inventive principles use
Convenient to use	33	28, 35, 30
Loss of time	25	15, 17, 13, 16

A count of the frequency with which individual inventive principles were suggested in this problem shows that Principle 28, Replacement of a Mechanical System, was cited four times. Those cited next in frequency were 13 (3), 15 (3), and 38 (3). The full description of Principle 28 is:

28. Replacement of a mechanical system
   a. Replace a mechanical system by an optical, acoustical, or odor system.
   b. Use an electrical, magnetic, or electromagnetic field for interaction with the object.
   c. Replace fields. Example: (1) stationary field change to rotating fields; (2) fixed fields become fields that change in time; (3) random fields change to structured ones.
   d. Use a field in conjunction with ferromagnetic particles.

Principle 28b suggests the creative solution of placing a magnet at the elbow to attract and hold a thin layer of powder that will serve to absorb the energy of particles navigating the 90° bend, thereby preventing erosion of the inside wall of the elbow. This solution will work, however, only if the metal particles are ferromagnetic so that they can be attracted to the pipe wall.

The algorithm of inventive problem solving (ARIZ) is the main analytical and solution tool of TRIZ. ARIZ places great emphasis on problem reformulation to

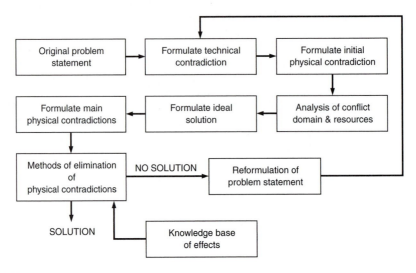

FIGURE 5.6
Main steps in the algorithm of inventive problem solving (ARIZ).

reveal the roots of the problem. Figure 5.6 shows the steps in the process in an abbreviated fashion. The latest version of ARIZ has about 100 discrete steps. Starting with the problem statement, we formulate a technical contradiction and proceed to search for concepts using the contradiction table, as illustrated above. If the contradiction table does not yield satisfactory solutions, we transform the problem to a physical contradiction. The process now focuses on the conflict between useful and harmful results, with emphasis on what resources can be identified to increase the useful zone. The next step is the formulation of the problem in terms of the ideal final result (IFR). The IFR is now transformed into a physical contradiction at a finer level of detail than initially. Elimination of the physical contradiction is based on one of three principles: (1) separation of opposing properties in time, (2) separation of opposing properties in space, or (3) separation of opposing properties between the system and its components. This process is aided by application of a knowledge base of physical effects.[1] If the ARIZ process has not solved the problem, then the problem should be reformulated, and the process is repeated.

5.6
CONCEPTUAL DECOMPOSITION

In solving any complex problem, a common tactic is to decompose the problem into smaller parts that are easier to manage. In breaking the system into subsystems, it is important to do it in such a way that the connections of elements in terms of structure and function within the "chunks" are stronger than those between the chunks. In prod-

---

1. The average engineer knows 50 to 100 physical and chemical laws, principles, and effects which can be used in the solution of design problems, but there are over 6000 such effects described in the scientific literature. Altshuller and his coworkers cataloged these with respect to function. They are available as part of the Invention Machine software.

uct design we typically start with a product or major subassembly and break this into subsidiary subassemblies and components. For example, an automobile would be decomposed into the major subassemblies of engine, drive train, suspension system, steering system, and body.

There are two chief approaches to conceptual decomposition. In *decomposition in the physical domain* the product or subassembly is decomposed directly into its subsidiary subassemblies and components. The second method is *functional decomposition.* Here the emphasis is on identifying *only* the subfunctions required to achieve the overall function. Then as a second step the embodiments (shape, force, motion, etc.) to achieve each subfunction are identified.

Most engineers instinctively follow the direct approach of physical decomposition by sketching a subassembly or part and by decomposing a design without thinking explicitly about the functions each component is created to perform. However, the approach of functional decomposition is inherently more basic. It builds on the principle that form follows function, but more importantly, because it does not initially impose a design, it allows more leeway for creativity and generates a wide variety of alternative solutions. The great advantage of functional decomposition is that the method facilitates the examination of options that most likely would not have been considered if the designer moved quickly to selecting specific physical principles or, even worse, selecting specific hardware.

5.6.1 Decomposition in the Physical Domain

Product conceptual design by physical decomposition starts with the product design specification (PDS), and possibly an industrial designer's preliminary concept of the size and shape. The first step is to decompose the product into those subassemblies and components that are essential for the overall functioning of the product. We need to be able to describe in qualitative physical terms how these design elements will work together to accomplish the required functions of the product. Thus, function is not ignored totally in physical decomposition. We need to be able to describe the principal functions of each subsidiary subassembly and component. We also need to understand the interactions or connections that each of these design elements has with each other, i.e., the *couplings.* Coupling among subsidiary subassemblies and components can be physical, as when they are physically connected, or can be energy or force connection. Another important requirement may be the allocation of some scarce resource like space, weight, or cost among the subsystems.

An important design consideration which is beginning to emerge at this point is *product architecture* (see Sec. 6.2). Product architecture is the scheme by which the functional elements of the product are arranged into physical building blocks (*chunks*). While product architecture is completely established in the phase of design called embodiment design, its origins start here.

Generally there is more than a single stage of conceptual decomposition for a product. Each of the subassemblies created at the first level of decomposition will be decomposed into its subassemblies and components, and each of these subassemblies will be decomposed into its subassemblies and components, until, ultimately, only components remain. For example, the automobile engine subassembly is decomposed into, among other things, an engine block and an ignition system, which is

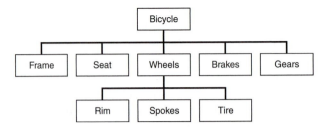

FIGURE 5.7
Direct decomposition of a bicycle into subassemblies and components.

decomposed into spark plugs, a wiring harness, and a fuel injection subassembly. Figure 5.7 shows the decomposition in the physical domain of a bicycle.

As the decomposition proceeds to lower and lower levels, the situation regarding design information changes. The information that is available about a subassembly that is created at the first level of decomposition is only slightly different from the state of information for the product as a whole. For a subassembly created two levels down, we not only have that information but additional specific information about the functions and couplings of the subassembly with the product. Note that the functions of subassemblies or components "down in the pile" are very different from the functions of the product. The functions of a spark plug are very specific and not even mentioned in the PDS of the product. The functions of subassemblies tend to be more specific and more technical than are the functions of a product, which are more closely aligned with the needs of the customer. These are the types of narrowly defined problems you have addressed in most of your design courses.

Another fact that should be clear by now is that large or complex products have a very large number of coupled subassemblies and components. The Boeing 777 air transport has one million parts, not counting rivets. Thus, information management in product development can be a very major and challenging problem.

5.6.2 Functional Decomposition

In functional decomposition the system's functions are described as a transformation between an initial state and a desired final state. We consider *function* to be in the nature of a *physical behavior* or *action*. Function tells us *what* the product must do, while its form or structure tells *how* it will accomplish it. Table 5.5 illustrates the functionality of several common devices.

The approach to concept generation by functional decomposition originated with the German school of design methodology.[1] From Table 5.5 we can see that the inputs and outputs to the functional devices are described in terms of either energy flow, material flow, or information flow (signals). Functions associated with a flow of energy are classified both by the type of energy and by its action on the system.

---

1. G. Pahl and W. Beitz, "Engineering Design," 2d ed., edited by Ken Wallace, Springer-Verlag, London, 1996; R. Koller, "Konstruktionslehre fuer den Maschinenbau," Springer-Verlag, Berlin, 1985.

TABLE 5.5
Functionality of some common devices

Device	Input	Function	Other effects	Output
Nozzle	Fluid flow	Increase velocity of fluid	Decrease pressure of fluid	Fluid flow
Motor	Electrical energy	Convert electrical energy to rotating mechanical energy	Thermal energy generated	Rotating mechanical energy
Pump	Fluid flow and mechanical energy	Increase pressure of fluid	Change direction of flow? Thermal energy	Fluid flow
Gear	Rotating mechanical energy	Change speed of rotation	Change direction of rotation	Rotating mechanical energy
Electric resistance element	Electrical energy	Convert to thermal energy	Element increases in temperature	Thermal energy
Pencil	Mechanical energy	Transfer graphite from pencil to paper		Graphite deposit on paper
Expansion valve in refrigeration system	Liquid refrigerant	Reduce temperature of refrigerant	Reduce pressure of refrigerant	Refrigerant (two-phase mixture)
Lever or wrench	Energy (force or torque in motion)	Increase magnitude of force or torque		Energy
Switch	Mechanical energy (force in motion)	Separate or join contacts	Flow of electricity enabled or stopped	Position of contacts moved
Room thermostat	Flow of room air	Separate or join contacts	Flow of electricity enabled or stopped	Position of contacts moved

J. R. Dixon and C. Poli, "Engineering Design and Design for Manufacturing," Field Stone Publishers, Conway, MA, 1995. Used with permission.

Normal energy types are mechanical, electrical, chemical, fluid, and thermal. The actions of a function are described by action verbs. A small initial set would be change, change back, enlarge, reduce, change in direction, conduct, insulate, connect, separate, join, divide, store, and destore. A much larger set of mechanical design action verbs is given in Table 5.6.

Material flow is divided into three main classes: (1) through-flow or material-conserving processes, in which material is manipulated to change its position or shape, (2) diverging flow, in which the material is divided into two or more parts, and (3) converging flow, in which material is joined or assembled.

Functions associated with information flow will be found in the form of mechanical or electrical signals, or software instructions. Usually, information flows are part of an automatic control system or the interface with a human being. For example, if you attached a bracket with screws, after you tighten the screws you would wiggle the bracket to see if it was securely attached. The observation of the system's response to wiggling is an information flow that confirms that the bracket is properly attached.

TABLE 5.6
Action verbs to be used with mechanical design

Absorb/remove	Drive	Rectify
Actuate	Grasp	Release
Amplify	Guide	Rotate
Assemble/disassemble	Hold or fasten	Secure
Attach	Increase/decrease	Separate
Change	Interrupt	Shield
Channel or guide	Join/separate	Start/stop
Clear or avoid	Lift	Steer
Collect	Limit	Store
Conduct	Locate	Supply
Control	Mix	Support
Convert	Move	Transform
Couple/interrupt	Orient	Translate
Direct	Position, relative to	Verify
Dissipate	Protect	

Source: D. G. Ullman, "The Mechanical Design Process," 2d ed., McGraw-Hill, New York, 1997.

The process of functional decomposition describes the design problem in terms of a flow of energy, material, and information. This creates a detailed understanding at the beginning of the process about *what* the product is expected to do. Some of this information is already contained in the QFD (see Sec 2.6). With the overall function defined by these flows, we decompose it into subfunctions expressed by a verb-noun combination. For example, in the function structure of a potato harvester we might use "separate leaves" (verb) (noun). The verb indicates what is being done by the product, and the noun indicates who or what is receiving the action. Next the subfunctions are reordered to give a logical order to achieve the overall objective of the design. Finally, each subfunction is examined to see whether it is decomposed as finely as possible. If not, then sub-subfunctions, and maybe even sub-sub-subfunctions, are created.

EXAMPLE. The design of a compact disc "jewel" case that was initiated in Chap. 2 is continued here. Turn to the QFD diagram (Fig. 2.5) to refresh your memory about the requirements of such a product.

1. *State the overall function that needs to be accomplished.* Put this one most important function inside of the system box (Fig. 5.8). Now the flows in and flows out need to be identified. Remember that energy and material must be conserved; i.e., whatever goes into the system must come out or be stored in the system. Information flow often is related to knowing whether the system is performing as expected. Also, shown at the top of the diagram are objects that interact or interface with the system. This "black box" model of the product shows the input and output flows for the primary high-level function of the design task.

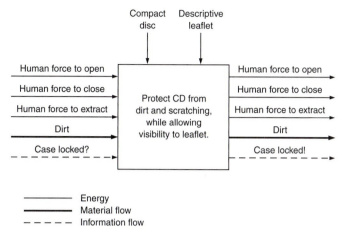

FIGURE 5.8
Overall system diagram for the CD jewel case.

2. *Create descriptions of subfunctions.* The objective of this step is to decompose the overall function in the black box into subfunctions described by verb-noun-modifier combinations. For example, *release disc from rosette.* (The rosette is the prickly plastic centerpiece that secures the CD in the case.) The customer needs (*whats*) in the QFD provide guidance about what functions the product should contain. Any structure-oriented *how* suggestions should be suppressed because they add detail too soon. It may be hard, at first, to think only in terms of function, since we conventionally think about functions through their physical embodiments.

Start with the overall function and break it down into separate subfunctions. Each subfunction represents a change or transformation in the flow of energy, material, or information. A useful way to do this is to trace the flow as it is transformed from its initial state to its final state when it leaves the product system boundary. Brainstorming or personal analogy to imagine you are actually in the flow or are in the role of the product may be useful in this process. Write the subfunctions in Post-It notes so they can be rearranged in step 3. Figure 5.9 shows several subfunctions generated for the CD case.

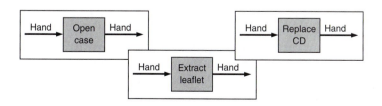

FIGURE 5.9
Several subfunctions generated from the overall function involving movement of the human hand.

FIGURE 5.10
Subfunctions required to open and close the CD case.

For the situation of a redesign problem, or where you are benchmarking a competitor's product, it is often possible to disassemble the product, one component at a time, to identify the subfunctions. Note what objects—features, components, assemblies, people—it interfaces with, and record the flow of energy, materials, and information between them.

3. *Arrange the subfunctions in logical order.* For some problems this is a trivial step, but for other design problems it can be major. The object is to order the functions and subfunctions found in step 2 to achieve the overall function in step 1. The QFD shows that one of the major customer needs is easier opening of the case and removing of the CD and the leaflet. Figure 5.10 shows the sequence of subfunctions to remove the CD and the leaflet.

4. *Refine the subfunctions.* The objective is to refine the subfunctions as finely as possible. Refinement stops when a subfunction can be fulfilled by a single solution that is an object or action, and the level of detail is sufficient to address the customer needs. For example, the subfunction 1.0 Open Case can be decomposed into 1.1 Hold and Grip Case, 1.2 Disengage Locking Mechanism, and 1.3 Expose CD.

Having now identified the functions and subfunctions, we are ready to generate design concepts. This is the subject of the next section.

Functional decomposition is not easy to implement in all situations. Dixon[1] suggests the compromise approach of starting with a physical decomposition and then identifying the functions that each subassembly and component fulfills. Then you can focus on ways of fulfilling the identified functions, and look for ways to separate and combine functions. Although this approach is not as general as functional decomposition, it is less abstract and therefore may be easier to implement.

5.7
GENERATING DESIGN CONCEPTS

Design concepts are the means for providing function. They are the *hows*. The exploration for ideas for concepts can take two general paths. First, as many as possible of the resources external to the design team that were described in Chap. 4 should be employed. Examples include reading patents, benchmarking related products, talking with consultants, and exploring the Web. The more information that is gathered about the problem the better will be the ideas leading to design concepts. Second, it is within the design team that the hard work of concept generation finally takes place. This

---

1. J. R. Dixon and C. Poli, "Engineering Design and Design for Manufacture," Field Stone Publishers, Conway, MA , pp. 6–8.

process occurs in two steps, a divergent process and a convergent process (see Fig. 5.14). The objective in the divergent step is to uncover as many concepts as possible that can provide for each function identified in the decomposition (Sec. 5.5). The convergent step combines these individual concepts into overall design concepts that satisfy all of the functional requirements.

5.7.1 Concept Development

Ultimately it is the storehouse of personal and team knowledge and creativity that generates design concepts. Studies of problem solving suggest that a group of people working as individuals for a period of time will generate more and better concepts than the same people working together for the same period.[1] This does not denigrate the importance of teams. Teams are vital for refining concepts, communication, and building consensus. The best procedure is for each team member to spend several hours working as an individual on some subset of the problem, such as how to satisfy some subfunction. The aids to creativity discussed in Sec. 5.3 should be employed. Then the team assembles to discuss and improve the concepts developed individually.[2]

As a result of the exploration for ideas, the team will have collected tens of fragments of design concepts.[3] These are solutions that provide *hows* for the various subfunctions.

5.7.2 Morphological Chart

The morphological chart arranges the functions and subfunctions in logical order, and for each subfunction lists the possible *hows*. The method was first proposed by Zwicky.[4] The word morphology means the study of shape or form; so morphological analysis is a way of creating new forms, i.e., design concepts. The purpose of the method is to uncover combinations of ideas that comprise design concepts that might not ordinarily be generated.

> **EXAMPLE.** We return to our now familiar problem of the design of the CD case. Table 5.7 shows the morphological chart built upon the earlier functional decomposition. We note that many ways of accomplishing the subfunctions are listed. No details of the possible embodiments are given except descriptive words or a very simple sketch. Note that for a few subfunctions only a single conceptual idea is given. This means that the designer has made a fundamental assumption. In this instance it is that the CD case will be opened by a human hand, and not by some other feasible but impractical method as with a robot

1. J. E. McGrath, "Groups: Interaction and Performance," Prentice-Hall, Englewood Cliffs, NJ, 1984.
2. K. T. Ulrich and S. D. Eppinger, op. cit., p. 90.
3. For many examples of how to apply the ideas in Table 5.2 to mechanical design see E. B. Magrab, "Integrated Product and Process Design and Development," CRC Press, Boca Raton, FL, 1997, pp. 114–126. Also see N. P. Chironis, "Mechanisms and Mechanical Devices Source Book," 2d ed., McGraw-Hill, New York, 1996.
4. F. Zwicky, "The Morphological Method of Analysis and Construction," Courant Anniversary Volume, pp. 461–470, Interscience Publishers, New York, 1948.

TABLE 5.7
Morphological chart for the design of a CD case

Subfunction	Concepts				
	(1)	**(2)**	**(3)**	**(4)**	**(5)**
1.0 Open case					
1.1 Hold and grip case	Flat box	Grooved box	Curved box	Case with handle	Rubber grab strips
1.2 Disengage lock	Friction lock	Inclined plane lock	Magnetic lock	Clamp lock	Clicking hinge lock
1.3 Expose CD	Conventional hinge	One-piece flex plastic hinge	Slide-out, like matchbox	Tilt like shampoo bottle top	
2.0 Extract CD					
2.1 Disengage from securing system	Conventional "rosetta"	Lift/lock device	Padded cradle		
2.2 Grasp CD and remove	Hand				
3.0 Extract leaflet					
3.1 Disengage from securing system	Tabs	Holding slot	Velcro straps	Tab that swivels	No securing system
3.2 Remove leaflet	Hand				
4.0 Replace CD					
4.1 Place CD in securing system	Hand				
4.2 Engage securing system	2-finger push	Whole hand			
5.0 Replace leaflet					
5.1 Place leaflet in securing system	Slide into position	Lay in position			
5.2 Engage securing system	Slide under tabs or in slot	Swivel tabs	Attach Velcro		
6.0 Close case					
6.1 Engage lock	Friction surfaces	Put magnets together	Slide platen into position		
7.0 Store case					
7.1 Place case in desired location	Put on table	Put on another CD	Put in special CD holder		

or your mouth. Another reason that only a single idea would be given could be that a physical embodiment was being given, or it could be that the design team is weak on ideas. We call this limited *domain knowledge*.

5.7.3 Combining Concepts

The next step is to combine concepts to arrive at a set of definitive design concepts. We note that the number of possible combinations is quite large. For the example given above there are $5 \times 5 \times 4 \times 3 \times 5 \times 2 \times 2 \times 3 \times 3 \times 3 = 162,000$ combinations, clearly too many to follow up in detail. We need to select one fragmentary concept for each subfunction. Some may be clearly infeasible or impractical and are not selected. However, care should be taken not to make this judgment too hurriedly. If in doubt, save an idea for the evaluation step (Sec. 5.9). Also, realize that some concepts will satisfy more than one subfunction. Likewise, not all subfunctions are independent, but rather they are coupled. This means that their solutions can be evaluated only in conjunction with the solutions for other subfunctions.

Do not rush into evaluation of design concepts. Outstanding designs often evolve out of several iterations of combining concept fragments from the morphological chart and working them into an integrated solution. This is a place where a smoothly functioning team pays off.

Although design concepts are quite abstract at this stage, it often is very helpful to utilize rough sketches. Sketches help us associate function with form, and they aid with our short-term memory as we work to assemble the pieces of a design. Moreover, sketches in a design notebook are an excellent way of documenting the development of a product for patent purposes.[1]

> **EXAMPLE.** Five design concepts for the CD case were evolved from the morphological chart in Table 5.7. The numbers after each feature ( ) represent the column in Table 5.7 where it is found.
>
> Concept 1. Conventional square box (1), with an inclined plane lock (2), and a slide-out matchbox for a hinge. The CD is secured with a conventional "rosetta" (1), while the leaflet is secured with tabs (1).
> Concept 2. A streamlined curved box to fit the hand (3), with a friction lock (2) and a conventional hinge (3). The CD is secured in padded elastomer cradle (3) and the CD cases are designed to stack flat (2).
> Concept 3. The box is grooved to the shape of the fingers (2), with a magnetic lock (3) and conventional hinges (1). A new lift/lock secures the CD (2). The leaflet fits in a slot in the top of the case (2).
> Concept 4. A standard square box (1) with magnetic lock (3) and conventional hinges (1). The CD is secured with a padded cradle (3), while the leaflet is secured with Velcro straps (3).
> Concept 5. A curved box (3) with inclined plane lock (2), with a slide-out match box (3). The CD is held by a rosetta (1) and the leaflet fits into a slot (2). The cases are designed to stack (2).

---

1. The notebook of the creative genius Leonardo da Vinci is available on CD-ROM from Corbis Corporation, 1996.

5.8
AXIOMATIC DESIGN

The methods discussed in the previous sections of this chapter represent the best thinking about how to regularize the design process. However, they are essentially empirical methods. There is little scientific basis for these methods. Rather, they represent a distillation of the best ideas about what works to enhance the design practice.

There is a natural desire to improve upon this situation by developing a theory of design. This would extend intuition and experience by providing a framework for evaluating and extending design concepts. A design theory would make it possible to answer such questions as: Is this a good design? Why is this design better than others? How many design parameters (DPs) do I need to satisfy the functional requirements (FRs)? Shall I abandon the idea or modify the concept? Professor Nam Suh[1] and his colleagues at MIT have developed such a theoretical basis for design which is focused around two design axioms, hence the name axiomatic design.

An axiom is a proposition which is assumed to be true without proof for the sake of studying the consequences that follow from it. An axiom must be general truth for which no exceptions or counterexamples can be found. Axioms stand accepted, based on weight of evidence, until otherwise shown to be faulty. Suh has proposed two conceptually simple design axioms.

> *Axiom 1: The independence axiom*
> Maintain the independence of functional requirements (FRs).
> *Axiom 2: The information axiom*
> Minimize the information content.

The meaning of these two axioms is explored at length in Nam Suh's book.

Fundamental to this theory of design is the idea of functional requirements (FRs) and design parameter (DPs). Suh views the engineering design process as a constant interplay between *what we want to achieve* and *how we want to achieve it*. The former objectives are always stated in the *functional domain,* while the latter (the physical solution) is always generated in the *physical domain*. The design procedure is concerned with linking these two domains at every hierarchical level of the design process (Fig. 5.11). The design objectives are defined in terms of specific requirements called *functional requirements* (FRs). In order to satisfy these functional

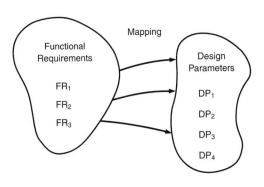

FIGURE 5.11
Suh's concept of design as the process of mapping from functional space to physical space to satisfy the functional requirements (FRs).

1. N. Suh, The Principles of Design," Oxford University Press, New York, 1990.

requirements a physical embodiment must be created in terms of *design parameters* (DPs). As Fig. 5.11 illustrates, the design process consists of mapping the FRs of the functional domain to the DPs of the physical domain to create a product, process, system, or organization that satisfies the perceived societal need. Note that this mapping process is not unique. Therefore, more than one design may result from the generation of the DPs that satisfy the FRs. Thus, the final outcome still depends on the designer's creativity. However, the design axioms provide the principles that the mapping techniques must satisfy to produce a good design, and they offer a basis for comparing and selecting designs.

There are hierarchies of FRs and DPs. Figure 5.12 shows the functional hierarchy for a metalcutting lathe. The hierarchical embodiment of these FRs in the physical domain is shown in Fig 5.13. FRs at the *i*th level cannot be decomposed into the next level of the FR hierarchy without first going over to the physical domain and developing a solution that satisfies the *i*th level FRs with all the corresponding DPs. For example, the FR concerning workpiece support and tool holder (Fig. 5.12) cannot be decomposed into the three FRs at the next lower level until it is decided in the physical domain (Fig. 5.13) that a tailstock will be used to satisfy the FRs. An experienced designer will take advantage of the hierarchical structure of FRs and DPs. By identifying the most important FRs at each level of the tree and ignoring the secondary factors from consideration at that level the designer manages to keep the work and information within bounds. Otherwise, the design process becomes too complex to manage. Remember that according to Axiom 1 each FR must be independent of the other FRs. This may be difficult to do on the first try; it is not unusual to expect that several iterations are required to get a proper set of FRs.

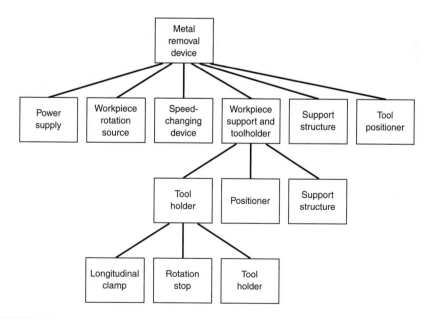

FIGURE 5.12

Hierarchy of functional requirements for a metalcutting lathe. (*From N. P. Suh, "The Principles of Design," copyright 1990 by Oxford University Press, Inc. Used by permission of Oxford University Press, Inc.*)

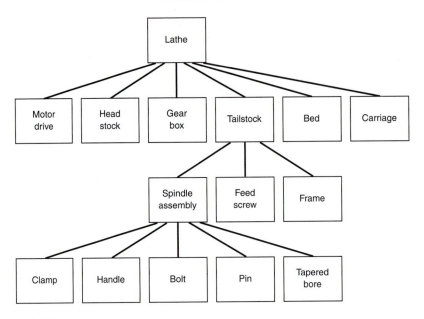

FIGURE 5.13
Hierarchy of lathe design in the physical domain. (*From N. P. Suh, "The Principles of Design," copyright 1990 by Oxford University Press, Inc. Used by permission of Oxford University Press, Inc.*)

Correspondingly, there can be many design solutions which satisfy a set of FRs. Also, when the set of FRs is changed a new design solution must be found. This new set of DPs must not be simply a modification of the DPs that were acceptable for the original FRs. Rather, a completely new solution should be sought.

Design constraints represent bounds on an acceptable solution. They can be input constraints or system constraints. The former are constraints in the design specifications, like weight, material strength, cost, or size, while the latter represent restrictions on capacity of machines, laws of nature, or shape. A constraint is different from an FR in that it does not have to be independent of other constraints or FRs. Moreover, constraints do not normally have tolerances associated with them, while FRs do. Note that it becomes easier to select FRs when a problem is highly constrained.

The two design axioms were given above in their most succinct form. We now restate them in a somewhat more descriptive form.[1]

Axiom 1: The independence axiom
   Alternative statement 1: An optimal design always maintains the independence of the FRs.
   Alternative statement 2: In an acceptable design the DPs and FRs are related in such a way that a specific DP can be adjusted to satisfy its corresponding FR without affecting other functional requirements.

---
1. N. P. Suh, op. cit, chap. 3.

Axiom 2: The information axiom

Alternative statement: The best design is a functionally uncoupled design that has the minimum information content.

Consider the design of the door for a conventional kitchen refrigerator. There are two FRs for the vertically hung door.

FR_1: provide access to the food in the refrigerator
FR_2: minimize the energy loss

In this design, when the door is opened to provide access to food (FR_1) the cold air escapes and warm air from the room rushes in. Thus, FR_2 is not satisfied and the vertical door is not a good design because the two FRs are coupled. A better uncoupled design is a door which is hinged horizontally and opens vertically, as in a freezer chest. With this design, when the door is opened to take out the food the cold air does not escape since cold air is heavier than the warm room air. Therefore, with this type of design the FRs are independent and the design satisfies the first axiom. We note that information content is related to complexity. While we aim to prevent functional coupling, physical coupling is desirable because the integration of more than one function in a single part reduces complexity, and according to the second axiom this is beneficial.

Seven important corollaries can be derived from the two basic axioms.[1] We can view these statements as design rules that can be useful in making design decisions.

Corollary 1: Decoupling of a coupled design

Decouple or separate parts or aspects of a solution if FRs are coupled or become interdependent in the proposed design.

Decoupling does not imply that a part has to be broken into two or more separate physical parts, or that a new element has to be added to the existing design.

Corollary 2: Minimize FRs

Minimize the number of FRs and constraints.

Increasing these elements of the design increases the information content. Do not try to produce a design that does more than was intended. A design that fulfills more functions than called for in the FRs will be more costly to operate and maintain, and may have lower reliability.

Corollary 3: Integration of physical parts

Integrate design features in a single physical part if FRs can be independently satisfied in the proposed solution.

Corollary 4: Use of standardization

Use standardized or interchangeable parts if the use of these parts is consistent with the FRs and constraints.

---

1. N. P. Suh, op, cit, pp. 52–54.

Corollary 5: *Use of symmetry*
Use symmetric shapes and/or arrangements if they are consistent with the FRs and constraints. Symmetrical parts require less information to manufacture and to orient in assembly.

Corollary 6: *Largest tolerance*
Specify the largest allowable tolerance in stating FRs.

Corollary 7: *Uncoupled design with less information*
Seek an uncoupled design that requires less information than coupled designs in satisfying a set of FRs. There is always an uncoupled design that involves less information than a coupled design. This corollary follows as a consequence of axioms 1 and 2, for if this corollary is not true then axioms 1 and 2 must be invalid. An implication of this corollary is that if a designer proposes an uncoupled design that has more information than a coupled design then the design should be started anew because a better design lies somewhere.

Unfortunately, space does not permit development of design principles, or *axiomatic design,* in greater detail. The reader is referred to Suh[1] for details of how to determine the independence of FRs, how to measure information content, and for a number of detailed examples of how to apply these techniques in design. Magrab[2] presents axiomatic design as the formal way of clarifying functional decomposition and the product goals, and gives many examples of its use.

5.9
EVALUATION METHODS

We have reached the point where we must choose which concept or small set of concepts to develop into finished designs. Unfortunately, we are forced to make this choice at a stage in the design process where we still have very little detailed information. Thus, we need initial evaluation methods that can be applied during the relatively unstructured process of concept development. Other evaluation methods will be useful later on in design when we must make choices based on more detailed technical information. To aid in learning about a full spectrum of evaluation methods, we shall temporarily abandon our practice of presenting new ideas in a sequence tied to the design process, and rather present a full spectrum of evaluation methods in this section.

Evaluation involves *comparison*, followed by *decision making*. To make a valid comparison the concepts must exist at the same level of abstraction. In an *absolute comparison* the concept is directly compared with some set of requirements. In a *relative comparison* the concepts are compared with each other.

---

1. N. P. Suh, op. cit.
2. E. B. Magrab, "Integrated Product and Process Design and Development," chap. 4, CRC Press, Boca Raton, FL, 1997.

5.9.1 Comparison Based on Absolute Criteria

As we saw in Fig. 5.2, our scheme begins by comparing the concepts to a series of *absolute filters.*[1]

1. *Evaluation based on judgment of feasibility of the design:* The initial screen is based on the overall evaluation of the design team as to the feasibility of each concept. Concepts should be placed into one of three categories:

   (*a*) It is not feasible (it will never work)? Before discarding an idea, ask "why is it not feasible?" If judged not feasible, will it provide new insight into the problem?

   (*b*) It is conditional—it might work if something else happens? The something else could be the development of a critical element of technology or the appearance in the market of a new microchip that enhances some function of the product.

   (*c*) Looks as if it will work! This is a concept that seems worth developing further. Obviously, the reliability of these judgments is strongly dependent on the expertise of the design team. When making this judgment, err on the side of accepting a concept unless there is strong evidence that it will not work.

2. *Evaluation based on assessment of technology readiness:* Except in unusual circumstances, the technology used in a design must be mature enough that it can be used in the product design without additional research effort. *Product design is not the appropriate place to do R & D.* Some indicators of technology maturity are:

   (*a*) Can the technology be manufactured with known processes?

   (*b*) Are the critical parameters that control the function identified?

   (*c*) Are the safe operating latitude and sensitivity of the parameters known?

   (*d*) Have the failure modes been identified?

   (*e*) Does hardware exist that demonstrates positive answers to the above four questions?

3. *Evaluation based on go-no-go screening of the customer requirements:* After a design concept has passed filters 1 and 2, the emphasis shifts to establishing whether it meets the customer requirements framed in the QFD. Each customer requirement must be transformed into a question to be addressed to each concept. The questions should be answerable as either yes (go), maybe (go), or no (no-go). The emphasis is not on a detailed examination (that comes below) but on eliminating any design concepts that clearly are not able to meet an important customer requirement.

   EXAMPLE.

   Question: In concept 5 (Sec. 5.7), is the CD case easy to open?

   Answer: Maybe (go).

   Question: Will the locking of the case be secure in concept 5?

   Answer: Yes (go).

---

1. D. G. Ullman, op. cit., pp. 155–160.

Proceed in this way through all of the customer requirements. Note that if a design concept shows mostly goes, but it has a few no-go responses, it should not be summarily discarded. The weak areas in the concept may be able to be fixed by borrowing ideas from another concept. Or the process of doing this go-no-go analysis may trigger a new idea.

5.9.2 Pugh's Concept Selection Method

A particularly good method for deciding on the most promising design concept at the concept stage is the *Pugh concept selection process.*[1] This method compares each concept relative to a reference or datum concept and for each criterion determines whether the concept in question is better than, poorer than, or about the same as the reference concept. Thus, it is a relative comparison technique. Remember that studies show that an individual is best at creating ideas—but a small group is better at selecting ideas. The concept selection method is done by the design team, usually in successive rounds of examination and deliberation. The steps in concept selection, as given by Clausing,[2] are:

1. *Choose the criteria by which the concepts will be evaluated:* The QFD is the starting place to develop the criteria. If the concept is well worked out, then the criteria will be based on the engineering characteristics listed in the columns of the House of Quality (Fig. 2.5). However, often the concepts have not been refined enough to be able to use the engineering characteristics, and then they must be based on the customer requirements listed in the rows of the QFD. Do not mix the two, since it is important to make comparisons at the same level of abstraction.

    A good way to arrive at the criteria is to ask each team member to create a list of 15 to 20 criteria, based on the QFD and functional analysis. Then as a team work session, the lists of criteria are merged, discussed, and prioritized. Note that by not just copying the criteria from the QFD it is possible to reduce the criteria to 15 to 20 items and to add important factors possibly not covered by the QFD like patent coverage, technical risk, and manufacturability. Also, in formulating the final list of criteria, it is important to consider the ability of each criterion to differentiate among concepts. A criterion may be very important, but if every concept satisfies it well, it will not help in selecting the final concept. Therefore, this concept should be left out of the concept selection matrix. Also, some teams want to determine a relative weight for each criterion. This should usually be avoided, since it adds a degree of detail that is not justified at the concept level of information. Instead, list the criteria in approximate decreasing order of priority.

2. *Formulate the decision matrix:* The criteria are entered into the matrix as the row headings. The concepts are the column headings of the matrix. Again, it is important that concepts to be compared are at the same level of abstraction. If a concept can be represented by a simple sketch, this should be used in the column heading. Otherwise, each concept is defined by a text list.

1. S. Pugh, "Total Design," Addison-Wesley, Reading, MA, 1991; S. Pugh, "Creating Innovative Products Using Total Design," Addison-Wesley, Reading, MA, 1996; D. Clausing, "Total Quality Development," ASME Press, New York, 1994.
2. D. Clausing, op. cit., pp. 153–164.

3. *Clarify the design concepts:* The goal of this step is to bring all members of the team to a common level of understanding about each concept. If done well, this will also develop team "ownership" in each concept. This is important, because if individual concepts remain associated with different team members the final team decision could be dominated by political negotiation.

   A good team discussion about the concepts often is a creative experience. New ideas often emerge and are used to improve concepts or to create entirely new concepts that are added to the list.

4. *Choose the datum concept:* One concept is selected by the team as a *datum* for the first round. This is the reference concept with which all other concepts are compared. In making this choice it is important to choose one of the better concepts. A poor datum would cause all of the concepts to be positive and would unnecessarily delay arriving at a solution. Generally the team members are asked for their ideas, and a majority vote prevails. It is not important which concept is chosen for the initial datum so long as it is a relatively good concept. For a redesign, the datum is the existing design reduced to the same level of abstraction as the other concepts. The column chosen as datum is marked accordingly, DATUM.

5. *Run the matrix:* It is now time to do the comparative evaluation. Each concept is compared with the datum for each criterion. The first criterion is applied to each concept, then the second, and so on. A three-level scale is used. At each comparison we ask the question, is this concept better ($+$), worse ($-$), or about the same (S) as the datum, and the appropriate symbol is placed in the cell of the matrix. Same (S) means that the concept is not clearly better or worse than the datum.

   Much more than filling in the scores occurs in a well-run concept selection meeting. There should be brief constructive discussion when scoring each cell of the matrix. Divergent opinions lead to greater team insight about the design problem. A good facilitator can keep the decision-making discussion to about 1 minute per cell. Long, drawn-out discussion usually results from insufficient information and should be terminated with an assignment to someone on the team to generate the needed information.

   Again, the team discussion often stimulates new ideas that lead to additional improved concepts. Someone will suddenly see that combining this idea from concept 3 solves a deficiency in concept 8, and a hybrid concept evolves. Another column is added for the new concept. A major advantage of the Pugh concept selection method is that it helps the team to develop better insights into the types of features that strongly satisfy the design requirements.

6. *Evaluate the ratings:* Once the comparison matrix is completed, the sum of the $+$, $-$, and S ratings is determined for each concept. Do not become too quantitative with these ratings. While it is appropriate to take a difference between the $+$ score and the $-$ scores, be careful about rejecting a concept with a high negative score without further examination. The few positive features in the concept may really be "gems" that could be picked up and used in another concept. For the highly rated concepts determine what their strengths are and what criteria they treat poorly. Look elsewhere in the set of concepts for ideas that may improve these low-rated criteria. Also, if most concepts get the same rating on a certain criterion examine it to see whether it is ambiguous or not uniformly evaluated from concept to concept. If this is an important criterion, then you will need to spend effort to generate better concepts or to clarify the criterion.

7. *Establish a new datum and rerun the matrix:* The next step is to establish a new datum, usually the concept that received the highest rating in the first round, and run the matrix again. Eliminate the lowest rating concepts from this second round. The main intent of this round is not to verify that the selection in round 1 is valid but to gain added insight to inspire further creativity. The use of a different datum will give a different perspective at each comparison that will help clarify relative strengths and weaknesses of the concepts.

8. *Plan further work:* Depending on the problem, the first round of concept selection will take between 1/2 and 1 day. For anything but the most elementary problem, this is not sufficient to arrive at an excellent solution. The team meeting may have uncovered missing pieces of information, calculations or experiments that need to be made, or experts that need to be consulted. Before the team leaves the initial meeting they make assignments to fill these gaps.

9. *Second working session:* Armed with the missing information, the team returns for a second session of concept development. About half of the concepts have been eliminated in the first session, so the objective is to improve upon the best that remain. Since team members have had time to think individually about the problem, and the problem has lain fallow in their subconscious minds, it is not surprising that new and creative concepts come forth and are added to the concept selection matrix.

   This session proceeds in the same way as described above to arrive at a smaller set of even better concepts. If the team reaches consensus on a final concept, the process stops. However, it may schedule another meeting in an attempt to inject new ideas and arrive at a better concept solution. As Fig. 5.14 shows, the Pugh concept selection method is an iterative process of convergent and divergent thinking which eventually arrives at a dominant concept.

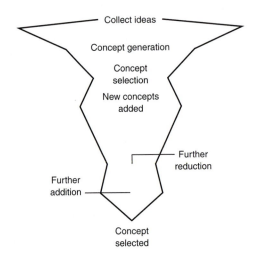

FIGURE 5.14

Concept generation and selection, viewed as alternating divergent and convergent processes.

EXAMPLE. Five concepts for a CD case were developed in Sec. 5.7. These are evaluated against the datum of the standard CD case with the Pugh concept selection method.

Criterion	Concept 1	Concept 2	Concept 3	Concept 4	Concept 5	Std. CD case
Manufacturing cost	S	−	S	−	S	
Easier opening	+	S	S	S	+	D
Easier to remove leaflet	S	S	−	+	−	A
Easier to remove CD	S	+	+	+	S	T
Hinge doesn't come apart	+	S	S	S	+	U
Stacking stability	S	S	S	S	+	M
More secure locking	+	S	+	+	+	
Fits hand better	S	+	+	S	+	
$\Sigma+$	3	2	3	3	5	
$\Sigma-$	0	1	1	1	1	
ΣS	5	5	4	4	2	

Study of the concept selection table shows that all concepts are highly positive, but concept 5 has the most pluses. Its only negative is in ease of removing the leaflet, since it might be difficult to get the leaflet out of the slot that contains it. Looking at the table shows that containing the leaflet with Velcro straps has an advantage in this function, and it is added to form a new concept 6. The only function that has not been improved in concept 5 is "ease of removal of the CD." The use of a urethane cradle will not work with the matchbox design, but the new "lift/lock" securing device might. We therefore create concept 7, which is the same as concept 6 with the substitution of the new "lift/lock" device. Both concepts 6 and 7 will be carried forward until a better idea of the manufacturing cost of the "lift/lock" device can be obtained.

Note that many of the criteria in the QFD, like crack resistance, scratch resistance, etc., were not included in the selection chart, because these issues would be rated equally for every concept.

5.9.3 Measurement Scales

Rating a design characteristic among several alternative designs is a measurement. Therefore, we need to have some idea of the various scales that can be used in such ranking.[1] A *nominal scale* is a named category or identifier like "thin" or "shiny." This is not a very quantitative scale of measurement, although data of this type can be counted and the mode determined. The next measurement scale is an *ordinal scale,* in

---

1. K. N. Otto, Measurement Methods for Product Evaluation, *Research in Engineering Design,* vol. 7, pp. 86–101, 1995.

which items are ranked. Each item in the set is deemed better than or worse than its counterparts, and is placed in order. This is called an ordinal scale because the elements are only ordered. This measurement scale says nothing about how far apart the elements are from each other. Both the mode and the median can be determined for data measured on this scale. Note that the Pugh concept selection method uses an ordinal scale.

A method of ranking alternatives on an ordinal scale is to use *pairwise comparison*. Each design objective or criterion is listed and is compared to every other objective, two at a time. In making the comparison the objective that is considered the more important of the two is given a 1 and the less important objective is given as a 0. The total number of possible comparisons is $N = n(n - 1)/2$, where n is the number of objectives under consideration.

Consider the case where there are five design objectives, A, B, C, D, and E. In comparing A vs. B we consider A to be more important, and give it a 1. In comparing A vs. C we feel C ranks higher and a 0 is recorded. Thus, the table is completed. The rank order established is B, D, A, E, C.

Design objectives	A	B	C	D	E	Row total
A	—	1	0	0	1	2
B	0	—	1	1	1	3
C	1	0	—	0	0	1
D	1	0	1	—	1	3
E	0	0	1	0	—	1
						—
						10

If we need information on how much worse A is relative to B, then an *interval scale* is needed. It is used when a quantitative scale is desired, but the objectives have no identifiable units of measure. An interval measurement scale has values with an arbitrary endpoint. The scale has no zero value. For example, we could distribute the results from the above example along a 1–10 scale to create an interval scale.

```
D                                      E
B                        A             C
10    9    8    7    6    5    4    3    2    1
```

The most important objectives have been given a value of 10, and the others have been given values relative to this.

A *ratio scale* is an interval scale in which a zero value is used to anchor the scale. In an interval scale some arbitrary value is assigned to the base point. A ratio scale is needed to establish meaningful weighting factors.

5.9.4 Weighted Decision Matrix

A decision matrix is a method of evaluating competing concepts by ranking the design criteria with weighting factors and scoring the degree to which each design concept

TABLE 5.8
Evaluation scheme for design objectives

11-point scale	Description	5-point scale	Description
0	Totally useless solution	0	Inadequate
1	Very inadequate solution		
2	Weak solution	1	Weak
3	Poor solution		
4	Tolerable solution	2	Satisfactory
5	Satisfactory solution		
6	Good solution with a few drawbacks		
7	Good solution	3	Good
8	Very good solution		
9	Excellent (exceeds the requirement)	4	Excellent
10	Ideal solution		

meets the criterion. In doing this it is necessary to convert the values obtained for different design criteria into a consistent set of values. The simplest way of dealing with design criteria expressed in a variety of ways is to use a point scale. A 5-point scale is used when the knowledge about the criteria is not very detailed. An 11-point scale (0–10) is used when the information is more complete (Table 5.8). It is best if several knowledgeable people participate in this evaluation.

EXAMPLE. A heavy steel crane hook, for use in supporting ladles filled with molten steel as they are transported through the steel mill, is being designed. Three concepts have been proposed: (1) built-up from steel plates, welded together; (2) built-up from steel plates, riveted together; (3) a monolithic cast-steel hook.

The first step is to identify the design criteria by which the concepts will be evaluated. The product design specification is a prime source of this information. The design criteria are identified as (1) material cost, (2) manufacturing cost, (3) time to produce another if one fails, (4) durability, (5) reliability, (6) reparability.

The next step is to determine the weighting factor for each of the design criteria. A good way to proceed is to construct a hierarchical objective tree (Fig. 5.15). The weights of the individual categories at each level of the tree must add to 1.0. We should note that this is a simplified problem. Many problems could have two or more additional levels of hierarchy. To get the weight of a factor on a lower level, multiply the weights as you go up the chain. Thus, the weighting factor for material cost, $O_{111} = 0.3 \times 0.6 \times 1.0 = 0.18$.

The decision matrix is given in Table 5.9. The weighting factors are calculated from Fig. 5.15. The score for each concept for each criterion is derived from Table 5.8. The rating for each concept at each design criterion is obtained by multiplying the score by the weighting factor. The overall rating for each concept is the sum of these ratings.

The weighted decision matrix indicates that the best overall design concept would be a crane hook made from elements cut from steel plate and fastened together with rivets.

The simplest procedure in comparing design alternatives is to add up the ratings for each concept and declare the concept with the highest score the winner. A better

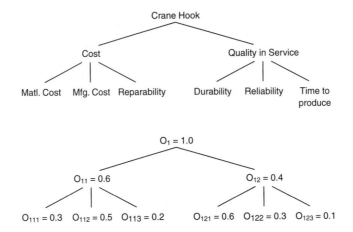

FIGURE 5.15
Objective tree for the design of a crane hook.

TABLE 5.9
Weighted decision matrix for a steel crane hook

Design criterion	Weight factor	Units	Built-up plates welded			Built-up plates riveted			Cast steel hook		
			Magnitude	Score	Rating	Magnitude	Score	Rating	Magnitude	Score	Rating
Material cost	0.18	¢/lb	60	8	1.44	60	8	1.44	50	9	1.62
Manufacturing cost	0.30	$	2500	7	2.10	2200	9	2.70	3000	4	1.20
Reparability	0.12	Experience	Good	7	0.84	Excellent	9	1.08	Fair	5	0.60
Durability	0.24	Experience	High	8	1.92	High	8	1.92	Good	6	1.44
Reliability	0.12	Experience	Good	7	0.84	Excellent	9	1.08	Fair	5	0.60
Time to produce	0.04	Hours	40	7	0.28	25	9	0.36	60	5	0.20
					7.42			8.58			5.66

way to use the decision matrix is to examine carefully the components that make up the rating to see what design factors influenced the result. This may suggest areas for further study or raise questions about the validity of the data or the quality of the individual decisions that went into the analysis. Pugh points out[1] that the outcome of a decision matrix depends heavily on the selection of the criteria. He worries that the method may instill an unfounded confidence in the user and that the designer will tend to treat the total ratings as being absolute.

---

1. S. Pugh, op. cit., pp. 92–99.

5.9.5 Analytic Hierarchy Process (AHP)

Saaty's analytic hierarchy process (AHP) is well suited for evaluation problems whose objectives have a hierarchical structure.[1] AHP is a multicriteria decision-making process that allows working with both numerical factors and those that are intangible and subjective. It provides a way to determine both the weights and values for each criterion in a consistent, methodologically correct, and intuitively acceptable manner.

Many evaluation problems in engineering design are framed in a hierarchy or system of stratified levels, each consisting of many elements or factors. The basic question to be answered is: "How strongly do the individual factors at the lowest level of the hierarchy influence its top factor, the overall objective of the design?" Figure 5.16 shows the hierarchical structure for the crane hook design problem introduced in the previous section.

AHP starts with pairwise comparison of the alternatives for each of the decision criteria to convert the verbal impression of importance into a numerical value. The criteria are arranged as the rows and columns of a matrix. Start with the first criterion in the first row and ask the question "How much more strongly does this criterion influence the outcome than the other criterion?" In answering this question, use Saaty's 9-point scale given in Table 5.10. This 9-point scale has been validated by statistical tests to give reproducibly accurate results. Moreover, it is a scale that people use instinctively.

> **EXAMPLE.** We shall use the AHP method to determine which design concept is best for the design of a crane hook (see previous example). The first task is to determine the weighting factors for each of the design criteria. This is done with a square matrix (Table 5.11). Start with the first row. Material cost is just as important as material cost in ranking the design criteria, so we put a 1.0 in the first cell. A value of 1.0 is put in each cell along the matrix diagonal because these cells compare one criterion with itself. Now we compare material cost with manufacturing cost and ask how much more important material cost is

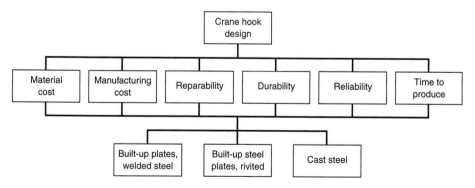

FIGURE 5.16
Hierarchical structure of the crane hook design.

---

1. T. L. Saaty, "The Analytic Hierarchy Process," McGraw-Hill, New York, 1980; T. L. Saaty, "Decision Making for Leaders," 3d ed., RWS Publications, Pittsburgh, PA, 1995.

TABLE 5.10
Saaty's fundamental scale for pairwise comparison

Intensity of importance	Definition	Description
1	Equal importance	Two activities contribute equally to the objective
3	Moderate importance	Judgment and experience slightly favor one activity over another
5	Strong importance	Judgment and experience strongly favor one activity over another
7	Very strong or demonstrated importance	An activity is favored very strongly over another; its dominance demonstrated in practice
9	Extreme importance	The evidence favoring one activity over another is of the highest possible
2, 4, 6, 8	These ratings are used to compromise between the above values	

TABLE 5.11
Square matrix to determine weighting factors

	Material cost	Manufacturing cost	Reparability	Durability	Reliability	Time to produce
Material cost	1	1/5 = 0.20	3	1/5 = 0.20	3	7
Manufacturing cost	5	1	7	3	3	7
Reparability	1/3 = 0.33	1/7 = 0.14	1	1/5 = 0.20	1/3 = 0.33	5
Durability	5	1/3 = 0.33	5	1	3	7
Reliability	1/3 = 0.33	1/3 = 0.33	3	1/3 = 0.33	1	7
Time to produce	1/7 = 0.14	1/7 = 0.14	1/5 = 0.20	1/7 = 0.14	1/7 = 0.14	1
Total	11.8	2.14	19.2	4.87	10.47	34.0

than manufacturing cost. Since we decide that manufacturing cost really is more important than material cost, we leave this cell blank for the moment and pass on. We ask how important is material cost compared with reparability, and decide from Table 5.10 that it is of moderate importance. Score a 3. We pass on along the first row in the manner just described. Now we are in the second row. We ask whether manufacturing cost is more important than material cost, and decide that it scores a 5 because it is of strong importance. Since manufacturing cost is considered strongly more important (5) than material cost, the importance of materials cost relative to manufacturing cost must be 1/5, or 0.20, and we enter this value in the second cell of row 1. Proceeding through the matrix of pairwise comparisons in this manner results in Table 5.11.

TABLE 5.12
Normalized values of Table 5.11, giving weighting factors

	Material cost	Manufacturing cost	Reparability	Durability	Reliability	Time to produce	Total	Weighting factor (AVG)
Material cost	0.085	0.093	0.156	0.041	0.286	0.206	0.867	0.144
Manufacturing cost	0.424	0.467	0.364	0.616	0.286	0.206	2.363	0.394
Reparability	0.028	0.065	0.052	0.041	0.031	0.147	0.364	0.061
Durability	0.424	0.154	0.260	0.205	0.286	0.206	1.535	0.256
Reliability	0.028	0.154	0.156	0.068	0.095	0.206	0.707	0.118
Time to produce	0.012	0.065	0.010	0.029	0.013	0.029	0.158	0.026
	1.000	1.000	1.000	1.000	1.000	1.000	6.000	1.000

Next we normalize the entries in each cell by dividing by the total for each column in Table 5.11. The average of the rows gives the weighting factor for each design criterion (Table 5.12). This is strictly true only if the pairwise comparisons are completely consistent. This is measured by the inconsistency ratio (IR). IR = 0 denotes complete consistency. IR ≤ 0.1 is considered to be acceptable. Consistency implies the transitivity relationship that if i is more important than j, and j is more important than k, then i is more important than k. Determining IR involves calculating the eigenvector of the matrix represented by Table 5.11. A much quicker way is to use the software developed for AHP.[1] Note that the values of the weighting factor have changed somewhat, compared with Table 5.9, because of the different evaluation approach used in the AHP method.

Now we will construct a decision matrix similar to Table 5.9. The magnitudes of the criterion factors for the three design concepts are given in that table. Let us convert the magnitudes for the manufacturing cost into ratings that can be used with AHP. Because a smaller number for cost is a "better number," we shall first convert the cost data by taking their reciprocals. Thus, a low cost will rate higher on this scale. The rating for each alternative is its fraction of the total.

	Built-up welded plates	Built-up riveted plates	Cast	
Manufacturing cost	2500	2200	3000	$ per crane hook
	400	454	333	reciprocal $\times 10^{-6}$
	0.34	0.38	0.28	fraction of total

Suppose we decide that the reparability of the crane hooks would be as given below. This ranking is based on expert judgment. It says that a crane hook built up by riveting cut steel plate is much easier to repair than a monolithic cast steel hook. The rating would be determined as follows.

1. Expert Choice®, version 9.0, available from Decision Support Software, Pittsburgh, PA.

	Built-up welded plates	Built-up riveted plates	Cast	
Reparability	6	10	1	Ranking
	0.35	0.59	0.06	Fraction of total

We can use AHP to create a ranking of the three design concepts with respect to the criterion durability. This is based on the judgment of the team applying Table 5.10 in the way illustrated above. Here we use a format that combines Tables 5.11 and 5.12 (see Table 5.13). It is important to realize rating values developed with the AHP are numbers on a ratio scale. Thus, we can multiply and divide these numbers without being concerned whether these operations are meaningful.

Now we are ready to combine the pieces of information into a decision matrix (Table 5.14). The rating values for the three design concepts come from the above small tables,

TABLE 5.13
AHP pairwise comparison to rank the design concepts with respect to durability

	Welded plate		Riveted plate		Cast		Total		Rating (average)
Welded plate	1.00	0.23	1/3 = 0.33	0.22	3.00	0.33		0.78	0.26
Riveted plate	3.00	0.69	1.00	0.65	5.00	0.56		1.90	0.63
Cast	1/3 = 0.33	0.08	1/5 = 0.20	0.13	1.00	0.11		0.32	0.11
Total	4.33	1.00	1.53	1.00	9.00	1.00		3.00	1.00

TABLE 5.14
Decision matrix for the crane hook problem

Design criterion	Weight factor	Welded plate	Riveted plate	Cast	Welded plate	Riveted plate	Cast
Material cost	0.14	0.31	0.31	0.38	0.043	0.043	0.053
Manufacturing cost	0.39	0.34	0.38	0.28	0.133	0.148	0.109
Reparability	0.06	0.35	0.59	0.06	0.021	0.035	0.004
Durability	0.25	0.26	0.63	0.11	0.065	0.157	0.027
Reliability	0.12	0.33	0.43	0.24	0.040	0.052	0.029
Time to produce	0.03	0.31	0.49	0.20	0.008	0.013	0.005
Total	1.00				0.31	0.45	0.23

plus others constructed in a similar way. In the same way as for the weighted decision matrix, each rating value for each concept is multiplied by the appropriate weighting factor for the design criterion. This gives the three columns to the right. The sum of each column is the relative importance of the design concept. Once again the riveted plate design is rated superior. Table 5.14 also shows how different design concepts rate with respect to various design criteria. For example, the cast hook rates rather low in ease of repair and durability.

We have carried out the AHP analysis step by step to give you a feel for what is involved. Using the Expert Choice® software would be a much quicker and better way of arriving at the results because it would allow the weighting factors to be adjusted readily to achieve an IR ≤ 0.1. The outcome comparable to Table 5.14, when the software is used, is shown in the table below.

Decision matrix for the selection of crane hook design

Welded plate	Riveted plate	Cast
0.32	0.50	0.17
IR $= 0.08$		

Analysis made with Expert Choice®

5.10
DECISION MAKING

In the previous section we presented a number of ways of evaluating alternative design concepts or problem solutions so as to establish the "best solution." These were analysis techniques that helped us establish a logical decision process. Another way to approach the problem is through the field of decision analysis, which is an important area of study in the larger field of *operations research*. The use of decision analysis methods to make decisions in engineering design is an active area of research.

5.10.1 Behavioral Aspects of Decision Making

Making a decision is a stressful situation for most people. This psychological stress arises from at least two sources.[1] First, decision makers are concerned about the material and social losses that will result from either course of action that is chosen. Second, they recognize that their reputations and self-esteem as a competent decision maker are at stake. Severe psychological stress brought on by decisional conflict can be a major cause of errors in decision making. There are five basic patterns by which people cope with the challenge of decision making.

---

1. I. L. Janis and L. Mann, *Am. Scientist,* November–December, 1976, pp. 657–667.

1. *Unconflicted adherence:* Decide to continue with current action and ignore information about risk of losses.
2. *Unconflicted change:* Uncritically adopt whichever course of action is most
   strongly recommended.
3. *Defensive avoidance:* Evade conflict by procrastinating, shifting responsibility to
   someone else, and remaining inattentive to corrective information.
4. *Hypervigilance:* Search frantically for an immediate problem solution.
5. *Vigilance:* Search painstakingly for relevant information that is assimilated in an
   unbiased manner and appraised carefully before a decision is made.

All of these patterns of decision making, except the last one, are defective.

   The quality of a decision does not depend on the particulars of the situation as
much as it does on the manner in which the decision-making process is carried out.
We will attempt to discuss the basic ingredients in a decision and the contribution
made by each.[1] The basic ingredients in every decision are listed in the accompanying table. That a substitution is made for one of them does not necessarily mean that
a bad decision will be reached, but it does mean that the foundation for the decision
is weakened.

Basic ingredients	Substitute for basics
Facts	Information
Knowledge	Advice
Experience	Experimentation
Analysis	Intuition
Judgment	None

   A decision is made on the basis of available facts. Great effort should be given to
evaluating possible bias and relevance of the facts. Emphasis should be on preventing
arrival at the right answer to the wrong question. It is important to ask the right questions to pinpoint the problem. When you are getting facts from subordinates, it is
important to guard against selectivity—the screening out of unfavorable results. The
status barrier between a superior and a subordinate can limit communication and
transmission of facts. The subordinate fears disapproval, and the superior is worried
about loss of prestige. Remember that the same set of facts may be open to more than
one interpretation. Of course, the interpretation of qualified experts should be
respected, but blind faith in expert opinion can lead to trouble.

   Facts must be carefully weighed in an attempt to extract the real meaning: knowledge. In the absence of real knowledge, we must seek advice. It is good practice to
check your opinions against the counsel of experienced associates. That should not be
interpreted as a sign of weakness. Remember, however, that even though you do make
wise use of associates, you cannot escape accountability for the results of your decisions. You cannot blame failures on bad advice; for the right to seek advice includes
the right to accept or reject it. Many people may contribute to a decision, but the decision maker bears the ultimate responsibility for its outcome. Also, advice must be

---

1. D. Fuller, *Machine Design,* July 22, 1976, pp. 64–68.

sought properly if it is to be good advice. Avoid putting the adviser on the spot; make it clear that you accept full responsibility for the final decision.

There is an old adage that there is no substitute for experience, but the experience does not have to be your own. You should try to benefit from the successes and failures of others. Unfortunately, failures rarely are recorded and reported widely. There is also a reluctance to properly record and document the experience base of people in a group. Some insecure people seek to make themselves indispensable by hoarding information that should be generally available. Disputes between departments in an organization often lead to restriction of the experience base. In a well-run organization someone in every department should have total access to the records and experience of every other department.

Before a decision can be made, the facts, the knowledge, and the experience must be brought together and evaluated in the context of the problem. Previous experience will suggest how the present situation differs from other situations that required decisions, and thus precedent will provide guidance. If time does not permit an adequate analysis, then the decision will be made on the basis of intuition, an instinctive feeling as to what is probably right (an educated guess). An important help in the evaluation process is discussion of the problem with peers and associates.

The last and most important ingredient in the decision process is judgment. Good judgment cannot be described, but it is an integration of a person's basic mental processes. Judgment is a highly desirable quality, as evidenced by the fact that it is one of the factors usually included in personal evaluation ratings. Judgment is particularly important because most decisional situations are shades of gray rather than either black or white. An important aspect of good judgment is to understand clearly the realities of the situation.

A decision usually leads to an *action*. A situation requiring action can be thought of as having four aspects:[1] should, actual, must, and want.

The *should aspect* identifies what ought to be done if there are no obstacles to the action. A should is the expected standard of performance if organizational objectives are to be obtained. The should is compared with the *actual,* the performance that is occurring at the present point in time. The *must* action draws the line between the acceptable and the unacceptable action. A must is a requirement that cannot be compromised. A *want* action is not a firm requirement but is subject to bargaining and negotiation. Want actions are usually ranked and weighted to give an order of priority. They do not set absolute limits but instead express relative desirability.

To summarize this discussion of the behavioral aspects of decision making, we list the sequence of steps that are taken in making a good decision.[2]

1. The objectives of a decision must be established first.
2. The objectives are classified as to importance. (Sort out the musts and the wants.)
3. Alternative actions are developed.
4. The alternatives are evaluated against the objectives.

---

1. C. H. Kepner and B. B. Tregoe, "The Rational Manager: A Systematic Approach to Problem Solving and Decision Making," Princeton Research Press, Princeton, NJ, 1976.
2. Ibid.

5. The choice of the alternative that holds the best promise of achieving all of the objectives represents the tentative decision.
6. The tentative decision is explored for future possible adverse consequences.
7. The effects of the final decision are controlled by taking other actions to prevent possible adverse consequences from becoming problems and by making sure that the actions decided on are carried out.

5.10.2 Decision Theory

An important area of activity within the broader subject field of operations research has been the development of a mathematically based theory of decisions.[1] Decision theory is based on utility theory, which develops values, and probability theory, which assesses our stage of knowledge. Decision theory has been applied more to business management situations than to engineering design decisions, but the potential for future applications in design appears strong. The purpose of this section is to acquaint the reader with the basic concepts of decision theory and point out references for future study.

A decision-making model contains the following six basic elements.

1. *Alternative courses of action* can be denoted as $a_1, a_2, \ldots, a_n$. As an example of alternative actions, the designer may wish to choose between the use of steel (a_1), aluminum (a_2), or fiber-reinforced polymer (a_3) in the design of an automotive fender.
2. *States of nature* are the environment of the decision model. Usually, these conditions are out of the control of the decision maker. If the part being designed is to withstand salt corrosion, then the state of nature might be expressed by $\theta_1 =$ no salt, $\theta_2 =$ weak salt concentration, etc.
3. *Outcome* is the result of a combination of an action and a state of nature.
4. *Objective* is the statement of what the decision maker wants to achieve.
5. *Utility* is the measure of satisfaction or value which the decision maker associates with each outcome.
6. *States of knowledge* is the degree of certainty that can be associated with the states of nature. This is expressed in terms of probabilities.

To carry out the simple design decision of selecting the best material to resist road salt corrosion in an automotive fender, we construct a table of the utilities for each outcome. A utility can be thought of as a generalized loss or gain all factors of which (cost of material, cost of manufacturing, corrosion resistance) have been converted to a common scale. We will discuss this complex problem later, but for the present consider that utility has been expressed on a scale of "losses." Table 5.15 shows the loss table for this material selection decision. Note that, alternatively, the utility could be expressed in terms of gains, and then the table would be called the payoff matrix.

_____

1. H. Raiffa, "Decision Analysis," Addison-Wesley, Reading, MA, 1968; S. R. Watson and D. M. Buede, "Decision Synthesis: The Principles and Practice of Decision Analysis," Cambridge University Press, Cambridge, 1987.

TABLE 5.15
Loss table for material selection decision

Courses of action	State of nature			
	θ_1	θ_2	θ_3	θ_4
a_1	1	4	10	15
a_2	3	2	4	6
a_3	5	4	3	2

Decision-making models usually are classified with respect to the state of knowledge.

1. *Decision under certainty:* Each action results in a known outcome that will occur with a probability of 1. Table 5.15 is based on this type of decision model.
2. *Decision under risk:* Each state of nature has an assigned probability of occurrence.
3. *Decision under uncertainty:* Each action can result in two or more outcomes, but the probabilities for the states of nature are unknown.
4. *Decision under conflict:* The states of nature are replaced by courses of action determined by an opponent who is trying to maximize his or her objective function. This type of decision theory usually is called *game theory.*

Decision making under risk: We can extend our design example to the situation of decision making under risk if we assume that the states of nature have the following probability of occurrence.

State of nature	θ_1	θ_2	θ_3	θ_4
Probability of occurrence	0.1	0.3	0.4	0.2

The expected value of an action a_i is given by

$$\text{Expected value of } a_i = E(a_i) = \sum_i P_i a_i$$

Thus, for the three materials in Table 5.15, the expected losses would be

$$(\text{steel}) \; E(a_1) = 0.1(1) + 0.3(4) + 0.4(10) + 0.2(15) = 8.3$$

$$(\text{aluminum}) \; E(a_2) = 0.1(3) + 0.3(2) + 0.4(4) + 0.2(6) = 3.7$$

$$(\text{FRP}) \; E(a_3) = 0.1(5) + 0.3(4) + 0.4(3) + 0.2(2) = 3.3$$

Therefore, for this example, we would select fiber-reinforced polymer (FRP) as the material that would minimize the loss in utility.

Decision making under uncertainty: The assumption in decision making under uncertainty is that the probabilities associated with the possible outcomes are not known. The approach used in this situation is to form a matrix of outcomes, usually expressed in terms of utilities, and base the decision on various decision rules. The *maximin decision rule* states that the decision maker should choose the alternative that maximizes the minimum payoff that can be obtained. This is a pessimistic approach, because it implies that the decision maker should expect the worst to happen.

In the case of losses, the selection is based on the course of action that minimizes the maximum loss, i.e., minimax. For example, in the *loss* table shown in Table 5.15 the greatest losses are

$$
\begin{array}{ccc}
a_1 & a_2 & a_3 \\
\hline
\theta_4 = 15 & \theta_4 = 6 & \theta_1 = 5
\end{array}
$$

The best choice among these is alternative 3, FRP, since it minimizes the maximum loss, i.e., minimax.

An opposite extreme in decision rules is the *maximax*. This rule states that the decision maker should select the alternative that maximizes the maximum value of the outcomes. This is an optimistic approach because it assumes the best of all possible worlds. For the loss table in Table 5.15 it would be the alternative with the smallest possible loss, and the decision based on a minimum criterion would be to select alternative 1, steel.

The use of the maximin decision rule implies that the decision maker is very averse to taking risks. In terms of a utility function, that implies perception of very little utility on any return above the minimum outcome (Fig. 5.17). On the other hand, the decision maker who adopts the maximax approach places little utility on values below the maximum. Neither decision rule is particularly logical.

Since the pessimist is too cautious and the optimist is too audacious, we would like to have an in-between decision rule. It can be had by combining the two rules. By using an index of optimism α, the decision maker can weight the relatively pessimistic and optimistic components. If we weight the decision as three-tenths optimistic, we get the results shown in Table 5.16. On that basis, FRP would be chosen as giving the lowest loss in utility.

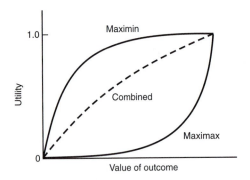

FIGURE 5.17
Utility functions implied by maximin and maximax decision rules.

TABLE 5.16
Combined decision criterion, $\alpha = 0.3$

Alternative	Optimistic	Pessimistic	Total
Steel	0.3(1)	+0.7(15)	= 10.8
Aluminum	0.3(2)	+0.7(6)	= 5.1
FRP	0.3(2)	+0.7(5)	= 4.1

5.10.3 Utility

All of the decision theory techniques discussed in this section presuppose the ability to determine the utility of each outcome. Utility is the intrinsic worth of an outcome. It is often expressed in monetary terms, but it has broader dimensions than just money. It is equal to the *value in use,* which is not necessarily equal to the *value in exchange* in the marketplace. In a general decision situation, decision makers show two types of preferences for outcomes: (1) a direct preference, as in the statement "I prefer outcome A to outcome C," and (2) an attitude toward risk, as in the statement "I prefer to play it safe and take the outcome that gives me $2000 with certainty to the strategy that gives me a 20 percent chance of losing $10,000 and an 80 percent chance of gaining $8000." Thus, the establishment of utilities is intimately concerned with the attitude toward risk. We have seen in Fig. 5.17 how different the utility functions are for a risk-averse and risk-taking individual.

Although utility is not solely expressed by monetary values, it is sufficient to illustrate the concept in monetary terms. In Table 5.17 are listed the probabilities associated with various outcomes related to the acceptance of two contracts that have been offered to a small R&D laboratory. Using expected values, we would accept contract I.

$$E(\text{I}) = 0.6(100{,}000) + 0.1(15{,}000) + 0.3(-40{,}000) = \$62{,}700$$

$$E(\text{II}) = 0.5(60{,}000) + 0.3(30{,}000) + 0.2(-10{,}000) = \$37{,}000$$

However, because contract I has a 30 percent chance of incurring a fairly large loss ($-\$40{,}000$), whereas contract II has only a 20 percent chance of a much smaller loss, our attitudes toward risk enter in and utility concepts most likely are important.

TABLE 5.17
Probabilities and outcomes to illustrate utility

Contract I		Contract II	
Outcome	Probability	Outcome	Probability
+100,000	0.6	+60,000	0.5
+15,000	0.1	+30,000	0.3
−40,000	0.3	−10,000	0.2

To establish the utility function we rank the outcomes in numerical order: +100,000, +60,000, +30,000, +15,000, 0, −10,000, −40,000. The value $0 is introduced to represent the situation in which we take neither contract. Because the scale of the utility function is wholly arbitrary, we set the upper and lower limits as:

$$U(+100{,}000) = 1.00 \quad U(-40{,}000) = 0$$

Note, however, that in the general case the utility function is not linear between these limits. To establish the utility associated with the outcome of +60,000, $U(+60{,}000)$, decision makers ask themselves a series of questions, as follows:

Q1: Which would you prefer?
    (*a*) Gaining $60,000 for certain
    (*b*) Having a 75 percent chance of gaining $100,000 and a 25 percent chance of
        losing $40,000

A1: I'd prefer (*a*) because (*b*) is too risky.

Q2: Now which would you prefer?
    (*a*) Gaining $60,000 for certain
    (*b*) Having a 95 percent chance of gaining $100,000 and a 5 percent chance of
        losing $40,000

A2: I'd prefer (*b*) with those odds.

Q3: How would you feel if the odds in (*b*) were 90 percent chance of gaining
    $100,000 and a 10 percent chance of losing $40,000?

A3: It would be a toss-up between (*a*) and (*b*) with those chances.

Therefore,

$$U(+60,000) = 0.9U(+100,000) + 0.1U(-40,000)$$

$$= 0.9(1.0) + 0.1(0) = 0.9$$

Thus, the technique is to vary the odds on the choices until the decision maker is indifferent to the choice between (*a*) and (*b*). The same procedure is repeated for each of the other values of outcome to establish the utility for those points. A difficulty with this procedure is that many people have difficulty in distinguishing between small differences in probability at the extremes, for example, 0.80 and 0.90 or 0.05 and 0.01.

    Nonmonetary values of outcome can be converted to utility in various ways. Clearly, quantitative aspects of a design performance, such as speed, efficiency, or horsepower, can be treated as dollars were in the above example. Qualitative performance indicators can be ranked on an ordinal scale, for example, 0 (worst) to 10 (best), and the desirability evaluated by a questioning procedure similar to the above.

    Two common types of utility functions that are found for the dependent variables important in engineering are shown in Fig. 5.18. The utility function shown in Fig. 5.18*a* is the most common. Above the design value the function shows diminishing marginal return for increasing the value of the outcome. The dependent variable (outcome) has a minimum design value set by specifications, and the utility drops sharply if the outcome falls below that value. The minimum pressure in a city water supply system and the rated life of a turbine engine are examples. For this type of utility function a reasonable design criterion would be to select the design with the maximum probability of exceeding the design value. The utility function sketched in Fig. 5.18*b* is typical of a high-performance situation. The variable under consideration is very dominant, and we are concerned with maximum performance. Although there is a minimum value below which the design is useless, the probability of going below the minimum value is considered to be very low.

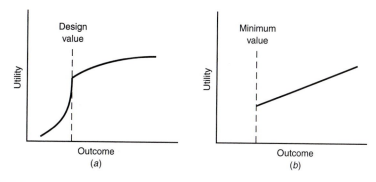

FIGURE 5.18
Common types of utility functions in engineering design.

In the usual engineering design situation more than one dependent variable is important to the design. This requires developing a *multiattribute utility function*.[1] These ideas, originally applied to problems in economics, have been developed into a design decision methodology called methodology for the evaluation of design alternatives (MEDA).[2] Using classical utility theory MEDA improves upon the methods discussed in Sec. 5.9 by providing a better measure of the worth of the performance levels of the attributes to the designer and more accurately quantifying the beneficial attribute trade-offs. The price is a considerable increase in the evaluation analysis.

5.10.4 Decision Trees

The construction of a decision tree is a useful technique when decisions must be made in succession into the future. Figure 5.19 shows the decision tree concerned with deciding whether an electronics firm should carry out R&D in order to develop a new product. The firm is a large conglomerate that has had extensive experience in electronics manufacture but no direct experience with the product in question. With the preliminary research done so far, the director of research estimates that a $4 million ($4M) R&D program conducted over 2 years would provide the knowledge to introduce the product to the marketplace.

A decision point in the decision tree is indicated by a square, and circles designate chance events (states of nature) that are outside the control of the decision maker. The length of line between nodes in the decision tree is not scaled with time, although the tree does depict precedence relations.

The first decision point is whether to proceed with the $4M research program or abandon it before it starts. We assume that the project will be carried out. At the end of the 2-year research effort the research director estimates there is a 50-50 chance of

---

1. J. Von Neumann and O. Morgenstern, "Theory of Games and Economic Behavior," 2d ed., Princeton University Press, Princeton, NJ, 1947; R. L. Keeney and H. Raiffa, "Decisions with Multiple Objectives: Preferences and Value Tradeoffs," Wiley, New York, 1976.
2. D. L. Thurston, "Research in Engineering Design," vol. 3, pp. 105–122, 1991.

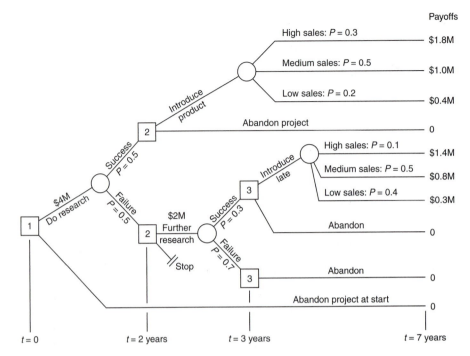

FIGURE 5.19
Decision tree for an R&D project.

being ready to introduce the product. If the product is introduced to the market, it is estimated to have a life of 5 years. If the research is a failure, it is estimated that an investment of an additional $2M would permit the R&D team to complete the work in an additional year. The chances of successfully completing the R&D in a further year are assessed at 3 in 10. Management feels that the project should be abandoned if a successful product is not developed in 3 years because there will be too much competition. On the other hand, if the product is ready for the marketplace after 3 years, it is given only a 1 in 10 chance of producing high sales.

The payoffs expected at the end are given to the far right at the end of each branch. The dollar amounts should be discounted back to the present time by using techniques of the time value of money (Chap. 13). Alternatively, the payoff could be expressed in terms of utility. As a decision rule we shall use the largest expected value of the payoff. Other decision rules, such as *maximin,* could be used.

The best place to start in this problem is at the end of the branches and work backward. The expected values for the chance events are:

$$E = 0.3(1.8) + 0.5(1.0) + 0.2(0.4) = 1.12 \text{ for the on-time project}$$

$$E = 0.1(1.4) + 0.5(0.8) + 0.4(0.3) = 0.66 \text{ for the delayed project at decision point 3}$$

$$E = 0.3(0.66) + 0.7(0) - 2 = -\$1.8M \text{ for the delayed project at decision point 2}$$

Thus, carrying the analysis for the delayed project backward to $\boxed{2}$ shows that to continue the project beyond that point results in a large negative expected payoff. The proper decision, therefore, is to abandon the research project if it is not successful in the first 2 years. Further, the calculation of the expected payoff for the on-time project at point $\boxed{1}$ is a large negative value.

$$E = 0.5(1.12) + 0.5(0) - 4.0 = -\$3.44$$

Thus, either the expected payoff is too modest or the R&D costs are too great to be warranted by the payoff. Therefore, based on the estimates of payoff, probabilities, and costs, this R&D project should not have been undertaken.

5.11
SUMMARY

This chapter deals with the heart of conceptual design—creating design concepts and deciding which among the alternatives is the best to carry forward in the design process. In presenting this subject we have tried to discuss both the attitudes with which you should approach these tasks and techniques which will help you find solutions.

We have tried to set the stage so that you will approach the task of generating design concepts in the most creative way possible. Keep an open mind and suspend judgment at the early stages of idea development. Learn to think with both hemispheres of your brain. Learn to brainstorm effectively, and to apply at least one other creativity tool such as the SCAMPER checklist, force fitting, or the mind map. Pay attention to and learn more about the theory of inventive problem solving (TRIZ). But above all, be persistent and realize that everyone has a degree of creativity, and the prize often goes to the person who works at the problem the hardest.

Concept development begins with decomposition of the problem. There are two majors ways to do this:

1. Physical decomposition breaks the design down into subassemblies and components. It continues down in the hierarchy until a level is reached where everything that has been decomposed is a component.
2. Functional decomposition is more basic, in that it breaks the design down in terms of the functions it must accomplish in terms of energy, material, and information flow. Functions are expressed by a verb-noun combination.

Generating concepts should start with individual effort using creativity aids and the information you have gathered using the sources discussed in Chap. 4. The functional decomposition serves as the roadmap, as the individuals work to generate concepts, not embodiments, to provide each function. Then, as a team they assemble these ideas into a morphological chart. Finally, the design team picks and chooses from the possible alternatives to develop a set of design concepts that look as if they meet the functional requirements.

Now the emphasis shifts to the evaluation of the design concepts to select the one, or few, which are carried forward to a final design. The first step in evaluation is comparing the concepts against a set of absolute criteria.

- Feasibility (will it work?)
- Assessment of technology readiness

Next, in an approximate way each concept is tested against a set of go-no-go questions based on the customer requirements framed in the QFD. The intent is to screen out obvious "losers." Be generous with awarding the "benefit of the doubt."

Those concepts deemed "possible winners" are passed to the next step. The evaluation tool most applicable to the level of detail usually available in conceptual design is Pugh's concept selection method. This method compares each concept relative to a reference concept and for each design criterion determines whether the concept is better than, poorer than, or about the same as the reference concept. This is an intensive team exercise, from which improved concepts often result.

For evaluation of designs for which more detail is available, typically at the subassembly or component level, a weighted decision matrix or a decision matrix based on the analytic hierarchy process are useful tools. Alternatively, the evaluation may be formulated in terms of the methods of decision analysis.

BIBLIOGRAPHY

Creativity

Bailey, R. L.: "Disciplined Creativity for Engineers," Ann Arbor Science Publishers, Ann Arbor, MI, 1978.
De Bono, E.: "Serious Creativity," HarperCollins, New York, 1992.
Lumsdaine, E., and M. Lumsdaine: "Creative Problem Solving," McGraw-Hill, New York, 1995.
Weisberg, R. W.: "Creativity: Beyond the Myth of Genius," W. H. Freeman, New York, 1993.

Conceptual Design Methods

Clausing, D.: "Total Quality Development," ASME Press, New York, 1994.
Cross, N.: "Engineering Design Methods," 2d ed., Wiley, New York, 1994.
French, M. J.: "Conceptual Design for Engineers," Springer-Verlag, New York, 1985.
Otto, K. P., and K. L. Wood: Conceptual and Configuration Design of Products and Assemblies, "ASM Handbook," vol. 20, pp. 15–32, ASM International, Materials Park, OH, 1997.
Pugh, S.: "Total Design," Addison-Wesley, Reading, MA, 1990.
Starkey, C. V.: "Engineering Design Decisions," Edward Arnold, London, 1992.
Ullman, D. G.: "The Mechanical Design Process," 2d ed., McGraw-Hill, New York, 1997.
Ulrich, K. T., and S. D. Eppinger: "Product Design and Development," McGraw-Hill, New York, 1995.

PROBLEMS AND EXERCISES

5.1. Select two pages at random from a large mail-order catalog. Select one item from each page and try to combine the two items into a useful innovation.

5.2. A technique for removing a blockage in the creative process is to apply transformation rules to an existing unsatisfactory solution. Some common transformation operators are

(1) put to other uses, (2) modify, (3) magnify, (4) diminish, (5) substitute, (6) rearrange, (7) reverse, (8) combine. A related technique is to use Roget's *Thesaurus* to suggest leads for alternative solutions. A key word from the existing solution is looked up in the thesaurus, and it provides a number of related and opposite words that stimulate new approaches.

Apply these techniques to the following problem. As a city engineer you are asked to suggest ways to eliminate puddles from pedestrian walkways after a rainstorm. Start with the obviously inadequate solution of waiting for the puddles to evaporate.

5.3. This problem will test your skills at lateral thinking. Join the pattern of dots with four straight lines. You cannot remove the pencil from the paper once you have started, so that the line is continuous.

5.4. As central station power plant operators consider reconverting from oil or gas to coal as the energy source, they sometimes find that there is not a suitable large land area near the plant that can be used for on-the-ground coal storage. Conduct a brainstorming session to propose alternative solutions to a conventional coal pile.

5.5. What are the questions that need to be asked and answered in order to prepare a problem statement? Develop a problem statement for the situation described in Prob. 5.4. Include the following elements in the problem statement: (1) need statement, (2) goals, (3) constraints and trade-offs, and (4) criteria for evaluating the design.

5.6. Use the idea of a morphological box (a three-dimensional morphological chart) to develop a new concept for personal transportation. Use as the three main factors (the axes of the cube) power source, media in which the vehicle operates, and method of passenger support.

5.7. Disassemble a small appliance or hand tool. Create a chart showing how the product is decomposed in the physical domain.

5.8. For the product used in Prob. 5.7, create a functional decomposition diagram.

5.9. Create a functional decomposition of an overhead projector.

5.10. Create a functional decomposition of the flip-lid trash can described in Prob. 2.9.

5.11. Using the results of Prob. 5.10, generate a number of design concepts for an improved flip-lid trash can. Evaluate the concepts using Pugh's concept selection method.

5.12. In the search for more environmentally friendly design, paper cups have replaced Styrofoam cups in some fast-food restaurants. These cups are not as good insulators, and the paper cups often get too hot for the hand. A design team is in search of a better disposable coffee cup. The designs to be evaluated are: (*a*) the current paper cup, (*b*) a standard Styrofoam cup, (c) a rigid injection-molded cup with a handle, (d) a double-wall disposable plastic cup, (e) a paper cup with a pull-out handle, and (f) a paper cup with a cellular wall. These design concepts are to be evaluated with the current paper cup as the datum.

The engineering characteristics on which the cups are evaluated are:

1. Temperature in the hand
2. Temperature of the outside of the cup
3. Material environmental impact
4. Indenting force of cup wall
5. Porosity of cup wall
6. Manufacturing complexity
7. Ease of stacking the cups
8. Ease of use by customer
9. Temperature loss of coffee over time

Using your knowledge of fast-food coffee cups, use the Pugh concept selection method to select the most promising design.

5.13. The following factors may be useful in deciding which brand of automobile to purchase: interior trim, exterior design, workmanship, initial cost, fuel economy, handling and steering, braking, ride, and comfort. To assist in developing the weighting factor for each of those attributes, group the attributes into four categories of body, cost, reliability, and performance and use a relevance tree to establish the individual weighting factors.

5.14. Four preliminary designs for sport-utility vehicles had the characteristics listed in the table below. First, see if you can get the same weighting factors as listed in the table. Using the weighted decision matrix, which design looks to be the most promising?

Characteristics	Parameter	Weight factor	Design A	Design B	Design C	Design D
Gas mileage	Miles per gal	0.175	20	16	15	20
Range	Miles	0.075	300	240	260	400
Ride comfort	Rating	0.40	Poor	Very good	Good	Fair
Ease to convert to 4-wheel drive	Rating	0.07	Very good	Good	Good	Poor
Load capacity	lb.	0.105	1000	700	1000	600
Cost of repair	Avg. of 5 parts	0.175	$700	$625	$600	$500

5.15. Repeat Prob. 5.14 using the AHP method.

5.16. Construct a simple personal decision tree over whether to take an umbrella when you go to work on a cloudy day.

5.17. This decision concerns whether to develop a microprocessor-controlled machine tool. The high-technology microprocessor-equipped machine costs $4 million to develop, and the low-technology machine costs $1.5 million to develop. The low-technology machine is less likely to receive wide customer acclaim ($P = 0.3$) vs. $P = 0.8$ for the microprocessor-equipped machine. The expected payoffs (present worth of all future profits) are as follows:

	Strong market acceptance	Minor market acceptance
High technology	$P = 0.8$	$P = 0.2$
	PW = $16 M	PW = $10M
Low technology	$P = 0.3$	$P = 0.7$
	PW = $12M	PW = 0

If the low-technology machine does not meet with strong market acceptance (there is a chance its low cost will be more attractive than its capability), it can be upgraded with microprocessor control at a cost of $3.2 million. It will then have an 80 percent chance of strong market acceptance and will bring in a total return of $10 million. The nonupgraded machine will have a net return of $3 million. Draw the decision tree and decide what you would do on the basis of various decision criteria.

6

EMBODIMENT DESIGN

6.1
INTRODUCTION

We have now brought the engineering design process to the point where a set of concepts has been developed and evaluated to give a single concept or small set of concepts for further development. It may be that some of the major dimensions have been established roughly, and the major components and materials have been approximately selected. At this point a *feasibility design review* is usually held to determine whether the resources should be committed to develop the design further.

The next stage of the design process is often called *embodiment design*. It is the stage where the design concept is invested with physical form, where we "put meat on the bones." Much of the activity at this stage is devoted to finalizing the product architecture, determining the form or shape of parts that will satisfy the required functions, and quantifying the important design parameters.

It is important to note that nomenclature becomes a bit fuzzy at this point. The term embodiment design comes from Pahl and Beitz,[1] and has been adopted by most European and British writers about design. Many U.S. writers divide the design process into three phases: conceptual design, preliminary (embodiment) design, and detail design. Others call embodiment design analytical design, because it is the design phase where most of the detailed analysis and calculation occurs. Some just lump everything beyond conceptual design into a broad phase called product design. To add further confusion to the nomenclature issue, there is a growing tendency in the United States to describe the phases of design that follow conceptual design as configuration design of parts,[2] parametric design, and detail design. Configuration design

---

1. G. Pahl and W. Beitz, "Engineering Design: A Systematic Approach," 2d English edition, Springer-Verlag, Berlin, 1996.
2. J. R. Dixon and C. Poli, "Engineering Design and Design for Manufacturing," Field Stone Publishers, Conway, MA, 1995.

is the selection of standard modules, like pumps, or the design of special-purpose parts, like a short, stubby box beam. Parametric design is the determination of the exact values, dimensions, or tolerances for the critical design parameters. We have chosen to incorporate both configuration design and parametric design, along with product architecture, into phase 2, embodiment design (see Fig. 6.1).

This text adopts the terminology conceptual design, embodiment design, and detail design because these words seem to be more descriptive of what takes place in each of these design phases. Doing this creates a problem of what is left in the design process for phase 3, detail design. Traditionally, detail design has been the stage where final dimensions and tolerances are established, and all information on the design is gathered into a set of "shop drawings" and bill of materials. However, moving the setting of dimensions and tolerances into embodiment design is in keeping with the current trend for utilizing computer-aided engineering so as to move the decision making as early as possible in the design process to compress the product development cycle. Not only does this save time, but it saves cost of rework compared to when errors are caught at the very end of the design process. It is still the purpose of detail design to provide whatever information is needed to describe the designed object fully and accurately in preparation for manufacturing. As will be shown in Chap. 16, detail design is becoming more involved in information management than just detailed drafting.

It is important to emphasize that Fig. 6.1 overly simplifies the design process in at least two major respects. In this figure the design process is represented as being sequential, with clear boundaries between each step. To be more realistic, this figure

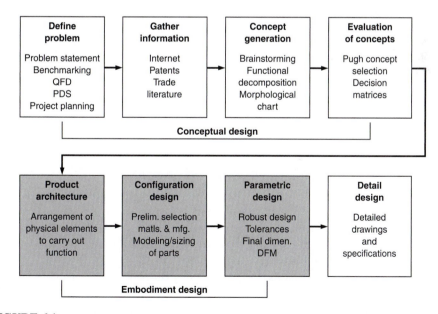

FIGURE 6.1
Steps in the design process, showing that embodiment design consists of establishing the product architecture and carrying out configuration design and parametric design.

should show arrows looping back from every step to those steps above it in the hierarchy. This represents the fact that design changes may be needed to accommodate decisions made further down in the hierarchy. For example, increases in weight brought about by the addition of heavier components demanded by a failure modes and effects analysis would require going back and beefing up support members and bracing.

The second simplification is that Fig. 6.1 implies that design is a serial process. For purposes of learning, we talk about the eight-step process in time sequence, whereas we learned in the discussion of *concurrent engineering* in Sec. 1.7 that *parallel design* is the key to shortening the product development cycle time. Thus, it is quite likely that one member of the design team is proof testing some subassembly that has been finished early, while other team members are still sizing the piping, while yet another member may be designing tooling to make another component. Different team members can be working on different design steps in parallel.

We need also to realize that not all engineering design is of the same type or level of difficulty.[1] Much of design is routine, where all possible solution types are known and often cataloged. Thus, in *routine design* the attributes that define the design and the strategies and methods for attaining them are well known. In *innovative design* not all attributes of the design may be known beforehand, but the knowledge base for creating the design is known. While no new knowledge is added, the solutions are novel, and new strategies and methods for attaining a solution may be required. In *creative design* neither the attributes of the design nor the strategies for achieving them are known ahead of time.

The conceptual design phase is most central to innovative design. At the opposite pole is *selection design* or *catalog design,* which is more central to routine design. Selection design involves choosing a standard component, like a bearing or a cooling fan, from a catalog listing similar items. While this may sound easy, it really can be quite complex owing to the presence of many different items with slightly different features and specifications. In this type of design the component is treated as a "black box" with specified properties, and the designer selects the item that will meet the requirements in the best way. In the case of dynamic components (motors, gearboxes, clutches, etc.) its characteristic curve and transfer function must be carefully considered.[2]

6.2
PRODUCT ARCHITECTURE

Product architecture is the arrangement of the physical elements of a product to carry out its required functions. The product architecture begins to emerge in the conceptual design phase with such things as diagrams of functions, rough sketches of concepts, and perhaps a proof-of-concept model. There the functional structure was determined as a way of generating design concepts. However, it is in the embodiment

1. M. B. Waldron and K. J. Waldron (eds.), "Mechanical Design: Theory and Methodology," Chap. 4, Springer-Verlag, Berlin, 1996.
2. J. F. Thorpe, "Mechanical System Components," Allyn and Bacon, Boston, 1989.

Let's Have a Slice of Computer

An interesting use of modular architecture can be found in the Hewlett-Packard Sojourn laptop computer. The laptop is designed as three modules or horizontal slices, like a layer cake. The top module, only 0.71 in thick, contains the keyboard and display. It is a very portable computer, weighing only 3.2 lb.

The next layer down is the multimedia module. It contains the CD-ROM and floppy drives, pop-out speakers, and printer port. When this slice is combined with the first module, it results in a full-feature laptop computer.

The third module, or base slice, is an extra battery pack that quadruples battery life to 6 h. It can be used with the keyboard module only or with both of the others. With all three modules the weight is 8.2 lb.

---

design phase that the layout and architecture of the product must be established by defining what the basic building blocks of the product should be in terms of what they do and what their interfaces will be between each other. Some organizations refer to this as *system-level design.*

Ulrich and Eppinger[1] designate the physical building blocks that the product is organized into as *chunks*. Other terminology is subsystem, subassembly, cluster, or module. Each chunk is made up of a collection of components that carry out the functions. Thus, the architecture of the product is given by the relationships among the components in the product and the functions the product performs. There are two entirely opposite styles of product architecture, modular and integral. In a *modular architecture* the chunks implement only one or a few functions and the interactions between chunks are well defined. An example would be an oscilloscope, where different measurement functions are obtained by plugging in different modules. In an *integral architecture* the implementation of functions uses only one or a few chunks, leading to poorly defined interactions between chunks. In integrated product architectures components perform multiple functions. Products designed with high performance as a paramount attribute often have an integral architecture. Of course, products are rarely strictly modular or integral, but they usually are a mixture of standard modules and customized components.

The interfaces between modules need to be given attention. Examples of interfaces are the crankshaft of an engine with a transmission or the connection between a computer monitor and the CPU. Interfaces should be designed so as to be as simple and stable as possible. Standard interfaces, which are well understood by designers and parts suppliers, should be used if possible. The personnel computer is an outstanding example of the use of standard interfaces, such that PCs can be customized, module by module, from parts supplied by many different suppliers.

---

1. K. T. Ulrich and S. D. Eppinger, "Product Design and Development," Chap. 7, McGraw-Hill, New York, 1995.

A modular design makes it easier to evolve the design over time, to adapt it to the needs of different customers, to replenish components as they wear out or are used up, and to reuse the product at the end of its useful life by remanufacture. Modular design may even be carried to the point of using the same component in multiple products, a *product family*. This *component standardization* allows the component to be manufactured in higher quantities than would otherwise be possible, with cost savings due to economy of scale. An excellent example is the Black & Decker rechargeable battery pack, which is used in many electrical hand tools and garden tools. Integral design is often adopted when constraints of weight, space, or cost require that performance be maximized. Thus, integral designs are sought where a single physical unit implements multiple functions, also called *function sharing*. One strong driver toward integration of components is the design for manufacturing (DFM) strategy which calls for minimizing the number of parts in a product (see Chap. 9). There is a natural tension between component integration to minimize costs and product architecture. The best approach is to consider integration of components only within a single chunk of the product architecture. Thus, product architecture has strong implications for manufacturing costs. DFM studies should begin early in design when the product architecture is being established to define these trade-offs.

A modular architecture tends to shorten the product-development cycle because modules can be developed independently provided there is not coupling of function between modules, and provided that interfaces are well laid out and understood.[1] Thus, a chunk is assigned to a single individual or small group to carry out because the decisions regarding interactions and constraints are confined within that chunk. Communication with other design groups is concerned only with the interfaces. However, if a function is implemented between two or more chunks, the interaction problem becomes much more severe and challenging. As a result, when a design is "farmed out" to an outside vendor or remote location within the corporation, it usually is a highly modular chunk, e.g., automotive seats.

Ulrich and Eppinger[2] propose a four-step process for establishing the product architecture.

- Create a schematic diagram of the product
- Cluster the elements of the schematic
- Create a rough geometric layout
- Identify the fundamental and incidental interactions

Because of the fundamental importance of the product architecture, it should be developed by a cross-functional product-development team.

---

1. One of the functions of systems engineering is *systems integration* in which hardware and software (modules) designed and made by different vendors are combined into a functioning system.
2. K. T. Ulrich and S. D. Eppinger, op. cit., pp. 138–148.

STEPS IN DEVELOPING PRODUCT ARCHITECTURE

6.2.1 Create the Schematic Diagram of the Product

The process of developing the product architecture will be illustrated with an example taken from Ulrich and Eppinger. It involves a machine for making plastic three-dimensional parts quickly and directly from computer-aided design (CAD) files. This is an example of a rapid prototyping process in which a smooth layer of plastic powder is selectively fused by a laser beam. The part is built up one layer at a time. The schematic diagram of the machine is shown in Fig. 6.2. We note that at this early stage in design some of the design elements are described by physical concepts, like the "part piston" that slowly retracts the part below the bed of powder, and physical components like the CO_2 laser. Yet other elements are described as functional elements

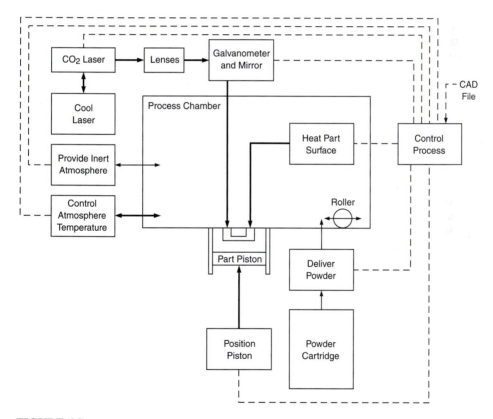

FIGURE 6.2

Schematic diagram of a laser-fusing rapid prototyping machine. Lines connecting the elements indicate a flow of force or energy (thick line), material (thin line), or signals (dashed line). (*From K. T. Ulrich and S. D. Eppinger, "Product Design and Development," McGraw-Hill, 1995, used with permission.*)

that have not been reduced to physical concepts or components, like "provide inert atmosphere" or "heat part surface."

Judgment should be used in deciding what level of detail to show on the schematic. Generally, no more than 30 elements should be used to establish the initial product architecture. Also, realize that the schematic is not unique. As in everything in design, the more iterations you investigate the better the chance of arriving at a best solution.

6.2.2 Cluster the Elements of the Schematic

The purpose of this step is to arrive at an arrangement of chunks by assigning each design element to a chunk. Looking at Fig. 6.3, we see that the following chunks have been established: (1) laser table; (2) process chamber; (3) powder engine; (4) atmos-

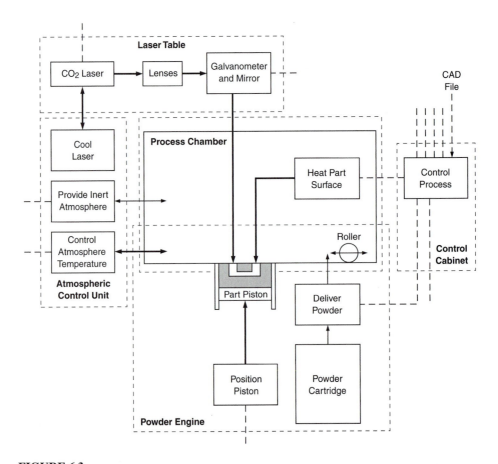

FIGURE 6.3
Design elements shown in Fig. 6.2 are clustered into chunks. (*From K. T. Ulrich and S. D. Eppinger, "Product Design and Development," McGraw-Hill, 1995, used with permission.*)

pheric control unit; (5) control cabinet. One way of deciding on chunks is to start with the assumption that each design element will be an independent chunk and then cluster the elements where there are advantages. Some of the reasons for clustering elements include elements requiring close geometric relationship or precise location, elements that can share a function, the desire to outsource part of the design, and the portability of interfaces; e.g., digital signals are much more portable and can be distributed more easily than mechanical motions.

6.2.3 Create a Rough Geometric Layout

This step determines whether there is likely to be geometrical, thermal, or electrical interference between elements and chunks. For some problems a two-dimensional drawing is adequate (Fig. 6.4), while for others a three-dimensional model (either physical or computer model) is required (see Chap. 7). Creating a geometric layout forces the team to decide whether the geometric interfaces between the chunks are feasible. For example, in Fig 6.4 the decision was made to locate the laser table at the top to remove it from the thermally active and powder areas. This introduced the design element of structurally rigid legs to accurately locate the laser relative to the part. They also introduced the key interface called the "reference plate."

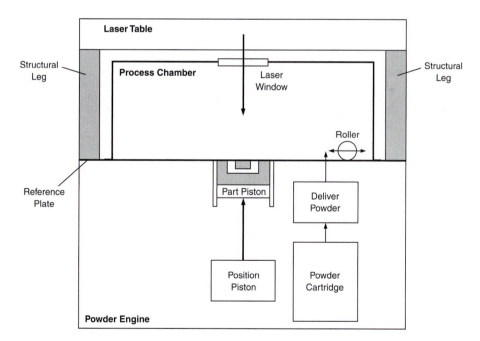

FIGURE 6.4
Geometric layout of the laser table, process chamber, and powder engine chunks. This is a vertical front view of the arrangement. Note that the control cabinet would be to the right side and the atmospheric control unit would be behind. (*From K. T. Ulrich and S. D. Eppinger, "Product Design and Development," McGraw-Hill, 1995, used with permission.*)

Note that sometimes it is not possible to arrive at a geometrically feasible layout, even after trying several alternatives. This means it is necessary to go back to the previous step and change the assignment of elements to chunks until an acceptable layout is achieved.

6.2.4 Identify the Fundamental and Incidental Interactions

For a quality design the interactions between the chunks should be known and controlled. The first category of interaction is *fundamental interactions.* These are given by the lines on the schematic diagram between chunks. For example, in Fig. 6.3 the inert atmosphere protecting the powder goes from the atmosphere control unit to the process chamber. This type of interaction usually is well understood and is carefully designed for. The second type of interaction, *incidental interactions,* arise from geometric arrangement of the chunks or as a result of the physical implementation of the functional elements. Although these interactions are not represented on the schematic drawing, they should be identified by the team and allowed for in the design. For a small number of interacting chunks this can be done graphically by listing the chunks and showing the interactions as arrows between each chunk. For a larger system an *interaction matrix*[1] is a useful tool.

6.3
CONFIGURATION DESIGN

In configuration design we establish the shape and general dimensions of components. Exact dimensions and tolerances are established in parametric design (Sec. 6.4). The term *component* is used in the generic sense to include special-purpose parts, standard parts, and standard assemblies or modules.[2] A *part* is a designed object that has no assembly operations in its manufacture. A *standard* part is one that has a generic function and is manufactured routinely without regard to a particular product. Examples are bolts, washers, rivets, and I-beams. A *special-purpose part* is designed and manufactured for a specific purpose in a specific product line. An *assembly* is a collection of two or more parts. A *subassembly* is an assembly that is included within another assembly or subassembly. A *standard assembly* or *standard module* is an assembly or subassembly which has a generic function and is manufactured routinely. Examples are electric motors, pumps, and gearboxes.

As already stated several times in previous chapters, the *form* or configuration develops from the *function.* However, the possible forms depend strongly on available *materials* and *production methods* used to generate the form from the material. Moreover, the possible configurations are dependent on the *spatial constraints* that define the envelope in which the product operates and the *product architecture.* This

---

1. T. U. Pimmler and S. D. Eppinger, Integration Analysis of Product Decompositions, *ASME Design Theory and Methodology Conference,* September 1994.
2. J. R. Dixon and C. Poli, "Engineering Design and Design for Manufacturing," pp. 1–8, Field Stone Publishers, Conway, MA, 1995.

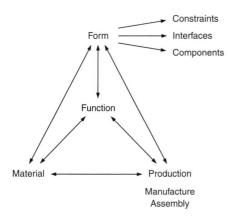

FIGURE 6.5
Schematic illustrating the close interrelationship between function and form and, in turn, its dependence on the material and the method of production. (*After Ullman.*)

set of close relationships is depicted in Fig. 6.5. Generally, decisions about the design of a component cannot proceed very far without making decisions about the material and the manufacturing process from which it will be made. These vital topics are considered in detail in Chaps. 8 and 9, respectively.

In starting configuration design we should follow these steps:[1]

1. Review the product design specification and any specifications developed for the particular subassembly to which the component belongs.
2. Establish the spatial constraints that pertain to the product or the subassembly being designed. Most of these will have been set by the product architecture (Sec. 6.2). In addition to physical spatial constraints, consider the constraints of a human working with the product (see Sec. 6.7) and constraints that pertain to the product's life cycle, such as the need to provide access for maintenance or repair or to dismantle it for recycling.
3. Create and refine the interfaces or connections between components. Again, the product architecture should give much guidance in this respect. Much design effort occurs at the connections between components. Identify and give special attention to the interfaces that carry the most critical functions.
4. In carrying out the design it is important to maintain functional independence in the design of an assembly or component. By this is meant that changing a critical dimension should affect only a single function.
5. Before spending much time on the design, answer the following questions:
   - Can the part be eliminated or combined with another part? Studies of DFM show that it is almost always less costly to make and assemble fewer, more complex parts than it is to design with a higher part count.
   - Can a standard part or module be used? While a standard part is generally less costly than a special-purpose part, two standard parts may not be less costly than one special-purpose part that replaces them.

---

1. J. R. Dixon and C. Poli, op. cit., Chap. 10; D. G. Ullman, "The Mechanical Design Process," 2d ed., Chap. 10, McGraw-Hill, 1997.

Generally, the best way to get started with configuration design is to just start sketching alternative configurations of a part. The importance of hand sketches and drawings[1] or solid CAD models should not be underestimated. Drawings are essential for communicating ideas between design engineers and between designers and manufacturing people. Sketches are an important aid in idea generation and a way for piecing together unconnected ideas into design concepts. Later as the drawings become scale drawings they provide a vehicle for providing missing data on dimensions and tolerances, for simulating the operation of the product (3-D solid modeling), and as a legal document for archiving the geometry and design intent.

The elements that comprise the configuration of a part are called *features.* Typical features in mechanical design are:

- Solid elements such as rods, cubes, and spheres
- Walls of various kinds (flat, curved, etc.)
- Add-ons to walls such as holes, bosses, notches, grooves, and ribs
- Intersections between walls, add-ons, and solid elements

Table 6.1 lists some common features that are designed into parts to improve manufacturing or to reduce material cost.

As modeling through drawing progresses, analysis based on strength of materials or machine design fundamentals, if it is a strength issue, or fluid flow or heat transfer, if it is a transport question, becomes appropriate. Mostly this can be done with hand calculators or PC-based equation solvers. Critical components will require the field-mapping capabilities of finite-element methods (see Chap. 7).

Designers have developed a variety of principles and practices to aid in the tasks of allocating space, selecting shapes and features, and roughing out dimensions. These are discussed in Sec. 6.5. One of the most important considerations in dealing with these issues is the ability to predict manufacturing costs at this early stage of design. This vital and topical subject is considered in Chaps. 9 and 14. Another important topic, design for safety, is considered in Chap. 12.

TABLE 6.1
Features designed into parts to aid manufacturing
or to reduce material cost

Function	Examples of part features
Aid manufacturing	Fillets, gussets, ribs, slots, holes
Add strength or stiffening	Ribs, fillets, gussets, rods
Reduce material use	Holes through walls, ribs that allow thinner walls
Provide a connection or contiguity	Walls, rods, gussets, tubes

From J. R. Dixon, "ASM Handbook," vol. 20: "Materials Selection and Design," p. 34, ASM International, Materials Park, OH, 1997. Used with permission.

---

1. J. M. Duff and W. A. Ross, "Freehand Sketching for Engineering Design," PWS Publishing Co., Boston, 1995.

With data developed on behavior, performance, and cost, it is possible to evaluate alternative designs using a weighted decision matrix or Pugh's selection method (Sec. 5.9). These methods will point to aspects of the design that may need improvement by a design iteration. Ullman[1] characterizes configuration design as refining and patching. *Refining* is the act of making an object less abstract. This is a natural process as we move through the design process. Figure 6.6 illustrates design refinement. At

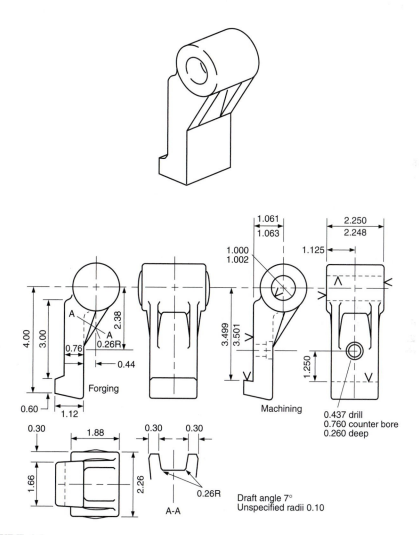

FIGURE 6.6
Sketches and engineering drawings illustrating refinement of a design. (*From D. G. Ullman, "The Mechanical Design Process," 2d ed., McGraw-Hill, New York, 1997. Used with permission.*)

---

1. D. G. Ullman, op. cit., pp. 195–199.

the top is an abstract sketch of a support bracket, while at the bottom is a detailed drawing showing the forged component, and the final configuration and dimensions after machining. *Patching* is the activity of changing a design without changing its level of abstraction. Patching occurs extensively in configuration design. For example, we may find the dimensions will not support the applied torque, or that a shelf will deflect excessively without stiffening, or that material needs to be removed to save on piece part cost, or that there is interference from another part. Detailed examples of configuration design are given by Ullman[1] and Dixon and Poli.[2]

A special but not infrequent type of situation in configuration design is selection design. By *selection design* we mean choosing a component from a list of similar standard parts in a catalog. Examples are the selection of bearings, pumps, and small motors. While configuration decisions are not needed in this case, considerable analysis may be required to be sure that the chosen component has the needed performance characteristics.

The final part configuration must satisfy a large number of considerations. Table 6.2 lists these factors and poses a set of questions which must be answered in order to reduce the design risk.

6.4
PARAMETRIC DESIGN

In configuration design the emphasis was on starting with the product architecture and then working out the best form for each component. Qualitative reasoning about physical principles and manufacturing processes played a major role. Dimensions and tolerances were set approximately, and while analysis was used to "size the parts" it generally was not highly detailed or sophisticated. Now the design moves into *parametric design,* the latter part of embodiment design.

In parametric design the attributes of parts identified in configuration design become the design variables for parametric design. A *design variable* is an attribute of a part whose value is under the control of the designer. This typically is a dimension or a tolerance, but it may be a material, heat treatment, or surface finish applied to the part. This aspect of design is much more analytical than concept or configuration design. The objective of parametric design is setting values for the design variables that will produce the best possible design considering both performance and manufacturability.

Distinguishing between configuration design and parametric design is of fairly recent origin. It has grown out of massive efforts by industry to improve the quality of their products, chiefly by improving robustness. *Robustness* means achieving excellent performance under the wide range of conditions that will be found in service. All products function reasonably well under ideal (laboratory) conditions, but robust designs continue to function well when the conditions to which they are exposed are far from ideal.

---

1. D. G. Ullman, op. cit., pp. 201–210.
2. J. R. Dixon and C. Poli, op. cit., pp. 10-23–10-28.

TABLE 6.2
Questions for revealing part configuration design risks

Factor	Questions
What are the most likely ways the part might fail in service?	
Excessive stress	Can the part be dimensioned to keep stresses below yield levels? Add ribs? Use stronger material?
Fatigue	If there will be cyclic loads, can the configuration be dimensioned so as to keep the internal stresses below the fatigue limit?
Stress concentrations	Can the part be dimensioned to keep local stress concentrations low?
Buckling	If buckling is a possibility, can the configuration be dimensioned to prevent it?
Unexpected shocks or loads	What unexpected dynamic loads might be encountered in service or in assembly? Can these be handled by the configuration?
What are the most likely ways the part might not meet its expected functionality?	
Tolerances	Is the configuration such that functionality will be especially sensitive to the actual tolerances that can be expected in a production situation? Are too many special (tight) tolerances required to make the part work well?
Creep	If creep is a possibility, will it result in loss of functionality?
Strain and deformation	If functional performance is sensitive to retention of size and shape, can the configuration be dimensioned to preserve the required integrity?
Thermal deformations	Might thermal expansion or contraction cause the configuration to deform so that function will be impaired?
Handling and assembly	Might there be unforeseen difficulties with handling and assembly?
Dimensions	Might the part end up being dimensioned so that assumptions about assembleability become invalid?
Tangling	Might the parts tangle if dimensioned in some way?
Will the available production machines be able to make the part?	
Production runs	Are the desired production runs consistent with the machines and expected costs?
Tooling wear	Is tooling wear or maintenance a possible problem that will impact part cost or performance?
Weld lines(a)	If the process is a flow process, can weld lines be located appropriately?
Other design and materials factors	
Geometric compatibility	Is the part geometrically compatible with its adjoining parts? What could go wrong in this regard? If there is a small change in this part, or in an adjoining part, can the configuration accommodate the change without major redesign? What about the effects of tolerances of the *adjoining* parts? Or on the assembly as a whole?
Materials	Is the material selected compatible with the configuration and the manufacturing process? Is surface finish properly accounted for? Will standard raw material supplies be of adequate quality? Has the material been thoroughly investigated for its use in *this* particular application? Are there previous uses in similar applications? Have experts on the properties and processing of the material been consulted? Is the material compatible with the rest of the product?
Designer and design team knowledge	Has every possible, unfortunate. unlikely, unlucky, even stupid "What if…" situation been considered? Are there aspects of the part design where the designer or design team is working without adequate knowledge? Where is the design based on insufficient knowledge of materials, forces, flows, temperatures, environment, etc.? Where are there guesses, hopes, fears, and assumptions instead of knowledge: Materials? Stresses? Fastening methods? Manufacturing process? Tolerances? Costs? Adjoining parts? Environmental conditions?

(a) A weld line is formed when a material flow must divide—say around a hole—and then rejoin. The weld lines tend to be weaker and more subject to fatigue failures.

From J. R. Dixon. "ASM Handbook," vol. 20: "Materials Selection and Design," p. 38, ASM International, Materials Park, OH, 1997.

There are basically three ways to improve the robustness of a design.[1]

- Improve the design concept that controls the function to be fixed
- Change the value of a critical parameter to an obvious feasible limit
- Carry out a systematic optimization of the design variables

Since an optimization of a complicated design is a major undertaking, the first two approaches usually should be taken first. Computer-based optimization methods can be useful when the design variables are all numerical and continuous functions. For the more general case, statistically designed experiments, especially the Taguchi approach, have met with increasing usage and success. These topics are covered in detail in Chap. 12.

Parametric design is about setting dimensions and tolerances so as to maximize quality and performance and minimize cost. Designing to maximize performance and quality involves the following design procedures.

6.4.1 Failure Modes and Effects Analysis (FMEA)

A *failure* is any aspect of the design or manufacturing process that renders a component, assembly, or system incapable of performing its intended function. FMEA is a methodology for determining all possible ways that components can fail and establishing the effect of failure on the system. To learn more about FMEA, see Sec. 11. 5.

6.4.2 Design for Reliability

Reliability is a measurement of the capacity of a system to operate without failure in the service environment. It is expressed as the probability of performing for a given time without failure. Chapter 11 gives considerable detail on methods for predicting and improving reliability.

6.4.3 Robust Design

One of the chief ways of assuring high quality in products is to reduce the variability in performance and manufacture over a wide range of operating conditions. Many manufacturers are adopting the statistically designed experimental methods of Genichi Taguchi to set values for the critical design variables. A process of robust design, as defined by Taguchi, has the following steps:

- *System design:* This corresponds to what we have called product architecture, where engineering principles are used to determine the basic configuration of the system.

---

1. D. Clausing, "Total Quality Development," p. 191, ASME, New York, 1994.

- *Parameter design:* Statistical methods are used to set nominal values of design variables that minimize variability from uncontrollable variables in the environment.
- *Tolerance design:* Further statistical methods are used to set the widest possible tolerances on the design variables without increasing variability.

The methodology for achieving robust design is considered in detail in Chap. 12. For additional examples see Dixon and Poli.[1]

6.4.4 Tolerances

Tolerances must be placed on dimensions of a part to limit the permissible variations in size because it is impossible to repeatedly manufacture a part exactly to a given dimension. A small tolerance results in greater ease of interchangeability of parts and less play or chance for vibrations in moving parts. However, this is achieved at an increased cost of manufacture.

Dimensions are used to specify size and location of features in a part. *Tolerance* is the acceptable variation to the ideal or *nominal* dimension. *Clearance* or *allowance* is the minimum space between mating parts. In modern engineering practice they are described and specified by a system of *geometric dimensioning and tolerancing* (GD&T) based on ASME standard Y14.5M-1994. Within this standard, tolerance is defined for geometric characteristics of form (straightness, flatness, etc.), profile tolerances, orientation tolerances (angularity, perpendicularity, parallelism), location tolerances (position, concentricity), and runout. Figure 6.7 shows the symbols used for these GD&T characteristics that are used on detail drawings. Understanding GD&T is rather complex, but straightforward. Space considerations do not begin to allow a thorough discussion. Instead, we present information on references[2] and training material[3] for further study. Any engineer involved in design or manufacturing will have to master this information.

The above discussion has dealt mainly with obtaining the best quality and performance from the design. A second goal of parametric design is minimizing manufacturing cost. This requires close attention to design for manufacturability, i.e., knowing how various design features affect manufacturing cost. This subject is discussed in detail in Chap. 9. As design features are patched in configuration design quantitative estimates can be made of only tooling costs for producing the part, because dimensions are known only approximately. However, at the parametric stage,

1. J. R. Dixon and C. Poli, op. cit., Chaps. 17, 18, 19.
2. "Tool and Manufacturing Engineers Handbook," vol. 4, "Quality Control and Assembly," Chap. 4, Society of Manufacturing Engineers, Dearborn, MI, 1987; L. W. Foster, "Geometrics II—The Application of Geometric Tolerancing Techniques," Addison-Wesley, Reading, MA, 1986; J. V. Liggett, "Dimensional Variation Management Handbook," Prentice-Hall, Englewood Cliffs, NJ, 1993; G. Henzold, "Handbook of Geometric Tolerancing," Wiley International, New York, 1995; C. M. Creveling, "Tolerance Design," Addison-Wesley, Reading, MA, 1997; H-C Zhang, "Advanced Tolerancing Techniques," Wiley, New York, 1997.
3. Geometric Dimensioning and Tolerancing video training program; 16 video tapes based on ASME Y14.5M, Society of Manufacturing Engineers.

SYMBOL FOR:	ANSI Y14.5	ISO
Straightness	—	—
Flatness	▱	▱
Circularity	○	○
Cylindricity	⌭	⌭
Profile of a line	⌒	⌒
Profile of a surface	⌓	⌓
All-around profile	⌾	NONE
Angularity	∠	∠
Perpendicularity	⊥	⊥
Parallelism	//	//
Position	⊕	⊕
Concentricity/coaxiality	◎	◎
Symmetry	NONE	≡
Circular runout	* ↗	↗
Total runout	* ↗↗	↗↗
At maximum material condition	Ⓜ	Ⓜ
At least material condition	Ⓛ	NONE
Regardless of feature size	Ⓢ	NONE
Projected tolerance zone	Ⓟ	Ⓟ
Diameter	⌀	⌀
Basic dimension	50	50
Reference dimension	(50)	(50)
Datum feature	– A –	* ⌱ OR * ⌱Ⓐ
Datum target	⌀6/A1	⌀6/A1
Target point	✕	✕

* May be filled in

FIGURE 6.7
Symbols for geometric dimensioning and tolerancing (GD&T) that are used on detail draw-ings. ANSI is American National Standards Institute. ISO is International Standards Organization.

final dimensions, tolerances and surface finish, and orientation and location of fea-tures are established. Then the relative cost to process the part can be made. The part cost is then the sum of the tooling cost, processing cost, and material cost, each on a per part basis. Examples of this analysis for parts produced by injection molding, die casting, and metal stamping are given by Dixon and Poli.[1]

By the conclusion of embodiment design a prototype of the product will have been constructed. This is a full-scale working model, technically and visually com-

1. J. R. Dixon and C. Poli, op. cit., Chaps. 15 and 16.

plete. The purpose of the prototype is to confirm that the design satisfies all customer requirements and performance criteria. Extensive testing will give information about the reliability and robustness of the design. Testing will also verify whether environmental, safety, and other legal requirements have been met. Although the prototype most likely will be built by hand, its construction will give confidence about the manufacturing operations that lie ahead, and it will help to determine the validity of the cost estimates.

6.5
BEST PRACTICES

It is much more difficult to give a prescribed set of methods for embodiment design than for conceptual design, because of the variety of issues that enter into the development of the configuration and performance of components. In essence, the rest of this text is about these issues, like selection of materials, design for manufacturability, and design for robustness. Nevertheless, many people have thought carefully about what constitutes the best practice of embodiment design. We record some of these insights here.

The general objectives of the embodiment phase of design are the fulfillment of the required technical function, at a cost that is economically feasible, and in a way that ensures safety to the user and to the environment. Pahl and Beitz[1] give the basic guidelines for embodiment design as *clarity, simplicity,* and *safety*. Clarity of function pertains to an unambiguous relationship between the various functions and the appropriate inputs and outputs of energy, material, and information flow. Simplicity refers to a design that is not complex and is easily understood and readily produced. Safety should be guaranteed by direct design, not by secondary methods such as guards or warning labels.

In many problems important insight can be gained by examining the effects of scale. For example, if all the dimensions of a pressure vessel (including wall thickness) are increased by the same scale factor, then the stresses are unchanged. The ideas of dimensional analysis and scaling are discussed in Secs. 7.3 and 7.4. A related aspect of design insight is proportion. A good designer can look at a drawing and tell where certain features are improperly sized.

Two important aspects of design that have been identified by French are matching and disposition. *Matching* refers to creating the proper interface between the separate components so that they can perform as an optimized system. *Disposition* is concerned with parceling out some constrained attribute, often space, between a number of functions in the best way. The concepts of matching and disposition cannot be readily discussed in the abstract. See French[2] for a detailed discussion.

---

1. G. Pahl and W. Beitz, "Engineering Design: A Systematic Approach," 2d ed. English translation by K. Wallace, Springer-Verlag, Berlin, 1996.
2. M. J. French, "Conceptual Design for Engineers," Springer-Verlag, New York, 1985, Chaps. 5 and 6.

6.5.1 Design Guidelines

It is difficult to list with confidence a set of rules of sufficient generality and enough concreteness to be useful and enduring. The guidelines laid out below have been given by French based on his extensive experience and study.

1. Avoid arbitrary decisions. Every design choice represents an opportunity to improve the design. Sometimes in design an arbitrary decision is made without recognizing it. Other times the designer is faced with a clear decision point but does not know how to take advantage to improve the design.
2. Search for alternatives. This imperative has been emphasized throughout this text. It behooves the designer to continually search for and catalog alternative solutions to design situations.
3. Solid models can be of very great help in many design problems because they often suggest ideas and alternative methods. Solid models produced by computer graphics (Sec. 7.6) may provide faster paths to this goal, with enhanced design results.
4. Increase the level of abstraction at which the problem is formulated. Moving the problem to a higher level of abstraction often suggests different solutions, as with functional analysis. For example, restricting a design study to the level of product features will produce entirely different ideas than if the study was conducted at the higher level of product alternatives.
5. Make tables of design functions and options and use them to develop competing design concepts.
6. In developing some design concept always pursue it to the limits, and then back off. The limits will be set by physical realizability or economic constraints.
7. Aim for clarity of function. There is a tendency in design to add features or devices to overcome problems that arise in the course of the design. The solution is to be willing to stop and start again when difficulties arise. The result will be a clearer and simpler design.
8. Exploit materials and manufacturing methods to the fullest. These topics are covered in Chaps. 8 and 9.
9. Develop a logical chain of reasoning for the design. A good design can be justified by a fairly tight line of reasoning. If a logical reasoning chain cannot be developed then it is difficult to have confidence in the design and more conceptual work is required. It is important to test the logic chain, link by link, to see whether there is some fault in the reasoning. If so, this points out where the design can be improved.
10. Ask questions. The designer should develop an attitude of incredulity. Is this part necessary? What would happen if this component failed? Why did we do it that way?

Among the extensive list of principles and guidelines for embodiment design, along with detailed examples, that are given by Pahl and Beitz,[1] four stand out for special mention.

---

1. G. Pahl and W. Beitz, op. cit., pp. 199–403.

- Force transmission
- Division of tasks
- Self-help
- Stability

Force transmission

In mechanical systems the function of many components is to transmit forces and moments between two points. In general, the force should be accommodated in such a way as to produce a uniformly distributed stress on the cross section. However, design conditions often impose nonuniform stress distributions because of space saving or geometry constraints. A method for visualizing how forces are transmitted through components and assemblies is to think of forces as flow lines, analogous to heat flow or magnetic flux.[1] In this model, the force will take the path of least resistance through the component. Figure 6.8 shows how the flow lines crowd together at the pin-connected ends, indicating a higher intensity of stress. Use sketches to trace out the path of the flow lines and label whether the major type of stress at a location is tension (T), compression (C), shear (S), or bending (B). Aim for parts that are stressed uniformly by avoiding sharp changes in cross section, sharp changes in the direction of forces, and changes in the flow-line density.

You should be aware that stiffness (resistance to deflection) more than stress determines the size of most components. Lack of rigidity or stiffness can cause interference between mating parts, produce misalignment of parts, and lead to undesirable wear. To achieve maximum stiffness, use the shortest and most direct force-transmission path possible. Avoid bending stresses by favoring geometries that result in symmetrical loading.

Mismatched deformations between related components can lead to uneven stress distributions and unwanted stress concentrations. Therefore, interacting components should be designed so that they deform in the same sense and by the same amount under load.

Realize that the main forces produced within a mechanical system as part of its function often give rise to secondary forces. The design should aim at balancing out these secondary forces by using symmetrical layouts and additional force-balancing elements.

FIGURE 6.8
Force "flow lines" in a bar loaded in tension with a force P.

---

1. D. G. Ullman, op. cit., pp. 192–194.

Division of tasks

The question of how rigorously to adhere to the principle of clarity of function is ever present. A component should be designed for a single function when the function is deemed critical and will be optimized for robustness. Assigning several functions to a single component (integral architecture) results in savings in weight, space, and cost but may compromise the performance of individual functions, and it may unnecessarily complicate the design.

Self-help

The idea of self-help concerns the improvement of a function by the way in which the components interact with each other. A *self-reinforcing* element is one in which the required effect increases with increasing need for the effect. An example is an O-ring seal which provides better sealing as the pressure increases. A *self-damaging* effect is the opposite. A *self-protecting* element is designed to survive in the event of an overload. One way to do this is to provide an additional force-transmission path that takes over at high loads, or a mechanical stop that limits deflection.

Stability

The stability of a design is concerned with whether the system will recover appropriately from a disturbance to the system. The ability of a ship to right itself in high seas is a classic example. Sometimes a design is purposely planned for instability. The toggle device on a light switch, where we want it to be either off or on and not at a neutral position, is an example.

Many factors need to be considered within the embodiment design phase. Table 6.3 provides a checklist for a design review at the end of embodiment design.[1]

6.6
INDUSTRIAL DESIGN

Industrial design, also often called just product design, is concerned with the visual appearance of the product and the way it interfaces with the customer. The terminology is not very precise in this area. Up until now, what we have called product design has dealt chiefly with the function of the design. However, in today's highly competitive market performance alone may not be sufficient to sell a product. The need to tailor the design for aesthetics and human usability has been appreciated for many years for consumer products, but today it is being given greater emphasis and is being applied more often to technically oriented industrial products.

Industrial design[2] deals chiefly with the aspects of a product that relate to the user. First and foremost is its *aesthetic* appeal. Aesthetics deal with the interaction of the product with the human senses—how it looks, feels, smells, or sounds. For most

---

1. G. Pahl and W. Beitz, op. cit., p. 206; C. Hales, "Managing Engineering Design," Longman Scientific & Technical, Harlow, England, 1993, p. 146.
2. R. Caplan, "By Design," St. Martin's Press, New York, 1982; C. H. Flurschein (ed.), "Industrial Design in Engineering," Design Council, London and Springer-Verlag, New York, 1983.

TABLE 6.3
Embodiment design checklist

Requirement	Issues	Where discussed in text
Function	Is function fulfilled? Will it work? Are other functions needed? Strength, stiffness, stability, fatigue, degradation, wear, corrosion	Sec. 5.6 Chap. 8
Layout	Product architecture	Sec. 6.2
Safety	Operational and human safety provided for?	Sec. 11.9
Quality	Designed for robustness? FMEA performed?	Chaps. 11 and 12
Manufacturing	Have DFM and DFA been applied? Are reliable suppliers available?	Chap. 9
Schedule	Will final design meet schedule? Any major problems with schedule?	Sec. 3.9
Cost	Are design and manufacturing costs meeting projections? Any new issues with operating costs?	Chap. 14
Ergonomics	Design reliable and easy to use? User-friendly controls and layout?	Sec. 6.7
Industrial design	Aesthetically pleasing? Is design subject to fashion?	Sec. 6.6
Environmental	Do materials and working fluids present problems in disposal? Meets all environmental regulations?	Sec. 6.8 Sec. 8.14
Life cycle	Quiet, vibration-free? Easy handling? Ease of inspection and maintenance? Recycling issues?	Chap. 16

products the visual appeal is most important. This has to do with whether the shape, proportion, balance, and color of the elements of the design create a pleasing whole. Often this goes under the rubric of *styling*. Proper attention to aesthetics in design can instill a pride of ownership and a feeling of quality and prestige in a product. Appropriate styling details can be used to achieve product differentiation in a line of similar products. Also, styling often is important in designing the packaging for a product. Finally, proper attention to industrial design is needed to develop and communicate to the public a corporate image about the products that it makes and sells. Many companies take this to the point where they have developed a *corporate style* that embodies their products, advertising, letterheads, etc.

The second major role of industrial design is in making sure that the product meets all requirements of the user human interface, a subject often called *ergonomics* or *usability*.[1] This activity deals with the user interactions with the product and making sure that it is easy to use and maintain. The human interface is discussed in Sec. 6.7.

The industrial designer is usually educated as an applied artist or architect. This is a decidedly different culture than in the education of the engineer. While engineers may see color, form, comfort, and convenience as necessary evils in the product design, the industrial designer is more likely to see these features as intrinsic in satisfying the needs of the user. The two groups have roughly opposite styles. Engineers work from the inside out. They are trained to think in terms of technical details.

---

1. A. March, Usability: The New Dimension of Product Design, *Harvard Business Review,* September–October 1994, pp. 144–149.

Industrial designers, on the other hand, work from the outside in. They start with a concept of a complete product as it would be used by a customer and work back into the details needed to make the concept work. Industrial designers often work in independent consulting firms, although large companies may have their own in-house staff. Regardless, it is important to have the industrial designers involved at the beginning of a project, for if they are called in after the details are worked out there may not be room to develop a proper concept. There currently is a trend for larger design firms that are staffed to handle a project from industrial design through to a functioning manufactured design.

6.6.1 Visual Aesthetics

Visual aesthetic values can be considered as a hierarchy[1] of human responses to visual stimuli. Aesthetics relate to our emotions. Since aesthetic emotions are spontaneous and develop beneath our level of consciousness, they satisfy one of our basic human needs. At the bottom level of the hierarchy is order of visual forms, their simplicity, and clarity. These values are related to our need to recognize and understand objects. We relate better to symmetric shapes with closed boundaries. Visual perception is enhanced by the repetition of visual elements related by similarity of shape, position, or color (rhythm). Another visual characteristic to enhance perception is homogeneity, or the standardization of shapes. For example, we relate much more readily to a square shape with its equal angles than to a trapezoid. Designing products so that they consist of well-recognized geometric shapes (geometrizing) greatly facilitates visual perception. Also, reducing the number of design elements and clumping them into more compact shapes aids recognition.

The base level of the visual aesthetics hierarchy deals with visual neatness. The second level is concerned with recognition of the functionality or utility of the design. Our everyday knowledge of the world around us gives us an understanding of the association between visual patterns and specific functions. For example, symmetrical shapes with broad bases suggest inertness or stability. Patterns showing a tendency toward visual separation from the base suggest a sense of mobility or action (see Fig. 6.9). A streamlined shape suggests speed. Looking around, you can observe many visual symbols of function.

The highest level of the hierarchy deals with the group of aesthetic values derived from the prevailing fashion, taste, or culture. These are the class of values usually associated with styling. There is a close link between these values and the state of available technology. For example, the advent of steel beams and columns made the high-rise building a possibility, and high-strength steel wire made possible the graceful suspension bridge. A strong driver of prevailing visual taste traditionally has been the influence of people in positions of power and wealth. In today's society this is most likely to be the influence of media stars. Another strong influence is the human need and search for newness.

---

1. Z. M. Lewalski, "Product Esthetics: An Interpretation for Designers," Design & Development Engineering Press, Carson City, NV, 1988.

FIGURE 6.9
Note how the design of this four-wheel-drive agricultural tractor projects rugged power. The clearly defined grid of straight lines conveys a sense of unity. The slight forward tilt of the vertical lines adds a perception of forward motion. (*From Z. M. Lewalski, Product Esthetics, Design & Development Engineering Press, Carson City, NV, used with permission.*)

6.6.2 The Process of Industrial Design

Just as engineers will follow an established process to generate and evaluate concepts for the technical functioning of a product, industrial designers will follow a generally similar methodology. There are two general situations: (1) the industrial designer is part of the integrated product-development team (IPDT) from its inception, and (2) industrial designers are brought in during embodiment design to provide styling and to ensure that human factors are given proper consideration. The first situation is preferred. For this case the steps in industrial design would be:[1]

- *Determine the customer needs:* Since industrial designers are skilled in recognizing issues involving user interaction, they can play a crucial role here.
- *Product conceptualization:* Industrial designers will concentrate on creating the product's form and user interfaces. This will take the form of simple sketches for each concept. Close coordination is needed to match these design concepts with the technical concepts of the engineer. The existence of designers on the IPDT who are comfortable in making sketches can be a great benefit.

1. K. T. Ulrich and S. D. Eppinger, op cit., chap. 8.

- *Preliminary refinement:* As the concepts are being evaluated it is important to have three-dimensional *soft models* made from plastic foam or foam-core board. The ability to touch and feel the product aids in concept evaluation.
- *Final concept selection:* Before the final product concept selection is made, it is usual to construct three-dimensional *hard models* made from metal or plastic and painted and textured to be close to the real product. Generally these are technically nonfunctional, and increasingly they are made by rapid prototyping processes (Sec. 7.9). Often three-dimensional drawings, known as *renderings,* are made. Today, these are often 3D-CAD models. Hard models and renderings are used frequently to get additional customer feedback and to sell the concept to the senior management of the organization.
- *Control drawings:* The completion of the industrial design process is the making of control drawings of the final concept. Control drawings document functionality, features, sizes, colors, surface finishes, and critical dimensions. While they contain much information about the product, they are not detailed engineering drawings suitable for manufacture of the components. Typically the control drawings are given to the part designers for reference. While this represents the end of the typical industrial design process, some industrial design firms offer comprehensive engineering services, including detail design, selection and management of outside suppliers of parts, and assembling the final product.

6.7
HUMAN FACTORS DESIGN

Human factors is the study of the interactions between people and the products and systems they use and the environments in which they work and live. This field also is described by the terms *human factors engineering* and *ergonomics.*[1] Human factors design applies information about human characteristics to the creation of human-made objects, facilities, and environments that people use. It goes beyond the issues of usability to consider design for ease of maintenance and for safety. Human factors expertise is found in industrial designers, who focus on ease of use of products, and in industrial engineers, who focus on design of production systems for productivity and freedom from accidents.

There are four ways that a human interacts with a product: (1) as an occupant of workspace (the cab of a tractor or a chair before a computer), (2) as a power source (usually muscle power), (3) as a sensor (looking for a warning light), or (4) acting as a controller (determining how much "pedal" to give to beat the red light with your car). Products that score high in human factors are generally regarded as quality products, since they are perceived to work well by the user (Table 6.4). Of course, the design must also be safe, both to the person and to property.

---

1. From the Greek words *ergon* (work) and *nomos* (study of).

TABLE 6.4
Correspondence between human factors characteristics and product performance

Product performance	Human factors characteristic
Comfortable to use	Good match between product and person in their workspace
Easy to use	Requires minimal human power; clarity of use
Operating condition easily sensed	Human sensing
Product is user-friendly	Control logic is natural to the human

6.7.1 Creating a User-Friendly Design[1]

1. *Fit the product to the user's physical attributes and knowledge:* The first step in designing the person for the workspace is to gather quantitative data on the dimensions of the human body. Such *anthropometric* data is available in many places.[2] Please remember that there is no such thing as an "average person," so that the design should accommodate 95 percent of the human population. Better yet, design in features so that the space can be adjusted to fit the user. Data on human-force generation is included in the study of *biomechanics.* Such information often is given along with anthropometric data.

    However, user-friendly design goes beyond the need to design for anthropometric data and incorporates Norman's concept[3] that the design must incorporate the general knowledge that many people in the population possess. For example, that a red light means stop, that the higher numbers of a dial should be in the clockwise direction, and that knobs tighten when turned in a clockwise direction. Be sure that you do not presume too much knowledge and skill on the part of the user.

2. *Simplify tasks:* Control operations should have a minimum number of operations and should be straightforward. The learning curve for users must be minimal. Incorporating technology (microelectronics) into the product may be used to simplify operation. The product should look simple to operate, with a minimum number of controls and indicators.

3. *Make the controls and their functions obvious:* Controls are the parts of the product such as knobs, levers, buttons, and slides that change the operational mode or

---

1. J. C. Bralla, "Design for Excellence," McGraw-Hill, 1996, Chap. 19.

2. G. Salvendy (ed.), "Handbook of Human Factors and Ergonomics," 2d ed., Wiley, New York, 1997; W. E. Woodson, "Human Factors Design Handbook," 2d ed., McGraw-Hill, New York, 1991; "Human Engineering Design Criteria for Military Systems and Facilities," MIL-STD 1472D; M. S. Sanderson and E. J. McCormick, "Human Factors in Engineering Design," 7th ed., McGraw-Hill, New York, 1992; J. H. Burgess, "Human Factors in Industrial Design," TAB Professional and Reference Books, Blue Ridge Summit, PA, 1989; "Human Factors Design Guide," http://www.tc.faa.gov/hfbranch/hfdg. Software for human factors design is available. For example, Job Evaluator Toolbox from ErgoWeb Inc, Salt Lake City, UT, HumanCAD, Melville, NY, and dV/Manikin, DIVISION, Ltd., Bristol, UK.

3. D. A. Norman, "The Design of Everyday Things," Doubleday, New York, 1988. This book is full of examples of good and poor ways to do human factor design. To learn by poor example, http://www.bad-designs.com/examples.html.

level of the product. Place the controls for a function adjacent to the device that is controlled. It may look nice to have all the buttons in a row, but it is not very user-friendly.

4. *Use mapping:* Make the control reflect, or map, the operation of the mechanism. For example, the seat position control in an automobile could have the shape of a car seat, and moving it up should move the seat up. The goal should be to make the operation clear enough that it is not necessary to refer to nameplates, stickers, or the operator's manual.

5. *Utilize constraints to prevent incorrect action:* Do not depend on the user always doing the correct thing. Controls should be designed so that an incorrect movement or sequence is not possible. An example is the automatic transmission that will not go into reverse when the car is moving forward.

6. *Provide feedback:* The product must provide the user with a clear, immediate response to any actions taken. This feedback can be provided by a light, a sound, or displayed information. The clicking sound and flashing dashboard light, in response to actuating an automobile turn signal, is a good example.

7. *Provide good displays:* The sensing characteristic of the human involves such *physiology factors* as the visual, tactile, and auditory senses. Most human-machine interfaces require that the human sense the state of the system and then control it based on the information received. Basic types of visual displays are shown in Fig. 6.10. The characteristics of these displays are summarized in Fig. 6.11. These are examples of the kinds of information to be found in the various human factors handbooks. In selecting the controller for a product, it is important

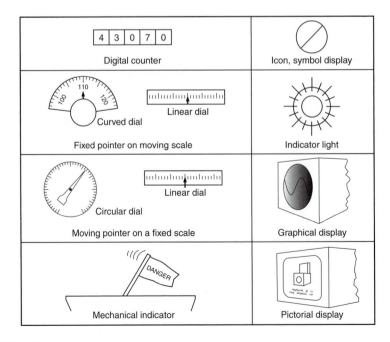

FIGURE 6.10
Types of visual displays. (*After Ullman.*)

	Exact value	Rate of change	Trend, direction of change	Discrete information	Adjusted to desired value
Digital counter	●	○	○	●	◐
Moving pointer on fixed scale	●	●	●	●	◐
Fixed pointer on moving scale	●	●	○	○	○
Mechanical indicator	○	○	○	●	○
Icon, symbol display	○	○	○	●	○
Indicator light	○	○	○	●	○
Graphical display	◐	◐	●	●	●
Pictorial display	◐	●	●	●	●

○ Not suitable   ◐ Acceptable   ● Recommended

FIGURE 6.11
Characteristics of common visual displays. (*After Ullman.*)

to match the intentions of the human with the actions required by the system. There should be a clear relationship between the human intent and the action that takes place on the system. The design should be such that when a person interacts with it there is only *one obviously correct* thing to do.

Displays should be clear, visible, large enough to read easily, and consistent in direction. Analog displays are preferred for quick reading and to show changing conditions. Digital displays provide more precise information. Locate the displays where viewing would be expected.

8. *Make controls easy to handle:* Shape knobs and handles of controls differently so they are distinguishable by look and by touch. Organize and group them to minimize complexity. Controls should not require a large force to actuate unless they are used only in emergencies. There are several strategies for the placement of controls: (1) left to right in the sequence they are used, (2) key controls located near the operator's right hand, (3) most commonly used controls near the operator's hand.

9. *Anticipate human errors:* Human errors are unavoidable. Furthermore, you cannot anticipate that the user will read the instruction manual, remember it, or follow it. The designer must anticipate possible errors and design to minimize them. Provide warnings if the user is in the process of making an error, and try to design so it is possible to reverse or undo an error easily.

10. *Avoid awkward and extreme motions for the user of the product:* Design the product so that twisting or lengthy arm movements, bending, and movements of the spine are not required, particularly if these motions will be repetitive. This can

lead to cumulative trauma disorders, where stresses cause nerve and other damage. Such situations will lead to operator fatigue and errors. Make sure that lighting and temperature control are adequate, or errors will increase.

11. *Standardize:* It pays to standardize on the arrangement and operation of controls because it increases the user knowledge. For example, the placement of the brake, clutch, and accelerator pedals in an automobile is arbitrary, but once standardized they become part of the user knowledge base and should not be changed.

6.7.2 Design for Serviceability

Serviceability is concerned with the ease with which maintenance can be performed on a product.[1] Many products require some form of *maintenance* or *service* to keep them functioning properly. Products often have parts that are subject to wear and that are expected to be replaced at periodic intervals. There are two general classes of maintenance. *Preventive maintenance* is routine service required to prevent operating failures, such as changing the oil in your car. *Breakdown maintenance* is the service that must take place after some failure or decline in function has occurred. It is important to anticipate the required service operations during the design of the product. Provision must be made for disassembly and assembly. Don't make a design like the automobile that requires the removal of a body panel to access the oil filter. Also, remember that service usually will be carried out in "the field" where special tools and fixtures used in factory assembly will not be available.

A concept closely related to serviceability is *testability*. This is concerned with the ease with which faults can be isolated in defective components and subassemblies. In complicated electronic and electromechanical products, testability must be designed into the product.

The best way to improve serviceability is to reduce the need for service by improving reliability. Reliability is the probability that a system or component will perform without failure for a specified period of time (see Sec. 11.3). Failing this, the product must be designed so that components that are prone to wear or failure, or which require periodic maintenance, are easily visible and accessible. It means making covers, panels, and housings easy to remove and replace. It means locating components that must be serviced in accessible locations. Avoid press fits, adhesive bonding, riveting, welding, or soldering for parts that must be removed for service. Modular design is a great boon to serviceability.

6.8
DESIGN FOR THE ENVIRONMENT

Protection of the earth's environment is high on the value scale of most citizens of the world's developed countries. Accordingly, most corporations realize that it is in their best interest to take a strong proenvironment attitude and approach to their business.

---

1. J. G. Bralla, op. cit., Chap. 16; M. A. Moss, "Designing for Minimum Maintenance Expense," Marcel Dekker, New York, 1985.

Moreover, there is near universal recognition that the most cost-effective way to improve the long-term environmental condition of our planet is through early and high-priority concern for the environment in product design—so-called green design.[1] A large proportion of our environmental problems are linked to the selection and use of particular technologies without previous adequate regard for the environmental consequences. Currently, and even more in the future, environmental impact will be considered in design along with function, appearance, cost, quality, and more traditional design factors.

Greater concern for the environment in product design places emphasis on *life-cycle design* through:

- Minimizing emissions and waste in the manufacturing process.
- Looking at all the ways that the product negatively impacts the environment. A polluting product is a defective product.
- Looking at ways to increase the useful life of the product, thereby prolonging the time when new material and energy resources need to be committed to a replacement product.

The useful life may be limited by degraded performance due to wear and corrosion, damage (either accidental or because of improper use), or environmental degradation. Other reasons to terminate the useful life not related to life-cycle issues are technological obsolescence (something better has come along) or styling obsolescence.

There are a variety of design strategies to extend a product's useful life.[2]

- *Design for durability:* Durability is the amount of use one gets from a product before it breaks down and replacement is preferable to repair.
- *Design for reliability:* Reliability is the ability of a product to neither malfunction nor fail within a specified time period.
- *Create an adaptable design:* A modular design allows for continuous improvement of the various functions.
- *Repair:* Feasibility of replacing nonfunctioning components to attain specified performance.
- *Remanufacture:* Worn parts are restored to like-new condition to attain specified performance.
- *Reuse:* Find additional use for the product or its components after the product has been retired from its original service.
- *Recyclability:* Reprocessing of the product to recover some or all of the materials from which it is made. This requires that the product can be disassembled cost-effectively, that the materials can be identified and have an economic value in excess of the cost (see Sec. 8.14).
- *Disposability:* All materials that are not recycled can be legally and safely disposed of.

---

1. U.S. Congress, Office of Technology Assessment, "Green Products by Design: Choices for a Cleaner Environment," OTA-E-541, Government Printing Office, Washington, DC, October 1992; S. B. Billatos and N. A. Basaly, "Green Technology and Design for the Environment," Taylor & Francis, Washington, DC, 1997.

2. E. B. Magrab, "Integrated Product and Process Design and Development," CRC Press, Boca Raton, FL, 1997, p. 251.

The accepted way of assessing the effects that products and processes have on the environment is with *life-cycle assessment* (LCA).[1] Figure 1.15 shows the life cycle for materials, and Fig. 6.12 shows a more general framework that is used for LCA. Life-cycle assessment proceeds in three stages:

- *Inventory analysis:* The flows of energy and materials to and from the product during its life are determined quantitatively.
- *Impact analysis:* Consideration of all potential environmental consequences of the flows cataloged above.
- *Improvement analysis:* Results of the above two steps are translated into specific actions that reduce the impact of the product or the process on the environment.

Detailed life-cycle assessments are often time-consuming and expensive to carry out in engineering systems, so other less rigorous scoring methods are used,[2] as described in Sec. 6.8.2.

It is generally believed that improvement of the environment is the joint responsibility of all citizens, joined by business and government. Government plays a crucial role, usually through regulation, to ensure that all businesses share equitably in the cost of an improved environment. Since these increased product costs often are passed on to the customer, it is the responsibility of government to use the tool of regulation prudently and wisely. Here the technical community can play an important role by providing fair and timely technical input to government. Finally, many vision-

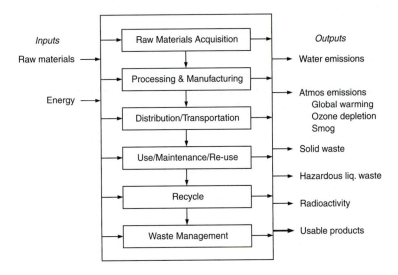

FIGURE 6.12
Framework for developing a life-cycle analysis.

1. T. E. Graedel and B. R. Allenby, "Design for Environment," Prentice-Hall, Upper Saddle River, NJ, 1996.
2. T. E. Graedel, "Streamlined Life-Cycle Assessment," Prentice-Hall, Upper Saddle River, NJ, 1998.

Hale
Huar
Pahl,

Sour

Beitz

Linga
"Mac

Pope,
Shigle
H

6.1. L
a

6.2. M

Fi

6.3. D
ho
ibl
an

—

12'

6.4. As
hanc

Diesel Engine Stages a Comeback?

The environmental pressure to reduce greenhouse gases is leading automotive engineers to take another look at small diesel engines. The diesel uses no spark plugs and burns fuel more efficiently than the gasoline engine. Because it gets greater fuel mileage the output of emission gases is less. Historical problems with diesel engines have been high noise and dirty exhaust.

Computerized fuel-control systems have quieted the engine and eliminated the "diesel bark." New catalytic converters have significantly improved NO_x emissions. The new diesel engines no longer belch clouds of black soot. The remaining environmental problem is microscopic particles of soot in the emissions, which can be removed with filters.

Another advantage of the diesel engine is its ability to run on a wider variety of fuels than the gasoline engine. Experimental engines have been run on nonfossil fuels and on a fuel made from natural gas.

---

aries see a future world based on *sustainable development* in which the world resources will no longer be depleted because the rate of resource consumption will be balanced by the rate of resource regeneration. Industry will operate with renewable energy and material resources and massive recycling. Designs will be based on many technologies different from those used today, e.g., the electric car vs. the gasoline internal-combustion engine.[1]

6.8.1 Design for Environment (DFE)[2]

We can divide DFE practices into two broad categories: those involving material recycling and remanufacture, and those avoiding the use of or production of hazardous substances.

1. *Design for material recovery and reuse:* See Sec. 8.14.
2. *Design for disassembly:* Provide for easy access and removal of components. Avoid embedding a part in an incompatible material. Minimize the use of adhesives and welds. Try to avoid the use of screws. Snap fits and spring clips make for easier disassembly.
3. *Design for product waste minimization:* An obvious way to minimize waste is to practice source reduction. Minimize the amount of material used by avoiding overdesign. This also saves cost. Realize which materials are incompatible in recycling and will have to be separated and segregated. Keep the number of different materials used in the design to a minimum to reduce recycling costs. Understand

---

1. Two design for environment web sites are: http://dfe.stanford.edu and http://www.flash.net/~rcade/dfe/index.html.
2. J. Fiksel (ed.), "Design for Environment," Chap. 8, McGraw-Hill, New York, 1996.

(*a*) Decide what performance characteristics and features you want the product to have.

(*b*) Operate the product for several hours and observe its performance. Does it meet the criteria you have established? Is it a quality product.

(*c*) Disassemble each product into its components. For each component, describe its intended function, the material from which it is made, and the most likely process used for its manufacture.

(*d*) What physical principles were used to achieve the function of each component? Refer to the TRIZ tables in Chap. 5 as an aid.

(*e*) Suggest how the design of the product could be improved with respect to performance, human factors design for the environment, or cost.

6.5. Take photographs of common consumer products, or tear pictures out of old magazines, to build a display of industrial designs that appeal to you, and designs that you consider need improvement. Be able to defend your decisions on the basis of aesthetic values.

6.6. Look at the web site http://www.baddesigns.com/examples.html for examples of poor user-friendly design. Then, from your everyday environment, identify five other examples. How would you change these designs to be more user-friendly?

6.7. Read the boxed example (Sec. 6.8) on the comeback of the diesel engine. Dig deeper into the story by learning more about diesel engines. What do you think is the likelihood of having a diesel engine in your sport-utility vehicle?

7

MODELING AND SIMULATION

7.1
THE ROLE OF MODELS IN ENGINEERING DESIGN

A model is an idealization of part of the real world that aids in the analysis of a problem. You have employed models in much of your education, and especially in the study of engineering you have learned to use and construct models such as the free-body diagram, electric circuit diagram, and the control volume in a thermodynamic system. Figure 7.1 illustrates some of the common types of conceptual models.

A model may be either descriptive or predictive. A *descriptive model* enables us to understand a real-world system or phenomenon; an example is a sectioned model of an aircraft gas turbine. Such a model serves as a device for communicating ideas and information. However, it does not help us to predict the behavior of the system. A *predictive model* is used primarily in engineering design because it helps us to both understand and predict the performance of the system.

We also can classify models as follows:[1] (1) static or dynamic, (2) deterministic or probabilistic, and (3) iconic-analog-symbolic. A *static model* is one whose properties do not change with time; a model in which time-varying effects are considered is *dynamic*. In the deterministic-probabilistic class of models there is differentiation between models that predict what will happen. A *deterministic model* describes the behavior of a system in which the outcome of an event occurs with certainty. In many real-world situations the outcome of an event is not known with certainty, and these must be treated with *probabilistic models*.

An *iconic model* is one that looks like the real thing. Examples are a scale model of an aircraft for wind tunnel test and an enlarged model of a polymer molecule. Iconic models are used primarily to describe the static characteristics of a system, and

---

1. W. J. Gajda and W. C. Biles, "Engineering Modeling and Computation," chap. 2, Houghton Mifflin Company, Boston, 1978.

specific states of the model have firmly established computer modeling and simulation as a powerful tool of engineering design.

Engineers use models for thinking, communicating, predicting, controlling, and training. Since many engineering problems deal with complex situations, a model often is an aid to visualizing and thinking about the problem. One of the results of an engineering education is that you develop a "menu of models" that are used instinctively as part of your thought process. Models are vital for communicating whether via the printed page, the computer screen, or oral presentation. Generally, we do not really understand a problem thoroughly until we have achieved a predictive capability. Engineers must make decisions concerning alternatives. The ability to simulate the operation of a system with a mathematical model is a great advantage in providing sound information, usually at lower cost and in less time than if experimentation had been required. Moreover, there are situations in which experimentation is impossible because of cost, safety, or time.

Use your modeling capability early and often in design. A simple mathematical model often serves to firm up a conceptual design. However, it is in embodiment design that a full-scale modeling approach becomes vital to success. Building a digital computer model (see Sec. 7.6) provides the platform for realistic determination of stresses, velocities, temperatures, etc., with modern methods of computer-aided engineering analysis.

7.1.1 The Prototype

The role of the prototype in product development is often overlooked. A prototype is a full-sized working model that is as close as possible to the design of the product that will be marketed. It rarely will be identical because the methods used in its manufacture most likely will be different from those used in making large numbers of product for sale. A prototype may evolve continuously once the design concept is firmed up, or it may be constructed at the end of the embodiment design phase. Either way, the purpose of the prototype is to reduce the risks incurred in putting a product into full production and then introducing it into the market.[1]

A prototype will be subjected to a variety of tests to see whether it performs as expected in a wide range of service conditions. When failure or underperformance occurs, the design must be modified. Testing will take place not only in the engineering department but with potential customers to see whether the industrial design and human factors aspects work well. Finally, after what could be considerable iteration, in the prototype and test cycle a decision is made to go into production.

While building, testing, and modifying the prototype is expensive and necessary, it certainly will not be the only place that modeling gets done during design. Experience shows that a good way to reduce the product development cycle time is to build and test many models for different aspects of the design. Combining computer analysis with working models is the way to go. For example, you may model the kine-

---

1. W. Bergwerk, "The Role of the Prototype in Managing Product Innovation Risks," *Proc. Inst. MechE.*, vol. 203, pp. 113–118, 1989.

matics of your design concept with a computer model and then validate it along with vibration analysis in an experimental model. At the same time the industrial designers will have built a set of styling models.

7.2
MATHEMATICAL MODELING

In mathematical modeling the components of a system are represented by idealized elements that have the essential characteristics of the real components and whose behavior can be described by mathematical equations. The first step is to devise a conceptual model that represents the real system to be analyzed. You have been exposed to many examples of simple mathematical models in your engineering courses, but modeling is a highly individualized art. A key issue is the assumptions, which determine on the one hand the degree of realism of the model and on the other hand the practicality of the model for achieving a numerical solution. Skill in modeling comes from the ability to devise simple yet meaningful models and to have sufficient breadth of knowledge and experience to know when the model may be leading to unrealistic results.

A generalized picture of a mathematical model for a system or component is shown in Fig. 7.2. The choice of the system that is modeled is an important factor in the success of the model. Engineering systems frequently are very complex. Progress often is better made by breaking the system into simpler components and modeling each of them. In doing that, allowance must be made for the interaction of the components with each other. Techniques for treating large and complex systems by isolating the critical components and modeling them are at the heart of the growing discipline called *systems engineering.*[1]

For example, in the gross sense we might consider the system to be a central-station power plant (Fig. 7.3). But on further reflection we realize that the basic elements of the plant are given by the block diagram in Fig. 7.4. However, each of these components is a complex piece of engineering equipment, so the total power plant system might be better modeled by Fig. 7.5.

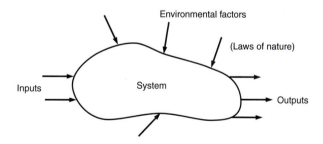

FIGURE 7.2
Characteristics of the modeling process.

---

1. H. Chestnut, "System Engineering Tools," Wiley, New York, 1965; ibid., "System Engineering Methods," Wiley, New York, 1967; A. P. Sage, "Systems Engineering," Wiley, New York, 1992.

FIGURE 7.3
Overall model for a steam power plant.

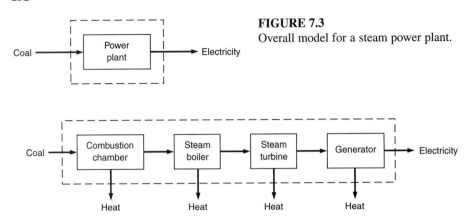

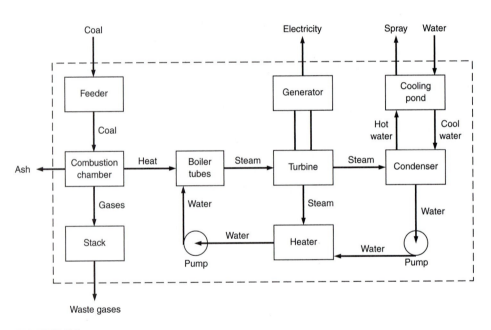

FIGURE 7.4
Block diagram of major components of power plant.

FIGURE 7.5
Detailed systems for a coal-fired power plant.

In developing a model we walk a fine line between simplification and realism. One way to achieve simplification is to minimize the number of physical quantities that must be considered. We do that routinely by neglecting small effects. Thus, we may assume a structural member is completely rigid when its elastic deformation is considered of little consequence to the problem, or we may assume a fluid is Newtonian viscous when it in fact shows a small deviation from ideality. We must be aware that factors that are neglected routinely because they are negligible in one environment may not be of minor consequence in a much different situation. Thus, we neglect surface tension effects when dealing with large objects in a fluid but must consider it when the particles are fine.

Another common assumption is that the environment is infinite in extent and therefore entirely uninfluenced by the system being modeled. In approximate models it also is common practice to assume that the physical and mechanical properties are constants that do not change with time or temperature. Generally, we start with two-dimensional models because they are more mathematically tractable.

Important simplification results when the distributed properties of physical quantities are replaced by their lumped equivalents. A system is said to have *lumped parameters* if it can be analyzed in terms of the behavior of the endpoints of a finite number of discrete elements. Lumped parameters have just single values, whereas distributed parameters have many values spread over a field in space. The mathematical model of a lumped-parameter system is expressed by differential equations, and a distributed-parameter system leads to partial differential equations. Systems that can be represented by linear models, i.e., linear differential equations, are much more easily solved than systems represented by nonlinear models. Thus, a common first step is to assume a linear model. However, since we live in a basically nonlinear world, this simplification often must be abandoned as greater realism is required. Likewise, the usual first step is to develop a deterministic model by neglecting the fluctuations or uncertainties that exist in the input values.

Once the chief components of the system have been identified, the next step is to list the important physical and chemical quantities that describe and determine the behavior of the system. As Fig. 7.2 indicates, they can be grouped into input and output parameters. Next, the various physical quantities are related to one another by the appropriate physical laws.[1] These are modified in ways appropriate to the model to transform the input quantities into the desired output. The relation that transforms the input quantities into output ones is called a *transfer function*. It may take the form of algebraic, differential, or integral equations. The solution of these equations, either analytically, numerically, or graphically, is the last step in the modeling process.

EXAMPLE. We illustrate mathematical modeling with an example taken from a classic book in engineering design.[2] This problem is typical of the kind of modeling one might do in configuration design in the embodiment design stage. Note that it is an application of the principles of mechanics that we all have studied in earlier courses.

We are faced with the selection of the motor to drive a conveyor belt to deliver sand at a flow rate Q lb/s. The design is shown in Fig. 7.6. We start the model from energy considerations, realizing that the energy supplied by the motor E_i must equal the energy stored in the sand plus the energy lost. The energy stored in the sand consists of its potential energy E_p and its kinetic energy E_k. The energy lost in friction due to sand sliding on the belt during acceleration as it exits the feed hopper is E_f. At constant velocity the kinetic energy stored in the sand equals the energy due to friction lost on the belt, E_f.

1. W. Hughes and E. Gaylord, "Basic Equations of Engineering Science," Schaum, New York, 1964; W. G. Reider and H. R. Busby, "Introductory Engineering Modeling," Wiley, New York, 1986; D. E. Thompson, "Design Analysis," Cambridge University Press, New York, 1999.
2. T. T. Woodson, "Introduction to Engineering Design," McGraw-Hill, New York, 1966, pp. 131–135.

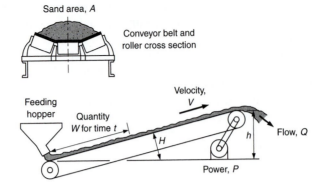

FIGURE 7.6
Schematic drawing for the modeling of the sand conveyor.

$$\text{Input} = \text{potential} + \text{kinetic} + \text{friction}$$

$$E_i = Wh + \frac{1}{2}\left(\frac{W}{g}\right)V^2 + \frac{1}{2}\left(\frac{W}{g}\right)V^2 = Wh + \frac{WV^2}{g} \qquad (7.1)$$

where $W =$ amount of sand, lb
$\quad\ \ g =$ gravitational constant 32 ft/s
$\quad\ \ V =$ belt velocity, ft/s

We convert energy to power by dividing both sides of Eq. (7.1) by time t. Also, we define the sand flow rate as $Q = W/t$.

$$P = \frac{E_i}{t} = Qh + Q\frac{V^2}{g} = Q\left(h + \frac{V^2}{g}\right)\frac{\text{ft-lb}}{\text{s}} \qquad (7.2)$$

Equation (7.2) gives us a mathematical model of the problem. Now we need to understand what it can tell us about the design.

First, let's find the condition at which power required is a minimum. When we differentiate P with respect to V and equate the result to zero, we find that minimum power occurs when $V=0$.

$$\frac{dP}{dV} = Q\left(0 + \frac{2V}{g}\right) = 0$$

$$\frac{2V}{g} = 0 \text{ and } V = 0 \qquad (7.3)$$

Equation (7.2) tells us that the more slowly the belt goes, the less power it will require, but we still don't have a design sense of what this could mean.

We input the following "trial values" of design parameters into Eq. (7.2) to see how power varies with belt velocity:

$Q =$ flow rate, 50 lb/s
$h =$ height, 55 ft

$A =$ sand area, 0.5 ft^2
$\rho =$ sand density, 100 lb/ft^3
$g =$ gravity, 32 ft/s^2

$$P = Q\left(h + \frac{V^2}{g}\right) = 50\left(55 + \frac{V^2}{32}\right) = 2750 + 1.563\,V^2\left(\frac{\text{ft-lb}}{\text{s}}\right)$$

Since 1 hp = 550 ft-lb/s

$$P = 5 + 0.003\,V^2\,(\text{hp}) \tag{7.4}$$

Thus, we have established that $P = 5$ hp at $V \cong 0$. Equation (7.4) is plotted in Fig. 7.7. It shows the portion of the power requirement that is due to potential energy and the portion due to kinetic energy. This plot shows that kinetic energy contributes very little to the power requirement. In fact, V must be 41 ft/s before it equals the potential energy portion of the total energy requirement.

We can examine the analysis for its *sensitivity* to the design variables and parameters. We see from Fig. 7.7 that power is insensitive to velocity up to about $V = 15$ ft/s. We also learned that power is proportional to Q. It also is proportional to h up to $V \cong 15$, where the power becomes influenced by V. When $V^2/g \gg h$ the power varies as V^2. The influence of changes in these variables on power requirement can be shown nicely with a spreadsheet or any of the equation-solver software tools.

What else can we learn from our model of the conveyor belt? We can express Q in terms of the design parameter A, the cross-sectional area of the sand on the belt, by $Q = AV\rho$. Using a design requirement of 100 tons/h (55.5 lb/s), we find that $V = 1.1$ ft/s when $A = 0.5$ ft.^{2} Thus, the power requirement will be satisfied with a 5-hp motor. Further, we can substitute for V in Eq. (7.2), to give

$$P = \frac{Q^3}{g\rho^2 A^2} + Qh\left(\frac{\text{ft-lb}}{\text{s}}\right) \tag{7.5}$$

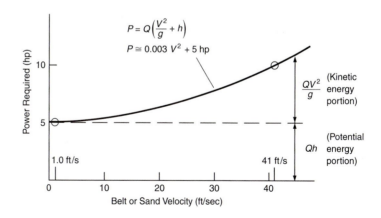

FIGURE 7.7
Characteristic curve for the power required for the sand conveyor. Plot of belt velocity vs. required driving power.

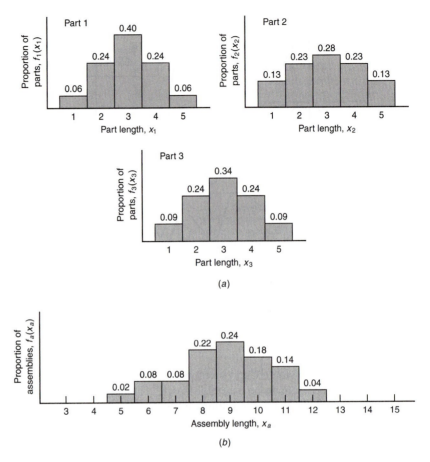

FIGURE 7.14
(*a*) Distribution of lengths of three parts. Part length is coded (see text). (*b*) Distribution of part lengths of 50 assemblies determined from a Monte Carlo simulation. (*M. F. Spotts, Machine Design, pp. 84, 85, Nov. 20, 1980.*)

TABLE 7.1
Assignment of Monte Carlo numbers

| \multicolumn{3}{c}{Part 1} | | | \multicolumn{3}{c}{Part 2} | | | \multicolumn{3}{c}{Part 3} | | |
Part 1			Part 2			Part 3		
x_1	$f_1(x_1)$	Monte Carlo No.	x_2	$f_2(x_2)$	Monte Carlo No.	x_3	$f_3(x_3)$	Monte Carlo No.
1	0.06	00 to 05	1	0.13	00 to 12	1	0.09	00 to 08
2	0.24	06 to 29	2	0.23	13 to 35	2	0.24	09 to 32
3	0.40	30 to 69	3	0.28	36 to 63	3	0.34	33 to 66
4	0.24	70 to 93	4	0.23	64 to 86	4	0.24	67 to 90
5	0.06	94 to 99	5	0.13	87 to 99	5	0.09	91 to 99

TABLE 7.2
A sample of random numbers

(1)	(2)	(3)	(4)	(5)	(6)
0095	8935	2939	3092	2496	0359
6657	0755	9685	4017	6581	7292
8875	8369	7868	0190	9278	1709
8899	6702	0586	6428	7985	2979
.	.	.	.	.	.
.	.	.	.	.	.
.	.	.	.	.	.

random numbers. We select the page of numbers by chance and then adopt a purely arbitrary selection scheme. In this case, we use only the first two digits of the four-digit numbers. Digits from columns 1, 2, and 3 are assigned to parts 1, 2, and 3, respectively, for the first 25 assemblies and columns 4, 5, and 6 for the last 25 assemblies. The total of the lengths for the three parts is the length of each assembly x_a. When all 50 assemblies are simulated, the number having length 6, 7, 8, etc., is determined, and the proportion with each length $f_a(x_a)$ is determined. Those numbers are plotted in Fig. 7.14b.

When we examine the histogram in Fig. 7.14b, we note that none of the assemblies is less than 5 units long or greater than 12. Based on the length distribution of the individual parts, it is possible for an assembly ($x_1 + x_2 + x_3 = x_a$) to be as short as 3 and as long as 15. However, the probability of selecting three successive parts from the left tails of the distribution or three successive parts from the right tails is very low. Therefore, the assembled lengths tend to bunch more than if the tolerances on the individual parts had simply been added.

7.6
GEOMETRIC MODELING ON THE COMPUTER

Geometric modeling on the computer has become the fastest-changing area of engineering design. When computer-aided design (CAD) was introduced in the late 1960s, it essentially provided an electronic drafting board for drawing in two dimensions. Through the 1970s CAD systems were improved to provide three-dimensional wireframe and surface models. By the mid-1980s nearly all CAD products had true solid modeling capabilities. More recently, tools for kinematic, stress, and thermal analysis have been seamlessly linked to the solid model so that analysis can be done along with modeling. In the beginning CAD required mainframe or midicomputers to support the software. Today, with the enhanced capabilities of personal computers, solid modeling software can be run on desktop machines.

From its initiation, CAD has promised five important benefits to the engineering design process.[1]

- Automation of routine design tasks to increase the productivity of designers and engineers and free them for more creative tasks.
- The ability to design in three dimensions to increase the designer's conceptual capacity, and hence the quality of the design.
- Design by solid modeling to create a digital geometric database which can be transferred downstream to permit engineering analysis and simulation, thereby minimizing the costly testing of prototypes.
- Electronic transfer of the design database to manufacturing (CAD/CAM) where it is used to generate NC tapes for machining on computerized machine tools, or for developing process plans. Alternatively, the database is transferred to a rapid prototyping system to generate 3D models of parts (see Sec. 7.9).
- A paperless design process is evolving, where digital databases rather than drawings are sent to customers and suppliers. This will reduce costs of producing, storing, and managing engineering drawings and will speed communications with customers and suppliers.

The first two benefits currently are within reach of any engineering organization or engineer. Seamless downstream data transfer for analysis, simulation, and computerized manufacture is a reality but has not yet been routinely achieved by many organizations. The design of the Boeing 777 transport as a completely paperless design is the best-known example.

A solid modeling system must unambiguously describe the three-dimensional shape of the object.[2] It must be able to distinguish between the inside and the outside of the object, and be able to support the calculation of mass properties such as weight and moment of inertia. In a two-dimensional wireframe model the object is created by drawing lines, arcs, circles, etc., on a plane in the manner of an electronic drafting board. The three-dimensional shape is conveyed with orthographic projections. This is little better in delivering information than a conventional hand-drawn portrayal. The advantage is in the ability to make easy drawing modifications and the ability to use predrawn modules. A three-dimensional wireframe model adds little additional information (see Fig. 7.27). The surfaces between the wireframe edges are not defined and the computer is unable to determine what is inside or outside of the object being drawn. Wireframe models often are confusing and difficult to interpret because of extraneous lines.

---

1. J. K. Liker, M. Fleischer, and D. Arnsdorf, Fulfilling the Promise of CAD, *Sloan Management Review,* Spring 1992, pp. 74–86.
2. J. MacKrell, Computer-Aided Design, "ASM Handbook," vol. 20, "Materials Selection and Design," ASM International, Materials Park, OH, 1997, pp. 155–165; D. LaCourse (ed.), "Handbook of Solid Modeling," McGraw-Hill, 1995; F. M. L. Amirouche, "Computer-Aided Design and Manufacturing," Prentice-Hall, Englewood Cliffs, NJ, 1993; J. D. Foley, A. Van Dam, S. K. Fenner, and J. F. Hughes, "Computer Graphics, Principles and Practice," 2d ed., Addison-Wesley, 1990; A. J. Medland, "The Computer-Based Design Process," 2d ed., Chapman & Hall, London, 1992.

7.6.1 Surface Modeling

The next evolution was to define the surfaces between the edges of the wireframe model. With such a *surface model* the definition of the object is more complete and the model can be used with finite-element analysis and numerical-controlled (NC) machining programs. Color-shaded images (renderings) may be created.

Ordinary first-degree analytical expressions can be used to define plane surfaces, but to define curved surfaces requires a third-degree polynomial expression. The *Bezier curve* is defined by the endpoints of the curve and two other control points which determine the degree of curvature. Figure 7.15 shows a Bezier surface patch with four end-points (*P*) and 12 control points (*Q*). The total surface is made up of many similar surface patches. The advantage of Bezier surface patches is that the surface can be easily shaped by manipulation of the control points, but they have the disadvantage that they do not provide local control. This means that a change in one control point affects the shape of the whole patch, and this could alter the edge which connects two patches together. A second disadvantage is that it is hard to calculate control points that will allow the patch to pass through already existing points in a model.

An improvement on the Bezier curve is the *bicubic spline* (B-spline). Patches constructed from B-splines allow local control by pulling a control point. This has the important feature that moving one control point does not affect the whole surface. Also, with B-splines it is much easier to create surfaces through predetermined points. The most advanced description of surfaces is with nonuniform rational B-splines (NURBS). This technique is called rational because for every point described on the surface there is an associated weighting function. Dividing the surface point function by the weight creates new surfaces that are not possible with nonrational surfaces like

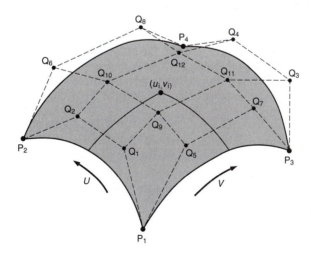

FIGURE 7.15

A Bezier surface patch with endpoints *P* and control points *Q*. (*G. R. Bertoline, E. N. Wiebe, C. L. Miller, and J. L. Mohler, "Technical Graphics Communication," 2d ed, Irwin, Chicago, 1997. Used with permission.*)

those created with B-splines. This means that NURBS can provide precise descriptions of basic geometric shapes, while B-spline surfaces can only approximate these shapes using many small patches. High-end CAD modelers that provide NURBS functionality can model both analytical and free-form surfaces.

7.6.2 Methods of Generating Solids

Solid models provide information about what is inside the 3D model as well as information about the surface of the object. The two basic techniques of solid modeling are *constructive solid geometry* (CSG) and *boundary representation* (B-rep) (Fig. 7.16). With CSG the solid is constructed in building-block fashion by combining primitive shapes like a cube, cylinder, cone, or sphere. These primitives may be scaled, rotated, and translated in space. They are then combined using the boolean operations of union, difference, and intersection (Fig. 7.17). The union operation (*a*) combines the two primitive shapes. The shape resulting from a difference operation resembles the original shape with the area of overlap removed. Note that a difference operation is sensitive to the order of operation. *A-B* gives a different shape (*b*) than *B-A* (Fig. 7.17c). With the intersection operation (*d*) only the volume of overlap remains.

In a boundary model (B-rep) solids are represented by the sets of faces that enclose them completely. To ensure that the boundary model is topologically valid, it must satisfy the Euler-Poincaré relationship between the number of faces (F), edges (E), vertices (V), inner loops of faces (L), bodies (B), and through holes ("genus").

$$F - E + V - L = 2(B - G) \qquad (7.24)$$

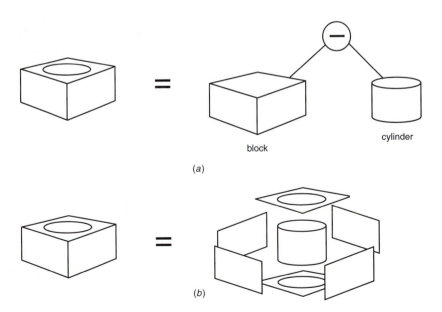

FIGURE 7.16
Two basic methods for representing solids in CAD: (*a*) constructive solid geometry (CSG); (*b*) boundary representation (B-rep).

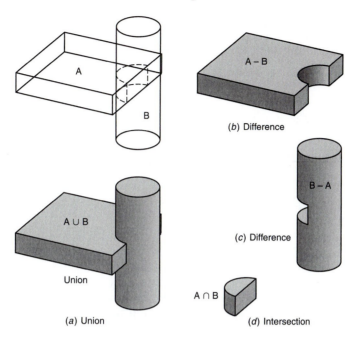

FIGURE 7.17
Three boolean operations involving primitive shapes *A* and *B*: (*a*) union; (*b*) difference *A-B*;
difference *B-A*; (*d*) intersection. (*G. R. Bertoline, E. N. Wiebe, C. L. Miller, and J. L. Mohler,
"Technical Graphics Communication," 2d ed., Irwin, Chicago, 1997. Used with permission.*)

For the B-rep shown in Fig. 7.16b, $7 - 17 + 10 - 0 = 2(1-1) = 0$. In the B-rep model,
the geometry is created and the topological validity is checked at the same time. The database must store information about the connectivity of the faces and the equations defining the geometry of the faces. A B-rep model is useful for complex parts that cannot be
modeled conveniently with primitive shapes. It defines the part topology and geometry
separately. After the topology has been defined, many different operations can be performed to adjust the geometry without changing the basic topology. A major disadvantage of B-rep modeling is that it is difficult to guarantee closure.

Sweeping is the creation of a solid shape by moving a line or a plane along a
defined trajectory (Fig. 7.18). The sweeping method is fast and easy to use. However,
it has the important limitation that it can only create shapes that have translational or
rotational symmetry. For example, a block with a hole through it is generated efficiently by sweeping, but a block with a hole only partway through the thickness cannot be generated by sweeping.

Spatial enumeration and *cell decomposition* use combinations of cubes to
approximate solid shapes. They are organized as three-dimensional binary trees called
octrees. Since modeling irregular surfaces requires almost an infinite number of tiny
cubes, the method is very memory-intensive. Nevertheless, the method is very good
at modeling irregular solids. In the variation called cell decomposition the cell may be
a distinctive shape, like a finite-element mesh. As computer memory and processing
power became even more plentiful it is likely that octree methods will become the
preferred method of solids modeling.

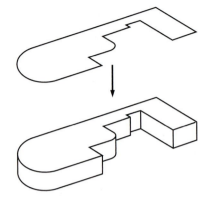

① A two-dimensional lamina is created with straight-line elements and arcs to represent the basic outline of the part base.

② A solid object is created by applying the linear sweep operator on the lamina.

FIGURE 7.18
Development of a solid model by sweeping.

Both types of modeling, CSG and B-rep, have their advantages. The CSG representation is easier to edit because changes to the way primitives are used produce major changes with minimal input. However, the B-rep data structure deals with the details like edges and faces that the designer will want to change, so it is more amenable to an interactive environment. Since no one method of solid modeling has a clear advantage, most solid modeling software contains a hybrid modeler which provides for conversion between the two systems as needed.

7.6.3 Constraint-Based Modelers and Features

Constraint-based solid modelers represent the current generation of CAD. They employ the methods of generating solids described above with important embellishments. Constraints are used to create a set of rules that control how changes can be made to a group of geometric elements such as lines or form features. The types of constraints in common use are classified as numeric, geometric, algebraic, or attributes. Numeric constraints provide the position of points, the length of lines, angular values, etc. Geometric constraints provide for parallelism, perpendicularity, tangency, etc. Algebraic constraints combine numeric and geometric constraints into simple equations (diameter of B = one-half the length of A) or complex sets of equations that include inequalities and IF-THEN-ELSE branches. Attribute constraints define characteristics of the part such as color, material, surface finish, maximum stress, etc.

There are two types of constraint modelers in general use: parametric and variational. The difference between them has to do with how the constraint equations are defined and solved. In a *parametric modeler* all of the constraint equations are captured and solved in the order in which they are created. For example, if points A and B are defined as to location, and point C is defined by reference to A and B, then the constraint equations for the system are A = point($x1$, $y1$, $z1$); B = point($x2$, $y2$, $z2$); $C = (A-B)/2$. In a parametric modeler points A and B can be moved and point C will automatically follow. In a *variational modeler* the set of equations is solved simultaneously so that the ordering of the equations is not so important. In the above example, points A and C can be moved and point B will automatically follow. For example, with a constraint modeler if a fillet is placed between a plane surface and a cylinder, the fillet radius will change if the cylinder diameter changes. In a traditional solid modeler the fillet would not change and the designer would have to remember to make this adjustment. In a constraint-based modeler the designer can countersink all of the holes in a part with a single command, rather than finding and changing every one. Parametric modelers impose order dependence, but they are faster than variational modelers. Most parametric modelers allow the user to redefine the parameterization and to reorder the parametric equations to allow different parameters to control the design. Variational modelers require more computing power to handle the more complex mathematics. However, they have the advantage that the designer does not have to give a lot of forethought to the hierarchy of the design constraints and can freely change the design without being limited by the order in which constraints were defined and features were added.

Constraint-based CAD systems combine constraints with form features. *Features* are higher-level constructs than lines, curves, and surfaces. Features such as slots, through holes, blind holes, etc., are predefined and can be called up with a few commands rather than having to be created each time with boolean operations. Features combine geometry with a set of dimensions and knowledge of how the feature combines with the solid object to which it is applied, i.e., *design intent.* They save the designer time because, since they are defined parametrically, the drawing can be quickly updated just by changing a dimension.

Constraint modelers also are relational modelers. Figure 7.19 shows a block with two features, a slot and a pair of holes. The slot is constrained by a *relation, $d_2 = 0.6d_1$,* such that whatever the value of d_1 the slot will always be 6/10 of the distance from the reference surface. The designer captures this design intent with the relational model.

Solid modeling software[1] provides several modes of operation. The part mode is used to model a part. Standard "shop drawings" using orthographic projection can be prepared from the solid model using the drawing mode. Figure 7.20 shows a solid model of a holder for the brushes of an electric motor, and the 2D orthographic projection

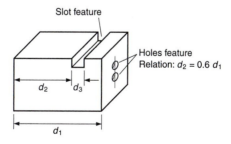

FIGURE 7.19
A part with two features: a slot and two holes.

---

1. There are many providers of solid modeling software. The high-end providers are Pro/ENGINEER from Parametric Technology Corp (www.ptc.com), CATIA-CADAM from IBM (www.catia.ibm.com), I-DEAS from SDRC (www.sdrc.com), and Unigraphics from Unigraphics Solutions (www.ugsolutions.com). Somewhat less expensive software that runs in a Windows environment is CADkey (www.cadkey.com), Mechanical Desktop (www. autodesk.com), Solid Edge (www.ugsolutions.com), Solid Works (www.solid works.com), and MicroStation (www.bentley.com).

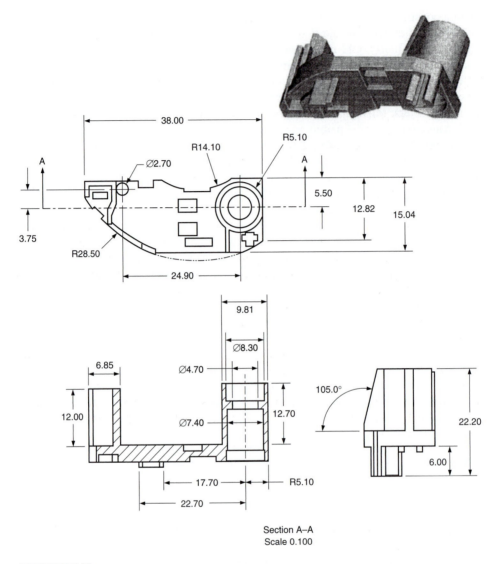

FIGURE 7.20
Digital geometric model of a brush holder for an electric motor (upper right) and the 2D drawing derived from the model. (*Courtesy of Professor Guangming Zhang, University of Maryland.*)

drawing derived from it. Solid models also can be used to create exploded assembly drawings using the assembly mode. An important capability is the ability to *associate* data between different modes. Thus, if a dimension is changed in the assembly mode, the software will automatically change the same dimension in the other modes.

Other capabilities of solid modeling software include the ability to make cutaway views, perspective views, and artistic renderings. Solid modeling CAD software can automatically generate complete meshes for finite-element analysis (FEA), and these can be linked with the FEA to perform design optimization.

7.6.4 Current Practice

An aspect of CAD modeling that is growing in importance is *data associativity,* the ability to share digital design data with other applications such as finite-element analysis or numerical control machining without each application having to translate or transfer the data. As we have seen, an important aspect of associativity is for the database of the application to be updated when a change is made in the basic CAD design data.

In order to integrate digital design data from design to manufacturing there must be a data format and transfer standard. The current standard, Standard for the Exchange of Product Data (STEP),[1] has been incorporated by all major CAD vendors. The STEP standards are organized into application protocols (APs), each tailored to a specific industry or process. For example, there are APs for the automotive, aerospace, and shipbuilding industries, as well as the design of sheet-metal die design. STEP also makes possible an open system of engineering information exchange using the Worldwide Web or private networks based on the Internet (*intranets*). Such a system will permit true concurrent engineering between company engineers and vendors working in different geographic locations. Facilitating this is the growing acceptance of the Windows NT® operating system,[2] with its ability to be truly multitasking and to share information running on multiple hardware platforms.

The trend also is to make the design software more productive by linking it with manufacturing specific design software (see Chap. 9) and enhancing its assembly modeling capability. Currently solids modeling software can handle large assemblies with thousands of parts. It will deal with the associativity of the parts, and manage the subsequent revisions to the parts. An increasing number of systems are providing top-down assembly modeling functions, where the basic assembly can be laid out and then populated later with parts.

7.7
FINITE-ELEMENT ANALYSIS

Finite-element analysis[3] (FEA) is a computer-based numerical method for solving problems in a wide range of engineering areas such as stress analysis, thermal analysis and fluid flow, diffusion, and magnetic field interactions. In finite-element analysis a continuum solid or fluid is considered to be built up of numerous tiny connected elements. Since the elements can be arranged in virtually any fashion, they can be

---

1. H. Baumgartner, A STEP to Improved CAD, *Mechanical Engineering,* February 1998, pp. 84–85. The current STEP standard only handles B-rep and feature-based models. Parametric and variational models are not allowed.
2. P. Dvorak, Windows NT makes CAD Hum, *Machine Design,* Jan. 10, 1994, pp. 46–52.
3. O. C. Zienkiewicz, "The Finite Element Method in Engineering Science," 4d ed., McGraw-Hill, New York, 1987; B. Szabo and I. Babuska, "Finite Element Analysis," Wiley, New York, 1989; K. H. Huebner, E. A. Thornton, and T. G. Byrom, "The Finite Element Method for Engineers," 3d ed., Wiley, New York, 1995; V. Adams and A. Askenazi, "Building Better Products with Finite Element Analyses," Onward Press, Santa Fe, NM, 1999.

used to model very complex shapes. Thus, it is no longer necessary to find an analytical solution that treats a close "idealized" model and guess at how the deviation from the model affects the prototype. As the finite-element method has developed, it has replaced a great deal of expensive preliminary cut-and-try development with quicker and cheaper computer modeling.

In contrast to the analytical methods that often require the use of higher-level mathematics, the finite-element method is based on simple algebraic equations. However, an FEA solution may require hundreds of simultaneous equations with hundreds of unknown terms. Therefore, the development of the technique required the availability of the high-speed digital computer for solving the equations efficiently by matrix methods. The rapid acceptance of finite-element analysis has been largely due to the increased availability of FEA software through interactive computer systems.

In the finite-element method the loaded structure is modeled with a mesh of separate elements (Fig. 7.21). We shall use triangular elements here for simplicity, but later we shall discuss other important shape elements. The elements are connected to one another at their corners, and the connecting points are called *nodes*. For a stress analysis, a solution is arrived at by using basic stress and strain equations to compute the deflections in each element by the system of forces transmitted from neighboring elements through the nodal points. The strain is determined from the deflection of the nodal points, and from the strain the stress is determined with the appropriate constitutive equation. However, the problem is more complex than first seen, because the force at each node depends on the force at every other node. The elements behave like a system of springs and deflect until all forces are in equilibrium. That leads to a complex system of simultaneous equations. Matrix algebra is needed to handle the cumbersome systems of equations.

The key piece of information is the *stiffness* matrix for each element. It can be thought of as a kind of spring constant that describes how much the nodal points are displaced under a system of applied forces. In matrix notation

$$\{f\} = [k]\{\delta\} \tag{7.25}$$

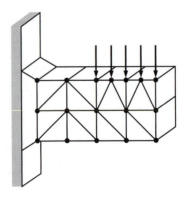

FIGURE 7.21
Simple finite-element representation of a beam.

where $\{f\}$ is the column matrix (vector) of the forces acting on the element, $[k]$ is the stiffness matrix for the element, and $\{\delta\}$ is the column matrix of the deflections of the nodes of the element. The stiffness matrix is constructed from the coordinate locations of the nodal points and the matrix of elastic constants of the material. A triangular element $[k]$ can be constructed from the principles of statics, but more complicated elements require the use of energy principles to derive $[k]$. When all the elements of the systems are assembled, the basic matrix equation is

$$\{F\} = [K]\{\delta\} \tag{7.26}$$

where $[K]$ = master stiffness matrix, assembled from the $[k]$ for all the elements
      $\{F\}$ = external forces at each node
      $\{\delta\}$ = displacements at each node

The force matrix is known because it consists of numerical values of loads and reactions computed prior to the start of the finite-element analysis. The displacements are the unknowns, and they are solved for by transposing the stiffness matrix in Eq. (7.26). This computer solution gives the displacements at all the nodes. When it is multiplied by the matrix of coordinate locations of the nodes $[B]$ and the matrix of elastic constants $[D]$, it gives the stress at every nodal point.

$$\{\sigma\} = [D][B]\{\delta\} \tag{7.27}$$

The computer analysis usually is carried through to compute the principal stresses and their directions throughout the part. The printed volume of computer output for complex models is enormous and difficult to handle. A graphics display to display final output data through stress contours, color graphics, etc., is very helpful, and now standard for most FEA software.

7.7.1 Types of Elements

Finite-element analysis was originally developed for two-dimensional (plane stress) situations. A three-dimensional solid causes orders of magnitude increase in the number of simultaneous equations that must be solved, but by using higher-order mesh elements and faster computers these issues are routinely handled in FEA. Broadly, there are two types of elements: *continuum* and *structural*. A continuum element is one whose geometry is completely defined by the coordinates of the nodes. Structural elements are elements that behave in accordance with assumptions as to how structures behave in strength of materials. Figure 7.22 shows a few of the elements available in FEA. Triangles and quadrilaterials, Fig. 7.22 *a* and *b,* are the simplest plane elements, with two degrees of freedom at each node. Adding additional nodes, either at the centroid or along the edges (Fig. 7.22c), provides for curved edges and faces. Three-dimensional models are best constructed from isoparametric elements with curved sides. Figure 7.22d is an isoparametric triangle, (*e*) is a tetrahedron (tet), and (*f*) is a hexahedron (hex). These elements are most useful when it is desirable to approximate curved boundaries with a minimum number of elements. The most common structural elements are the beam element (*g*), the plate element (*h*), and the shell element (*i*).

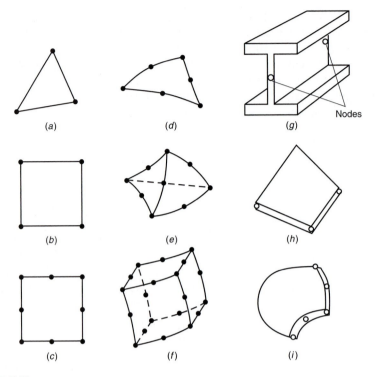

FIGURE 7.22
Some common elements used in finite-element analysis.

The components of the displacement δ in a 2D triangular element, with three nodes and two displacements per node, are

$$u = a_1 + a_2x + a_3y$$

$$v = a_4 + a_5x + a_6y \tag{7.28}$$

Since the strain in the x-direction is $\varepsilon_x = \dfrac{\partial u}{\partial x} = a_2$, we see that for this first-order element the strain is constant throughout the element. For a 2D quadrilateral, with four nodes and two displacements per node, the components of the displacement are

$$u = a_1 + a_2x + a_3y + a_4xy$$

$$v = a_5 + a_6x + a_7y + a_8xy \tag{7.29}$$

Now the strain in the x-direction is $\varepsilon_x = \dfrac{\partial u}{\partial x} = a_2 + a_4y$, which provides for a strain gradient in this higher-order element. Elements with additional nodes, like Fig. 7.22c, lead to still higher-order polynomials to express the strain more accurately within the element and to more accurately represent curved boundaries.

Since FEA creates a model of elements that aims to predict the behavior of a continuum, the selection of type of element and its size is very important. For example, using straight-sided triangular elements to model the region around a hole in a plate would lead to a poor approximation to the circular hole unless the size of the elements is very small. This is an example of a type of modeling error called *discretization.* *Formulation* errors arise from using elements that do not exactly duplicate the way the physical part would behave under the loading. If we think that displacements change linearly over the meshed region, then a linear element would be appropriate. However, if the displacements vary quadratically, then there would be a formulation error, and a higher-order element should be chosen. To create a good element mesh, the stress distribution, not the stresses, should be understood beforehand.

FEA software generally provides the capability for automatically meshing a solid with triangles, quadrilaterals, tetrahedrons, and hexahedrons. The accuracy of the model will be determined by its *convergence error.* This is the percent difference between the results of one run and the next iteration as either the element size or the nature of the element are changed. There are two ways that FEA software approaches this problem. In the *h-element method,* the size of the element (*h* is the element size) is reduced. In the *p-element method* (*p* for polynomial), the software increases the element's polynomial order without changing the original mesh.

Figure 7.23 illustrates these approaches. At the top left, we see the original mesh of first-order elements in an *h*-element approach. After the first run the analysis shows

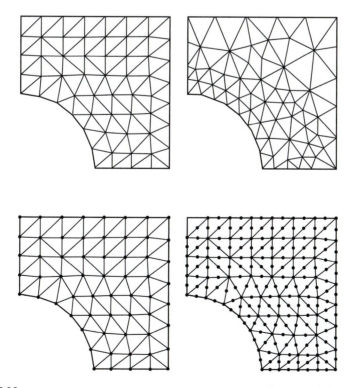

FIGURE 7.23
Top: *h*-element approach to improving accuracy in remeshing. Bottom: *p*-element approach.

the stresses to be highest in the curved region, and the automatic mesher decreases the size of the elements in this region to account for the steep stress gradient (top right). A third iteration, with still smaller elements in the critical region, is run to see whether the stress is converging. The bottom of Fig. 7.23 illustrates the *p*-element approach. The initial mesh (bottom left) is changed by the automatic mesher to second-order elements without changing the mesh size. There is controversy as to which is the best approach. The *p*-element method gives better representation of curved surfaces and is better suited where stress gradients are high. However, it requires much greater computational resource and it cannot be used with nonlinear material models. Moreover, if the element is too large in the *p*-element method, it can have a major impact on accuracy. Many FEA software programs provide for both approaches. As to choice of elements, most 3D work in FEA modeling is done with the parabolic elements shown in Fig. 7.22 *e* and *f*. The 10-node tetrahedron element and the 20-node hexahedron element provide good stress results for reasonable mesh size with a similar number of nodes. However, the 10-node tetrahedron elements provide accuracy comparable to the 20-node hexahedron at less computation time.[1]

7.7.2 Steps in FEA Process

The steps in finite-element modeling are shown in Fig. 7.24. The *preprocessing stage* starts with identification of the geometry and the input of the material behavior through appropriate constitutive equations. Ideally, the geometry can be imported from the CAD model. However, because solid models contain great detail, they often must be simplified by deleting small nonstructural features and taking advantage of symmetry to reduce computation time. The constitutive equation for the material (linear, nonlinear, etc.) that relates displacement to stress must be established.

Dividing the geometry into elements, often called *meshing,* is the most important and most difficult step in FEA. The problem is knowing which elements to use and

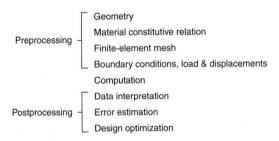

FIGURE 7.24
The steps in constructing a finite-element model.

---

1. A. M. Niazy, *Machine Design,* Nov. 6, 1997, pp. 54–58.

building a mesh that will provide a solution with the needed accuracy and efficiency. Most FEA software provides a means for automatically meshing the geometry. The finite-element mesh is applied in one of two ways: *structured* (mapped) *mesh* or *unstructured* (free) *mesh*. Structured meshes have a clear structure of triangles or quadrilateral elements (for 2D) or tets or hexes (for 3D) that are produced by rule-based mapping techniques. Grid points can be distributed along lines with effective spacing, and well-graded grids can be constructed. This approach is effective when the geometry is relatively simple. With complex geometries a multiblock approach is used, in which the geometry is filled with an assemblage of meshed cubes. This presents the problem of setting up the connections between the blocks. Unstructured meshing does not show structure in the placement of the elements. The advancing front method[1] starts at the surface of the model and builds up a mesh of tet elements as it moves into the solid. With the Delaunay method the boundary elements are first used to form a coarse triangular mesh, and then additional points are inserted into the domain by a refinement strategy. Regardless of the meshing technique, loads and boundary conditions should be defined relative to the geometry, rather than the mesh. This allows for remeshing of the solid without the need to redefine loads and boundary conditions. It is typical for preprocessing to consume about 80 percent of the time spent in FEA.

A finite-element solution could easily contain thousands of field values. Therefore, *postprocessing* operations are needed to interpret the numbers efficiently. Typically the output data are plotted as color contours. Mathematical operations may have to be performed on the data by the FEA software before it is displayed, e.g., determining the Von Mises stress. The FEA program also should calculate error indicators. If the error is too large, the problem will have to be remeshed. Also, the FEA software might be combined with an optimization package and used in iterative calculations to optimize a critical dimension or shape.

7.7.3 Current Practice

The key to practical utilization of finite-element modeling is the model itself. To minimize cost the model should contain the smallest number of elements to produce the needed accuracy. The best procedure is to use an iterative modeling strategy whereby coarse meshes with few elements are increasingly refined in critical areas of the model. Coarse models can be constructed with beam and plate structural models, ignoring details like holes and flanges. Once the overall structural characteristics have been found with the coarse model, a fine-mesh model is used, with many more elements constructed in regions where stress and deflection must be determined more accurately. Accuracy increases rapidly as a function of the number of degrees of freedom (DOF), defined as the product of the number of nodes times the number of unknowns per node. However, cost increases exponentially with DOF.

---

1. K-J. Bathe, Current Directions in Meshing, *Mechanical Engineering,* July 1998, pp. 70 –71.

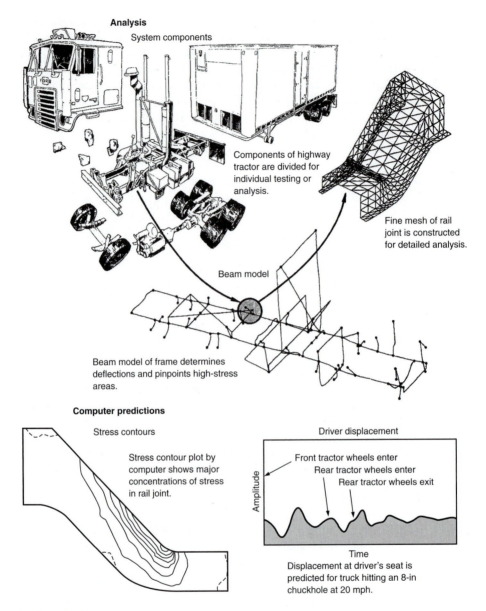

Analysis

System components

Components of highway tractor are divided for individual testing or analysis.

Fine mesh of rail joint is constructed for detailed analysis.

Beam model

Beam model of frame determines deflections and pinpoints high-stress areas.

Computer predictions

Stress contours

Stress contour plot by computer shows major concentrations of stress in rail joint.

Driver displacement

Front tractor wheels enter
Rear tractor wheels enter
Rear tractor wheels exit

Amplitude

Time

Displacement at driver's seat is predicted for truck hitting an 8-in chuckhole at 20 mph.

FIGURE 7.25
Example of application of FEA in design. (*From Machine Design.*)

The application of FEA to the complex problem of a truck frame is illustrated in Fig. 7.25. A "stick figure" or beam model of the frame is constructed first to find the deflections and locate the high-stress areas. Once the critical stresses are found, a fine-mesh model is constructed to get detailed analysis. The results are a computer-generated drawing of the part with the stresses plotted as contours.

FEA is becoming widely available for use on high-end PCs, and it comes as part of the package of most CAD modeling software. FEA software has become much more user-friendly.[1] Now, the emphasis is shifting to more effectively sharing information from FEA between users. Thus, FEA has finally become an everyday engineering tool. With this comes the danger of misuse.[2] It is easy to get wrong answers by blindly applying the analysis without really understanding the fine points of meshing, error analysis, and data interpretation. Engineers should use common sense and sound engineering judgment when evaluating, interpreting, and verifying results.

7.8
COMPUTER VISUALIZATION

The rapidly growing power and speed of the computer has greatly increased its use in carrying out simulations of all kinds. The goal is to substitute analytical simulation for physical tests on actual models and prototypes, with the objective of reducing design time and cost. Two major classes of computer simulations stand out in design. The first deals with rapidly occurring events in solids[3] and fluids.[4] The other concerns the fast-growing field of the simulation of interactive product design. These forms of simulation are so realistic, and bring the human viewer so much into the modeling world, that they are becoming known as *computer visualization.*

7.8.1 Dynamic Analysis

Figure 7.26 shows the computer simulation of the deformation of the bottom dome of an aluminum can after it strikes the ground. For the simulation the computer uses a dynamical FEA that combines the equations that describe the complicated interplay of mass, velocity, acceleration, pressure of the fluid, and mechanical behavior of the can material to calculate and display the distortion of the can at 1-ms intervals after impact. Previous to using the computer simulation a conventional development approach was used in which experimental designs were fabricated and tested. Such a program would take 6 months to a year and cost around $100,000, whereas computer simulation took less than 2 weeks and cost about $2000.[5]

Computer simulation is evolving from a specialized analysis done with supercomputers to a more conventional engineering analysis done on a computer workstation or PC.[6] *Event simulation* software that combines kinematic nonlinear simulation

1. FEA software divides naturally into stand-alone programs and those offered as modules with CAD software. Among the better-known stand-alone packages are Adina (www.adina.com), Algor (www.algor.com), Ansys (www.ansys.com), MSC/PATRAN (www.macsch.com). Typical of FEA software that is offered as part of a CAD modeler is Pro/MECHANICA (www.ptc.com), I-DEAS Simulation and Modeling Set (www.sdrc.com), CATIA Analysis and Simulation Solution (www.catia.ibm.com), UG/Scenario for FEA (www.ug.eds.com).

2. P. Kurowski, Avoiding Pitfalls in FEA, *Machine Design,* Nov. 7, 1994, pp. 78–85.

3. J. E. Crosheck, Mechanism Dynamics and Simulation, "ASM Handbook," vol. 20: "Materials Selection and Design," ASM International, Materials Park, OH, 1997, pp. 166–175.

4. P. J. O'Rourke, D. C. Haworth, and R. Ranganathan, Computational Fluid Dynamics, "ASM Handbook," vol. 20: "Materials Selection and Design," ASM International, Materials Park, OH, 1997, pp. 186–203.

5. R. Pool, *Science,* vol. 244, pp. 1438–1440, June 23, 1989.

6. M. Bussler, How Engineering Will Be Done in the 21st Century, *Machine Design,* Oct. 23, 1997, pp. 58–61.

FIGURE 7.26
Computer simulation of bottom dome of aluminum
can striking the ground. Only one-quarter of the
bottom of the can is included in the simulation.
Note that the bottom dome pops downward after the
can hits the ground. (*Courtesy of Alcoa.*)

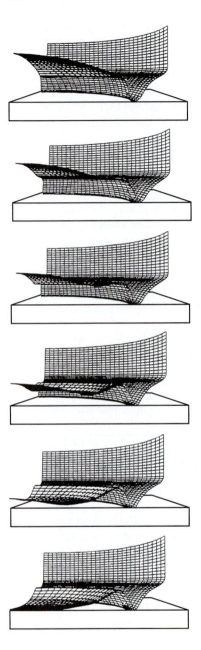

with nonlinear static and dynamic FEA provides a visual three-dimensional image of
the part in the deformed state caused by a dynamical event. For example, the designer
of a coffee pot could simulate dropping it from a certain height to a surface of specified hardness. The computer screen shows the trajectory and bounce of the pot,
whether it shatters or dents, and the stress contours at any time during the impact or
bounce. The computer simulation is in effect a prototype test.

7.8.2 Interactive Product Simulation

Computer simulation is being integrated into a total software package that creates visual databases from CAD, CAE, CAM, and industrial design geometries that is called *interactive product simulation* (IPS). IPS is a real-time visualization and interaction system.[1] IPS software has two main capabilities: large-scale assembly visualization and the ability to simulate the product's functionality. The benefits inherent in IPS are:

- Earlier access to "virtual" prototypes and fast updates as changes are made to the design
- Distribution of design information in easy-to-understand format to all members of the product development team, regardless of geographical location
- Ability for engineers at remote locations to communicate in real time about the design, even extending to being able to pick up objects independently and move around on the computer screen

Some additional capabilities of IPS software are the ability to visualize the assembly and disassembly of the product (for repair and recycling), the real-time functional simulation, including the human engineering of the product with simulated mannequins. The motion simulation checks out part paths and sequences and checks for collision and clearances. The use of IPS models in early focus group sessions with potential users can yield valuable feedback at the conceptual design stage. The ultimate goal is to use the IPS software in a *virtual reality* environment where the user feels that they are actually using the product.

Clearly computer simulation has much to offer engineering design. Developments are moving fast, but a word of caution is in order. Do not become so enamored with the glitz of simulation that you lose sight of practicality. While simulation can greatly reduce the time and cost to arrive at a product, realize that in the end a real working design is required. Until you have considerable experience and confidence with computer simulation, do not totally abandon physical prototypes.

7.9
RAPID PROTOTYPING

Rapid prototyping (RP) is the name given to a variety of processes that produce prototype parts directly from CAD models. Some people refer to it as three-dimensional printing or desktop manufacturing. The objective is to produce "fit and feel" models to verify geometry and alignment of parts, and in some instances the part function, in a matter of hours instead of the days or weeks that might be required for producing them by conventional methods. The prototyped object can also be used as a pattern for producing tooling to manufacture the part by casting.

---

1. S. Ghee, The Virtues of Virtual Products, *Mechanical Engineering,* June 1998, pp. 60–63.

A rapid prototyping process[1] begins by receiving an electronic representation of the solid CAD model of the part. The model is decomposed into parallel cross sections, with each layer between 0.05 and 0.2 mm thick. The prototype part is then made by building layer by layer, with each layer bonded together. Figure 6.2 shows a schematic drawing of an RP machine in which polymer powder is formed into a solid by laser fusing. The movement of the laser is driven by the CAD file from the solid model. The prototype part is built up, layer by layer, as the solid part is fused in the powder mass, and then slowly withdrawn from the bed of powder.

Many variants of this process have been developed to form the solid body, but they all take a three-dimensional digital model, convert it into a series of two-dimensional layers, and then generate a thin solid layer from a "formless" material. The original RP process, stereolithography, uses a liquid vat of photopolymer resin. Where the liquid polymer is exposed to a laser light, it solidifies. Modifications of this use a bed of metal or ceramic powder coated with polymer binder. Other RP processes form the solid by extruding polymer from a tiny orifice or build up the solid by depositing droplets of polymer binder, like from an inkjet printer head, into a bed of powder. Finally, there is a class of RP machines that build the solid by cutting the cross section from sheets of paper, polymer, or composite material, and stacking and laminating them into a solid. This is known as *laminated object manufacturing* (LOM).

The main use of RP has been to produce models that are used by engineers to visualize their designs and make necessary adjustments before committing to production. More recently, interest has focused on using RP in rapid manufacturing of tooling durable enough for limited production and for making fully functional prototypes.[2] Mostly, these processes start with the polymer RP model and use it as the pattern for making a mold for precision casting. In another approach, the CAD model is used to create a prototype by numerically controlled (NC) three-axis milling of aluminum. The analogy with other RP methods is that the CAD model is digitally divided into slablike layers to greatly reduce the machining time, and then the slabs are combined into a working prototype by vacuum brazing.

EXAMPLE OF USE OF CAD AND RP.[3] Itech Sport Products, a Montreal-based producer of athletic equipment, used concurrent engineering tools to take a new hockey helmet from concept to hard tooling in about 6 months. The helmet design started with a digital database of head sizes and shapes created by laser scanning human heads. These data were used to create a 3D CAD model of the interior helmet contour. From this a 3D wireframe model of a two-part shell design was made (Fig. 7.27). Full-scale 2D cross sections of the solid model were made and used to hand-fabricate a plastic foam model for use in marketing studies.

The modifications made for this physical model led to a final 3D CAD model. The reverse engineering capabilities of a coordinate measuring machine were used to update the CAD model from the physical model. Then the digital design data were transferred to

1. C. L. Thomas, Rapid Prototyping, "ASM Handbook," vol. 20: "Materials Selection and Design," ASM International, Materials Park, OH, 1997, pp. 231–239; P. Jacobs, "Stereolithography and Other RP&M Technologies," Society of Manufacturing Engineers, Dearborn, MI, 1996.
2. S. Ashley, *Mechanical Engineering,* July 1998, pp. 64–67.
3. *Plastics Technology,* July 1997, p. 35.

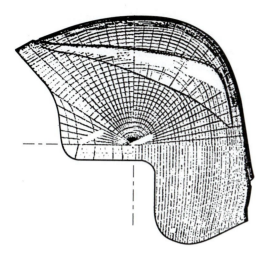

FIGURE 7.27
A CAD wireframe model of the helmet shell. (*Plastics Technology. Used with permission of Bill Communications.*)

the toolmaker's CAM system for NC tool-path generation to machine the molding dies for the nylon shell.

A rapid prototyping model was made by the selective laser sintering (SLS) process to check out the telescoping features of the helmet which allow it to adjust to different size heads. The SLS process was chosen because it can make a model from nylon powder similar to the actual helmet material. Modifications made on the prototype were electronically transmitted to the CAM system used for machining the steel mold for the nylon shell.

This is an example of a project where three separate design centers (CAD, CAM, and SLS) in three different locations electronically communicated critical engineering information with multiple iterations at each site. A true example of concurrent engineering.

7.10
SUMMARY

Models play an important role in engineering design. They can be either descriptive or predictive, static or dynamic, deterministic or probabilistic. We can identify three classes of engineering models:

- *Iconic models:* Models that look like the real thing. Examples are a model airplane for a wind-tunnel test or a shaded solid computer model of a new car.
- *Analog models:* Models that behave like the real thing. Examples are electric circuits that model mechanical systems.
- *Symbolic models:* Models based on abstraction, usually mathematics. The wide use of digital computers in engineering design has greatly enhanced and increased the use of mathematical modeling.

Physical models (iconic models) play an important role in design, from the proof-of-concept model used in conceptual design to the full-scale working prototype that is usually developed by the end of embodiment design.

Simulation involves subjecting models (of all types) to various inputs or experimental conditions to observe how they behave, and thus explore how the real-world design might behave under similar conditions. The testing of physical models in this way is called *simulated service testing*. Exercising a mathematical computer model in this way is *computer simulation.*

The great advances in the speed and power of computers, and in computer software, has greatly enhanced the engineer's ability to model designs. Nowhere in design has progress been more rapid than in our ability to increase productivity in drafting, construct computer solid models, perform static and dynamic analysis in solids and fluids with finite-element analysis, create a rapid prototype from a CAD file, and transfer the final design results to computer-aided manufacturing. While knowledge and practice in each of these areas has advanced to different degrees, there is no question that the practice of design has been greatly transformed.

A few guidelines for mathematical modeling are suggested, lest we place too much confidence in mathematical modeling as opposed to physical modeling.

- Do not build a complicated model when a simple one will do.
- Beware of shaping the problem to fit the modeling technique.
- The model cannot be any better than the information that goes into it. (GI-GO).
- Models should be validated before implementation.
- A model should never be pressed to do that for which it was never intended.
- A model cannot replace decision makers.

BIBLIOGRAPHY

General Engineering Modeling

Starfield, A. M., K. A. Smith, and A. L. Bleloch: "How to Model It," McGraw-Hill, New York, 1990.

Graphical Modeling

Bertoline, G. R., E. N. Wiebe, C. L. Miller, and J. L. Mohler: "Technical Graphics Communication," 2d ed., Irwin, Chicago, 1997.
Rodriquez, W.: "The Modeling of Ideas: Graphics and Visualization Ideas for Engineers," McGraw-Hill, New York, 1992.
Steidel, R. E., and J. M. Henderson: "The Graphic Languages of Engineering," Wiley, New York, 1983.

Mathematical Modeling

Cellier, F. E.: "Continuous System Modeling," Springer-Verlag, Berlin, 1991.
Law, A. M., and W. D. Kelton: "Simulation Modeling and Analysis," McGraw-Hill, New York, 1991.
Smith, D. L.: "Introduction to Dynamic Systems Modeling for Design," Prentice-Hall, Upper Saddle River, NJ, 1994.
Svobodny, T. P.: "Mathematical Modeling for Industry and Engineering," Prentice-Hall, Upper Saddle River, NJ, 1998.

Computer Modeling

Akin, J. E.: "Computer-Assisted Mechanical Design," Prentice-Hall, Englewood Cliffs, NJ, 1990.

Anand, V. B.: "Computer Graphics and Geometric Modeling for Engineers," Wiley, New York, 1993.

Lewis, J. W.: "Modeling Engineering Systems: PC-based Techniques and Design Tools," HighText Publications, Solana Beach, CA, 1994.

Mantyla, M.: "Introduction to Solid Modeling," Computer Science Press, Rockville, MD, 1988.

Mortenson, M. E.: "Geometric Modeling," 2d ed., Wiley, New York, 1997.

PROBLEMS AND EXERCISES

7.1. The following is a generally accepted list of the category of models used in engineering practice:

- Proof-of-concept model
- Scale model
- Experimental model
- Prototype model

Going down in this hierarchy, the model increases in cost, complexity, and completeness. Define each category of model in some detail.

7.2. Classify each of the following models as iconic, analogic, or symbolic. Give your reasons.
   (*a*) The front view and left side view of a new fuel-efficient automobile when the scale is 1 in = 1 ft.
   (*b*) The relation between the flow rate Q through a packed bed in terms of area A, pressure drop Δp, and height of the bed is given by

$$Q = \frac{K_D A \Delta p}{L}$$

   (*c*) A strip chart recording showing the temperature-time profile for a carburizing cycle.
   (*d*) A flowchart showing movement of a cylinder block through a machine shop.
   (*e*) A free body diagram like Fig. 7.1*a*.
   (*f*) A set of N outcomes from a random experiment, represented by $A = a_1, a_2, \ldots, a_n$.

7.3. With the slowdown in the growth in the use of electric power, many utilities have been forced to take older coal-fired power plants off base-load operation (steady state) and use them only to supply daily peak demands. That means the plants will be used in a cycling mode of operation, being fired to full power only at the daily period of peak demand. By using the model for a coal-fired power plant in Fig. 7.5, identify a number of materials-related problems that would be expected to result from a cycling type of operation.

7.4. A novel idea for absorbing energy in an automotive crash is a "mechanical fuse" (see M. C. Shaw, *Mechanical Engineering,* pp. 22–29, April 1972). The idea is to use the energy absorbed by metal cutting to dissipate the kinetic energy of the moving vehicle in a bumper energy absorber. In the concept, a round steel bar is pushed through a circular cutting tool, thereby creating a chip that absorbs the energy of the impact. In effect, the bar is "skinned" by the cutting tool.

Develop a mathematical model for the mechanical fuse. It is appropriate to assume that the metal cutting force is independent of velocity and displacement. Assume that the specific cutting energy (300,000 lbf/in^3 for steel) is not affected by the circular geometry of the tool or by the impact load.

7.5. Use dimensional analysis to determine the dimensionless groups that describe the forced-convection heat transfer of a fluid in a tube. Experience shows that the heat-transfer coefficient h is given by

$$h = f(\bar{V}, \rho, k, \eta, C_{p,} D)$$

where $\bar{V}$ = mean velocity
ρ = mean density
k = thermal conductivity
η = viscosity
C_p = specific heat at constant pressure
D = diameter

7.6. Use dimensional analysis to determine relationships for a cantilever beam loaded with a concentrated force F at its end. The relevant variables are deflection δ, force F, moment of inertia I, modulus of elasticity E, and length L. The units are either Newtons (N) or meters (m). Work the problem in this system of units, not M, L, T.

7.7. Use dimensional analysis to develop Griffith's equation for the fracture strength of a brittle material. This will be an equation similar to Eq. (8.14).

7.8. The following table gives the number of defects that are found to occur in parts of type A and B.

Defects	No. of times occurring in part A	No. of times occurring in part B
0	5	2
1	5	3
2	15	5
3	30	10
4	20	20
5	10	40
6	5	10
7	5	5
8	3	3
9	2	2
	100	100

Find with the help of a Monte Carlo simulation the expected numbers of defects in the final assembly C, which is made up of parts A and B. The first three random numbers generated in the simulation are 14 15; 58 20; 82 14.

7.9. Show the boolean operations with primitive shapes required to produce the shape given below.

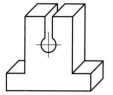

8

MATERIALS SELECTION AND
MATERIALS IN DESIGN

8.1
INTRODUCTION

The selection of the correct materials for a design is a key step in the process because it is the crucial decision that links computer calculations and lines on an engineering drawing with a working design. Materials, and the manufacturing processes which convert the material into a useful part, underpin all of engineering design. The enormity of the decision task in materials selection is given by the fact that there are well over 100,000 engineering materials to choose from. On a more practical level, the typical design engineer should have ready access to information on 50 to 80 materials, depending on the range of applications he or she deals with.

The recognition of the importance of materials selection in design has increased in recent years. The adoption of concurrent engineering methods has brought materials engineers into the design process at an earlier stage, and the importance given to manufacturing in present-day product design has reinforced the fact that materials and manufacturing are closely linked in determining final product performance. Moreover, world pressures of competitiveness have increased the level of automation in manufacturing to the point where materials costs comprise 50 percent or more of the cost for most products. Finally, the extensive activity in materials science worldwide has created a variety of new materials and focused our attention on the competition between six broad classes of materials: metals, polymers, elastomers, ceramics, glasses, and composites. Thus, the range of materials available to the engineer is much broader than ever before. This presents the opportunity for innovation in design by utilizing these materials to provide greater performance at lower cost. To achieve this requires a rational process for materials selection.

8.1.1 Relation of Materials Selection to Design

An incorrectly chosen material can lead not only to failure of the part but also to unnecessary life-cycle cost. Selecting the best material for a part involves more than

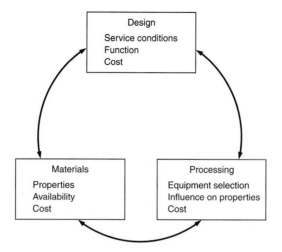

FIGURE 8.1
Interrelations of design, materials, and processing to produce a product.

selecting a material that has the properties to provide the necessary performance in service; it is also intimately connected with the processing of the material into the finished part (Fig. 8.1). A poorly chosen material can add to manufacturing cost and unnecessarily increase the cost of the part. Also, the properties of the material can be enhanced or diminished by processing, and that may affect the service performance of the part. Chapter 9 focuses on the relationship between materials processing and manufacturing and design.

Faced with the enormous combination of materials and processes from which to choose, the materials selection task can only be done by introducing simplification and systemization. As design proceeds from concept design, to configuration and parametric design (embodiment design), to detail design, the material and process selection becomes more detailed.[1] Figure 8.2 compares the design methods and tools used at each design stage with the materials and processes selection. At the concept level of design essentially all materials and processes are considered rather broadly. The materials selection methodology and charts developed by Ashby[2] are highly appropriate at this stage (see Sec. 8.3). The decision is to determine whether each design concept will be made from metal, plastics, ceramic, composite, or wood, and to narrow it to a group of materials. The precision of property data needed is rather low. Note that if an innovative choice of material is to be made it must be done at the conceptual design phase because later in the design process too many decisions have been made to allow for a radical change. At the embodiment level of design the emphasis is on determining the shape and approximate size of a part using engineering analysis. Now the designer will have decided on a class of materials and processes, e.g., a range of aluminum alloys, wrought and cast. The material properties must be known to a greater level of precision. At the parametric design level the decision will have narrowed to a single material and

1. M. F. Ashby, *Met. Mat. Trans.,* vol. 26A, pp. 3057–3064, 1995.
2. M. F. Ashby, "Materials Selection in Mechanical Design," Pergamon Press, 1992.

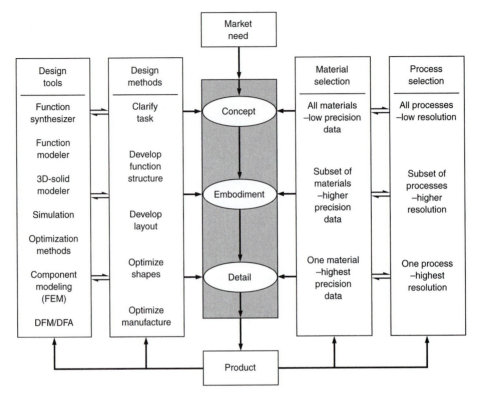

FIGURE 8.2
Schematic of the design process, with design tools shown on the left and materials and process selection on the right. Shows the narrowing-down process used to arrive at the best combination of material and manufacturing process. (*M. F. Ashby, Materials, Bicycles, and Design, Met. Mat. Trans., vol. 26A, p. 3057, 1995.*)

only a few manufacturing processes. Here the emphasis will be on deciding on critical tolerances, optimizing for robust design (see Chap. 12) and selecting the best manufacturing process using quality engineering and cost modeling methodologies. Depending on the criticality of the part, materials properties may need to be known to a high level of precision. At the extreme this requires the development of a detailed database from an extensive materials testing program. Thus, material and process selection is a progressive process of narrowing from a large universe of possibilities to a specific material and process selection (Fig. 8.2).

8.1.2 General Criteria for Selection

Materials are selected on the basis of four general criteria.

- Performance characteristics (properties)
- Processing characteristics

- Environmental profile
- Business considerations

Selection on the basis of performance characteristics is the process of matching values of the properties of the material with the requirements and constraints imposed by the design. Most of this chapter deals with this issue.

Selection on the basis of processing characteristics deals with finding the process that will form the material into the required shape with a minimum of defects at the least cost. Chapter 9 is devoted exclusively to this topic.

Selection on the basis of an environmental profile is concerned with the impact of the material throughout its life cycle on the environment. As discussed in Sec. 6.8, environmental considerations are growing in importance because of the dual pressures of greater consumer awareness and governmental regulation. Design for recycling is discussed in Sec. 8.14.

The chief business consideration that affects materials selection is the cost of the part that is made from the material. This considers both the purchase cost of the material and the cost to process it into a part. A more rational basis for selection is life-cycle cost, which includes the cost of replacing failed parts and the cost of disposing of the material at the end of its useful life. Chapter 14 discusses cost evaluation in detail.

8.2
PERFORMANCE CHARACTERISTICS OF MATERIALS

The performance or functional requirements of a material usually is expressed in terms of physical, mechanical, thermal, electrical, or chemical properties. Material properties are the link between the basic structure and composition of the material and the service performance of the part (Fig. 8.3).

We can divide structural engineering materials into metals, ceramics, and polymers. Further division leads to the categories of elastomers, glasses, and composites. Finally, there is the technology driving class of electronic, magnetic, and semiconductor materials. The chief characteristics of metals, ceramics, and polymers are given in Table 8.1.

Not too long ago metals dominated mechanical design so that it was possible to ignore other classes of materials. Today the range of materials available to the engineer

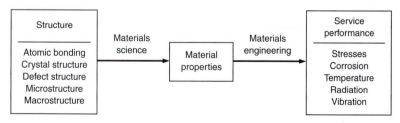

FIGURE 8.3
Material properties, the link between structure and performance.

TABLE 8.1
Property characteristics of material classes

Metals	Ceramics	Polymers
Strong	Strong	Weak
Stiff	Stiff	Compliant
Tough	Brittle	Durable
Electrically conducting	Electrically insulating	Electrically insulating
High thermal conductivity	Low thermal conductivity	Temperature sensitive

is much larger and growing rapidly. It is important to be cognizant of the opportunities for innovation and product improvement that new materials provide.

The ultimate goal of materials science is to predict how to improve the properties of engineering materials by understanding how to control the various aspects of structure. Structure can vary from atomic dimensions to the dimensions of a macroscopic crack in a fillet weld. Figure 8.4 relates various dimensions of structure with typical structural elements. The chief methods of altering structure[1] are through composition control (alloying), heat treatment, and deformation processing. A general background in the way structure controls the properties of solid materials usually is obtained from a course in materials science or fundamentals of engineering materials.[2]

The first task in materials selection is to determine which material properties are relevant to the situation. We look for material properties that are easy and cheap to measure, are reproducible, and are associated with a material behavior that is well defined. However, for reasons of technological convenience we often determine something other than the most fundamental material property. Thus, the elastic limit measures the first significant deviation from elastic behavior; but it is tedious to measure, so we substitute the easier and more reproducible 0.2 percent offset yield strength. That, however, requires a carefully machined test specimen, so the yield stress may be approximated by the exceedingly cheap and rapid hardness test.

Figure 8.5 shows the relations between some common failure modes and the mechanical properties most closely related to the failures. Note that in most modes of failure, two or more mechanical properties interact to control the material behavior. In addition, the service conditions met by materials in general are more complex than the test conditions used to measure material properties. Thus, the stress level is not likely to be a constant value; instead, it is apt to fluctuate with time in a nonsinusoidal way. Or the service condition consists of a complex superposition of environments, such as a fluctuating stress (fatigue) at high temperature (creep) in a highly oxidizing atmosphere (corrosion). Specialized service simulation tests are developed to "screen materials" for complex service conditions. Finally, the best candidate materials must

---

1. R. W. Cahn, *J. Metals,* pp. 28–37, February 1973.
2. L. E. Van Vlack, "Elements of Materials Science and Engineering," 6th ed., Addison-Wesley, Reading, MA, 1989; D. R. Askeland, "The Science and Engineering of Materials," 3d ed., Wadsworth, Belmont, CA, 1994; W. D. Callister, "Materials Science and Engineering," 3d ed., Wiley, New York, 1993; J. F. Shackelford, "Introduction to Materials Science for Engineers," 3d ed., Macmillan, New York, 1992; W. F. Smith, "Principles and Materials Science and Engineering," 3d ed., McGraw-Hill, New York, 1997.

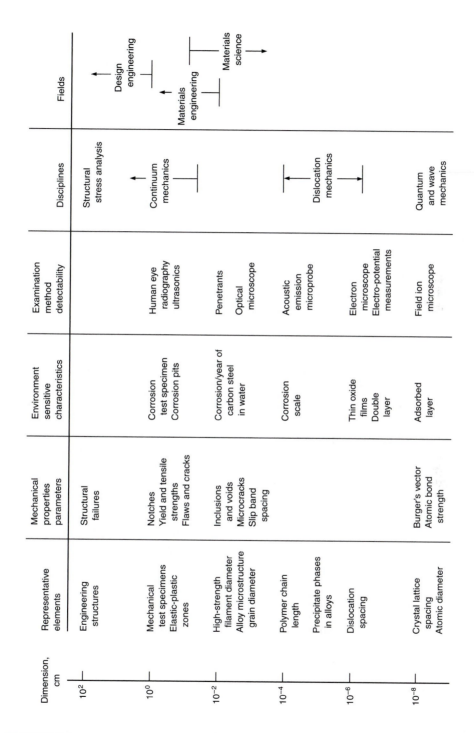

Dimension, cm	Representative elements	Mechanical properties parameters	Environment sensitive characteristics	Examination method detectability	Disciplines	Fields
10^2	Engineering structures	Structural failures			Structural stress analysis	Design engineering
10^0	Mechanical test specimens Elastic-plastic zones	Notches Yield and tensile strengths Flaws and cracks	Corrosion test specimen Corrosion pits	Human eye radiography ultrasonics	Continuum mechanics	Materials engineering
10^{-2}	High-strength filament diameter Alloy microstructure grain diameter	Inclusions and voids Microcracks Slip band spacing	Corrosion/year of carbon steel in water	Penetrants Optical microscope		Materials science
10^{-4}	Polymer chain length Precipitate phases in alloys		Corrosion scale	Acoustic emission microprobe	Dislocation mechanics	
10^{-6}	Dislocation spacing		Thin oxide films Double layer	Electron microscope Electro-potential measurements		
10^{-8}	Crystal lattice spacing Atomic diameter	Burger's vector Atomic bond strength	Adsorbed layer	Field ion microscope	Quantum and wave mechanics	

FIGURE 8.4
Dimensions of material structure.

Figure 8.5 — Relations between failure modes and mechanical properties.

Material property (● = shaded block indicating that a particular material property is influential in controlling a particular failure mode)

Failure mode	Ultimate tensile strength	Yield strength	Compressive yield strength	Shear yield strength	Fatigue properties	Ductility	Impact energy	Transition temperature	Modulus of elasticity	Creep rate	K_{Ic}	K_{Iscc}	Electrochemical potential	Hardness	Coefficient of expansion
Gross yielding		●		●											
Buckling			●						●						
Creep										●					
Brittle fracture							●	●			●				
Fatigue, low cycle					●	●									
Fatigue, high cycle	●				●										
Contact fatigue			●												
Fretting													●		
Corrosion	●												●		
Stress-corrosion cracking												●	●		
Galvanic corrosion													●		
Hydrogen embrittlement	●														
Wear	●													●	
Thermal fatigue										●					●
Corrosion fatigue					●								●		

Shaded block at intersection of material property and failure mode indicates that a particular material property is influential in controlling a particular failure mode.

FIGURE 8.5

Relations between failure modes and mechanical properties. (*C. O. Smith and B. E. Boardman: "Metals Handbook," 9th ed., vol. I, p. 828, American Society for Metals, Metals Park, Ohio, copyright American Society for Metals, 1980.*)

be evaluated in prototype tests or field trials to evaluate their performance under actual service conditions.

The material properties usually are formalized through specifications. There are two types of specification: performance specifications and product specifications. *Performance specifications* delineate the basic functional requirements of the product and set out the basic parameters from which the design can be developed. They are based on the need the product is intended to satisfy and an evaluation of the likely risk and consequences of failure. The *product design specification* defines conditions under which the components of the designs are purchased or manufactured. Material properties are an important part of product specifications.

Table 8.2 provides a fairly complete listing of material performance characteristics. It can serve as a checklist in selecting materials to assure that no important properties are overlooked. The crucial subject of producibility or fabrication properties is considered only briefly in Table 8.2. This subject is considered in much greater detail in Chap. 9.

TABLE 8.2
Material performance characteristics

Physical properties:
Crystal structure
Density
Melting point
Vapor pressure
Viscosity
Porosity
Permeability
Reflectivity
Transparency
Optical properties
Dimensional stability

Electrical properties:
Conductivity
Dielectric constant
Coercive force
Hysteresis

Nuclear properties:
Half-life
Cross section
Stability

Mechanical properties:
Hardness
Modulus of elasticity
  Tension
  Compression
Poisson's ratio
Stress-strain curve
Yield strength
  Tension
  Compression
  Shear
Ultimate strength
  Tension
  Shear
  Bearing
Fatigue properties
  Smooth
  Notched
  Corrosion fatigue
  Rolling contact
  Fretting
Charpy transition temp.
Fracture toughness (K_{Ic})
High-temperature
  Creep
  Stress rupture
Damping properties
Wear properties
  Galling
  Abrasion
  Erosion
  Cavitation
  Spalling
  Ballistic impact

Thermal properties:
Conductivity
Specific heat
Coef. of expansion
Emissivity
Absorptivity
Ablation rate
Fire resistance

Chemical properties:
Position in
  electromotive series
Corrosion and degradation
  Atmospheric
  Salt water
  Acids
  Hot gases
  Ultraviolet
Oxidation
Thermal stability
Biological stability
Stress corrosion
Hydrogen embrittlement
Hydraulic permeability

Fabrication properties:
Castability
Heat treatability
Hardenability
Formability
Machinability
Weldability

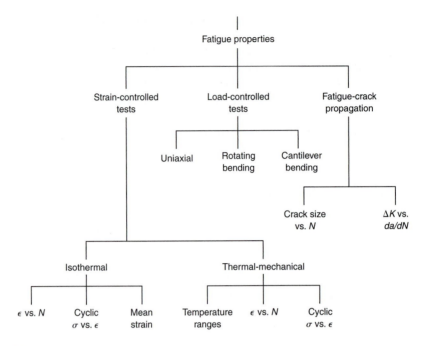

FIGURE 8.6
Generic tree for fatigue properties.

The subject of material properties can quickly become rather complex. A consideration of any one of the properties listed in Table 8.2 can be expanded to include the type of test environment, stress state, or even specimen configuration. Figure 8.6 illustrates the generic tree that is developed by expanding the category of fatigue properties.

There is a whole set of important factors in materials selection that are not covered by Table 8.2. The economic and practical considerations, which are discussed in detail in Sec. 8.5, are as follows:

1. Availability.
   a. Are there multiple sources of supply?
   b. What is likelihood of availability in future?
   c. Is the material available in the forms needed (tubes, wide sheet, etc.)?
2. Size limitations and tolerances on available material.
3. Variability in properties.
4. Environmental impact, including ability to recycle the material.
5. Cost. Materials selection comes down to buying properties at best available price.

8.3
THE MATERIALS SELECTION PROCESS

The selection of materials on a purely rational basis is far from easy. The problem is not only often made difficult by insufficient or inaccurate property data but is typi-

Consider the question of materials selection for an automotive exhaust system. The **product design specification** states that it must provide the following functions:

- Conduct engine exhaust gases away from the engine
- Prevent noxious fumes from entering the car
- Cool the exhaust gases
- Reduce the engine noise
- Reduce the exposure of automobile body parts to exhaust gases
- Affect the engine performance as little as possible
- Help control unwanted exhaust emissions
- Have an acceptably long service life
- Have a reasonable cost, both as original equipment and as a replacement part

The basic **system configuration** is a series of tubes that collect the gases at the engine and convey them to the rear of the automobile. The size of the tubes is determined by the volume of gases to be carried away and the extent to which the exhaust system can be permitted to impede the flow of gases from the engine (back pressure). In addition, a muffler is required for noise reduction and a catalytic converter to change polluting gases to less harmful emissions.

Material requirements for an automotive exhaust system.

Mechanical property requirements not overly severe.

- Suitable rigidity to prevent excessive vibration
- Moderate fatigue resistance
- Good creep resistance in hot parts

Limiting property: corrosion resistance, especially in the cold end where gases condense to form corrosive liquids.

Properties of unique interest: the requirements are so special that only a few materials meet them regardless of cost.

- Pt-base catalysts in catalytic converter
- Special ceramic carrier that supports the catalyst

Previous materials used: Low-carbon steel with corrosion-resistant coatings. Material is relatively inexpensive, readily formed and welded. Life of tailpipe and muffler is limited.

Newer materials used: With greater emphasis on automotive quality many producers have moved to specially developed stainless steels with improved corrosion and creep properties. Ferritic 11% Cr alloys are used in the cold end components and 17 to 20% Cr ferritic alloys and austenitic Cr-Ni alloys in the hot end of the system.

cally one of decision making in the face of multiple constraints without a clear-cut objective function. A problem of materials selection usually involves one of two different situations.

1. Selection of the materials for a new product or design.
2. Reevaluation of an existing product or design to reduce cost, increase reliability, improve performance, etc.

It generally is not possible to realize the full potential of a new material unless the product is redesigned to exploit both the properties and the manufacturing characteristics of the material. In other words, *a simple substitution of a new material without changing the design rarely provides optimum utilization of the material.* Most often the essence of the materials selection process is *not* that one material competes against another for adoption; rather, it is that the processes associated with the production or fabrication of one material compete with the processes associated with the other. For example, the pressure die casting of a zinc-based alloy may compete with the injection molding of a polymer. Or a steel forging may be replaced by sheet metal because of improvements in welding sheet-metal components into an engineering part.

Materials selection for a new product or new design. In this situation the steps that must be followed are:

1. Define the functions that the design must perform and translate these into required materials properties such as stiffness, strength, corrosion resistance, etc., and such business factors as the cost and availability of the material.
2. Define the manufacturing parameters such as the number of parts to be produced, the size and complexity of the part, its required tolerance and surface finish, general quality level, and overall fabricability of the material.
3. Compare the needed properties and parameters with a large materials property database (most likely computerized) to select a few materials that look promising for the application. In this initial screening process it is helpful to establish several screening properties. A *screening property* is any material property for which an absolute lower (or upper) limit can be established. No trade-off beyond this limit is allowable. It is a *go-no-go* situation. The idea of the screening phase in materials selection is to ask the question: "Should this material be evaluated further for this application?"
4. Investigate the candidate materials in more detail, particularly in terms of trade-offs in product performance, cost, fabricability, and availability in the grades and sizes needed for the application. Material property tests and computer simulation often is done in this step. The objective is to narrow the material selection down to a single material and to have a small number of possible manufacturing processes.
5. Develop design data and/or a design specification. *Design data properties* are the properties of the selected material in its *fabricated state* that must be known with sufficient confidence to permit the part to function with a specified level of reliability. Step 4 results in the selection of a single material for the design and a suggested process for manufacturing the part. In most cases this results in establishing the minimum properties through defining the material with a generic material standard such as ASTM, SAE, ANSI, or a MIL spec. The extent to which step 5 is pursued depends on the nature of the application. In many product areas, service conditions are not severe and commercial specifications such as ASTM may be used without adopting an extensive testing program. In other applications, such as the aerospace and nuclear areas, it may be necessary to conduct an extensive testing program to develop design data which are statistically reliable.

Materials substitution in an existing design. In this situation the following steps pertain.

1. Characterize the currently used material in terms of performance, manufacturing requirements, and cost.
2. Determine which characteristics must be improved for enhanced product function. Often failure analysis reports play a critical role in this step (see Sec. 11.7).
3. Search for alternative materials and/or manufacturing routes. Use the idea of screening properties to good advantage.
4. Compile a short list of materials and processing routes and use these to estimate the costs of manufactured parts. Note that a method of engineering analysis called *value engineering* has proved useful for this purpose (see Sec. 8.11). Value engineering is a problem-solving methodology that focuses on identifying the key function(s) of a design so that unnecessary costs can be removed without compromising the quality of the design.
5. Evaluate the results of step 4 and make a recommendation for a replacement material. Define the critical properties with specifications or testing, as in step 5 of the previous section.

8.3.1 Design Process and Materials Selection

There are two approaches[1] to determining the material-process combination for a part. First, in the *material first approach* the designer begins by selecting a material class and narrowing it down as described above. Then manufacturing processes consistent with the selected material are considered and evaluated. Chief among the factors to consider are production volume and information about the size, shape, and complexity of the part. Second, with the *process first approach* the designer begins by selecting the manufacturing process, guided by the same factors. Then materials consistent with the selected process are considered and evaluated, guided by the performance requirements of the part. Both approaches end up at the same decision point. Most design engineers and materials engineers instinctively use the materials first approach, since it is the method taught in strength of materials and machine design courses. Manufacturing engineers and those heavily involved with process engineering gravitate toward the other approach.

While materials selection issues arise at every stage in the design process, the opportunity for greatest innovation in materials selection occurs at the conceptual design stage. At this stage, where all options are open, the designer requires approximate data on the broadest possible range of materials. A metal may be the best material for one design concept while a polymer is best for a different concept, even though the two concepts provide the same function.

---

1. J. R. Dixon and C. Poli, "Engineering Design and Design for Manufacturing," Field Stone Publishers, Conway, MA, 1995.

8.3.2 Ashby Charts

Ashby[1] has created useful materials selection charts that are very helpful in comparing a large number of materials at the concept design phase. Two of many available charts are shown in Fig. 8.7. Figure 8.7a plots the elastic modulus of polymers, metals, ceramics, and composites against density, while Fig. 8.7b plots strength against density. Note how the classes of materials group into common regions.

A common design criterion is to minimize cost or weight. Lines of constant slope are drawn on the diagram. Depending upon the geometry and the type of loading, lines of different slope apply (see Table 8.4 and Sec. 8.8). For simple axial loading the

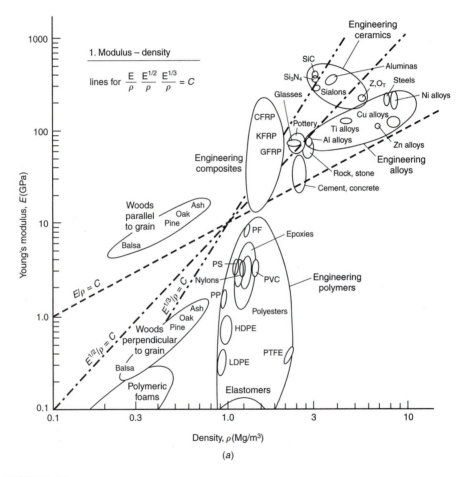

FIGURE 8.7a
Ashby materials selection chart: Young's modulus vs. density. (*M. F. Ashby, Mater. Sci. and Tech. vol. 5, p. 521, 1989.*)

1. M. F. Ashby, op. cit.

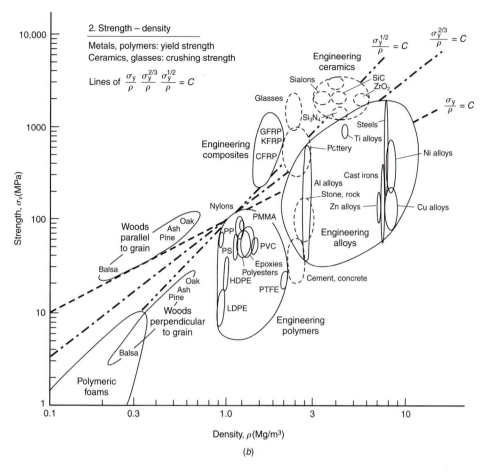

FIGURE 8.7b
Ashby materials selection chart: strength vs. density. (*M. F. Ashby, Mater Sci. and Eng., vol. 5, p. 522, 1989.*)

relationship is $E/\rho = C$ or $\sigma/\rho = C$. For buckling of a slender column, $E^{1/2}/\rho = C$ applies, and for the bending of a plate it is $E^{1/3}/\rho = C$. For example, if we are trying to determine which materials would be suitable for a column in compression, we would lay a straightedge with the slope $E^{1/2}/\rho = C$. Start at the lower right-hand corner and move it toward the upper left-hand corner. All of the materials which lie on the line will perform equally well when loaded as a column in compression. Those materials which lie above the line are better, and those the farthest from the line are the best. A constraint of the design might be that $E \geq 12$ GPa and $\rho \leq 2$ Mg/m^3. We would draw horizontal and vertical lines to conform with these constraints, and the search would be confined to the upper left-hand corner of the chart. The Ashby chart would tell us that the only materials that could meet this condition would be a fiber-reinforced composite material, such as a graphite-fiber-reinforced polymer (GFRP) or a Kevlar™ fiber-reinforced polymer (KFRP).

8.3.3 Materials Selection in Embodiment Design

Detailed materials selection is typically carried out in the embodiment design phase using the process shown in Fig. 8.8. At the beginning there are parallel materials selection and component design paths. A small set of tentative materials is chosen based on the Ashby charts and sources of data described in Sec. 8.4. At the same time a tentative component design is developed which fulfills the functional requirements, and using the material properties an approximate stress analysis is carried out to calculate stresses and stress concentrations. The two paths merge in an examination of whether the best material, fabricated into the component by its expected manufacturing process, can bear the loads, moments, and torques that the component is expected to withstand. Often the information is inadequate to make this decision with confidence and finite-element modeling or some other computer-aided design tool is used to gain the needed knowledge. Alternatively, a prototype component is made and subjected to a simulated-service test. Sometimes it becomes clear that the initial selections of materials are just inadequate and the process iterates back to the top and the selection process starts over.

When the material-process selection is deemed adequate for the requirements, the process passes to a detailed specification of the material and the design. This is the *parametric design* step discussed in Chap. 6. In this design step an attempt should be made to optimize the critical dimensions and tolerances to achieve a component that is robust to its service environment, using an approach such as the Taguchi robust design methodology (see Chap. 12). The next step is to finalize the choice of the production method. This is based chiefly on a detailed calculation of the cost to manufacture the component (Chap. 14). The material cost and the inherent workability and formability of the material, to reduce scrapped parts, are a major part of this determination. Another important consideration is the quality of the manufactured component, again strongly influenced by the choice of material. Yet other considerations are the heat treatment, surface finishing, and joining operations that will be required.

Once the component goes into production, the early runs will be used to fine-tune the manufacturing process and to gauge the market receptivity to the product. If this is satisfactory, then full-scale production is established. However, it is important to follow the service experience of the product to determine any weak or potential points of failure, to identify parts of the design that could be improved by a redesign, or to determine ways to reduce cost using value analysis (Sec. 8.11).

8.4
SOURCES OF INFORMATION ON MATERIALS PROPERTIES

Most practicing engineers develop a file of trade literature, technical articles, and company reports. Material property data comprise an important part of this personal data system. In addition, many large corporations and government agencies develop their own compendiums of data on materials properties.

The purpose of this section is to provide a guide to material property data that are readily available in the published technical literature. There are several factors to have clearly in mind when using property data in handbooks and other open-literature sources. Usually a single value is given for a property, and it must be assumed that the

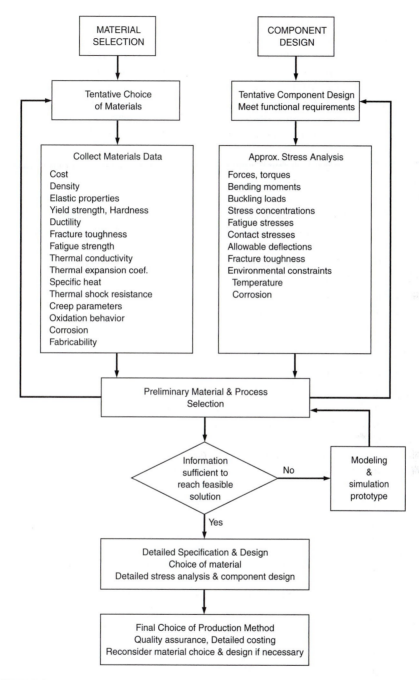

FIGURE 8.8
Steps in materials selection at the embodiment (configuration) design phase.

value is typical. When scatter or variability of results is considerable, the fact may be indicated in a table of property values by a range of values (i.e., the largest and smallest values) or be shown graphically by scatter bands. However, it is rare to find a property data presented in a proper statistical manner by a mean value and the standard deviation (see Chap. 10). Obviously, for critical applications in which reliability is of great importance, it is necessary to determine the frequency distribution of both the material property and the parameter that describes the service behavior. Figure 8.9 shows that when the two frequency distributions overlap, there will be a statistically predictable number of failures.

It is important to realize that a new material cannot be used in a design unless the engineer has access to reliable material properties and cost data. This is a major reason why the tried and true materials are used repeatedly for designs even though better designs could be achieved with advanced materials. At the start of the design process low-precision but all-inclusive data is needed. At the end of the design process, data is needed for only a single material, but it must be accurate and very detailed. Listed below is information on some widely available sources of information on materials properties.[1] A significant start has been made in developing computerized materials databases, and in converting handbook data to CD-ROM for easier searching and retrieval.

8.4.1 Conceptual Design

Materials Selector, special annual issue of *Materials Engineering* magazine giving tabular data
      for a broad range of metals, ceramics, plastics, and composites.
"The Materials Selector," 2d ed., N. A. Waterman and M. F. Ashby, eds., Chapman and Hall,
      New York, 1966. Comprehensive 3-volume, 2500-page compendium of engineered materials and related methods of manufacture.
"ASM Metals Handbook Desk Edition," 2d ed., ASM International, Materials Park, OH 1998.
      A compact compilation of metals, alloys, and processes.

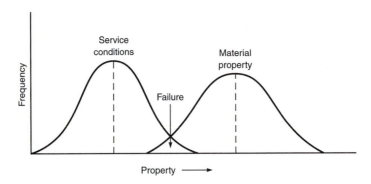

FIGURE 8.9
Overlapping distribution of material property and service requirement.

---

1. For extensive information on sources of materials property data see J. H. Westbrook, in "ASM Handbook," vol. 20: "Materials Selection and Design," pp. 491–506, ASM International, Materials Park, OH, 1997.

"ASM Engineered Materials Reference Book," 2d ed., ASM International, Materials Park, OH, 1994. A compact compilation of data for ceramics and polymers.

"ASM Metals Reference Book," 3d ed., ASM International, Materials Park, OH, 1993. Gives mechanical and physical properties of standard industrial alloys.

Cambridge Materials Selector, V2.0 for Windows, Granta Design Ltd., Cambridge, UK. This software implements the Ashby materials selection scheme for PC-based computers. http:/www.granta.com.uk.

8.4.2 Embodiment Design

At this phase of design decisions are being made on the layout and size of parts and components. The design calculations require materials properties for a narrower class of materials but specific to a particular heat treatment or manufacturing process. These data are typically found in handbooks and computer databases, and in data sheets published by trade associations of materials producers.

Metals

"ASM Handbook," vol. 1, "Properties and Selection: Irons, Steels, and High-Performance Alloys," ASM International, 1990.

"ASM Handbook," vol. 2, "Properties and Selection: Nonferrous Alloys and Special-Purpose Alloys," ASM International, 1991.

"SAE Handbook," Part 1, "Materials, Parts, and Components," Society of Automotive Engineers, Warrendale, PA, published annually.

"Metallic Materials and Elements for Aerospace Vehicle Structures," Military Standardization Handbook, MIL-HDBK-5D, U.S. Department of Defense. Updated every 6 months and reissued every 3 years.

"Woldman's Engineering Alloys," 8th ed., L. Frick (ed.), ASM International, 1994. References approx. 55,000 alloys. Use this to track down information on an alloy if you know only the trade name. Available in electronic form.

Alloy Digest on CD-ROM, ASM International, 1995. Comprehensive mechanical, physical, and processing data on over 3500 alloys.

Mat. DB, ASM International, A materials database, in PC format, of properties and processing for metals, and some polymers. Use with MAPP 1.1 to provide a Windows/Macintosh interface. http://www.asm-intl.org.

Ceramics

"ASM Engineered Materials Handbook," vol. 4, "Ceramics and Glasses, ASM International," 1991.

R. Morrell, "Handbook of Properties of Technical and Engineering Ceramics," HMSO, London, Part 1, 1985, Part 2, 1987.

M. Schwartz, "Handbook of Structural Ceramics," McGraw-Hill, 1991.

Polymers

"ASM Engineered Materials Handbook," vol. 2, Engineered Plastics, ASM International, 1988.

"Modern Plastics Encyclopedia," published yearly by *Modern Plastics* magazine. McGraw-Hill, New York.

"Polymers and Composite Materials for Aerospace Vehicle Structures," MIL-HDBK-17, U.S. Department of Defense.

"Handbook of Plastics, Elastomers, and Composites," 3d ed., C. A. Harper (ed.), McGraw-Hill, 1997.

"PDL Electronic Handbooks," Plastics Design Library, Norwich, NY: http://www.norwich.net/~wai001.

Plascams, Rapra Technology, Ltd., Shrewsbury, UK. A searchable computer database for over 350 thermoplastic, thermosets, composites, and elastomers.

Composites

"ASM Engineered Materials Handbook," vol. 1, ASM International, 1987.

Electronic materials

"Handbook of Materials and Processes for Electronics," C. A. Harper (ed.), McGraw-Hill, New York, 1970.

"Electronic Materials Handbook," vol. 1; "Packaging," ASM International, 1989.

Thermal properties

"Thermophysical Properties of High Temperature Solid Materials," vols. 1 to 9, Y. S. Touloukian (ed.), Macmillan, New York, 1967.

Chemical properties

"NACE Corrosion Engineer's Reference Book," 2d ed.," R. S. Tressler (ed.), National Association of Corrosion Engineers, Houston, TX, 1991.

"ASM Handbook," vol. 13, "Corrosion," ASM International, 1987.

P. A. Schwitzer, "Corrosion Resistance Tables," 4th ed., Marcel Dekker, New York, 1995.

Several useful references are given below to aid in the identification of alloys and producers.

"Guide to Engineering Materials Producers," J. C. Bittence (ed.), ASM International, 1993. A comprehensive source of manufacturers of metallic and nonmetallic engineered materials.

"Metals and Alloys in the Unified Numbering System," 7th ed., ASM International, 1996. Provides a cross reference to many different alloy numbering systems, including ASME, ASTM, AISI, SAE, federal, and military.

8.4.3 Detail Design

At the detail design stage very precise data is required. These data are best found in data sheets issued by materials suppliers or by generation internally. This is particularly true for polymers, whose properties vary considerably depending upon how they are manufactured. For all materials, with critical parts, tests on the actual material from which they will be made are a requirement.

There is a wide range of material information that may be needed at the detail design stage. This goes beyond just material properties to include information on manufacturability, including final surface finish and tolerances, cost, the experience in using the material in other applications (failure reports), availability in the sizes and forms (sheet, plate, wire, etc.) needed, and issues of repeatability of properties and quality assurance. Two often overlooked factors are whether the manufacturing process will produce different properties in different directions in the part, and whether the part will contain a detrimental state of residual stress after manufacture. These and other issues that influence the cost of the manufactured part are considered in detail in Chap. 9.

8.5
ECONOMICS OF MATERIALS

Ultimately the decision on a particular design will come down to a trade-off between performance and cost. There is a continuous spectrum of applications, varying from those where performance is paramount (aerospace and defense are good examples) to those where cost clearly predominates (household appliances and low-end consumer electronics are typical examples). In the latter type of application the manufacturer does not have to provide the highest level of performance that is technically feasible. Rather the manufacturer must provide a value to cost ratio that is no worse, and preferably better, than the competition. By *value* we mean the extent to which the performance criteria appropriate to the application are satisfied. *Cost* is what must be paid to achieve that level of value.

Because cost is such an overpowering consideration in many materials selection situations, we need to give this factor additional attention. The basic cost of a material depends upon (1) scarcity, as determined by either the concentration of the metal in the ore or the cost of the feedstock, (2) the cost and amount of energy required to process the material, and (3) the basic supply and demand for the material. In general, large-volume-usage materials like stone and cement have very low prices, while scarce materials, like industrial diamonds, have high prices.

As is true of any commodity, as more work is invested in the processing of a material, the cost increases (value is added). Table 8.3 shows how the relative price of various steel products increases with processing.

TABLE 8.3
Relative price of various steel products

Product	Price relative to pig iron
Pig iron	1.0
Billets, blooms, and slabs	1.4
Hot-rolled carbon steel bars	2.3
Cold-finished carbon steel bars	4.0
Hot-rolled carbon steel plate	3.2
Hot-rolled sheet	2.6
Cold-rolled sheet	3.3
Galvanized sheet	3.7

Increases in properties, like yield strength, beyond those of the basic material are produced by changes in structure brought about by compositional changes and additional processing steps. For example, changes in the strength of steel are promoted by expensive alloy additions such as nickel or by heat treatment such as quenching and tempering. However, the cost of an alloy may not simply be the weighted average of the cost of the constituent elements that make up the alloy. Often, a high percentage of the cost of an alloy is the need to control one or more impurities to very low levels. That could mean extra refining steps or the use of expensive high-purity raw materials.

Because most engineering materials are produced from depletable mineral resources, there is a continuous upward trend of cost with time. Although some materials, e.g., copper, have fluctuated in price considerably because of temporary oversupply, the costs of materials will rise at a rate greater than the costs of goods and services in general. Therefore, wise use of materials is increasingly important. The general price level for metals and ceramics can be obtained from the following sources.

Metals: *American Metal Market/Metalworking News*
Ceramics: *Ceramic Industry Magazine*
Polymers: *Plastics Technology*

However, the pricing structure of many engineering materials is quite complex, and true prices can be obtained only through quotations from vendors. The reference sources listed above give only the nominal, baseline price. The actual price depends upon a variety of price extras in addition to the base price (very much as when a new car is purchased). The actual situation varies from material to material, but the situation for steel products is a good illustration.[1]

Price extras are assessed for the following situations.

Metallurgical requirements:
   Grade extra. Each AISI grade has an extra over the cost of the generic type of steel, i.e., hot-rolled bar, hot-rolled plate, etc.
   Chemistry extra. Nonstandard chemical composition for the grade of steel.
   Quality extra, e.g., vacuum melting or degassing.
   Inspection and testing. A charge is made for anything other than routine tensile tests and chemical analysis.
   Special specifications.

Dimensions:
   Size and form. Special shapes, or sizes.
   Length. Precise requirements on length are costly.
   Cutting. Sheared edge, machined edge, flame-cut edge, etc.
   Tolerances. Tighter tolerances on OD or thickness cost extra.

Processing:
   Thermal treatment, e.g., normalizing or spheroidizing.
   Surface treatment, e.g., pickling or oil dip.

---

1. R. F. Kern and M. E. Suess, "Steel Selection," Wiley, New York, 1979.

Quantity:

Purchases in less than heat lots (50 to 300 tons) are an extra.

Pack, mark, load:

Packing, wrapping, boxing, etc.

Marking. Other than stamped numbers may be an extra.

Loading. Special blocking for freight cars, etc.

From this detailed listing of price extras we can see how inadvertent decisions of the designer can significantly influence cost. Standard chemical compositions should be used whenever possible, and the number of alloy grades should be standardized[1] to reduce the cost of stocking many grades of steel. Manufacturers whose production rates do not justify purchasing in heat lots should try to limit their material use to grades that are stocked by local steel service centers. Special section sizes and tolerances should be avoided unless a detailed economic analysis shows that the cost extras are really justified.

8.6
METHODS OF MATERIALS SELECTION

There is no method or small number of methods of materials selection that has evolved to a position of prominence. Partly, this is due to the complexity of the comparisons and trade-offs that must be made. Often the properties we are comparing cannot be placed on comparable terms so a clear decision can be made. Partly it is due to the fact that little research and scholarly effort have been devoted to the problem.

A variety of approaches to materials selection are followed by designers and materials engineers. A common path is to examine critically the service of designs in environments similar to the one of the new design. Information on service failures can be very helpful. The results of accelerated laboratory screening tests or short-time experience with a pilot plant can provide valuable input. Often a minimum innovation path is followed and the material is selected on the basis of what worked before or what is used in the competitor's product.

Some of the more common and more analytical methods of materials selection are:

1. Selection with computer-aided databases (8.7)
2. Performance indices (8.8)
3. Decision matrices (8.9)
   Pugh selection method
   Weighted property index
4. Selection with expert systems (8.10)
5. Value analysis (8.12)
6. Failure analysis (11.8)
7. Benefit-cost analysis (13.11)

---

1. "Metals Handbook," 8th ed., vol. 1, pp. 281–289, American Society for Metals, Metals Park, OH, 1961.

Because cost is so important in selecting materials, it is logical to consider cost at the outset of the materials selection process. Considerable effort is being given to developing computer-based methods of estimating manufacturing cost that can be employed in the early stages of design. If this is not available it is usually possible to set a target cost and eliminate the materials that are too expensive. Since the final choice is a trade-off between cost and performance (properties), it is logical to attempt to express that relation as carefully as possible. Figure 8.10 shows the costs of substituting lightweight materials to achieve weight saving (fuel economy) in automobiles. The horizontal axis shows the weight reduction made possible by each substitution and the vertical axis shows the cost of the lightweight material relative to its conventional counterpart. In most cases the lightweight materials lie above the breakeven curve where the cost of the substitute part equals the cost of the conventional part, because less material by weight needs to be purchased for the substitute. The exception is high-strength steel substituted for mild steel. Note that this plot does not consider possible savings in processing and assembly for the lightweight material.

It is important to realize that the cost of a material expressed in dollars per pound

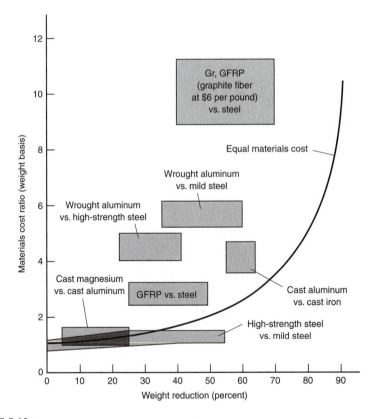

FIGURE 8.10

Cost of substituting a lightweight material in automotive applications. Note: GFRP is glass fiber-reinforced polymer. Gr, GFRP is GFRP containing graphite fibers. (*W. D. Compton and N. A. Gjostein, Scientific American, vol. 255, p. 98, October 1986.*)

may not always be the most valid criterion. Often materials provide more of a space-filling function than a load-bearing function, and then dollars per cubic foot is a more appropriate criterion. For example, the cost of plastics usually is justified on a cost per unit volume, rather than a cost per unit weight, basis. It is also important to emphasize that there are many ways to compute costs (see Chap. 14). *Total life-cycle cost* is the most appropriate cost to consider. It consists of the initial material costs plus the cost of manufacturing and installation plus the costs of operation and maintenance plus the cost of recycling or remanufacture. Consideration of factors beyond just the initial materials cost leads to relations like the relation shown in Fig. 8.11. Part costs (curve *A*) can increase with strength because of greater difficulty in fabricating the higher-strength parts. However, the number of parts needed, or the weight per part, will decrease with increasing strength, or the service life might increase with strength, etc., to lead to curve *B*. The total cost is the sum of *A* and *B,* and the optimum value of the property occurs at the minimum cost.

A classic situation regarding cost is the choice between two or more materials with different initial costs and different expected lives. That is a standard problem in the field of engineering economy. It is discussed in detail in Chap. 13.

With the growing access to computer-aided materials databases more engineers are finding the materials they need with a computerized search. Section 8.7 discusses this popular method of materials selection, suggests some cautions that should be observed, and gives an important merit parameter that can be used when comparing a number of materials.

A rational way to compare materials is by using the material performance index (Sec. 8.8). This is an important adjunct to the use of the Ashby selection charts during the initial screening in the conceptual design phase.

Various types of decision matrices were introduced in Chap. 5 to evaluate design concepts. These can be used to good advantage to compare materials when it is necessary to satisfy more than one performance requirement. The weighted property index, considered in Sec. 8.9, is the most comprehensive of these methods.

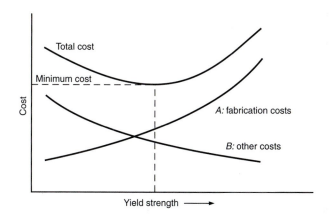

FIGURE 8.11
Relations between cost factors and a material property.

Value analysis is an organized method of finding the least expensive way to make a product without compromising quality or reliability. It is a general problem-solving methodology that is much broader than its application just to materials selection. However, because it can be a valuable materials selection tool, it is presented in Secs. 8.11 and 14.9.

A rational way to select materials is to determine the way in which actual production parts, or parts similar to a new design, fail in service. Then, on the basis of that knowledge, materials that are unlikely to fail are selected. The general methodology of failure analysis is considered in Chap. 11.

Regardless of how well a material has been characterized and how definitive the performance requirements and the program schedule are, there will always be a degree of uncertainty about the ability of the material to perform. The major risk areas encountered in any complex design and manufacturing program include: (1) program schedule, (2) cost, (3) producibility, (4) performance, (5) maintainability, (6) repairability, and (7) survivability. For high-performance systems such that the consequences of failure can be very severe, material selection based on risk analysis can be very important. Some of the ideas of risk analysis are discussed in Chap. 11.

8.7
SELECTION WITH COMPUTER-AIDED DATABASES

The use of computer-aided tools allows the engineer to minimize the materials selection information overload. A computerized materials search can accomplish in minutes what may take hours or days by a manual search. Over 100 materials databases are available worldwide. However, the data contained in most of them is limited to numerical values and text. Image information (micrographs, fractographs, etc.) can be provided through CD-ROM. All materials property databases allow the user to search for a material match by comparing four or more property parameters, each of which can be specified as below, above, or within a stated range of values. Some databases have the ability to weight the significance of the various properties. The most advanced databases allow the materials property data to be transmitted directly to a design software package, such as finite-element analysis, so that the effect of changing material properties on the geometry and dimensions of a part can be directly observed on the computer monitor. However, this capability is generally limited to a narrow homogeneous group of materials.

Most existing databases provide numerical material properties. Usually mechanical and corrosion properties are well covered, with less extensive coverage of magnetic, electrical, and thermal properties. Since it is unlikely that any database will be sufficiently comprehensive for a specific user, it is vital that the system be designed so that users may easily add their own data, and subsequently search, manipulate, and compare these values along with the entire collection of data.

The validity of the data in a database is an important issue. In order to increase the reliability and accuracy of data in a computerized database, the ASTM recommends that the following descriptors be used:

- Source (name of handbook, government agency, producer, etc.)
- Statistical basis of the data
- Status of material (production, experimental, obsolete)
- Evaluation status (if so, by whom and how)
- Validation status (if so, by whom and how)
- Certification status (if so, by whom and how)

To compare different materials using a computerized database it may be useful to employ the *limits on properties* criterion.[1] In this method of materials selection the performance requirements are divided into three categories: (1) lower limit properties, (2) upper limit properties, and (3) target properties.

For example, if it is required to have a stiff light material, we would put a lower limit on Young's modulus and an upper limit on density. This is the approach used when screening a large number of materials with a computer database. After screening, the remaining materials are those whose properties are above the lower limits, below the upper limits, and within the target values of the specified requirements.

To arrive at a merit parameter for each material the properties are first assigned weighting factors using pairwise comparison (see Sec. 8.9). A merit parameter p is then calculated for each material using the relationship

$$ p = \left[\sum_{i=1}^{n_l} w_i \frac{Y_i}{X_i} \right]_l + \left[\sum_{j=1}^{n_u} w_j \frac{X_j}{Y_j} \right]_u + \left[\sum_{k=1}^{n_t} w_k \left(\frac{X_k}{Y_k} - 1 \right) \right]_t \tag{8.1} $$

where l, u, and t stand for lower-limit, upper-limit, and target properties, respectively.

n_l, n_u, and n_t are the numbers of lower-limit, upper-limit, and target value properties.

w_i, w_j, and w_k are the weighting factors on lower-limit, upper-limit, and target-value properties.

X_i, X_j, and X_k are the candidate material lower-limit, upper-limit, and target value properties.

Y_i, Y_j, and Y_k are the specified lower limits, upper limits, and target values.

Based on Eq. (8.1), the lower the value of the merit parameter p the more suitable the material. Cost can be treated as an upper-limit property and given the appropriate weight.

8.8
MATERIALS PERFORMANCE INDICES

A materials performance index is a group of material properties which governs some aspect of the performance of a component.[2] If the performance index is maximized, it gives the best solution to the design requirement. Consider the tubular frame of a

1. M. M. Farag, "Selection of Materials and Manufacturing Processes for Engineering Design," Prentice-Hall, Englewood Cliffs, NJ, 1989, p. 430.
2. M. F. Ashby, *Acta Met.*, vol. 37, p. 1273, 1989.

bicycle.[1] The design requirement calls for a light, strong tubular beam of fixed outer diameter. Its *function* is to carry bending moments. The *objective* is to minimize the mass m of the frame. The mass per unit length m/L can expressed by

$$\frac{m}{L} = 2\pi rt\rho \tag{8.2}$$

where r is the outer tube radius, t is the wall thickness, and ρ is the density of the material from which it is made. Equation (8.2) is the *objective function*, the quantity to be minimized. This optimization must be done subject to several *constraints*. The first constraint is that the tube strength must be sufficient so it will not fail. Failure could occur by buckling, fast fracture, plastic collapse, or fatigue caused by repeated cyclic loads. If fatigue is the likely cause, then the cyclic bending moment M_b the tube can withstand with infinite life is

$$M_b = \frac{I\sigma_e}{r} \tag{8.3}$$

where σ_e is the endurance limit and $I = \pi r^3 t$ is the second moment of inertia for a thin-walled tube. The second constraint is that r is fixed. However, the wall thickness is free and this should be chosen so that it will just support M_b. Substituting Eq. (8.3) into Eq. (8.2) gives the mass per unit length in terms of the design parameters and material properties.

$$\frac{m}{L} = \frac{2M_b}{r}\left[\frac{\rho}{\sigma_e}\right] \tag{8.4}$$

The lightest tube which performs the function and meets the constraints is one that is made from the material with the greatest value of the material performance index M_1.

$$M_1 = \frac{\sigma_e}{\rho} \tag{8.5}$$

A change of function, objective, or constraints changes the materials performance index. If the function were to select a tube for carrying a torsional load with the constraint of stiffness rather than strength, but still with the objective of minimum weight, then $M_1 = G^{1/2}/\rho$, where G is the shear modulus. For the bicycle beam in bending, if the constraint changes from strength to stiffness, the index becomes

$$M_2 = \frac{E}{\rho} \tag{8.6}$$

Next, if the other constraint, a fixed tube radius, is relaxed and replaced with a fixed tube shape (r/t fixed) the materials performance index for strength becomes

$$M_3 = \frac{\sigma_e^{2/3}}{\rho} \tag{8.7}$$

---

1. M. F. Ashby, *Met. Mat. Trans.*, vol. 26A, pp. 3057–3064, 1995.

TABLE 8.4
Performance indices

Design objective: minimum weight for different shapes and loadings	To maximize strength	To maximize stiffness
Bar in tension: load, stiffness, length are fixed; section area is variable	σ_f/ρ	E/ρ
Torsion bar: torque, stiffness, length are fixed; section area is variable	$\sigma_f^{2/3}/\rho$	$G^{1/2}/\rho$
Beam in bending: loaded with external forces or self-weight; stiffness, length fixed; section area free	$\sigma_f^{2/3}/\rho$	$E^{1/2}/\rho$
Plate in bending: loaded by external forces or self-weight; stiffness, length, width fixed; thickness free	$\sigma_f^{1/2}/\rho$	$E^{1/3}/\rho$
Cylindrical vessel with internal pressure: elastic distortion, pressure, and radius fixed; wall thickness free	σ_f/ρ	E/ρ
Other design objectives, as stated below	Maximize	
Thermal insulation: minimize heat flux at steady state; thickness given	$1/\kappa$	
Thermal insulation: minimum temperature after specified time; thickness given	$C_p\rho/\kappa$	
Minimize thermal distortion	κ/α	
Maximize thermal shock resistance	$\sigma_f/E\alpha$	

σ_f = failure strength (yield or fracture stress as appropriate to problem); E = Young's modulus; G = shear modulus; ρ = density; C_p = specific heat capacity; α = thermal expansion coefficient; κ = thermal conductivity.

and for stiffness it becomes

$$M_4 = \frac{E^{1/2}}{\rho} \tag{8.8}$$

Table 8.4 lists materials performance indices for many common design situations.[1]

A commonly used objective is to minimize cost rather than weight. The performance index can be readily transformed to this objective by replacing the density ρ with $C_m\rho$, where C_m is the cost of a unit mass of material.

> **EXAMPLE.** We wish to select the optimum material for a light, stiff beam. The beam is b cm on a side and L long. The mass of the beam is $m = AL\rho$ where A is the area of the beam cross section and ρ is the density of the material. For a simply supported beam with a load P at the midlength, the central deflection is
>
> $$\delta = \frac{PL^3}{48EI} \qquad \text{where the moment of inertia is } I = \frac{b^4}{12} = \frac{A^2}{12}$$

1. M. F. Ashby, "Materials Selection in Mechanical Design," Pergamon Press, New York, 1992.

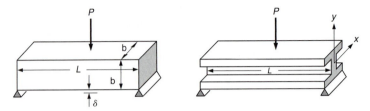

The beam stiffness is

$$S = \frac{P}{\delta} = \frac{48EI}{L^3} = \frac{4EA^2}{L^3}$$

Eliminating A between this equation and the equation for m gives

$$m = \left(\frac{S^{1/2}}{2}\right)(L^{5/2})\left(\frac{\rho}{E^{1/2}}\right)$$

Note that the equation describing the mass of the beam consists of three terms. The first describes the required performance of the part, in this case expressed by the constraint of delivering a prescribed stiffness S. The second term is the geometrical constraint of length L. The last term is the material performance index

$$M = \frac{E^{1/2}}{\rho}$$

By selecting a material that maximizes M, the mass of the beam is minimized. We note that M represents the slope of the lines drawn on the Ashby chart to determine the feasible set of materials for a design (see Fig. 8.7a).

The cross-sectional shapes of beams used in practice are rarely simple squares but are more often the irregular shape typical of the I-beam shown above to the right. Engineers have long understood that such shapes maximize performance while minimizing the use of materials. Ashby[1] has shown that shape can be included in the performance index by introducing a dimensionless shape factor ϕ defined by

$$\phi = \frac{4\pi I}{A^2}$$

Then the best selection of material and section shape for a light stiff beam is that which maximizes

$$M = \frac{(\phi E)^{1/2}}{\rho}$$

The form of the shape factor will be different depending upon whether the part is stressed in bending or torsion and whether the objective is to maximize stiffness or strength.[2]

1. M. F. Ashby, *Acta Metal. Mater.,* vol. 39, no. 6, pp. 1025–1039, 1991.
2. M. F. Ashby, "Materials Selection in Mechanical Design," Pergamon Press, New York, 1992.

8.9
DECISION MATRICES

In most applications it is necessary that a selected material satisfy more than one performance requirement. In other words, compromise is needed in materials selection. We can separate the requirements into three groups (1) go-no-go parameters, (2) nondiscriminating parameters, and (3) discriminating parameters. Go-no-go parameters are those requirements which must meet a certain fixed minimum value. Any merit in exceeding the fixed value will not make up for a deficiency in another parameter. Examples of go-no-go parameters are corrosion resistance or machinability. Nondiscriminating parameters are requirements that must be met if the material is to be used at all. Examples are availability or general level of ductility. Like the previous category these parameters do not permit comparison or quantitative discrimination. Discriminating parameters are those requirements to which quantitative values can be assigned.

The decision matrix methods that were introduced in Chap. 5 are very useful in materials selection. They organize and clarify the selection task, provide a written record of the selection process (which can be useful in redesign), and improve the understanding of the interactions between alternative solutions.

Three important factors in any formalized decision-making process are the alternatives, the criteria, and the relative weight of the criteria. In materials selection, each candidate material, or material-process pair, is an alternative. The criteria are the material properties or factors that are deemed essential to satisfy the functional requirements. The weighting factors are the numerical representations of the relative importance of each criterion. As we saw in Chap. 5, it is usual practice to select the weighting factors so that their sum equals unity.

8.9.1 Pugh Selection Method

The simplest decision method discussed in Chap. 5 is the Pugh concept selection method. This method involves qualitative comparison of each alternative to a reference or datum alternative, criterion by criterion. It is useful in conceptual design because it requires the least amount of detailed information. It is also useful in redesign, where the current material serves automatically as the datum.

EXAMPLE. The Pugh decision method is used to select a replacement material for a helical steel spring in a wind-up toy train.[1] The alternatives to the ASTM A227 class I hard-drawn steel wire are the same material in a different design geometry, ASTM A228 music spring-quality steel wire, and ASTM A229 class I steel wire, quenched and oil tempered.

In the decision matrix, if an alternative is judged better than the datum, it is given a +, if it is poorer it gets a −, and if it is about the same it is awarded an S. The +, −, and S responses are then totaled.

---

1. D. L. Bourell, Decision Matrices in Materials Selection, in "ASM Handbook," vol. 20, "Materials Selection and Design," ASM International, Materials Park, OH, 1997.

Pugh Decision Matrix for Helical Spring Redesign

	Alternative 1 present material hard-drawn steel ASTM A227	Alternative 2 hard-drawn steel class I ASTM A227	Alternative 3 music wire quality steel ASTM A228	Alternative 4 oil-tempered steel class I ASTM A229
Wire diameter, mm	1.4	1.2	1.12	1.18
Coil diameter, mm	19	18	18	18
Number of coils	16	12	12	12
Relative material cost	1	1	2.0	1.3
Tensile strength, MPa	1750	1750	2200	1850
Spring constant	D	−	−	−
Durability	A	S	+	+
Weight	T	+	+	+
Size	U	+	+	+
Fatigue resistance	M	−	+	S
Stored energy		−	+	+
Material cost (for one spring)		+	S	S
Manufacturing cost		S	+	−
$\Sigma+$		3	6	4
ΣS		2	1	2
$\Sigma-$		3	1	2

Both the music spring-quality steel wire and the oil-tempered steel wire are superior to the original material selection. The music wire is selected because it ranks highest, especially with regard to manufacturing cost. Note that the Pugh method weights each criterion equally.

8.9.2 Weighted Property Index

The weighted decision matrix that was introduced in Sec. 5.9 is well suited to materials selection with discriminating parameters.[1] In this method each material property is assigned a certain weight depending on its importance to the required service performance. Techniques for assigning weighting factors are considered in Sec. 5.9. Since different properties are expressed in different units the best procedure is to normalize these differences by using a scaling factor. The scaling is a simple technique to bring all the different properties within one numerical range. Since different properties have widely different numerical values, each property must be so scaled that the largest value does not exceed 100.

---

1. M. M. Farag, "Materials and Process Selection in Engineering," Chap. 13, Applied Science Publ., Ltd., London, 1979.

$$\beta = \text{scaled property} = \frac{\text{numerical value of property}}{\text{largest value under consideration}}\ 100 \qquad (8.9)$$

For properties such that it is more desirable to have low values, e.g., density, corrosion loss, cost, and electrical resistance, the scale factor is formulated as follows:

$$\beta = \text{scaled property} = \frac{\text{lowest value under consideration}}{\text{numerical value of property}}\ 100 \qquad (8.10)$$

For properties that are not readily expressed in numerical values, e.g., weldability and wear resistance, some kind of subjective rating is required. For example, use Table 5.8, or the scheme given below.

Property	Alternative materials			
	A	B	C	D
Weldability	Excellent	Good	Good	Fair
Relative rating	5	3	3	1
Scaled property	100	60	60	20

The weighted property index γ is

$$\gamma = \Sigma \beta_i w_i \qquad (8.11)$$

where β_i is summed over all the properties (criteria) and w_i is the weighting factor for the ith property.

There are two ways to treat cost in this analysis. First, cost can be considered as one of the properties, usually with a high weighting factor. Or the weighted property index can be modified as follows:

$$\gamma' = \frac{\gamma}{C_m \rho} \qquad (8.12)$$

where C_m is the cost of a unit mass of material and ρ is the density.

EXAMPLE. The material selection for a cryogenic storage vessel for liquefied propane gas is being evaluated on the basis of the following properties: (1) low-temperature fracture toughness, (2) fatigue strength, (3) stiffness, (4) thermal expansion, and (5) cost.

Determine the weighting factors for these properties using pairwise comparison. There are $N = 5(5 - 1)/2 = 10$ possible comparisons of pairs.

By comparing each pair of properties, and deciding which is paramount for that comparison (1), and which is less important (0), we get the following table.

Possible design combinations

Property	1 (1) (2)	2 (1) (3)	3 (1) (4)	4 (1) (5)	5 (2) (3)	6 (2) (4)
1. Fracture toughness	1	1	1	1		
2. Fatigue strength	0				0	1
3. Stiffness		0			1	
4. Thermal expansion			0			0
5. Cost				0		

Property	7 (2) (5)	8 (3) (4)	9 (3) (5)	10 (4) (5)	Positive decisions	Weighting factor w_i
1. Fracture toughness					4	0.4
2. Fatigue strength	0				1	0.1
3. Stiffness		0	0		1	0.1
4. Thermal expansion		1		0	1	0.1
5. Cost	1		1	1	3	0.3
					10	1.0

The chart for selecting a material based on the weighted property index is shown in Table 8.5. Several go-no-go screening parameters are included. It is assumed that the aluminum alloy will not be available in the required plate thickness, so that material is dropped from further consideration. The body of the table shows both the raw data and the data in scaled form. The β values for toughness, fatigue strength, and stiffness were determined from Eq. (8.9). The β values for thermal expansion and cost were determined from Eq. (8.10) because for these properties a smaller value is better. Since no comparable fracture toughness data was available for the candidate materials, a relative scale 1 to 5 was used. The best material among these choices for the application is the 9 percent nickel steel, which has the largest value of weighted property index.

8.10
MATERIALS SELECTION BY EXPERT SYSTEMS

Material selection represents an obvious application for expert systems (see Sec. 4.9). While broad comprehensive knowledge-based systems for materials selection still are under development,[1] expert systems that deal with more restrictive situations currently exist. An example is ESCORT (Expert Software for Corrosion Technology),[2] an expert system for materials selection in the chemical process industry. It consists of chemical process industry. It consists of five knowledge bases: materials, equipment (type of heat exchanger, packed columns, pumps, etc.), environmental characteristics, types of corrosion, and preventive measures.

In starting the expert system the user is shown a *Corrosion Engineering Worksheet* consisting of three windows: (1) industry, process, or operation; (2) environment, and (3) equipment. These windows permit the user to specify the problem in considerable detail. The more detailed the input the more specific will be the final material selection. By specifying the kind of industry or operation in the first window the program will make assumptions concerning certain process conditions. This enables the expert system to ask for additional specific information from the user. Window (2) allows the first selection of the process parameters. Window (3) calls for the specification of the process equipment in which the materials are to be used.

1. "Computer Aided Materials Selection During Structural Design," National Materials Advisory Board, NMAB-467, National Academy Press, Washington, DC, 1995.
2. M. J. S. Vancoille, W. F. Bogaerts, and M. J. Rijckaert, Nat. Assoc. Corrosion Engrs. 88 Conference, Paper 121, 1988.

TABLE 8.5
Weighted property index chart for selection of a material for a cryogenic storage vessel

| Material | Go-no-go screening | | | Toughness (0.4) | | Fatigue strength (0.1) | | Stiffness (0.1) | | Thermal expansion (0.1) | | Cost (0.3) | | Weighted property index |
	Corrosion	Weldability	Available in thick plate	Rel. Scale	β	ksi	β	10^6 psi	β	μin/in °F	β	$/lb	β	γ
304 stainless	S	S	S	5	100	30	60	28.0	93	9.6	80	3.00	50	78.3
9% Ni steel	S	S	S	5	100	50	100	29.1	97	7.7	100	1.80	83	94.6
3% Ni steel	S	S	S	4	80	35	70	30.0	100	8.2	94	1.50	100	88.4
Aluminum alloy	S	S	U											

S = satisfactory
U = unsatisfactory
Relative scale: 5 = excellent, 4 = very good
Sample calculation: 304 stainless steel: $\gamma = 0.4(100) + 0.1(60) + 0.1(93) + 0.1(80) + 0.3(50) = 78.3$

World 1, World 2, and *Merge Worlds* are features used while designing process equipment. Frequently the fluids inside and outside of heat-transfer tubing are different. The process stream inside the tube might be highly corrosive acid while the outside of the tube might be cooled with water containing chlorides. Using *World 1* the expert system would select all materials which are suitable for the inside of the tubing (acid) and *World 2* would select all materials suitable for the outside of the tubing (water). By using *Merge Worlds* the program will identify the materials compatible with both the inside and outside of the tubing. The command *Input Complete* will cause the system to begin the reasoning process and proceed to the next worksheet. Often the system will produce questions that ask for more information than given in the input to the first corrosion engineering worksheet. *Call NACE-NBS* introduces the option of calling upon external databases, such as the NACE-NBS(NIST) corrosion database. In this way it is not necessary to burden the expert system with too much data. A section called *Requirements* prompts the user to specify the classes of materials he or she prefers. These decisions could be based on previous experience, availability, cost, or codes and specifications.

When the input process has been completed the expert system starts the main reasoning operation using if-then production rules. The system uses two types of rules: shallow rules and deep rules. The shallow rules are based on readily available relationships, like those discussed earlier in this chapter, and on the expertise of individual specialists. The deep rules are even less easily defined, and comprise things like availability, economic factors, company policies, etc. The deep rules further narrow down the material selection. When the expert system completes its task, it will present a list of materials that meet the input specifications. The user can ask why a certain material was chosen and the program will show the rules it applied to make this selection. The user can also ask for additional information about a material, e.g., how should it be welded?

The usefulness and power of expert systems in material selection should be obvious from the above brief scenario. Considerable development is expected in this area of knowledge-based engineering in the near future.

8.11
VALUE ANALYSIS

Value analysis, or value engineering, is a team problem-solving process to optimize the value of a product for the customer.[1] It involves breaking a product down into its component parts and determining the value of these design elements relative to the importance of the functions which they provide. As a result it sometimes is called functional cost analysis.

Value analysis has much broader applicability to engineering design than in just the materials selection problem.[2] It is often used as a first step in the redesign of a

1. C. Fallon, "Value Analysis to Improve Productivity," Wiley, New York, 1971; D. Miles, "Techniques of Value Analysis in Engineering," 2d ed., McGraw-Hill, New York, 1972; T. C. Fowler, "Value Analysis in Design," Van Nostrand Reinhold, New York, 1990.
2. T. C. Fowler, L. Parrot, and V. Thompson, Value Analysis in the Design Process, "ASM Handbook," vol. 20, "Materials Selection and Design," ASM International, Materials Park, OH, 1997.

product, or in the production-planning step of design. However, its greatest potential is when used early in the process before design details have been set too firmly.

Success with value analysis depends heavily on understanding the relationships between each design feature of a component and the function that it is to provide. This important aspect of design was stressed in Chap. 5 as the foundation stone for conceptual design.

The concept of value, as used in value analysis methodology, is given by

$$\text{Value} = \frac{\text{worth of a feature, component, or assembly}}{\text{cost}} \tag{8.13}$$

Worth is the value of something measured by its qualities or the esteem in which it is held. Considerable skill is needed to determine worth in terms of dollars. Focus groups of customers may be used to gain their insight into the product. Another way to determine the worth of a function is to ask yourself what would be a reasonable amount to pay if it were to come from your own pocket.

Value analysis asks the following questions: How can a given function of a design system be performed at minimum cost? What is the value of the contribution that each feature of the design makes to the specific function that the part must fulfill? Value analysis usually is carried out by a team of engineers and managers possessing different backgrounds and viewpoints, so that the problem gets looked at from many aspects. However, a value analysis needs the support and endorsement of top management if it is to be most successful.

The value analysis approach seeks the development of answers to the following questions.

Can we do without the part?
Does the part do more than is required?
Does the part cost more than it is worth?
Is there something that does the job better?
Is there a less costly way to make the part?
Can a standard item be used in place of the part?
Can an outside supplier provide the part at less cost without affecting dependability?

The value analysis job plan consists of those tasks or functions necessary to perform the study (Fig. 8.12). A structured plan like the one shown assures that consideration is given to all important aspects, provides for a logical separation of the study into convenient units, and provides a convenient basis for maintaining a written record of progress in the value analysis study.

A key step in value analysis is evaluation of the function of the design or system. It has the same importance that definition of the problem has to more generalized problem solving. In value analysis, function is expressed by two words, a verb and its noun object. The verb answers the question what does it do; the noun answers the question what does it do it to. It might, for example, transmit torque, conduct current, or maintain records. As in preparing a problem statement, the function should be identified so as not to limit the ways in which it could be performed.

Once the functions have been established, the next step is to establish a dollar value or worth for each function. Next we determine the cost of the method used to carry out the function. Cost applies to the actual design that is used, whereas worth applies to the function. An important feature of value analysis is to identify high-cost

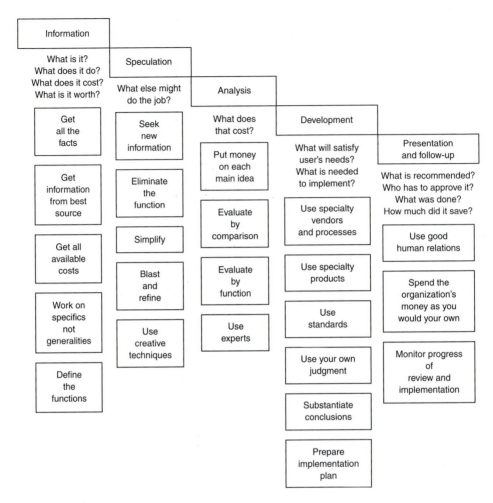

FIGURE 8.12
Components of the value analysis job plan.

elements of the design and focus attention on them. In that regard we should be aware of Pareto's law (Fig. 8.13), which states that about 80 percent of the total effect of any group will come from only 20 percent of the components of that group. Thus, about 20 percent of the elements of a design contribute 80 percent of the costs. Obviously, attention should be given to this small but important part of the distribution. In analyzing costs, the total unit cost is broken down into material, labor, and overhead costs (see Chap. 14). Since the decisions made will depend on the reliability of the data, great care should be used to establish the cost data.

An example of functional cost analysis is given in Fig. 8.14. This shows how cost can be assigned to the parts of an aircraft air valve based on function.[1] Note that many parts contribute to more than one function, so that estimates of the cost need to be dis-

---

1. N. Cross, "Engineering Design Methods," 2d ed., Wiley, Chichester, UK, 1994.

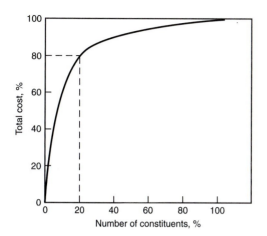

FIGURE 8.13
Pareto's law of distribution of costs.

Parts	Functions										$ total cost	%
	Stop air	Sense ram air	Sense servo air	Sense cabin air	Connect parts	Provide mounting	Resist corrosion	Provide support	Provide interchangeability	No function		
Banjo assembly			0.2		0.4				0.47		1.07	5.5
Valve body	0.4	1.0			2.82	0.8	0.2	0.8		0.6	6.62	34.0
Spring										0.39	0.39	2.0
Diaphragm assembly	0.6	0.1	0.1	0.1	0.94		0.2	0.1			2.14	11.0
Cover			0.4		1.2	0.1	0.1	0.34	0.1		2.24	11.5
Lug										0.1	0.1	0.5
Nuts, bolts, and washers					2.14		0.1		0.1		2.34	12.0
Assembly					4.58						4.58	23.5
Total	1.0	1.1	0.7	0.1	12.08	0.9	0.6	1.24	0.67	1.09	19.48	100.0
% total	5.1	5.7	3.4	0.5	62.0	4.6	3.1	6.4	3.4	5.6		
High or low					H					H		

FIGURE 8.14
Example of a functional cost analysis used in value analysis. (*Reprinted by permission of John Wiley & Sons, Ltd.*)

tributed over several functions. This analysis showed that the valve body is the most costly part but that the "connect parts" function is by far the most expensive. It also identified parts for which no function is apparent. A redesign reduced the number of parts and the overall component cost was decreased by over 60 percent.

EXAMPLE. The design under study was the housing for a hydraulic valve (Fig. 8.15). The housing was made of nodular cast iron and was heat-treated to a fully pearlitic matrix

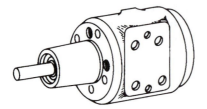

FIGURE 8.15
Hydraulic valve housing.

to provide the necessary wear and resistance for the valve stem. The objective of the valve analysis was to reduce the cost of manufacture of the part. Analysis of the costs for each step in the manufacture showed that machining was the preponderant cost driver. It was readily appreciated that machining time (and cost) could be reduced substantially if the valve housings were heat-treated to a fully ferritic matrix. However, wear resistance of the softer condition was not adequate.

Functional analysis showed that the valve housing had two basic functions: (1) to provide an enclosure for the hydraulic fluid and (2) to provide a wear-resistant surface for the piston-like valve stem. In the speculation phase it was realized that those two basic functions were coupled in the original design and that hence the heat treatment of the nodular iron was constrained by the need for high hardness to provide wear resistance. Other suggestions for wear resistance were the use of a tool-steel sleeve in a soft valve housing and the chrome plating of the cylinder bore. The analysis phase further substantiated the cost savings to be achieved by improving the machinability of the nodular iron castings and rejected the new ideas as probably more costly. In the development phase extensive tests were made on the wear characteristics of the valves. These studies showed that wear was significant only during the first 500 cycles of operation as the piston, machined to close tolerances, was running in. This suggested that a low-friction surface coating that would eventually wear away might be a suitable alternative. The tests also pointed out that surface finish of the cylinder bore was a critical operational parameter that could not be achieved in dead-soft ferritic nodular iron.

The final solution was to heat-treat the valve housings to a mixed microstructure of ferrite and spheroidized pearlite. This was a compromise metallurgical condition between the hard wear-resistant condition and the soft condition that would have improved machinability but poor surface finish. Wear during the running-in period was accommodated by a phosphate coating on the cylinder bore that provided lubrication during the critical period.

8.12
DESIGN EXAMPLE—MATERIALS SYSTEMS

Engineered systems contain many components, and for each a material must be selected. The automobile is our most familiar engineering system and one that is undergoing a major change in the materials used for its construction. These trends in materials selection reflect the great effort that is being made to decrease the fuel consumption of cars by downsizing the designs and adopting weight-saving materials. Prior to 1975, steel and cast iron comprised about 78 percent of the weight of a car, with aluminum and plastics each at slightly less than 5 percent. Today, ferrous materials comprise about 65 percent of the total weight, with plastics at about 15 percent and aluminum at about 8 percent. Aluminum is in an ongoing battle with steel to take over the structural frame and part of the sheet panels.

Frequently, complex and severe service conditions can be met economically only by combining several materials into a single component. The surface hardening of gears and other automotive components by carburizing or nitriding[1] is a good example. Here the high hardness, strength, and wear resistance of a high-carbon steel is produced in the surface layers of a ductile and tougher low-carbon steel.

An excellent example of a complex materials system used in a difficult environment is the exhaust valve in an internal-combustion engine.[2] Valve materials must have excellent corrosion- and oxidation-resistance properties to resist "burning" in the temperature range 1350 to 1700°F. They must have (1) sufficient high-temperature fatigue strength and creep resistance to resist failure and (2) suitable hot hardness to resist wear and abrasion.

The critical failure regions in an exhaust valve are shown in Fig. 8.16. Maximum operating temperature occurs in areas A and C. Corrosion and oxidation resistance are especially critical there. The underhead area of the valve, area C, experiences cyclic

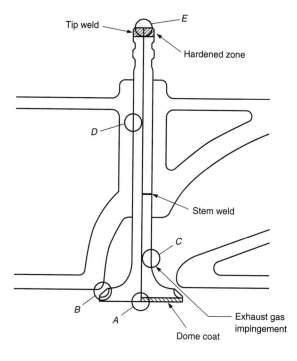

FIGURE 8.16
Typical exhaust valve showing critical regions of failure.

1. Surface Hardening of Steel, "Metals Handbook," 9th ed., vol. I, "Properties and Selection of Irons and Steels," American Society for Metals, Metals Park, OH, 1978.
2. J. M. Cherrie and E. T. Vitcha, *Metal Prog.*, pp. 58–62, September 1971; J. F. Kocis and W. M. Matlock, ibid. pp. 58–62, August 1975.

loading, and because of the mild stress concentrations, fatigue failure may occur at that point. The valve face, area *B,* operates at a somewhat lower temperature because of heat conduction into the valve seat. However, if an insulating deposit builds up on the valve face, it can lead to burning. Also, the valve seat can be damaged by indentation by abrasive fuel ash deposits. The valve stem is cooler than the valve head. However, wear resistance is needed. Surface wear of the valve stem, area *D,* can lead to scuffing, which will cause the valve to stick open and burn. Wear at the valve tip, area *E,* where the valve contacts the rocker arm, will cause valve lash and cause the valve to seat with higher than normal forces. Eventually, that will cause failure.

The basic valve material for passenger car application, where T_{max} = 1300°F, is an austenitic stainless steel that obtains its good high-temperature properties from a dispersion of precipitates. This alloy, 21-2N, contains 20.35 percent chromium for oxidation and corrosion resistance. It has good PbO corrosion resistance, and its high-temperature fatigue strength is exceeded only by that of the more expensive nickel-base superalloys (Table 8.6).

The entire body of one-piece valves is 21-2N, except for a hard-steel tip at *E* and a hard chromium plate in area *D.* However, it is generally more economical to use a two-piece valve in which 21-2N is replaced in the cooler stem portions by a cheaper alloy steel such as SAE 3140 or 4140. Either steel will have sufficient wear resistance, and the lower stem does not need the high oxidation and corrosion resistance of the high-chromium, high-nickel steel. The two materials are joined by welding, as shown in Fig. 8.16. Burning of the valve face, area *B,* is generally avoided by coating the valve surface with aluminum to produce an Fe–Al alloy or, in severe cases, by hard-facing the valve seat with one of the Co–C–Cr–W Stellite alloys.

With the removal of lead compounds from gasoline has come an easing of the high-temperature corrosion problems that cause burning, but a new problem has arisen. The combustion products of lead-free gasoline lack the lubricity characteristics of those of fuels that contain lead. As a result, the wear rates at the valve seating surfaces have increased significantly. The solution has been to harden the cast-iron cylinder head to improve its wear resistance or install a wear-resistant insert. Unless one or the other is done, the wear debris will weld to the valve seat and cause extensive damage. With increased use of hard-alloy inserts, it is becoming more common to hard-face the valve seat with a Stellite alloy.

TABLE 8.6
Some properties of valve materials

Alloy	C	Mn	Cr	Ni	Other	Strength at 1350°F		
						Table strength, psi	Creep strength*	Fatigue limit†
21-2N	0.55	8.25	20.35	2.10	0.3N; bal. Fe	57,700	34,000	26,000
DV2A	0.53	11.50	20.5	—	2W; 1Cb; bal. Fe	78,000	34,000	30,000
Inconel 751	0.10 max	1.0 max	15.5	Bal.	2.3Ti; 1.2Al	82,000	45,000	45,000

*Stress to produce 1 percent elongation in 100 h
†Stress to produce failure in 10^8 cycles

8.13
DESIGN EXAMPLE—MATERIALS SUBSTITUTION

This design example illustrates the common problem of substituting a new material for one that has been used for some time. It illustrates that material substitution should not be undertaken unless appropriate design changes are made. Also, it illustrates some of the practical steps that must be taken to ensure that the new material and design will perform adequately in service.

Aluminum alloys have been substituted for gray cast iron[1] in the external supporting parts of integral-horsepower induction motors (Fig. 8.17). The change in materials was brought about by increasing cost and decreasing availability of gray-iron castings. There has been a substantial reduction in gray-iron foundries, partly because of increased costs resulting from the more stringent environmental pollution and safety regulations imposed in recent years by governmental agencies. The availability of aluminum casting has increased owing to new technology and the lesser problem of operating a foundry at a temperature much lower than that required for cast iron.

There are a variety of aluminum casting alloys.[2] Among the service requirements for this application, strength and corrosion resistance were paramount. The need to provide good corrosion resistance to water vapor introduced the requirement to limit the copper content to an amount just sufficient to achieve the necessary strength. Actual alloy selection was dependent on the manufacturing processes used to make the part. That in turn depended chiefly (see Chap. 9) on the shape and the required quantity of parts. Table 8.7 gives details on the alloys selected for this application.

Since the motor frame and end-shield assemblies have been made successfully from gray cast iron for many years, a comparison of the mechanical properties of the aluminum alloys with cast iron is important (Table 8.8).

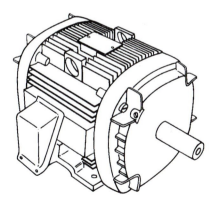

FIGURE 8.17
Horizontal aluminum alloy motor. (*Courtesy of General Electric Company.*)

1. T. C. Johnson and W. R. Morton, IEEE Conference Record 76CH1109-8-IA, Paper PCI-76-14, available from General Electric Company as Report GER-3007.
2. "Metals Handbook," 9th ed., vol. 2, pp. 140–179, American Society for Metals, Metals Park, OH, 1979.

TABLE 8.7
Aluminum alloys used in external parts of motors

| Part | Alloy | Composition | | | Casting process |
		Cu	Mg	Si	
Motor frame	356	0.2 max	0.35	7.0	Permanent mold
End shields	356	0.2 max	0.35	7.0	Permanent mold
Fan casing	356	0.2 max	0.35	7.0	Permanent mold
Conduit box	360	0.6 max	0.50	9.5	Die casting

TABLE 8.8
Comparison of typical mechanical properties

Material	Yield strength, ksi	Ultimate tensile strength, ksi	Shear strength, ksi	Elongation in 2 in, percent
Gray cast iron	18	22	20	0.5
Alloy 356 (as cast)	15	26	18	3.5
Alloy 360 (as cast)	25	26	45	3.5
Alloy 356-T61	28	38		5
(solution heat-treated and artificially aged)				

The strength properties for the aluminum alloys are approximately equal to or exceed those of gray cast iron. If the slightly lower yield strength for alloy 356 cannot be tolerated, it can be increased appreciably by a solution heat treatment and aging (T6 condition) at a slight penalty in cost and corrosion resistance. Since the yield and shear strength of the aluminum alloys and gray cast iron are about equal, the section thickness of aluminum to withstand the loads would be the same. However, since the density of aluminum is about one-third that of cast iron, there will be appreciable weight saving. The complete aluminum motor frame is 40 percent lighter than the equivalent cast-iron design. Moreover, gray cast iron is essentially a brittle material, whereas the cast-aluminum alloys have enough malleability that bent cooling fins can be straightened without breaking them.

One area in which the aluminum alloys are inferior to cast iron is compressive strength. In aluminum, as with most alloys, the compressive strength is about equal to the tensile strength, but in cast iron the compressive strength is several times the tensile strength. That becomes important at bearing support, where, if the load is unbalanced, the bearing can put an appreciable compressive load on the material surrounding and supporting it. With an aluminum alloy end shield, that leads to excessive wear. To minimize the problem, a steel insert ring is set into the aluminum alloy end shield when it is cast. The design eliminates any clearance fit between the steel and aluminum, and the steel insert resists wear from the motion of the bearing just as the cast iron always did.

The greater ease of casting aluminum alloys permits the use of cooling fins thinner and in greater number than in cast iron. Also, the thermal conductivity of aluminum is about three times greater than that of cast iron. Those factors result in more uniform temperature throughout the motor, and the results are longer life and reliability. Because of the higher thermal conductivity and larger surface area of cooling fins, less cooling air is needed. With the air requirements reduced, a smaller fan can be used, and as a result there is a small reduction of noise.

The coefficient of expansion of aluminum is greater than that of cast iron, and that makes it easier to ensure a tight fit of motor frame to core. Only a moderate temperature is needed to expand the aluminum frame sufficiently to insert the core, and on cooling the frame contracts to make a tight bond with the core. That results in a tighter fit between the aluminum frame and the core and better heat transfer to the cooling fins. Complete design calculations need to be made when aluminum is substituted for cast iron to be sure that clearances and interferences from thermal expansion are proper.

Since a motor design that had many years of successful service was being changed in a major way, it was important to subject the redesigned motor to a variety of simulated service tests. The following were used:

- Vibration test
- Navy shock test (MIL-S-901)
- Salt fog test (ASTM B117-57T)
- Axial and transverse strength of end shield
- Strength of integral cast lifting lugs
- Tests for galvanic corrosion between aluminum alloy parts and steel bolts

This example illustrates the importance of considering design and manufacturing in a material-substitution situation. Although the use of aluminum in this situation is very favorable, it must be recognized that this is an application in which the much reduced elastic stiffness of aluminum is not of great importance.

8.14
RECYCLING AND MATERIALS SELECTION

The heightened public awareness of environmental issues has resulted in significant legislation and regulation that provide new constraints on design. For example, in Germany a law requires manufacturers to take back all packaging used in transport and distribution of a product, and in the Netherlands old or broken appliances must be returned to the manufacturer for recycling. Traditionally the recycling and reuse of materials was dictated completely by economics. Those materials like steel, and more recently aluminum cans, that can be collected and reprocessed at a profit were recycled. However, with widespread popular support for improving the environment other benefits of recycling are being recognized.

The complete life cycle of a material is shown in Fig. 1.15. The *materials cycle* starts with the mining of a mineral or the drilling for oil or the harvesting of an agricultural fiber like cotton. These raw materials must be processed to refine or extract bulk material (e.g., a steel ingot) that is further processed into a finished engineering material (e.g., a steel sheet). At this stage an engineer designs a product that is manufactured from the material, and the product is put into useful service. Eventually the product wears out or is made obsolete because a better product comes on the market. At this stage our tendency has been to junk the product and dispose of it in some way, like a landfill, that eventually returns the material to the earth. However, society is more and more mindful of the dangers to the environment of such haphazard practices. As a result, more emphasis is being placed on recycling of the material back into the materials cycle. Figure 8.18 shows a flow diagram that can be used to determine whether or not a product can be designed for disposal or reuse.

The obvious benefits of materials recycling are the contribution to the supply of materials, with corresponding reduction in consumption of natural resources, and the reduction in the volume of solid waste that must be disposed of, mostly in a landfill. Moreover, recycling contributes to environmental improvement through the amount of energy saved by producing the material from recycled (secondary) material rather than primary sources (ore or chemical feedstock). Recycled aluminum requires only 5 percent of the energy required to produce it from the ore. Between 10 and 15 percent of the total energy used in the United States is devoted to the production of steel, aluminum, plastics, or paper. Since most of this is generated from fossil fuels, the saving from carbon dioxide and particulate emissions due to recycling is appreciable. Recycling of materials also causes reduced pollution directly. For example, the use of steel scrap in making steel bypasses the blast furnace (a considerable economic benefit), and this eliminates the pollution associated with this process and with the production of coke, which is an essential ingredient for blast furnace smelting.

An alternative to recycling is *remanufacturing,* where instead of the product's being disassembled to recycle the materials it is restored to near new condition by cleaning and replacement of worn parts. We have long been familiar with rebuilt automotive parts, like alternators and carburetors, but now remanufacture has moved to appliances such as copiers and printers.

Special terms are used to designate where secondary materials are generated in the materials processing and utilization cycle. *Home scrap* is the term for residual material from primary material production, which can be returned directly to the production process. Examples would be cropped tops of ingots or the sheared edges of plate. *Prompt industrial scrap* or new scrap is that generated during the manufacture of products, for example, compressed bundles of turnings or stamping discard. *Old scrap* is scrap generated from a product which has completed its useful life, e.g., a scrapped automobile or refrigerator. This is also called obsolete or postconsumer scrap.

Recyclability refers to the ease with which a material can be recovered from a given product. The steps in recycling a material are (1) collection and transport, (2) separation, (3) identification and sorting, (4) reprocessing, and (5) marketing.[1] Collection for recycling from consumers can be an expensive proposition. For example, plastic containers represent a high volume of material for little weight. Collection methods include curbside pickup, buy-back centers (for some containers), and resource recovery centers where solid waste is sorted for recyclables and the waste is burned for energy. Separation typically follows one of two paths. In the first, selective dismantling takes place. Toxic materials like engine oil or lead batteries are removed first and given special treatment. High-value materials like gold and copper are removed and segregated appropriately. Dismantling leads naturally to sorting of materials into like cate-

1. M. E. Henstock, "Design for Recyclability," Institute of Metals, London, 1988.

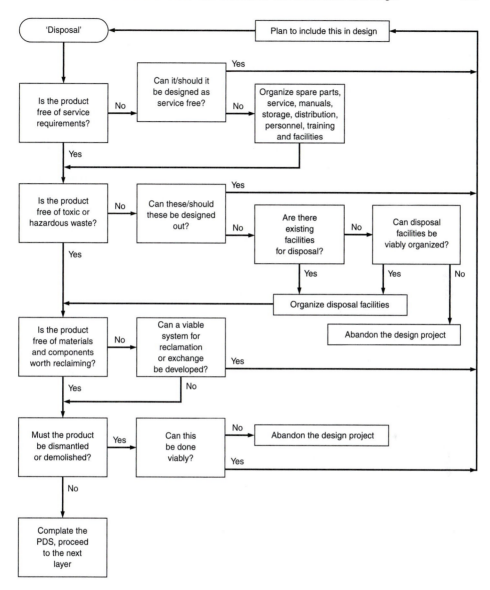

FIGURE 8.18

A flow diagram to help decide whether or not a product can be designed for disposal or reuse. (*From B. Hollins and S. Pugh, "Successful Product Design," Butterworths, London, 1990. Used with permission of Heinemann Publishers, Oxford, U.K.*)

gories. The second separation route is shredding, in which the product is subjected to multiple high-energy impacts to batter it into small irregular pieces. Automobile hulks are routinely processed by shredding. Shredded material requires sorting into different materials using techniques like magnetic separation, vibratory sieving, air classification, and wet flotation. Identification of materials may be a problem, especially for plastics. A solution for this is to mold or cast into the surface of the part the standard identification symbols of the SAE or Society for the Plastic Industry. However, a better solution may be a recently developed device which can identify the chemical composition of plastic parts at rates of more than 100 pieces per second. Using the principles of Raman spectroscopy, it reads the light scattered off the plastic surface from an incident laser beam, much like a bar-code scanner. After sorting, the recycled material is sold to a secondary materials producer. Metals are remelted into ingots; plastics are ground and processed into pellets. These are then introduced into the materials stream by selling the recycled material to part manufacturers.

It is important to know if the properties of the secondary material have been degraded by the recycling. Generally this is less of a problem with metals than with plastics. In metallic alloys, there is generally a limit on "'tramp elements" which influence critical properties or workability or castability. Some alloys are restricted to the use of "virgin material." In steel more than 0.20% copper or 0.06% tin limits the hot workability. As more and more steel is recycled through scrap, there is concern with a buildup in these critical elements. Degradation of plastics is more critical than that of metals, since it is often difficult to ensure that different types of polymers were not mixed together. Only thermoplastic polymers can be recycled. Often recycled material is used for a less critical application than its original use. There is an intensive effort to improve the recycling of plastics, and it is claimed that under the best of conditions engineered plastics can be recycled three or four times without losing more than 5 to 10 percent of their original strength. Thermosets, which are degraded by high temperature, cannot be recycled. Other materials that may not be economically recycled are zinc-coated steel (galvanized), ceramic materials (except glass), and parts with glued identification labels made from a different material than the part. Composite materials consisting of mixtures of glass and polymer represent an extreme problem in recycling.

There are several steps that the designer can take to enhance the recyclability of a product.

- Make it easier for the manual disassembly of the product.
- Minimize the number of different materials in the product.
- Choose materials that are compatible and do not require separation before recycling.
- Identify the material that the part is made from right on the part.

These guidelines can lead to contradictions that require serious trade-offs. Minimizing the number of materials may require a compromise on performance by selecting a material with less than optimum properties. Heavily galvanized steel can lead to unacceptable zinc buildup in steel scrap, yet the galvanized undercoat has greatly reduced corrosion on automobile bodies. A clad metal sheet or chromium-plated metal provides the wanted attractive surface at a reasonable cost, yet it cannot be readily recycled. In the past, decisions of this type would be made exclusively on the basis of cost. Today, we are moving toward a situation where the customer may be willing to pay extra for a recyclable design, or the recyclable design may be mandated by government regulations.

EXAMPLE. Terne-coated steel (8 percent tin-lead coating) has been the traditional material selection for automotive gas tanks.[1] A number of federal laws have mandated radical changes in automotive design. Chief among these is the act which mandates the Corporate Average Fuel Economy, which places continual pressure on weight reduction to increase gas mileage. The Alternative Motor Fuels Act of 1988 and the Clear Air Act Amendments of 1990 created a need to prepare for the wider use of alternative-fuel vehicles to reduce oil imports and increase the use of U.S. sourced renewable fuels such as methanol and ethanol. Tests have shown that neither painted nor bare terne-coated steel will resist the corrosion effects of alcohol for the 10-year expected life of the fuel tank. In addition, the EPA has introduced fuel-permeation standards that challenge the designs and materials used in automotive fuel tanks.

In the selection of a material for an automotive fuel tank the following factors are most important:

Manufacturability/cost/weight/corrosion/permeability/recyclability

Another critical factor is safety and the ability to meet crash requirements.

Two new competing materials have emerged to replace terne steel: electrocoated zinc-nickel steel sheet and high-density polyethylene (HDPE). The steel sheet is painted on both sides with an aluminum-rich epoxy. The epoxy is needed to provide exterior protection from road-induced corrosion. Stainless steel performs admirably for this application but at a nearly $5\times$ cost premium. HDPE is readily formed by blow-molding into the necessary shape, and has long-term structural stability, but it will not meet the permeability requirement. Two approaches have been used to overcome this problem. The first is multilayer technology, in which an inner layer of HDPE is adhesively joined to a barrier layer of polyamide. The second approach is a barrier technology which involves treating the HDPE with fluorine.

The first step in arriving at a material selection is to make a competitive analysis, laying out the advantages and disadvantages of each candidate.

Steel

Terne-coated steel

Advantages: low cost at high volumes, modest material cost, meets permeability requirement

Disadvantages: shape flexibility, poor corrosion protection from alcohol fuels, lead containing coating gives problems with recycling or disposal.

Electrocoated Zn-Ni steel

Advantages: low cost at high volumes, effective corrosion protection, material cost, meets permeability requirement

Disadvantages: weldability, shape flexibility

Stainless steel

Advantages: corrosion, recyclable, meets permeability requirements

Disadvantages: cost at all volumes, formability, weldability

---

1. P. J. Alvarado, Steel vs. Plastics: The Competition for Light-Vehicle Fuel Tanks, *JOM,* July, 1996, pp. 22–25.

Plastics

HDPE

  Advantages: shape flexibility, low tooling costs at low volumes, weight, corrosion resistance

  Disadvantages: high tooling costs at high volumes, high material cost, does not meet permeability and recyclability requirements

Multilayer and barrier HDPE

  Advantages: same as HDPE plus meets permeability requirement

  Disadvantages: higher tooling costs at high volume, higher material cost, hard to recycle

The next step would be to use one of the matrix methods discussed in Sec. 8.9 to arrive at the selection. As of 1996, two of the big-three automobile producers are changing to some variant of the Zn-Ni coated steel. The other is going with HDPE fuel tanks.

8.15
MATERIALS IN DESIGN

This chapter is titled Materials Selection and Materials in Design. So far we have focused on the issues of materials selection. The rest of the chapter deals with topics of material performance that are not usually covered in courses in strength of materials or machine design, but with which the mechanical designer needs to be familiar. Specifically we consider the following topics:

- Design for Brittle Fracture (Sec. 8.16)
- Design for Fatigue Failure (Sec. 8.17)
- Design for Corrosion Resistance (Sec. 8.18)
- Designing with Plastics (Sec. 8.19)
- Designing with Brittle Materials (Sec. 8.20)

While the remainder of this chapter on materials in design is quite comprehensive, it does not consider all of the ailments that can befall materials and cause them to fail. The most prominent omissions are high temperature creep and rupture,[1] oxidation,[2] abrasion, erosion, and wear,[3] and a variety of embrittling mechanisms.[4]

8.16
DESIGN FOR BRITTLE FRACTURE

An important advance in engineering knowledge has been the ability to predict the influence of cracks and cracklike defects on the brittle fracture of materials through

---

1. D. A. Woodford, Design for High-Temperature Applications, "ASM Handbook," vol. 20, pp. 573–588, ASM International, Materials Park, OH, 1997.
2. J. L. Smialek, C. A. Barrett, and J. C. Schaeffer, Design for Oxidation Resistance, "ASM Handbook," vol. 20, pp. 589–602.
3. R. G. Bayer, Design for Wear Resistance, "ASM Handbook," vol. 20, pp. 603–614; Friction, Lubrication , and Wear Technology, "ASM Handbook," vol. 18, 1992.
4. G. H. Koch, Stress-Corrosion Cracking and Hydrogen Embrittlement, "ASM Handbook," vol. 19, pp. 483–506, 1996.

the science of fracture mechanics.[1] Fracture mechanics had its origin in the ideas of A. A. Griffiths, who showed that the fracture strength of a brittle material, like glass, is inversely proportional to the square root of the crack length. G. R. Irwin proposed that fracture occurs at a fracture stress corresponding to a critical value of crack-extension force G_c, according to

$$\sigma_f = \left(\frac{EG_c}{\pi a} \right)^{1/2}$$

(8.14)

where G_c = crack extension force, in, lb/in^2
     E = elastic modulus of material, lb/in^2
     a = length of the crack, in

An important conceptualization was that the elastic stresses in the vicinity of a crack tip (Fig. 8.19) could be expressed entirely by a stress field parameter K called the stress intensity factor.

The equations for the stress field at the end of the crack can be written

$$\sigma_x = \frac{K}{\sqrt{2\pi r}} \left[\cos \frac{\theta}{2} \left(1 - \sin \frac{\theta}{2} \sin \frac{3\theta}{5} \right) \right]$$

$$\sigma_y = \frac{K}{\sqrt{2\pi r}} \left[\cos \frac{\theta}{2} \left(1 + \sin \frac{\theta}{2} \sin \frac{3\theta}{2} \right) \right]$$

$$\tau_{xy} = \frac{K}{\sqrt{2\pi r}} \left(\sin \frac{\theta}{2} \cos \frac{\theta}{2} \cos \frac{3\theta}{2} \right)$$

(8.15)

Equations (8.15) show that the elastic normal and elastic shear stresses in the vicinity of the crack tip depend only on the radial distance from the tip r, the orientation θ, and K. Thus, the magnitudes of these stresses at a given point are dependent

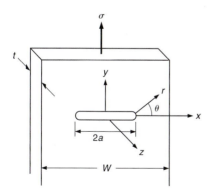

FIGURE 8.19
Model for equations for stress at a point near a crack.

1. J. F. Knott, "Fundamentals of Fracture Mechanics," Wiley, New York, 1973; S. T. Rolfe and J. M. Barsom, "Fracture and Fatigue Control in Structures," 2d ed., Prentice-Hall, Englewood Cliffs, NJ, 1987; T. L. Anderson, "Fracture Mechanics Fundamentals and Applications," CRC Press, Boca Raton, FL, 1991; D. P. Miannay, "Fracture Mechanics," Springer-Verlag, New York, 1998.

completely on the stress intensity factor K. However, the value of K depends on the nature of the loading, the configuration of the stressed body, and the mode of crack displacement, i.e., mode I (opening mode), mode II (shearing), or mode III (tearing).

For a center crack of length $2a$ in an infinite thin plate subjected to a uniform tensile stress σ the stress intensity factor K is given by

$$K = \sigma\sqrt{\pi a} = GE \tag{8.16}$$

where K is in ksi$\sqrt{\text{in}}$ or MPa$\sqrt{\text{m}}$. Values of K have been determined for a variety of situations by using the theory of elasticity, often combined with numerical methods and experimental techniques.[1] For a given type of loading, Eq. (8.16) usually is written as

$$K = \alpha\sigma\sqrt{\pi a} \tag{8.17}$$

where α is a parameter that depends on the specimen and crack geometry. For example, for a plate of width w containing a central crack of length $2a$

$$K = \sigma\sqrt{\pi a}\left(\frac{w}{\pi a}\tan\frac{\pi a}{w}\right)^{1/2} \tag{8.18}$$

Since the crack tip stresses can be described by the stress intensity factor K, a critical value of K can be used to define the conditions that produce brittle fracture. The tests usually used subject the specimen to the crack opening mode of loading (mode I) under a condition of plane strain at the crack front. The critical value of K that produces fracture is K_{Ic}, the plane-strain fracture toughness. If a_c is the critical crack length at which failure occurs, then

$$K_{Ic} = \alpha\sigma\sqrt{\pi a_c} \tag{8.19}$$

K_{Ic} is a basic material parameter called fracture toughness. If K_{Ic} is known, then it is possible to compute the maximum allowable stress to prevent brittle fracture for a given flaw size. As Fig. 8.20 illustrates, the allowable stress in the presence of a crack of a given size is directly proportional to K_{Ic}, and the allowable crack size for a given stress is proportional to the square of the fracture toughness. Therefore, increasing K_{Ic} has a much larger influence on allowable crack size than on allowable stress. For increasing load in a cracked part, a higher fracture toughness results in a larger allowable crack size or larger allowable stresses at fracture.

Although K_{Ic} is a basic material property, in the same sense as yield strength, it changes with important variables such as temperature and strain rate. The K_{Ic} of materials with a strong temperature and strain-rate dependence usually decreases with

1. A. S. Kobayashi (ed.), "Experimental Techniques in Fracture Mechanics," Society for Experimental Stress Analysis, Westport, CT, 1973; G. G. Sih, "Handbook of Stress Intensity Factors," Institute of Fracture and Solid Mechanics, Lehigh University, Bethlehem, PA, 1973; P. C. Paris and G. C. Sih, ASTM Spec. Tech. Publ. 381, pp. 30–83, 1965; Y. Murakami et al., (eds.), "Stress Intensity Factors Handbook," (2 vols.), Pergamon Press, New York, 1987; A. Liu, Summary of Stress-Intensity Factors, "ASM Handbook," vol. 19, pp. 980–1000, ASM International, Materials Park, OH, 1996.

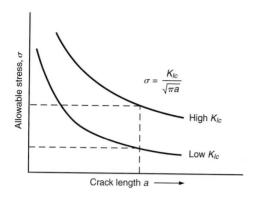

FIGURE 8.20
Relation between fracture toughness and allowable stress and crack size.

$$\sigma = \frac{K_{Ic}}{\sqrt{\pi a}}$$

decreased temperature and increased strain rate. The K_{Ic} of a given alloy is strongly dependent on such variables as heat treatment, texture, melting practice, impurity level, and inclusion content.

The fracture toughness measured under plane-strain conditions is obtained under maximum constraint or material brittleness. The plane-strain fracture toughness is designated K_{Ic} and is a true material property. Figure 8.21 shows how the measured fracture stress varies with specimen thickness B. A mixed-mode, ductile brittle fracture with 45° shear lips is obtained for thin specimens. Once the specimen has the critical thickness for the toughness of the material, the fracture surface is flat and the fracture stress is constant with increasing specimen thickness. The minimum thickness to achieve plane-strain conditions and valid K_{Ic} measurement is

$$B = 2.5\left(\frac{K_{Ic}}{\sigma_0}\right) \tag{8.20}$$

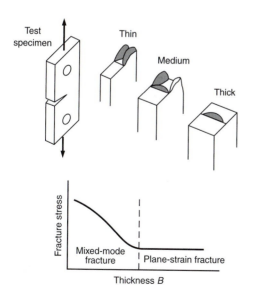

FIGURE 8.21
Effect of specimen thickness on stress and mode of fracture. Sketches depict appearance of specimen fracture surface.

where σ_0 is the 0.2 percent offset yield strength. Standardized test specimens and test procedures for fracture toughness testing have been developed by ASTM.[1]

The fracture mechanics concept is strictly correct only for linear elastic materials, i.e., under conditions in which no yielding occurs. However, reference to Eq. (8.15) shows that as r approaches zero the stress at the crack tip approaches infinity. Thus, in all but the most brittle material, local yielding occurs at the crack tip and the elastic solution should be modified to account for crack tip plasticity. However, if the plastic zone size r_y at the crack tip is small relative to the local geometry, for example, r_y/t or $r_y/a < 0.1$, crack tip plasticity has little effect on the stress intensity factor. That limits the strict use of fracture mechanics to high-strength materials. Moreover, the width restriction to obtaining valid measurements of K_{Ic}, as described in Eq. (8.20), makes the use of linear elastic fracture mechanics (LEFM) impractical for low-strength materials. Extensive research to extend LEFM to low-strength materials is underway.[2]

DESIGN EXAMPLE. We shall illustrate the use of LEFM in design with the example of a disk spinning at high speed.[3] Rotating disks are important components in steam and gas turbines. The maximum stress is a circumferential tensile stress that occurs at the surface of the bore of the disk.[4]

$$\sigma_{max} = \frac{3 + v}{4} \rho \omega^2 c^2 \left[1 + \frac{1 - v}{3 + v} \left(\frac{b}{c} \right)^2 \right] \tag{8.21}$$

where ρ = mass density
v = Poisson's ratio
ω = $2\pi N/60$, the angular velocity

and c and b are as defined in Fig. 8.22. The stress intensity factor for a crack of length a located at the bore of a disk is given by[5] Eq. (8.22) if $b/a > 10$.

$$K_{Ic} = 1.12\sigma_{max}\sqrt{\pi a} \tag{8.22}$$

Combining Eqs. (8.21) and (8.22) results in an equation for the critical speed, in rpm, of the disk in terms of the fracture toughness and the crack length.

$$N_c = 13.560 \frac{\sqrt{K_{Ic}}}{a^{1/4}\sqrt{\rho[(3 + v)c^2 + (1 - v)b^2]}} \tag{8.23}$$

Critical speed is plotted vs. crack length in Fig. 8.22 by using material parameters of a high-strength 4340 steel.

It is important to note that even in this relatively simple example the use of fracture mechanics lends important realism to the calculation. If $a = 0.1$ in and $c/b = 2$, the critical speed, from Fig. 8.22, is 4000 rpm. In the conventional design approach, which ignores the presence of a crack, N_c is determined by setting σ_{max} in Eq. (8.21) equal to the ultimate tensile strength. This oversimplification leads to a critical speed of 8400 rpm.

1. ASTM Standards, Part 31, Designation E399-70T.
2. A. Saxena, "Nonlinear Fracture Mechanics for Engineers," CRC Press, Boca Raton, FL, 1998.
3. G. C. Sih, *Trans. ASME, Ser. B., J. Eng. Ind.,* vol. 98, pp. 1243–1249, 1976.
4. R. J. Roark and W. C. Young, "Formulas for Stress and Strain," 6th ed., McGraw-Hill, New York, 1989.
5. G. C. Sih, op. cit. For other design applications see A. Blake, "Practical Fracture Mechanics in Design," Marcel Dekker, New York, 1996.

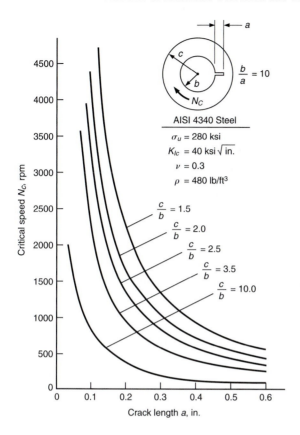

FIGURE 8.22
Critical speed as a function of crack length. (*From G. C. Sih, Trans. ASME, sec. B., vol. 98, p. 1245, 1976.*)

8.17
DESIGN FOR FATIGUE FAILURE

Materials subjected to a repetitive or fluctuating stress will fail at a stress much lower than that required to cause fracture on a single application of load. Failures occurring under conditions of fluctuating stresses or strains are called fatigue failure.[1] Fatigue accounts for the majority of mechanical failures in machinery.

A fatigue failure is a localized failure that starts in a limited region and propagates with increasing cycles of stress or strain until the crack is so large that the part cannot withstand the applied load, and it fractures. Plastic deformation processes are involved in fatigue, but they are highly localized.[2] Therefore, fatigue failure occurs without the warning of gross plastic deformation. Failure usually initiates at regions of local high stress or strain because of abrupt changes in geometry (stress concentration), temperature differentials, residual stresses, or material imperfections. Much

1. G. E. Dieter, "Mechanical Metallurgy," 3d ed., Chap. 12, McGraw-Hill, New York, 1986; S. Suresh, "Fatigue of Materials," Cambridge University Press, Cambridge, 1991; N. E. Dowling, "Mechanical Behavior of Materials," 2d ed., Prentice-Hall, Englewood Cliffs, NJ, 1999; Fatigue and Fracture, "ASM Handbook," vol. 19, ASM International, Materials Park, OH, 1996.
2. "Fatigue and Microstructure," American Society for Metals, Metals Park, OH, 1979; "ASM Handbook," vol. 19, pp. 63–109.

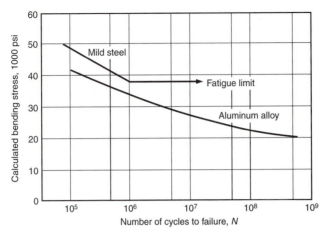

FIGURE 8.23
Typical fatigue curves for ferrous and nonferrous metals.

basic information has been obtained about the mechanism of fatigue failure, but at present the chief opportunities for preventing fatigue lie at the engineering design level. Fatigue prevention is achieved by proper choice of material, control of residual stress, and minimization of stress concentrations through careful design.

Basic fatigue data are presented in the *S-N* curve, a plot of stress *S* vs. the number of cycles to failure *N*. Figure 8.22 shows the two typical types of behavior. The *S-N* curve is chiefly concerned with fatigue failure at high numbers of cycles ($N > 10^5$ cycles). Under these conditions the gross stress is elastic, although fatigue failure results from highly localized plastic deformation. As can be seen from Fig. 8.23, the number of cycles of stress that a material can withstand before failure increases with decreasing stress. For most materials, e.g., aluminum alloys, the *S-N* curve slopes continuously downward with decreasing stress. At any stress level there is some large number of cycles that ultimately causes failure. For steels in the absence of a corrosive environment, however, the *S-N* curve becomes horizontal at a certain limiting stress. Below that stress, called the fatigue limit, the steel can withstand an infinite number of cycles.

8.17.1 Fatigue Design Criteria

There are several distinct philosophies concerning design for fatigue that must be understood to put this vast subject into proper perspective.

1. *Infinite-life design:* This design criterion is based on keeping the stresses at some fraction of the fatigue limit of the steel. This is the oldest fatigue design philosophy. It has largely been supplanted by the other criteria discussed below. However, for situations in which the part is subjected to very large cycles of uniform stress it is a valid design criterion.
2. *Safe-life design:* Safe-life design is based on the assumption that the part is initially flaw-free and has a finite life in which to develop a critical crack. In this approach to design one must consider that fatigue life at a constant stress is sub-

ject to large amounts of statistical scatter. For example, the Air Force historically designed aircraft to a safe life that was one-fourth of the life demonstrated in full-scale fatigue tests of production aircraft. The factor of 4 was used to account for environmental effects, material property variations, and variations in as-manufactured quality. Bearings are another good example of parts that are designed to a safe-life criterion. For example, the bearing may be rated by specifying the load at which 90 percent of all bearings are expected to withstand a given lifetime. Safe-life design also is common in pressure vessel and jet engine design.

3. *Fail-safe design:* In fail-safe design the view is that fatigue cracks may occur: therefore, the structure is designed so that cracks will not lead to failure before they can be detected and repaired. This design philosophy developed in the aircraft industry, where the weight penalty of using large safety factors could not be tolerated but neither could the danger to life from very small safety factors be tolerated. Fail-safe designs employ multiple-load paths and crack stoppers built into the structure along with rigid regulations and criteria for inspection and detection of cracks.

4. *Damage-tolerant design:* The latest design philosophy is an extension of the fail-safe design philosophy. In damage-tolerant design the assumption is that fatigue cracks will exist in an engineering structure. The techniques of fracture mechanics are used to determine whether the cracks will grow large enough to cause failure before they are sure to be detected during a periodic inspection. The emphasis in this design approach is on using materials with high fracture toughness and slow crack growth. The success of the design approach depends upon having a reliable nondestructive evaluation (NDE) program and in being able to identify the damage critical areas in the design.

Although much progress has been made in designing for fatigue, especially through the merger of fracture mechanics and fatigue, the interaction of many variables that is typical of real fatigue situations makes it inadvisable to depend on a design based solely on analysis. Simulated service testing[1] should be part of all critical fatigue applications. The failure areas not recognized in design will be detected by these tests. Simulating the actual service loads requires great skill and experience. Often it is necessary to accelerate the test, but doing so may produce misleading results. For example, when time is compressed in that way, the full influence of corrosion or fretting is not measured, or the overload stress may appreciably alter the residual stresses. It is common practice to eliminate many small load cycles from the load spectrum, but they may have an important influence on fatigue crack propagation.

The following are some of the most important engineering factors that determine fatigue performance.

Stress cycle
   Repeated or random
   Mean stress
Combined stress state

---

1. R. M. Wetzel (ed.), "Fatigue Under Complex Loading," Society of Automotive Engineers, Warrendale, PA, 1977.

Stress concentration
    Fatigue notch factor
    Fatigue notch sensitivity
Statistical variation in fatigue life and fatigue limit
Size effect
Surface finish
Surface treatment
Residual stress
Corrosion fatigue
Fretting fatigue
Cumulative fatigue damage

Space precludes a discussion of these factors. The reader is referred to the basic references that have been cited for details on how the factors control fatigue. There is a considerable literature on design methods to prevent fatigue failure.

Fatigue design for infinite life is considered in:

R. C. Juvinall, "Engineering Consideration of Stress, Strain, and Strength," McGraw-Hill, New York, 1967. Chapters 11 to 16 cover fatigue design in considerable detail.
L. Sors, "Fatigue Design of Machine Components," Pergamon Press, New York, 1971. Translated from the German, this presents a good summary of European fatigue design practice.
C. Ruiz and F. Koenigsberger, "Design for Strength and Production," Gordon & Breach Science Publishers, New York, 1970. Pages 106 to 120 give a concise discussion of fatigue design procedures.

Detailed information on stress concentration factors and the design of machine details to minimize stress can be found in:

W. D. Pilkey, "Peterson's Stress Concentration Factors," 2d ed., Wiley, New York, 1997.
R. B. Heywood, "Designing Against Fatigue of Metals," Reinhold, New York, 1967.
C. C. Osgood, "Fatigue Design," 2d ed., Pergamon Press, New York, 1982.

The most complete books on fatigue design, which considers the more modern work on safe-life design and damage-tolerant design are:

H. O. Fuchs and R. I. Stephens, "Metal Fatigue in Engineering," Wiley, New York, 1980.
J. E. Bannantine, J. J. Comer, and J. L. Handrock, "Fundamentals of Metal Fatigue Analysis," Prentice-Hall, Englewood Cliffs, NJ, 1990.
"Fatigue Design Handbook," 3d ed., Society of Automotive Engineers, Warrendale, PA, 1997.
E. Zahavi, "Fatigue Design," CRC Press, Boca Raton, FL, 1996.

We will now attempt to bring this huge subject of fatigue into focus by presenting several design examples.

DESIGN EXAMPLE—INFINITE-LIFE DESIGN.[1] A steel shaft heat-treated to a Brinell hardness of 200 has a major diameter of 1.5 in and a small diameter of 1.0 in. There is a 0.10-in radius at the shoulder between the diameters. The shaft is subjected to completely reversed cycles of stress of pure bending. The fatigue limit determined on pol-

---

1. Based on design procedures described by R. C. Juvinall, "Engineering Consideration of Stress, Strain, and Strength," McGraw-Hill, New York, 1967.

ished specimens of 0.2-in diameter is 42,000 psi. The shaft is produced by machining from bar stock. What is the best estimate of the fatigue limit of the shaft?

Since an experimental value for fatigue limit is known, we start with it, recognizing that tests on small unnotched polished specimens represent an unrealistically high value of the fatigue limit of the actual part.[1] The procedure, then, is to factor down the idealized value.

We start with the stress concentration (notch) produced at the shoulder between two diameters of the shaft. A shaft with a fillet in bending is a standard situation covered in all machine design books. If $D = 1.5$, $d = 1.0$, and $r = 0.10$, the important ratios are $D/d = 1.5$ and $r/d = 0.1$. Then, from standard curves, the theoretical stress concentration factor is $K_t = 1.68$. However, K_t is determined for a brittle elastic solid, and most materials exhibit a lesser value of stress concentration when subjected to fatigue. The extent to which the plasticity of the material reduces K_t is given by the fatigue notch sensitivity q.

$$q = \frac{K_f - 1}{K_t - 1} \tag{8.24}$$

where K_t = theoretical stress concentration factor

$$K_f = \text{fatigue notch factor} = \frac{\text{fatigue limit unnotched}}{\text{fatigue limit notched}}$$

For a BHN 200 steel, $q = 0.8$ and $K_f = 1.54$.

Returning to the fatigue limit for a small polished specimen, $S'_\text{fl} = 42,000$ psi, we need to reduce this value because of size effect, surface finish, and type of loading and for statistical scatter

$$S_\text{fl} = S'_\text{fl} C_S C_F C_L C_Z \tag{8.25}$$

where C_S = factor for size effect
$\quad\quad\ C_F$ = factor for surface finish
$\quad\quad\ C_L$ = factor for type of loading
$\quad\quad\ C_Z$ = factor for statistical scatter

Increasing the specimen size increases the probability of surface defects, and hence the fatigue limit decreases with increasing size. Typical values of C_S are given in Table 8.9. In this example we use $C_S = 0.9$.

TABLE 8.9
Fatigue reduction factor
due to size effect

Diameter, in	C_S
$D \le 0.4$	1.0
$0.4 \le D \le 2.0$	0.9
$2.0 \le D \le 9.0$	$1 - \dfrac{D - 0.03}{15}$

---

1. If fatigue data are not given, they must be determined from the published literature or estimated from other mechanical properties of the material; see H. O. Fuchs and R. I. Stephens, "Metal Fatigue in Engineering," pp. 156–160, Wiley, New York, 1980; "ASM Handbook," vol. 19, 1996.

TABLE 8.10 Loading factor for fatigue tests	
Loading type	C_L
Bending	1.0
Torsion	0.58
Axial	0.9

TABLE 8.11 Statistical factor for fatigue limit	
Reliability, percent	C_Z
50	1.0
99	0.814
99.9	0.752

Curves for the reduction in fatigue limit due to various surface finishes are available in standard sources.[1] For a standard machined finish in a steel of BHN 200, $C_F = 0.8$.

Laboratory fatigue data (as opposed to simulated service fatigue tests) commonly are determined in a reversed bending loading mode. Other types of loading, e.g., axial and torsional, generate different stress gradients and stress distributions and do not produce the same fatigue limit for the same material. Thus, fatigue data generated in reversed bending must be corrected by a load factor C_L, if the data are to be used in a different loading mode. Table 8.10 gives typical values. Since the bending fatigue data are used for an application involving bending, $C_L = 1.0$.

Fatigue tests show considerable scatter in results. Fatigue limit values are normally distributed with a standard deviation that can be 8 percent of the mean value. If the test or literature value is taken as the mean value of fatigue limit (which in itself is a big assumption), then this value is reduced by a statistical factor[2] according to the reliability level that is desired (Table 8.11).

If we assume that a 99 percent reliability level is acceptable, then $C_Z = 0.814$. Therefore, the unnotched fatigue limit corrected for these factors is

$$S = S'_{fl} C_S C_F C_L C_Z$$

$$= 42{,}000(0.9)(0.8)(1.0)(0.81) = 24{,}494 \text{ psi}$$

Since $K_f = 1.54$, the fatigue limit of the notched shaft is estimated to be

$$S_{fl\,shaft} = \frac{24.494}{1.54} = 15{,}900 \text{ psi}$$

We note that the working stress is 38 percent of the laboratory value of fatigue limit.

This example is fairly realistic, but it has not included the important situation in which the mean stress is other than zero.[3] This also permits consideration of fatigue strengthening with compressive residual stresses.

DESIGN EXAMPLE—SAFE-LIFE DESIGN. Safe-life design based on failure at a finite number of cycles as influenced by mean stress can be treated by Haigh diagram as presented by Fuchs.[4] However, a more general and rapidly growing viewpoint is to con-

1. R. C. Juvinall, op. cit., p. 234.
2. G. Castleberry, *Machine Design,* pp. 108–110, Feb. 23, 1978.
3. See, for example, R. C. Juvinall, op. cit., chap. 14.
4. H. O. Fuchs, *Trans. ASME,* Ser. D, *J. Basic Eng.,* vol. 87, pp. 333–343, 1965; H. O. Fuchs and R. I. Stephens, op. cit., pp. 148–160.

sider design based on strain-life curves, which are often called low-cycle fatigue curves because much of the data is obtained at less than 10^5 cycles.

When fatigue occurs at a relatively low number of cycles, the stresses that produce failure often exceed the yield strength. Even when the gross stress remains elastic, the localized stress at a notch is inelastic. Under these conditions it is better to carry out fatigue tests under fixed amplitude of strain (strain control) rather than fixed amplitude of stress. Figure 8.24 shows the stress-strain loop that is produced by strain control cycling.

The strain-life curve is expressed in terms of the total strain amplitude vs. the number of strain reversals to failure (Fig. 8.25). At given life the total strain is the sum of the elastic and plastic strains.

$$\frac{\Delta\varepsilon}{2} = \frac{\Delta\varepsilon_e}{2} + \frac{\Delta\varepsilon_p}{2} \tag{8.26}$$

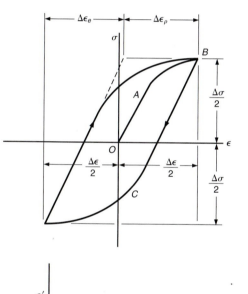

FIGURE 8.24
Typical cyclic stress-strain loop.

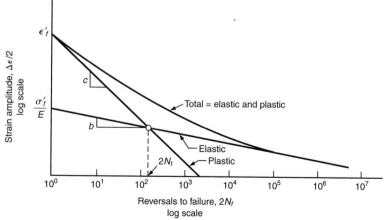

FIGURE 8.25
Typical strain-life curve for mild steel.

Both the elastic and plastic curves are approximated as straight lines. At small strains or long lives and the elastic strain predominates, and at large strains or short lives the plastic strain is predominant. The plastic curve has a negative slope of c and an intercept at $2N = 1$ of ε'_f. The elastic curve has a negative slope of b and an intercept of σ'_f/E. Substituting into Eq. (8.26) yields

$$\frac{\Delta\varepsilon}{2} = \frac{\sigma'_f}{E}(2N)^b + \varepsilon'_f(2N)^c \qquad (8.27)$$

The exponent b ranges from about -0.06 to -0.14 for different materials, and a typical value is -0.1. The exponent c ranges from about -0.5 to -0.7, and -0.6 is a representative value. The term ε'_f in Eq. (8.27) called the fatigue ductility coefficient, is some fraction (0.35 to 1.0) of the true fracture strain measured in the tension test. Likewise, the fatigue strength coefficient σ'_f is approximated by the true fracture stress. Manson[1] has simplified Eq. (8.27) with his method of universal slopes to give

$$\Delta\varepsilon = 3.5\frac{S_u}{E}N^{-0.12} + \varepsilon_f^{0.6}N^{-0.6} \qquad (8.28)$$

where S_u = ultimate tensile strength
$\quad\;\; \varepsilon_f$ = true strain at fracture in tension
$\quad\;\; E$ = elastic modulus in the tension test

Equation (8.28) can be used as the first approximation for the strain-life curve for fully reversed fatigue cycles in an unnotched specimen. Strain-life curves give the number of cycles to the formation of a detectable crack with a length between 0.25 and 5 mm, i.e., the life to crack initiation.

An important use of the low-cycle fatigue approach is to predict the life to crack initiation at notches in machine parts where the nominal stresses are elastic but the local stresses and strain at the notch root are inelastic. When there is plastic deformation, both a strain concentration K_ε and a stress concentration K_σ must be considered. Neuber's rule relates these by

$$K_f = (K_\sigma K_\varepsilon)^{1/2} \qquad (8.29)$$

The situation is described in Fig. 8.26a, where ΔS and Δe are the elastic stress and strain increments at a location remote from the notch and $\Delta\sigma$ and $\Delta\varepsilon$ are the local stress and strain at the root of the notch.

$$K_\sigma = \frac{\Delta\sigma}{\Delta S} \qquad \text{and} \qquad K_\varepsilon = \frac{\Delta\varepsilon}{\Delta e}$$

$$K_f = \left(\frac{\Delta\sigma\Delta\varepsilon}{\Delta S\,\Delta e}\right)^{1/2} = \left(\frac{\Delta\sigma\;\Delta\varepsilon E}{\Delta S\,\Delta\varepsilon E}\right)^{1/2} \qquad (8.30)$$

and $\qquad\qquad K_f(\Delta S\,\Delta eE)^{1/2} = (\Delta\sigma\;\Delta\varepsilon E)^{1/2} \qquad (8.31)$

For nominally elastic loading, $\Delta S = \Delta eE$, and

1. S. S. Manson, *Exp. Mechanics*, vol. 5, no. 7, p. 193, 1965.

$$K_f \, \Delta S = (\Delta \sigma \, \Delta \varepsilon E)^{1/2} \tag{8.32}$$

Thus, Eq. (8.32) allows stresses and strains remotely measured from the notch to be used to predict notch behavior. Rearranging Eq. (8.32) gives

$$\Delta \sigma \, \Delta \varepsilon = \frac{(K_f \, \Delta S)^2}{E} = \text{const} \tag{8.33}$$

which is the equation of a rectangular hyperbola (Fig. 8.26c). If a nominal stress S_1 is applied to the notched specimen, (Fig. 8.26b,) then the right side of Eq. (8.33) is known provided K_f is known. The cyclic stress-strain curve also is plotted in Fig. 8.26 (solid curve), and its intersection with Eq. (8.33) gives the local stress and strain at the notch root. If the low-cycle fatigue curve (Fig. 8.25) is entered with the value of $\Delta \varepsilon$ at the notch root, an estimate of the fatigue life is obtained.

Actual application of this analysis technique is often much more complex than this simple example.[1] However, the strain-based technique has shown great promise for fatigue analysis of structural components. The technique[2] utilizes the cyclic stress-strain curve and low-cycle fatigue curve. The complex strain-time history that the part experiences is determined experimentally, and this strain signal is separated into individual cycles by a cycle-counting routine such as the rainflow method. The method separates the strain history into stress-strain hysteresis loops that are comparable with those found in constant-amplitude strain-controlled cycling. Then the notch analysis described above is used in a computer-aided procedure to determine the stresses and strains at the notch for each cycle and the fatigue life is assessed with a cumulative damage rule. Although the design procedures still are evolving, they show great promise of placing fatigue life prediction on a firm basis.

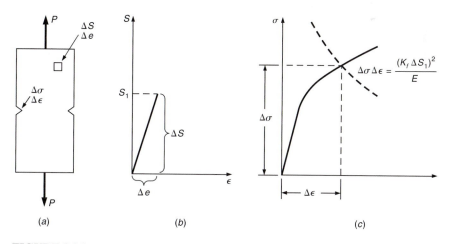

(a)   (b)   (c)

FIGURE 8.26
Notch stress analysis based on Neuber's analysis.

---

1. H. O. Fuchs and R. I. Stephens, op. cit., pp. 161–165.
2. R. W. Landgraf and N. R. LaPointe, Society of Automotive Engineers, Paper 740280, 1974; A. R. Michetti, *Exp. Mechanics,* pp. 69–76, February 1977.

DESIGN EXAMPLE—DAMAGE-TOLERANT DESIGN. Damage-tolerant design starts with the premise that the part contains a fatigue crack of known dimensions and geometry and predicts how many cycles of service are available before the crack will propagate to a catastrophic size that will cause failure. Thus, emphasis is on fatigue crack growth.

Figure 8.27 shows the process of crack propagation from an initial crack of length a_0 to a crack of critical flaw size a_{cr}. The crack growth rate da/dN increases with the cycles of repeated load. An important advance in fatigue design was the realization that there is a generalization in fatigue growth data if the crack growth rate da/dN, is plotted against the stress intensity factor range, $\Delta K = K_{max} - K_{min}$ for the fatigue cycle. Since the stress intensity factor $K = \alpha\sigma\sqrt{\pi a}$ is undefined in compression, K_{min} is taken as zero if σ_{min} is compression in the fatigue cycle.

Figure 8.28 shows a typical plot of rate of crack growth versus ΔK. The typical curve is sigmoidal in shape with three distinct regions. Region I contains the threshold value ΔK_{th} below which there is no observable crack growth. Below K_{th} fatigue cracks behave as nonpropagating cracks. The threshold starts at crack propagation rates of around 10^{-8} in/cycle and at very low values of ΔK, for example 8 ksi$\sqrt{in}$ for stress. Region II exhibits essentially a linear relation between log da/dN and log ΔK, which results in

$$\frac{da}{dN} = A(\Delta K)^n \tag{8.34}$$

where $n =$ slope of the curve in region II
$A =$ coefficient found by extending the line to $\Delta K = 1$ ksi$\sqrt{in}$

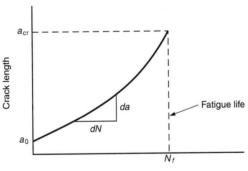

FIGURE 8.27
Process of crack propagation (schematic).

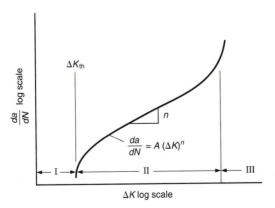

FIGURE 8.28
Schematic fatigue crack growth vs. ΔK curve.

Region III is a region of rapid crack growth that is soon followed by failure. This region is controlled primarily by the fracture toughness K_{Ic}.

The relation between fatigue crack growth and ΔK expressed by Eq. (8.34) ties together fatigue design[1] and linear elastic fracture mechanics (LEFM). The elastic stress intensity factor is applicable to fatigue crack growth even in low-strength, high-ductility materials because the K values needed to cause fatigue crack growth are very low and the plastic zone sizes at the tip are small enough to permit an LEFM approach. By correlating crack growth and stress intensity factor, it is possible to use data generated under constant-amplitude conditions with simple specimens for a broad range of design situations in which K can be calculated. When K is known for the component under relevant loading conditions, the fatigue crack growth life of the *component* can be obtained by integrating Eq. (8.34) between the limits of initial crack size and final crack size. From Sec. 8.16, recall that

$$\Delta K = \alpha \sigma \sqrt{\pi a} \tag{8.35}$$

$$\frac{da}{dN} = A(\Delta K)^n = A(\Delta \sigma \sqrt{\pi a} \alpha)^n \tag{8.36}$$

$$= A(\Delta \sigma)^n (\pi a)^{n/2} \alpha^n$$

Now, if the final crack length a_f results in fracture,

$$\sigma_f = \frac{1}{\pi} \left(\frac{K_{Ic}}{\sigma_{max}\alpha} \right)^2 \tag{8.37}$$

$$N_f = \int_0^{N_f} dN = \int_{a_0}^{a_f} \frac{da}{A(\Delta \sigma)^n (\pi a)^{n/2} \alpha^n} \tag{8.38}$$

$$N_f = \frac{a_f^{(-n/2)+1} - a_0^{(-n/2)+1}}{(-n/2+1)A(\Delta \sigma)^n (\pi)^{n/2} \alpha^n} \tag{8.39}$$

Equation (8.39) is the integration of Eq. (8.36) for the special case in which α is not a function of a and n 2. Other cases require numerical integration.

Working with Eq. (8.39) shows that the fatigue life is much more dependent on the small initial crack size than on K_{Ic}. Thus, refined NDE techniques to detect small flaws are an important part of a fracture control program.

8.18
DESIGN FOR CORROSION RESISTANCE

Failure of metal components by corrosion is as common as failure due to mechanical causes, such as brittle fracture and fatigue. The National Institute of Standards and Technology estimates that corrosion annually costs the United States $70 billion, of

---

1. R. P. Wei, *Trans. ASME*, Ser. H., *J. Eng. Materials Tech.*, vol. 100, pp. 113–120, 1978; A. F. Liu, "Structural Life Assessment Methods," ASM International, Materials Park, OH, 1998.

which at least $10 billion could be prevented by better selection of materials and design procedures. Although corrosion failures are minimized by proper materials selection and careful attention to control of metallurgical structure through heat treatment and processing, many corrosion-related failures can be minimized by proper understanding of the interrelation of the fundamental causes of corrosion and design details.[1]

Corrosion of metals is driven by the basic thermodynamic force of a metal to return to the oxide or sulfide form, but it is more related to the electrochemistry of the reactions of a metal in an electrolytic solution. There are eight basic forms of corrosion.[2]

Uniform attack. The most common form of corrosion is uniform attack. It is characterized by a chemical or electrochemical reaction that proceeds uniformly over the entire exposed surface area. The metal becomes thinner and eventually fails.

Galvanic corrosion. The potential difference that exists when two dissimilar metals are immersed in a corrosive or conductive solution is responsible for galvanic corrosion. The less-resistant (anodic) metal is corroded relative to the cathodic metal. Table 8.12 gives a brief galvanic series for some commercial alloys immersed in seawater. Note that the relative position in a galvanic series depends on the electrolytic environment as well as the metal's surface chemistry (passivity).

To minimize galvanic corrosion, use pairs of metals that are close together in the galvanic series and avoid situations in which a small anode metal is connected to a larger surface area of more noble metal. If two metals far apart in the series must be used in contact, they should be insulated electrically from each other. Do not coat the anodic surface to protect it, because most coatings are susceptible to pinholes. The coated surface would corrode rapidly in contact with a large cathodic area. When a galvanic couple is unavoidable, consider utilizing a third metal that is anodic and sacrificial to both of the other metals.

Crevice corrosion. An intense localized corrosion frequently occurs within crevices and other shielded areas on metal surfaces exposed to corrosive attack. This type usually is associated with small volumes of stagnant liquid at design details such as holes, gasket surfaces, lap joints, and crevices under bolt and rivet heads.

Pitting. Pitting is a form of extremely localized attack that produces holes in the metal. It is an especially insidious form of corrosion because it causes equipment to fail after only a small percentage of the designed-for weight loss.

Intergranular corrosion. Localized attack along the grain boundaries with only slight attack of the grain faces is called intergranular corrosion. It is especially common in austenitic stainless steel that has been sensitized by heating to the temperature range 950 to 1450°F. It can occur either during heat treatment for stress relief or during welding, when it is known as *weld decay*.

---

1. V. P. Pludek, "Design and Corrosion Control," Wiley, New York, 1977.
2. M. G. Fontana and N. D. Greene, "Corrosion Engineering," 3d ed., McGraw-Hill, New York, 1986.

TABLE 8.12
**A brief galvanic series for commercial
metals and alloys**

Noble (cathodic)	Platinum
	Gold
	Titanium
	Silver
	316 stainless steel
	304 stainless steel
	410 stainless steel
	Nickel
	Monel
	Cupronickel
	Cu-Sn bronze
	Copper
	Cast iron
	Steel
Active (anodic)	Aluminum
	Zinc
	Magnesium

Selective leaching. The removal of one element from a solid-solution alloy by corrosion processes is called selective leaching. The most common example of it is the selective removal of zinc from brass (dezincification), but aluminum, iron, cobalt, and chromium also can be removed. When selective leaching occurs, the alloy is left in a weakened, porous condition.

Erosion-corrosion. Deterioration at an accelerated rate is caused by relative movement between a corrosive fluid and a metal surface; it is called erosion-corrosion. Generally the fluid velocity is high and mechanical wear and abrasion may be involved, especially when the fluid contains suspended solids. Erosion destroys protective surface films and enhances chemical attack. Design can play an important role in erosion control in such areas as reducing fluid velocity, eliminating situations in which direct impingement occurs, and minimizing abrupt changes in the direction of flow. Some erosion situations are so aggressive that neither a suitable material nor design can ameliorate the problem. Here the role of design is to provide for easy detection of damage and for quick replacement of damaged components.

A special kind of erosion-corrosion is *cavitation,* which arises from the formation and collapse of vapor bubbles near the metal surface. Rapid bubble collapse can produce shock waves that cause local deformation of the metal surface.

Another special form of erosion-corrosion is fretting corrosion. It occurs between two surfaces under load that are subjected to cycles of relative motion. Fretting produces breakdown of the surface into an oxide debris and results in surface pits and cracks that usually lead to fatigue cracks.

Stress-corrosion cracking. Cracking caused by the simultaneous action of a tensile stress and a specific corrosive medium is called stress-corrosion cracking. The stress may be a result of applied loads or "locked-in" residual stress. Only specific combinations of alloys and chemical environment lead to stress-corrosion cracking. However, may are of common occurrence, such as aluminum alloys and seawater, copper alloys and ammonia, mild steel and caustic soda, and austenitic steel and salt-water.[1] Over 80 combinations of alloys and corrosive environments are known to cause stress-corrosion cracking. Design against stress-corrosion cracking involves selecting an alloy that is not susceptible to cracking in the service environment; but if that is not possible, then the stress level should be kept low. The concepts of fracture mechanics have been applied to SCC.

General precautions. Some of the more obvious design rules for preventing corrosion failure have been discussed above. In addition, tanks and containers should be designed for easy draining and easy cleaning. Welded rather than riveted tanks will provide less opportunity for crevice corrosion. When possible, design to exclude air; if oxygen is eliminated, corrosion can often be reduced or prevented. Exceptions to that rule are titanium and stainless steel, which are more resistant to acids that contain oxidizers than to those that do not. Many examples of design details for corrosion prevention are given by Pludek[2] and Elliott.[3]

8.19
DESIGN WITH PLASTICS

Most mechanical design is taught with the implicit assumption that the part will be made from a metal. However, plastics are increasingly finding their way into design applications because of their light weight, attractive appearance, and freedom from corrosion and the ease with which many parts may be manufactured from polymers.

Polymers are sufficiently different from metals to require special attention in design.[4] With respect to mechanical properties, steel has about 100 times the Young's modulus of a polymer and about 10 times the tensile strength. Also, the strength properties of polymers are time-dependent at or near room temperature, which imposes a different way of looking at allowable strength limits. On the other hand, polymers are 1/7th as dense as metals, but their thermal conductivity is 1/200th of steel and their

1. G. H. Koch, "ASM Handbook," vol. 19, pp. 483–506, ASM International, Materials Park, OH, 1996.
2. V. R. Pludek, op. cit.
3. P. Elliott, Design Details to Minimize Corrosion in "Metals Handbook," 9th ed., vol. 13, "Corrosion," pp. 338–343, ASM International, Metals Park, OH, 1987.
4. R. D. Beck, "Plastic Product Design," 2d ed., Van Nostrand Reinhold, New York, 1980; S. Levy and J. H. DuBois, "Plastic Product Design Engineering Handbook," 2d ed., Chapman and Hall, 1985; Engineering Plastics, "Engineered Materials Handbook," vol. 2, ASM International, Materials Park, OH, 1988; M. L. Berins, ed., "Plastics Engineering Handbook of the Society of Plastics Industry," 5th ed., Van Nostrand Reinhold, New York, 1991; E. A. Muccio, "Plastic Part Technology," ASM International, 1991; D. V. Rosato, D. P. DiMattia, and D. V. Rosato, Jr., "Designing with Plastics and Composites," Van Nostrand Reinhold, New York, 1991.

thermal expansion is 7 times greater. These last two properties influence their processing (see Sec. 9.16). Therefore, these differences in properties must be allowed for in design with plastics.

8.19.1 Properties of Plastics

The majority of plastics are synthetic materials characterized by having a carbon-carbon backbone modified by other organic side groups. Plastics are made up of tens of thousands of small molecular units (mers) that are *polymerized* into long-chain macromolecules, hence the scientific term *polymers*. The polymer chains, depending on composition and processing, can take many configurations (coiled, cross-linked, crystalline), to change the polymer properties.[1] Polymers are divided into two general classes: thermoplastics (TP) and thermosets (TS). The difference between these two classes of plastics lies in the nature of bonding and their response to increases in temperature. When a TP is heated to a sufficiently high temperature it will become viscous and pliable. In this condition it can be formed into useful shapes and will retain these shapes when cooled to room temperature. If reheated to the temperature, it becomes viscous and then can be reshaped and retains the shape when cooled. When a TS polymer is heated, or a catalyst is added, covalent bonding occurs between the polymer chains, resulting in a rigid, cross-linked structure. This structure is "set in place," so that if the TS is reheated from a cooled state it will not return to the fluid viscous state but instead degrades and chars on continued heating.

Few polymers are used in their pure form. Copolymers are made by polymerizing two or more different mers so that the repeating unit in the polymer chain is not just a single mer. Styrene-acrylonitrile copolymer (SAN) and acrylonitrile-butadiene-styrene copolymer (ABS) are common examples. Blends are combinations of polymers that are mechanically mixed. They do not depend on chemical bonding, like copolymers, but they need chemical compatibilizer additions to keep the components of the blend from separating. Copolymers and blends are developed to improve upon one or more property of a polymer, like impact resistance, without degrading its other properties.

About three-quarters of the plastics sold are thermoplastics. Therefore, we will concentrate on this category of plastics. For commercial purposes TP plastics can be divided into commodity plastics and engineering plastics. Commodity plastics are generally used in low-load bearing applications like packaging, house siding, water pipe, toys, and furniture. Polyethylene (PE), polystyrene (PS), polyvinyl chloride (PVC), and polypropylene (PP) are good examples. These plastics generally compete with glass, paper, and wood. Engineering TPs compete with metals, since they can be designed to carry significant loads for a long period of time. Examples are polyoxymethylene or acetal (POM), polyamides or nylon (PA), polyamide-imide (PAI), polycarbonates (PC), polyethylene terephthalate (PET), and polyetheretherketones (PEEK).

1. A. B. Strong, "Plastics: Materials and Processing," Prentice-Hall, Englewood Cliffs, NJ, 1996; G. Gruenwald, "Plastics: How Structure Determine Properties," Hanser Publishers, New York, 1993; C. P. MacDermott and A. V. Shenoy, "Selecting Thermoplastics for Engineering Applications," 2d ed., Marcel Dekker, New York, 1997.

Figure 8.29 compares the tensile stress-strain curve for polycarbonate with a soft mild steel and a high-carbon steel. We note that the level of strength is much lower and that yielding and fracture occur at much larger strains. The level of the stress-strain curve is strongly dependent on the strain rate (rate of loading). Increasing the strain-rate raises the curve and decreases the ductility. Table 8.13 lists some short-time mechanical properties of metals and plastics at room temperature. Because many polymers do not have a truly linear initial portion of the stress-strain curve, it may be difficult to specify a yield strength; so the tensile strength usually is reported. Some polymers are brittle at room temperature because their glass-transition temperature is above room temperature. Brittleness is measured by the impact test. Note the marked improvement by introducing glass fibers. The data in Table 8.13 are aimed at illustrating the difference in properties between metals and polymers. They should not be used for design purposes. Moreover, it should be recognized that the mechanical properties of plastics are more subject than metals to variations due to blending and processing.

Tests other than the tensile and impact tests are frequently made on plastics and reported in the producer's literature as an aid in selecting polymers. One is a flexure test, in which the plastic is bent as a beam until failure. The load at failure, divided by the cross-sectional area, gives the flexural strength. Heat resistance is measured by the *heat-deflection temperature* (HDT). A plastic bar is bent under a low constant stress of 264 psi as the temperature of a surrounding oil bath is raised slowly from room temperature. The HDT is the temperature at which the sample shows a deflection of 0.010 in. While useful in ranking materials, such a test is of no value in predicting structural performance of a plastic at a given temperature and stress.

While plastics do not corrode like metals, because they are insulators, they are susceptible to various types of environmental degradation. Some plastics are attacked

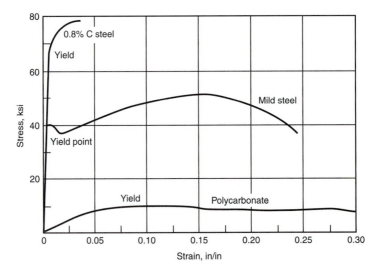

FIGURE 8.29
Comparison of engineering stress-strain curves for a thermoplastic polymer (polycarbonate) with 0.2% carbon mild steel and 0.8% carbon steel.

TABLE 8.13
Typical mechanical properties of metals and polymers

Material	Young's modulus, psi $\times 10^{-6}$	Tensile strength, ksi	Impact ft-lb/in	Specific gravity, g/cm^3
Aluminum alloys	10	20–60		2.7
Steel	30.	40–200		7.9
Polyethylene	0.08–0.15	3–6	1–12	0.94
Polystyrene	0.35–0.60	5–9	0.2–0.5	1.1
Polycarbonate	0.31–0.35	8–10	12–16	1.2
Polyacetal	0.40–0.45	9–10	1.2–1.8	1.4
Polyester-glass reinforced	1.5–2.5	20–30	10–20	1.7

by organic solvents and gasoline. Some are highly susceptible to the absorption of water vapor, which degrades mechanical properties and causes swelling. Many plastics are affected by ultraviolet radiation, causing cracking, fading of the color, or loss of transparency.

Many additives are compounded with the polymer to improve upon its properties. Fillers like wood flour, silica flour, or fine clay are sometimes added as extenders to decrease the volume of plastic resin used, and therefore the cost, without severely degrading properties. Chopped glass fiber is commonly added to increase the stiffness and strength of plastics. Plasticizers are used to enhance flexibility and melt flow during processing. Flame retardants are added to reduce the flammability of plastics, but some of these are proving to be toxic. Colorants such as pigments and dyes are used to impart color. Ultraviolet light absorbers are used to stabilize the color and lengthen the life of the product when exposed to sunlight. Antistatic agents are used to reduce the buildup of electrostatic charges on an insulating plastic surface.

A category of plastics that is growing in importance as a material for engineered structures is plastic matrix composite materials. In these materials high-modulus, strong but often brittle fibers are embedded in a matrix of thermosetting plastic.[1] The fibers most often used are graphite or glass. Because of the high cost of materials and the complexity of fabrication, composite structures are used where a high premium is placed on lightweight, strong structures. Composite structures represent the ultimate in materials design in that the structure is laid up layer upon layer. Because the strength and stiffness reside in the fibers, and these are highly directional, great attention must be paid to directional properties in the design.[2]

Figure 8.30 depicts the interrelationship between part performance requirements, material properties, and manufacturing considerations in selecting the material and in designing the part. While this diagram pertains to issues specific to plastics, it can be applied to any material system.

---

1. Composites, "Engineered Materials Handbook," vol. 1, ASM International, Materials Park, OH, 1987; D. Hull, "An Introduction to Composite Materials," Cambridge University Press, Cambridge, 1981.
2. R. J. Diefendorf, Design with Composites, "ASM Handbook," vol. 20, pp. 648–665, ASM International, Materials Park, OH, 1997.

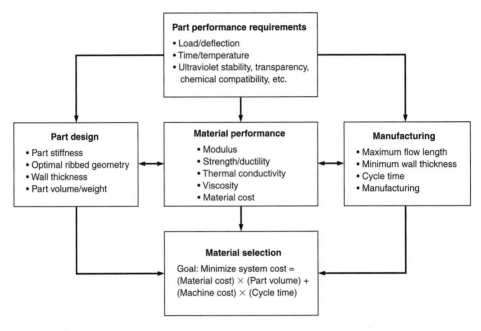

FIGURE 8.30
Interrelationship between materials selection and manufacturing in part design. (*Source:* *"ASM Handbook," vol. 20, used with permission of ASM International, Materials Park, OH.*)

8.19.2 Design for Stiffness

Since Young's modulus is low for plastics compared with metals, resistance to deflection (stiffness) is often a concern regarding the use of plastics. The stiffness of a structure is dependent upon the elastic modulus of the material and the part geometry. For example, the flexural rigidity of a flat plate is given by[1]

$$D = \frac{Et^3}{12(1 - v^2)} \tag{8.40}$$

where $E =$ Young's modulus
$v =$ Poisson's ratio
$t =$ plate thickness

If a polymer has $E = 300,000$ psi and a metal $E = 10,000,000$ psi, the polymer plate will have to be about three times thicker than the metal plate to have the same flexural rigidity. That thickness is probably cost-prohibitive. However, if we increase the thickness of the structure by having two plastic plates of thickness $t/2$ separated by a distance $3t$, then the moment of inertia of the split plate is 37 times that of the single plate with the same total thickness.

The separation of plates may be achieved by a thin honeycomb structure or a lightweight foamed plastic core (Fig 8.31). The structural rigidity of the double plate

1. J. H. Faupel, "Engineering Design," Wiley, New York, 1964.

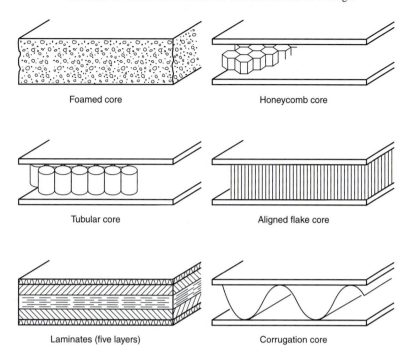

Foamed core   Honeycomb core

Tubular core   Aligned flake core

Laminates (five layers)   Corrugation core

FIGURE 8.31

Examples of sandwich and laminate structures to achieve increased structural rigidity. (*From W. W. Chow, "Cost Reduction in Product Design," p. 211, Van Nostrand Reinhold, New York, 1978; reprinted by permission of author.*)

described above exceeds that of a metal plate of equivalent thickness having an elastic modulus 30 times greater. However, it should be noted that the improved performance of a sandwich panel is limited chiefly to bending resistance. The shear strength of the core must be great enough to resist the shear stresses generated by bending and the interface between the skin and core must be integrally bonded. For further details on stiffness analysis with FEA, see Trantina and Nimmer.[1]

Increases in stiffness may be achieved somewhat more cheaply by stiffened skin approaches that use ribs and corrugations (Fig. 8.32). The ease with which ribs and corrugations can be molded and thermoformed in plastics enhances this approach. The stiffening effect of ribs and corrugations arises from the fact that the geometry results in an increased distance of material from the neutral axis of bending so that the moment of inertia is increased. The stiffening factor may be calculated from the geometry[2] and is expressed by

$$\text{SF} = \text{stiffening factor} = \frac{\text{I with ribs}}{\text{I without ribs}} \tag{8.41}$$

1. G. Trantina and R. Nimmer, "Structural Analysis of Thermoplastic Components," McGraw-Hill, New York, 1994.
2. W. W. Chow, "Cost Reduction in Product Design," chap. 4, Van Nostrand Reinhold, New York, 1978.

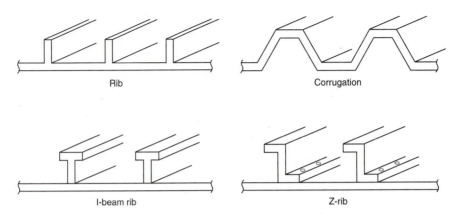

FIGURE 8.32
Examples of stiffened skin designs. (*From W. W. Chow, "Cost Reduction in Production Design,"
p. 211, Van Nostrand Reinhold, New York, 1978; reprinted by permission of author.*)

When a polymer is substituted for a metal, equivalent stiffness is achieved if the total
stiffening factor is unity.

$$\text{Total stiffening factor} = \text{SF}\left(\frac{t_p}{t_s}\right)^3 \frac{E_p}{E_s} \qquad (8.42)$$

8.19.3 Time-Dependent Part Performance

The mechanical properties of plastics are *viscoelastic*.[1] This means that they vary with
time under load, the rate of loading, and the temperature. This material behavior
shows itself most prominently in the phenomena of *creep* and *stress relaxation*. Creep
is the permanent deformation of a material over time, under constant load, and at constant temperature. Stress relaxation is the decreasing stress required to cause a constant strain at a constant temperature. For example, the loosening of a snap fit with
time, or the loosening of a self-tapping screw, is a result of stress relaxation.

In creep a plastic part undergoes an initial deformation when the load is first
applied, and then continues to deform at a slower rate as long as the load is applied.
This is in contrast to a metal part at room temperature where design is based on elastic behavior, and the deformation is not time-dependent. However, if a metal part is
required to operate at temperatures in excess of one-half of its melting point, then
creep must be considered.

Let us consider a simple beam of polyethylene terephthalate (PET) reinforced by
30 volume percent of short glass fibers that is statically loaded with 10 lb in the mid-

1. J. G. Williams, "Stress Analysis of Polymers," 2d ed., John Wiley—Halsted Press, New York, 1980.

dle of its 8-in length.[1] The beam is 1 in thick and has a moment of inertia of $I = 0.0025$ in^4. For a beam loaded at midspan the maximum bending moment is

$$M = \frac{PL}{4} = \frac{10 \times 8}{4} = 20 \text{ in-lb}$$

The bending stress on the outer fiber of the beam is

$$\sigma = \frac{Mc}{I} = \frac{20 \times 0.5}{0.0025} = 4000 \text{ psi}$$

Figure 8.33 shows the flexural creep curves for 30 percent GF PET. Since we are interested in long-term behavior, we look for the creep strain at 1000h under the environmental conditions of 4000 psi (27.6 MPa ≈ 28 MPa) and room temperature 73°F (23°C). The curve shows a strain of 0.7 percent or 0.007 in/in. Then, the *apparent modulus* is

$$E_a = \frac{\sigma}{\varepsilon} = \frac{4000}{0.007} = 0.572 \times 10^6 \, psi$$

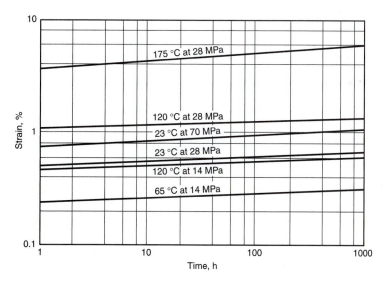

FIGURE 8.33

Creep curve in flexure for 30 percent GF PET. (*"Engineered Materials Handbook," vol. 2, p. 173, ASM International, Materials Park, OH, 1988. Used with permission.*)

---

1. Example taken from "Engineered Materials Handbook," vol. 2, p. 174, ASM International, Materials Park, OH, 1988.

For a simply supported beam the maximum deflection under the load at midspan is given by

$$y = \frac{PL^3}{48E_aI} = \frac{10 \times 8^3}{48\,(0.572 \times 10^6)(0.0025)} = 0.075 \text{ in}$$

If the operating temperature is raised to 250°F (120°C) the creep strain at 1000 h is 1.2 percent or 0.012 in/in. E_a changes to 0.333×10^6 and the deflection under the load is 0.13 in.

Note that the static elastic modulus for this material at room temperature is 1.3×10^6 psi. If this value is used in the deflection equation without considering creep, a severe underestimate of 0.033 in results.

8.20
DESIGN WITH BRITTLE MATERIALS

Brittle materials are those materials which fracture without substantial yielding. Such a material has a linear stress-strain curve up to fracture. When gross yielding does not occur, it means that there is no plastic flow around flaws and defects to relieve the buildup of local stress concentrations. Ceramic materials are brittle materials. This means that the attractive properties of ceramics, high strength-to-weight ratio and strength and oxidation resistance at very high temperatures, cannot be utilized without employing extremely precise design and manufacturing processes.

The brittleness, or very low strain tolerance, of ceramics means that the design analysis must be refined enough to locate and eliminate all high-stress areas. The second issue is that because the mechanical properties of brittle materials are controlled by a statistical population of flaws, the strength data show large variability. We address this issue next.

8.20.1 Strength Properties of Brittle Materials

The large scatter in mechanical properties that is characteristic of brittle materials requires a statistical analysis. The usual procedure is to analyze the data with a Weibull frequency distribution (see Sec. 10.7). If the probability of failure at a series of stress levels is plotted on Weibull probability paper, the data will plot as a straight line (Fig. 8.34). The mathematical equation for the lines is

$$\ln \ln \frac{1}{1-P} = m \ln \sigma - m \ln \sigma_0 + \ln V \qquad (8.43)$$

where P = probability of failure at a stress σ
  m = Weibull modulus, a measure of the scatter
  V = volume of material
  σ_0 = a constant called the *characteristic strength,* the strength with a probability of failure of 63.2 percent

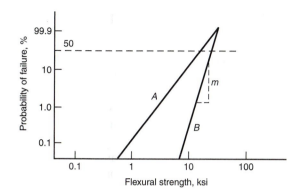

FIGURE 8.34
Typical Weibull plots for brittle materials. Material B has a higher Weibull modulus, and thus less scatter.

The Weibull modulus m is an important parameter of a brittle material. A high value of m indicates less scatter in strength values. Note from Fig. 8.34 that the mean failure stress (the stress at $P = 0.50$) has little significance in the design of brittle materials. Rather, brittle materials must be designed on the basis of some low probability of failure, for example, $P = 0.01$. The volume of material enters into the Weibull plot because brittle materials exhibit a strong size effect. Since failure is initiated at flaws and the likelihood of a critical-size flaw increases with the volume of the part, the strength is reduced in large section sizes. The relationship between these design factors is illustrated by the data[1] given in Table 8.14

This statistical approach is used two ways in design. The first, as shown above, is to characterize the mechanical properties of a brittle material accurately. The second is to replace the factor-of-safety concept by a more precise strength-reliability relation. Figure 8.35, which is derived directly from Weibull's theory,[2] shows the probability of failure vs. relative fracture strength (mean fracture strength/maximum stress) for a material with a Weibull modulus of 7 and for large and small stressed volumes. This plot shows that $\sigma_{max} = \sigma_f/1.5$ at $P = 0.5$, but it changes to $\sigma_{max} = \sigma_f/5$ for $P = 0.0001$. Even greater reductions in working stress are required for the large stressed volume.

TABLE 8.14
Allowable stress, ksi

Failure Probability	$m = 8$		$m = 32$	
	V_1	$V_2 = 100V_1$	V_1	$V_2 = 100V_1$
0.50	50	28.1	50	43.3
0.05	36.4	20.3	46.1	39.9
0.01	29.4	16.5	43.8	37.9

1. W. Duckworth and G. Bansal, *ASME Jnl. Eng. for Power,* vol. 100, pp. 260–266, 1978.
2. G. G. Trantina and H. G. de Lorenzi, *Trans. ASME,* ser. A, vol. 99, pp. 559–566, 1977.

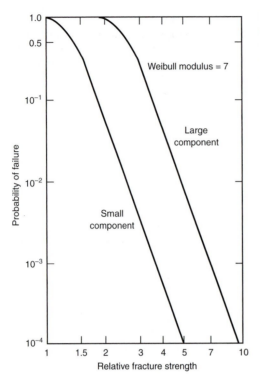

FIGURE 8.35
Probability of failure vs. relative fracture strength (fracture strength/maximum stress in structure). (*From Trantina and de Lorenzi, 1977.*)

Tensile testing of brittle materials is difficult because of the problem of machining specimens from hard brittle material, and because the alignment and concentricity must be very good since ductility is not present to even out small misalignments. Therefore, bend tests, in which the material is flexed as a beam until fracture, are often used to evaluate mechanical properties. The stress in the tensile surface is calculated from the standard beam formula and is known as the *modulus of rupture* (MOR). The value of MOR will vary depending on the volume of the highly stressed region.

8.20.2 Design Approach

Design with brittle materials starts with the hypothesis that the material contains a population of flaws, many of which are too small to be detected with nondestructive evaluation techniques (NDT). These flaws grow by a subcritical crack growth phenomenon, chiefly fatigue, until they reach a critical length, defined by fracture mechanics, and catastrophic crack propagation ensues. Therefore, the design of brittle materials builds on the understanding of brittle fracture and fatigue discussed in Secs. 8.16 and 8.17. This requires detailed mapping of the stress and temperature fields with FEA techniques. Probabilistic design methods are needed because of the nature of the flaws from which cracks grow. All of this has been woven together in a comprehensive design software called Ceramics Analysis and Reliability Evaluation of Structures Life Prediction (CARES/LIFE) by workers at the NASA Lewis Research Center.[1] While the all-ceramic engine has not entered the marketplace, there are components such as turbocharger rotors, valves, piston pins, and liners that are being used in engines.[2]

8.21
SUMMARY

This chapter has shown that there are no magic formulas for materials selection. Rather, the solution of a materials selection problem is every bit as challenging as any other aspect of the design process and follows the same general approach of problem solving. Successful materials selection depends on the answers to the following questions.

1. Have performance requirements and service environments been properly and completely defined?
2. Is there a good correlation between the performance requirements and the material properties used in evaluating the candidate materials?
3. Has the relation between properties and their modification by subsequent manufacturing process been fully considered?
4. Is the material available in the shapes and configurations required and at an acceptable price?

The steps in materials selection are:

1. Define the functions that the design must perform and translate these into required materials properties, and to business factors such as cost and availability.
2. Define the manufacturing parameters such as number of parts required, size and complexity of the part, tolerances, quality level, and fabricability of the material.
3. Compare the needed properties and process parameters with a large materials database to select a few materials that look promising for the application. Use several screening properties to identify the candidate materials.
4. Investigate the candidate materials in greater detail, particularly in terms of trade-offs in performance, cost, and fabricability. Make a final selection of material.
5. Develop design data and a design specification.

Materials selection can never be totally separated from the consideration of how the part will be manufactured. This large topic is covered in Chap. 9.

The Ashby charts are very useful for screening a wide number of materials at the conceptual design stage, and should be employed with materials performance indices. Computer screening of materials databases is widely employed in embodiment

---

1. N. N. Nemeth, L. M. Powers, L. A. Janosik, and J. P. Gyekenyes, *American Ceramic Soc. Bull.,* vol. 72, December 1993, pp. 59–69; the software is available from the Computer Software Management and Information Center, University of Georgia.
2. Ceramics and Glasses, "Engineered Materials Handbook," vol. 4, Sec. 10, ASM International, Materials Park, OH, 1991.

design. Many of the evaluation methods that were introduced in Chap. 5 are readily applied to narrowing down the materials selection. The Pugh selection method and weighted decision matrix are most applicable. Failure analysis (see Chap. 11) is an important input to materials selection when a design is modified. The value analysis technique has broad implications in design, but it is especially useful for materials selection when a design review on a new product is being conducted or an existing product is being redesigned. Life-cycle issues should always be considered, especially those having to do with recycling and disposal of materials.

A second thrust of this chapter deals with the use of materials in design. Brief introductions are given about how the engineer can design to avoid failure from brittle fracture, fatigue, and corrosion. Also, the special design considerations in working with plastics and brittle materials, like ceramics, are described in some detail.

BIBLIOGRAPHY

Materials Selection

Ashby, M. F.: "Materials Selection in Mechanical Design," Pergamon Press, New York, 1992.
"ASM Handbook," vol. 20, "Materials Selection and Design," ASM International, Materials Park, OH, 1997.
Budinski, K. G.: "Engineering Materials: Properties and Selection," 5th ed., Prentice-Hall, Upper Saddle River, NJ, 1996.
Charles, J. A., F. A. A. Crane, and J. A. G. Furness: "Selection and Use of Engineering Materials," 3d ed., Butterworth Heinemann, Boston, 1997.
Farag, M. M.: "Materials Selection for Engineering Design," Prentice-Hall, London, 1997.
Kern, R. F., and M. E. Suess: "Steel Selection," Wiley, New York, 1979.
Lewis, G.: "Selection of Engineering Materials," Prentice-Hall, Englewood Cliffs, NJ, 1990.
Mangonon, P. L.: "The Principles of Materials Selection for Engineering Design," Prentice-Hall, Upper Saddle River, NJ, 1999.
Shackelford, J. F., and J. S. Park: "CRC Practical Handbook of Materials Selection," CRC Press, Boca Raton, FL, 1995.

Materials in Design

"ASM Handbook," vol. 20, "Materials Selection and Design," ASM International, Materials Park, OH, 1997.
Chawla, K. K., and M. A. Meyers: "Mechanical Behavior of Materials," Prentice-Hall, Upper Saddle River, NJ, 1998.
Derby, B., D. A. Hills, and C. Ruiz: "Materials for Engineering: A Fundamental Design Approach," Wiley, New York, 1992.
Dieter, G. E.: "Mechanical Metallurgy," 3d ed., McGraw-Hill, New York, 1986.
Dowling, N. E.: "Mechanical Behavior of Materials," 2d ed., Prentice-Hall, Englewood Cliffs, NJ, 1999.
Edwards, K. S., and R. B. McKee: "Fundamentals of Mechanical Component Design," McGraw-Hill, New York, 1991.
Jones, D. R. H.: "Engineering Materials 3: Materials Failure Analysis," Pergamon Press, Oxford, 1993.

PROBLEMS AND EXERCISES

8.1. Think about why books are printed on paper. Suggest a number of alternative materials that could be used. Under what conditions (costs, availability, etc.) would the alternative materials be most attractive?

8.2. Consider a soft drink can as a materials system. List all the components in the system and consider alternative materials for each component.

8.3. Select a tool material for thread-rolling mild-steel bolts. In your analysis of the problem you should consider the following points: (1) functional requirements of a good tool material, (2) critical properties of a good tool material, (3) screening process for candidate materials, and (4) selection process.

8.4. Rank-order the following materials for use as an automobile radiator: copper, stainless steel, brass, aluminum, ABS, galvanized steel.

8.5. Discuss the use of aluminum vs. steel for electric power transmission line towers.

8.6. Classify the common stainless-steel alloys into a "family of alloys" and develop a procedure for selecting among them.

8.7. Chromium is a highly strategic metal whose supply could be shut off completely by a war in southern Africa. Discuss the reasons for the strategic nature of chromium, and develop available corporate strategies to deal with the problem. What potential substitutes for chromium are available?

8.8. In an aerospace application, total cost of a component can be expressed by

$$C_t = C_f + C_m W + PW$$

where C_f = cost of fabrication
   C_m = material cost, $/lb
   W = weight of the component, lb
   P = penalty factor by which performance is jeopardized, $/lb

(a) Discuss the various strategies available to minimize C_t.
(b) Determine whether material B can be economically substituted for material A if (1) there is no weight penalty factor and (2) the weight penalty is $100 per lb.

Material	C_m, $/lb	W, lb
A	1	90
B	100	2

8.9. Titanium alloys are being considered for automotive engine valves. What advantages would the titanium provide? Suggest alloys for both the inlet and outlet valves.

8.10. The wider use of many materials in automobiles has led to problems in recycling because of the difficulty in separating such a wide range of materials. For example, a midsize car may contain up to 250 lb of as many as 20 varieties of plastics. Propose a solution for this problem.

8.11. Select a group of candidate materials for an energy-storing flywheel using the Ashby charts. First develop the performance index M from the following information about flywheels.

Consider the flywheel to be a solid disk of radius R and thickness t, rotating with an angular velocity ω. The energy stored in the flywheel is

$$U = \frac{1}{2} J\omega^2 = \frac{1}{2}\left(\frac{\pi}{2}\rho R^4 t\right)\omega^2$$

The quantity to be maximized is the kinetic energy per unit mass. The maximum centrifugal stress in the spinning disk is given by

$$\sigma_{max} = \left(\frac{3+v}{8}\right)\rho R^2 \omega^2$$

8.12. Use the information in the example in Sec. 8.14 to construct a Pugh concept selection matrix to aid in deciding which material to select.

8.13. Two materials are being considered for an application in which electrical conductivity is important

Material	Working strength MN/m²	Electrical conductance %
A	500	50
B	1000	40

The weighting factor on strength is 3 and 10 for conductance. Which material is preferred based on the weighted property index?

8.14. An aircraft windshield is rated according to the following material characteristics. The weighting factors are shown in parentheses.

Resistance to shattering (10)   The candidate materials are:
Fabricability (2)
Weight (8)   A   plate glass
Scratch resistance (9)   B   PMMA
Thermal expansion (5)   C   tempered glass
                                      D   a special polymer laminate

The properties are evaluated by a panel of technical experts, and they are expressed as percentages of maximum achievable values.

Property	Candidate material			
	A	*B*	*C*	*D*
Resistance to shattering	0	100	90	90
Fabricability	50	100	10	30
Weight	45	100	45	90
Scratch resistance	100	5	100	90
Thermal expansion	100	10	100	30

Use the weighted property index to select the best material.

8.15. A cantilever beam is loaded with force P at its free end to produce a deflection $\delta = PL^3/3EI$. If the beam has a circular cross section, $I = \pi r^4/4$. Develop a figure of merit for selecting a material that minimizes the weight of a beam for a given stiffness (P/δ). By using the material properties given below, select the best material (a) on the basis of performance and (b) on the basis of cost and performance.

Material	E		ρ, Mgm^{-3}	Approx. Cost, $/ton (1980)
	GNm^{-2}	ksi		
Steel	200	29×10^3	7.8	450
Wood	9–16	1.7×10^3	0.4–0.8	450
Concrete	50	7.3×10^3	2.4–2.8	300
Aluminum	69	10×10^3	2.7	2,000
Carbon-fiber-reinforced plastic (CFRP)	70–200	15×10^3	1.5–1.6	200,000

8.16. Select the most economical steel plate to construct a spherical pressure vessel in which to store gaseous nitrogen at a design pressure of 100 psi at ambient weather conditions down to a minimum of $-20°F$. The pressure vessel has a radius of 138 in. Your selection should be based on the steels listed below and expressed in terms of cost per square foot of material. Use a value of 489 lb/ft^3 for steel.

ASTM spec.	Grade	Allowable stress, psi	Pricing, ¢/lb (estimated 1997 prices)						
			Base	Special grade	Quality extra	Width extra	Testing	Heat-treat	Total
A-36		12,650	29.1	0.40	—	3.0	—	—	32.5
A-285	C	13,750	29.1	4.00	—	3.0	—	—	36.1
A-442	60	15,000	29.1	—	4.0	4.0	0.70	—	37.8
A-533	B	20,000	40.0	15.60	3.20	6.2	0.70	18.2	83.9
A-157	B	28,750	40.0	11.70	3.20	8.2	3.00	18.2	84.3

8.17. Using the example of the hydraulic valve housing given in Sec. 8.11, put it into the framework of the solution steps for value analysis shown in Fig. 8.12.

8.18. Compare steels A and B for the construction of a pressure vessel 30 in inside diameter and 12 ft long. The pressure vessel must withstand an internal pressure of 5000 psi. Design against a radial internal flaw of length $2a$. Use a factor of safety of 2. For each steel, determine (a) critical flaw size and (b) flaw size for a leak-before-break condition.

Steel	Yield strength, psi	K_{Ic}, ksi $\sqrt{\text{in}}$
A	260	80
B	110	170

8.19. A high-strength steel has a yield strength of 100 ksi and a fracture toughness $K_{Ic} = 150$ ksi $\sqrt{\text{in}}$. By use of a certain nondestructive evaluation technique, the smallest size flaw that can be detected routinely is 0.3 in. Assume that the most dangerous crack geometry in the structure is a single-edge notch so that $K_{Ic} = 1.12\sigma \sqrt{\pi a}$. The structure is subjected to cyclic fatigue loading in which $\sigma_{max} = 45$ ksi and $\sigma_{min} = 25$ ksi. The crack growth rate for the steel is given by da/dN $0.66 \times 10^{-8}(\Delta K)^{2.25}$. Estimate the number of cycles of fatigue stress that the structure can withstand.

8.20. A 3-in-diameter steel shaft is rotating under a steady bending load. The shaft is reduced in diameter at each end to fit in the bearings, which gives a theoretical stress concentration factor at this radius of 1.7. The fatigue limit of the steel in a rotating beam test is 48,000 psi. Its fatigue notch sensitivity factor is 0.8. Since the fatigue limit is close to a 50 percent survivability value, we want to design this beam to a 99 percent reliability. What point in the beam will be the likely point for a failure, and what is the maximum allowable stress at that point?

8.21. A plastic beam (6-in diameter), used to haul small loads into a loft, extends 8 ft in the horizontal plane from the building. If the beam is made from glass-reinforced PET, how much would the beam deflect on a 23°C day when hauling up a 900-lb load? Hint: Use the appropriate information in Prob. 8.15 and data in Fig. 8.33.

8.22. Stress-corrosion failures occur in the 304 and 316 stainless-steel recirculation piping of boiling-water nuclear reactors (BWR). As of April 1981, a total of 254 incidents of cracking had been reported. What are the three conditions necessary for stress-corrosion cracking? Suggest remedies for the cracking.

8.23. What are the chief advantages and disadvantages of plastic gears? Discuss how material structure and processing are utilized to improve performance.

9

MATERIALS PROCESSING AND DESIGN

9.1
ROLE OF PROCESSING IN DESIGN

Producing the design is a critical link in the chain of events that starts with a creative idea and ends with a successful product in the marketplace. In modern technology the function of production no longer is a routine activity. Rather, design, materials selection, and processing are inseparable, as shown in Fig. 8.1.

There is a confusion of terminology concerning the engineering function we have called processing. Materials engineers may talk about materials processing to refer to the conversion of semifinished products, like steel blooms or billets, into finished products, like cold-rolled sheet or hot-rolled bar. A mechanical, industrial, or manufacturing engineer is more likely to refer to the conversion of the sheet into an automotive body panel as manufacturing. *Processing* is the more generic term, but *manufacturing* is the more common terminology. *Production engineering* is a term used in Europe to describe what we call manufacturing in the United States.

The first half of the twentieth century saw the maturation of manufacturing operations in the western world. Increases in the scale and speed of operations brought about increases in productivity, and manufacturing costs dropped while wages and the standard of living rose. There was a great proliferation of available materials as basic substances were tailor-made to have selectively improved properties. One of the major achievements of this era was the development of the production line for mass-producing automobiles, appliances, and other consumer goods. Because of the pre-eminence in manufacturing that developed in the United States, there has been a tendency in the last half of the century to take the production function for granted. Manufacturing has been downplayed, or even ignored completely, in the education of engineers. Manufacturing positions in industry have been considered routine and not challenging, and as a result they have not attracted their share of the most talented engineering graduates. Fortunately, this situation is improving as the importance of the manufacturing function is being rediscovered and the nature of manufacturing is being changed by automation and computer-aided manufacturing.

Peter Drucker, the prominent social scientist and management expert, has termed the current manufacturing situation "the third industrial revolution." The second industrial revolution began roughly a century ago when machines were first driven directly by fractional-horsepower motors. The use of power, whether generated by falling water or a steam engine, was, of course, the first industrial revolution. However, before machines were direct-driven by electric motors, they were driven by belts and pulleys, which meant that the equipment had to be very close to the source of power. Thus, the second industrial revolution gave flexibility and economy to manufacturing. The third industrial revolution, in which information processing is becoming part of the machine or tool, is converting production from a manual into a knowledge-based operation.

A serious problem has been the tendency to separate the design and manufacturing functions into separate organizational units. Barriers between design and manufacturing can inhibit the close interaction that the two engineering functions should have, as discussed previously under concurrent engineering (Sec. 1.7). When technology is sophisticated and fast-changing, a close coupling between the people in research, design, and manufacturing is very necessary. That has been demonstrated best in the area of solid-state electronic devices. As semiconductor devices replaced vacuum tubes, it became apparent that design and processing could no longer be independent and separable functions. Vacuum-tube technology was essentially a linear situation in which specialists in materials passed on their input to specialists in components who passed on their input to circuit designers who, in turn, communicated with system designers. With the advent of transistors the materials, device construction, and circuit design functions became closely coupled. Then, with the microelectronics revolution of large-scale integrated circuits, the entire operation from materials to system design became interwoven and manufacturing became inseparable from design. The result was a situation of rapid technical advance requiring engineers of great creativity, flexibility, and breadth. The payoff in making the minicomputer a reality has been huge. Nowhere has productivity been enhanced as rapidly as in the microelectronics revolution. That should serve as a model of the great payoff that can be achieved by closer coupling of research, design, and manufacturing.

More conventional manufacturing is divided into (1) process engineering, (2) tool engineering (3) work standards, (4) plant engineering, and (5) administration and control. *Process engineering* is the development of a step-by-step sequence of operations for production. The overall product is subdivided into its components and subassemblies, and the steps required to produce each component are arranged in logical sequence. An important part of process engineering is to specify the related tooling. Vital parameters in process engineering are the rate of production and the cost of manufacturing. *Tool engineering* is concerned with the design of tools, jigs, fixtures, and gauges to produce the part. Jigs both hold the part and guide the tool during manufacture, while fixtures hold a part to be joined, assembled, or machined. Tools do the machining or forming; gauges determine whether the dimensions of the part are within specification. *Work standards* are time values associated with each manufacturing operation. Other standards that need to be developed in manufacturing are tool standards and materials standards. *Plant engineering* is concerned with providing the plant facilities (utilities, space, transportation, storage, etc.) needed to carry out the manufacturing process. *Administration and control* is production planning, scheduling, and supervising to assure that materials, tools, machines, and people are available at the right time and in the quantities needed to produce the part.

We ordinarily think of modern manufacturing in terms of the automotive assembly line, but mass production manufacturing systems account for less than 25 percent of metal parts manufactured. In fact, 75 percent of the parts manufactured are produced in lots of fewer than 50 pieces. About 40 percent of the employment in manufacturing is in such job-shop operations. Studies of batch-type metal-cutting production shops have shown that, on the average, a workpiece in such a shop is on a machine tool being productively processed *only 5 percent of the time.* During all the rest of the time the workpiece is in the shop, it is waiting in an inventory of unfinished parts. Moreover, of the very small fraction of time the part is being worked on, it is being cut by the tool only about 30 percent of the time. The remaining 70 percent of the time is taken up by loading and unloading, positioning the workpiece, gauging the part, etc.

Thus, there is a major opportunity for greatly increasing manufacturing productivity in small-lot manufacture. Computer-automated machine tool systems, which include industrial robots and computer software for scheduling and inventory control, have a demonstrated potential for increasing machine utilization time from an average of 5 percent to as much as 90 percent. The introduction of computer-controlled machining centers that can perform many operations greatly increases the productivity of the machine tool. We are moving toward the computer-automated factory in which all steps in parts manufacturing will be optimized by computer software systems, the machine tools will be under computer control, and at least half of the machines will be part of a versatile manufacturing system featuring multiple machining capability and automatic parts handling between work stations. This automated factory differs from the automotive transfer line in that it will be a flexible manufacturing system capable of producing a wide variety of parts under computer control. This broadbased effort throughout industry to link computers into all aspects of manufacturing is called *computer-integrated manufacturing* (CIM). Figure 9.1 shows the complete spectrum of activities which are encompassed by manufacturing.

9.2
CLASSIFICATION OF MANUFACTURING PROCESSES

It is not a simple task to classify the tremendous variety of manufacturing processes. We start this analysis through the hierarchical classification of business and industry shown in Fig. 9.2. The service industries consist of enterprises, such as banking, education, insurance, and communication, that provide important services to modern society but do not create wealth by converting raw materials. The producing industries acquire raw materials (minerals, natural products, or petroleum) and process them,

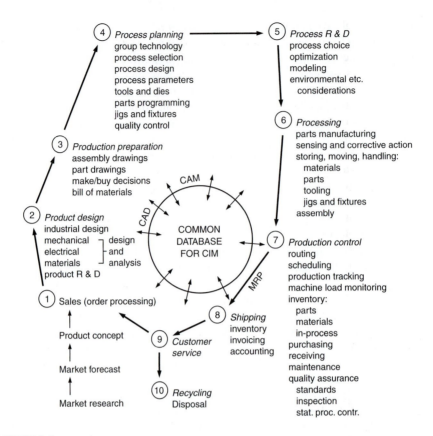

FIGURE 9.1
Spectrum of activities which are encompassed by manufacturing. (*From J. A. Schey, "Introduction to Manufacturing Processes," 2d ed., McGraw-Hill, 1987. Used with permission of McGraw-Hill.*)

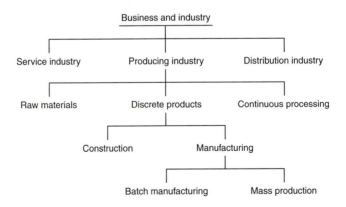

FIGURE 9.2
A hierarchical classification of business and industry.

through the use of energy, machinery, and brainpower, into products that serve the needs of society. The distribution industries, such as merchandising and transportation, make those products available to the general public. A characteristic of modern industrialized society is that an increasingly smaller percentage of the population produces the wealth that makes our affluent society possible. The producing industries account for about 20 percent and the service industries about 70 percent of our gross national product (GNP). Just as the past century saw the United States change from a predominantly agrarian society to a nation in which only 4 percent of the population works in agriculture, so we are becoming a nation in which an ever-decreasing percentage of the work force is engaged in manufacturing. In 1947 about 30 percent of the work force were in manufacturing; in 1980 it was about 22 percent. By the year 2000 less than 10 percent of U.S. workers will be engaged in manufacturing.

The producing industries can be divided conveniently into raw materials producers (mining, petroleum, agriculture), producers of discrete products (autos, television, etc.), and industries engaged in continuous processing (gasoline, wood products, steel, chemicals, etc.). Two major divisions of the discrete products are construction (buildings, dams, etc.) and manufacturing. Under manufacturing we recognize batch (low-volume) manufacturing and mass production.

9.2.1 Types of Manufacturing Processes

A manufacturing process converts a material into a finished part or product. The changes that take place occur with respect to external part geometry, or they can affect the internal microstructure and properties of the material. For example, a sheet of brass that is being drawn into the cylindrical shape of a cartridge case is also being hardened and reduced in ductility by the process of dislocation glide on slip planes.

We will recall from Chap. 5 that the functional decomposition of a design was described initially in terms of energy, material, and information flows. These same three factors are present in manufacturing. Thus, a manufacturing process requires an energy flow to cause the material flow that brings about changes in shape. The information flow, which consists of both shape and material property information, depends on the type of material, the process used, i.e., whether mechanical, chemical, or thermal, the characteristics of the tooling used, and the pattern of movement of the material relative to the tooling.

A natural division among the hundreds of manufacturing processes is whether the process is *mass conserving* or *mass reducing.* In a mass conserving process the mass of the starting material is approximately equal to the mass of the final part. A *shape-replication* process is a mass conserving process in which the part replicates the information stored in the tooling by being forced to assume the shape of the surface of the tool cavity. Casting, injection molding, and closed-die forging are examples. In a mass reducing process the mass of the starting material is greater than the mass of the final part. Such processes are *shape-generation* processes because the part shape is produced by the relative motion between the tool and the workpiece. Material removal is caused by controlled fracture, melting, or chemical reaction. A machining process, such as milling, is an example of controlled fracture.

We can classify the great number of processes used in manufacture into the following eight categories.

1. *Solidification (casting) processes:* Molten metal, plastic, or glass is cast into a mold and solidified into a shape.
2. *Deformation processes:* A material, usually metal, is plastically deformed hot or cold to give it improved properties and change its shape. Typical processes of this type are forging, rolling, extrusion, and wiredrawing. Sheet-metal forming is a special category in which the deformation occurs in a two-dimensional stress state.
3. *Material removal or cutting (machining) processes:* Material is removed from a workpiece with a sharp tool by a variety of methods such as turning, milling, grinding, shaving, polishing, and lapping.
4. *Polymer processing:* The special properties of polymers have brought about the development of processes, such as injection molding and thermoforming, that are not duplicated in the above categories, which are more oriented to metals processing.
5. *Powder processing:* This rapidly developing area includes the consolidation of particles of metal, ceramics, or polymers by pressing and sintering, hot compaction, or plastic deformation. It also includes the processing of composite materials.
6. *Joining processing:* Included in joining processing are all categories of welding, brazing, soldering, diffusion bonding, riveting, bolting, and adhesive bonding.
7. *Heat treatment and surface treatment:* This category includes the improvement of mechanical properties by thermal heat treatment processes as well as the improvement of surface properties by diffusion processes like carburizing and nitriding or by alternative means such as sprayed or hot-dip coatings, electroplating, and painting. The category also includes the cleaning of surfaces preparatory to surface treatment.
8. *Assembly processes:* In this, usually the final, step in manufacturing a number of parts are brought together and combined into a subassembly or finished part.

9.2.2 Sources of Information on Manufacturing Processes

In this book we cannot describe in any detail the many processes used in modern manufacturing. The general references listed in the bibliography are suggested for those who have never had a course in manufacturing processes and therefore are unfamiliar with terminology and the general functioning of the processes. The most important reference sources are "Tool and Manufacturing Engineers Handbook," 4th ed., published in nine volumes by the Society of Manufacturing Engineers, and various volumes of "ASM Handbook" published by ASM International devoted to specific manufacturing processes. Books dealing with each of the eight categories of manufacturing processes are listed below.

Solidification Processes

M. C. Flemings, "Solidification Processing," McGraw-Hill, New York, 1974.
"Casting, ASM Handbook," vol. 15, ASM International, Materials Park, OH, 1988.

M. Blair and T. L. Stevens, eds., "Steel Castings Handbook," 6th ed., ASM International, 1995.

S. P. Thomas, ed., "Design and Procurement of High-Strength Structural Aluminum Castings," American Foundrymen's Society, Cleveland, 1995.

Deformation Processes

W. A. Backofen, "Deformation Processing," Addison-Wesley, Reading, MA, 1972.

Z. Marciniak and J. L. Duncan, "The Mechanics of Sheet Metal Forming," Edward Arnold, London, 1992.

E. Mielnik, "Metalworking Science and Engineering," McGraw-Hill, New York, 1991.

T. G. Breyer, ed., "Forging Handbook," Forging Industry Association, Cleveland, 1985.

"Forming, Tool and Manufacturing Engineers Handbook," vol. 2, 4th ed., Society of Manufacturing Engineers, Dearborn, MI, 1984.

"Forming and Forging, ASM Handbook," vol. 14, ASM International, Materials Park, OH, 1988.

K. Lange, ed., "Handbook of Metal Forming," Society of Manufacturing Engineers, Dearborn, MI, 1985.

R. Pearce, "Sheet Metal Forming," Adam Hilger, 1991.

Material Removal Processes

G. Boothroyd and W. W. Knight, "Fundamentals of Machining and Machine Tools," 2d ed., Marcel Dekker, New York, 1989.

S. Malkin, "Grinding Technology: Theory and Applications," Ellis Horwood, 1989.

M. C. Shaw, "Metal Cutting Principles," 4th ed., Oxford University Press, New York, 1984.

D. A. Stephenson and J. S. Agapiov, "Metal Cutting: Theory and Practice," Marcel Dekker, New York, 1996.

"Machining, Tool and Manufacturing Engineers Handbook," vol. 1, 4th ed., Society of Manufacturing Engineers, Dearborn, MI, 1983.

"Machining, ASM Handbook," vol. 16, ASM International, Materials Park, OH, 1989.

Polymer Processing

J. F. Agassant, P. Avenas, J. Sergent, and P. J. Carreau, "Polymer Processing: Principles and Modeling," Hanser Gardner Publications, Munich, 1991.

E. A. Muccio, "Plastics Processing Technology," ASM International, Materials Park, OH, 1994.

A. B. Strong, "Plastics: Materials and Processing," Prentice-Hall, Englewood Cliffs, NJ, 1996.

"Plastics Parts Manufacturing, Tool and Manufacturing Engineers Handbook," vol. 8, 4th ed., Society of Manufacturing Engineers, Dearborn, MI, 1995.

"Engineering Plastics, Engineered Materials Handbook," vol. 2, ASM International, Materials Park, OH, 1988.

Powder Processing

R. M. German, "Powder Metallurgy Science," Metal Powder Industries Federation, Princeton, NJ, 1985.

R. M. German, "Powder Metallurgy of Iron and Steel," Wiley, New York, 1998.

"Powder Metal Technologies and Applications, ASM Handbook," vol. 7, ASM International, Materials Park, OH, 1998.

"Powder Metallurgy Design Manual," 2d ed., Metal Powder Industries Federation, Princeton, NJ, 1995.

Joining Processes

S. Kuo, "Welding Metallurgy," Wiley, New York, 1987.
"Adhesives and Sealants, Engineered Materials Handbook," vol. 3, ASM International, Materials Park, OH, 1990.
R. O. Parmley, ed., "Standard Handbook for Fastening and Joining," 3d ed., McGraw-Hill, New York, 1997.
"Welding, Brazing, and Soldering, ASM Handbook," vol. 6, ASM International, Materials Park, OH, 1993.
"Welding Handbook," 8th ed., American Welding Society, Miami, FL, 1996.

Heat Treatment and Surface Treatment

"Heat Treating, ASM Handbook," vol. 4, ASM International, Materials Park, OH, 1991.
"Surface Engineering, ASM Handbook," vol. 5, ASM International, Materials Park, OH, 1994.
"Materials, Finishing, and Coating, Tool and Manufacturing Engineers Handbook," vol. 3, 4th ed., Society of Manufacturing Engineers, Dearborn, MI, 1985.

Assembly Processes

G. Boothroyd, C. Poli, and L. E. Murch, "Automatic Assembly," Marcel Dekker Inc., New York, 1982.
E. K. Henriksen, "Jig and Fixture Design," Industrial Press, New York, 1973.
E. Hoffman, "Fundamentals of Tool Design," 3d ed., Society of Manufacturing Engineers, Dearborn, MI, 1991.
"Assembly Processes, Tool and Manufacturing Engineers Handbook," vol. 9, 4th ed., Society of Manufacturing Engineers, Dearborn, MI, 1998.

9.2.3 Types of Process Systems

We can identify four types of process systems: job shop, batch, assembly line, and continuous flow. The characteristics of these production systems are listed in Table 9.1. The job shop is characterized by small batches of a large number of different part types every year. There is no regular work flow and work-in-process must often wait in queues for its turn on the machine. Hence, it is difficult to specify job shop capacity because it is highly dependent on the product mix. Batch flow, or decoupled flow line, is used when the product line is relatively stable and produced in periodic batches, but the volume for an individual product is not sufficient to warrant the cost of specialized, dedicated equipment. Examples are heavy equipment or ready-to-wear clothing. With assembly-line production the equipment is laid out in the sequence of usage. The large number of assembly tasks is divided into small subsets to be performed at successive workstations. Examples are the production of automobiles or power hand tools. Finally, a continuous flow process is the most specialized type. The equipment is highly specialized, laid out in a circuit, and usually automated. The material flows continuously from input to output. Examples are a gasoline refinery or a paper mill.

TABLE 9.1
Characteristics of production systems

Characteristic	Job shop	Batch flow	Assembly line	Continuous flow
Equipment and physical layout				
Batch size	Low (1–100 units)	Moderate (100–10,000 units)	Large (10,000–millions/year)	Large. Measured in tons, gals., etc.
Process flow	Few dominant flow patterns	Some flow patterns	Rigid flow patterns	Well defined and inflexible
Equipment	General-purpose	Mixed	Specialized	Specialized
Setups	Frequent	Occasional	Few and costly	Rare and expensive
Process changes for new products	Incremental	Often incremental	Varies	Often radical
Information and Control				
Production information requirements	High	Varies	Moderate	Low
Scheduling	Uncertain, frequent change	Varies	Fixed schedule	Inflexible
Raw material inventory	Small	Moderate	Varies; frequent deliveries	Large
Work-in-process	Large	Moderate	Small	Very small

A process is said to be *mechanized* when it is being carried out by powered machinery and not by hand. Nearly all manufacturing processes in developed countries are mechanized. A process is *automated* when the steps in the process, along with the movement of material and inspection of the parts, are automatically performed or controlled by self-operating devices. Automation involves mechanization plus sensing and controlling (programmable logic controllers and PCs). Hard automation is hard-linked and hard-wired, while *flexible automation* includes the added capability of being reprogrammed to meet changed conditions.

Design for production (DFP) evaluates a design with respect to a product's capacity requirements (number of units per year) and manufacturing cycle time (seconds per unit). This is distinct from the more common design for manufacturing (DFM) and design for assembly (DFA) methodology discussed in Secs. 9.4 and 9.5. A DFP analysis may identify a product feature that creates a production "bottleneck" (an operation which causes an increase in cycle time), because there is insufficient in-house machine capacity. Or maybe there is no in-house capacity and the parts will have to be taken to the shop of a subcontractor to do this operation. Study of the DFP aspects of the design[1] early in the design process can avoid embarrassing and costly problems at the manufacturing stage.

---

1. A. Kusiak and W. He, Design of Components for Schedulability, *European J. of Operational Research,* vol. 76, pp. 49–59, 1994.

9.2.4 Computer-Integrated Manufacturing (CIM)

Computer-integrated manufacturing (CIM) is the term used to describe the vision in which digital data is linked seamlessly from the design of a product to the automated manufacture of the product. This involves the steps of design, including analysis and simulation, documentation, manufacturing planning and control, including material and machine scheduling, and factory automation, including materials processing, inspection, assembly, and materials handling. Coupled to this technical chain are the business functions of product strategic planning, marketing, finance, and human resource management. While examples of true CIM exist, it is by no means the current standard in industry due to the large investments in plant and human capital needed to make CIM work effectively. Also, the standards for interchange of digital data between computers and machines have not yet reached the point where they are as widely adopted as will be necessary for widespread use of CIM.

9.3
FACTORS DETERMINING PROCESS SELECTION

The factors that influence the selection of a process to make a part are:

- Cost of manufacture and life cycle cost
- Quantity of parts required
- Complexity—shape, form, size
- Material
- Quality of part
- Availability, lead time, and delivery schedule

9.3.1 Cost of Manufacture

The manufacturing cost of the finished product is the most important factor determining the selection of the manufacturing process and the material. The computer-based tools for DFM and DFA that are discussed in Sec. 9.7 are based on calculating the cost of a part. Strictly speaking, we should be focusing on the life cycle cost of the part, allowing for such costs as maintenance and disposal (see Sec. 14.15). A basic expression for the unit cost of a part C is

$$C = C_m + \frac{C_C}{n} + \frac{C_L}{\dot{n}} \tag{9.1}$$

where $C_m =$ material cost per unit
$\quad C_C =$ capital cost of plant, machinery, and tooling required to make the part
$\quad C_L =$ labor cost per unit time
$\quad n =$ annual number of parts produced
$\quad \dot{n} =$ the production rate, number of parts produced per unit time

The least expensive material that provides the properties needed for the design should be chosen. In making this decision, you need to be aware of cost extras, as discussed in Sec. 8.5.

9.3.2 Quantity of Parts Required

All manufacturing processes have a minimum number of pieces (volume) that must be made to justify its use. Some processes, like an automatic screw machine, are inherently high-volume processes, in that the setup time is long relative to the time to produce a single part. Others, like the hand lay-up of a fiberglass polymer boat, are low-volume processes. This leads to the concept of an *economical lot size,* the break-even volume at which one process with higher tooling costs becomes less expensive per unit than a process with lower tooling costs (see Fig. 9.5).

Related to this concept is the *flexibility* of the process. Flexibility in manufacturing is the ease with which a process can be adapted to produce different products or variations of the same product. It is influenced greatly by the time to change and set up tooling. At a time when product customization is becoming more important, this process attribute has gained importance.

> **EXAMPLE.** With the drive to reduce the weight of automobiles, a plastic bumper looks attractive. Such a bumper must have good rigidity to maintain dimensional limits, low-temperature impact resistance (for crash-worthiness), and dimensional stability over the operating range of temperature.[1] In addition, it must have the ability to be finished to match the adjoining painted metal parts.
>
> With these performance requirements of chief importance, four polymeric materials were chosen from the large number of engineering plastics.
>
> - Polyester reinforced with chopped-glass fiber to improve toughness
> - Polyurethane with glass-flake filler to increase stiffness
> - Rubber-modified polypropylene to decrease the ductile-brittle transition to below $-30°C$
> - A polymer blend of polyester and polycarbonate to combine the excellent solvent resistance of the former with the high toughness of the latter
>
> Four polymer processes are under consideration for making the bumpers from these polymers. Each works well with the engineered plastics chosen, but they vary greatly in tooling costs and flexibility.

Process	Mold cost	Labor input/unit
Injection molding	$450,000	3 min = $1
Reaction injection molding	$90,000	6 min = $2
Compression molding	$55,000	6 min = $2
Contact molding	$20,000	1 h = $20

---

1. L. Edwards and M. Endean, eds., *Manufacturing with Materials,* Butterworths, Boston, 1990.

Then, the part cost is the sum of the mold cost per part plus the labor input, neglecting the material cost, which is roughly the same for each.

	Cost per part			
Process	1000 parts	10,000 parts	100,000 parts	1,000,000 parts
Injection molding	$451	$46	$5.50	$1.45
Reaction injection molding	$92	$11	$2.90	$2.09
Compression molding	$57	$7.50	$2.55	$2.06
Contact molding	$40	$22	$20.20	$20.02

Note how the unit part cost varies greatly with quantity of parts required. The hand lay-up process of contact molding is the least expensive for a low part volume, while the low-cycle-time injection molding process excels at the highest part volume. Assuming that the material cost for the bumper is $30 per part, we see how material cost represents the largest fraction of the total cost as the part volume increases.

9.3.3 Complexity

The complexity of a part refers to its shape, size, and type and number of features that it contains. Most mechanical parts have a three-dimensional shape, although sheet-metal fabrications are basically two-dimensional. Shapes may be further divided into symmetrical and nonsymmetrical. Figure 9.13, p. 407, shows some basic shapes, classified according to increasing complexity. Different manufacturing processes vary in their limitations on producing complex shapes. For example, many processes will not allow the manufacture of parts with undercuts. Extrusion requires a part with cylindrical symmetry, and powder metallurgy has geometric restrictions because the weak powder compact must be ejected from the die. Therefore, the complexity of a part will determine whether a process is a candidate to make the part.

One way of expressing the complexity of a part is through its information content I.[1]

$$I = n \log_2 \left(\frac{\bar{l}}{\overline{\Delta l}} \right) \tag{9.2}$$

where n = number of dimensions of the part

$\bar{l} = (l_1 \cdot l_2 \cdot l_3 \dots l_n)^{1/n}$ is the geometric mean dimension

$\overline{\Delta l} = (\Delta l_1 \cdot \Delta l_2 \cdot \Delta l_3 \dots \Delta l_n)^{1/n}$ is the geometric mean of the tolerance (the $\pm$ dimension)

$\log_2(x) = \dfrac{\log_{10}(x)}{\log_{10}(2)}$

Simple shapes contain only a few bits of information. Complex shapes, like integrated circuits, contain many. A cast engine block might have 10^3 bits of information, but after machining the complexity increases by both adding new dimensions (n) and

---

1. M. F. Ashby, "Materials Selection in Mechanical Design," Pergamon Press, New York, 1992.

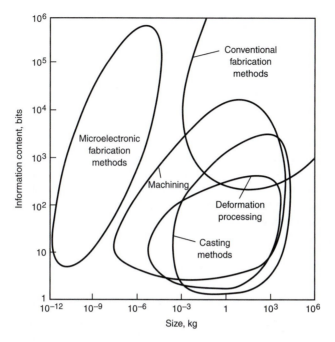

FIGURE 9.3

An example of an Ashby process selection chart. This chart plots the information content of the part against the part size. It delineates those processes that are most applicable in various regimes. (*From "ASM Handbook," vol. 20, p. 249. Used with permission.*)

improving the precision (reducing $\overline{\Delta}\,\bar{l}$). Ashby has shown[1] the relationship between complexity factor and size for different manufacturing processes (Fig. 9.3). Some processes are inherently small-scale, while others like casting, machining, and forging can produce parts with a large range of sizes, depending on the scale of available equipment. Fabrication methods involving welding and bolting produce large engineering structures like bridges and buildings.

9.3.4 Materials

Just as the shape limits the available selection of processes, the selection of a material places certain restrictions on the available manufacturing processes. The melting point of the material and its level of deformation resistance and ductility are the chief factors. The melting point of the material determines the casting processes that can be employed. Low-melting-point metals can be used with a wide number of casting processes, but as the melting point rises problems with mold reaction and atmosphere contamination limit the available processes. Some materials may be too brittle for shape creation by deformation processes, while others are too reactive to have good weldability.

1. M. F. Ashby, op. cit.

Figure 9.4 shows a matrix laying out the manufacturing processes generally used with the most common classes of engineering materials.[1] The table is further divided with respect to the quantity of parts needed for economical production. This is an excellent way to narrow down the possibilities to a manageable few processes.

Steels, aluminum alloys, and other metallic alloys can be purchased from the supplier in a variety of metallurgical conditions other than the annealed (soft) state. Examples are quenched and tempered steel bars, solution-treated and cold-worked and aged aluminum alloys, or cold-drawn and stress-relieved brass rod. It may be more economical to have the metallurgical strengthening produced in the workpiece by the material supplier than to heat-treat each part separately after it has been manufactured.

When parts have very simple geometric shapes, as straight shafts and bolts have, the form in which the material is obtained and the method of manufacture are readily apparent. However, as the part becomes more complex in shape, it becomes possible to make it from several forms of material and by a variety of manufacturing methods. For example, a small gear may be machined from bar stock or, perhaps more economically, from a precision-forged gear blank. The selection of one of several alternatives is based on overall cost of a finished part (see Chap. 14 for details of cost evaluation). Generally, the production quantity is an important factor in cost comparisons (Fig. 9.5). There will be a break-even point beyond which it is more economical to invest in precision-forged preforms in order to produce a gear with a lower unit cost. As the production quantity increases, it becomes easier economically to justify a larger initial investment in tooling or special machinery to lower the unit cost.

9.3.5 Required Quality of the Part

The quality of the part is defined by three related areas: (1) freedom from external and internal defects, (2) surface finish, and (3) dimensional accuracy and tolerance. To a high degree, the achievement of high quality in these areas is influenced by the *workability* or formability of the material.[2] While different materials exhibit different workability in a given process, the same material may show different workability in different processes. For example, in deformation processing, the workability increases with the extent that the process provides a condition of hydrostatic compression. Thus, steel has greater workability in extrusion than in forging, and even less in drawing.

Defects

Defects may be internal to the part or concentrated mainly at the surface. Internal defects are such things as voids, porosity, cracks, or regions of different chemical

1. K. G. Swift and J. D. Booker, "Process Selection: From Design to Manufacture," Arnold, London, 1997. Also available from Wiley, New York.
2. G. E. Dieter, Introduction to Workability, "Forming and Forging," vol. 14, "ASM Handbook," ASM International, 1988, pp. 363–372. For castability, weldability, machinability, moldability, etc., consult reference on those processes.

Note: The PRIMA selection matrix cannot be regarded as comprehensive and should not be taken as such. It represents the main common industrial practice but there will always be exceptions at this level of detail. Also, the order in which the PRIMAs are listed in the nodes of the matrix has no significance in terms of preference.

QUANTITY \ MATERIAL	IRONS	STEEL (carbon)	STEEL (tool, alloy)	STAINLESS STEEL	COPPER & ALLOYS	ALUMINUM & ALLOYS	MAGNESIUM & ALLOYS	ZINC & ALLOYS	TIN & ALLOYS	LEAD & ALLOYS	NICKEL & ALLOYS	TITANIUM & ALLOYS	THERMOPLASTICS	THERMOSETS	FR COMPOSITES	CERAMICS	REACTIVE METALS	PRECIOUS METALS
VERY SMALL 1 TO 100	[1.5][1.6] [1.7][4.M]	[1.5][1.7] [3.6][4.M] [5.1][5.5] [5.6]	[1.1][1.7] [3.6][4.M] [5.1][5.5] [5.6]	[1.7][3.6] [4.M][5.1] [5.5][5.6]	[1.5][1.7] [3.6][4.M] [5.1]	[1.5][1.7] [3.6][4.M] [5.1][5.5]	[1.6][1.7] [3.6][4.M] [5.1][5.5]	[1.1][1.7] [3.6][4.M] [5.5]	[1.1][1.7] [3.6][4.M] [5.5]	[1.1][3.6] [4.M][5.5]	[1.5][1.7] [3.6][4.M] [5.1][5.5][5.6]	[1.1][1.6] [4.M][5.1] [5.5][5.6]	[2.3] [2.5]		[2.6]	[5.6]		[5.5]
SMALL 100 TO 1,000	[1.2][1.3] [1.5][1.6] [1.7][4.M] [5.3][5.4]	[1.2][1.3][1.5] [1.7][3.6] [4.M][5.1] [5.3][5.4][5.5]	[1.1][1.7][3.6] [4.M][5.1] [5.3][5.5][5.6]	[1.2][1.7] [3.6][4.M] [5.1][5.5] [5.4][5.5]	[1.2][1.3][1.5] [1.7][1.8][3.3] [4.M][5.1] [5.3][5.4]	[1.2][1.3][1.5] [1.7][1.8][3.6] [4.M][5.3] [5.4][5.5]	[1.3][1.6] [1.7][1.8] [3.6][4.M] [5.5]	[1.1][1.3] [1.7][1.8] [3.6][4.M] [5.5]	[1.1][1.3] [1.7][1.8] [3.6][4.M] [5.5]	[1.1][1.3] [1.8][3.6] [4.M][5.5]	[1.2][1.3][1.5] [1.7][3.6] [4.M][5.1] [5.3][5.4][5.5]	[1.1][1.6] [4.M][5.3] [5.4][5.5] [5.6]	[2.2][2.3] [2.5]	[2.2]	[2.2][2.6]	[5.6]		[5.5]
SMALL TO MEDIUM 1,000 TO 10,000	[1.2][1.3] [1.5][1.6] [1.7][3.7] [4.A][5.2]	[1.2][1.3][1.5] [1.7][3.1][3.2] [3.7][4.A][5.2] [5.3][5.4][5.5]	[1.2][1.7][3.1] [3.2][3.7][4.A] [5.2][5.3] [5.4][5.5]	[1.2][1.7][3.1] [3.2][3.7][4.A] [5.2][5.3] [5.4][5.5]	[1.2][1.3][1.4] [1.5][1.8][3.1] [3.2][3.7][4.A] [5.3][5.4][5.5]	[1.2][1.3][1.4] [1.5][1.8][3.1] [3.2][3.7][4.A] [5.3][5.4][5.5]	[1.3][1.4] [1.6][1.8] [3.1][3.2] [4.A][5.5]	[1.3][1.4] [1.8][3.2] [4.A][5.5]	[1.3][1.4] [3.2]	[1.3][1.4] [3.2]	[1.2][1.3][1.5] [1.7][3.1][3.2] [3.7][4.A][5.2] [5.3][5.4][5.5]	[3.1][3.7] [4.A][5.3] [5.4][5.5] [5.5]	[2.1][2.2] [2.3][2.4]	[2.2]	[2.2]	[5.2]		[5.5]
MEDIUM TO HIGH 10,000 TO 100,000	[1.2][1.3] [3.7][4.A]	[3.1][3.2] [3.3][3.7] [3.8][4.A] [5.5]	[3.2][3.3] [3.8][4.A] [5.2]	[3.1][3.2] [3.3][3.7] [3.8][4.A]	[1.2][1.4] [3.1][3.2] [3.3][3.7] [3.8][4.A]	[1.2][1.4] [3.1][3.2][3.3] [3.7][3.8] [4.A][5.5]	[1.3][1.4] [3.1][3.2] [3.3][3.8] [4.A]	[1.4][3.2] [3.8][4.A]	[1.4][3.2] [3.8]	[1.4][3.2] [3.3][3.8] [4.A]	[3.3][3.7] [3.8][4.A] [5.2][5.5]	[3.7][3.8] [4.A][5.2] [5.5]	[2.1][2.3] [2.4][2.7]	[2.1][2.2] [2.7]	[2.2]	[3.7]		[3.3]
HIGH 100,000+	[1.2][1.3] [3.7]	[3.1][3.2] [3.3][3.8] [4.A]				[1.2][1.3] [1.4][3.2] [3.3][3.8] [4.A]	[1.3][1.4] [3.1][3.8] [4.A]	[1.4][3.2]		[1.4][3.2]			[2.1][2.4] [2.7]	[2.1][2.2] [2.7]	[2.2]	[3.7]		
ALL QUANTITIES	[1.1]	[1.1][1.6] [3.4][3.5]	[1.6]	[1.1][1.6] [3.4][3.5]	[1.1][1.6] [3.4][3.5] [5.5]	[1.1][1.6] [3.4][3.5]	[1.1][3.4] [3.5]	[3.4][3.5]			[1.1][1.6] [3.4][3.5]	[3.4][3.5]				[5.5]		[1.6]

KEY TO MATRIX:

[1.1] SAND CASTING
[1.2] SHELL MOLDING
[1.3] GRAVITY DIE CASTING
[1.4] PRESSURE DIE CASTING
[1.5] CENTRIFUGAL CASTING
[1.6] INVESTMENT CASTING
[1.7] CERAMIC MOLD CASTING
[1.8] PLASTER MOLD CASTING

[2.1] INJECTION MOLDING
[2.2] COMPRESSION MOLDING
[2.3] VACUUM FORMING
[2.4] BLOW MOLDING
[2.5] ROTATIONAL MOLDING
[2.6] CONTACT MOLDING
[2.7] CONTINUOUS EXTRUSION (PLASTICS)

[3.1] CLOSED DIE FORGING/ UPSET FORGING
[3.2] COLD FORMING
[3.3] COLD HEADING
[3.4] SHEET METAL SHEARING
[3.5] SHEET METAL FORMING
[3.6] SPINNING
[3.7] POWDER METALLURGY
[3.8] CONTINUOUS EXTRUSION (METALS)

[4.A] AUTOMATIC MACHINING
[4.M] MANUAL MACHINING
(THE ABOVE HEADINGS COVER A BROAD RANGE OF MACHINING PROCESSES AND LEVELS OF CONTROL TECHNOLOGY. FOR MORE DETAIL, THE READER IS REFERRED TO REFERENCES ON THE INDIVIDUAL PROCESSES.)

[5.1] ELECTRICAL DISCHARGE MACHINING
[5.2] ELECTROCHEMICAL MACHINING
[5.3] ELECTRON BEAM MACHINING
[5.4] LASER BEAM MACHINING
[5.5] CHEMICAL MACHINING
[5.6] ULTRASONIC MACHINING

FIGURE 9.4
Prima selection matrix showing which materials and processes are usually used in practice. (*From K. G. Swift and J. D. Booker, "Process Selection," p. 18, Arnold, London, 1997. Used with permission.*)

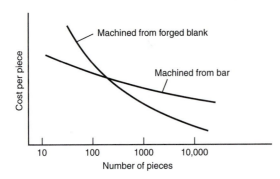

FIGURE 9.5

Comparison of costs with different forms of starting material.

composition (segregation). Surface defects can be surface cracks, rolled-in oxide, extreme roughness, or surface discoloration or attack. The amount of material used to make the part should be just enough larger than the final part to allow for removal of surface defects by machining or another surface conditioning method. Thus, extra material in a casting may be needed to permit machining the surface to a specified finish, or a heat-treated steel part may be made oversized to allow for the removal of a decarburized layer.

Often the manufacturing process dictates the use of extra material, such as sprues and risers in castings and flash in forgings and moldings. At other times extra material must be provided for purposes of handling, positioning, or testing the part. Even though extra material removal is costly, it usually is cheaper to purchase a slightly larger workpiece than to pay for a scrapped part.

Process modeling is being used effectively to investigate the design of tooling and the flow of material to minimize defect formation. Also, improved nondestructive inspection methods make more certain the detection of defects before a part is placed into service. Defects can often be eliminated by subjecting the part to a high hydrostatic pressure, e.g., 15,000 psi, at elevated temperature, in a process called hot-isostatic pressing (HIP).[1] HIPing has been used effectively in investment casting to replace parts previously made by forging.

Surface finish

The surface finish of a part determines its appearance, affects the assembly of the part with other parts, and may determine its resistance to corrosion. The surface roughness of a part must be specified and controlled because of its influence on fatigue failure, friction and wear, and assembly with other parts.

No surface is smooth and flat like the straight line we make on an engineering drawing. On a highly magnified scale it is rough, as sketched in Fig. 9.6. Several parameters are used to describe the state of surface roughness.[2]

---

1. H. V. Atkinson and B. A. Rickinson, "Hot Isostatic Pressing," Adam Hilger, 1991.
2. See Surface Texture, ANSI Std. B46.1, ASME, 1985.

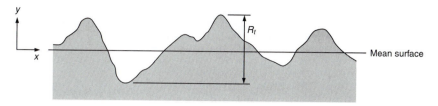

FIGURE 9.6
Cross-sectional profile of surface roughness with vertical direction magnified.

R_t is the height from maximum peak to deepest trough.
R_a is the centerline average (CLA), the arithmetic average based on the deviation from the mean surface

$$R_a = \frac{y_1 + y_2 + y_3 + \cdots + y_n}{n}$$

R_{rms} is the root-mean-square value of height.

$$R_{rms} = \left(\frac{y_1^2 + y_2^2 + y_3^2 + \ldots + y^2}{n} \right)^{1/2}$$

Surface roughness measurements typically are expressed in microinches (1 μin = 0.025 μm = 0.000001 in). The arithmetic average is most widely used for reporting surface roughness. The rms value will always be greater than the arithmetic average because the larger deviations will figure more strongly in its calculation.

There are other important characteristics of a surface besides the height of the roughness. Surfaces may exhibit a directionality characteristic called *lay.* Surfaces may have a strong directional lay (e.g., from machining grooves), a random lay, or a circular pattern of marks. Another characteristic of the surface is its *waviness,* which occurs over a longer distance than the peaks and valleys of roughness. All these surface characteristics are specified on the engineering drawing by the scheme shown in Fig. 9.7.

It is important to realize that specifying a surface by average roughness height is not an ideal approach. Two surfaces can have the same value of R_a (or R_{rms}) and vary considerably in the details of surface profile.

Surface texture does not completely describe a surface. For example, there is an altered layer just below the surface texture layer. This layer is characteristic of the nature and amount of energy that has been put into creating the surface. It can contain cracks, residual stresses, hardness differences, and other alterations.[1] Control of this surface and subsurface layer as it is influenced by processing is called *surface integrity.*[2]

---

1. Surface Integrity, ANSI Std. B211.1-1986, Society of Manufacturing Engineers.
2. A. R. Marder, Effects of Surface Treatments on Materials Performance, "ASM Handbook," vol. 20, pp. 470–490, 1997; E. W. Brooman, Design for Surface Finishing, "ASM Handbook," vol. 20, pp. 820–827, ASM International, Materials Park, OH, 1997.

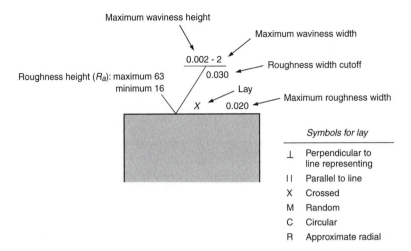

FIGURE 9.7
Symbols used to specify finish characteristics. Roughness given in microinches.

Dimensional accuracy and tolerances

Processes differ in their ability to meet close tolerances. Generally, good workability materials can be held to closer tolerances. Achieving dimensional accuracy depends on both the nature of the material and the process. Solidification processes must allow for the shrinkage that occurs when a molten metal solidifies. Polymer processes must allow for the much higher thermal expansion of polymers than metals, and hot working processes with metals must allow for oxidation of the surface.

Each manufacturing process has the capability of producing a part to a certain surface finish and tolerance range without incurring extra cost. Figure 9.8 shows this general relationship. The tolerances apply to a 1-in dimension and are not necessarily scalable to larger and smaller dimensions for all processes. For economical design, the loosest possible tolerances and coarsest surface finish that will fulfill the function of the design should be specified. As Fig. 9.9, p. 396, shows, processing cost increases nearly exponentially as the requirements for tolerances and surface finish are made more stringent.

We noted above that the tolerances in Fig. 9.8 applied only to a 1-in (25-mm) dimension. Information on the way the tolerance varies with dimension for many processes is given in the manufacturing process information maps (PRIMA) given by Swift and Booker.[1] Figure 9.10 shows the way that tolerance changes with dimension for sand casting. Three regions are delineated: the region where the process does not normally operate (white), the normal working capability of the process (light gray), and the region where the tolerance can be achieved only at high cost (black).

9.3.6 Availability, Lead Time, and Delivery

Next to cost, a paramount business factor in selecting a manufacturing process is the availability of the production equipment, the lead time required to make tooling, and the

---

1. K. G. Swift and J. D. Booker, "Process Selection: From Design to Manufacture," Arnold, London, 1997.

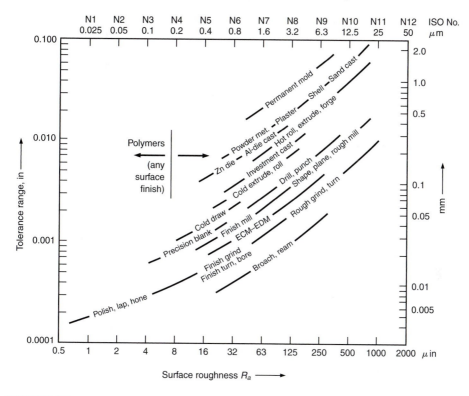

FIGURE 9.8

Approximate values of surface roughness and tolerance on dimensions typically obtained with different manufacturing processes. (*From J. A. Schey, "Introduction to Manufacturing Processes," p. 670, McGraw-Hill Book Company, New York, 1987; used with permission of McGraw-Hill Book Co.*)

reliability of the expected delivery date for the parts. Large structural parts, such as rotors for electrical generators, or the main structural forgings for military aircraft, can be made in only a few factories in the world because of limitations of equipment. Careful scheduling with the design cycle may be needed to mesh with the production schedule. Complex forging dies and polymer injection molding dies can have lead times of a year. These kinds of issues clearly affect the choice of the manufacturing process.

9.3.7 Further Information for Process Selection

For further details on manufacturing processes see the references at the end of this chapter. The book by Schey[1] and the handbook chapter by the same author[2] are particularly helpful in the way they compare a large spectrum of processes.

1. J. A. Schey, "Introduction to Manufacturing Processes," 2d ed., McGraw-Hill, New York, 1987.
2. J. A. Schey, Manufacturing Processes and Their Selection, "ASM Handbook," vol. 20, pp. 687–704, ASM International, Materials Park, OH, 1997.

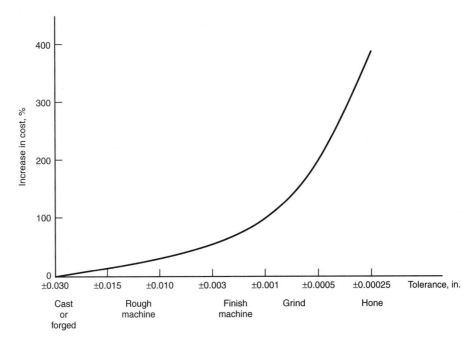

FIGURE 9.9
Influence of tolerance on processing costs (schematic).

A comparison of the manufacturing processes is given in Table 9.2. This is based on a series of data cards published by the Open University.[1] A useful feature of this table is the references to the extensive series of ASM Handbooks (AHB) and Engineered Materials Handbooks (EMH), which give many practical details on the processes.

The Manufacturing Process Information Maps (PRIMA) give much information useful for an initial selection of process.[2] The PRIMA selection matrix (Fig. 9.4) gives a set of 5 to 10 possible processes for different combinations of material and quantity of parts. Each PRIMA then gives the following information, which is a good summary of the information needed to make an intelligent decision on manufacturing process.

- Process description.
- Materials: materials typically used with the process.
- Process variations: common variants of the basic process.
- Economic factors: cycle time, minimum production quantity, material utilization, tooling costs, labor costs, lead times, energy costs, equipment costs, etc.
- Typical applications: examples of parts commonly made with this process.
- Design aspects: general information on shape complexity, size range, minimum thickness, draft angles, undercuts, and limitations on other features.

1. Data cards to accompany L. Edwards and M. Endean, op. cit.
2. K. G. Swift and J. D. Booker, op. cit.

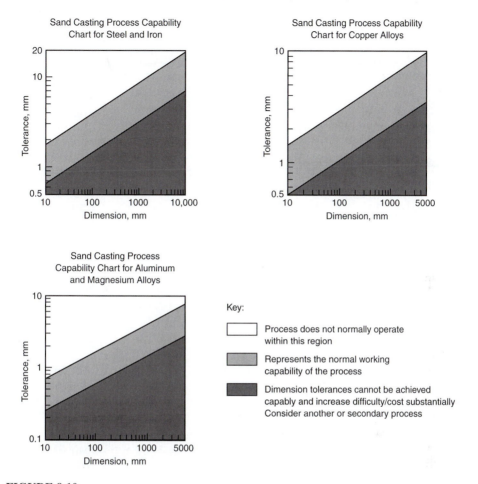

FIGURE 9.10

Process information map for tolerance capability in the sand casting process. (*From K. G. Swift and J. D. Booker, "Process Selection," p. 24, Arnold, London, 1997; used with permission of publisher.*)

- Quality issues: describes defects to watch out for, expected range of surface finish, and process capability charts showing dimensional tolerances as a function of dimension.

This 200-page book is a goldmine of information for the serious design engineer.

9.4
DESIGN FOR MANUFACTURABILITY (DFM)

The past decade has seen a large amount of effort devoted to the integration of design and manufacture, with the goals of reducing manufacturing cost and improving

TABLE 9.2
Rating of characteristics of common manufacturing processes

Process	Shape	Cycle time	Flexibility	Material utilization	Quality	Equipment tooling costs	Handbook reference
Casting							
Sandcasting	3D	2	5	2	2	1	AHB, vol. 15
Evaporative foam	3D	1	5	2	2	4	AHB, vol. 15, p. 230
Investment casting	3D	2	4	4	4	3	AHB, vol. 15, p. 253
Permanent mold casting	3D	4	2	2	3	2	AHB, vol. 15, p. 275
Pressure die casting	3D solid	5	1	4	2	1	AHB, vol. 15, p. 285
Squeeze casting	3D	3	1	5	4	1	AHB, vol. 15, p. 323
Centrifugal casting	3D hollow	2	3	5	3	3	AHB, vol. 15, p. 296
Injection molding	3D	4	1	4	3	1	EMH, vol. 2, p. 308
Reaction injection molding (RIM)	3D	3	2	4	2	2	EMH, vol. 2, p. 344
Compression molding	3D	3	4	4	2	3	EMH, vol. 2, p. 324
Rotational molding	3D hollow	2	4	5	2	4	EMH, vol. 2, p. 360
Monomer casting contact molding	3D	1	4	4	2	4	EMH, vol. 2, p. 338
Forming							
Forging, open die	3D solid	2	4	3	2	2	AHB, vol. 14, p. 61
Forging, hot closed die	3D solid	4	1	3	3	2	AHB, vol. 14, p. 75
Sheet metal forming	3D	3	1	3	4	1	AHB, vol. 14, p. 445
Rolling	2D	5	3	4	3	2	AHB, vol. 14, p. 343
Extrusion	2D	5	3	4	3	2	AHB, vol. 14, p. 315
Superplastic forming	3D	1	1	5	4	1	AHB, vol. 14, p. 852
Thermoforming	3D	3	2	3	2	3	EMH, vol. 2, p. 399
Blow molding	3D hollow	4	2	4	4	2	EMH, vol. 2, p. 352
Pressing and sintering	3D solid	2	2	5	2	2	AHB, vol. 7
Isostatic pressing	3D	1	3	5	2	1	AHB, vol. 14, p. 419
Slip casting	3D	1	5	5	2	4	EMH, vol. 4, p. 153
Machining							
Single-point cutting	3D	2	5	1	5	5	AHB, vol. 16
Multiple-point cutting	3D	3	5	1	5	4	AHB, vol. 16
Grinding	3D	2	5	1	5	4	AHB, vol. 16, p. 421
Electrical discharge machining	3D	1	4	1	5	1	AHB vol. 16, p. 557
Joining							
Fusion welding	All	2	5	5	2	4	AHB, vol. 6
Brazing/soldering	All	2	5	5	3	4	AHB, vol. 6, p. 929
Adhesive bonding	All	2	5	5	3	5	EMH, vol. 3
Fasteners	3D	4	5	4	4	5	. . .
Surface treatment							
Shot peening	All	2	5	5	4	5	AHB, vol. 5, p. 138
Surface hardening	All	2	4	5	4	4	AHB, vol. 4
CVD/PVD	All	1	5	5	4	3	AHB, vol. 13, p. 456

Rating scheme: 1, poorest; 5, best. From "ASM Handbook," vol. 20, p. 299, ASM International. Used with permission.

product quality. The processes and procedures that have been developed have become known as *design for manufacturability* or *design for manufacture* (DFM). Associated with this is the closely related area of *design for assembly* (DFA). The field is often simply described by DFM/DFA.

Design for manufacture represents an awareness of the importance of design as the first manufacturing step. To achieve the goals of DFM requires a concurrent engineering team approach (Sec. 1.7) in which appropriate representatives from manufacturing, including outside suppliers, are members of the design team from the start.

9.4.1 DFM Guidelines

DFM guidelines are statements of good design practice that have been empirically derived from years of experience.[1] Using these guidelines helps narrow the range of possibilities so that the mass of detail that must be considered is within the capability of the designer.

1. *Minimize total number of parts:* Eliminating parts results in great savings. A part that is eliminated costs nothing to make, assemble, move, store, clean, inspect, rework, or service. A part is a good candidate for elimination if there is no need for relative motion, no need for subsequent adjustment between parts, and no need for materials to be different. However, part reduction should not go too far so that it adds cost because the remaining parts become too heavy or complex.

   The best way to eliminate parts is to make minimum part count a functional requirement of the design at the conceptual stage of design. Combining two or more parts into an integral design is another approach. Plastic parts are particularly well suited for integral design.[2]

   As an example of this principle, a cash register manufacturer designed a new electronic cash register that has 85 percent fewer parts than the model it replaced, and takes only 25 percent as much time to assemble. A chief factor was the elimination of all screws by snap-fastened plastic parts. Heavy use was made of CAD, and through concurrent engineering the product was introduced in the market just 24 months after development began.

2. *Standardize components:* Costs are minimized and quality is enhanced when standard commercially available components are used in design. The life and reliability of these components have already been established, and cost reduction comes through quantity discounts, elimination of design effort, avoidance of equipment and tooling costs, and better inventory control.

3. *Use common parts across product lines:* It is good business sense to use parts in more than one product. Specify the same materials, parts, and subassemblies in each product as much as possible. This provides economies of scale that drive down per unit cost and simplify operator training and process control.

4. *Design parts to be multifunctional:* A good way to minimize part count is to design such that parts can fulfill more than one function. For example, a part might serve

1. H. W. Stoll, *Appl. Mech. Rev.,* vol. 39, no. 9, pp. 1356–1364, 1986.
2. W. Chow, "Cost Reduction in Product Design," chap. 5, Van Nostrand Reinhold, New York, 1978.

as both a structural member and a spring. The part might be designed to provide a guiding, aligning, or self-fixturing feature in assembly.

5. *Design parts for ease of fabrication:* Most of this chapter addresses this guideline. As discussed in Chap. 8, the least costly material that satisfies the functional requirements should be chosen. In determining this cost, the cost of manufacture must be considered. However, in many products like automobiles 50 to 60 percent of the total cost is attributable to materials. Since machining processes tend to be costly, manufacturing processes that produce the part to near net shape are preferred whenever possible.

6. *Avoid too tight tolerances:* Tolerances must be set with great care. Specifying tolerances that are tighter than needed results in increased cost. These come about from the need for secondary finishing operations like grinding, honing, and lapping, from the cost of building extra precision into the tooling, from longer operating cycles, and from the need for more skilled workers.

7. *Avoid secondary operations:* Eliminate or minimize secondary manufacturing operations such as deburring, heat treatment, polishing, painting, and plating. Use only when there is a functional reason for doing this. Machine a surface only when the functionality requires it or if it is needed for aesthetic purposes.

8. *Utilize the special characteristics of processes:* Be alert to the special design features that many processes provide. For example, molded polymers can be provided with "built-in" color, as opposed to metals that need to be painted or plated. Aluminum extrusions can be made in intricate cross sections that can then be cut to short lengths to provide parts. Powder-metal parts can be made with controlled porosity that provide self-lubricating bearings.

9.4.2 Specific Design Rules

A number of rules for design, more specific than those given in the previous section have been developed.[1]

1. Space holes in machined, cast, molded, or stamped parts so they can be made in one operation without tooling weakness. This means that there is a limit on how close holes may be spaced due to strength in the thin section between holes.

2. Avoid generalized statements on drawings, like "polish this surface" or "toolmarks not permitted" which are difficult for manufacturing personnel to interpret. Notes on engineering drawings must be specific and unambiguous.

3. Dimensions should be made from specific surfaces or points on the part, not from points in space. This greatly facilitates the making of gauges and fixtures.

4. Dimensions should all be from a single datum line rather than from a variety of points to avoid overlap of tolerances.

5. The design should aim for minimum weight consistent with strength and stiffness requirements. While material costs are minimized by this criterion, there also will usually be a reduction in labor and tooling costs.

---

1. J. G. Bralla (ed.), "Handbook of Product Design for Manufacture," McGraw-Hill, New York, 1986.

6. Whenever possible, design to use general-purpose tooling rather than special dies, form cutters, etc. An exception is high-volume production where special tooling may be more cost-effective.
7. Use generous fillets and radii on castings, molded, formed, and machined parts.
8. Parts should be designed so that as many operations as possible can be performed without requiring repositioning. This promotes accuracy and minimizes handling.

9.5
DESIGN FOR ASSEMBLY (DFA)

Once parts are manufactured, they need to be assembled into subassemblies and products. The assembly process consists of two operations, *handling* followed by *insertion.* There are three types of assembly, classified by the level of automation. In *manual assembly* a human operator at a workstation reaches and grasps a part from a tray, and then moves, orients, and prepositions the part for insertion. The operator then places the parts together and fastens them, often with a power tool. In *automatic assembly* handling is accomplished with a parts feeder, like a vibratory bowl, that feeds the parts correctly oriented for insertion to an automatic workhead, which in turn inserts the part.[1] In *robotic assembly* the handling and insertion of the part is done by a robot arm under computer control.

The cost of assembly is determined by the number of parts in the assembly and the ease with which the parts can be handled and inserted. Design can have a strong influence in both areas. Reduction in the number of parts can be achieved by elimination of parts, e.g., replacing screws and washers with snap or press fits, and by combining several parts into a single component. Ease of handling and insertion is achieved by designing so that the parts cannot become tangled or nested in each other, and by designing with symmetry in mind. Parts that do not require end-to-end orientation prior to insertion, as a screw does, should be used if possible. Parts with complete rotational symmetry around the axis of insertion, like a washer, are best. For automatic handling it is better to make a part highly asymmetric if it cannot be made symmetrical. For ease of insertion a part should be made with chamfers or recesses for ease of alignment, and clearances should be generous to reduce the resistance to assembly. Self-locating features are important, as is providing unobstructed vision and room for hand access. Figure 9.11 illustrates some of these points. Complete information on design details is given by Boothroyd and Dewhurst.[2]

A list of DFA guidelines is given below.

1. *Minimize the total number of parts:* Clearly a part that is not required by the design is a part that does not need to be assembled. Go through the list of parts in the

1. G. Boothroyd, C. Poli, and L. E. Murch, "Automatic Assembly," Marcel Dekker, New York, 1982; G. Boothroyd, "Assembly Automation and Product Design," Marcel Dekker, New York, 1992; Quality Control and Assembly, "Tool and Manufacturing Engineers Handbook," vol. 4, Society of Manufacturing Engineers, Dearborn, MI, 1987.
2. G. Boothroyd and P. Dewhurst, "Product Design for Assembly," Boothroyd Dewhurst Inc., Wakefield, RI, 1989.

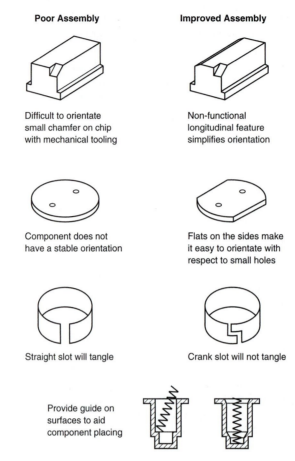

Poor Assembly

Difficult to orientate
small chamfer on chip
with mechanical tooling

Component does not
have a stable orientation

Straight slot will tangle

Provide guide on
surfaces to aid
component placing

Improved Assembly

Non-functional
longitudinal feature
simplifies orientation

Flats on the sides make
it easy to orientate with
respect to small holes

Crank slot will not tangle

FIGURE 9.11
Some design features that improve assembly.

assembly and identify those parts that are essential for the proper functioning of the product. All others are candidates for elimination. The criteria for an essential part are:

- The part must exhibit motion relative to another part that is declared essential.
- There is a fundamental reason that the part be made from a material different from all other parts.
- It would not be possible to assemble or disassemble the other parts unless this part is separate.
- Parts used only for fastening or connecting other parts are prime candidates for elimination.

Designs can be evaluated for efficiency of assembly from Eq. (9.3),

$$\text{Design efficiency} = \frac{3 \times \text{minimum number of parts}}{\text{total assembly time}} \tag{9.3}$$

where the time taken to assemble a "theoretical" part is 3 s.

2. *Minimize the assembly surfaces:* Simplify the design so that fewer surfaces need to be prepared in processing, and all work on one surface is completed before moving to the next one.

3. *Avoid separate fasteners:* The use of screws in assembly is expensive. Snap fits should be used whenever possible. Where screws must be used, quality risks can be reduced by minimizing the number, size, and variations of fasteners and by using standard fasteners.

4. *Minimize assembly direction:* All parts should be designed so that they can be assembled from one direction. The need to rotate in assembly requires extra time and motion and may require additional transfer stations and fixtures. The best situation in assembly is when parts are added in a top-down manner to create a *z-axis stack.*

5. *Maximize compliance in assembly:*[1] Excessive assembly force may be required when parts are not identical or perfectly made. Allowance for this should be made in the product design. Designed-in compliance features include the use of generous tapers, chamfers, and radii. If possible, one of the components of the product can be designed as the part to which other parts are added (part base) and as the assembly fixture. This may require design features which are not needed for the product function.

6. *Minimize handling in assembly:* Parts should be designed to make the required position easy to achieve. Since the number of positions required in assembly equates to increased equipment expense and greater risk of defects, quality parts should be made as symmetrical as function will allow. Orientation can be assisted by design features which help to guide and locate parts in the proper position. Parts that are to be handled by robots should have a flat, smooth top surface for vacuum grippers, or an inner hole for spearing, or a cylindrical outer surface for gripper pickup.

9.6
EARLY ESTIMATION OF MANUFACTURING COST

The decisions about materials, shape, features, and tolerances that are made in the embodiment phase of design determine the manufacturing cost of the product. It is not often possible to get large cost reductions once production has begun because of the high cost of change at this stage of the product life cycle. Therefore, we need a way of identifying costly designs as early as possible in the design process.

Of course, one way to achieve this goal is to include knowledgeable manufacturing personnel on the product design team. The importance of this is unassailable, but it is not always possible from a practical standpoint due to conflicts in time commitments, or even because the design and manufacturing personnel may not be in the same location. Also, there is the conservative tendency of manufacturing people to go with what they know works, thereby limiting opportunity for improvement.

---

1. J. L. Nevins and D. E. Whitney, "Concurrent Design of Products and Processes," McGraw-Hill, New York, 1989, chaps. 5, 6, and 7.

One system that is applicable to an early design stage where not much detail has been yet established was developed at the University of Hull.[1] It is based on data obtained from British automotive, aerospace, and light manufacturing companies. The process starts with the identification of a small number of materials and processes to consider (Fig. 9.4). Comparison between alternatives is based on estimated manufacturing cost.

$$C = VC_{mv} + P_c \cdot R_c \tag{9.4}$$

where C_{mv} = cost of material per unit volume
$\quad\quad V$ = volume of material input to the process
$\quad\quad P_c$ = basic processing cost for an ideal part
$\quad\quad R_c$ = cost coefficient for the part design that takes into account shape complexity, material workability, section thickness, surface finish, and tolerances.

The basic processing cost is given by

$$P_c = \alpha T + \beta/N \tag{9.5}$$

where α = cost of setting up and operating a specific process
$\quad\quad T$ = cycle time in seconds to produce an ideal part
$\quad\quad \beta$ = total tooling cost for an ideal part
$\quad\quad N$ = annual production quantity for the part

Figure 9.12 shows average data for Eq. (9.5) for eight common processes based on one shift per day and a 2-year payback for equipment and tooling.

The design-dependent factors are included in the R_c term.

$$R_c = C_{mp} \cdot C_c \cdot C_s \cdot C_{ft} \tag{9.6}$$

where C_{mp} = relative cost associated with material-process suitability (workability or fabricability)
$\quad\quad C_c$ = relative cost associated with shape complexity
$\quad\quad C_s$ = relative cost associated with achieving minimum section thickness
$\quad\quad C_{ft}$ = the higher of the cost in achieving a specified surface finish or tolerance, but not both

For the ideal part design each of the coefficients in Eq. (9.6) is unity. As the design moves away from this situation, one or more of the coefficients will increase in value, thereby changing the manufacturing cost in Eq. (9.4).

Table 9.3 shows the suitability rating of the material-process combinations that give C_{mp}. Note that many of the combinations are inadmissible for technical or economic reasons.

---

1. K. G. Swift and J. D. Booker, op. cit.; also A. J. Allen and K. G. Swift, *Proc. Instn. Mech. Engrs.*, vol. 204, pp. 143–148, 1990.

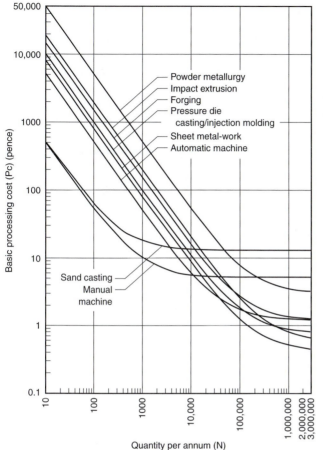

FIGURE 9.12
Average curves of P_c vs. N for selected manufacturing processes. (*From K. G. Swift and J. D. Booker, "Process Selection," p. 177, Arnold, London. Used with permission.*)

TABLE 9.3
Relative cost of material-process suitability, C_{mp}

Process Material	Impact extrusion	Sand casting	Pressure die casting	Forging	Sheet metal-working	Machining	Powder metallurgy	Injection molding
Cast iron		1				1.2	1.6	
Low carbon steel	1.3	1.2		1	1.2	1.4	1.2	
Alloy steel	2	1.3		2	1.5	2.5	1.1	
Stainless steel	2	1.5		2	1.5	4	1.1	
Copper alloy	1	1	3	1	1	1.1	1	
Aluminum alloy	1	1	1.5	1	1	1	1	
Zinc alloy	1	1	1.2	1	1	1.1	1	
Thermoplastic						1.1		1
Thermoset						1.2		1
Elastomers						1.1		1.5

From K. G. Swift and J. D. Booker, "Process Selection," p. 178, Arnold, London, 1997. (Used with permission.)

The shape complexity factor C_c is based on the classification system shown in Fig. 9.13. This system divides shapes into three basic categories: (A) solids of revolution, (B) prismatic solids, and (C) flat or thin-wall section components. Within each category the complexity increases from 1 to 5. Shape classification is based on the finished shape of the part. C_c is the most important coefficient in Eq. (9.6). Therefore, careful study of the category definitions is important. Once the shape subcategory has been established, a set of curves for the process is entered (Fig. 9.14) and the coefficient is picked off. Figure 9.15 shows the curves for obtaining the section coefficient C_s. If the required section thickness falls to the left of the black vertical line, additional secondary processing (machining or grinding) is likely to be necessary. This cost is included in the value of the coefficient.

Curves for tolerances C_t and surface finish C_f are given in Fig. 9.16, p. 409. In using these charts, identify the most severe tolerance requirement in the part, and note whether it applies on one or more than one plane. The same applies for the surface finish requirements. Note that the value for either C_t or C_f is used, depending on which coefficient is larger. Costs estimated this way have been shown to be within at least 15 percent of the actual cost.

> **EXAMPLE.** We wish to estimate the cost of manufacturing the part shown at the top of A3 in Fig. 9.13. Its dimensions are as follows: large diameter 1 in, small diameter 0.25 in, length of long cylinder 2 in, length of short cylinder 1 in. The diameter of the cross-bore is such that the wall thickness is 5 mm. The tolerance on this bore diameter is 0.005 in. The surface finish is 5 μm. We expect to need 10,000 parts per year. For strength reasons the part will be made from a quenched and tempered medium alloy steel.
>
> Figure 9.4 suggests we look at whether the part should be made by sand casting, machining, or powder metallurgy. Using the charts reproduced in this section we estimate the coefficients for Eq. (9.6) to be:

Process	C_{mp}	C_c	C_s	C_t	C_f	R_c	P_c	$P_c R_c$
Sand casting	1.3	1.3	1.2	1.9	1.9	3.85	10.5p	40p
Machining	2.5	3.0	1.0	1.0	1.0	7.50	5p	38p
Powder metallurgy	1.1	2.4	1.0	1.0	1.0	2.64	60p	158p

> Multiplying coefficients according to Eq. (9.6) gives R_c. Note how machining carries the greatest penalty because of the poor machinability of alloy steel and the geometry complexity. However, the basic processing cost P_c is relatively low. The required number of parts is not sufficient to pay back the tooling costs of powder metallurgy processing.
>
> Turning now to Eq. (9.4), the volume of the part is 0.834 in³. The cost of alloy steel is about $1.25 per lb. Since the density of steel is 0.283 lb/in³, C_{mv} is $0.354 per in³ and VC_{mv} is $0.30. The rest of Eq. (9.4) is in pence (p). There are 100 pence to the pound (£), and 1£ ≈ 1.65$. Therefore,

$$\frac{0.30\$}{1.65\frac{\$}{£}} = 0.182£ \cdot 100\frac{p}{£} = 18p$$

> Thus the estimated manufacturing cost is
>
> | | |
> |---|---|
> | Sand casting | 58p |
> | Machining | 56p |
> | Powder metallurgy | 176p |

A ✎ Part Envelope is Largely a Solid of Revolution

Single/Primary Axis		Secondary Axes: Straight line features parallel and/or perpendicular to primary axis		Complex Forms
Basic rotational features only	Regular secondary/ repetitive features	Internal	Internal and/or external features	Irregular and/or complex forms.
A 1	A 2	A 3	A 4	A 5
Category includes: Rotationally symmetrical/ grooves, undercuts, steps, chamfers, tapers and holes along the primary axis/centre line	Internal/external threads, knurling and simple contours through flats/splines/keyways on/around the primary axis/centre line	Holes/threads/ counterbores and other internal features not on the primary axis	Projections, complex features, blind flats, splines, keyways on secondary axes	Complex contoured surfaces, and/or series of features which are not represented in previous categories

B ▱ Part Envelope is Largely a Rectangular or Cubic Prism

Single Axis/Plane		Multiple Axes		Complex Forms
Basic features only	Regular secondary/ repetitive features	Orthogonal/straight line based features	Simple curved features on a single plane	Irregular and/or contoured forms
B 1	B 2	B 3	B 4	B 5
Category includes: Through steps, chamfers and grooves/channels/slots and holes/threads on a single axis	Regular through features, T-slots and racks/plain gear sections etc. Repetitive holes/threads/counter bores on a single plane	Regular orthogonal/straight line based pockets and/or projections on one or more axis. Angled holes/threads/ counter bores	Curves on internal and/or external surfaces	Complex 3-D contoured surfaces/geometries which cannot be assigned to previous categories

C ⬐ Flat Or Thin Wall Section Components

Single Axis	Secondary/Repetitive Regular Features		Regular Forms	Complex Forms
Basic features only	Uniform section/ wall thickness	Non-uniform section/ wall thickness	Cup, cone and box-type parts	Non-uniform and/or contoured forms
C 1	C 2	C 3	C 4	C 5
Category includes: Blanks, washers, simple bends, forms and through features on or parallel to primary axis	Plain cogs/gears, multiple or continuous bends and forms	Component section changes not made up of multiple bends or forms. Steps, tapers and blind features	Components may involve changes in section thickness	Complex or irregular features or series of features which are not represented in previous categories

FIGURE 9.13

The shape classification system used to determine C_c. (*From K. G. Swift and J. D. Booker, "Process Selection," p. 180, Arnold, London, 1987. Used with permission.*)

407

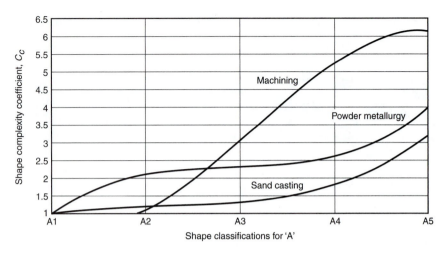

FIGURE 9.14
Curves for determining C_c for shape classification A. (*From K. G. Swift and J. D. Booker, "Process Selection," p. 181, Arnold, London, 1997. Used with permission.*)

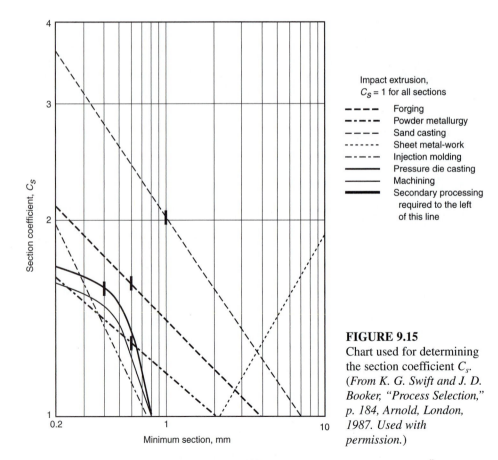

FIGURE 9.15
Chart used for determining the section coefficient C_s. (*From K. G. Swift and J. D. Booker, "Process Selection," p. 184, Arnold, London, 1987. Used with permission.*)

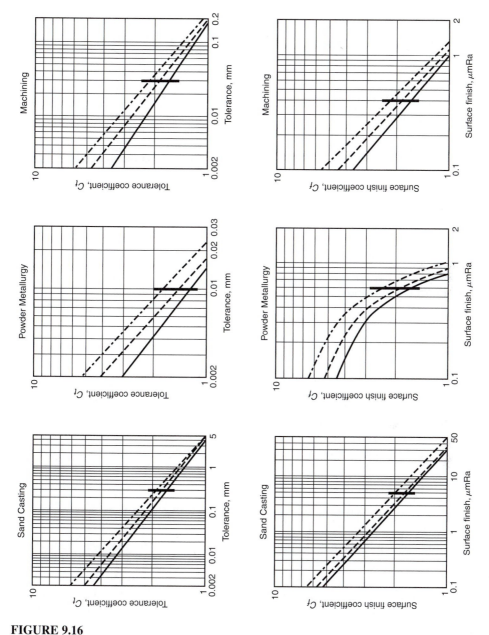

FIGURE 9.16
Charts used to determine C_t (top row) and C_f (bottom row). (*From K. G. Swift and J. D. Booker, "Process Selection," Arnold, London, 1997. Used with permission.*)
—— 1 plane   - - - 2 planes   - · - 3 planes

For this lot size machining is the preferred process over sand casting because the surfaces and tolerances will be better. At this number of parts powder metallurgy is ruled out because of high tooling costs. However, experienced designers would have already ruled it out because they would have recognized that molding in a cross-bore is not economical

TABLE 9.4
Sources of information on how manufacturing cost changes with design details

Process	Reference
Assembly	
Manual assembly	B. D. K., chap. 3
Automatic and robotic assembly	B. D. K., chap. 5
Injection molding	B. D. K., chap. 8
	D.&P., chaps. 11 and 15
Die casting	B. D. K., chap. 10
	D.&P., chaps. 11 and 15
Sheet metal forming	B. D. K., chap. 9
	D.&P., chaps. 12 and 16
Forging	Knight and Poli, *Machine Design,* Jan. 24, 1985
Powder metallurgy	B. D. K., chap. 11
Machining	B. D. K., chap. 7
Printed circuit board design and assembly	B. D. K., chap. 6

Abbreviations: B. D. K.: G. Boothroyd, P. Dewhurst, and W. Knight, "Product Design for Manufacture and Assembly," Marcel Dekker, New York, 1994.
D. & P.: J. R. Dixon and C. Poli, "Engineering Design and Design for Manufacturing," Field Stone Publishers, Conway, MA, 1995.

in P/M. This illustrates the need to know something about the design rules that are associated with each process. Much of the rest of the chapter is devoted to this topic.

The method just presented can be used in the early stage of configuration design, maybe even at the end of a detailed conceptual design, because it depends on only a few details of the design. Other methods that allow the designer to choose on the basis of cost between alternate details of component design in configuration and parametric design are available. Table 9.4 summarizes the sources of information.

9.7
COMPUTER METHODS FOR DFM AND DFA

The simple example given in the previous section demonstrates that a large amount of information must be made available to the designer to do the job of design for manufacture (DFM) and design for assembly (DFA). In addition to the data correlations given by the various curves, there are many design rules specific to particular processes and materials. Thus, it is not surprising that design software has been developed to ease the designer's task and make DFM and DFA more of a reality. The objective of this software is to put relevant manufacturing information in the hands of the designer so that manufacturing decisions can be more accurately balanced against functional decisions during the early stages of the design process.[1]

The DFA methodology and software was developed prior to DFM. The focus of DFA on the entire assembly and its emphasis on reducing part count serves as a driver

---

1. For examples of using these methods, see http://www.munroassoc.com and http://www.sintef.no/units/matek/projects/dfm.

for DFM. DFM, by nature, focuses more on the individual part and how to make it more cheaply. The objective is to quickly provide information on costs while the design is still fluid. While DFM and DFA methods can be done manually on paper, the use of a computerized method greatly aids the designer by providing prompts and help screens, and making it easy to quickly see the effect of design changes. Whatever the method, a major benefit from using a DFA/DFM analysis is that the rigor of using a formal analysis scheme invariably leads to better questions, and ultimately to better solutions.

EXAMPLE.[1] A motor-drive assembly that moves vertically on two steel guiderails is to be designed. The motor must be fully enclosed and have a removable cover for access to the sensor. The chief functional requirement is that there be a rigid base which supports the motor and the sensor, and move up and down on the rails.

Figure 9.17 shows the initial design of the motor-drive assembly. Two brass bushings are pressed in the base to provide suitable friction and wear characteristics. An end plate

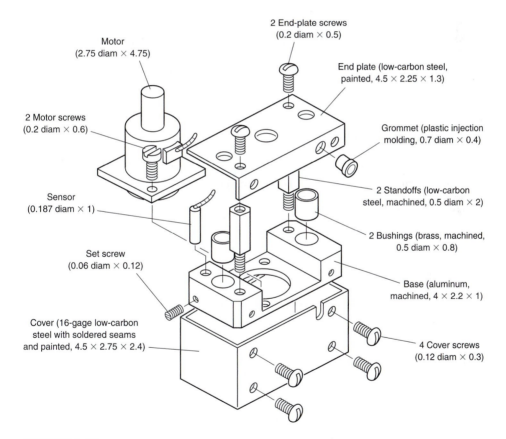

FIGURE 9.17
Initial design of the motor-drive assembly. (*"ASM Handbook," vol. 20, p. 680, ASM International, Materials Park, OH, 1997. Used with permission.*)

1. G. Boothroyd, Design for Manufacture and Assembly, "ASM Handbook," vol. 20, pp. 676–686, ASM International, Materials Park, OH, 1997.

is screwed to two standoffs, which are, in turn, screwed into the base. Finally, a box-shaped cover slides over the assembly from below and is held in place with four cover screws. The entire assembly requires 19 parts and has an assembly time of 160 s (Table 9.5).

Next we review the DFA Guidelines given in Sec. 9.5 to determine which parts are theoretically needed. All of the screws are separate fasteners and theoretically can be eliminated. The base is the main structural part and cannot be eliminated. The motor and the sensor are purchased subassemblies and cannot be eliminated. That leaves two covers, two standoffs, and two bushings. The standoffs and one of the covers could be eliminated by redesign of the base. The bushings are a separate material from the aluminum base and on first examination would have a fundamental reason to remain as separate parts. However, if the material of the base was changed to a material with good frictional properties like nylon, then they also could be eliminated. Table 9.5 shows that the theoretical part count could be as low as four. The design efficiency, based on Eq. (9.3), is 7.5 percent.

In the redesign process we decide, for reliability reasons, to keep the two screws to secure the motor to the base and the setscrew to hold the sensor. The redesigned motor drive is shown in Fig. 9.18. Table 9.6 shows that the assembly time has been reduced to 46 s and the design efficiency increased to 26 percent.

The DFM software calculates the cost of making the individual parts. For the original design most of the parts are either machined or purchased. For the redesign the base is machined from nylon, but the snap-on cover is injection-molded plastic. The tooling cost for this is recovered over 1000 parts. Table 9.7 compares the part costs for the initial design and the redesign. The purchased costs of the motor, sensor, and wiring are not included. This shows a reduction in part cost of 37 percent due to elimination of parts and machining operations. In addition there is a saving in assembly cost of $0.95 per unit.

TABLE 9.5
Results of DFA analysis for the motor-drive assembly (initial design)

Part	No.	Theoretical part count	Assembly time, s	Assembly cost, ¢
Base	1	1	3.5	2.9
Bushing	2	0	12.3	10.2
Motor subassembly	1	1	9.5	7.9
Motor screw	2	0	21.0	17.5
Sensor subassembly	1	1	8.5	7.1
Setscrew	1	0	10.6	8.8
Standoff	2	0	16.0	13.3
End plate	1	1	8.4	7.0
End-plate screw	2	0	16.6	13.8
Plastic bushing	1	0	3.5	2.9
Thread leads	. . .	. . .	5.0	4.2
Reorient	. . .	. . .	4.5	3.8
Cover	1	0	9.4	7.9
Cover screw	4	0	31.2	26.0
Totals	19	4	160.0	133.0

Design efficiency for assembly = (4 × 3)/160 = 7.5%

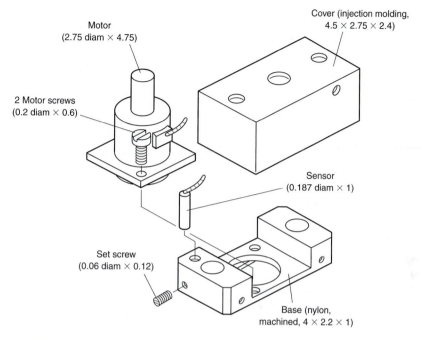

FIGURE 9.18
Redesign of motor-drive assembly based on DFA analysis. (*"ASM Handbook," vol. 20, p. 681, ASM International, Materials Park, OH, 1997. Used with permission.*)

TABLE 9.6
Results of DFA analysis for motor-drive assembly after redesign

Part	No.	Theoretical part count	Assembly time, s	Assembly cost, ¢
Base	1	1	3.5	2.9
Motor subassembly	1	1	4.5	3.8
Motor screw	2	0	12.0	10.0
Sensor subassembly	1	1	8.5	7.1
Setscrew	1	0	8.5	7.1
Thread leads	. . .	. . .	5.0	4.2
Plastic cover	1	1	4.0	3.3
Totals	7	4	46.0	38.4

Design efficiency for assembly = $(4 \times 3)/46 = 26\%$

9.7.1 DFA and DFM Software

In this section we discuss briefly the capability of the main DFA and DFM software currently being marketed. This design methodology is focused on mechanical based assemblies of a size like automotive water pumps or video recorders. Specialized software will be mentioned in sections devoted to specific manufacturing processes.

TABLE 9.7

Part cost for motor-drive assembly based on initial design and redesign with DFA. Costs arrived at with Boothroyd Dewhurst DFMA software

Item	Cost, $
Initial design (Fig. 9.17)	
Base (aluminum)	15.29
Bushing (2)	3.06
Motor screw (2)	0.20(a)
Setscrew	0.10(a)
Standoff (2)	9.74
End plate	2.26
End-plate screw (2)	0.20(a)
Plastic bushing	0.10(a)
Cover	3.73
Cover screw (4)	0.40
Total (initial design)	**35.08**
Redesign (Fig. 9.18)	
Base (nylon)	13.04
Motor screw (2)	0.20(a)
Setscrew	0.10(a)
Plastic cover	8.66(b)
Total (redesign)	**22.00**

(*a*) Purchased in quantity. (*b*) Includes tooling cost for plastic cover ($8000).

The most widely used software of this type in U.S. industry is the Boothroyd Dewhurst, Inc.,[1] software DFMA. This starts with the DFA analysis to minimize part count through redesign, and then goes into DFM analysis to provide cost estimates of individual parts. Modules are available for early cost estimation for machining, sheet metalworking, and injection molding. Other modules consider the life-cycle issues of design for service and design for environment.

The Lucas DFA software came about as a result of a collaboration between Lucas Engineering Systems[2] and the University of Hull in England. It has been combined with the manufacturing process analysis described in Sec. 9.6, and other tools like QFD and FMEA, into an integrated suite of concurrent engineering tools[3] called ToolSET.

HKB, from the German for manufacturing cost calculation, is a rule-based expert system.[4] The system has the capability for determining a complete process plan, while

1. Boothroyd Dewhurst, Inc., 138 Main Street, Wakefield, RI, 02879; http://www.dfma.com.
2. B. L. Miles, *Proc. Instn. Mech. Engrs.,* vol. 203, pp. 29–38, 1989.
3. TeamSet, CSC, PO Box 52, Shirley, Solihull, B90 4JJ, Great Britain; http://www.teamset.com.
4. Mirakon, Gallen, Switzerland. Available in either German or English versions.

for the designer it can do comparative exercises to see how differences in materials or design details will affect the part cost.[1]

A different kind of software provides a variation simulation analysis (VSA) to determine whether the statistical variations that arise from the manufacturing tolerance on dimensions will result in problems in assembly.[2] The analysis starts by defining the three-dimensional features of the components in the assembly by the geometric dimensioning and tolerancing methodology. Then each feature is allowed to vary within its defined zone to determine dimensional stackup and interferences. The software also can simulate the effects of assembly order and fixtures.

In the following sections of this chapter we consider the most common manufacturing processes and discuss how they are related to design. Wherever possible, we give references to design software that is being used to optimize a design for a particular process. There is a strong trend to integrate this kind of software with solid CAD modelers, often in a less sophisticated and easier to use version, so that DFM can be implemented more readily in the design process.

There is no question that DFM and DFA software has greatly advanced the application of these methods in engineering design. Boothroyd Dewhurst have surveyed their users and found[3] a 51 percent reduction in part count, a 37 percent reduction in part cost, a 62 percent reduction in assembly time, and a 57 percent reduction in manufacturing cycle time. However, before we leave this topic, it is important to inject a word of caution about blind acceptance of design rules and design methodologies that are embedded in the software. The overall product view must be paramount. In the drive to reduce part count, it is quite possible to increase cost because of greater geometric complexity.

9.8
DESIGN FOR CASTINGS

One of the shortest routes from raw material to finished part is casting. In a casting, a molten metal is poured into a mold or cavity that approximates the shape of the finished part (Fig. 9.19). Heat is extracted through the mold (in this case a sand mold), and the molten metal solidifies into the final solid shape. This seemingly simple process can be quite complex metallurgically, since the metal undergoes a complete transition from the superheated molten state to the solid state. Liquid metal shrinks on solidification. Thus, the casting and mold must be so designed that a supply of molten metal is available to compensate for the shrinkage. The supply is furnished by introducing feeder heads (risers) that supply molten metal but must be removed from the final casting (Fig. 9.19). Allowance for shrinkage and thermal contraction must be provided in the design. Also, since the solubility of dissolved gases in the liquid

1. P. Ferreirinha, V. Hubka, and W. E. Eder, "Design for Manufacturability," DE-52, ASME, New York, 1993, pp. 97–104.
2. M. Craig, Dimensional Management and Tolerance, "ASM Handbook," vol. 20, pp. 219–221, ASM International, Materials Park, OH, 1997.
3. S. Ashley, *Mechanical Engineering,* March 1995, pp. 74–77.

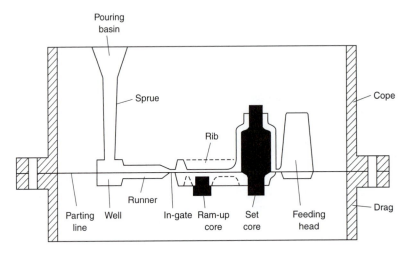

FIGURE 9.19
Parts of a conventional sand casting process. (*From J. A. Schey, "Introduction to Manufacturing Processes," 2d ed., p. 160, McGraw-Hill Book Company, 1987, used with permission of McGraw-Hill Book Company.*)

decreases suddenly as the metal solidifies, castings are subject to the formation of gas bubbles and porosity unless proper preventive steps are taken.

The mechanical properties of a casting are determined during solidification and subsequent heat treatment. The grain structure of the casting, and thus its properties, is determined by how fast each part of the casting freezes. This cooling rate is roughly proportional to the ratio of the square of the surface area of the casting to the square of its volume. Thus, bulky castings freeze much more slowly than thin castings and have lower properties. A sphere of a given volume will freeze more slowly than a thin plate of the same volume because the plate has much more surface area to transfer heat into the mold. Moreover, the casting must be designed so that the flow of molten metal is not impeded by solidified metal before the entire mold cavity fills with molten metal. The casting should freeze progressively, with the region farthest from the source of molten metal freezing first so that the risers can supply liquid metal to feed shrinkage that occurs during solidification. Designing the needed solidification pattern can be achieved with finite-element modeling to construct temperature distributions as a function of time.[1] The FEA can predict shrinkage regions due to lack of feeding and grain size (property) distribution in the casting.[2] This process is called solidification modeling. When such modeling methods are employed in design, and the casting is subjected to hot isostatic pressing to close up any residual porosity, it is possible to produce high-integrity castings that are as reliable as forgings.

1. Commercially available software is ProCAST from UES, Inc., Dayton, OH, or Rapid/CAST from Concurrent Technologies Corp., Johnstown, PA.
2. M. J. Beffel, K. O. Yu, M. Robinson, and K. R. Schneider, *JOM*, February 1989, pp. 27–30.

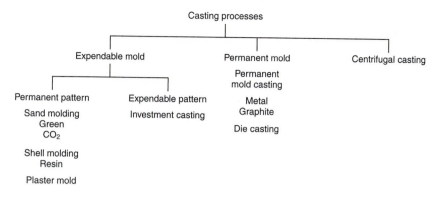

FIGURE 9.20
Classification of casting processes.

There are a large number of casting processes, which can be classified best with respect to the type of mold that is employed (Fig. 9.20). A brief description of each process is given in Table 9.8. The selection of the proper casting process depends on the following factors.

1. Complexity of the shape:
   a. External and internal shape
   b. Types of core required
   c. Minimum wall thickness
2. Cost of the pattern or die
3. Quantity of parts required
4. Tolerances required
5. Surface finish required
6. Strength
7. Weight
8. Overall quality required

Proper attention to design details can minimize casting problems and lead to lower costs.[1] Therefore, close collaboration between the designer and the foundry engineer is important. The use of solidification modeling in this design collaboration is recommended.

The chief consideration is that the shape of the casting should allow for orderly solidification by which the solidification front progresses from the remotest parts toward the points where molten metal is fed in. Whenever possible, section thickness should be uniform. Large masses of metal lead to hot spots, where freezing is delayed, and a shrinkage cavity is produced when the surrounding metal freezes first.

---

1. "Casting Design Handbook," American Society for Metals, 1962; "Metals Handbook," 9th ed.; vol. 15, American Society for Metals, Metals Park, OH, 1988; T. S. Piwonka, Design for Casting, "ASM Handbook," vol. 20, pp. 723–729, ASM International, Materials Park, OH, 1997.

TABLE 9.8

Advantages and disadvantages of the common casting processes

Form	The process	Advantages	Limitations
Sand castings	Green sand. Moist, bonded sand is packed around a wood or metal pattern, the pattern removed, and molten metal poured into the cavity; when metal solidifies, mold is broken and casting removed	Almost any metal can be used; almost no limit on size and shape of part; extreme complexity possible; low tool cost; most direct route from pattern to casting	Some machining always necessary; large castings have rough surface finish; close tolerances difficult to achieve; long, thin projections not practical; some alloys develop defects
	Dry sand. Same as above except: core boxes used instead of patterns, sand bonded with a setting binder, and core baked in an oven	Same as above plus ability to handle long, thin projections	Usually limited to smaller parts than possible with green sand
Shell-mold castings	Sand coated with a thermosetting plastic resin is dropped onto a heated metal pattern (which cures resin); shell halves are stripped off and assembled. When poured metal solidifies, shell is broken away from finished casting	Rapid production rate; high dimensional accuracy; smooth surfaces; uniform grain structure; minimized finishing operations	Some metals cannot be cast; requires expensive patterns; equipment, and resin binder; size of part limited
Full-mold castings	Sand casting process using foamed plastic such as polystyrene for pattern and one-piece mold. Pattern is vaporized during casting. One piece or multipiece patterns can be used depending on complexity	Most metals can be cast; almost no limit on shape, size; useful for complex shapes. Plastic patterns easily handled; no draft required; no flash	Pattern costs can be high for low quantities; some limitations imposed by low strength of pattern material
Permanent-mold castings	Mold cavities are machined into metal die blocks designed for repetitive use; molten metal is gravity-fed to cavity (pressure sometimes applied after pouring). Mold consists of two or more parts and is hinged and clamped for easy removal of castings	Good surface finish and grain structure; high dimensional accuracy; repeated use of molds (up to 25,000); rapid production rate; low scrap loss; low porosity	High initial mold costs; shape, size and intricacy limited; high-melting metals such as steel unsuitable
Die castings	Molten metal is poured into closed steel die under pressures varying from 1500 to 25,000 psi; when the metal solidifies, the die is opened and the casting ejected	Extremely smooth surfaces; excellent dimensional accuracy; rapid production rate	High initial die costs; limited to nonferrous metals; size of part limited

TABLE 9.8 (continued)

Plaster-mold castings	Slurry of special gypsum plaster, water, and other ingredients is poured over pattern and allowed to set; pattern is removed and the mold baked. When poured metal cools, mold is broken for removal of casting	High dimensional accuracy; smooth surfaces; almost unlimited intricacy; low porosity	Limited to nonferrous metals; limited to relatively small parts; mold-making time is relatively long
Ceramic-mold castings	Precision technique using stable ceramic powders, binder, and gelling agent for mold. Mold can be ceramic or ceramic facing with sand backup	Intricate, close tolerance parts with smooth finishes can be cast	Some limit to maximum size
Investment castings	Refractory slurry is cast around (or dipped on) a pattern formed from wax or plastic; when slurry hardens, pattern is melted out and mold is baked. When poured metal solidifies, mold is broken away from casting	High dimensional accuracy; excellent surface finish; almost unlimited intricacy; almost any metal can be used; no flash to remove; no parting line tolerances	Size of part limited; requires expensive patterns and molds; high labor costs
Centrifugal castings	Sand, metal, or graphite mold is rotated in a horizontal or vertical plane (true centrifugal method); molten metal introduced into the revolving mold is thrown to mold wall, where it is held by centrifugal force until solidified	Good dimensional accuracy; rapid production rate; good soundness and cleanliness of castings; ability to produce extremely large cylindrical parts	Shape of part limited; spinning equipment expensive

From "Materials Selector," *Materials Engineering Magazine*, Penton/IPC, Cleveland, OH.

Figure 9.21 illustrates some design features that can alleviate the shrinkage cavity problem. A transition between two sections of different thicknesses should be made gradually (*a*). As a rule of thumb, the difference in thickness of adjoining sections should not exceed 2 to 1. Wedge-shaped changes in wall thickness should not have a taper exceeding 1 to 4. The thickness of a boss or pad (*b*) should be less than the thickness of the section the boss adjoins, and the transition should be gradual. The local heavy section caused by omitting the outer radius at a corner (*c*) should be eliminated. The radius for good shrinkage control should be from one-half to one-third of the section thickness. A strong hot spot is produced when two ribs cross each other (*d*); it can be eliminated by offsetting the ribs as shown. A good way to evaluate the hot spot brought about by a large mass of molten metal is to inscribe the largest circle possible in the cross section of the part. The larger the diameter of the circle the greater the thermal mass effect.

Castings must be so designed as to ensure that the pattern can be removed from the mold and the casting from the permanent mold. A draft, or taper, of less than 3° is required on vertical surfaces so the pattern can be removed from the mold. Projecting details or undercuts should be avoided. Allowance for shrinkage and thermal contraction must be provided on the pattern. Molds made with extensive cores cost more money, so castings should be designed to minimize the use of cores. Also, provisions must be made for placing cores in the mold cavity and holding them in place. Locating points for machining is important. When considerable machining is to be done on a casting, a boss should be provided to serve as a reference point for subsequent machining. A finish allowance for machining should be added to the allowances for shrinkage.

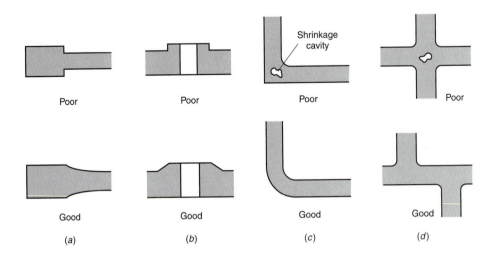

FIGURE 9.21
Some design details to prevent shrinkage cavity.

9.9
DESIGN FOR FORGINGS

Forging processes are among the most important means of producing parts for high-performance applications. Forging is typical of a group of bulk deformation processes in which a solid billet is forced under high pressure to undergo extensive plastic deformation into a final near-to-finished shape. Forging usually is carried out on a hot workpiece, but other deformation processes such as cold extrusion or impact extrusion may be conducted cold, depending upon the material. Because of the extensive plastic deformation that occurs in forging, the metal undergoes metallurgical changes. Any porosity is closed up, and the grain structure and second phases are deformed and elongated in the principal directions of working, creating a "fiber structure." The forging billet has an axial fiber structure due to hot working, but this is redistributed depending upon the geometry of the forging (Fig. 9.22).

The mechanical fibering due to the preferred alignment of inclusions, voids, segregation, and second-phase particles in the direction of working introduces a directionality to structure-sensitive properties such as ductility, fatigue strength, and fracture toughness. The principal direction of working (such as the long axis of a bar) is defined as the *longitudinal direction*. The *short-transverse direction* is the minimum dimensions of the forging, such as the thickness of a platelike shape. The *long-transverse direction* is perpendicular to both the longtudinal and the short-transverse direction. The variation of reduction of area in the tensile test (the most sensitive measure of ductility) with the angle that the specimen axis makes with the forging axis is shown in Fig. 9.23.

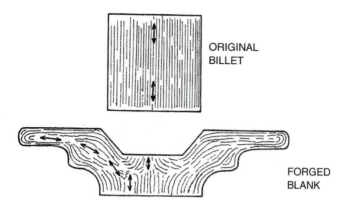

ORIGINAL BILLET

FORGED BLANK

FIGURE 9.22
The redistribution of the fiber structure direction during the forging of a part.

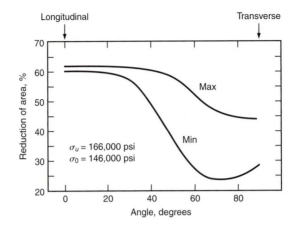

FIGURE 9.23

Relation between reduction of area and orientation within the forging. (*From G. E. Dieter, "Mechanical Metallurgy," McGraw-Hill Book Company, 1976; used with permission of McGraw-Hill Book Company.*)

Forgings often are classified as to whether they are made on open, or flat, dies, or in closed dies. Open dies are used to impose localized forces for deforming billets progressively into simple shapes, much as the blacksmith does with his hammer and anvil. Closed die forging or impression die forging uses presses or hammers to force the metal to flow into a closed cavity to produce complex shapes to close dimensional tolerances. A wide variety of shapes, sizes, and materials can be utilized in forging. Table 9.9 describes the advantages of the common forging processes. With proper forging die design, grain flow is controlled to give the best properties at the critically stressed regions.

Closed-die forgings rarely are done in a single step. The billet, often a piece of a bar stock, must be shaped in blocker dies to place the material properly so it will flow to fill the cavity of the finishing die completely (Fig. 9.24). To ensure complete filling of the die cavity, a slight excess of material is used. It escapes into the flash and is trimmed off from the finished forging.

There are a number of factors that must be considered in the design[1] if the forging is to be made economically and defect-free. As with a casting, vertical surfaces of a forging must be tapered to permit removal of the forging from the die cavity. The normal draft angle on external surfaces is 5 to 7°, and for internal surfaces it is 7 to 10°. The maximum flash thickness should not be greater than $\frac{1}{4}$ in or less than $\frac{1}{32}$ in on average.

---

1. "Forging Design Handbook," American Society for Metals, Metals Park, OH, 1972; "Metals Handbook," 9th ed., vol. 14, 1988; "Forging Industry Handbook," Forging Industry Association, Cleveland, OH; A. Thomas, "Die Design," Drop Forging Research Association, Sheffield, England, 1980; W. A. Knight and C. Poli, *Machine Design,* Jan. 24, 1985, pp. 94–99; C. Poli and W. A. Knight, "Design for Forging Handbook," University of Massachusetts, 1984; B. L. Ferguson, Design for Deformation Processes, "ASM Handbook," vol. 20, pp. 730–744, ASM International, Materials Park, OH, 1997.

TABLE 9.9

Advantages and disadvantages of the common forging process

Form	The process	Advantages	Limitations
Open-die forgings	Compressive forces (produced by hand tools or mechanical hammers) are applied locally to heated metal stock; little or no lateral confinement is involved. Desired shape is achieved by turning and manipulating work-piece between blows	Simple, inexpensive tools; useful for small quantities; wide range of sizes available; good strength characteristics	Limited to simple shapes; difficult to hold close tolerances; machining to final shape necessary; slow production rate; relatively poor utilization of material; high degree of skill required
Closed-die forgings	Compressive forces (produced by a mechanical hammer in a mechanical or hydraulic press) are applied over the entire surface of heated metal stock, forcing metal into a die cavity of desired shape. There are several types of closed-die forgings	Relatively good utilization of material; generally better properties than open-die forgings; good dimensional accuracy; rapid production rate; good reproducibility	High tool cost for small quantities; machining often necessary
	Blocker type. Uses single-impression dies and produces parts with somewhat generalized contours	Low tool costs; high production rates	Machining to final shape necessary; thick webs and large fillets necessary
	Conventional type. Uses preblocked workpiece and multiple-impression dies	Requires much less machining than blocker type; rapid production rates; good utilization of material	Somewhat higher tool cost than blocker type
	Precision type. Uses minimum draft (often 0°)	Close tolerances; machining often unnecessary; excellent material utilization; very thin webs and flanges possible	Requires intricate tooling and elaborate provision for removing forging from tools
Upset forgings	Heated metal stock is gripped by dies (which also form the impression) and pressed into desired shape	Fair amount of intricacy possible; good dimensional accuracy; rapid production rate	Limited to cylindrical shapes; finish not as good as with other forgings; size of part limited; high die costs
Cold-headed parts	Similar to upset forging except metal is cold. Wire up to about 1 mm diam is fed to die in punch press and positioned with one end protruding; this end mushrooms out under force of punch and is formed between die and punch face	Good surface strength; alloys used are generally tough, ductile and crack resistant; excellent surface finish; no scrap loss; rapid production rate	Head volume and shape limited; internal stresses may be left at critical points; size of part limited

From "Materials Selector," *Materials Engineering Magazine*, Penton/IPC, Cleveland, OH.

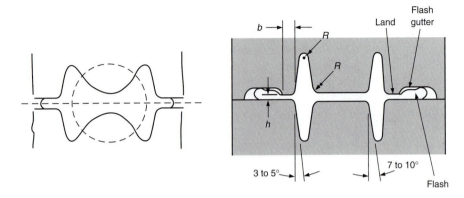

FIGURE 9.24
Schematic of closed-die forging. (*a*) Blocker die; (*b*) finishing die (*After Schey.*)

The parting line, where the die halves meet, is an important design consideration because its location helps to influence grain flow, die costs, and die wear. For optimum economy it should be kept to a single plane if at all possible, since that will make die sinking, forging, and trimming less costly. Because the forging fiber is unavoidably cut through when the flash is trimmed, the parting line is best placed where the minimum stresses arise in the service performance of the forging. Figure 9.25 illustrates a variety of simple shapes and the correct and incorrect parting line locations.

In case 1 the preferred orientation of the part avoids a deep impression that would require high forging pressure for complete filling and might lead to die breakages. Cases 2 and 3 represent nonsymmetrical parts in which the parting line (PL) and the forging plane are no longer coincident. The forging plane always is perpendicular to the direction of ram travel. The preferred locations of the parts in the die orient the parts so that the transverse forces balance out and there is no resultant side thrust on the die. When the PL is not in a single plane, as in cases 2 and 3, die construction is more costly. Sometimes the most economical solution to producing a nonsymmetrical part is to build a die with two cavities in mirror-image positions and in that way balance out side forces without tilting the PL. A good rule in forging die design is to locate the PL near the central height of the part. That avoids deep impressions in either the top or bottom die. However, when parts are dished or hollow, that may not produce the best strength because a centrally located PL interrupts the grain flow, cases 4 and 5. In case 4, the satisfactory location of the PL provides the least expensive design because only the top half of the die requires a machined impression. However, the most desirable grain flow pattern is produced when the parting line is at the top of the dish. In case 5 the location of the PL also is based on grain flow considerations. Placing the PL in the most desirable location often introduces manufacturing problems and is used only when grain flow is an extremely critical factor in design.

Whenever possible in the design of forgings, as in the design of castings, it is desirable to maintain all adjacent sections as uniform as possible. Rapid changes in section thickness should be avoided. Laps and cracks are most likely where metal flow changes because of large differences in the bulk of the sections. To prevent these defects, generous radii must be provided at those locations.

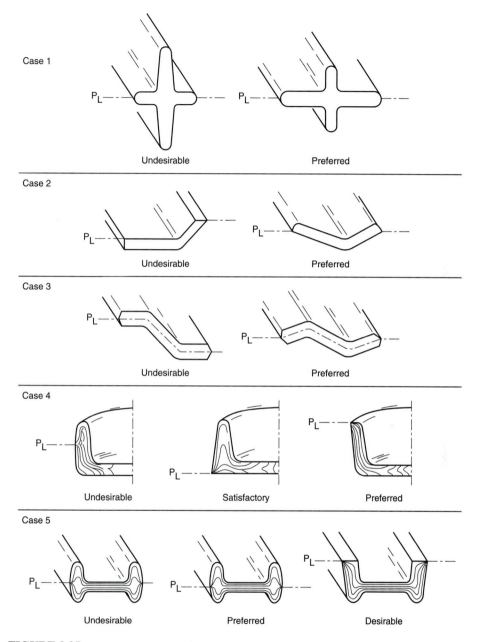

FIGURE 9.25
Examples of desirable and undesirable location of parting line. (*"Forging Industry Association Handbook."*)

The *machining envelope* is the excess metal that must be removed to bring the forging to the finished size. The ultimate in precision forging is the net-shape forging, in which the machining allowance is zero. Generally, however, allowance must be made for removing surface scale (oxide), correcting for warpage and mismatch (where the upper and lower dies shift parallel to the parting plane), and for dimensional mistakes due to thermal contraction or die wear.

The large deformation in forging and other metal deformation processes can be modeled with a FEA code called ALPID.[1] Strain distribution, free surface profiles, heat transfer from the workpiece, and stress in the tooling are among the significant parameters that can be determined.

9.10
DESIGN FOR SHEET-METAL FORMING

Sheet metal is widely used for industrial and consumer parts because of its capacity for being bent and formed into intricate shapes. Sheet-metal parts comprise a large fraction of automotive, agricultural machinery, and aircraft components and of consumer appliances. Successful sheet-metal forming depends on the selection of a material with adequate formability, the proper design of the part and the tooling, the surface condition of the sheet, selection and application of lubricants, and the speed of the forming press.

The cold stamping of a strip or sheet of metal with dies can be classified as either a cutting or forming operation.[2] Cutting operations are designed to punch holes or to separate entire parts from sheets by blanking. A blanked shape may be either a finished part or the first stage in a forming operation in which the shape is created by plastic deformation.

The sheared edge that is produced when sheet metal is punched or blanked is neither perfectly smooth nor perpendicular to the sheet surface. Since the die cost depends upon the length and the intricacy of the contour of the blank, simple blank contours should be used whenever possible. It may be less expensive to construct a component from several simple parts than to make an intricate blanked part. Blanks with sharp corners are expensive to produce.

The layout of the blanks on the sheet should be such as to minimize scrap loss. As Fig. 9.26 illustrates, a simple change in design can often greatly improve the mate-

---

1. R. Duggirala, *JOM,* February 1990, pp. 24–27. Commercially available modeling systems are DEFORM from Scientific Forming Technologies, Columbus, OH (www.deform.com) and MARC/AutoForge (www.marc.com).

2. For examples see "Metals Handbook," 9th ed., vol. 14, American Society for Metals, Metals Park, OH, 1988; G. Sachs, "Principles and Methods of Sheet Metal Fabrication," 2d ed., Reinhold, New York, 1966; F. Strasser, "Functional Design of Metal Stampings," Society of Manufacturing Engineers, Dearborn, MI, 1971; I. Suchy, "Handbook of Die Design," McGraw-Hill, New York, 1998.

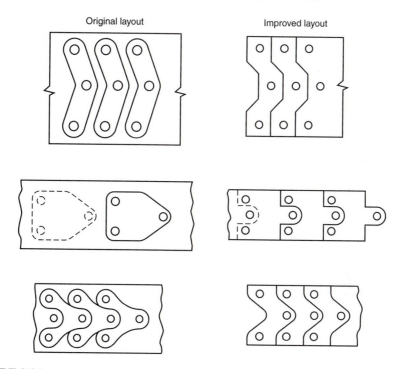

FIGURE 9.26
Changes in design to minimize scrap loss in blanking. (*"Tool and Manufacturing Engineers Handbook," 3d ed., pp. 15–18.*)

rial utilization. Notching a blank along one edge results in an unbalanced force that makes it difficult to control dimensions as accurately as with blanking round the entire contour. The usual tolerances on blanked parts are ±0.003 in.

When holes are punched in metal sheet, only part of the metal thickness is sheared cleanly; that is, a hole with tapered sides is created. If the hole is to be used as a bearing surface, then a subsequent operation will be required to obtain parallel walls. Diameters of punched holes should not be less than the thickness of the sheet or a minimum of 0.025 in. Smaller holes result in excessive punch breakage and should be drilled. The minimum distance between holes, or between a hole and the edge of the sheet, should be at least equal to the sheet thickness. If holes are to be threaded, the sheet thickness must be at least one-half the thread diameter.

Bending is the simplest sheet-forming operation. The greatest formability in bending is obtained when the bend is made across the "metal grain" (i.e., the line of the bend is perpendicular to the rolling direction of the sheet). The largest possible bend radius should be used, and the bend radius should not be less than the sheet thickness t. The bendability of sheet is usually expressed in multiples of the sheet thickness; thus a $2t$ material has a greater formability than a sheet metal whose minimum bend radius is $4t$. The total length of metal required for bending is the sum of the two legs of the bend plus the bend allowance. The *bend allowance* depends upon how much the metal stretches on bending, which is a function of the angle of bend and the bend radius.

Cost can sometimes be reduced by using sheet metal that is thinner than normal if the strength and rigidity are increased by bending and forming the sheet into ribs, corrugations, and beads.

During forming, the contour of the part matches that of the dies; but upon release of the load, the elastic forces are released. Consequently, the bent material springs back and both the angle of the bend and the bend radius increase. Therefore, to compensate for springback, the metal must be bent to a smaller angle and sharper radius so that when the metal springs back, it is at the desired values. Springback becomes more severe with increasing yield strength and section thickness.

Most sheet forming operations consist of a combination of stretching and deep drawing. In stretching, the limit of deformation is the formation of a localized region of thinning (necking) in the sheet. This behavior is governed by the uniform elongation of the material in a tension test. The greater the capacity of the material to undergo strain hardening the greater its resistance to necking in stretching.

The classic example of deep drawing is the formation of a cup. In deep drawing, the blank is drawn with a punch into a die. The circumference of the blank is decreased when the blank is forced to conform to the smaller diameter of the punch. The resulting circumferential compressive stresses cause the blank to thicken and also to wrinkle unless a sufficient hold-down pressure is applied. However, as the metal is drawn into the die over the die radius, it is bent and then straightened while being subjected to tension. That results in substantial thinning of the sheet in the region between the punch and the die wall. The deformation conditions in deep drawing are substantially different from those in stretching. Success in deep drawing is enhanced by factors that restrict sheet thinning: a die radius about 10 times the sheet thickness, a liberal punch radius, and adequate clearance between the punch and die. Of considerable importance is the crystallographic texture of the sheet. If the texture is such that the slip mechanisms favor deformation in the width direction over slip in the thickness direction of the sheet, then deep drawing is facilitated. This property of the material can be measured in tension test on the sheet from the *plastic strain ratio r.*

$$r = \frac{\text{strain in width direction}}{\text{strain in thickness direction}} = \frac{\varepsilon_w}{\varepsilon_t} \tag{9.7}$$

The best deep-drawing sheet steels have an *r* of about 2.0.

An important tool in developing sheet-forming operations is the Keeler-Goodman forming limit diagram (Fig. 9.27). It is experimentally determined for each sheet material by placing a grid of circles on the sheet before deformation. When the sheet is deformed, the circles distort into ellipses. The major and minor axes of an ellipse represent the two principal strain directions in the stamping. The strains are measured at points of failure for different stampings with different geometries to fill out the diagram. Strain states above the curve cause failure, and those below do not cause failure. The tension-tension sector is essentially stretching, whereas the tension-compression sector is closer to deep drawing. As an example of how to use the diagram, suppose point *A* represents the critical strains in a particular stamping. This failure could be

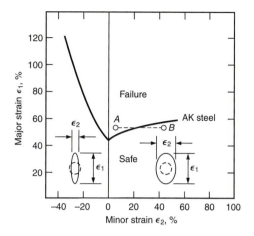

FIGURE 9.27
Keeler-Goodwin forming limit diagram.

eliminated by moving the strain state to B by increasing the die radius. Alternatively, a material of greater formability could be substituted.

Design guidelines for metal stampings and fabrications have been published.[1] Also, several computer-aided design tools are available for designing parts to be made by sheet metal forming. Most DFM packages contain sheetforming modules. SheetAdviser assists with the design of relatively flat stampings consisting mostly of bends and punchings.[2] Other software generates an FEA model and compares the calculated strain distributions in deep-drawn parts with the forming limit diagram of the material.[3] Thus the formability of complex stampings can be analyzed before any tooling is made.

9.11
DESIGN FOR MACHINING

Machining operations represent the most versatile and most common manufacturing processes. Practically every part is subjected to some kind of machining operation in its final finishing stages of manufacture. Parts that are machined may have started out as castings or forgings, or they may be machined completely from bar stock or plate.

---

1. PMA Design Guidelines, Precision Metalforming Association, Richmond Heights, OH; C. Poli, P. Dastidar, P. Mahajan, and R. J. Graves, *Jnl of Mechanical Design,* ASME, vol. 115, pp. 735–743, 1993.
2. Developed by Hewlett-Packard, SheetAdvisor is available from CoCreate Software, Ft. Collins, CO.
3. For example, see PamStamp from Pam Systems International, ESI-Group Software, Rungis, France, FAST_FORM3D from Forming Technologies Inc., Oakville, Canada (www.forming.com), and Optris and HyperForm from Altair Computing (www.altair.com).

There is a wide variety of machining processes with which the design engineer should be familiar.[1] We can break them down into three broad classes: metal-cutting processes, grinding processes, and unconventional or electrical effects. Conventional machining processes, which represent the greatest concentration of effort, can be categorized by whether the tool translates or rotates or is stationary while the workpiece rotates. The classification of machining processes based on this system is shown in Fig. 9.28. The operations and machines that can be used to generate flat surfaces are

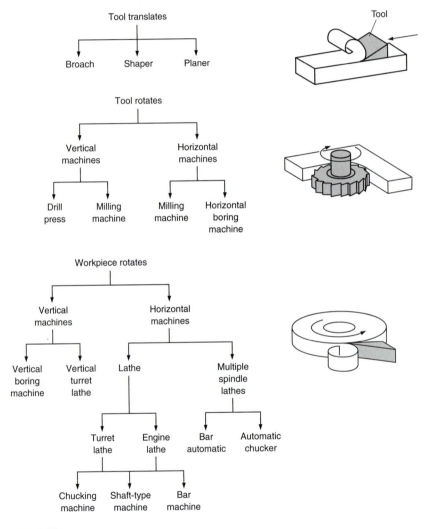

FIGURE 9.28
Classification of metal-cutting processes.

---

1. For examples see "ASM Handbook," vol. 16, ASM International, Materials Park, OH, 1989; E. P. DeGarmo, J. T. Black, and R. Kohser, "Materials and Processes in Manufacturing," 8th ed., Prentice-Hall, Saddle River, NJ, 1997.

shown in Fig. 9.29. The operations and machines for machining external cylindrical surfaces are shown in Fig. 9.30, and the common operations and machines for machining internal cylindrical surfaces are shown in Fig. 9.31.

Most metals and plastics can be machined, but they vary a great deal in their ease of machining, or *machinability*. Machinability is a complex technological property that is difficult to define precisely. A material has good machinability if the tool wear

Operation	Block diagram	Most commonly used machines	Machines less frequently used	Machines seldom used
Shaping	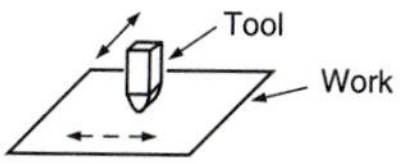	Horizontal shaper	Vertical shaper	
Planing	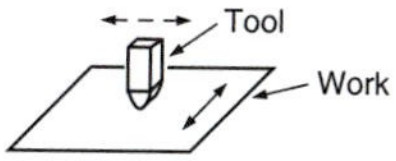	Planer		
Milling	Slab milling 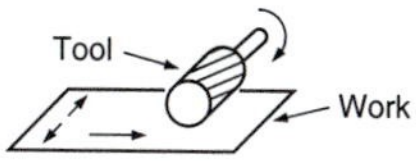 / Face milling 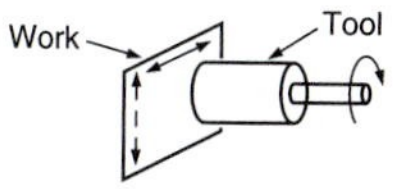	Milling machine		Lathe (with special attachment)
Facing	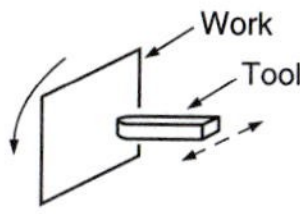	Lathe	Boring mill	
Broaching	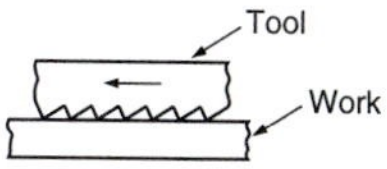	Broaching machine		
Grinding	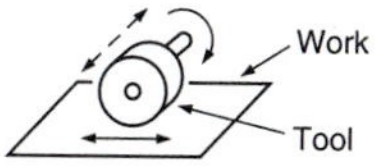	Surface grinder		Lathe (with special attachment)
Sawing	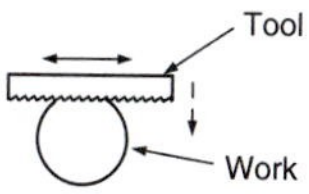	Cutoff saw	Contour saw	

⟶ Tool and work motion   ⟵--⟶ Feed only

FIGURE 9.29

Operations and machines for machining flat surfaces. (*From E. P. DeGarmo, "Materials and Processing in Manufacturing," 5th ed., Macmillan Publishing Co., New York, 1979; copyright 1979 by Darvic Associates, Inc.*)

Operation	Block diagram	Most commonly used machines	Machines less frequently used	Machines seldom used
Turning	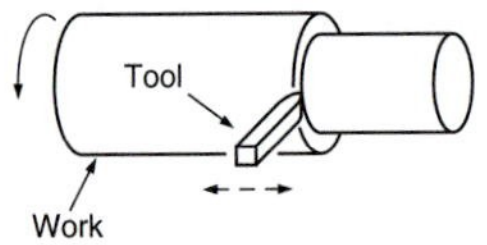	Lathe	Boring mill	Vertical shaper Milling machine
Grinding	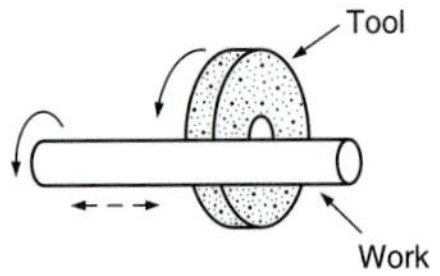	Cylindrical grinder		Lathe (with special attachment)
Sawing	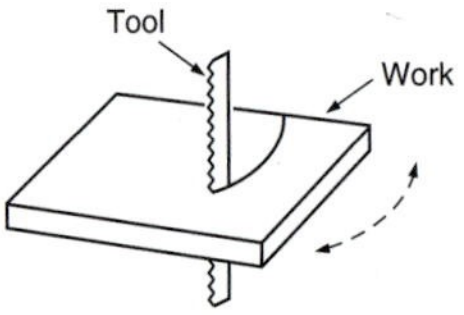	Contour saw		

FIGURE 9.30

Operating and machines for machining external cylindrical surfaces. (*From E. P. DeGarmo, "Materials and Processing in Manufacturing," 5th ed., Macmillan Publishing Co., New York, 1979; copyright 1979 by Darvic Associates, Inc.*)

is low, the cutting forces are low, the chips break into small pieces instead of forming long snarls, and the surface finish is acceptable. Machinability is a system property that depends on the workpiece material, the cutting tool material and its geometry, the type of machining operation, and its operating conditions.[1] Table 9.10 lists common classes of metallic materials and machining processes, in decreasing order of machinability. Nothing has greater impact on machining costs and quality of machined parts than the machinability of the work material. The one generalization that can be applied to machinability is that the higher the hardness of the workpiece material the poorer the machinability.

An important factor for economy in machining is to specify a machined surface only when it is needed for the functioning of the part. Two design examples for reducing the amount of machined area are shown in Fig. 9.32.

In designing a part, the sequence by which the part would be machined must be kept in mind so the design details that make machining easy are incorporated.[2] The

---

1. D. A. Stephenson and J. S. Agapiou, "Metal Cutting Theory and Practice," Chaps. 2, 11, and 13, Marcel Dekker, New York, 1996.
2. There are a large number of vendors for computer-numerical control software for machining operations. See B. B. Beckert, *Computer-Aided Engineering,* December, 1998, pp. 40–46.

Operation	Block diagram	Most commonly used machines	Machines less frequently used	Machines seldom used
Drilling	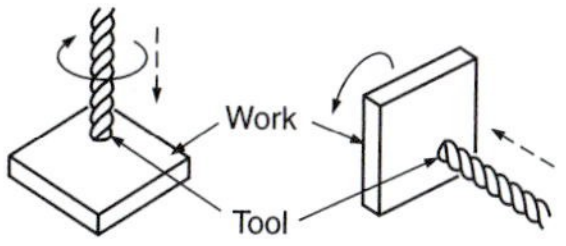	Drill press	Lathe	Milling machine Boring mill Horizontal boring machine
Boring	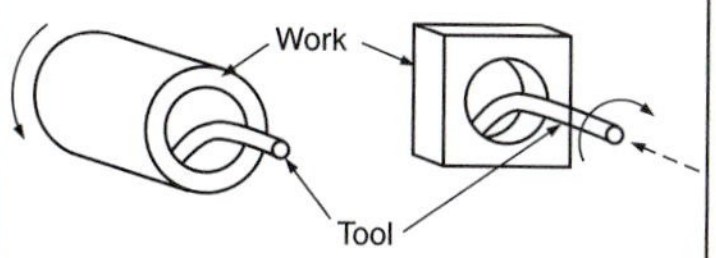	Lathe Boring mill Horizontal boring machine		Milling machine Drill press
Reaming	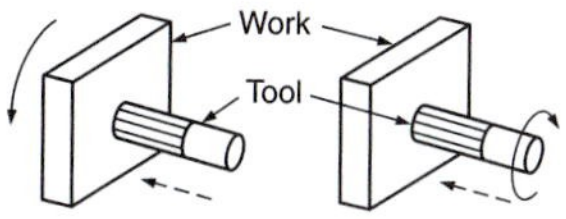	Lathe Drill press Boring mill Horizontal boring machine	Milling machine	
Grinding	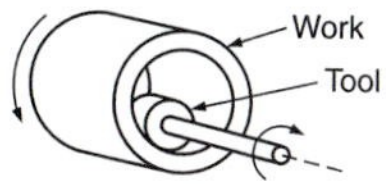	Cylindrical grinder		Lathe (with special attachment)
Sawing	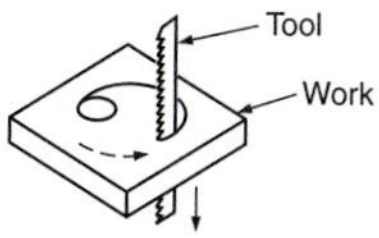	Contour saw		
Broaching	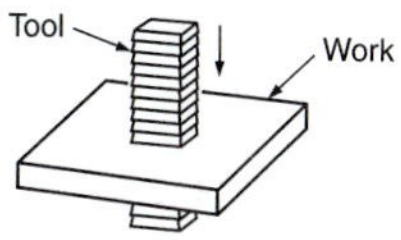	Broaching machine		

FIGURE 9.31

Operations and machines for machining internal cylindrical surfaces. (*From E. P. DeGarmo, "Materials and Processing in Manufacturing," 5th ed., Macmillan Publishing Co., New York, 1979; copyright 1979 by Darvic Associates, Inc.*)

workpiece must have a reference surface that is suitable for holding it on the machine tool or in a fixture. A surface with three-point support is better than a large flat surface because the workpiece is then less likely to rock. Sometimes a supporting foot or tab must be added to the rough casting for support purposes, and be removed from the final machined part. When possible, the design should permit all the machining to be done without reclamping the workpiece (Fig. 9.33a). If the part needs to be clamped

TABLE 9.10
Classes of metals and machining processes, listed in decreasing order of machinability

Classes of metals	Machining processes
Magnesium alloys	Grinding
Aluminum alloys	Sawing
Copper alloys	Turning with single-point tools
Gray cast iron	Drilling
Nodular cast iron	Milling
Carbon steels	High-speed, light feed, screw machine work
Low-alloy steels	Screw machining with form tools
Stainless steels	Boring
Hardened and high-alloy steels	Generation of gear teeth
Nickel-base superalloys	Tapping
Titanium alloys	Broaching

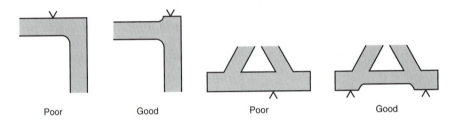

Poor   Good   Poor   Good

FIGURE 9.32
Examples of design details that minimize the area of the machined surface.

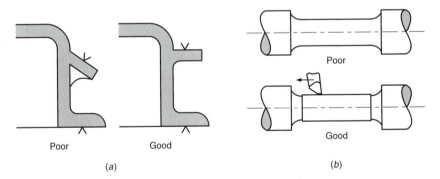

Poor   Good   Poor

                                       Good

(a)   (b)

FIGURE 9.33
Design details for improved machining.

in a second, different position, one of the already machined surfaces should be used as the reference surface. Whenever possible, the design should be such that existing tools can be used in production. When possible, the radius of the part should be the same as the radius of the tool. Also when possible, design should permit the use of the larger tools, which are stronger and can cut at higher speeds. Remember that a cutting tool often requires a runout space because the tool cannot be retracted instantaneously

(Fig. 9.33*b*). Many more examples of design details that enhance machining are cataloged by Singh.[1]

In drilling, the cost of a hole increases proportionately with depth; but when the depth exceeds three times the diameter, the cost increases more rapidly. When a drill is cutting, it should meet equal resistance on all cutting edges. It will if the entry and exit surfaces it encounters are perpendicular to its axis. Holes should not be placed too near the edge of the workpiece. If the workpiece material is weak and brittle, like cast iron, it will break away. Steel, on the other hand, will deflect at the thin section and will spring back afterward to produce a hole that is out of round.

Some additional design guidelines for machining processes are given below.[2] This reference also describes a procedure for the early cost estimation of machined parts.

1. Try to design the part so that it can be machined on one machine tool only.
2. Try to design the part so that machining is not needed on the unexposed surfaces of the workpiece when the part is gripped in the work-holding device.
3. Avoid specifying machined features that the company shop is not equipped to carry out.
4. Design the part so that the workpiece, when gripped in the work-holding device, is sufficiently rigid to withstand the machining forces.
5. Make sure that when the part is machined the tool, tool holder, workpiece, and work-holding device do not interfere with one another.
6. Make sure that auxiliary holes or main bores are cylindrical and have L/D ratios that make it possible to machine them with standard drills or boring bars. For dimensions of standard tools see "Machinery's Handbook."[3]
7. Make sure that auxiliary holes are parallel or normal to the axis of the workpiece and related by a logical drilling pattern.
8. Make sure that the ends of blind holes are conical and, for the case of a tapped blind hole, that the thread does not continue to the bottom of the hole.

For components with rotational symmetry:
9. Try to make sure that cylindrical surfaces are concentric and plane surfaces are normal to the part axis.
10. For internal corners on the part, specify radii equal to the radius of the rounded tool corner.
11. Avoid internal features in long parts.
12. Avoid parts with very large or very small L/D ratios.

For components with nonrotational symmetry:
13. Design in a base for work holding and reference.
14. If possible, make sure that the exposed surfaces of the part consist of a series of mutually perpendicular plane surfaces parallel to and normal to the base.

1. K. Singh, "Mechanical Design Principles: Applications, Techniques and Guidelines for Manufacture," Nantel Publications, Melbourne, 1996.
2. G. Boothroyd and W. A. Knight, "Fundamental of Machining and Machine Tools," 2d ed., Chap. 13, Marcel Dekker, New York, 1989; Simplifying Machining in the Design Stage, "Tool and Manufacturing Engineers Handbook," vol. 6, "Design for Manufacturability," SME, Dearborn, MI, 1992.
3. E. Oberg, F. D. Jones, and H. L. Horton, "Machinery's Handbook," 23d ed., Industrial Press, New York, 1988.

15. Make sure that internal corners normal to the base have a radius equal to the tool radius and that internal corners in machined pockets have as large a radius as possible.
16. If possible, restrict plane surface machining (slots, grooves, etc.) to one surface of the part.
17. Avoid cylindrical bores in long parts.
18. Avoid machined surfaces on long parts by using work material preformed to the required cross section, e.g., extrusions.
19. Avoid extremely long or thin parts.

Assembly:
20. Make sure that it is possible to assemble the parts into the component.
21. Make sure that each operating machined surface on the part has a corresponding machined surface on the mating part.
22. Make sure that internal corners do not interfere with a corresponding external corner on the mating part.

Surface finish and accuracy:
23. Design for the widest tolerances and the roughest surface that will give acceptable performance for operating surfaces.
24. Make sure that surfaces to be finished ground are raised and never intersect to form internal corners.

9.12
DESIGN FOR POWDER METALLURGY

Powder metallurgy (PM) produces parts to final shape and with little or no machining required. In the conventional PM process a finely divided metallic powder is compressed in a die to produce a porous shape. The green compact is sintered in an atmosphere (usually nonoxidizing or reducing) at elevated temperature to close up the porosity of the as-pressed compact. Generally, the sintered part contains from 4 to 10 percent porosity. The porosity can be decreased by coining or restriking the sintered compact. This step also improves dimensional tolerance. Still further densification and strength improvement can be achieved by resintering the part after it has been coined.

Powder metallurgy processing has found greatest acceptance for small parts (under 1 lb) in automotive and appliance applications in which the ability to produce to final shape with a minimum of machining provides a strong economic advantage. The requirements for mechanical strength must not be too severe for pressed and sintered PM parts.

One of the chief advantages of powder metallurgy is versatility. Metals that can be combined in no other way can be produced by powder metallurgy. Some examples are copper combined with carbon for electrical brushes and cobalt and tungsten carbide for cutting tools. The density (porosity) of the part can be controlled over wide limits to fabricate products with special features such as self-lubricating bearings, metallic filters, and parts with unusual damping properties.

However, the chief trend in powder metallurgy is in the production of full-density, high-strength parts with fewer processing steps than the competing processes of casting or forging. One of the approaches to this is PM hot forging, whereby a sintered PM preform is completely densified and hot-forged to finished shape in a single operation. Another exciting approach is hot isostatic pressing (HIP). In this process, the powder is sealed in a metal or ceramic container that has the shape of the desired part. The container is placed in a special pressure vessel that has the capability of simultaneously heating the container and subjecting it to a hydrostatic argon gas pressure. The powder is compacted, densified, and sintered in one step. The HIP process is particularly suited to producing parts from high-temperature alloys that are difficult to forge and machine. A 35 percent saving in the material needed to make a gas turbine disk has been reported for the HIP process.

Several design rules[1] must be considered to make economical parts by the conventional press and sinter PM process.

1. The design must be such that the part can be ejected from the die. Parts with straight walls are preferred. No draft is required for ejection from a lubricated die. Parts with undercuts or holes at right angles to the direction of pressing cannot be made.
2. In designing the part, consideration should be given to the need for the powder to flow properly into all parts of the die. Therefore, do not design for thin walls, narrow splines, or sharp corners. In general, sidewalls should be thicker than 0.03 in. Abrupt changes in wall thickness should be avoided, since they lead to distortion after sintering.
3. The shape of the part should permit the construction of strong tooling. Dies and punches should have no sharp edges. There should be a reasonable clearance between top and bottom punches during pressing.
4. Since pressure is not transmitted uniformly through a deep bed of powder, the length of a die-pressed part should not exceed about two and a half times the diameter.
5. Keep the part shape simple. The part should be designed with as few levels (diameters) and axial variations as possible.
6. Provide wide dimensional tolerances whenever possible. Wide tolerances mean lower piece-part cost and longer tool life.
7. PM parts may be bonded by assembling in the green (as-pressed) condition and then sintering together to form a bonded assembly.

9.13
DESIGN FOR WELDING

Welding is the most prominent process for joining large components into complex assemblies or structures. Smaller assemblies are typically joined by bolts, screws, studs, rivets, snap fasteners, soldering, brazing, and adhesive bonding. These fastening

---

1. "Powder Metallurgy Design Manual," Metal Power Industries Federation, 2d ed., Princeton, NJ, 1995; H. I. Sanderow, Design for Powder Metallurgy, "ASM Handbook," vol. 20, pp. 745–753, ASM International, Materials Park, OH, 1997.

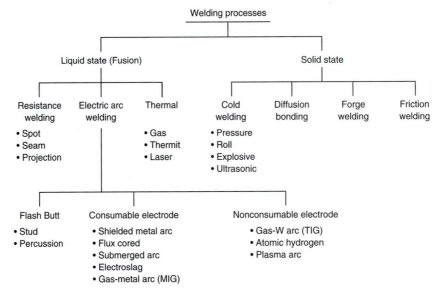

FIGURE 9.34
Classification of welding processes.

methods are covered in books on machine design.[1] In this section we deal more with the metallurgical aspects of the broad spectrum of welding processes[2] (Fig. 9.34).

Welding is a process in which two materials are joined permanently by coalescence. It involves some combination of temperature, pressure, and surface condition. A necessary condition for welding is that the two surfaces to be joined must be brought into intimate contact. When fusion takes place in the welding process, the contact is achieved by a molten metal. When melting does not occur, contact is achieved by overcoming the surface roughness of the mating surfaces by producing plastic deformation of the surface asperities and by removing any oxide or other layer that may be on the surfaces.

9.13.1 Welding Processes

Probably the oldest welding process is the solid-state method called forge welding. It is the technique, used by the blacksmith, in which two pieces of steel or iron are heated and forced together under point contact. Slag and oxides are squeezed out, and

---

1. J. E. Shigley and C. R. Mischke, "Standard Handbook of Machine Design," 2d ed., McGraw-Hill, New York, 1997.
2. For detailed descriptions of these processes and examples of their use, see "Metals Handbook," 9th ed., vol. 6, "Welding, Brazing, and Soldering," American Society for Metals, Metals Park, OH, 1983, "Welding Handbook," 7th ed., published in six sections by American Welding Society, Miami, FL, 1976; R. A. Lindberg and N. R. Braton, "Welding and Other Joining Processes," Allyn & Bacon, Inc., Boston, 1976.

interatomic bonding of the metal results. In the modern version of forge welding, steel pipe is heated by induction or electrical resistance and butt-welded under axial force. The pipe is produced by forming sheet into a cylinder and welding the edges together by forge-seam welding, in which either the sheet is pulled through a conical die or the hot strip is passed between shaped rolls.

As the name implies, cold-welding processes are carried out at room temperature without any external heating of the metal. The surfaces must be clean, and the local pressure must be high to produce substantial cold-working. The harmful effect of interface films is minimized when there is considerable relative movement of the surfaces to be joined. The movement is achieved by passing the metal through a rolling mill or subjecting the interface to tangential ultrasonic vibration. In explosive bonding there is extensive vorticity at the interfaces. Diffusion bonding takes place at a temperature high enough for diffusion to occur readily across the bond zone. Hot roll bonding is a combination of diffusion bonding and roll bonding.

Friction welding (inertia welding) utilizes the frictional heat generated when two bodies slide over each other. In the usual way of doing friction welding, one part is held fixed and the other part (usually a shaft or cylinder) is rotated and, at the same time, forced axially against the stationary part. The friction quickly heats the abutting surfaces and, as soon as the proper temperature is reached, the rotation is stopped and the pressure is maintained until the weld is complete. The impurities are squeezed out into a flash, but essentially no melting takes place. The heated zone is very narrow, and therefore dissimilar metals are easily joined.

In the majority of welding applications the interatomic bond is produced by melting. In welding, the workpiece materials and the filler material in the joint have similar compositions and melting points. By contrast, in soldering and brazing, the filler material has a much different composition that is selected to have lower melting point than the workpiece materials.

Resistance welding utilizes the heat generated at the interface between two metal parts when a high current is passed through the parts. Spot welding is used extensively to join metal sheets at discrete points (spots). Rather than produce a series of spots, an electrode in the form of a roller often is used to produce a seam weld. If the part to be welded contains small embossed dimples or projections, they are easily softened under the electrode and pushed back to produce the weld nugget.

Other sources of heat for welding are chemical sources or high-energy beams. Gas welding, especially the reaction between oxygen and acetylene to produce an intense flame, has been used for many years. Thermit welding uses the reaction between Fe_2O_3 and Al, which produces Fe and an intense heat. The process is used to weld heavy sections such as rails. Energy from a laser beam is being used to produce welds in sheet metal. Its advantage over an electron beam is that a vacuum is not required. Each form of energy is limited in power, but it can be carefully controlled. Laser beam and electron beam welding lend themselves to welding thin gauges of hardened or high-temperature materials.

The thermal energy produced from an electric arc has been utilized extensively in welding. In flash butt welding, an arc is created between two surfaces, typically tubes or bars, which, after they reach temperature, are forced together axially to squeeze out a radial flash. Stud welding is a variant of the process in which a threaded stud is fastened to a flat surface by arc welding. Percussion welding is a process in which two workpieces are brought together at a rapid rate such that, just before the pieces meet, an arc melts both of the colliding surfaces. Percussion welding is particularly good for joining small-diameter wires or dissimilar materials. In all of these flash welding processes, a true electric arc is generated at the welded interface.

Most electric arc welding is done with an arc struck between a consumable electrode (the filler rod) and the workpiece. A coating is applied to the outside surface of the metal electrode to provide a protective atmosphere around the weld pool. The electrode coating also acts as a flux to remove impurities from the molten metal and as an ionizing agent to help stabilize the arc. This is the commonly used shielded metal arc process. Since the electrode coating is brittle, only straight stick electrodes can be used. That restricts the process to a slow hand operation. If the flux coating is placed inside a long tube, the electrode can be coiled, and then the shielded arc process can be made continuous and automatic. In the submerged arc process the consumable electrode is a bare filler wire and the flux is supplied from a separate hopper in a thick layer that covers the arc. In the electroslag process the electrode wire is fed into a pool of molten slag that sits on top of the molten weld pool. Metal transfer is from the electrode to the weld pool through the molten slag. This process is used for welding thick plates and can be automated. In the gas metal arc process the consumable metal electrode is shielded by an inert gas such as argon or helium. Because there is no flux coating, there is no need to remove the slag deposit from the weld bead after each pass.

In nonconsumable electrode welding an inert tungsten electrode is used. Depending on the weld design, a filler rod may be required. In gas tungsten arc welding (TIG welding), argon or helium is used. The process produces high-quality welds in almost any material, especially in thinner-gauge sheet. In the atomic hydrogen process molecular hydrogen is passed through an arc maintained by two tungsten electrodes. The hydrogen dissociates to atomic hydrogen in the arc and then reassociates to diatomic hydrogen molecules at the weld surface. This chemical reaction produces an arc with very high temperature. In plasma arc welding two separate gas flows are used. Gas from the central system surrounds the electrode and becomes ionized as it passes through the arc. The flow of this ionized gas is constricted by a small orifice directly below the point of the electrode, which increases the ionization and generates a plasma. The central column of plasma is surrounded by a second cooler sheath of shielding gas. The plasma arc generates the highest temperature of any welding system.

To design a weldment properly, consideration must be given to the selection of materials, the joint design, the selection of the welding process, and the stresses generated by the design. The welding process subjects the workpiece at the joint to a temperature excursion that may exceed the melting point of the material. Heat is applied rapidly and locally. We have a miniature casting in the weld pool, which usually is repeated as successive weld beads are laid down. The base metal next to the weld bead, the heat-affected zone, is subjected to rapid heating and cooling, so that there the original microstructure and properties of the base metal are changed in a nonequilibrium way. Thus, considerable opportunity for defects exists unless the weld processing is properly designed.

Since fusion welding is a melting process, controls appropriate to producing quality castings must be applied. Reactions with the atmosphere are prevented by sealing off the molten pool with an inert gas or a slag or by carrying out the welding in a vacuum chamber. The surfaces of the weld joint should be cleaned of scale or grease before welding is undertaken. The thermal expansion of the weld structure on heating, followed by solidification shrinkage, can lead to high internal tensile stresses that can produce cracking and/or distortion. Rapid cooling of alloy steels in welding can result in brittle martensite formation and consequent crack problems. As a result, it is common to limit welding to carbon steels with less than 0.3 percent carbon or to alloy steels in which the carbon equivalent[1] is less than 0.3 percent carbon. When steels with 0.3 to 0.6 percent carbon must be used because their high strength and high toughness are required, welding without martensite cracking can be performed if the weld joint is preheated before welding and postheated after the weld bead has been deposited. These thermal treatments decrease the rate of cooling of the weld and heat-affected zone, and they thereby reduce the likelihood of martensite formation.

The chief factors in the design of a weldment are (1) the selection of the material, (2) the design of the joint, (3) selection of the welding process, and (4) design of the welded joint so it will withstand the applied stresses.

Material selection for welding involves choosing a material with high weldability. *Weldability,* like machinability, is a complex technological property that combines many more basic properties. The melting point of the material, together with the specific heat and latent heat of fusion, will determine the heat input necessary to produce fusion. A high thermal conductivity allows the heat to dissipate and therefore requires a higher *rate* of heat input. Also, metals with higher thermal conductivity result in more rapid cooling and more problems with weld cracking. Greater distortion results from a high thermal expansion, with higher residual stresses and greater danger of weld cracking. There is no absolute rating of weldability of metals because different welding processes impose a variety of conditions that can affect the way a material responds.

9.13.2 Weld Joint Design

The basic types of welded joint are shown in Fig. 9.35. Many variations of these basic designs are possible, depending on the type of edge preparation that is used. A square-edged butt joint requires a minimum of edge preparation. However, an important parameter in controlling weld cracking is the ratio of the width of the weld bead to the depth of the weld. It should be close to unity. Since narrow joints with deep weld pools are susceptible to cracking, the most economical solution is to spend machining money to shape the edges of the plate to produce a joint design with a more acceptable width-to-depth ratio. Ideally, a butt weld should be a full-penetration weld that fills the joint completely throughout its depth. When the gap in a butt joint is wide, a backing strip is used at the bottom of the joint.

---

1. $C_{eq} = C + \dfrac{Mn}{6} + \dfrac{Cr + Mo + V}{5} + \dfrac{Ni + Cu}{15}$

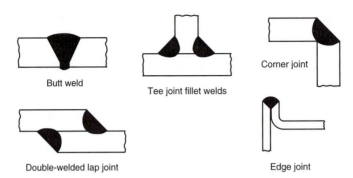

FIGURE 9.35
Basic type of welded joints.

Welding electrodes are specified with a code such as E60XX. The last two digits indicate the type of welding application, and the two digits immediately following E indicate the minimum tensile strength of the weld metal in kips per square inch. For example, E7024 has a 70-ksi tensile strength and is intended for ac or dc electric arc welding of steel fillet welds in the horizontal or flat position. The load-carrying ability of a full-penetration butt weld made with E60XX electrodes (50-ksi yield strength) would be

$$\frac{P}{L} = 50t$$

where P/L is joint strength, in kips per inch, and t is the plate thickness. Welds frequently are made with weld metal "reinforcement" that extends above or below the surface of the base metal plate. Some designers believe this increases the strength of the joint and compensates for any weld imperfection. However, such a joint design would serve as a stress concentration under fatigue loading. Therefore, reinforcements on welds should not be used when the welds are subject to fatigue.

Fillet welds are the welds most commonly used in structural design. They are inherently weaker than full-penetration butt welds (Fig. 9.36). A fillet weld fails in shear at the weld throat, given by 0.707h, and the American Welding Society code allows a shearing yield strength of 30 percent of the tensile strength designation of the electrode. Thus, for an E60XX electrode the load-carrying capacity of a fillet weld is $P/L = 0.30(60)(0.707h)$ kips/in.

The selection of the appropriate welding process depends on the required heat input, the availability of equipment, and the economics of the process.[1] Since welding involves the rapid application of heat to a localized area, followed by the rapid removal of the heat, distortion is ever-present. One of the best ways to eliminate welding distortion is to design the welding sequence with thermal distortion in mind. If, because of the geometry, distortion cannot be avoided, then the forces that produce the shrinkage distortion should be balanced with other forces provided by fixtures and

---

1. R. A. Lindberg and N. R. Braton, op. cit., chap. 13.

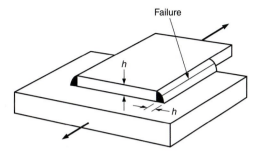

FIGURE 9.36
Fillet weld in lap joint.

clamps. Also, the shrinkage forces can be removed after welding by postwelding annealing and stress-relief operations. It should be kept in mind that distortion arises from welding, per se, so that the design should call for only the amount of weld metal that is absolutely required. Overwelding adds not only to the shrinkage forces but also to the costs.

Obviously, the design of weldments[1] calls for much expertise that is beyond the scope of this chapter. However, we can suggest some general design guidelines.

1. Welded designs should reflect the flexibility and economy inherent in the welding process. Do not copy designs based on casting or forging.
2. In the design of welded joints, try to provide for a straight-line force pattern. Avoid the use of welded straps, laps, and stiffeners except as required for strength.
3. Use the minimum number of welds.
4. Whenever possible, weld together parts of equal thickness.
5. Locate the welds at areas in the design where stresses and/or deflections are least critical.
6. Carefully consider the sequence with which parts should be welded together and include that information as part of the design drawing.
7. Make sure that the welder or welding machine has unobstructed access to the joint so that a quality weld can be produced. Whenever possible, the design should provide for welding in the flat or horizontal position, not overhead.

9.14
RESIDUAL STRESSES IN DESIGN

Residual stresses are the system of stresses that can exist in a part when the part is free from external forces, and they are sometimes referred to as internal stresses or locked-

1. "Design of Weldments," The James F. Lincoln Arc Welding Foundation, Cleveland, OH, 1963; O. W. Blodgett, "Design of Welded Structures," ibid., 1966; T. G. F. Gray and J. Spencer, "Rational Welding Design," 2d ed., Butterworths, London, 1982.

in stresses.[1] They arise from nonuniform plastic deformation of a body, chiefly as a result of inhomogeneous changes in volume or shape.

9.14.1 Origin of Residual Stresses

For example, consider a metal sheet that is being rolled under conditions such that plastic flow occurs only near the surfaces of the sheet (Fig. 9.37*a*). The surface fibers of the sheet are cold-worked and tend to elongate while the center of the sheet is unchanged. Since the sheet must remain a continuous whole, its surface and center must undergo strain accommodation. The center fibers tend to restrain the surface fibers from elongating, and the surface fibers seek to stretch the central fibers of the sheet. The result is a residual stress pattern in the sheet that consists of a high compressive stress at the surface and a tensile residual stress at the center (Fig. 9.37*b*). In general, the sign of the residual stress that is produced by inhomogeneous deformation will be opposite the sign of the plastic strain that produced the residual stress. Thus, for the case of the rolled sheet, the surface fibers that were elongated in the longitudinal direction by rolling are left in a state of compressive residual stress when the external load is removed.

The residual stress system existing in a body must be in static equilibrium. Thus, the total force acting on any plane through the body and the total moment of forces on any plane must be zero. For the longitudinal stress pattern in Fig. 9.37*b* this means that the area under the curve subjected to compressive residual stresses must balance the area subjected to tensile residual stresses. Actually, the situation is not quite so

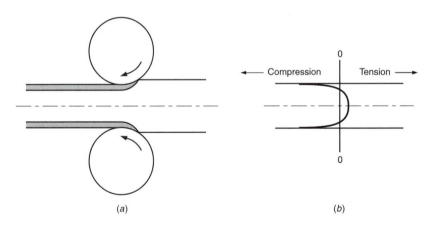

(a)   (b)

FIGURE 9.37
(*a*) Inhomogenous deformation in rolling of sheet; (*b*) resulting distribution of longitudinal residual stress over thickness of sheet (schematic).

1. W. B. Young (ed.), "Residual Stresses in Design, Process and Materials Selection," ASM International, Materials Park, OH, 1987; U. Chandra, Control of Residual Stresses, "ASM Handbook," vol. 20, pp. 811–819, ASM International, Materials Park, OH, 1997.

simple as is pictured in Fig. 9.37. For a complete analysis, the residual stresses acting across the width and thickness of the sheet should be considered, and the state of residual stress at any point is a combined stress derived from the residual stresses in the three principal directions. Frequently, because of symmetry, only the residual stress in one direction need be considered. A complete determination of the state of residual stress in three dimensions is a considerable undertaking.

Residual stresses are to be considered as only elastic stresses. The maximum value that the residual stress can reach is the elastic limit of the material. A stress in excess of that value, with no external force to oppose it, will relieve itself by plastic deformation until it reaches the value of the yield stress.

Residual and applied stress add algebraically, so long as their sum does not exceed the elastic limit of the material. For example, if the maximum applied stress due to service loads is 60,000 psi tension and the part already contains a tensile residual stress of 40,000 psi, the total stress at the critically stressed region is 100,000 psi. However if the residual stress is a compressive 40,000 psi produced by shot peening, then the actual stress is 20,000 psi.

Any process, whether mechanical, thermal, or chemical, that produces a permanent nonuniform change in shape or volume creates a residual stress pattern. Practically, all cold-working operations develop residual stresses because of nonuniform plastic flow. In surface-working operations, such as shot peening, surface rolling, or polishing, the penetration of the deformation is very shallow. The distended surface layer is held in compression by the less-worked interior. A surface compressive residual stress pattern is highly desirable in reducing the incidence of fatigue failure.

Residual stresses arising from thermal processes may be classified as those due to a thermal gradient alone or to a thermal gradient in conjunction with a phase transformation, as in heat-treating steel. These situations arise most frequently in quenching, casting, and welding.

9.14.2 Quenching Stresses

The situation of greatest practical interest involves the residual stresses developed during the quenching of steel for hardening. In this case, however, the residual stress pattern is due to thermal volume changes plus volume changes resulting from the transformation of austenite to martensite. The simpler situation, in which the stresses are due only to thermal volume changes, will be considered first. This is the situation encountered in the quenching of a metal that does not undergo a phase change on cooling. It is also the situation encountered when steel is quenched from a tempering temperature below the A_1 critical temperature.

The distribution of residual stress over the diameter of a quenched bar in the longitudinal, tangential, and radial directions is shown in Fig. 9.38a for the usual case of a metal that contracts on cooling. Figure 9.38c shows that the opposite residual stress distribution is obtained if the metal expands on cooling (this occurs for only a few situations). The development of the stress pattern shown in Fig. 9.38a can be visualized as follows: The relatively cool surface of the bar tends to contract into a ring that is both shorter and smaller in diameter than it was originally. This tends to extrude the hotter, more plastic center into a cylinder that is longer and thinner than it was originally. If

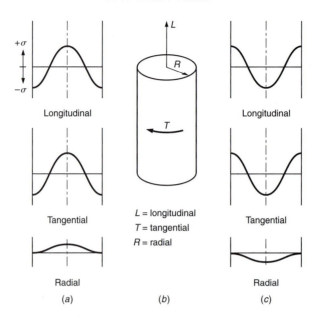

FIGURE 9.38
Residual stress patterns found in quenched bars and due to thermal strains (schematic).
(*a*) For metal which contracts on cooling; (*b*) orientation of directions; (*c*) for metal which
expands on cooling.

the inner core were free to change shape independently of the outer ring, it would
change dimensions to a shorter and thinner cylinder on cooling. However, continuity
must be maintained throughout the bar so that the outer ring is drawn in (compressed)
in the longitudinal, tangential, and radial directions at the same time the inner core is
extended in the same directions. The stress pattern shown in Fig. 9.38 results.

The magnitude of the residual stresses produced by quenching depends on the
stress-strain relations for the metal and the degree of strain mismatch produced by the
quenching operation. For a given strain mismatch, the higher the elastic modulus of
the metal the higher the residual stress. Further, since the residual stress cannot exceed
the yield stress, the higher the yield stress the higher the possible residual stress. The
yield stress-temperature curve for the metal also is important. If the yield stress
decreases rapidly with increasing temperature, the strain mismatch will be small at
high temperature because the metal can accommodate to thermally produced volume
changes by plastic flow. On the other hand, metals that have a high yield strength at
elevated temperatures, like superalloys, will develop large residual stresses from
quenching.

The following combinations of physical properties will lead to high mismatch
strains on quenching:

- Low thermal conductivity k
- High specific heat c
- High coefficient of thermal expansion α
- High density ρ

These factors can be combined into the thermal diffusivity $D_t = k/\rho c$. Low values of thermal diffusivity lead to high strain mismatch. Other factors that produce an increase in the temperature difference between the surface and center of the bar promote high quenching stresses. They are (1) a large diameter of the cylinder, (2) a large temperature difference between the initial temperature and the temperature of the quenching bath, and (3) a high severity of quench.

The control of residual stresses starts with understanding the fundamental source of the stress and identifying the parameters in the manufacturing process that influence the stress. Then, experiments are performed in varying the process parameters to produce the desired level of stress. Recently, FEA modeling has been used effectively in predicting how residual stresses can be reduced.[1]

In the quenching of steels austenite begins to transform to martensite whenever the local temperature of the bar reaches the M_s temperature. Since an increase in volume accompanies the transformation, the metal expands as the martensite reaction proceeds on cooling from the M_s to M_f temperature.[2] This produces a residual stress distribution of the type shown in Fig. 9.38c. The residual stress distribution in a quenched steel bar is the resultant of the competing processes of thermal contraction and volume expansion due to martensite formation. Transformation of austenite to bainite or pearlite also produces a volume expansion, but of lesser magnitude. The resulting stress pattern depends upon the transformation characteristics of the steel, as determined chiefly by composition and hardenability, and the heat-transfer characteristics of the system, and the severity of the quench.

Figure 9.39 illustrates some of the possible residual stress patterns that can be produced by quenching steel bars. On the left side of the figure is a typical isothermal transformation diagram for the decomposition of austenite. The cooling rates of the outside, midradius, and center of the bar are indicated on the diagram by the curves marked *o, m,* and *c.* In Fig. 9.39a the quenching rate is rapid enough to convert the entire bar to martensite. By the time the center of the bar reaches the M_s temperature, the transformation has been essentially completed at the surface. The surface layers try to contract against the expanding central core, and the result is tensile residual stresses at the surface and compressive stresses at the center of the bar (Fig. 9.39b). However, if the bar diameter is rather small and the bar has been drastically quenched in brine so that the surface and center transform at about the same time, the surface will arrive at room temperature with compressive residual stresses. If the bar is slack-quenched so that the outside transforms to martensite while the middle and center transform to pearlite (Fig. 9.39c), there is little restraint offered by the hot, soft core during the time when martensite is forming on the surface, and the core readily accommodates to the expansion of the outer layers. The middle and center pearlite regions then contract on cooling in the usual manner and produce a residual stress pattern consisting of compression on the surface and tension at the center (Fig. 9.39d).

---

1. U. Chandra, op. cit.

2. M_s and M_f are the temperature at which martensite starts to form and finishes forming, respectively, on quenching.

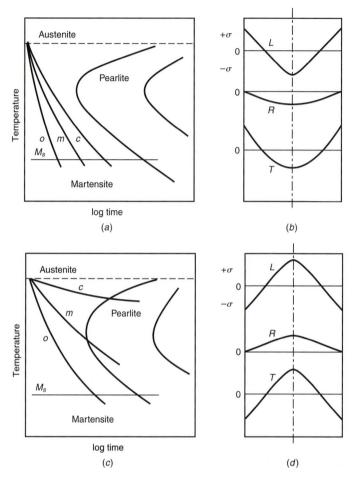

FIGURE 9.39
Transformation characteristics of a steel (*a* and *c*), and resulting residual stress distributions
(*b* and *d*).

9.14.3 Other Sources of Residual Stresses

To a first approximation the residual stresses in castings are modeled by a quenched
cylinder. However, the situation in castings is made more complicated by the fact that
the mold offers a mechanical restraint to the shrinking casting. Moreover, the casting
design may produce greatly different cooling rates at different locations that are due
to variations in section size and the introduction of chills, which produce an artifi-
cially rapid cooling rate.

Appreciable residual stresses are developed in welding, even in the absence of a
phase transformation. Figure 9.40 shows the residual stresses developed in the longitudinal direction (parallel to the weld joint). As the weld metal and heat-affected zone

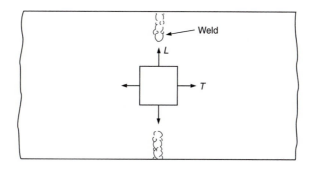

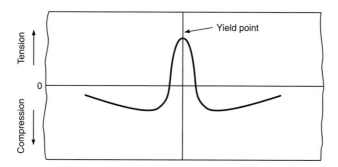

FIGURE 9.40
Longitudinal residual stresses in a butt-welded plate.

shrink on cooling, they are restricted by the cool surrounding plate. The weld contains tensile residual stresses, which are balanced by compressive stresses in the adjacent region. Because thermal gradients tend to be steep in welding, the residual stress gradients also tend to be steep.

Chemical processes such as oxidation, corrosion, and electroplating can generate large surface residual stresses if the new surface material retains coherency with the underlying metal surface. Other surface chemical treatments such as carburizing and nitriding cause local volume changes by the diffusion of an atomic species into the surface.

Residual stresses are measured by either destructive or nondestructive methods.[1] The destructive methods relax the locked-in stress by removing a layer from the body. The stress existing before the cut was made is calculated from the deformation produced by relaxing the stress. The nondestructive method depends on the fact that the spacing of atomic planes in a crystalline material is altered by stress. This change can be measured very precisely with a diffracted x-ray beam. The x-ray method is nondestructive, but it gives only the value of residual surface stress.

1. A. A. Denton, *Met. Rev.,* vol. 11, pp. 1–22, 1966; C. O. Ruud, *J. Metals,* pp. 35–40, July 1981.

9.14.4 Relief of Residual Stresses

The removal or reduction in the intensity of residual stress is known as stress relief. Stress relief may be accomplished either by heating or by mechanical working operations. Although residual stresses will disappear slowly at room temperature, the process is very greatly accelerated by heating to an elevated temperature. The stress relief that comes from a stress-relief anneal is due to two effects. First, since the residual stress cannot exceed the yield stress, plastic flow will reduce the residual stress to the value of the yield stress at the stress-relief temperature. Only the residual stress in excess of the yield stress at the stress-relief temperature can be eliminated by immediate plastic flow. Generally, most of the residual stress will be relieved by time-dependent stress relaxation. Since the process is extremely temperature-dependent, the time for nearly complete elimination of stress can be greatly reduced by increasing the temperature. Often a compromise must be made between the use of a temperature high enough for the relief of stress in a reasonable length of time and the annealing of the effects of cold-working.

The differential strains that produce high residual stresses also can be eliminated by plastic deformation at room temperature. For example, products such as sheet, plate, and extrusions are often stretched several percent beyond the yield stress to relieve differential strains by yielding. In other cases the residual stress distribution that is characteristic of a particular working operation may be superimposed on the residual stress pattern initially present in the material. A surface that contains tensile residual stresses may have the stress distribution converted into beneficial compressive stresses by a surface-working process like rolling or shot peening. However, it is important in using this method of stress relief to select surface-working conditions that will completely cancel the initial stress distribution. For example, it is conceivable that, if only very light surface rolling were used on a surface that initially contained tensile stresses, only the tensile stresses at the surface would be reduced. Dangerously high tensile stresses could still exist below the surface.

9.15
DESIGN FOR HEAT TREATMENT

One of the reasons why steel is such an important engineering material is that its metallurgical structure, and hence its properties, can be varied over wide limits by heat treatment. Precipitation (aging) reactions are important strengthening heat treatments in aluminum and nickel alloys. Also, annealing heat treatments are important in removing the damaging effects of cold-working so that metal forming can be carried well beyond the point at which fracture would ordinarily occur.

These are only some of the more important examples of the role of thermal treatment in controlling the properties of metals. However, processing by heat treatment requires energy. In addition, it requires a protective atmosphere or protective surface coating to prevent the metal part from oxidizing or otherwise reacting with the furnace atmosphere. Metal parts soften, creep, and eventually sag upon long exposure at elevated temperatures. Therefore, parts may require special fixtures to support them during heat treatment. Since heat treatment is a special processing step, it would be advantageous to eliminate it if possible. Sometimes a cold-worked sheet or bar can be substituted for a heat-treated part, but generally the flexibility and/or superior properties that result from heat treatment will be wanted.

The steel with the best combination of high strength and high toughness is produced by quenching from within the austenite temperature region (1400 to 1650°F depending on composition) and cooling rapidly enough that hard and brittle martensite is formed. The part is then reheated below the austenite region to allow the martensite to break down (temper) into a fine precipitation of carbides in a soft ferrite matrix. The formation of a proper *quenched and tempered* microstructure depends on cooling fast enough that pearlite or other nonmartensitic phases are not formed. This requires a balance between the heat transfer from the part (as determined chiefly by geometry), the cooling power of the quenching medium (water, oil, or air), and the transformation kinetics of the steel (as controlled by the alloy chemistry). These factors are interrelated by the property called hardenability.[1]

In heating for austenitization care should be taken to subject the parts to uniform temperature in the furnace. Long thin parts are especially prone to distortion from nonuniform temperature. Also, parts containing residual stress from previous processing operations may distort on heating.

Quenching is a severe condition to impose upon a piece of steel.[2] In quenching, the part suddenly is cooled at the surface. The part must shrink rapidly because of thermal contraction (steel is at least 0.125 in/ft larger before quenching from the austenitizing temperature), but it also undergoes a volume increase when it transforms to martensite at a comparatively low temperature. As discussed in Sec. 9.14 and Fig. 9.39, this can produce a condition of high residual (internal) stresses. The value of the local tensile stresses may be high enough to produce fractures called quench cracks. Also, local plastic deformation can occur in quenching even if cracks do not form, and that causes warping and distortion.

Problems with quench cracks and distortion chiefly are caused by nonuniformity of temperature distribution and, in turn, by geometry as influenced by the design. Thus, many of the heat-treatment problems can be prevented by proper design. The most important feature is to make the cross sections of the part as uniform as possible. In the ideal design for heat treatment all sections should have equal ability to absorb or give up heat. Unfortunately, designing for uniform thickness or sectional area usually interferes with the functions of the design. A sphere is the ideal geometry for uniform heat transmission, but obviously only a limited number of parts can utilize this shape.

A typical heat-treating problem is illustrated by the gear blank shown in Fig. 9.41. The region *A,* where the teeth will be machined, is thinner than the hub region *B.*

1. Heat Treating, "ASM Handbook," vol. 4, ASM International, Materials Park, OH, 1991; C. A. Siebert, D. V. Doane, and D. H. Breen, "The Hardenability of Steels," ASM International, 1977.
2. A. J. Fletcher, "Thermal Stress and Strain Generation in Heat Treatment," Elsevier Applied Science, New York, 1989.

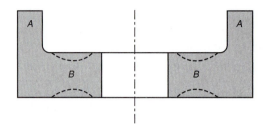

FIGURE 9.41
Gear blank geometry.

When the part was heated to the austenitizing temperature, region A increased in temperature faster than region B. As it tried to expand, it was attached to a thicker hub region B whose average temperature was a bit lower. Since the hot metal in A had to go somewhere, local hot upsetting of the surface occurred in region A. On quenching, region A cooled faster, so that it was through the temperature region of martensite formation before the thicker hub region B was. Since there were local temperature variations, there were local distortions that resulted in residual stresses. If there were a sharp fillet radius or other type of stress concentration, it could result in quench cracks in the brittle martensite. Thus, nonuniform temperatures, brought on by nonuniform geometry, lead to distortion and possibly fracture. A typical solution is to remove unneeded material in an attempt to produce more uniform heat transfer. In the gear blank of Fig. 9.41 this could be done by redesigning the hub region to remove metal along the dashed lines. In other cases, however, the approach toward uniformity might be to add metal.

Design details that minimize stress concentrations, as would be good design practice to prevent fatigue, also minimize quench cracking. Distortion in heat treatment is minimized by designs that are symmetrical. A single keyway in a shaft is a particularly difficult situation. A part with a special distortion problem may have to be quenched in a special fixture that restrains it from distorting beyond tolerances. Another consideration is to so design the part that the quenching fluid has access to all critical regions that must be hardened. Since the quenching fluid produces a vapor blanket when it hits the hot steel surface, it may be necessary to design for special venting or access holes for the quenching fluid.

A process approach to minimizing the difficulties with heat treatment of steel is called marquenching. In it large thermal gradients are eliminated by initially quenching to a temperature just above the M_s temperature and holding long enough to equalize the temperature of the part but not long enough to transform the austenite (Fig. 9.39). Then the parts are cooled slowly to below the M_f temperature to produce martensite. While seemingly an ideal way to heat-treat steel, marquenching may be difficult to carry out on a practical basis. It generally is more economical to select a steel that can be heat-treated by direct quenching.

9.16
DESIGN FOR PLASTICS PROCESSING

Many manufacturing processes used with plastics take advantage of the unique flow properties of polymers. Compared with metals, this means that the flow stress is much lower and highly strain rate dependent, the viscosity is much higher, and the formability is much greater. See Sec. 8.19 for a consideration of how the properties of plastics affect their use in design. We learned in Sec. 8.19 that plastics divide into (1) thermoplastic polymers (TP) that soften on heating and harden when cooled and can be remelted repeatedly; (2) thermosetting polymers (TS) that set or cross-link upon heating in an irreversible way; and (3) polymer composites that have either a TS or TP matrix reinforced with fibers of glass or graphite. TP polymers are polymerized in their primary manufacturing step and enter plastics processing as a granule or pellet resin. TS polymers are polymerized during the processing step, usually by the addition of a catalyst or simply by the addition of heat.

The plastic manufacturing processes considered in this section are:

- Injection molding (mostly TP)
- Extrusion (TP)
- Blow molding (TP)
- Rotational molding (TP)
- Thermoforming (TP)
- Compression molding (mostly TS)
- Casting (mostly TS)
- Composite processing (mostly TS)

Plastic manufacturing processes excel in producing parts with good surface finish and fine detail.[1] By adding dyes and colorants, the part can be given a color that eliminates a secondary painting operation. Depending on the process, the cycle time can vary from 10 s to 10 h. Generally polymer manufacturing is the preferred method for producing small to medium-sized parts for consumer and electronic products where mechanical stresses are not too high.

9.16.1 Injection Molding

Injection molding is a process in which plastic granules are heated and forced under pressure into a die cavity. It is a fast process (10 to 60 s cycle time) that is economical for production runs in excess of 10,000 parts. It is well suited for producing true three-dimensional shapes which require fine details like holes, snaps, and surface details.

---

1. Plastic Part Manufacturing, "Tool and Manufacturing Engineers Handbook," vol. 8, Society of Manufacturing Engineers, Dearborn, MI, 1996; Engineering Plastics, "Engineered Materials Handbook," vol. 2, sec. 3, ASM International, Materials Park, OH, 1988; E. A. Muccio, Design for Plastics Processing, "ASM Handbook," vol. 20, pp. 793–803, 1997.

Design of the gating and feed system for the die is crucial to ensure complete die fill.[1] As in design for casting, it is important to design the molding so that solidification does not prevent complete mold fill. The design and location of the gates for entry of polymer into the die is a crucial design detail. In large parts there may need to be more than one gate through which resin will flow in two or more streams into the mold. These will meet inside of the mold to create a fusion line. This may be a source of weakness or a surface blemish.

It is obvious that the mold must be designed so that the solid part can be ejected without distortion. Thus, the direction of mold closure, the parting surface between the two halves of the mold, and the part design must be considered together. By proper considerations of orientation at the beginning, it may be possible to avoid expensive mold costs like side cores. If at all possible, design the part so that it can be ejected in the direction of mold closure.

In addition to the economics of the process, the main concerns deal with the ability to achieve the required dimensional tolerances.[2] Mostly this is concerned with shrinkage, which is much larger in plastics than in metals. As the polymer cools from a plastic melt to a solid, the volume decreases (the density increases). Different plastics show different amounts of shrinkage. To minimize shrinkage, fillers, like glass fiber, wood flour, or natural fibers, are added during molding. Also, shrinkage can be influenced by the rate and direction of injecting the melt into the mold. It is best to have any shrinkage occur while the part is confined by the mold. However, with some plastics and part geometries postmold shrinkage can occur. This is related to the generation of high residual stresses during the molding process, and their gradual relief over hours, days, or weeks at room temperature. The creation of these residual stresses is a function of the mold design and the operating conditions of temperature and cooling rate in the process.

An important variation of the injection molding process is *reaction injection molding* (RIM). In the RIM process two liquid monomers are mixed as they enter the mold, where they react to form the polymer. Compared with injection molding, RIM offers lower pressure requirements, tooling costs, and capital investment. The process is used chiefly with urethanes, which can be made into parts with a load-bearing structural skin and a lightweight, rigid, cellular core. Large, complex parts can be made without residual stresses.

9.16.2 Extrusion

Extrusion is one of the few continuous plastic processes. It is used to produce sheet (>0.010 in thick), film, long lengths with a profiled cross section, and fiber. The chief concerns with the process are *die swell* and *orientation.* In die swell the extrudate swells to a size greater than the die from which it just exited. Thus, the design must

---

1. Software to aid in mold design and provide practical advice on manufacturing constraints is available as an add-on module with most 3-D CAD software. The most common software is Moldflow (www.mold flow.com) and C-MOLD (www.cmold.com).
2. R. A. Malloy, "Plastic Part Design for Injection Molding," Hanser Publishers, New York, 1994.

compensate for the swell. Polymer molecules become highly oriented in one or two directions as a result of the strongly oriented flow inherent in the extrusion process. When orientation is controlled, it can improve the properties of the material.

9.16.3 Blow Molding

Blow molding produces hollow products. A heated thermoplastic tube (called a parison) is held inside a mold and is expanded under air pressure to match the contours of the mold. The part cools, hardens, and is ejected from the mold. The process produces a part that is dimensionally defined on its external dimensions, but the interior surfaces are not controlled. Examples are milk bottles and automotive fuel tanks. The process does not lend itself to incorporating design details such as holes, sharp corners, or narrow ribs.

9.16.4 Rotational Molding

Like blow molding, rotational molding produces a hollow part. Rotational molding uses a fine TP powder that is placed inside a hollow heated metal mold that is slowly rotated about two perpendicular axes. Gravity rather than centrifugal force causes even coating of the mold surface. Then, still rotating, the mold is cooled and the part solidifies and hardens. Rotational molding can produce large parts, up to 500 gal capacity. Since it is a low-pressure process and the plastic is not forced through narrow channels, as in injection molding or extrusion, the process does not induce a significant amount of residual stress. Therefore, parts made by rotational molding exhibit a high degree of dimensional stability.

9.16.5 Thermoforming

Thermoforming, or vacuum forming, is a sheet forming process in which a TP sheet is clamped to a mold, softened, and a vacuum is applied to draw the sheet into the contour of the mold. When the sheet cools, it will retain the shape of the mold when the mold is removed. Traditionally, thermoforming is done with only a single mold, but for more precise control of dimensions two matching mold halves are used.

9.16.6 Compression Molding

Compression molding, the oldest plastics process, is similar to powder metallurgy. A preform of polymer, usually TS, is placed in a heated mold cavity and a plunger applies pressure to force the polymer to fill the mold cavity. The plastic is allowed to cure and is then ejected from the mold. Because the amount of flow is much less than in injection molding or extrusion, the level of residual stress in the part is low.

A variation of compression molding is *transfer molding*. In this process the plastic is preheated in a transfer mold and then "shot" into the mold as a viscous liquid with a transfer ram. The ram holds the plastic under pressure until it begins to cure. Then the ram retracts, and the part completes its cure cycle and is ejected. Starting with a heated plastic reduces the cycle time from 40 to 300 s for compression molding to from 30 to 150 s. Also, because a liquid plastic enters the mold it is possible to mold in inserts or to encapsulate parts. However, parts made this way have sprues and runners which must be trimmed and which result in lower yield.

9.16.7 Casting

Plastics are cast much less frequently than metals. The oldest applications are the casting of sheets and rods of acrylics and the "potting" of electrical components in epoxy. The development of a wider range of casting resins has led to consideration of casting as a way to make prototypes and low-volume production parts. Casting produces parts with low residual stress and a high degree of dimensional stability.

9.16.8 Composite Processing

The most common composite materials are plastics reinforced with glass, metal, or carbon fibers.[1] The reinforcement may be in the form of long, continuous filaments, short fibers, or flakes. TS polymers are the most common matrix materials. Except for filament winding, as in making a rocket motor case, the fiber and the matrix are combined in some preliminary form prior to processing. Molding compounds consist of TS resin with short randomly dispersed fibers. *Sheet molding compound* (SMC) is a combination of TS resin and chopped fibers rolled into a sheet about 1/4 in thick. *Bulk molding compound* (BMC) consists of the same ingredients made in billet form instead of sheet. SMC is used in the lay-up of large structures. BMC is used in compression molding. *Prepreg* consists of long fibers in partially cured TS resin. Prepregs are available as tape or cross-plied sheets or fabrics.

Composites are made by either open-mold or closed-mold processes. Hand layup, in which successive layers of resin and fiber are applied to the mold by hand, with the resin being rolled into the fiber, is the simplest process. An alternative is an open-mold process in which the liquid resin and chopped glass fibers are sprayed into the surface of the mold. In bag molding a plastic sheet or elastomer bag is clamped over the mold and pressure is applied either by drawing a vacuum or with compressed air.

Closed-mold composite processing follows closely the compression molding process. Variations have evolved to better place and orient the fibers in the composite. In resin transfer molding (RTM) a glass preform or mat is placed in the mold and a TS resin is transferred into the cavity under moderate pressure to impregnate the mat. A newer process is structural reaction injection molding (SRIM), a combination of RIM and RTM. SRIM is finding increasing application in the automotive industry for large structural parts, such as door panels, instrument panels, and rear window decks, in quantities under about 10,000 parts per year.

---

1. Composites, "Engineered Materials Handbook," vol. 1, ASM International, Materials Park, OH, 1987.

9.16.9 Design Issues

The issues of plastic design brought about because of their lower strength and stiffness than metals should be reviewed (Sec. 8.19).

The *wall* is the most important design feature of the plastic part. The wall thickness should not vary greatly. The nominal wall thickness will vary from about 0.015 to 0.160 in depending on the plastic. The rate of change of the thickness of the nominal wall should be gradual to ensure mold filling. Avoid thick walls. They require more plastic, but more importantly, they reduce the cycle time by requiring longer time to cool.

The typical projections from a molded wall are ribs, webs, and bosses. Ribs and webs are used to increase stiffness rather than increasing wall thickness. A rib is a piece of reinforcing material between two other features that are more or less perpendicular. A web is a piece of bracing material between two features that are more or less parallel. A boss is a short block of material protruding from a wall which is used to drive a screw through or to support something in the design.

It is important to design as many features, such as pilot holes, countersinks to receive fasteners, snap fits, and living hinges, as are needed into the molding rather than adding them as secondary operations. A big part of the attraction of plastic manufacture is that it minimizes the need for secondary operations.

Part design and process selection affects the residual stresses in the part. These stresses arise from inhomogeneous flow as the polymer molecules flow through the passages of the mold. Thus, generous radii, higher melt temperatures (which result in longer cycle time), and processes which minimize polymer flow result in lower residual stresses. Lower stresses lead to better dimensional stability.

9.17
SUMMARY

This chapter completes the core theme that design, materials selection, and processing are inseparable. Moreover, decisions concerning the manufacturing of parts should be made as early as possible in the design process—certainly in embodiment design. We recognize that there is a great deal of information that the designer needs to intelligently make these decisions. To aid in this the chapter provides:

- An overview of the most commonly used manufacturing processes, with emphasis on the factors which need to be considered most prominently
- References to a carefully selected set of books and handbooks that will provide both in-depth understanding of how the processes work, and detailed data needed for design
- An introduction to a simple methodology for ranking manufacturing processes on a unit cost basis which can be used early in the design process
- Reference to some of the rapidly growing collection of computer tools for design for assembly and design for manufacturing

A material and a process for making a part must be chosen in concert. The overall factor in deciding on the material and the manufacturing process is the cost to make a quality part. When making a decision on the material, the following factors must be considered:

- Material composition: grade of alloy or plastic
- Form of material: bar, tube, wire, strip, plate, pellet, powder, etc.
- Size: dimensions and tolerance
- Heat-treated condition
- Directionality of mechanical properties (anisotropy)
- Quality level: control of impurities, inclusions, microstructure, etc.
- Ease of manufacture: workability, weldability, castability, etc.
- Ease of recyclability
- Cost of material

The decision on the manufacturing process will be based on the following factors:

- Unit cost of manufacture
- Life cycle cost per unit
- Quantity of parts required
- Complexity of the part, with respect to shape, features, and size
- Compatibility of the process for use with candidate materials
- Ability to make a defect-free part
- Economically achievable surface finish
- Economically achievable dimensional accuracy and tolerances
- Availability of equipment
- Lead time for delivery of tooling
- Make-buy decision. Should we make the part in-house or purchase from a supplier?

Design can decisively influence manufacturing cost. That is why we must adopt methods to bring manufacturing knowledge into the embodiment design. An integrated product design team which contains experienced manufacturing people is a very good way of doing this. Design for manufacturability guidelines is another way. Some DFM guidelines are:

- Minimize total number of parts in a design
- Standardize components
- Use common parts across product lines
- Design parts to be multifunctional
- Design parts for ease of fabrication
- Avoid too-tight tolerances
- Avoid secondary manufacturing operations
- Utilize the special characteristics of a process

Experience has shown that a good way to proceed is to first do a rigorous design for assembly (DFA) analysis in an attempt to reduce part count. This will trigger a process of critical examination that can be followed up by "what if" exercises on critical parts to drive down manufacturing cost.

BIBLIOGRAPHY

Manufacturing Processes

DeGarmo, E. P., J. T. Black, and R. Kohser: "Materials and Processes in Manufacturing," 8th ed., Prentice-Hall, Upper Saddle River, NJ, 1997.
Groover, M. P.: "Fundamentals of Modern Manufacturing," Prentice-Hall, Upper Saddle River, NJ, 1996.
Kalpakjian, S.: "Manufacturing Processes for Engineering Materials," 3d ed., Addison-Wesley, Reading, MA, 1997.
Koshal, D.: "Manufacturing Engineer's Reference Book," Butterworth-Heinemann, Oxford, 1993.
Schey, J. A.: "Introduction to Manufacturing Processes," 3d ed., McGraw-Hill, New York, 1999.
Walker, J. M. (ed.): "Handbook of Manufacturing Engineering," Marcel Dekker, New York, 1996.

Design for Manufacturing

Boothroyd, G., P. Dewhurst, and W. Knight: "Product Design for Manufacture and Assembly," Marcel Dekker, New York, 1994.
Bralla, J. G. (ed.): "Design for Manufacturability Handbook," McGraw-Hill, New York, 1998.
Corbett, J., M. Dooner, J. Meleka, and C. Pym: "Design for Manufacture," Addison-Wesley, Reading, MA, 1991.
Design for Manufacturability, "Tool and Manufacturing Engineers Handbook," vol. 6, Society of Manufacturing Engineers, Dearborn, MI, 1992.
Ettlie, J. E., and H. W. Stoll: "Managing the Design-Manufacturing Process," McGraw-Hill, New York, 1990.
Hundal, M. S.: "Systematic Mechanical Design," ASME Press, New York, 1997.
Materials Selection and Design, "ASM Handbook," vol. 20, ASM International, Materials Park, OH, 1997.
Singh, K.: "Mechanical Design Principles: Applications, Techniques, and Guidelines for Manufacture," Nantel Publications, Melbourne, Australia, 1996.

PROBLEMS AND EXERCISES

9.1. Classify the following manufacturing processes as to shape-replication or shape-generative: (*a*) honing the bore of a cylinder, (*b*) powder metallurgy gear, (*c*) rough turning a cast roll, (*d*) extrusion of vinyl house siding.

9.2. One evolution in the design of automobile engines has been the change from in-line long-stroke engines to compact four- and six-cylinder engines. As a result, the crankshaft material has been changed from quenched and tempered steel forgings to cast crankshafts made from pearlitic malleable cast iron or nodular iron. Discuss this change in materials and processing in terms of the service performance and the properties of the materials.

9.3. A small hardware fitting is made from free-machining brass. For simplicity consider that the production cost is the sum of three terms: (1) material cost, (2) labor costs, and (3) overhead costs. Assume that the fitting is made in production lots of 500, 50,000, and 5×10^6 pieces by using, respectively, an engine lathe, a tracer lathe, and an automatic screw machine. By using the cost per part as an indicator, schematically plot the relative distribution of the cost due to materials, labor, and overhead for each of the production quantities.

9.4. Difficult-to-work materials mean that complex shapes often can be produced only by machining operations that convert a large part of the workpiece into chips. Let α be the fraction of the workpiece that is converted into chips to make the part. These chips will be cleaned, reprocessed, and melted and rolled to produce useful material and sold back to the manufacturer. If C is the cost of the material, this will go through the same cycle over and over to produce a real cost of material $C(1 + \alpha + \alpha^2 + \alpha^3 + \cdots) = C/(1 - \alpha)$. If the chips are sold at a ratio of β to the price of the material, determine a relation for the true cost of the material under a condition in which scrap (chips) is involved. If titanium alloys cost $10,000 per ton and $\alpha = 0.90$ and $\beta = 0.1$, what is the real cost of the workpiece material?

9.5. Explain why alloys designed for casting generally are not used for forgings, and vice versa.

9.6. You are concerned with the cast nodular-iron crankshafts. What design factors determine the manufacturing cost? Which of the costs are determined by the foundry and which by the purchaser?

9.7. Describe the manufacturing steps to produce an automobile rear-axle housing from a tubular blank.

9.8. Titanium alloys are more difficult to form into sheet-metal structural members than aluminum alloys in aircraft construction. Discuss the reasons for this difference and also discuss manufacturing methods that have been developed to overcome the difficulties.

9.9. A machine shaft has a diameter of 1.75 in and is 3.5 ft long. It must withstand a maximum bending stress of 80,000 psi. Since other parts of the same machine are made from 4140H steel, we would like to use that material for the shaft if it has sufficient hardenability. Use data available in the ASM "Metals Handbook" to determine whether 4140H steel is acceptable for the application.

9.10. Make an early design stage estimate of the manufacturing cost of the cylinder with the conical end shown in category A1 in Fig. 9.13. This part must be resistant to mild sulfuric acid attack to a temperature of 200°F. It must withstand 90 psi internal pressure. The wall thickness is 0.80 ±0.01 mm. The surface finish is 0.5 μm. A total of 1000 parts is needed annually.

9.11. Make a brief literature study of hot isostatic processing (HIP). Discuss the mechanics of the process, its advantages and disadvantages. Think broadly about how HIPing can improve more conventional processes, and how it can impact on design.

9.12. Product cycle time is the time it takes for raw materials to be transformed into a finished product. A firm makes 1000 products per day. Before it is sold, each product represents $200 in materials and labor. (*a*) If the cycle time is 12 days, how many dollars are tied up with in-process inventory? If the company's internal interest rate is 10 percent, what is the annual cost due to in-process inventory? (*b*) If the cycle time is reduced to 8 days as a result of process improvement, what is the annual cost saving?

9.13. A tubular structural member must transmit fluid, while at the same time resisting a 1200 Nm bending moment. It must sustain 6 MPa internal pressure and a fiber stress of 95 MPA. The ID of the tube is 60 mm. The original design was a steel tube, but it is now important to reduce weight by a new design. Think creatively about combining a change of material, with its processing capability, and a geometry change, to accomplish these design objectives.

9.14. A wheel spindle must be capable of developing RC 50 minimum when quenched to 90 percent martensite at the critical section. The critical section is 2.0 in in the final machined condition, but 0.20 in must be allowed for removal by machining from the forged and heat-treated surface. A steel with a 0.40 percent carbon content has been specified. The Jominy equivalent cooling rate at 0.20 in below the surface is found to be J9.5.
(*a*) What is the ideal critical diameter needed for this application?
(*b*) Suggest and prove out a more economical approach than using a medium-carbon alloy steel and an oil quench.

10

ENGINEERING STATISTICS

10.1
STATISTICS AND DESIGN

Probability and statistics have become working tools of the engineer in many areas of engineering practice. Since in engineering design we typically deal with poorly defined situations or are forced to use data that have low precision, it is easy to appreciate how the proper application of statistical analysis can help greatly with engineering design. This fact is recognized in some engineering curricula by requiring extensive course work in statistics. Unfortunately, that is far from the usual situation, and many engineers are forced to acquire their statistical background in bits and pieces obtained from several courses or from independent reading.

Thus, in no other area within this book is there greater disparity in prior background. Those who have taken a complete course in engineering statistics will have received a far richer background than can be imparted by this chapter. You lucky ones should use the chapter as a review while paying particular attention to the examples. For those of you who have not had previous instruction in engineering statistics, the chapter aims at providing a basic background. However, emphasis is on the application of statistical methods rather than the mathematical concepts underlying the methods. Many references to in-depth texts and technical papers are provided for those who wish to pursue the topics further. Also, the discussion of many topics is keyed to the excellent statistics handbook by Natrella[1] and text by Montgomery and Runger.[2]

Statistical techniques are important in engineering decision making. The underlying philosophy is that we use observed samples to estimate the properties of the sta-

---

1. M. G. Natrella, "Experimental Statistics," Nat. Bur. Standards Handbook 91, Government Printing Office, 1963; also available from John Wiley & Sons, New York, 1983.
2. D. C. Montgomery and G. C. Runger, "Applied Statistics and Probability for Engineers," Wiley, New York, 1994.

tistical population. The known properties of the sample distribution provide the basis for decision making.

In making physical measurements we are concerned with their precision and accuracy. The *precision* of an instrument indicates the instrument's capacity to reproduce a certain reading time after time. The fact that the same reading is not always obtained is due to the existence of *random error.* The reading also may contain systematic error or *bias,* such that the readings are consistently high or low. The presence of both random error and systematic error affects the accuracy of the reading. The *accuracy* is the deviation of the reading from the true value.

At least four major aspects of statistical analysis are important in engineering design. At the most basic level we are interested in the analysis of experimental data, so that the results can be unequivocally described by appropriate statistical parameters. Next we are concerned with *statistical inference,* which uses statistics to make reliable decisions utilizing tests of hypotheses and confidence limits. *Hypotheses tests* are used to determine whether there is a significant difference between the characteristics of an observed set of data and a proposed mathematical model of the data. *Confidence limits* allow us to determine a range in which the true characteristic of a population is likely to lie. A third important area, *analysis of variance,* is a test for the equality of means and/or variances of groups of observations, such as the means resulting from different experimental procedures. Finally, these concepts lead to consideration of the most efficient way to collect data through the *statistical design of experiments.* They also permit us to determine empirical relations between variables with regression analysis and response surface methodology.

10.2
PROBABILITY

The concept of probability is part of our general base of knowledge. When asked the probability that a tossed coin will come up heads, most people will answer "one-half," or when asked the probability that a six-sided die will come up a 4, they will answer "one-sixth." Thus, they recognize the concept of a probability scale in which a nonevent has a probability of zero [$P(A) = 0$] and an assured event has a probability of one [$P(A) = 1$]. If there are N possible outcomes, then the probability that an event A will occur is

$$P(A) = \frac{\text{number of ways in } N \text{ that produce A}}{N} = \frac{n}{N} \tag{10.1}$$

The probability that an event will not occur is $P(\overline{A})$, and

$$P(\overline{A}) = 1 - P(A) = \frac{N - n}{N} \tag{10.2}$$

Since the event A will either occur or not occur,

$$P(A) + P(\overline{A}) = 1.0 \tag{10.3}$$

The odds of event A occurring are given by $P(A)/P(\overline{A})$.

A basic underlying assumption of probability theory is that it deals with random events. A *random event* is one in which the conditions are such that each member of the population N has an equal chance of being chosen.

10.2.1 Notation of Probability

A special and precise system of language and notation is used in probability theory. Two events A and B are said to be *independent* if the occurrence of either one has no effect on the occurrence of the other. Two events that have no elements in common are said to be *mutually exclusive* events.

1. If A and B are independent events, then the probability of *both* A and B occurring (joint probabilities) is

$$P(\text{A and B}) = P(\text{AB}) = P(\text{A})P(\text{B}) \tag{10.4}$$

If A and B are *mutually exclusive* events,

$$P(\text{AB}) = 0 \tag{10.5}$$

2. The probability of A or B occurring (for non-mutually-exclusive events) is

$$P(\text{A or B}) = P(\text{A + B}) = P(\text{A}) + P(\text{B}) - P(\text{AB}) \tag{10.6}$$

If A and B are mutually exclusive events,

$$P(\text{A + B}) = P(\text{A}) + P(\text{B}) \tag{10.7}$$

3. A conditional probability is one in which the probability of the event depends upon whether the other event has occurred.

$P(\text{A}\,|\,\text{B}) = $ probability that A will occur given that B has occurred

$$P(\text{A}\,|\,\text{B}) = \frac{P(\text{AB})}{P(\text{B})} \tag{10.8}$$

or

$$P(\text{B}\,|\,\text{A}) = \frac{P(\text{AB})}{P(\text{A})} \tag{10.9}$$

4. Bayes' theorem allows us to modify a probability estimate as additional information becomes available. From Eqs. (10.8) and (10.9), $P(\text{AB}) = P(\text{A}\,|\,\text{B})P(\text{B}) = P(\text{B}\,|\,\text{A})P(\text{A})$, since $\text{AB} = \text{BA}$, and

$$P(\text{A}\,|\,\text{B}) = \frac{P(\text{B}\,|\,\text{A})P(\text{A})}{P(\text{B})} \tag{10.10}$$

Since $P(\text{A}) + P(\overline{\text{A}}) = 1$, it follows that event B must occur jointly with either A or $\overline{\text{A}}$

$$P(\text{B}) = P(\text{AB}) + P(\overline{\text{A}}B)$$

and from Eq. (10.9)

$$P(B) = P(A)P(B \mid A) + P(\bar{A})P(B \mid \bar{A}) \qquad (10.11)$$

Substituting Eq. (10.11) into Eq. (10.10) gives

$$P(A \mid B) = \frac{P(A)P(B \mid A)}{P(A)P(B \mid A) + P(\bar{A})P(B \mid \bar{A})} \qquad (10.12)$$

If event A has more than two available alternatives, then Eq. (10.12) is expressed as

$$P(A_i \mid B) = \frac{P(A_i)P(B \mid A_i)}{\sum_i P(A_i)P(B \mid A_i)} \qquad (10.13)$$

EXAMPLE 10.1. The number of defective and acceptable parts received from vendor 1 (V_1) and vendor 2 (V_2) is:

	V_1	V_2	Total
Defective	300	750	1,050
Acceptable	9,700	6,250	15,950
	10,000	7,000	17,000

Let A_1 = event "part from vendor 1"
$\quad A_2$ = event "part from vendor 2"
$\quad B_1$ = event "acceptable part"
$\quad B_2$ = event "defective part"
$B_1 \mid A_1$ = event "acceptable part from vendor 1"

$$P(A_1) = \frac{10,000}{17,000} = 0.59 \qquad P(A_2) = \frac{7,000}{17,000} = 0.41$$

$$P(B_1 \mid A_1) = 0.97 \qquad P(B_2 \mid A_1) = 0.03 \qquad P(B_2 \mid A_2) = 0.11$$

$$P(B_1 \mid A_2) = 0.89 \qquad P(B_2 \mid A_2) = 0.11$$

We wish to know the probability of selecting a part that is made by vendor 1 and is also defective.

Solution. From Eq. (10.8)

$$P(A_1 B_2) = P(B_2 A_1) = P(A_1)P(B_2 \mid A_1)$$

$$= 0.59(0.03) = 0.018$$

In a similar way, the probability of selecting a defective part that is made by vendor 2 is

$$P(A_2 B_2) = P(A_2)P(B_2 \mid A_2)$$

$$= 0.41(0.11) = 0.045$$

Because the selection of a defective part made by vendor 1 and vendor 2 are mutually exclusive events, the probability of selecting a defective part, irrespective of which vendor provided it, is given by

$$P(B_2) = P(A_1B_2) + P(A_2B_2)$$

$$= 0.018 + 0.045 = 0.063$$

EXAMPLE 10.2. Referring to Example 10.1, if we find a part to be defective, we can then ask, what is the *posterior probability* that it came from vendor 1?

Solution. This uses Bayes' theorem.

$$P(A_1 \mid B_2) = \frac{P(A_1)P(B_2 \mid A_1)}{P(A_1)P(B_2 \mid A_1) + P(A_2)P(B_2 \mid A_2)}$$

$$= \frac{0.59(0.03)}{0.59(0.03) + 0.41(0.11)} = 0.28$$

Thus, using the new information that the part selected did fail, the posterior probability that it came from vendor 1 is much less than the prior probability 0.59.

10.3
ERRORS AND SAMPLES[1]

The act of making any type of experimental observation involves two types of errors: systematic errors (which exert a nonrandom bias) and experimental, or random, errors. Systematic errors arise because of faulty control of the environment. Experimental errors are due to limitations of the measuring equipment or to inherent variability in the material being tested. As an example, in the measurement of the reduction of area of a fractured tensile specimen, a systematic error could be introduced if an improperly zeroed micrometer were used for measuring the diameter, whereas random errors would result from slight differences in fitting together the two halves of the tensile specimen and from the inherent variability of reduction-of-area measurements on metals. By averaging a number of observations, the random error will tend to cancel out. The systematic error, however, will not cancel upon averaging. One of the major objectives of statistical analysis is to deal quantitatively with random error.

When a tensile specimen is cut from a steel forging and the reduction of area is determined for it, the observation represents a sample of the *population* from which it was drawn. The population, in this case, is the collection of all possible tensile specimens that could be cut from the forging or from all other forgings. As more and more tensile specimens are cut from the forging and reduction-of-area values are measured, the sample estimate of the population values becomes more accurate. However, it is obviously impractical to sample and test the entire forging. Therefore, one of the main

---

1. Sections 10.3 to 10.6 are from G. E. Dieter, "Mechanical Metallurgy," McGraw-Hill, New York, 1961.

purposes of statistical techniques is to determine the best *estimate* of the population parameters from a randomly selected sample. The approach that is taken is to postulate that, for each sample, the population has fixed and invariant parameters. However, the corresponding parameters calculated from samples contain random errors, and, therefore, the sample provides only an estimate of the population parameters. It is for this reason that statistical methods lead to conclusions having a given *probability* of being correct.

10.4
FREQUENCY DISTRIBUTION

When a large number of observations are made from a random sample, a method is needed to characterize the data. The most common method is to arrange the observations into a number of equal-valued *class intervals* and determine the frequency of the observations falling within each class interval. In Table 10.1, out of a total sample of 449 measurements of yield strength, 4 observations fell between 114,000 and 115,900

TABLE 10.1
Frequency tabulation of yield strength of steel*

(1) Yield strength, 1000 psi, class interval	(2) Class midpoint x_i	(3) Frequency f_i	(4) $f_i x_i$	(5) Relative frequency, % of total	(6) Cumulative frequency	(7) Cumulative frequency, %
114–115.9	115	4	460	0.9	4	0.9
116–117.9	117	6	702	1.3	10	2.2
118–119.9	119	8	952	1.6	18	3.8
120–121.9	121	26	3146	5.8	44	9.6
122–123.9	123	29	3657	6.5	73	16.1
124–125.9	125	44	5500	9.8	117	25.9
126–127.9	127	47	5969	10.5	164	36.4
128–129.9	129	59	7611	13.1	223	49.5
130–131.9	131	67	8777	15.0	290	64.5
132–133.9	133	45	5985	10.0	335	74.5
134–135.9	135	49	6615	10.9	384	85.4
136–137.9	137	29	3973	6.5	413	91.9
138–139.9	139	17	2363	3.8	430	95.7
140–141.9	141	9	1269	2.0	439	97.7
142–143.9	143	6	858	1.3	445	99.0
144–145.9	145	4	580	0.9	449	99.9

$$\sum f_i = 449 \qquad \sum f_i x_i = 58{,}417$$

$$\bar{x} = 58{,}417/449 = 130{,}000 \text{ psi}$$

*Data from F. B. Stulen, W. C. Schulte, and H. N. Cummings *in* D. E. Hardenbergh (ed.), "Statistical Methods in Materials Research," Pennsylvania State University, University Park, PA, 1956.

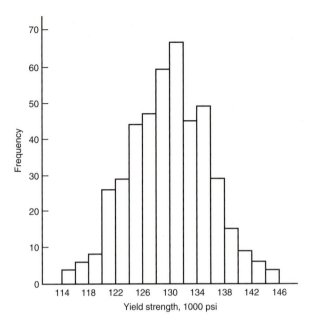

FIGURE 10.1
Frequency histogram of data in Table 10.1.

psi, 26 fell between 120,000 and 121,900 psi, etc. An estimate of the frequency distribution of the observations can be obtained by plotting the frequency of observations against the class intervals of the yield strength measurements (Fig. 10.1). This type of bar diagram is known as a *histogram.* As the number of observations increases, the size of the class interval can be reduced until we obtain the limiting curve, which represents the *frequency distribution* of the sample (Fig. 10.2). Note that most of the values of yield strength fall within the interval from 126,000 to 134,000 psi.

If the frequency of observations in each class interval is expressed as a percentage of the total number of observations, the area under a frequency-distribution curve, which is plotted on this basis, is equal to unity (Fig. 10.3).

The probability that a single random measurement of yield strength will be between the value x_1 and a slightly higher value $x_1 + \Delta x$ is given by the area under the frequency-distribution curve bounded by those two limits. Similarly, the probability that a single observation will be greater than some value x_2 is given by the area under the curve to the right of x_2, while the probability that the single observation will be less than x_2 is given by the area to the left of x_2. Before inferring probabilities from the frequency distribution, we need to fit the experimental results to a standard statistical distribution such as the normal or lognormal distribution (Sec. 10.6).

Another way to present these data is to arrange the frequency in a cumulative manner. In Table 10.1, column 6, the frequency for each class interval is accumulated with the values of frequency for the class intervals below it to indicate the total number of observations with a value of yield strength less than or equal to the value of the upper limit of the class interval. As an example, for the sample described in Table 10.1, 164 out of 449 observations of yield strength have a value of 127,900 psi or less. If the cumulative frequency is expressed as the percentage of the total (Table 10.1,

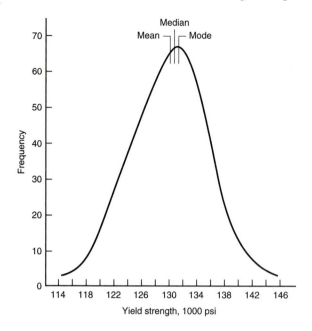

FIGURE 10.2
Frequency distribution of data in Table 10.1.

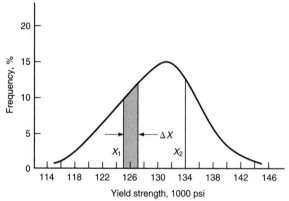

FIGURE 10.3
Frequency distribution based on relative frequency.

column 7), the values represent the probability that the yield strength will be less than or equal to the value of the observation. Figure 10.4 shows the cumulative frequency distribution plotted on that basis. The presentation of data as a cumulative distribution is sometimes preferred to a frequency distribution because it is much less sensitive than the frequency distribution to the choice of class intervals.

10.5
MEASURES OF CENTRAL TENDENCY AND DISPERSION

A frequency distribution such as that of Fig. 10.2 can be described with numbers that indicate the central location of the distribution and how the observations are spread

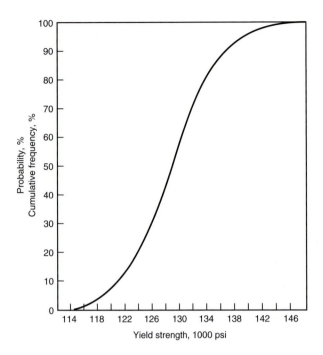

FIGURE 10.4
Cumulative frequency
distribution.

out from the central location (dispersion). The most common and important measure of the central value of an array of data is the *arithmetic mean,* or *average.* The mean of $x_1, x_2, \ldots, x_n$ observations is denoted by $\bar{x}$ and is given by

$$\bar{x} = \frac{\sum\limits_{i=1}^{n} x_i}{n} \qquad (10.14)$$

The arithmetic mean is equal to the summation of the individual observations divided by the total number of observations n. If the data are arranged in a frequency table, as in Table 10.1, the mean can be most conveniently determined from

$$\bar{x} = \frac{\sum\limits_{i=1}^{k} f_i x_i}{\Sigma f_i} \qquad (10.15)$$

where f_i is the frequency of observations in a particular class interval with midpoint x_i. The summation is taken over all class intervals (see Table 10.1, column 4).

Two other common measures of central tendency are the mode and the median. The *mode* is the value of the observations that occurs most frequently. The *median* is the middle value of a group of observations. For a set of discrete data the median can be obtained by arranging the observations in numerical sequence and determining which value falls in the middle. For a frequency distribution the median is the value that divides the area under the curve into two equal parts. The mean and the median are frequently close together; but if there are extreme values in the observations (either high or low), the mean will be more influenced by these values than the median. As an extreme example, the situation sometimes arises in fatigue testing that out of a group of specimens tested at a certain stress a few do not fail in the time allotted for the test and therefore presumably have infinite lives. These extreme values could not be grouped with the failed specimens to calculate a mean fatigue life; yet they could be considered in determining the median. The positions of the mean, median, and mode are indicated in Fig. 10.2.

The most important measure of the dispersion of a sample is given by the unbiased estimate of the *variance* s^2.

$$s^2 = \frac{\sum\limits_{i=1}^{n}(x_i - \bar{x})^2}{n-1} \tag{10.16}$$

The term $x_i - \bar{x}$ is the deviation of each observation x_i from the arithmetic mean $\bar{x}$ of the n observations. The quantity $n - 1$ in the denominator is called the number of degrees of freedom and is equal to the number of observations minus the number of linear relations between the observations. Since the mean represents one such relation, the number of degrees of freedom for the variance about the mean is $n - 1$. For computational purposes it is often convenient to calculate the variance from the following relation:

$$s^2 = \frac{n\sum\limits_{i=1}^{n} x_1^2 - \left(\sum\limits_{i=1}^{n} x_i\right)^2}{n(n-1)} \tag{10.17}$$

Equation (10.16) is preferable to Eq. (10.17) when the number of significant digits is important. When the data are arranged in a frequency table, the variance can be most readily computed from the following equation:

$$s^2 = \frac{\sum\limits_{i=1}^{k} f_i x_i^2 - \dfrac{\left(\sum\limits_{i=1}^{k} f_i x_i\right)^2}{n}}{n-1} \tag{10.18}$$

In dealing with the dispersion of data it is usual practice to work with the *standard deviation s,* which is defined as the positive square root of the variance.

$$s = \left[\frac{\sum\limits_{i=1}^{n}(x_i - \bar{x})^2}{n-1}\right]^{1/2} \tag{10.19}$$

Sometimes it is desirable to describe the variability relative to the average. The coefficient of variation v is used for this purpose.

$$v = \frac{s}{\bar{x}} \tag{10.20}$$

A measure of dispersion that is sometimes used because of its extreme simplicity is the *range*. It is simply the difference between the largest and smallest observation. The range does not provide as precise estimates of dispersion as does the standard deviation.

The need to distinguish between the sample and the population has been emphasized above. Standard notation is adopted in statistics to make this distinction between the characteristics of the sample and the population.

	Sample	Population
Arithmetic mean	$\bar{x}$	μ
Standard deviation	s	σ

10.6
THE NORMAL DISTRIBUTION

Many physical measurements follow the symmetrical, bell-shaped curve of the normal, or Gaussian, frequency distribution; repeated measurements of the length or diameter of a bar would closely approximate it. The distributions of yield strength, tensile strength, and reduction of area from the tension test have been found to follow the normal curve to a suitable degree of approximation. The equation of the normal curve is

$$f(x) = \frac{1}{\sigma\sqrt{2\pi}} \exp\left[-\frac{1}{2}\left(\frac{x - \mu}{\sigma}\right)^2 \right] \tag{10.21}$$

where $f(x)$ is the height of the frequency curve corresponding to an assigned value x, μ is the mean of the population, and σ is the standard deviation of the population.[1]

The properties of the normal distribution are well known and are readily available in tabular form. The central limit theorem of statistics states that the distribution of the mean of N independent observations from *any* frequency distribution with a finite mean and variance approaches the normal distribution as N approaches infinity. When a random variable represents the total effect of a number of independent small causes, the central limit theorem leads one to expect that the distribution of the measured variable will be normal. However, because of the prominence of the normal distribution, there is a tendency to assume that every experimental variable is normally distributed unless proved otherwise. This is wrong.

The normal distribution extends from $x = -\infty$ to $x = +\infty$ and is symmetrical about the population mean μ. The existence of negative values and long "tails" makes the normal distribution a poor model in certain engineering problems.

---

1. It should be noted that σ is used for normal stress in other chapters of this book.

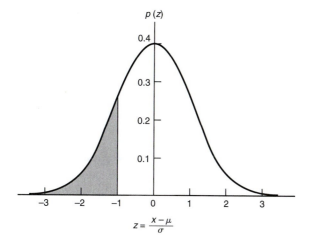

FIGURE 10.5
Standardized normal frequency distribution.

In order to place all normal distributions on a common basis in a standardized way, the normal curve frequently is expressed in terms of the standard normal variable z.

$$z = \frac{x - \mu}{\sigma} \tag{10.22}$$

Now, the equation of the standard normal curve becomes

$$f(z) = \frac{1}{\sqrt{2\pi}} \exp\left(-\frac{z^2}{2}\right) \tag{10.23}$$

Figure 10.5 shows this standardized normal curve, where $\mu = 0$ and $\sigma = 1$. The total area under the curve is unity. The relative frequency of a value of z falling between $z = -\infty$ and a specified value of z is given by the total area under the curve. Some typical values are listed in Table 10.2.[1]

EXAMPLE 10.3. A highly automated factory is producing ball bearings. The average ball diameter is 0.2152 in and the standard deviation is 0.0125 in. These dimensions are normally distributed.
(a) What percentage of the parts can be expected to have a diameter less than 0.2500 in? Determining the standard normal variable

$$z = \frac{x - \mu}{\sigma} \approx \frac{x - \bar{x}}{s} = \frac{0.2500 - 0.2512}{0.0125} = \frac{-0.0012}{0.0125} = -0.096$$

From Table B-1, $P(z < -0.09) = 0.4641$ and $P(z < 0.10) = 0.4602$. Interpolating, the area under the z distribution curve at $z = -0.096$ is 0.4618. Therefore, 46.18 percent of the ball bearings are below 0.2500 in diameter.
(b) What percentage of the balls are between 0.2574 and 0.2512 in?

---

1. For complete values see M. G. Natrella, op. cit., and Table B-1 in appendix, or any other statistics text.

TABLE 10.2
Areas under standardized normal frequency curve

$z = x - \mu/\sigma$	Area	z	Area
-3.0	0.0013	-3.090	0.001
-2.0	0.0228	-2.576	0.005
-1.0	0.1587	-2.326	0.010
-0.5	0.3085	-1.960	0.025
0.0	0.5000	-1.645	0.050
$+0.5$	0.6915	1.645	0.950
$+1.0$	0.8413	1.960	0.975
$+2.0$	0.9772	2.326	0.990
$+3.0$	0.9987	2.576	0.995
		3.090	0.999

$$z = \frac{0.2512 - 0.2512}{0.0125} = 0.0 \quad \text{Area under curve from } -\infty \text{ to } z = 0 \text{ is } 0.5000.$$

$$z = \frac{0.2574 - 0.2512}{0.0125} = \frac{0.0062}{0.0125} = +0.50 \quad \begin{array}{l} \text{Area under curve from} \\ -\infty \text{ to } z = 0.5 \text{ is } 0.6915 \end{array}$$

Therefore, percentage of ball diameters in interval 0.2512 to 0.2574 is $0.6915 - 0.5000 = 0.1915$ or 19.15 percent.

10.6.1 Lognormal Distribution

The lognormal distribution is used as a model for lifetime data. For example, the life to failure of fatigue tests at a constant stress follows a normal distribution if the logarithm of fatigue life is plotted instead of the fatigue life. Unlike the normal distribution, the lognormal distribution has no negative values and thus it lends itself to life data.[1] It is an asymmetric distribution that is positively skewed with a long right tail.

10.6.2 Tests for Normal Distribution

In a convenient method for determining whether a sample frequency distribution approximates a normal distribution a cumulative frequency plot on normal probability paper is used. First the sample mean and standard deviation are computed. Plot $\bar{x}$

---

1. C. H. Lipson and N. J. Sheth, "Statistical Design and Analysis of Engineering Experiments," pp. 32–36, McGraw-Hill, New York, 1973; "ASM Metals Handbook," 9th ed., vol. 8, pp. 630–632, ASM International, Metals Park, OH, 1985; J. Aitchison and J. A. C. Brown, "The Lognormal Distribution," Cambridge University Press, Cambridge, 1966.

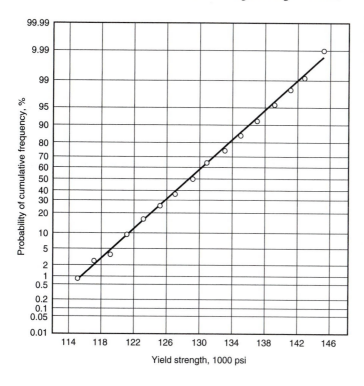

FIGURE 10.6
Normal probability plot of data in Table 10.1.

at $P = 0.5$ and $\bar{x} + s$ at $P = 0.84$. A straight line is then drawn through those points. The individual points are then plotted on the normal probability paper. For small sample populations the probability is obtained by rank-ordering the data from smallest to largest value. The probability of an event occurring is

$$P_m = \frac{m}{n + 1} \qquad (10.24)$$

where $m =$ rank number of the value
$n =$ total number of data points

The degree of fit to the normal distribution is determined by how well the data group around the straight line (Fig. 10.6).

The statistical advantages of the normal distribution are quite considerable; so that when the data do not fit the normal curve, it is worthwhile to search for a simple transformation that will normalize them. The most common transformations are $x' = \log x$ and $x' = x^{1/2}$.

10.7
WEIBULL AND OTHER FREQUENCY DISTRIBUTIONS

The normal distribution is an unbounded symmetrical distribution with long tails extending to $-\infty$ and to $+\infty$. However, many random variables follow a bounded, nonsymmetrical distribution. A good example is a distribution describing the life of a component for which all values are positive (there are no negative lives) and for which there are occasional long-lived results.

10.7.1 Weibull Distribution

The Weibull distribution[1] is widely used for many engineering problems because of its versatility. Originally proposed for describing fatigue life, it is widely used to describe the life of parts or components, such as ball bearings, gears, and electronic components. The *two-parameter* Weibull function is described by

$$ f(x) = \frac{m}{\theta} \left(\frac{x}{\theta} \right)^{m-1} \exp\left[-\left(\frac{x}{\theta} \right)^{m} \right] \qquad x > 0 \qquad (10.25) $$

where $f(x) =$ frequency distribution of the random variable x
$\quad\quad m =$ shape parameter, which is sometimes referred to as the Weibull modulus
$\quad\quad \theta =$ scale parameter, sometimes called the characteristic value

The change in the Weibull distribution for various values of shape parameter is shown in Fig. 10.7. The mean of a Weibull population is given by

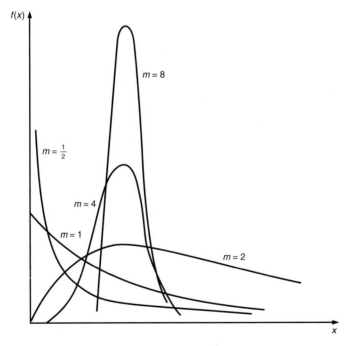

FIGURE 10.7
The Weibull distribution for $\theta = 1$ and different values of m.

1. W. Weibull, *J. Appl. Mech.*, vol. 18, pp. 293–297, 1951; *Materials Research and Stds.*, pp. 405–411, May 1962; "ASM Metals Handbook," op. cit., pp. 632–634; C. R. Mischke, *Jnl. Mech. Design*, vol. 114, pp. 29–34, 1992.

$$\bar{x} = \theta \Gamma\left(1 + \frac{1}{m}\right) \tag{10.26}$$

where Γ is the standard gamma function (see next under gamma distribution). The variance of a Weibull population is given by

$$\text{Variance} = \theta^2\left[\Gamma\left(1 + \frac{2}{m}\right) - \Gamma^2\left(1 + \frac{1}{m}\right)\right] \tag{10.27}$$

The cumulative frequency distribution is given by

$$F(x) = 1 - \exp\left[-\left(\frac{x}{\theta}\right)^m\right] \quad x > 0 \tag{10.28}$$

Rewriting Eq. (10.28) as

$$\frac{1}{1 - F(x)} = \exp\left(\frac{x}{\theta}\right)^m$$

$$\ln\frac{1}{1 - F(x)} = \left(\frac{x}{\theta}\right)^m$$

$$\ln\left(\ln\frac{1}{1 - F(x)}\right) = m\ln x - m\ln\theta = m(\ln x - \ln\theta) \tag{10.29}$$

which is a straight line of the form $Y = mx + C$. Special Weibull probability paper is available to assist in the analysis according to Eq. (10.29). When the probability of failure is plotted against x (life) on Weibull paper a straight line is obtained (Fig. 10.8). The slope is the Weibull modulus m. The greater the slope the smaller the scatter in the random variable x.

θ is called the characteristic value of the Weibull distribution. If $x = \theta$, then

$$F(x) = 1 - \exp\left[-\left(\frac{\theta}{\theta}\right)^m\right] = 1 - e^{-1} = 1 - \frac{1}{e}$$

Therefore,

$$F(x) = 1 - \frac{1}{2.718} = 0.632$$

Thus, for any Weibull distribution the probability of being less than or equal to the characteristic value is 0.632. Therefore, θ will divide the area under the probability distribution function into 0.632 and 0.368 for all values of m.

If the data do not plot as a straight line on Weibull graph paper, then either the sample was not taken from a population with a Weibull distribution or it may be that

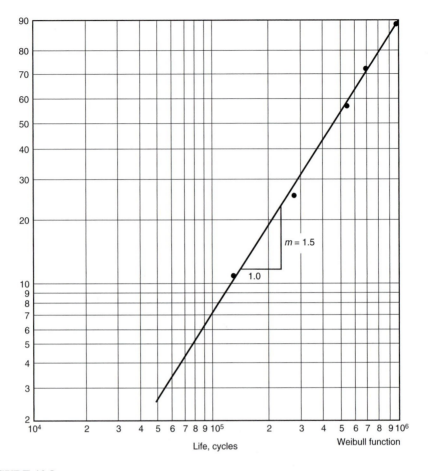

FIGURE 10.8
Weibull plot for life of ball bearings. (*C. Lipman and N. J. Sheth, "Statistical Design and Analysis of Engineering Experiments," p. 41. Copyright © 1973 by McGraw-Hill. Used with the permission of McGraw-Hill Book Company.*)

the Weibull distribution has a minimum value x_0 which is greater than $x_0 = 0$. This leads to the three-parameter Weibull distribution.

$$F(x) = 1 - \exp\left[-\left(\frac{x - x_0}{\theta - x_0}\right)^m \right] \tag{10.30}$$

For example, in the distribution of fatigue life at a constant stress, it is unrealistic to expect a minimum life of $x_0 = 0$. The easiest procedure for finding x_0 is to use the Weibull probability plot. First, plot the data as in the two-parameter case where $x_0 = 0$. Now, pick a value of x_0 between 0 and the lowest value of x_i and subtract it from each of the observed x. Continue adjusting x_0 and plotting $x - x_0$ until a straight line is obtained on the Weibull graph paper.

10.7.2 Gamma Distribution

The two-parameter gamma density function is used to describe random variables that are bounded at one end.[1] It has the form:

$$f(x) = \frac{\lambda^\eta x^{\eta-1} e^{-\lambda x}}{\Gamma(\eta)} \qquad \text{for } x > 0; \lambda > 0; \eta > 0 \qquad (10.31)$$

where $\Gamma(\eta) = $ gamma function $= \displaystyle\int_0^\infty x^{\eta-1} e^{-x}\, dx$
$\qquad \eta = $ shape factor
$\qquad \lambda = $ scale factor

Figure 10.9 illustrates the wide variety of shaped distributions that can be described by the gamma distribution. The gamma distribution models the time required for a total of η independent events to take place if the events occur at a constant rate λ. For example, in a complex engineering system, if failure occurs when η subfailures have taken place at a rate λ, the time for system failure is distributed according to the gamma distribution. The gamma distribution also is important in queuing theory (the analysis of waiting lines). It has been used as an empirical distribution to describe such

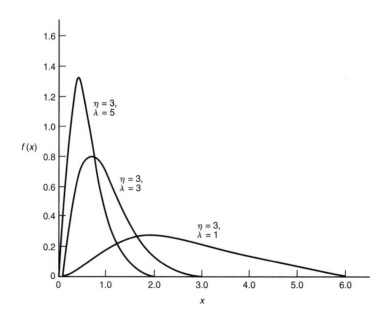

FIGURE 10.9
Gamma frequency distributions with $\eta = 3$ and various values of λ.

1. G. J. Hahn and S. S. Shapiro, "Statistical Models in Engineering," Wiley, New York, 1967.

factors as the distribution of family income and the time to failure for capacitors. The chi-square and exponential distributions are special cases of the gamma distribution.

10.7.3 Exponential Distribution

The exponential distribution is a special case of the gamma distribution for $\eta = 1$.

$$f(x) = \lambda e^{-\lambda x} \qquad \text{for } x > 0 \text{ and } \lambda > 0 \tag{10.32}$$

We also can consider the exponential distribution to be a special case of the Weibull distribution when $m = 1$, and $x_0 = 0$.

$$f(x) = \frac{1}{\theta} e^{-x/\theta} \tag{10.33}$$

The Weibull distribution models the probability of failure of a component like a bearing or a shaft, but for the complete system, such as a pump, the exponential distribution is a better model.

10.7.4 Distributions for Discrete Variables

The normal distribution and the other distributions discussed in this section deal with continuous random variables, which can take on any value over a considerable range. However, there are important engineering problems in which the random variable takes on only discrete values. Such a situation occurs at a quality-control station on a production line, where a part is either accepted or rejected.

A random variable follows a binomial distribution when the following four characteristics occur: (1) there are n trials; (2) the trials are independent; (3) each trial has only two possible outcomes; and (4) the probability of the event remains constant from trial to trial. Let the overall probability of an unsuccessful occurrence be p and the probability of a success be $q = 1 - p$. The probability of occurrence of any value of $r = 0, 1, 2, \ldots, n$ in n trials is given by

$$P(r) = \frac{n!}{r!\,(n-r)!} p^r q^{n-r} \tag{10.34}$$

The mean of a binomial distribution is given by $n \cdot p$. The variance $= npq$.

The hypergeometric distribution is related to the binomial distribution. It applies when samples are randomly drawn from a finite population without replacement, so that the probability of the event cannot remain constant from trial to trial. As with the binomial distribution, each trial has only two possible outcomes, p and q.

$$P(\text{number of defects } d) = \frac{\dbinom{N_p}{d}\dbinom{N_q}{n-d}}{\dbinom{N}{n}} \tag{10.35}$$

where N = lot size
n = sample size
N_p = number of failures in the lot
N_q = number of successes in the lot
d = number of failures in a sample

and
$$\binom{a}{b} = \frac{a!}{b!\,(a-b)!} \qquad \text{for the generalized case}$$

Note that $0! = 1$.

EXAMPLE 10.4. A parts supplier receives 12 transmissions and sells 9 of them. Later he is informed that 3 of the 12 transmissions were defective, but which three parts were defective was not specifically known. What is the probability that all three of the defective parts were sold to customers?

Solution.

$$N = 12$$
$$p = \text{fraction defective} = \tfrac{3}{12} = 0.25$$
$$q = 1 - p = 0.75$$
$$N_p = 12(0.25) = 3$$
$$N_q = 12(0.75) = 9$$
$$n = 9$$
$$d = 3$$

$$P(d=3) = \frac{\binom{3}{3}\binom{9}{6}}{\binom{12}{9}} = \frac{3!}{3!(3-3)!} \times \frac{9!}{6!(9-6)!} \times \frac{9!\,(12-9)!}{12!}$$

$$= \frac{3!}{3!\,0!} \times \frac{9!}{6!\,3!} \times \frac{9!\,3!}{12!} = 0.382$$

With both the binomial and hypergeometric distributions, we know the number of failures and successes. There are situations in which we know the number of times the event occurred but it is not practical to observe the number of times an event did not occur. The distribution of isolated events in a continuum of events is described by the Poisson distribution. If y represents the average number of occurrences of an event, the probability of observing the occurrence of r events is given by

$$P(r) = \frac{e^{-y}y^r}{r!} \qquad \text{for } r = 0, 1, 2, \dots \tag{10.36}$$

EXAMPLE 10.5. The following data were obtained for the number of breakdowns of sheet-forming presses in a stamping shop. Records were taken for 350 consecutive days. Determine the probability of having zero, one, two, three, four, or five breakdowns per day.

Number of breakdowns per day	Number of days breakdowns occurred	Number of breakdowns		
0	80	0×80	=	0
1	134	1×134	=	134
2	62	2×62	=	124
3	41	3×41	=	123
4	23	4×23	=	92
5	6	5×6	=	30
6	3	6×3	=	18
7	1	7×1	=	7
	350			528

Solution. Average breakdown per day $= \frac{528}{350} = 1.51 = y$

$$e^{-y} = e^{-1.51} = 0.22$$

$$P(0) = 0.22 \frac{1.51^{0}}{0!} = 0.22$$

$$P(1) = 0.22 \frac{1.51^{1}}{1!} = 0.33$$

$$P(2) = 0.22 \frac{1.51^{2}}{2!} = 0.25$$

$$P(3) = 0.22 \frac{1.51^{3}}{3!} = 0.13$$

$$P(4) = 0.22 \frac{1.51^{4}}{4!} = 0.05$$

$$P(5) = 0.22 \frac{1.51^{5}}{5!} = 0.01$$

10.8
SAMPLING DISTRIBUTIONS

We have already emphasized that the central problem in statistics is relating the population and the samples that are drawn from it. This problem is viewed from two perspectives: (1) what does the population tell us about the behavior of the samples drawn from it and (2) what does a sample or series of samples tell us about the population from which the sample came? The first question is the subject of this section. Section 10.9 is concerned with the second question.

Suppose we have a population consisting of 300 items. We take all possible samples of $n = 10$ and determine the mean of each sample $\bar{x}$. This would give a population of $\bar{x}$'s, one for each sample of 10. The frequency distribution of this population is called the sampling distribution of $\bar{x}$ for samples of $n = 10$. Similarly, we could have sampling distributions for s^2 or for any other sample statistic. Such sampling distributions have been worked out by exact mathematical methods for random samples from any normal distribution, but comparable information from nonnormal populations still is often incomplete.

10.8.1 Distribution of Sample Means

The mean of a sample provides an unbiased estimate of the mean of the population from which the sample was drawn. The sample values of the mean will be normally distributed about the population mean μ. The mean of the distribution of $\bar{x}$, $\mu_{\bar{x}}$ is equal to the mean of the population of x's if the population is very large. The standard deviation of the sampling distribution is called the standard error of the mean, and it is equal to

$$\sigma_{\bar{x}} = \frac{\sigma}{\sqrt{n}} \tag{10.37}$$

If, for example, $n = 36$, the scatter of $\bar{x}$ about $\mu_{\bar{x}}$ is only one-sixth what it would be for the distribution of x about μ.

We can use the above to establish the standardized variable z, which is normally distributed with a mean 0 and standard deviation 1.

$$z = \frac{\bar{x} - \mu}{\sigma_{\bar{x}}} = \frac{\bar{x} - \mu}{\sigma/\sqrt{n}} \tag{10.38}$$

10.8.2 t Distribution

It is a common situation not to know the standard deviation of the population. When we do not know it, we cannot substitute s for σ in Eq. (10.38) and assume that the statistic will be normally distributed, unless n is greater than about 30. However, if x is normally distributed, the sampling distribution for the mean is the t distribution:

$$t = \frac{\bar{x} - \mu}{s/\sqrt{n}} \qquad \text{for } v = n - 1 \text{ degrees of freedom} \tag{10.39}$$

Values of the t distribution are given in the Appendix Table B-2. The distribution is symmetrical, with a different curve for each value of v. As v increases, the distribution of t approaches the normal distribution.

10.8.3 Distribution of Sample Variances

The distribution of sample variances s^2 from a normal population with a variance σ^2 is given by the *chi-square distribution*.

TABLE 10.3
Summary of statistical tests of hypotheses

Hypothesis	Conditions	Test statistic	Distribution of test statistic	Alternative hypothesis	Reject null hypothesis
1. $\mu = \mu_0$	σ known	$z = \dfrac{\bar{x} - \mu_0}{\sigma}\sqrt{n}$	Normal (0, 1)	$\mu \neq \mu_0$	$z < -z_{\alpha/2}$ $z > z_{\alpha/2}$
				$\mu < \mu_0$	$z < -z_\alpha$
				$\mu > \mu_0$	$z > z_\alpha$
2. $\mu = \mu_0$	σ unknown	$t = \dfrac{\bar{x} - \mu_0}{s/\sqrt{n}}$	t distribution $v = n - 1$	$\mu \neq \mu_0$	$t < -t_{\alpha/2}$ $t > t_{\alpha/2}$
				$\mu < \mu_0$	$t < -t_\alpha$
				$\mu > \mu_0$	$t > t_\alpha$
3. $\sigma = \sigma_0$	μ unknown	$\chi^2 = (n-1)s^2/\sigma_0^2$	chi-square $v = n - 1$	$\sigma \neq \sigma_0$	$\chi^2 < \chi^2_{\alpha/2}$ $\chi^2 > \chi^2_{1-\alpha/2}$
				$\sigma < \sigma_0$	$\chi^2 < \chi^2_\alpha$
				$\sigma > \sigma_0$	$\chi^2 > \chi^2_{1-\alpha}$
4. $\mu_1 - \mu_2 = \delta$	σ_1 and σ_2 known	$z = \dfrac{(\bar{x}_1 - \bar{x}_2) - \delta}{\sigma_{\bar{x}_1 - \bar{x}_2}}$ $\sigma_{\bar{x}_1 - \bar{x}_2} = \left(\dfrac{\sigma_1^2}{n_1} + \dfrac{\sigma_2^2}{n_2}\right)^{1/2}$	Normal (0, 1)	$(\mu_1 - \mu_2) \neq \delta$	$z < -z_{1-\alpha/2}$ $z > z_{1-\alpha/2}$
				$(\mu_1 - \mu_2) < \delta$	$z < z_{1-\alpha}$
				$(\mu_1 - \mu_2) > \delta$	$z > z_{1-\alpha}$
5. $\sigma_1 = \sigma_2$	x's normal and independent $F < F_{1-\alpha/2}$	$F = s_1^2/s_2^2$	F_{v_1, v_2}	$\sigma_1 \neq \sigma_1$	$F > F_{1-\alpha/2}$
				$\sigma_1 > \sigma_2$	$F > F_{1-\alpha}$

6. $\mu_1 - \mu_2 = \delta$ σ unknown but $\sigma_1 = \sigma_2$

$$t = \frac{(\bar{x}_1 - \bar{x}_2) - \delta}{s_{\bar{x}_1 - \bar{x}_2}}$$

$$s_{\bar{x}_1 - \bar{x}_2} = \left[s_p^2 \left(\frac{1}{n_1} + \frac{1}{n_2} \right) \right]^{1/2}$$

$$s_p^2 = \frac{(n_1 - 1)s_1^2 + (n_2 - 1)s_2^2}{n_1 + n_2 - 2}$$

t distribution
$v = n_1 + n_2 - 2$

$(\mu_1 - \mu_2) \neq \delta$	$t < -t_{\alpha/2}$ $t > t_{\alpha/2}$
$(\mu_1 - \mu_2) > \delta$	$t > t_\alpha$
$(\mu_1 - \mu_2) < \delta$	$t < -t_\alpha$

---

7. $\mu_1 - \mu_2 = \delta$ σ unknown and $\sigma_1 \neq \sigma_2$

$$t = \frac{(\bar{x}_1 - \bar{x}_2) - \delta}{\left(\dfrac{s_1^2}{n_1} + \dfrac{s_2^2}{n_2} \right)^{1/2}}$$

t distribution

$$v = \frac{1}{\left(\dfrac{k^2}{v_1} \right) + \dfrac{(1-k)^2}{v_2}}$$

$$k = \frac{s_1^2/n_1}{s_1^2/n_1 + s_2^2/n_2}$$

$(\mu_1 - \mu_2) \neq \delta$	$t < -t_{\alpha/2}$ $t > t_{\alpha/2}$
$(\mu_1 - \mu_2) < \delta$	$t < -t_\alpha$
$(\mu_1 - \mu_2) > \delta$	$t > t_\alpha$

---

8. $\mu_{1j} - \mu_{2j} = \mu_{dj} = 0$ Matched pairs; equal variability

$$t = \frac{\bar{d}}{s_d/\sqrt{n}} \qquad \bar{d} = \frac{\sum d}{n}$$

$$s_d = \frac{\sum d^2 - (\sum d)^2/n}{n-1}$$

t distribution
$v = n - 1$, where n is the number of pairs

$\mu_{dj} = \mu_d \neq 0$	$t > t_{\alpha/2}$ $t < -t_{\alpha/2}$
$\mu_{dj} = \mu_d > 0$	$t > t_\alpha$

H_0: $\mu = \mu_0 = 10{,}000$ $\alpha = 0.05$

H_1: $\mu < \mu_0 = 10{,}000$

The critical region for rejecting H_0 is $t < -t_{0.05}$, or from Appendix B, if $t < 1.90$ for $8 - 1$ $= 7$ degrees of freedom,

$$t = \frac{\bar{x} - \mu_0}{s/\sqrt{n}} = \frac{9250 - 10{,}000}{110\sqrt{8}} = -19.3$$

Therefore, we reject H_0 and conclude that the sample did not come from a population with $\mu = 10{,}000$ lb.

EXAMPLE 10.8. The manager of a quality-control lab is concerned with whether two stress rupture machines are giving reliable test results. Six specimens of the same material are run on each machine at the same temperature and stress level. The logarithm of the rupture time is used as the test response. The results for the two machines are:

Machine 1	Machine 2
$\bar{x}_1 = 0.91$	$\bar{x}_2 = 1.78$
$s_1 = 0.538$	$s_2 = 0.40$
$n_1 = 6$	$n_2 = 6$

Does the manager have cause for concern?

Solution. Since σ is not known, we must first establish whether there is a significant difference in the variances of the two machines

$$F = \frac{s_1^2}{s_2^2} = \frac{0.538^2}{0.40^2} = \frac{0.289}{0.160} = 1.88$$

From a table of the F distribution for $\nu_1 = \nu_2 = 5$, we find the critical region is $F > 11.0$ for $\alpha = 0.01$. Therefore, $\sigma_1 = \sigma_2$ and we can follow entry 6 in Table 10.3.

H_0: $\mu_1 - \mu_2 = \delta = 0$

H_1: $\mu_1 - \mu_2 \neq 0$ $\alpha = 0.05$

The critical region is $t < -t_{\alpha/2}$ and $t > t_{\alpha/2}$, which for $\nu = n_1 + n_2 - 2$ is $t < -2.23$ and $t > 2.23$.

$$t = \frac{(\bar{x}_1 - \bar{x}_2) - \delta}{s_{\bar{x}_1 - \bar{x}_2}} \qquad s_{\bar{x}_1 - \bar{x}_2} = \left[s_p^2 \left(\frac{1}{n_1} + \frac{1}{n_2} \right) \right]^{1/2}$$

$$s_p^2 = \frac{(n_1 - 1)s_1^2 + (n_2 - 1)s_2^2}{n_1 + n_2 - 2} = \frac{5(0.289) + 5(0.16)}{10} = 0.225$$

$$s_{\bar{x}_1 - \bar{x}_2} = \left[0.225 \left(\frac{1}{6} + \frac{1}{6} \right) \right]^{1/2} = [0.225(0.333)]^{1/2} = 0.274$$

$$t = \frac{(0.91 - 1.78) - 0}{0.274} = \frac{-0.87}{0.274} = -3.17$$

Since the calculated value of t is well into the critical region, we reject the null hypothesis and conclude that the two stress rupture machines do not give comparable results.

10.10
STATISTICAL INTERVALS

Interval estimation is commonly used to make probability statements about the population from which a sample has been drawn or to predict the results of a future sample from the same population. A common method is to determine the confidence limits for a parameter so that we can have a specified degree of confidence that the parameter lies within the interval. For example, if we have determined the 90 percent confidence limits of the mean, then, in the long run, the true mean of the population will lie within the limits 90 percent of the time. A number of statistical intervals have been established.

10.10.1 Confidence Interval

The confidence interval on the mean gives limits that can be claimed with a $(1 - \alpha)$ 100 percent degree of confidence to contain the unknown value of the population mean. When σ is known, the two-sided confidence interval is

$$\bar{x} - z_{\alpha/2} \frac{\sigma}{\sqrt{n}} < \mu < \bar{x} + z_{\alpha/2} \frac{\sigma}{\sqrt{n}} \tag{10.42}$$

For the usual case in which the population standard deviation is unknown, we must use the t distribution with $n - 1$ degrees of freedom.

$$\bar{x} - t_{\alpha/2} \frac{s}{\sqrt{n}} < \mu < \bar{x} + t_{\alpha/2} \frac{s}{\sqrt{n}} \tag{10.43}$$

The confidence interval on the variance, from Sec. 10.8, is given by

$$(n - 1) \frac{s^2}{\chi^2_{1-\alpha/2}} < \sigma^2 < (n - 1) \frac{s^2}{\chi^2_{\alpha/2}} \tag{10.44}$$

10.10.2 Tolerance Interval

An interval that can be claimed to contain at least a specified proportion p of the population with a confidence $(1 - \alpha)$ 100 percent is known as a statistical tolerance interval. This type of interval is especially useful in quality-control situations in mass production. For example, if we have obtained data on a random sample of 30 ball bearings, a tolerance interval will provide limits to specify a given proportion, for example, $p = 0.8$, of the population of bearings from which the sample was selected.

10.10.3 Prediction Interval

A *prediction interval* will contain all of k future observations with a confidence level $(1 - \alpha)$ 100 percent. Alternatively, the prediction interval could be selected to contain the mean of k future observations. Prediction intervals often are of interest to manufacturers of equipment that is produced in small lots. Thus, a prediction interval that contains all future k items of production would be required to establish limits on performance for a small shipment of equipment when the satisfactory performance of all units is to be guaranteed.

The statistical interval can be determined from

$$\bar{x} \pm cs \tag{10.45}$$

Details on how to compute the appropriate value of c are given by Hahn.[1] The appropriate factors for two-sided 95 percent intervals are given in Fig. 10.11.

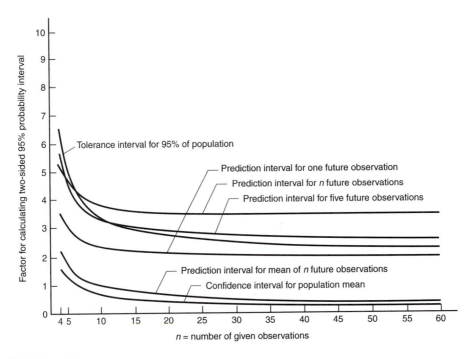

FIGURE 10.11
Comparison of c factor for calculating some two-sided 95 percent probability intervals. (*G. J. Hahn, J. Qual Tech., vol. 2 no. 3, p. 117, 1970.*)

---

1. G. J. Hahn, *J. Qual. Tech.*, part I, vol. 2, no. 3, pp. 115–125, 1970; part II, vol. 2, no. 4, pp. 195–206, 1970.

TABLE 10.4
Rejection of outliers by Chauvenet's criterion

Number of observations n	Ratio of max. acceptable deviation to standard deviation d_{max}/σ
2	1.15
3	1.38
4	1.54
5	1.65
6	1.73
7	1.80
10	1.96
15	2.13
25	2.33
50	2.57
100	2.81

10.10.4 Rejection of Outliers

It is not uncommon to find a bad data point in a set of observations, i.e., a point that "looks out of place" when compared with the bulk of the data. It is incorrect just to discard the outlier, but statistics can lend assistance on this point.[1]

If we take n measurements, and n is large enough that we can expect a Gaussian error distribution, then we can use the distribution to compute the probability that a given point will deviate a certain amount from the mean. It would be unlikely to expect the outlier to occur with a probability less than $1/n$. Thus, if the probability for the observed deviation of the outlier is less than $1/n$, it confirms our suspicions that this an extreme observation. There are a number of statistical criteria for rejecting outliers. The *Chauvenet criterion* specifies that a reading may be rejected if the probability of obtaining the particular deviation from the mean is less than $1/2n$. Table 10.4 gives the value of the deviation from the point to the mean that must be exceeded in order to reject the point according to the Chauvenet criterion. Once all the outliers are rejected, a new mean and standard deviation are calculated for the sample.

10.11
ANALYSIS OF VARIANCE

Section 10.9 provided methods for making statistically based decisions between two samples or treatments. Now we shall consider the situation in which we have three or more treatments. The statistical procedure is called the analysis of variance (ANOVA). An understanding of ANOVA will be important later when we consider the important topic of design of experiments. With ANOVA we determine:

---

1. See M. G. Natrella, op. cit., chap. 17.

1. The total spread of results *between* the different treatments
2. The spread of results *within* each treatment

The variability between treatments is compared with the variability within each treatment.

10.11.1 One-Way Classification

Consider data for a single factor observed at k levels or treatments, each containing n observations. Let the jth observation of the ith treatment be y_{ij}

Treatment	Observations	Totals	Mean
1	$y_{11}\, y_{12} \cdots y_{1j} \cdots y_{1n}$	$y_1.$	$\bar{y}_1.$
2	$y_{21}\, y_{22} \cdots y_{2j} \cdots y_{2n}$		
i	$y_{i1}\, y_{i2} \cdots y_{ij} \cdots y_{in}$	$y_i.$	$\bar{y}_i.$
k	$y_{k1}\, y_{k2} \cdots y_{kj} \cdots y_{kn}$	$y_k.$	$\bar{y}_k.$

$$y_{i.} = \sum_{j=1}^{n} y_{ij} \quad \bar{y}_{i.} = y_{i.}/n \quad i = 1, 2, \ldots k$$

Thus, the "dot" subscript notation implies summation over the subscript it replaces. Therefore, the grand mean $\bar{y}..$ is given by

$$y.. = \sum_{i=1}^{k} \sum_{j=1}^{n} y_{ij} \text{ and } \bar{y}.. = y../kn$$

To test the null hypothesis that the means of the k treatments are equal, we compare the variance based on the variation *between* the treatment means with the variance based on variation *within* the treatments. The total variability of the observations is described by the *total sum of squares.*

$$SST = \sum_{i=1}^{k} \sum_{j=1}^{n} (y_{ij} - \bar{y}..)^2$$

The total sum of squares can be partitioned into a *treatment sum of squares* (SSTr) and an *error sum of squares* (SSE).

$$SST = SSTr + SSE$$

where $$SSTr = n\sum_{i=1}^{k} (\bar{y}_{i.} - \bar{y}..)^2 \text{ and } SSE = \sum_{i=1}^{k} \sum_{j=1}^{n} (y_{ij} - \bar{y}_{i.})^2$$

We see that SSTr is based on the difference between the treatment means and the grand mean, while SSE is based on the difference between an observation within a treatment and the treatment mean.

The null hypothesis is tested with the F statistic using the *mean square*, which is the variance divided by the appropriate number of degrees of freedom.

$$F = \frac{MS_{Tr}}{MS_E} = \frac{SSTr/(k-1)}{SSE/k(n-1)}$$

If the ratio given by the above equation is larger than the F statistic for $F_{\alpha, k-1, k(n-1)}$ then we reject the null hypothesis and conclude that the variable y is influenced by the level of treatment.

10.11.2 Two-Way Classification

Consider $1, 2, \ldots, a$ treatments divided between $1, 2, \ldots, b$ blocks. A block is a portion of an experimental design that is expected to be more homogeneous than the total design space. Greater precision is obtained by confining treatment comparisons to within blocks.

	B_1	B_2	$\cdots$	B_j	$\cdots$	B_b	Mean
Treatment 1:	y_{11}	y_{12}	$\cdots$	y_{1j}	$\cdots$	y_{1b}	$\bar{y}_{1.}$
Treatment 2:	y_{21}	y_{22}	$\cdots$	y_{2j}	$\cdots$	y_{2b}	$\bar{y}_{2.}$
Treatment i:	y_{i1}	y_{i2}	$\cdots$	y_{ij}	$\cdots$	y_{ib}	$\bar{y}_{i.}$
Treatment a:	y_{a1}	y_{a2}	$\cdots$	y_{aj}	$\cdots$	y_{ab}	$\bar{y}_{a.}$
Mean:	$\bar{y}_{.1}$	$\bar{y}_{.2}$		$\bar{y}_{.j}$		$\bar{y}_{.b}$	$\bar{y}_{..}$

The sum of the squares is given by

$$\underset{\text{SST}}{\sum_{i=1}^{a}\sum_{j=1}^{b}(y_{ij}-\bar{y}_{..})^2} = \underset{\text{SSE}}{\sum_{i=1}^{a}\sum_{j=i}^{b}(y_{ij}-\bar{y}_{i.}-\bar{y}_{.j}+\bar{y}_{..})^2} + \underset{\text{SS(Tr)}}{b\sum_{i=1}^{a}(y_{i.}-\bar{y}_{..})^2}$$

$$+ \underset{\text{SS(Bl)}}{a\sum_{j=1}^{b}(y_{.j}-\bar{y}_{..})^2}$$

EXAMPLE 10.9. A sensitive measure of the quality of a steel forging is the reduction of area of a tensile specimen that is cut from a direction normal to the forging fiber, i.e., the transverse reduction of area, abbreviated RAT. Table 10.5 gives the values of RAT that were determined from forgings with 10:1 and 30:1 reduction from the ingot, with and without a homogenization heat treatment. We are interested in answering the following questions.

1. Is there a significant difference in RAT for 10 to 1 and 30 to 1 forging reductions?
2. Is there a significant difference in RAT between two forgings one of which was given a homogenization treatment and the other was not?
3. Is there a significant interaction[1] between forging reduction and homogenization that affects the level of RAT?

---

1. An interaction is present if the observed values for different levels of one factor are altered by the presence of another factor.

TABLE 10.5
Effect of forging reduction and homogenization treatment on
transverse reduction of area (RAT)

Forging reduction	Homogenization		
	None	2400°F, 50 h	
	25.8	26.4	
	30.4	35.6	
	28.6	30.2	
10:1	35.6	21.4	
	20.3	27.2	
	$T = 140.7$	$T = 140.8$	$T_{i.} = 281.5$
	22.8	30.6	
	19.5	35.3	
	30.6	27.9	
30:1	28.5	28.2	
	25.5	31.5	
	$T = 126.9$	$T = 153.5$	$T_{i.} = 280.4$
	$T_{.j} = 267.8$	$T_{.j} = 294.3$	$T_{..} = 561.9$

The experiment outlined in Table 10.5 is a two-way classification with two columns and two rows and with five replicate measurements of RAT for each combination of forging reduction and homogenization condition. Therefore, $a = 2$, $b = 2$, $r = 5$, and $n = 2 \times 2 \times 5 = 20$.

$$\text{SST} = \Sigma \, y_{ij}^2 - \frac{T_{...}^2}{abr} = 16{,}208.67 - \frac{561.9^2}{2(2)(5)} = 422.09$$

The treatment sum of squares (between rows) is

$$\text{SS(Tr)} = \Sigma \, \frac{T_{i.}^2}{br} - \frac{T_{...}^2}{abr} = \frac{281.5^2 + 280.4^2}{2(5)} - \frac{561.9^2}{2(2)(5)} = 0.06$$

The blocks sum of squares (between columns) is

$$\text{SS(Bl)} = \Sigma \, \frac{T_{.j}^2}{ar} - \frac{T_{...}^2}{abr} = \frac{267.8^2 + 294.3^2}{2(5)} - \frac{561.9^2}{2(2)(5)} = 46.35$$

By introducing replication, we are able to separate out the variation due to interaction effects from that due to random error. The sum of the squares due to random error within each test condition (within cells) is determined by first determining the subtotal sum of squares (SSS).

$$\text{SSS} = \Sigma \, \frac{T^2}{r} - \frac{T_{...}^2}{abr} = \frac{140.7^2 + 140.8^2 + 126.9^2 + 153.5^2}{5} - \frac{561.9^2}{2(2)(5)} = 70.82$$

Then, the within-cells sum of squares is

TABLE 10.6
Analysis of variance table

Source of variability	Sum of squares	Degrees of freedom	Mean square
Forging reduction (between rows)	0.06	1	0.06
Homogenization (between columns)	46.35	1	46.35
Interaction	24.41	1	24.41
Within cells (error)	351.27	16	21.95
Total		19	

$$\text{SSWC} = \text{SST} - \text{SSS} = 422.09 - 70.82 = 351.27$$

Finally, the interaction sum of squares is given by

$$\text{SSI} = \text{SSS} - \text{SS(Tr)} - \text{SS}(Bl) = 70.82 - 0.06 - 46.35 = 24.41$$

Once these computations are made, the results are entered into an analysis of variance table (ANOVA table), Table 10.6.

The values of mean square are obtained by dividing the values of sum of the squares by their respective degrees of freedom. We use the mean squares in the F test to determine whether there are significant differences. For example, to determine whether there is a significant effect of homogenization on the level of RAT, the following F ratio is determined.

$$F = \frac{\text{between columns mean square}}{\text{within cells (error) mean square}} = \frac{46.35}{21.95} = 2.11$$

$$F_{0.05,1,16} = 4.45$$

Since the calculated F ratio does not exceed 4.45, the effect of homogenization is not significant at the 95 percent level.

To determine whether there is a significant interaction between forging reduction and homogenization treatment, the following F ratio is determined.

$$F = \frac{\text{interaction mean square}}{\text{error mean square}} = \frac{24.41}{21.95} = 1.09$$

Once again the effect is not significant at the 5 percent level.

10.12
STATISTICAL DESIGN OF EXPERIMENTS (DOE)

The greatest benefit can be gained from statistical analysis when the experiments are planned in advance so that data are taken in a way that will provide the most unbiased and precise results commensurate with the desired expenditure of time and money. This can best be done through the combined efforts of a statistician and the engineer during the planning stage of the project.

Probably the most important benefit from statistically designed experiments is that more information per experiment will be obtained that way than with unplanned experimentation. A second benefit is that statistical design results in an organized approach to the collection and analysis of information. Conclusions from a statistically designed experiment very often are evident without extensive statistical analysis, whereas with a haphazard approach the results often are difficult to extract from the experiment even after detailed statistical analysis. Still another advantage of statistical planning is the credibility that is given to the conclusions of an experimental program when the variability and sources of experimental error are made clear by statistical analysis. Finally, an important benefit of statistical design is the ability to discover interactions between experimental variables.

A major impetus to the use of experimental design has been given by the strong emphasis being shown in design for quality. The techniques developed in Japan by Genichi Taguchi employ experimental design methods in a unique way to produce *robust designs*. This chapter sets the statistical basis for considering this and other concepts of quality engineering in Chap. 12.

In any experimental program involving a large number of tests, it is important to randomize the order in which the specimens are selected for testing. By randomization we permit any one of the many specimens involved in the experiment to have an equal chance of being selected for a given test. In this way, bias due to uncontrolled second-order variables is minimized. For example, in any extended testing program errors can arise over a period of time owing to subtle changes in the characteristics of the testing equipment or in the proficiency of the operator of the test. In taking metal specimens from large forgings or ingots the possibility of the variation of properties with position in the forging must be considered. If the objective of the test is to measure the average properties of the entire forging, randomization of the test specimens will minimize variability due to position in the forging.

One way to randomize a batch of specimens is to assign a number to each specimen, put a set of numbered tags corresponding to the specimen numbers in a jar, mix the tags thoroughly, and then withdraw the tags from the jar. Each tag should be placed back in the jar after it is withdrawn to allow an equal probability of selecting that number. The considerable labor involved in this procedure can be minimized by using a table of random numbers.

We have previously considered the various stages in the design process. In the same way, any experimental investigation starts with preliminary familiarization experiments. These may take the form of trying to duplicate earlier results or learning to operate new equipment. Much of the work at this stage is mainly intuitive and aimed at defining the problem; generally, statistical design is not very applicable. However, after the goals of the investigation are better defined, the next step is to reduce the large number of possible variables to the few most important ones. Here statistical design can be very useful. As the experimentation moves into the optimization stage, the number of variables has been reduced to only a few. Once again statistical design can be effective in finding the optimum values for the variables. Finally, statistical analysis can assist in choosing between possible models for the process or phenomenon under study.

The response variables are the data we obtain from an experimental run. Responses can be classified as quantitative, qualitative, or quantal. A quantitative response, which is measured by a continuous scale, is the most common and easiest to work with in statistical analysis. Qualitative responses, like luster or odor, can be ranked on an ordinal scale, for example, 0 (worst) to 10 (best). Quantal or binary responses produce one of two values, go or no-go, pass or fail. An example is fatigue specimens tested near the fatigue limit that either fail or run out. The case of the quantal response is treated by special statistical techniques.[1]

Factors are experimental variables that are controlled by the investigator. An important part of planning an experimental program is identifying the important variables that affect the response and deciding how to exploit them in the experiment. This is best done jointly by the engineer and the statistician. However, identifying the variables, and assuring that no important factors are forgotten, is the responsibility of the engineer. Frequently the scientific model of the problem is examined for important variables. Previous experience should always be called upon. We often take advantage of dimensional analysis in establishing the factors.

Factors may be *independent* in the sense that the level of one factor is independent of the level of the other factors. However, two or more factors may *interact* with one another, i.e., the effect on the response of one variable depends upon the levels of the other variables. Figure 10.12 illustrates different types of behavior between the

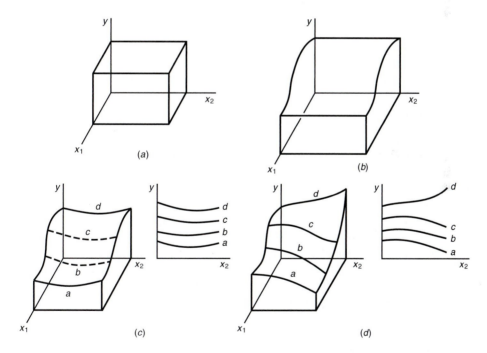

FIGURE 10.12
Different behavior of response y as a function of the factors x_1 and x_2. (*a*) No effect of x_1 and x_2 on y. (*b*) Main effect of x_1 on y. No effect of x_2 on y. (*c*) Effect of x_1 and x_2 on y but no $x_1 - x_2$ interaction. (*d*) Main effects of x_1 and x_2. Interaction between x_1 and x_2.

1. D. J. Finney, "Probit Analysis," 3d ed., Cambridge University Press, New York, 1971; "Metals Handbook," 9th ed., vol. 8, pp. 695–720, ASM International, Metals Park, OH, 1985.

factors and the response. In this case the response y is the yield strength of an alloy as it is influenced by two factors, temperature x_1 and aging time x_2. Interactions between the factors are determined by varying factors simultaneously under statistical control rather than one at a time.

In addition to the primary variables that are under the control of the experimenter, there are other variables, which may not be under strict experimental control. An example would be subtle differences in the way different machine operators run an experiment or carry out a test or slight differences in humidity or other environmental factors. The use of randomization of the test runs so as to remove unconscious bias in results has been discussed above. Also, the use of an experimental block to produce relatively homogeneous test conditions has been mentioned in Sec. 10.9. When an experiment is run with blocking, the effect of a background variable is removed from the experimental error, but randomization alone does not usually produce this effect.

It is important to realize that not all primary factors may be capable of variation with equal facility. Thus, completely randomizing the sequence of testing may be impractical. Often the final experimental plan is a compromise between the information that can be obtained and the cost of the information. In developing the experimental design, physical reality should take precedence over slavish adherence to statistics. Whenever faced with this decision, the experimental design should be modified to accommodate the real-world situation, not the reverse.

It is important to make some initial estimates of the overall repeatability before embarking on a major experimental test program. Often this information is available from previous experiments; but if it is not, it would be wise to conduct some preliminary experiments under identical conditions over a reasonable time period. If these experiments show large variability in response, then the primary variables may not have been properly identified.

It is not necessary to conduct a statistically designed experiment all the way through to completion. In fact, there are advantages to conducting the experimental program in stages. That permits changes to be made in later tests based on the information gained from early results. Conducting the experiment in stages is attractive when the experimenter is searching for an optimum response, since it takes investigation closer to the optimum at each stage. On the other hand, if there are large start-up costs for each stage or if there is a long time delay between preparing the samples and measuring their performance (as in agricultural research), then carrying the designed experiment straight through to completion may be preferred.

In general, there are three classes of statistically designed experiments.

1. Blocking designs use blocking techniques to remove the effect of background variables from the experimental error. The most common designs are the randomized block plan and the balanced incomplete block,[1] which remove the effect of a single extraneous variable. The Latin square and the Youden square designs remove the effects of two extraneous variables. Graeco-Latin square and hyper-Latin square designs remove the effects of three or more extraneous variables.

---

1. M. G. Natrella, op. cit., pp. 13-1 to 13-29.

2. Factorial designs are experiments in which all levels of each factor in an experiment are combined with all levels of every other factor. This important class of experimental design is discussed in Sec. 10.13.
3. Response surface designs are used to determine the empirical functional relation between the factors (independent variables) and the response (performance variable). The central composite design and rotatable designs are frequently used for this purpose. This is discussed in Sec. 10.15.

We emphasize that a working partnership between the engineer and the statistician leads to the best results. Frequently the best experimental design is specially made for the problem and deviates significantly from the standard textbook designs. By getting involved in the project at an early stage, the statistician can help evolve an experimental design to minimize testing effort and maximize information content.

10.13
FACTORIAL DESIGN

A factorial experiment is one in which we control several factors and investigate their effects at each of two or more levels. The experimental design consists of making an observation at each of all possible combinations that can be formed for the different levels of the factors. Each different combination is called a treatment combination. The approach in a factorial experiment is much different than in the traditional experiment, in which all factors but one are held constant.

The simplest, and most common type of factorial design is one that uses two levels, i.e., a 2^n factorial design. Not only does it reduce the number of experimental conditions but convenient computational methods exist for the 2^n case. A disadvantage is that with only two levels it is not possible to distinguish between linear and higher-order effects.

In a 2^n experimental design, factors that are set at the low level are indicated $(-)$ and those at the high level $(+)$. The particular combination of treatments can be easily determined by using the special Yates notation. Each factor is represented by a lowercase letter arranged in a standard order. When a lowercase letter representing a factor is missing, it indicates that the factor is at a lower level. Otherwise the factor is at the higher level. Table 10.7 illustrates this for a three-factor, 2^3, design.

The main effect of a factor is the change in response produced by a change in level of the factor.

$$\text{Main effect of } x_i = \frac{\Sigma \text{ responses at high } x_i - \Sigma \text{ responses at low } x_i}{\text{half the number of runs in experiment}}$$

Thus, the main effect of factor x_B is

$$\text{Main effect } x_B = \frac{(y_b + y_{ab} + y_{bc} + y_{abc}) - (y_1 + y_a + y_c + y_{ac})}{4}$$

To find the interaction effects, extend Table 10.7 as shown in Table 10.8. Then, the effect for the BC interaction is written as:

$$BC \text{ interaction} = \frac{(y_1 + y_a + y_{bc} + y_{abc}) - (y_b + y_{ab} + y_c + y_{bc})}{4}$$

The main effects and interactions for a 2^3 factorial design are given by the following equations, where $a = y_a$, $b = y_b$, etc.

$$x_A = \frac{1}{4}(a - 1)(b + 1)(c + 1)$$

$$x_B = \frac{1}{4}(a + 1)(b - 1)(c + 1)$$

$$x_C = \frac{1}{4}(a + 1)(b + 1)(c - 1)$$

$$x_{AB} = \frac{1}{4}(a - 1)(b - 1)(c + 1) \qquad (10.46)$$

$$x_{AC} = \frac{1}{4}(a - 1)(b + 1)(c - 1)$$

$$x_{BC} = \frac{1}{4}(a + 1)(b - 1)(c - 1)$$

$$x_{ABC} = \frac{1}{4}(a - 1)(b - 1)(c - 1)$$

TABLE 10.7
Three-factor, 2^3, a factorial design

Yates std. order	Run no.	Level of factor		
		x_A	x_B	x_C
1	1	−	−	−
a	2	+	−	−
b	3	−	+	−
ab*	4	+	+	−
c	5	−	−	+
ac*	6	+	−	+
bc*	7	−	+	+
abc†	8	+	+	+

*First-order interaction.
†Second-order interaction.

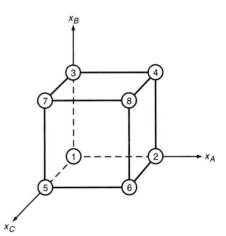

TABLE 10.8
Interactions for 2^3 factorial design

Yates std. order	Level of factor						
	x_A	x_B	x_C	$x_A x_B$	$x_A x_C$	$x_B x_C$	x_{ABC}
(1)	−	−	−	+	+	+	−
a	+	−	−	−	−	+	+
b	−	+	−	−	+	−	+
ab	+	+	−	+	−	−	−
c	−	−	+	+	−	−	+
ac	+	−	+	−	+	−	−
bc	−	+	+	−	−	+	−
abc	+	+	+	+	+	+	+

These equations can be generalized for a 2^n factorial design with $A, B, C, \ldots, Q$ factors as

$$x_A = \frac{1}{2}^{n-1}(a-1)(b+1)(c+1)\cdots(q+1)$$

$$x_{AB} = \frac{1}{2}^{n-1}(a-1)(b-1)(c+1)\cdots(q+1)$$

$$x_{ABC\cdots Q} = \frac{1}{2}^{n-1}(a-1)(b-1)(c-1)\cdots(q-1)$$

EXAMPLE 10.10. The strength of steel oil well drill pipe is controlled by the austenitization temperature A, the temperature at which it enters the tube mill B, and the amount of deformation produced in the mill C. A 2^3 factorial design is developed. The tensile strength of the steel tube, in 1000 psi, is the measured response y.

Treatment	Response UTS (y)	x_A	x_B	x_C	x_{AB}	x_{AC}	x_{BC}	x_{ABC}
(1)	151	−	−	−	+	+	+	−
a	147	+	−	−	−	−	+	+
b	147	−	+	−	−	+	−	+
ab	101	+	+	−	+	−	−	−
c	182	−	−	+	+	−	−	+
ac	159	+	−	+	−	+	−	−
bc	173	−	+	+	−	−	+	−
abc	158	+	+	+	+	+	+	+

The main effect of factor A

$$A = \frac{(147 + 101 + 159 + 158) - (151 + 147 + 182 + 173)}{4}$$

$$= 141 - 163 = -22$$

The minus sign signifies that the lower level of austenization temperature (factor A) resulted in higher values of the response.

We can compare the mean value of factor A at its high level with the mean at its low level by using the paired t test, Table 10.3, Entry 8.

Low A	High A	Difference	Deviation from $\bar{d}$
151	147	4	−18
147	101	46	24
182	159	23	1
173	158	15	−7
		$\bar{d} = \dfrac{88}{4} = 22$	

$$s^2 = \frac{\Sigma(deviation)^2}{n-1} = \frac{(-18)^2 + (24)^2 + (1)^2 + (-7)^2}{3} = 316.67$$

$$s = 17.79$$

$H_0: \quad \mu = 0 \qquad \text{for } \bar{d} = \bar{x}_L - \bar{x}_H \qquad t = \frac{\bar{d}}{s/\sqrt{n}} = \frac{22}{17.79/\sqrt{4}} = 2.47$

$H_1: \quad \mu \neq 0 \qquad\qquad\qquad\qquad\qquad\qquad v = 4 - 1 = 3$

For the two-sided t distribution (Appendix B) $t_{0.10} = 2.35$ and $t_{0.05} = 3.18$. Thus, the null hypothesis is rejected at a 10 percent level of significance.

To increase the precision of the experiment, we perform duplicate observations for each treatment combination. This is called replicating the experiment. For each of the eight treatments there are duplicate observations of the response, y_1 and y_2. The standard deviation of these duplicates about their own mean is $s = [(y_1 - y_2)/2]^{1/2}$. These individual deviations are combined into a pooled standard deviation s_p.

$$s_p = \left[\frac{\dfrac{(y_1-y_2)_1^2}{2} + \dfrac{(y_1-y_2)_2^2}{2} + \dfrac{(y_1-y_2)_3^2}{2} + \cdots + \dfrac{(y_1-y_2)_8^2}{2}}{2 + 2 + 2 + \cdots + 2} \right]^{1/2}$$

and in general, for n duplicates

$$s_p = \left[\frac{\Sigma\,(y_1-y_2)^2}{2n} \right]^{1/2} \tag{10.47}$$

When duplicate observations are made for the 2^3 experiment (data not given), $x_p = 4.8$. The mean difference between the low and high levels of factor A has changed slightly from

$\bar{x}_L - \bar{x}_H = 22$ to 20. We observe a mean difference of 20 ksi based on two groups of tests each with eight observations. The standard error of the mean difference is given by

$$s_{\bar{x}-\bar{x}} = \left(\frac{4.8^2}{8} + \frac{4.8^2}{8}\right)^{1/2} = 2.4$$

Now, applying the null hypothesis for paired observations,

$$t = \frac{\bar{d}}{s_{\bar{x}-\bar{x}}} = \frac{20}{2.4} = 8.35 \quad v = 8 - 1 = 7$$

$t_{0.01} = 3.50$, so the null hypothesis is rejected with a very high (0.999) level of confidence. The main effect of austenitization temperature is shown to be highly significant. This illustrates how replication greatly increases the precision of the statistical analysis.

Let us now look at the interaction between austenitization temperature A and the temperature of working in the tube mill B. The AB interaction is given by

$$AB \text{ interaction} = \frac{(151 + 101 + 182 + 158) - (147 + 147 + 159 + 173)}{4}$$

$$= \frac{592 - 626}{4} = \frac{-34}{4} = -8.5$$

Applying the null hypothesis with the paired t test

$$t = \frac{\bar{d}}{s_{\bar{x}-\bar{x}}} = \frac{-8.5 - 0}{2.4} = 3.54 \quad v = 7$$

$$t_{0.01} = 3.50$$

so the AB interaction is highly statistically significant.

10.13.1 Size of the Experiment

In a 2^n design all of the treatment combinations are used in each estimation of main effect and interaction. However, it is important to have some way to estimate the total number of runs N required to achieve a desired level of precision.[1]

We first need to decide on the minimum change in the response that will be of practical interest Δ. Then σ is our best estimate of the standard deviation. The relation between N and Δ/σ for Δ to be declared significant with a high degree of assurance is given in Fig. 10.13. We see that for a given, Δ, if σ is large, the number of runs is correspondingly large. If N is larger than the number of runs that will be provided by an experimental design, replication of the design is required. This was illustrated by Example 10.10. Although its primary purpose is to increase the precision of the estimate, replication works in another way to give a more precise estimate of σ.

1. T. D. Murphy, Jr., *Chem. Eng.*, pp. 168–182, June 6, 1977.

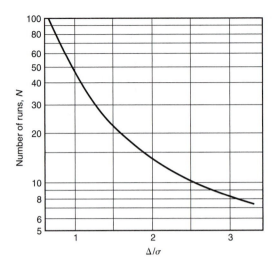

FIGURE 10.13
Number of experimental runs required
to achieve a given level of precision.

10.13.2 Fractional Factorial Designs

The number of treatment combinations in a factorial design increases rapidly with an increase in the number of factors. Thus, if $n = 4$, $2^n = 16$, but if $n = 8$, $2^n = 256$. Since engineering experimentation can easily involve 6 to 10 factors, the number of experiments required can rapidly become prohibitive in cost.

Consider a 2^6 factorial experiment. There are:

6 main effects	n
15 first-order interactions	$n(n - 1)/2$
20 second-order interactions	$n(n - 1)(n - 2)/2 \times 3$
15 third-order interactions	
6 fourth-order interactions	
1 fifth-order interaction	

Since we usually are not interested in higher-order interactions, we are collecting a great deal of extraneous information if we perform all 64 treatment combinations. This permits us to design an experiment that is some fraction of the total factorial design.

As a simple example, we consider a one-half replicate of a 2^3 factorial design.

$$N = r\,(2^n) = \frac{1}{2}(2^3) = \frac{8}{2} = 4$$

We split the treatment combinations into two blocks (Table 10.9). The main effect of factor A in block I is

$$x_A = (ab + ac) - [(1) + bc]$$

Also, in block I the BC interaction is

$$x_{BC} = [(1) + bc] - (ab + ac)$$

TABLE 10.9
Fractional factorial design for 2^3 experiment

	Treatment combinations	Effects						
		x_A	x_B	x_C	x_{AB}	x_{AC}	x_{BC}	x_{ABC}
	(1)	−	−	−	+	+	+	−
	ab	+	+	−	+	−	−	−
Block I	ac	+	−	+	−	+	−	−
	bc	−	+	+	−	−	+	−
	a	+	−	−	−	−	+	+
Block II	b	−	+	−	−	+	−	+
	c	−	−	+	+	−	−	+
	abc	+	+	+	+	+	+	+

Therefore, we see that $x_A = -x_{BC}$, and we can also show that $x_B = -x_{AC}$ and $x_C = -x_{AB}$. We say that BC, AC, and AB are *aliased* with A, B, and C, respectively. Therefore, by working with just one-half of the total factorial experiment, i.e., block I, we can determine the main effects for A, B, C if we assume there are no interactions.

The half replicate of the 2^3 factorial is the smallest practical fractional design. Fractional factorial designs usually are used when four or more factors must be considered. For example, in the half replicate of a 2^5 factorial design no first-order interactions are aliased with each other. Sometimes one-quarter replicates are used. A great many fractional factorial designs are given in Natrella.[1]

10.14
REGRESSION ANALYSIS

It is a frequent experience in engineering design and experimentation to be interested in determining whether there is a relation between two or more variables. Regression analysis is the statistical technique for establishing such relations. There are two different but related situations. The first determines a functional relation between variables when one or more variables are set at fixed values (independent variables) and the response (dependent variable) is established; here the emphasis is on *prediction*. In the second situation two or more variables are assumed to be *associated* with each other, i.e., there is a correlation between the variables, which vary jointly. Although the mathematical relations for regression and correlation analysis are identical, the concepts are related but distinct.

---

1. M. G. Natrella, op. cit., pp. 12–14 to 12–21; Montgomery and Runger, op. cit., pp. 757–776.

10.14.1 Method of Least Squares

The simplest case of regression analysis is a linear relation between an independent variable x and the dependent variable y. The mathematical model for the population is

$$y_i = \alpha + \beta x_i + \varepsilon_i \tag{10.48}$$

in which y_i and x_i are the ith observations of the dependent and independent variables, respectively. α and β are the population values of the intercept and slope regression coefficients. ε_i is the error.

The sample equation is

$$\hat{y} = a + bx \tag{10.49}$$

in which $\hat{y}$ is the predicted value of the dependent variable and a and b are the sample estimates of the regression coefficients. The assumptions that underlie the method of least squares are:

1. The errors are normally distributed with a mean of zero and a constant variance (σ_e^2).
2. The errors are independent of each other.

The method of least squares establishes the line through the data such that the sum of the squares of the vertical deviations of observations from the line is smaller than for any other line that could be drawn through the data.

$$\varepsilon_i = a + bx_i - y_i$$

and
$$\Sigma \, \varepsilon_i^2 = \Sigma \, (a + bx_i - y_i)^2 \tag{10.50}$$

We want to establish the values of the parameters a and b that result in the least value of $\Sigma \varepsilon_i^2$. This is done by taking the partial derivatives of Eq. (10.50) with respect to a and b and setting the resulting equations equal to zero.

$$\frac{\partial \varepsilon_i^2}{\partial a} = 2 \sum_{i=1}^{n} (a + bx_i - y_i) = 0$$

$$\frac{\partial \varepsilon_i^2}{\partial b} = 2 \sum_{i=1}^{n} (a + bx_i - y_i)x_i = 0 \tag{10.51}$$

If the terms within the parentheses are separated, the two derivatives can be rearranged to provide two simultaneous equations, which are called the normal equations. Note that all summations extend over the observations in the sample of size n.

$$\Sigma \, y_i = an + b \, \Sigma \, x_i$$

$$\Sigma \, x_i y_i = a \, \Sigma \, x_i + b \, \Sigma \, x_i^2 \tag{10.52}$$

When the normal equations are solved simultaneously, we get the equations for the parameters a and b.

$$b = \frac{n \Sigma xy - \Sigma x \Sigma y}{n \Sigma x^2 - (\Sigma x)^2} = \frac{\Sigma (x - \bar{x})(y - \bar{y})}{\Sigma (x - \bar{x})^2} \tag{10.53}$$

$$a = \frac{\Sigma y - b \Sigma x}{n} = \bar{y} - b\bar{x} \tag{10.54}$$

The sample *correlation coefficient* is given by

$$r = \frac{n \Sigma xy - \Sigma x \Sigma y}{\sqrt{[n \Sigma x^2 - (\Sigma x)^2][n \Sigma y^2 - (\Sigma y)^2]}} \tag{10.55}$$

The value of r varies between -1 and $+1$. When $r = \pm 1$, the experimental points fit the linear model $y = a + bx$ perfectly. When r approaches zero, there is very low correlation between the data, and the linear model does not fit the data. The procedure for establishing the confidence interval for the regression line is given by Natrella.[1]

10.14.2 Linear Multiple Regression Analysis

The engineering design situation usually involves more than one independent variable. A frequently used model has the form:

$$y = b_0 + b_1 x_1 + b_2 x_2 + \cdots + b_m x_m \tag{10.56}$$

The bivariate model of Eq. (10.48) is a special case of Eq. (10.56) in which $m = 1$. To solve Eq. (10.56), the same procedure as for when $m = 1$ is used; but matrix methods and digital computer techniques are used to solve the $m + 1$ linear equations.[2]

An example of a linear multiple regression equation without interaction effects is

YS (1000 psi) = 13.29 + 5.90 Mn + 10.21 Si + 0.220 (% pearlite) + 0.476$d^{-1/2}$

This equation predicts the yield strength of hot-rolled ferritic-pearlitic steel based on chemical composition, volume fraction of pearlite, and the grain size. The composition range was 0 to 0.3 wt% carbon, 0 to 1.60% manganese, and 0 to 0.8% silicon. The term d is the mean linear ferrite intercept grain size, in inches. The correlation coefficient is 0.89 and the 95 percent confidence limit for prediction of yield strength is ± 3800 psi.

10.14.3 Nonlinear Regression Analysis

We are well aware that nonlinearity is encountered in many engineering problems. A second-order model with two independent variables is given by:

---

1. M. G. Natrella, op. cit., pp. 5–36 to 5–46; Montgomery and Runger, op. cit., pp. 498–503.
2. C. Daniel and F. S. Wood, "Computer Analysis of Multifactor Data," 2d ed., Wiley, New York, 1980; N. R. Draper and H. Smith, "Applied Regression Analysis," 3d ed., ibid., 1998; D. C. Montgomery and G. C. Runger, op. cit., chap. 10.

$$y = b_0 + b_1x_1 + b_2x_2 + b_{11}x_1^2 + b_{12}x_1x_2 + b_{22}x_2^2 \qquad (10.57)$$

Such a model is needed to detect nonlinearity and second-order effects. Standardized computer programs are usually used to determine the parameters of this type of equation.[1]

The reader is cautioned against indiscriminate use of "canned" statistical programs in computer-based data reduction. There is no question that the availability of these programs in almost every computer center, together with the proliferation of home computers and programmable hand calculators, has increased the use of regression analysis, but it is important to understand what the statistical package really does. The user should take the time and trouble to read the backup documentation for the program and, in the event of failing to understand everything, seek the advice of a statistical consultant.

10.14.4 Linearization Transformations

The computational difficulties associated with nonlinear regression analysis sometimes can be avoided by using simple transformations that convert a problem that is nonlinear into one that can be handled by simple linear regression analysis. The most common transformations are given in Table 10.10. The reader needs to be aware that logarithmic transformation can introduce bias in the prediction of the response variable.[2]

TABLE 10.10
Some common linearization transformation, $W = c + dV$

		Linearized variables	
Nonlinear equation	Linearized equation	W	V
1. $y = a + bx$ (linear)	$y = a + bx$	y	x
2. $y = ax^b$ (logarithmic)	$\ln y = \ln a + b \ln x$	$\ln y$	$\ln x$
3. $y = ae^{bx}$ (exponential)	$\ln y = \ln a + bx$	$\ln y$	x
4. $y = 1 - e^{-bx}$ (exponential)	$\ln\dfrac{1}{1-y} = bx$	$\ln\dfrac{1}{1-y}$	x
5. $y = a + b\sqrt{x}$ (square root)	$y = a + bx$	y	$\sqrt{x}$
6. $y = a + \dfrac{b}{x}$ (inverse)	$y = a + bx$	y	$\dfrac{1}{x}$
7. $y = 1 - \exp -\left(\dfrac{x - x_0}{\theta - x_0}\right)^m$ (Weibull)	$\ln \ln\dfrac{1}{1-y} =$ $- m \ln (\theta - x_0)$ $+ m(x - x_0)$	$\ln \ln y$	$\ln (x - x_0)$

---

1. C. Daniel and F. S. Wood, op. cit.
2. R. W. McCuen, R. B. Leahy, and P. A. Johnson, *J. Hydraulic Engr,* vol. 11b, March 1990, pp. 414–428.

10.15
RESPONSE SURFACE METHODOLOGY

In the preceding section we saw that complex engineering problems can be analyzed to give a regression equation that describes either a linear or a nonlinear model. The equations give the response in terms of the several independent variables of the problem. If the response is plotted as a function of x_1, x_2, etc., we obtain a response surface. A powerful statistical procedure that employs factorial analysis and regression analysis has been developed for the determination of the optimum operating condition on a response surface.[1]

Response surface methodology (RSM) has two objectives:

1. To determine with one experiment where to move in the next experiment so as to continually seek out the optimal point on the response surface.
2. To determine the equation of the response surface near the optimal point.

Response surface methodology (RSM) uses a two-step procedure aimed at rapid movement from the current operating position into the central region of the optimum. This is followed by the characterization of the response surface in the vicinity of the optimum by a mathematical model. The basic tools used in RSM are two-level factorial designs and the method of least squares model and its simpler polynomial forms.

Initially, a small factorial experiment is run over a small area of the response surface where the surface may be considered to be planar, point A, Fig. 10.14. This is

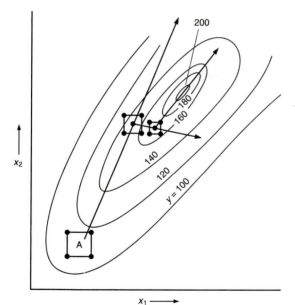

FIGURE 10.14
Concept of response surface methodology.

1. G. E. P. Box and K. B. Wilson, *J. Roy. Stat. Soc.,* sec. B., vol. 13, pp. 1–45, 1951; R. H. Meyers, "Response Surface Methodology," Allyn & Bacon, Inc., Boston, 1971; O. L. Davies and P. L. Goldsmith (eds.), "Statistical Methods in Research and Production Methodology," 4th ed., Longman Group Ltd., London, 1976; G. E. P. Box and N. R. Draper, "Empirical Model-Building and Response Surfaces," Wiley, New York, 1987; D. C. Montgomery and G. C. Runger, op. cit., pp. 778–791.

TABLE 11.3
Fatality rate

Cause of fatality	Fatality per person per year
Smoking (20 per day)	5×10^{-3}
Cancer, in general	3×10^{-3}
Race car driving	1×10^{-3}
Motor vehicle driving	3×10^{-4}
Fires	4×10^{-5}
Poison	2×10^{-5}
Industrial machinery	1×10^{-5}
Air travel	9×10^{-6}
Railway travel	4×10^{-6}
California earthquake	2×10^{-6}
Lightning	5×10^{-7}

industry), (2) health information, and (3) accident statistics. Usually the data are differentiated between fatalities and injuries. Risk is usually expressed as the probability of the risk of a fatality per person per year. A risk that exceeds 10^{-3} fatalities per person per year is generally considered unacceptable while a rate that is less than 10^{-5} is not of concern to the average person.[1] The range 10^{-3} to 10^{-5} is the tolerable range. However, an individual's perception of risk depends upon the circumstances. If the risk is voluntarily assumed, like smoking or driving a car, then there is a greater acceptance of the risk than if the risk was assumed involuntarily, as with traveling in a train or breathing secondhand smoke. There is a large difference between individual risk and societal risk. While 10^{-3} is the boundary where unacceptable risk begins for a single voluntary death, the acceptable fatality rate shifts to about 10^{-5} if 100 deaths are involved. Table 11.3 gives some generally accepted fatality rates for a variety of risks.

The most common approach in risk assessment is a benefit-cost analysis (Sec. 13.11). One difficulty with this approach is that benefits are counted in lives saved while costs are in monetary terms. This requires the often distasteful task of placing a value on a human life. One way to do this is the human capital approach, which estimates the lost earnings potential of the victim, with allowance for the pain and suffering of the victim's loved ones. This is about $2.6 million for a life lost in a highway accident.[2] Another approach is the "willingness to pay" method, in which people are asked directly what they would be willing to pay to avoid danger or harm.

11.2
PROBABILISTIC APPROACH TO DESIGN

Conventional engineering design uses a deterministic approach. It disregards the fact that material properties, the dimensions of the components, and the externally applied

1. D. J. Smith, "Reliability, Maintainability, and Risk," 5th ed., Butterworth-Heinemann, Oxford, 1997.
2. *The Economist,* December 4, 1993, p. 74.

loads are stochastic in nature. In conventional design these uncertainties are handled by applying a factor of safety. In critical design situations, such as aircraft, space, and nuclear applications, however, there is a growing trend toward using a probabilistic approach to better quantify uncertainty and thereby increase reliability.[1]

There are three typical approaches for incorporating probabilistic effects in design. In order of increasing sophistication they are: the use of a factor of safety, the use of the absolute worst case design, and the use of probability in design. We will present the probabilistic approach first, and then show how the other approaches can be folded in.

Consider a structural member subjected to a static load that develops a stress σ. The variation in load or sectional area results in the distribution of stress shown in Fig. 11.1, where the mean is $\bar{\sigma}$ and the standard deviation[2] of the sample of stress values is s. The yield strength of the material S_y has a distribution of values given by $\bar{S}_y$ and s_y. However, the two frequency distributions overlap and it is possible for $\sigma > S_y$, which is the condition for failure. The probability of failure is given by

$$P_f = P(\sigma > S_y) \tag{11.1}$$

The reliability R is defined as

$$R = 1 - P_f \tag{11.2}$$

If we subtract the stress distribution from the strength distribution, we get the distribution $Q = S_y - \sigma$ shown at the left in Fig. 11.1.

We now need to be able to determine the mean and standard deviation of the distribution Q constructed by performing algebraic operations on two independent random variables x and y, that is, $Q = x \pm y$. Without going into statistical details,[3] the results are as given in Table 11.4. Referring now to Fig. 11.1, and using the results in Table 11.4, we see that the distribution $Q = S_y - \sigma$ has a mean value $\bar{Q} = 40 - 30 = 10$ and $s_Q = \sqrt{6^2 + 8^2} = 10$. The part of the distribution to the left of $Q = 0$

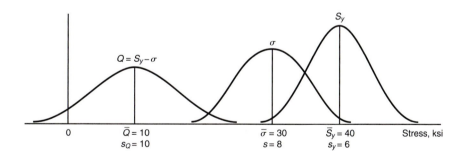

FIGURE 11.1
Distributions of yield strength S_y and stress.

1. E. B. Haugen, "Probabilistic Mechanical Design," Wiley-Interscience, New York, 1980; J. N. Siddal, "Probabilistic Engineering Design," Marcel Dekker, New York, 1983.
2. Note that probabilistic design is at the intersection of two engineering disciplines: mechanical design and engineering statistics. Thus, confusion in notation is a problem.
3. E. B. Haugen, op. cit., pp. 26–56.

Design Stage	Design Activity
Conceptual design	Problem definition:
	Estimate reliability requirement
	Determine likely service environment
Embodiment design	Configuration design:
	Investigate redundancy
	Provide accessibility for maintenance
	Parametric design:
	Select highly reliable components
	Build and test physical and computer prototypes
	Full environmental tests
	Establish failure modes/FMEA
	Estimate MTBF
	User trials/modification
Detail Design	Produce & test preproduction prototype
	Final estimate of reliability
Production	Production models:
	Further environmental tests
	Establish quality assurance program
Service	Deliver to customer:
	Feedback field failures and MTBFs to designers
	Repair and replace
	Retirement from service

FIGURE 11.8
Reliability activities throughout design, production, and service.

11.4.1 Causes of Unreliability

The malfunctions that an engineering system can experience can be classified into five general categories.[1]

1. *Design mistakes:* Among the common design errors are failure to include all important operating factors, incomplete information on loads and environmental conditions, erroneous calculations, and poor selection of materials.
2. *Manufacturing defects:* Although the design may be free from error, defects introduced at some stage in manufacturing may degrade it. Some common examples are (1) poor surface finish or sharp edges (burrs) that lead to fatigue cracks and (2) decarburization or quench cracks in heat-treated steel. Elimination of defects in manufacturing is a key responsibility of the manufacturing engineering staff,

---

1. W. Hammer, "Product Safety Management and Engineering." chap. 8, Prentice-Hall, Englewood Cliffs, NJ, 1980.

but a strong relationship with the R&D function may be required to achieve it. Manufacturing errors produced by the production work force are due to such factors as lack of proper instructions or specifications, insufficient supervision, poor working environment, unrealistic production quota, inadequate training, and poor motivation.

3. *Maintenance:* Most engineering systems are designed on the assumption they will receive adequate maintenance at specified periods. When maintenance is neglected or is improperly performed, service life will suffer. Since many consumer products do not receive proper maintenance by their owners, a good design strategy is to make the products maintenance-free.

4. *Exceeding design limits:* If the operator exceeds the limits of temperature, speed, etc., for which it was designed, the equipment is likely to fail.

5. *Environmental factors:* Subjecting equipment to environmental conditions for which it was not designed, e.g., rain, high humidity, and ice, usually greatly shortens its service life.

11.4.2 Minimizing Failure

A variety of methods are used in engineering design practice to improve reliability. We generally aim at a probability of failure of $P_f < 10^{-6}$ for structural applications and $10^{-4} < P_f < 10^{-3}$ for unstressed applications.

Margin of safety

We saw in Sec. 11.2 that variability in the strength properties of materials and in loading conditions (stress) leads to a situation in which the overlapping statistical distributions can result in failures. In Fig. 11.5 we saw that the variability in strength has a major impact on the probability of failure, so that failure can be reduced with no change in the mean value if the variability of the strength can be reduced.

Derating

The analogy to using a factor of safety in structural design is derating electrical, electronic, and mechanical equipment. The reliability of such equipment is increased if the maximum operating conditions (power, temperature, etc.) are derated below their nameplate values. As the load factor of equipment is reduced, so is the failure rate. Conversely, when equipment is operated in excess of rated conditions, failure will ensue rapidly.

Redundancy

One of the most effective ways to increase reliability is with redundancy (see Sec. 11.3). In parallel redundant designs the same system functions are performed at the same time by two or more components even though the combined outputs are not required. The existence of parallel paths may result in load sharing so that each component is derated and has its life increased by a longer than normal time.

Another method of increasing redundancy is to have inoperative or idling standby units that cut in and take over when an operating unit fails. The standby unit wears out much more slowly than the operating unit does. Therefore, the operating strategy

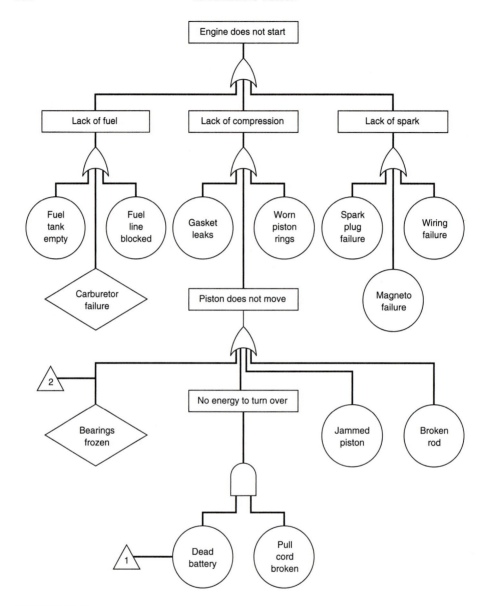

FIGURE 11.12
Fault tree for the failure of a lawn mower engine to start.

The AND gate ⌂ is used to represent a logic condition in which all inputs below the gate must be present for the output (at the top of the gate) to occur.

The OR gate ⌂ is used to represent the situation in which any of the input events will lead to the output.

These symbols are employed in Fig. 11.12, which is a fault tree for the failure of a lawn mower engine. The bottom branch of the tree consists of failures or initiating events that are errors. They point out events where FEMA should be done. Starting with the top event and moving down the branches, we enter a hierarchy of causes. Each event branches into other events. In constructing the tree, it is important to proceed slowly and deliberately down from the top event and list every direct immediate cause before going on to consider the next level of causes. The description used for each event must be carefully chosen to indicate the cause.

The fault tree clearly indicates where corrective action should be taken. For example, the need for disassembling the piston and/or bearings is indicated. Also, the need for preventive maintenance and inspection of the battery or pull cord is called for.

In summary, FTA points out the critical areas in a complex system where further study in failure mode analysis and reliability engineering is required. It also can be used as a tool in troubleshooting a piece of equipment that will not operate. FTA is very useful in accident investigations and the analysis of failures. Since preparation of a fault tree requires considerable detail about the system, it is possible only after the design has been completed. However, it can point the way to improved design for reliability and safety.

11.7
DEFECTS AND FAILURE MODES

Failures of engineering designs and systems are a result of deficiencies in four broad categories.

- Hardware failure—failure of a component to function as designed.
- Software failure—failure of the computer software to function as designed.
- Human failure—failure of human operators to follow instructions or respond adequately to emergency situations.
- Organizational failure—failure of the organization to properly support the system. Examples might be overlooking defective components, slowness to bring corrective action, ignoring bad news, etc.

11.7.1 Causes of Hardware Failure

Failures are caused by design errors or deficiencies in one or more of the following categories:

1. Design deficiencies
   Failure to adequately consider the effect of notches
   Inadequate knowledge of service loads and environment
   Difficulty of stress analysis in complex parts and loadings
2. Deficiency in selection of material
   Poor match between service conditions and selection criteria
   Inadequate data on material
   Too much emphasis given to cost and not enough to quality

3. Imperfection in material due to manufacturing
4. Overload and other abuses in service
5. Inadequate maintenance and repair
6. Environmental factors
   Conditions beyond those allowed for in design
   Deterioration of properties with time of exposure to environment

Deficiencies in the design process, or defects in the material or its processing, can be classified in the following ways. At the lowest level is a lack of conformance to a stated specification. An example would be a dimension "out of spec" or a strength property below specification. Next in severity is a lack of satisfaction by the customer or user. This may be caused by a critical performance characteristic set at an improper value, or it may be a system problem caused by rapid deterioration. The ultimate defect is one which causes failure of the product. Failure may be an actual fracture or disruption of physical continuity of the part, or failure may be inability of the part to function properly.

11.7.2 Failure Modes

The specific modes of failure of engineering components can usually be grouped into four general classes:

1. Excessive elastic deformation
2. Excessive plastic deformation
3. Fracture
4. Loss of required part geometry through corrosion or wear

The most common failure modes are listed in Table 11.11. Some of these failure modes are directly related to a standard mechanical property test, but most are more complex, and failure prediction requires using a combination of two or more properties (see Fig. 6.5).

11.7.3 Importance of Failure

It is a human tendency to be reluctant to talk about failure or to publish much information about failures. Spectacular system failures, like the Tacoma Narrows bridge or the O-ring seal on the space shuttle *Challenger* solid rocket booster, have caught the public's attention, but most failures go unheralded.[1] This is a shame, because much learning in engineering occurs by studying failures. Simulated service testing and proof-testing of preproduction prototypes are important steps at arriving at a success-

----

1. See http://dol1.eng.sunysb.edu/disaster.

TABLE 11.11
Failure modes for mechanical components

1. Elastic deformation	9. Impact
2. Yielding	*a.* Impact fracture
3. Brinelling	*b.* Impact deformation
4. Ductile failure	*c.* Impact wear
5. Brittle fracture	*d.* Impact fretting
6. Fatigue	*e.* Impact fatigue
a. High-cycle fatigue	10. Fretting
b. Low-cycle fatigue	*a.* Fretting fatigue
c. Thermal fatigue	*b.* Fretting wear
d. Surface fatigue	*c.* Fretting corrosion
e. Impact fatigue	11. Galling and seizure
f. Corrosion fatigue	12. Scoring
g. Fretting fatigue	13. Creep
7. Corrosion	14. Stress rupture
a. Direct chemical attack	15. Thermal shock
b. Galvanic corrosion	16. Thermal relaxation
c. Crevice corrosion	17. Combined creep and fatigue
d. Pitting corrosion	18. Buckling
e. Intergranular corrosion	19. Creep buckling
f. Selective leaching	20. Oxidation
g. Erosion-corrosion	21. Radiation damage
h. Cavitation	22. Bonding failure
i. Hydrogen damage	23. Delamination
j. Biological corrosion	24. Erosion
k. Stress corrosion	
8. Wear	
a. Adhesive wear	
b. Abrasive wear	
c. Corrosive wear	
d. Surface fatigue wear	
e. Deformation wear	
f. Impact wear	
g. Fretting wear	

ful product. While the literature on engineering failures is not extensive, there are several useful books on the subject.[1]

11.8
TECHNIQUES OF FAILURE ANALYSIS

When the problem of determining the cause of a failure and proposing corrective action must be faced, there is a definite procedure for conducting the failure analysis.[2]

---

1. "Case Histories in Failure Analysis," ASM International, Materials Park OH, 1979; H. Petrowski, "Design Paradigms: Case Histories of Error and Judgment in Engineering," Cambridge University Press, 1994; C. F. Jones, "Accidents May Happen: Fifty Inventions Discovered by Mistake," Delacorte Press, 1996; R. R. Whyte (ed.), "Engineering Progress Through Trouble," The Institution of Mechanical Engineers, London, 1975.
2. G. F. Vander Voort, *Metals Eng. Quart.,* pp. 31–36, May 1975; R. Roberts and A. W. Pense, *Civil Eng.,* pp. 64–67, May 1980; pp. 60–62, July 1980.

Frequently a failure analysis requires the efforts of a team of people, including experts in materials behavior, stress analysis, vibrations, and sophisticated structural and analysis techniques.

11.8.1 Field Inspection of the Failure

The most useful first approach is to inspect the failure at the site of the accident *as soon as possible* after the failure occurs. This site visit should be lavishly documented with photographs, for very soon the accident will be cleared away and repair begun. It is best to take photographs in color. Start taking pictures at a distance and move up to the site of the failure. Shoot pictures from several angles. Careful sketches and detailed notes help to orient the photographs and allow you to completely reconstruct the scene months or years later when you are in a design review or a courtroom.

The following critical pieces of information should be obtained during the field inspection.

1. Location of all broken pieces relative to each other
2. Identification of the origin of failure
3. Orientation and magnitude of the stresses
4. Direction of crack propagation and sequence of failure
5. Presence of obvious material defects, stress concentrations, etc.
6. Presence of oxidation, temper colors, or corrosion products
7. Presence of secondary damage not related to the main failure

It is important to interview operating and maintenance personnel to get their version of what happened and learn about any unusual operating history, such as unusual vibration or noise prior to failure. Whenever possible, the failure should be brought back to the laboratory for more detailed analysis. Any cutting that is required should be done well away from the fracture surface so as not to alter that surface. Whenever possible, samples should be obtained from identical material or components that did not fail. Samples of process fluids, lubricants, etc., should be obtained for corrosion-related failures. Be sure to label all pieces and key their identification to your notes.

Great care should be exercised in preserving the fracture surface. Never touch the fractured surfaces, and do not attempt to fit them back together. Avoid washing a fracture surface with water unless it has been contaminated with seawater or fire-extinguisher fluids. To prevent corrosion of a fracture surface, dry the surface with a jet of water-free compressed air and place the part in a desiccator or pack it with a suitable desiccant.

When the failure surface cannot be removed from the field for investigation in the laboratory, it is necessary to take the laboratory into the field. A portable metallographic laboratory has been developed for such a situation.[1]

---

1. H. Crowder, "Metals Handbook," 8th ed., vol. 10, pp. 26–69.

11.8.2 Background History and Information

A complete case history on the component that failed should be developed as soon as possible. Ideally, most of this information should be obtained before making the site visit, since more intelligent questions and observations will result. The following is a list of data that need to be assembled.

1. Name of item, identifying numbers, owner, user, manufacturer or fabricator
2. Function of item
3. Data on service history, including inspection of operating logs and records
4. Discussion with operating personnel and witnesses concerning any unusual conditions or events prior to failure
5. Documentation on materials used in the item
6. Information on manufacturing and fabrication methods used, including any codes or standards
7. Documentation on inspection standards and techniques that were applied
8. Date and time of failure; temperature and environmental conditions
9. Documentation on design standards and calculations performed in the design
10. A set of shop drawings, including any modifications made to the design during manufacturing or installation

11.8.3 Macroscopic Examination

A macroscopic examination is made at magnifications ranging from $1\times$ to about $100\times$. Certainly this type of examination should occur at the site of the failure, but it is better repeated back in the laboratory where the lighting and other conditions are more favorable. The purpose of macroscopic examination is to observe the gross features of the fracture, the presence or absence of cracks, the presence of any gross defects, and the presence of corrosion or oxidation products. Most examination is done in the $1\times$ to $10\times$ range. An illuminated $10\times$ magnifier is a good tool for this type of study. Working in that magnification range, you should try to make an initial assessment of the origin of fracture and thus narrow down the region of the fracture for further study at higher magnification.

Often it is possible to identify the type of fracture from macroscopic examination. If there is gross deformation near the break and a dark fibrous texture to the fracture surface, you most likely have a ductile rupture. If the fracture surface is flat, with shiny flat grain facets visible in the surface, a brittle fracture is indicated. Often the surface of a brittle fracture has "chevron markings" pointing back to the origin of the fracture (Fig. 11.13a). Sometimes a brittle fracture has shear lips, regions of local deformation. In macroexamination, fatigue failures often show concentric rings or beach marks emanating from the origin of the fracture (Fig. 11.13b). The surface of a fatigue fracture generally is flat, with no shear lips, and is oriented normal to the largest tensile stress. A fracture surface that is heavily corroded or oxidized is usually evidence that the crack existed for a long time before finally propagating to failure. A large collection of macrophotographs of failures can be found in "Metals Handbook," 8th ed., vols. 9 and 10.

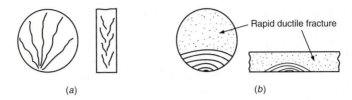

(a)   *(b)*

FIGURE 11.13
Schematic drawings of fracture surfaces. (*a*) Brittle fracture of shaft and a plate; (*b*) fatigue
fracture of a shaft and a plate.

11.8.4 Microscopic Examination

A microscopic examination is made at magnifications greater than $100\times$. The term
covers the use of such instruments as the metallurgical (reflected light) microscope,
the scanning electron microscope (SEM), the transmission electron microscope
(TEM), and the x-ray microprobe analyzer. All these instruments usually will be
available in a modern materials laboratory. When used properly, they provide highly
valuable diagnostic information.

Metallographic examination with the metallurgical microscope requires a small
section of material to be cut out, mounted, polished, and etched.[1] This type of examination is used to determine the microstructure of the material. The presence, size, and
arrangement of phases is important documentation of the thermal and mechanical history of the metal.[2] Microstructural analysis will identify such features as grain size,
inclusion size, and distribution of second phases. The technique also can be used to
follow crack growth through the microstructure. For example, it can be used to determine whether a crack propagates in a transgranular or intergranular manner, whether
a hard brittle phase cracks to initiate the fracture, or whether some microconstituent
serves to impede crack propagation.

The scanning electron microscope (SEM) examines the actual surface of the fracture with a beam of electrons in an evacuated chamber (typically 1 by 2 by 5 in). A
back-scattered image is recorded on a CRT display. Magnifications from $1000\times$ to
$40,000\times$ are available. The image has great depth of field and a three-dimensional
character. This makes SEM outstandingly useful for the examination of fractures. The
crack propagation associated with a particular fracture mode leaves a characteristic
appearance on the fracture surface. These *fractographs* are directly revealed by the
SEM and provide an identification of the fracture mode[3] (see Fig. 11.14). When the
SEM is equipped with an energy-dispersive x-ray probe, elements higher than atomic
number 10 can be detected and analyzed quantitatively. This microanalysis capability
is useful for identifying inclusions, second phases, and corrosion products.

---

1. "Metals Handbook," 9th ed., vol. 10, American Society for Metals, Metals Park, OH, 1986.
2. See "Metals Handbook," 9th ed, vol. 9, "Metallography and Microstructures," for a large collection of
microstructures with extensive explanation and interpretation.
3. See "Metals Handbook," 9th ed., vol. 12, "Fractography."

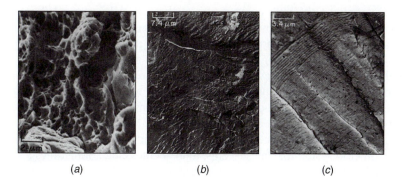

(a)   (b)   (c)

FIGURE 11.14
Examples of fractographs made with SEM. (*a*) Ductile rupture in 4340 steel; (*b*) cleavage
type features (river pattern) in steel; (*c*) striations from fatigue crack growth in aluminum
alloy. Note: (*b*) and (*c*) were made from plastic carbon replicas of the fracture surface. (*From
"Metals Handbook," 8th ed., vol. 9, ASM International, Materials Park, OH. Copyright ASM
International, 1974.*)

The transmission electron microscope (TEM) is used to examine the microstructure and defect structure at magnifications up to 1,000,000× (on a few special instruments). Since the electron beam must pass through it, the specimen must either be a
very thin foil or a plastic replica of the fracture surface. The TEM is an important
research instrument for microstructure analysis; with selected area electron diffraction
accessories, it is possible to determine crystallographic relations between phases and
to study solid-state transformations and the interplay of dislocations with structural
features. However, because of its great depth of field and the ease of specimen preparation, the SEM has much greater application in fracture analysis.

Other sophisticated analytical instrumentation often is helpful in fracture analysis. In electron microprobe analysis the chemical composition of surface layers, precipitates, segregated regions, etc., is quantitatively determined. It is the same function
that is carried out with the energy-dispersive probe on many SEM units. The scanning
Auger electron spectrometer is a very sensitive tool for analyzing the first few atomic
layers of a surface. Auger spectroscopy has been used very effectively in detecting
trace elements such as bismuth, arsenic, and antimony, which cause embrittlement in
steels.

X-ray diffraction techniques also may be useful in fracture analysis. X-ray methods can be used for the qualitative and quantitative identification of phases, the determination of crystallographic orientation, the measurement of residual stresses, and the
characterization of texture or preferred orientation.

11.8.5 Additional Tests

It usually is necessary to obtain a number of other types of experimental data in order
to put together the pieces of the puzzle that lead to the identification of the cause of

failure. Determining the bulk composition of the material is part of the process of identifying the material and finding whether it meets the specification. Some failures are caused simply by a mixup of materials during manufacturing or maintenance. Quick spot tests for identifying materials in the field are available.[1]

It also is important to measure the mechanical properties of an unused specimen of the material that failed. Sometimes not enough unused material is available to machine specimens for tensile or impact tests, and it may then be necessary to "infer" the properties from the results of a hardness test.

It is important to probe the failed part with various nondestructive testing techniques in order to search for flaws, seams, hidden cracks, etc. Sometimes a section of the material is deep-etched with acid to reveal defects such as segregation, hydrogen flakes, decarburized layer, or soft, spongy spots.

11.8.6 Analysis of the Data

The critical step is assembling the facts and pieces of data into a coherent picture of the cause of the failure. Early in the process it is common to develop a hypothesis of the cause. All data should be cross-checked against the hypothesis, and any contradictions should be run down and either confirmed or discarded as spurious. When firm contradictions exist, they usually result in refining the hypothesis until gradually all pieces fit together. An experienced failure analyst not only considers the available data but will take note of the absence of features that experience suggests should be present. It is common for a failure to be caused by more than one factor. Therefore, developing a plausible hypothesis usually is not just a straightforward procedure.

Often when it becomes apparent that a critical piece of data is missing it will be necessary to reexamine the site of the failure and/or to start over again in analyzing the failure. At this stage it may be very helpful to assemble a team of experts from different disciplines so that as many varied viewpoints as possible can be brought to bear on the analysis.

Sometimes the failure testing of a model, or even a full-size duplicate of the failed unit, can be very illuminating. Computer models of the stress or temperature distribution, fatigue crack propagation rate, etc., may be needed in complex fracture problems.

Most failure analyses are failure-mechanism analyses which describe *what* failed and *how* the failure occurred. The question of *why* the failure occurred is often left to speculation or skipped over. To answer this important issue requires a *root-cause* failure analysis.[2] In a root-cause failure analysis every fact and item of evidence relating to the failed component and system is examined. Operating and maintenance people are interviewed in detail.

1. "Identification of Metals and Alloys," bulletin published by International Nickel Co., New York; "Symposium on Rapid Methods for the Identification of Metals," ASTM Spec. Tech. Publ. No. 98, ASTM, Philadelphia, 1950; "Metals Handbook," 8th ed., vol, pp. 270–286.
2. C. M. Jackson and R. D. Buchheit, *Mechanical Engineering,* July 1984, pp. 32–37.

11.8.7 Report of Failure

The report of the analysis of a failure is one of the most difficult written technical communications because a failure is often a matter of great sensitivity that may be fraught with legal implications. The best procedure is to stick to the hard facts, refrain from conjecture, and keep the technical jargon to a minimum. Additional tips on written technical communication will be found in Chap. 17.

11.9
DESIGN FOR SAFETY

Safety may well be the paramount issue in product design.[1] Normally we take safety for granted, but the recall of an unsafe product can be very costly in terms of product liability suits, replaced product, or tarnished reputation. The product must be safe to manufacture, to use, and to dispose of after use.[2]

A safe product is one that does not cause injury or property loss. Also included under safety is injury to the environment. Achieving safety is no accident. It comes from a conscious focus on safety during design, and in knowing and following some basic rules. There are three aspects to design for safety.

1. Make the product safe, i.e., design all hazards out of the product.
2. If it is not possible to make the product inherently safe, then design in protective devices like guards, automatic cutoff switches, pressure-relief valves, to mitigate the hazard.
3. If step 2 cannot remove all hazards, then warn the user of the product with appropriate warnings like labels, flashing lights, and loud sounds.

A *fail-safe design* seeks to ensure that a failure will either not affect the product or change it to a state in which no injury or damage will occur. There are three variants of fail-safe designs.

- Fail-passive design. When a failure occurs, the system is reduced to its lowest-energy state and the product will not operate until corrective action is taken. A circuit breaker is an example of a fail-passive device.
- Fail-active design. When failure occurs, the system remains energized and in a safe operating mode. A redundant system kept on standby is an example.
- Fail-operational design. The design is such that the device continues to provide its critical function even though a part has failed. A valve that is designed so that it will remain in the open position if it fails is an example.

---

1 C. O. Smith, Safety in Design, "ASM Handbook," vol. 20, pp. 139–145, ASM International, Materials Park, OH, 1997.
2. For a comprehensive safety web site see http://www.safetyline.net.

11.9.1 Potential Dangers

We list below some of the general categories of safety hazards that need to be considered in design.

Acceleration/deceleration—falling objects, whiplash, impact damage
Chemical contamination—human exposure or material degradation
Electrical—shock, burns, surges, electromagnetic radiation, power outage
Environment—fog, humidity, lightning, sleet, temperature extremes, wind
Ergonomic—fatigue, faulty labeling, inaccessibility, inadequate controls
Explosions—dust, explosive liquids, gases, vapors, finely powdered materials
Fire—combustible material, fuel and oxidizer under pressure, ignition source
Human factors—failure to follow instructions, operator error
Leaks or spills
Life cycle factors—frequent startup and shutdown, poor maintenance
Materials—corrosion, weathering, breakdown of lubrication
Mechanical—fracture, misalignment, sharp edges, stability, vibrations
Physiological—carcinogens, human fatigue, irritants, noise, pathogens
Pressure/vacuum—dynamic loading, implosion, vessel rupture, pipe whip
Radiation—ionizing (alpha, beta, gamma, x-ray), laser, microwave, thermal
Structural—aerodynamic or acoustic loads, cracks, stress concentrations
Temperature—changes in material properties, burns, flammability, volatility

Product hazards are often controlled by government regulation. The U.S. Consumer Products Safety Commission is charged with this responsibility.[1] Products designed for use by children are held to much higher safety standards than products intended to be used by adults. The designer must also be cognizant that in addition to providing a safe product for the customer, it must be safe to manufacture, sell, install, and service.

In our society, products that cause harm invariably result in court suits for damages under the product liability laws. Design engineers must understand the consequences of these laws and how they must practice to minimize safety issues and the threat of litigation. This topic is covered in Chap. 15.

11.9.2 Guidelines for Design for Safety[2]

1. Recognize and identify the actual or potential hazards, and then design the product so they will not affect its functioning.
2. Thoroughly test prototypes of the product to reveal any hazards overlooked in the initial design.

---

1. See the CPSC web site, gopher://cpsc.gov.
2. C. O. Smith, op. cit.; J. G. Bralla, "Design for Excellence," McGraw-Hill, New York, chap. 17.

3. Design the product so it is easier to use safely than unsafely.
4. If field experience turns up a safety problem, determine the root cause and redesign to eliminate the hazard.
5. Realize that humans will do foolish things, and allow for it in your design. More product safety problems arise from improper product use than from product defects. A user-friendly product is usually a safe product.
6. There is a close correspondence between good ergonomic design and a safe design. For example:
   - Arrange the controls so that the operator does not have to move to manipulate them.
   - Make sure that fingers cannot be pinched by levers, etc.
   - Avoid sharp edges and corners.
   - Point-of-operation guards should not interfere with the operator's movement.
   - Products that require heavy or prolonged use should be designed to avoid cumulative trauma disorders like carpal tunnel syndrome. This means avoiding awkward positions of the hand, wrist, and arm and avoiding repetitive motions and vibration.
7. Minimize the use of flammable materials, including packaging materials.
8. Paint and other surface finishing materials should be chosen to comply with EPA and OSHA regulations for toxicity to the user and when they are burned, recycled, or discarded.
9. Think about the need for repair, service, or maintenance. Provide adequate access without pinch or puncture hazards to the repairer.
10. Electrical products should be properly grounded to prevent shock. Provide electrical interlocks so that high-voltage circuits will not be energized unless a guard is in the proper position.

11.9.3 Warning Labels

With rapidly escalating costs of product liability manufacturers have responded by plastering their products with warning labels. Warnings should supplement the safety-related design features by indicating how to avoid injury or damage from the hazards which could not be feasibly designed out of the product without seriously compromising its utility. The purpose of the warning label is to alert the user to a hazard and tell how to avoid injury from it.

For a warning label to be effective the user must receive the message, understand it, and act on it. The engineer must properly design the label with respect to the first two issues to achieve the third. The label must be prominently located on the product. Most warning labels are printed in two colors on a tough wear-resistant material, and fastened to the product with an adhesive. Attention is achieved by printing *Danger, Warning,* or *Caution* depending on the degree of the hazard. The message to be communicated by the warning must be carefully composed to convey the nature of the hazard and the action to be taken. It should be written at the sixth-grade level, with no long words and technical terms. For products that will be used in different countries the warning label must be in the local language.

11.10
SUMMARY

Modern society places strong emphasis on avoiding risk, while insisting on products that last longer and require less service or repair. This requires greater attention to risk assessment in the concept of a design, in using methods for deciding on potential modes of failure, and in adopting design techniques that increase the reliability of engineered systems.

A *hazard* is the potential for damage. *Risk* is the likelihood of a hazard materializing. *Danger* is the unacceptable combination of hazard and risk. *Safety* is freedom from danger. Thus, we see that the engineer must be able to identify hazards to the design, evaluate the risk in adopting a technology or course of action, and understand when conditions constitute a danger. Design methods which mitigate a danger lead to safe design. One of the common ways this is achieved is by designing with respect to accepted codes and standards.

Reliability is the probability that a system or component will perform without failure for a specified time. Most systems follow a three-stage failure curve: (1) an early burn-in or break-in period, in which the failure rate decreases rapidly with time, (2) a long period of nearly constant failure rate (useful life), and (3) a final wear-out period of rapidly increasing failure rate. The failure rate is usually expressed as the number of failures per 1000 h, or by its reciprocal, the mean time between failures (MTBF). System reliability is determined by the arrangement of its components, i.e., in series or parallel.

System reliability is heavily influenced by design. The product design specification should contain a reliability requirement. The configuration of the design determines the degree of redundancy. The design details determine the level of defects. Early estimation of potential failure modes by FMEA and FTA lead to more reliable designs. Other methods to increase the reliability of the design are use of highly durable materials and components, derating of components, reduction in part count and simplicity of the design, and adoption of a damage-tolerant design coupled with ready inspection. Extensive testing of preproduction prototypes to "work the bugs out" is a method that works well. Methods for carrying out a root cause analysis of the reasons for a failure are an important means of improving the reliability of designs.

A safe design is one that instills confidence in the customer. It is a design that will not incur product liability costs. In developing a safe design the primary objective should be to identify potential hazards and then produce a design that is free from the hazards. If this cannot been done without compromising the functionality of the design, the next best approach is to provide protective devices that prevent the human from coming in contact with the hazard. Finally, if this cannot be done, then warning labels, lights, or buzzers must be used.

BIBLIOGRAPHY

Risk Assessment

Haimes, Y. Y.: "Risk Modeling, Assessment, and Management," Wiley, New York, 1998.
Michaels, J. V.: "Technical Risk Management," Prentice-Hall, Upper Saddle River, NJ, 1996.

Schwing, R. C., and W. A. Alpers, Jr. (eds.): "Societal Risk Assessment: How Safe Is Enough?"
    Plenum Publishing Co., New York, 1980.

Reliability Engineering

Bentley, J. P.: "An Introduction to Reliability and Quality," Wiley, New York, 1993.
Ireson, W. G.(ed.): "Handbook of Reliability Engineering and Management," 2d ed., McGraw-
    Hill, New York, 1996.
Lewis, E. E.: "Introduction to Reliability Engineering," 2d ed., Wiley, New York, 1995.
O'Connor, P. D. T.: "Practical Reliability Engineering," 3d ed., Wiley, New York, 1996.
Rao, S. S., "Reliability-Based Design," McGraw-Hill, New York, 1992.
Smith, D. J.: "Reliability, Maintainability, and Risk," 5th ed., Butterworth-Heinemann, Oxford,
    1997.

Safety Engineering

Bahr, N. J.: "System Safety Engineering and Risk Assessment," Taylor & Francis, Washington,
    DC, 1997.
Covan, J.: "Safety Engineering," Wiley, New York, 1995.
Hunter, T. A.: "Engineering Design for Safety," McGraw-Hill, New York, 1992.
Seiden, R. M.: "Product Safety Engineering for Managers," Prentice-Hall, Upper Saddle River,
    NJ, 1984.

PROBLEMS AND EXERCISES

11.1. Assume you are part of a federal commission established in 1910 to consider the risk
to society of the expected widespread use of the motor car powered with highly flammable gasoline. Without the benefit of hindsight, what potential dangers can you con-
template? Use a worst-case scenario. Now, taking advantage of hindsight, what lesson
can you draw about evaluating the hazards of future technologies? Do this as a team
exercise.

11.2. Comment on the following statistics in terms of the societal perception of risk.

According to the Consumer Product Safety Commission, in 1974 there were
about 40 serious injuries from fires started in wood stoves or fireplaces. By 1979 the
number had jumped to 400.

In 1980 the National Fire Protection Association counted 26 fires started by wood
stoves or fireplaces in which 3 or more people died.

11.3. A steel tensile link has a mean yield strength of $\bar{S}_y = 27,000$ psi and a standard deviation on strength of $S_y = 4000$ psi. The variable applied stress has a mean value of
$\bar{\sigma} = 13,000$ psi and a standard deviation $s = 3000$ psi.
   (*a*) What is the probability of failure taking place? Show the situation with carefully
        drawn frequency distributions.
   (*b*) The factor of safety is the ratio of the mean material strength divided by the mean
        applied stress. What factor of safety is required if the allowable failure rate is 5
        percent?
   (*c*) If absolutely no failures can be tolerated, what is the lowest value of the factor of
        safety?

11.4. A machine component has average life of 120 h. Assuming an exponential failure distribution, what is the probability of the component operating for at least 200 h before failing?

11.5. A nonreplacement test was carried out on 100 electronic components with a known constant failure rate. The history of failures was as follows:

1st failure after	93 h
2nd failure after	1,010 h
3rd failure after	5,000 h
4th failure after	28,000 h
5th failure after	63,000 h

The testing was discontinued after the fifth failure. If we can assume that the test gives an accurate estimate of the failure rate, determine the probability that one of the components would last for (*a*) 10^5 h and (*b*) 10^6 h.

11.6. The failure of a group of mechanical components follows a Weibull distribution, where $\theta = 10^5$ h, $m = 4$, and $t_0 = 0$. What is the probability that one of these components will have a life of 2×10^4 h?

11.7. A complex system consists of 550 components in a series configuration. Tests on a sample of 100 components showed that 2 failures occurred after 1000 h. If the failure rate can be assumed to be constant, what is the reliability of the system to operate for 1000 h? If an overall system reliability of 0.98 in 1000 h is required, what would the failure rate of each component have to be?

11.8. A system has a unit with MTBF = 30,000 h and a standby unit (MTBF = 20,000 h). If the system must operate for 10,000 h, what would be the MTBF of a single unit (constant failure rate) that, without standby, would have the same reliability as the standby system?

11.9. A reliability block diagram for an engineering system is given in Fig. 11.15. Determine the overall system reliability.

11.10. Make a failure modes and effects analysis for a ball-point pen.

11.11. Construct a qualitative fault tree diagram for a coal miner who is injured by a falling mine roof.

11.12. List a number of reasons why the determination of product life is important in engineering design.

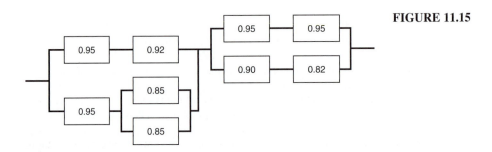

FIGURE 11.15

11.13. There has been growing societal concern about oil spills from tankers. What major changes in design would you propose to significantly change the risk of a major environmental accident?

11.14. By reading the literature, classify the macroappearance of fatigue failures with respect to type of loading (bending, torsion, axial) and presence of stress concentrations.

11.15. Using the principles of mechanical metallurgy, what would a torsion failure look like in a ductile material and a brittle material?

11.16. What steps of failure analysis would you undertake to determine what caused the crack in the Liberty Bell?

11.17. Read one of the following detailed accounts of a failure analysis.
   (a) C. O. Smith, Failure of a Twistdrill, *Trans. ASME, J. Eng. Materials Tech.,* vol. 96, pp. 88–90, April 1974.
   (b) C. O. Smith, Failure of a Welded Blower Fan Assembly, ibid., vol. 99, pp. 83–85, January 1977.
   (c) R. F. Wagner and D. R. McIntyre, Brittle Fracture of a Steel Heat Exchanger Shell, ibid., vol. 102, pp. 384–387, October 1980.

11.18. The following case study deals with forensic engineering, the combination of technical analysis and legal discovery.

Background

A manufacturer of modular housing units provided units to be used by the U.S. Navy as modular air traffic control towers. The original Navy specifications called for painted steel siding, but the supplier induced the Navy to substitute panels made from aluminum bonded on one side to plywood. These panels were used for both interior and exterior sheathing.

The supplier began to manufacture panels in February 1976, and delivery to the Navy was completed in November 1976. However, in October 1976 the Navy notified the supplier that the panels were deteriorating. Deterioration was in the form of blistering of the surface paint, delamination of the aluminum from the plywood, and bubbling and cracks in the aluminum skin. These phenomena occurred in all towers constructed, in both interior and exterior panels.

The Dialog

The Navy demanded the panels be repaired, charging they were defective in manufacture. The supplier maintained that, because the panels had been stored in an open field and then improperly assembled by the Navy, moisture in the form of rainwater had been allowed to enter and cause the observed damage.

The Failure Analysis

Macro- and microexamination of the delaminated and cracked panels showed that the aluminum was failing because of corrosion at the plywood-aluminum interface.

Problem Analysis

(a) Give three possible sources of the corrosion.
(b) List six detailed questions about plywood manufacture and bonding to plywood that must be answered. The answers to the questions point the finger toward a fire retardant in the plywood as the only possible source of corrosion.
(c) What simple tests would you perform to check out the retardant hypothesis? The tests showed an unusually high concentration of phosphorus and strongly

suggested that a fire retardant was involved. Checking the records of the supplier revealed that the panels had been painted on the plywood face and edges with a fire retardant paint containing phosphorus. In addition, x-ray fluorescence of the powdery white corrosion product with an electron beam microprobe showed the same elements as in the 3006 aluminum alloy, with the one exception that chlorine was present. Moreover, it was established from the manufacturer that the fire retardant paint had no history of causing corrosion when applied to bare aluminum.

(d) What are your tentative conclusions? A wood technologist expert in adhesives and plywood manufacture was employed to assist the metallurgist in this investigation. He quickly noticed the previously overlooked fact that the corrosion was worse in the center of the panels than at the edges. He also focused on the source of water to act as a leachant for chloride. He pointed out that plywood has a high initial water content (8 to 12 percent), which can be increased by the water-base glue used in its manufacture. His expertise caused the emphasis to be focused on the manufacture of the panels.

(e) Based on this new technical input, what is your analysis of the cause of failure? How could this be backed up by tests? For the trial litigation and the final outcome of this case, read the discussion of this problem by W. G. Dobson, N. J. Dilloff, and H. B. Gatslick, *Metal Progress,* pp. 61–68, August 1980. Do not read the article until you have tried to answer the preceding questions.

11.19. Consult the home page of the Consumer Products Safety Commission to determine what products have recently received rulings. Divide the work up between teams, and together, prepare a set of detailed design guidelines for safe product design.

11.20. Discuss the practice of using consumer complaints to establish that a product is hazardous and should recalled.

12

ROBUST AND QUALITY DESIGN

12.1
THE CONCEPT OF TOTAL QUALITY

In the 1980s many manufacturers around the world became threatened by the high quality of products produced by Japan. Not only were these products of high quality but they were competitively priced. The competitive threat forced a frantic search for the "magic bullet" that enabled Japanese manufacturers to capture market share. However, what the investigators found was a system of continuous quality improvement, *kaizen,* using simple statistical tools, emphasizing working in teams, and focused on delighting the customer. We have introduced many of these concepts throughout this text, starting with quality function deployment (QFD) in Chap. 2 and covering team methods and most of the problem-solving tools in Chap. 3. The concepts learned from the Japanese became known as *total quality management* (TQM) in the western world.

An important lesson learned from Japan is that the best way to achieve high quality in a product is to design it into the product from the beginning, and then to assure that it is maintained throughout the manufacturing stage. A further lesson, advanced by Dr. Genichi Taguchi, is that the enemy of quality is variability in performance of a product and in its manufacture. A *robust design* is one that has been created with a system of design tools that reduce product or process variability, while simultaneously guiding the performance toward an optimal setting. A product that is robustly designed will provide customer satisfaction even when subjected to extreme conditions on the manufacturing floor or the service environment.

12.1.1 Definition of Quality

Quality is a concept that has many meanings depending upon your perspective. On the one hand, quality implies the ability of a product or service to satisfy a stated or implied need. On the other hand, a quality product or service is free from defects or deficiencies. In Sec. 2.5 we defined quality in terms of Garvin's eight basic dimensions of a manufactured product.

In another basic paper, Garvin[1] identifies five distinct approaches toward the concept of quality.

1. The transcendent approach. This is a philosophical approach that holds that quality is some absolute and uncompromising high standard that we learn to recognize only through experience.
2. Product-based approach. This is completely opposite from the above, and views quality as a precise and measurable parameter. A typical parameter of quality might be the number of product features, or its expected life.
3. Manufacturing-based approach. In this view quality is defined by conformance to requirements or specifications. High quality is equated with "doing it right the first time."
4. Value-based approach. In this view quality is defined in terms of costs and prices. A quality product is one that provides performance at an acceptable price. This approach equates quality (excellence) with value (worth).
5. User-based approach. This approach views quality as "being in the eyes of the beholder." Each individual is considered to have a highly personal and subjective view of quality.

The phrase *total quality* used in the heading of this section denotes a broader concept of quality[2] than simply checking the parts for defects as they come off the production line. The idea of preventing defects by improved design, manufacturing, and process control plays a big role in total quality. We refer to the first aspect as off-line quality control, while the latter is on-line quality control. In order for total quality to be achieved it must be made the number one priority of the organization. This is rooted in the belief that quality is the best way to assure long-term profitability. In a study in which companies were ranked by an index of perceived quality, the firms in the top third showed an average return on assets of 30 percent while the firms in the bottom third showed an average return of 5 percent.

For total quality to be achieved requires a customer focus. Quality is meeting customer requirements consistently. To do this we must know who our customers are and what they require. This attitude should not be limited to external customers. Within our organization we should consider those we interact with as our customers. This means that a manufacturing unit providing parts to another unit for further processing should be just as concerned about defects as if the parts were shipped directly to the customer.

Total quality is achieved by the use of facts and data to guide in decision making. Thus, data should be used to identify problems and to help determine when and if action should be taken. Because of the complex nature of the work environment this requires a considerable skill in data acquisition and analysis with statistical methods.

1. D. A. Garvin, What Does Product Quality Really Mean, *Sloan Management Review,* Fall, 1984, pp. 25–44.
2. A. V. Feigenbaum, "Total Quality Control," 3d ed., McGraw-Hill, New York, 1983.

Finally, the quest for quality must be continuous and requires total involvement of the organization. Total quality will not be achieved by spurts or campaigns for quality. Also, it must involve all employees, especially those engaged in production.

12.1.2 Deming's 14 Points

Work by Walter Shewhart, Edwards Deming, and Joseph Juran in the 1920s and 1930s pioneered the use of statistics for the control of quality in production. These quality control methods were mandated by the War Department in World War II for all ordnance production and were found to be very effective. After the war, with a pent-up demand for civilian goods and relatively cheap labor and materials costs, these statistical quality control (SQC) methods were largely abandoned as unnecessary and an added expense. However, in Japan, whose industry had been largely destroyed by aerial bombing, it was a different story. The Japanese Union of Scientists and Engineers invited Dr. W. Edwards Deming to Japan in 1950 to teach them SQC. His message was enthusiastically received and SQC became an integral part of the rebuilding of Japanese industry. An important difference between how Americans and Japanese were introduced to SQC is that in Japan the first people converted were top management while in America it was largely engineers who adopted it. The Japanese have continued to be strong advocates of SQC methods and have extended it and developed new adaptations. Today the world looks to Japanese products as a standard of quality and is rapidly adopting statistical methods in design and production. In Japan, the national award for industrial quality, a very prestigious award, is called the Deming Prize.

Dr. Deming views quality in a broader philosophy of management.[1] This is expressed by his fourteen points.

1. Create a constancy and consistency of purpose toward improvement of product and service. Aim to become competitive and to stay in business and to provide jobs.
2. Adopt the philosophy that we are in a new economic age. Western management must awaken to the challenge, must learn their responsibilities, and take on the leadership of change.
3. Stop depending on inspection to achieve quality. Eliminate the need for production line inspection by building quality into the product's design.
4. Stop the practice of awarding business only on the basis of price. The goal should be to minimize total cost, not just acquisition cost. Move toward a single supplier for any one item. Create a relationship of loyalty and trust with your suppliers.
5. Search continually for problems in the system and seek ways to improve it.
6. Institute modern methods of training on the job. Management and workers alike should know statistics.
7. The aim of supervision should be to help people and machines to do a better job. Provide the tools and techniques for people to have pride of workmanship.

---

1. W. E. Deming, "Out of Crisis," MIT Center for Advanced Engineering Study, Cambridge, MA, 1986; M. Tribus, Mechanical Engineering, January 1988, pp. 26–30.

8. Eliminate fear, so that everyone may work effectively for the company. Encourage two-way communication.

9. Break down barriers between departments. Research, design, sales, and production must work as a team.

10. Eliminate the use of numerical goals, slogans, posters for the workforce. 80 to 85 percent of the causes of low quality and low productivity are the fault of the system, 15 to 20 percent are because of the workers.

11. Eliminate work standards (quotas) on the factory floor and substitute leadership. Eliminate management by objective, management by numbers, and substitute leadership.

12. Remove barriers to the pride of workmanship.

13. Institute a vigorous program of education and training to keep people abreast of new developments in materials, methods, and technology.

14. Put everyone in the company working to accomplish this transformation. This is not just a management responsibility—it is everybody's job.

12.2
QUALITY CONTROL AND ASSURANCE

"Quality control"[1] refers to the actions taken throughout the engineering and manufacturing of a product to prevent and detect product deficiencies and product safety hazards. The American Society for Quality (ASQ) defines *quality* as the totality of features and characteristics of a product or service that bear on ability to satisfy a given need. In a narrower sense, "quality control" refers to the statistical techniques employed in sampling production and monitoring the variability of the product. *Quality assurance* refers to those systematic actions vital to provide satisfactory confidence that an item or service will fulfil defined requirements.

Quality control received its initial impetus in the United States in World War II when war production was facilitated and controlled with QC methods. The traditional role of quality control has been to control the quality of raw materials, control the dimensions of parts during production, eliminate imperfect parts from the production line, and assure functional performance of the product. With increased emphasis on tighter tolerance levels, slimmer profit margins, and stricter interpretation of liability laws by the courts (see Sec. 15.6), there has been even greater emphasis on quality control. More recently the heavy competition for U.S. markets from overseas producers who have emphasized quality in the extreme has placed even more emphasis on QC by U.S. producers.

---

1. J. M. Juran and F. M. Gryna (eds.), "Juran's Quality Control Handbook," 4th ed., McGraw-Hill, New York, 1988; J. M. Juran and F. M. Gryna, "Quality Planning and Analysis," 2d ed., McGraw-Hill, New York, 1980.

12.2.1 Fitness for Use

An appropriate engineering viewpoint of quality is to consider that it is *fitness for use.* The consumer may confuse quality with luxury, but in an engineering context quality has to do with how well a product meets its design and performance specifications. The majority of product failures can be traced back to the design process. It has been found that 75 percent of defects originate in product development and planning, and that 80 percent of these remain undetected until the final product test or in service.[1]

The particular technology used in manufacturing has an important influence on quality. We saw in Chap. 9 that each manufacturing process has an inherent capability for maintaining tolerances, generating a shape, and producing a surface finish. This has been codified into a methodology called conformability analysis.[2] This technique aims, for a given design, to identify the potential process capability problems in component manufacture and assembly and to estimate the level of potential failure costs.

As computer-aided applications pervade manufacturing, there is a growing trend toward automated inspection. This permits a higher volume of part inspection and removes human variability from the inspection process. An important aspect of QC is the design of inspection fixtures and gauging.[3]

The skill and attitude of production workers can have a great deal to do with quality. Where there is pride in the quality of the product there is greater concern for quality on the production floor. A technique used successfully in Japan and meeting with growing acceptance in the United States is the quality circle, in which small groups of production workers meet regularly to suggest quality improvements in the production process.

Management must be solidly behind it, or total quality will not be achieved. There is an inherent conflict between achieving quality and wanting to meet production schedules at minimum cost. This is another manifestation of the perennial conflict between short- and long-term goals. There is general agreement that the greater the autonomy of the quality function in the management structure the higher the level of quality in the product. Most often the quality control and manufacturing departments are separate and both the QC manager and the production manager report to the plant manager.

The concept of availability was discussed in Chap. 11. Reliability should be designed into the product; maintainability is a function of design plus manufacturing quality. Logistic support includes the cost and ease of obtaining the parts needed to keep the equipment operational.

Field service comprises all the services provided by the manufacturer after the product has been delivered to the customer: equipment installation, operator training,

1. K. G. Swift and A. J. Allen, Product Variability, Risks, and Robust Design, *Proc. Instn. Mech. Engrs.,* vol. 208, pp. 9–19, 1994.
2. K. G. Swift, M. Raines, and J. D. Booker, Design Capability and the Costs of Failure, *Proc. Instn. Mech. Engrs.,* vol. 211, Part B, pp. 409–423, 1997.
3. C. W. Kennedy and D. E. Andrews, "Inspection and Gaging," 6th ed., Industrial Press, Inc., New York, 1987.

repair service, warranty service, and claim adjustment. The level of field service is an important factor in establishing the value of the product to the customer, so that it is a real part of the fitness-for-use, concept of quality control. Customer contact by field service engineers is one of the major sources of input about the quality level of the product. Information from the field "closes the loop" of quality assurance and provides needed information for redesign of the product.

12.2.2 Quality-Control Concepts

A basic tenet of quality control is that variability is inherent in any manufactured product. Someplace there is an economic balance between reducing the variability and the cost of manufacture.[1] Statistical quality control considers that part of the variability is inherent in the materials and process and can be changed only by changing those factors. The remainder of the variability is due to assignable causes that can be reduced or eliminated if they can be identified.

The basic questions in establishing a QC policy for a part are four in number: (1) What do we inspect? (2) How do we inspect? (3) When do we inspect? (4) Where do we inspect?

What to inspect

The objective should be to focus on a few critical characteristics of the product that are good indicators of performance. This is chiefly a technically based decision. Another decision is whether to emphasize nondestructive or destructive inspection. Obviously, the chief value of an NDI technique is that it allows the manufacturer to inspect a part that will actually be sold. Also, the customer can inspect the same part before it is used. Destructive tests, like tensile tests, are done with the assumption that the results derived from the test are typical of the population from which the test samples were taken. Often it is necessary to use destructive tests to verify that the nondestructive test is measuring the desired characteristic.

How to inspect

The basic decision is whether the characteristic of the product to be monitored will be measured on a continuous scale (inspection by variables) or whether the part passes or fails some go-no-go test. The latter situation is known as measurement by attribute. Inspection by variables uses the normal, lognormal, or some similar frequency distribution. Inspection by attributes uses the binomial and Poisson distributions.

When to inspect

The decision on when to inspect determines the QC method that will be employed. Inspection can occur either while the process is going on (process control) or after it has been completed (acceptance sampling). A process control approach is used when the inspection can be done nondestructively at low unit cost. An important benefit of

---

1. J. L. Plunkett and B. G. Dale, *Int. J. Prod. Res.,* vol. 26, pp. 1713–1726, 1988.

process control is that the manufacturing conditions can be continuously adjusted on the basis of the inspection data to reduce the percent defectives. Acceptance sampling often involves destructive inspection at a high unit cost. Since not all parts are inspected, it must be expected that a small percentage of defective parts will be passed by the inspection process. The development of sampling plans[1] for various acceptance sampling schemes is an important aspect of statistical quality control.

Where to inspect

This decision has to do with the number and location of the inspection steps in the manufacturing process. There is an economic balance between the cost of inspection and the cost of passing defective parts to the later stages of the production sequence or to the customer. The number of inspection stations will be optimal when the marginal cost of another inspection exceeds the marginal cost of passing on some defective parts. Inspection operations should be conducted before production operations that are irreversible, i.e., operations that are very costly or where rework is impossible. Inspection of incoming raw material to a production process is one such place. Steps in the process that are most likely to generate flaws should be followed by an inspection. In a new process, inspection operations might take place after every process step; but as experience is gathered, the inspection would be maintained only after steps that have been shown to be critical.

12.2.3 New Ideas

The success of the Japanese in designing and producing quality products has led to new ideas about quality control. Rather than flooding the receiving dock with inspectors who establish the quality of incoming raw material and parts, it is cheaper and faster to require the supplier to provide statistical documentation that the incoming material meets quality standards. This can only work where the buyer and seller work in an environment of cooperation and trust.

In traditional QC an inspector makes the rounds every hour, picks up a few parts, takes them back to the inspection area, and checks them out. By the time the results of the inspection are available it is possible that bad parts have been manufactured and it is likely that these parts have either made their way into the production stream or have been placed in a bin along with good parts. If the latter happens, the QC staff will have to perform a 100 percent inspection to separate good parts from bad. We end up with four grades of product—first quality, second quality, rework, and scrap. To achieve close to real-time control inspection must be an integral part of the manufacturing process. Ideally, those responsible for making the parts should also be responsible for acquiring the process performance data so that they make appropriate adjustments. There is a trend to using electronic data collectors to eliminate human error and to speed up analysis of data.

---

1. See MIL-STD-105D and MIL-STD-414.

12.2.4 Quality Assurance

Quality assurance is concerned with all corporate activities that affect customer satisfaction with the quality of the product. There must be a quality assurance department with sufficient independence from manufacturing to act to maintain quality. This group is responsible for interpreting national and international codes and standards in terms of each purchase order and for developing written rules of operating practice. Emphasis should be on clear and concise written procedures. A purchase order will generate a great amount of in-plant documentation, which must be accurate and be delivered promptly to each workstation. Much of this paper flow has been computerized, but there must be a system by which it gets on time to the people who need it. There must also be procedures for maintaining the identity and traceability of materials and semifinished parts while in the various stages of processing. Definite policies and procedures for dealing with defective material and parts must be in place. There must be a way to decide when parts should be scrapped, reworked, or downgraded to a lower quality level. A quality assurance system must identify which records should be kept, and establish procedures for accessing those records as required.

Quality control is not something that can be put in place and then forgotten. There must be procedures for training, qualifying, and certifying inspectors and other QC personnel. Funds must be available for updating inspection and laboratory equipment and for the frequent calibration of instruments and gauges.

12.2.5 ISO 9000

An important aspect of quality assurance is the audit of the quality system against written standards.[1] The most prevalent quality standard is ISO 9000, and its companion standards, that are issued by the International Organization for Standards (ISO). ISO 9000 has become required by companies doing business in the European Community, and since it is a worldwide marketplace, companies around the world have been scrambling to become ISO 9000 certified. Certification to ISO 9000 is accomplished by submitting to an audit by an accredited ISO registrar. Over 100,000 certificates have been issued worldwide.

To achieve ISO 9000 certification a company must take the following steps:

- Assess current conformance to ISO standards; identify areas that need to be improved.
- Prepare and implement procedures to fix areas of noncompliance; train employees on how to use the procedures.
- Submit quality manual, procedure manual, and work instructions to the auditing body for off-site review.

---

1. D. Hoyle, "ISO 9000: Quality System Assessment Handbook," Butterworth-Heinemann, Oxford, 1996.

- The auditing body will make an on-site visit to ensure that procedures are being followed and that all employees have been properly trained in their use.
- Following the on-site review, the auditing group will submit a report to the certification committee for review and approval.
- If the report is favorable, ISO certification will be granted. If deficiencies are noted, the company is given a period of time in which to correct them. If all deficiencies are corrected, ISO certification will be granted.

The system of ISO 9000 standards is listed in Table 12.1. ISO 9001 is the most complete since it extends from design to field service. Clause 4.4, Design Control, lays out many of the issues discussed in this text (Table 12.2). The big three automotive producers in the United States have advanced their own standard, QS 9000, similar to ISO 9000 which must be followed by their suppliers.

A related standard to ISO 9000 is ISO 14000, which deals with environmental management guidelines. Currently there is a move toward making these two standards very compatible. It is expected that ISO 14000 will have even more impact on companies than ISO 9000.

TABLE 12.1
ISO 9000 Standards

Standard	Subject
IS0 9000	Guidelines for Selection & Use
ISO 9001	Quality Assurance in Design, Production, Installation, and Servicing
ISO 9002	Quality Assurance in Production, Installation, and Servicing
ISO 9003	Quality Assurance in Final Inspection
ISO 9004	Guidelines for Implementation

TABLE 12.2
Topics covered in ISO 9001, Clause 4.4,
Design Control

Subclause	Topic
4.4.1	General
4.4.2	Design and development planning
4.4.3	Organizational and technical interfaces
4.4.4	Design input
4.4.5	Design output
4.4.6	Design review
4.4.7	Design verification
4.4.8	Design validation
4.4.9	Design changes

12.3
QUALITY IMPROVEMENT

Four basic costs are associated with quality.

Prevention—those costs incurred in planning, implementing, and maintaining a quality system. Included are the extra expense in design and manufacturing to ensure the highest quality product.

Appraisal—costs incurred in determining the degree of conformance to the quality requirements. The cost of inspection is the major contributor.

Internal failure—costs incurred when materials, parts, and components fail to meet the quality requirements for shipping to the customer. These parts are either scrapped or reworked.

External failure—costs incurred when products fail to meet customer expectation. These result in warranty claims, ill will, or product liability suits.

To simply collect statistics on defective parts and weed them out of the assembly line is not sufficient for quality improvement and cost reduction. A proactive effort must be made to determine the root causes of the problem so that permanent corrections can be made. Two commonly used techniques in this area of problem solving are the *Pareto diagram* and *cause-and-effect analysis.*

12.3.1 Pareto diagram

In 1897 an Italian economist, Vilfred Pareto, studied the distribution of wealth in Italy and found that a large percentage of the wealth was concentrated in about 10 percent of the population. This was published and became known as Pareto's law. Shortly after World War II, inventory control analysts observed that about 20 percent of the items in the inventory accounted for about 80 percent of the dollar value. In 1954 Joseph Juran generalized Pareto's law as the "80/20 rule," e.g., 80 percent of sales are generated by 20 percent of the customers, 80 percent of the product defects are caused by 20 percent of the parts, etc. While there is no widespread validation of the 80/20 rule it is widely quoted as a useful axiom. Certainly Juran's admonition "to concentrate on the vital few and not the trivial many" is excellent advice in quality improvement, as in other aspects of life.

12.3.2 Cause-and-Effect Analysis

Cause-and-effect analysis uses the "fishbone diagram" or Ishikawa diagram,[1] Fig. 12.3, to identify possible causes of a problem. The poor quality is associated with four categories of causes: operator (man), machine, method, and material. The likely

---

1. K. Ishikawa. "Guide to Quality Control," 2d ed., UNIPUB, New York, 1982.

causes of the problem are listed on the diagram under these four main categories. Suggested causes of the problem are generated by the manufacturing engineers, technicians, and production workers meeting to discuss the problem. The use of the cause-and-effect diagram provides an orderly, step-by-step approach to improving manufacturing quality. Manufacturing management should develop a cause-and-effect diagram for each manufacturing process and update it as more knowledge about the process becomes available.

EXAMPLE 12.1. A manufacturing plant is producing injection molded automobile grilles.[1] The process was newly installed and the parts produced had a number of defects. Therefore, a quality improvement team consisting of operators, setup people, manufacturing engineers, production supervisors, quality control staff, and statisticians was assembled to improve the situation. The first task was to agree on what the defects were and how to specify them. Then a sampling of 25 grilles was examined for defects. Figure 12.1a shows the control chart (see Sec. 12.4 for more details) for the grilles produced by the process. It shows a mean of 4.5 defects per part. The pattern is typical of a process out of control.

A Pareto diagram was prepared to portray the relative frequency of the various types of defects, Fig. 12.2. This was based on the data in Fig. 12.1a. It shows that black spots (degraded polymer patches on the surface) are the most prevalent type of defect. Therefore, it was decided to focus attention initially on this defect.

Focusing on the causes of the black spots resulted in the "fishbone" diagram shown in Fig. 12.3. The causes are grouped under the four-Ms of manufacturing. Note that for some items, like the injector screw, the level of detail is greater. The group decided that the screw had been worn through too much use and needed to be replaced.

When the screw was changed the black spots completely disappeared (see control chart in Fig. 12.1b). Then after a few days the black spots reappeared to about the same level of intensity as before. Thus, it must be concluded that the root cause of black spots had not been identified. The quality team continued to meet to discuss the black spot problem. It was noted that the design of the vent tube on the barrel of the injection molding machine was subject to clogging and was difficult to clean. It was hypothesized that either polymer accumulated in the vent tube port, became overheated and periodically broke free and continued down the barrel, or was pushed back into the barrel during cleaning. A new vent tube design which minimized these possibilities was designed and constructed, and when installed the black spots disappeared, Fig. 12.1c.

Having solved the most prevalent defect problem the team turned their attention to scratches, the defect with the second highest frequency. A press operator proposed that the scratches were caused by the hot plastic parts falling on the metal lacings of the conveyor belt. He proposed using a continuous belt without metal lacings. However, this type of belt cost twice as much. Therefore, an experiment was proposed in which the metal lacings were covered with a soft latex coating. When this was done the scratches disappeared, but after time they reappeared as the latex coating wore away. With the evidence from this experiment the belt with metal lacings was replaced by a continuous vulcanized belt, not only on the machine under study but for all the machines in the shop.

---

1. This example is based on "Tool and Manufacturing Engineer's Handbook," 4th ed., vol. 4, pp. 2-20 to 2-24, Society of Manufacturing Engineers, Dearbon, MI, 1987.

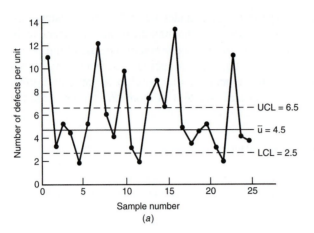

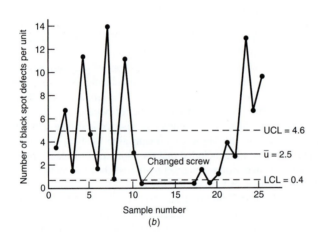

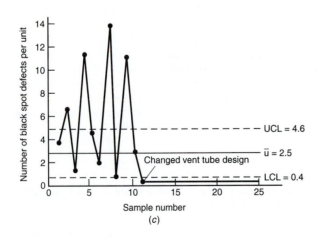

FIGURE 12.1
Control chart for the number of defects for injection molded grilles: (*a*) process out of control; (*b*) process after injection screw was changed; (*c*) process after new vent system was installed. (*From "Tool and Manufacturing Engineers Handbook," 4th ed., vol. 4, p. 2-22, 1987, courtesy of Society of Manufacturing Engineers, Dearborn, MI.*)

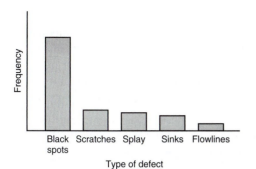

FIGURE 12.2
Pareto diagram for defects in automotive grille. (*From "Tool and Manufacturing Engineers Handbook," 4th ed., vol. 4, p. 2-22, 1987, courtesy of Society of Manufacturing Engineers, Dearborn, MI.*)

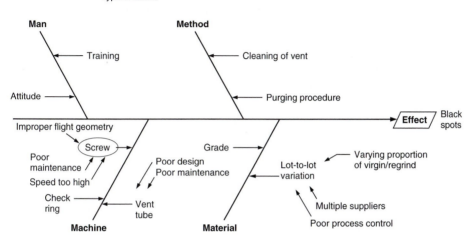

FIGURE 12.3
Cause-and-effect (Ishikawa) diagram for black spot defects on automobile grille. (*From "Tool and Manufacturing Engineers Handbook," 4th ed., vol. 4, p. 2-23, 1987, courtesy of Society of Manufacturing Engineers, Dearborn, MI.*)

12.4
STATISTICAL PROCESS CONTROL

Collecting manufacturing performance data and keeping charts on this data is common practice in industrial plants. William Shewhart[1] showed that such data could be interpreted and made useful through a simple but statistically sound method called a control chart.

12.4.1 Control Charts

The use of the control chart is based on the viewpoint that every manufacturing process is subject to two source of variation: (1) chance variation, also called common causes

---

1. W. A. Shewhart, "Economic Control of Quality in Manufacture Product," Van Nostrand Reinhold Co., New York, 1931.

of variation, and (2) assignable variation, or that due to special causes. Chance variation arises from numerous factors which are individually of small importance. Generally it is not feasible to detect or identify them individually. An assignable variation is due to a cause like poorly trained operators or worn production machines, which it is possible and important to detect and identify. The control chart is one of the chief methods used in the important branch of applied statistics known as quality control.[1]

> **EXAMPLE 12.2.** Consider a commercial heat-treating operation in which bearing races are being quenched and tempered in a conveyor-type furnace on a continuous 24-h basis. Every 2 h the Rockwell hardness is measured in 10 bearing races to determine whether the product conforms to the specifications. The mean of each sample $\bar{x}$ is computed, and the dispersion is determined by computing the range R. Separate control charts are kept for the mean and the range (Fig. 12.4).
>
> If the population parameters are known, then the $\pm 3\sigma$ limits represent the upper control limit (UCL) and the lower control limit (LCL). In the more usual case, in which the population parameters are unknown, it is necessary to estimate the parameters from preliminary samples. If k samples, each of size n, are used, then we compile the mean for each sample $\bar{x}_i$ and determine the grand mean from

$$\bar{\bar{x}} = \frac{1}{k} \sum_{i=1}^{k} \bar{x}_i \tag{12.1}$$

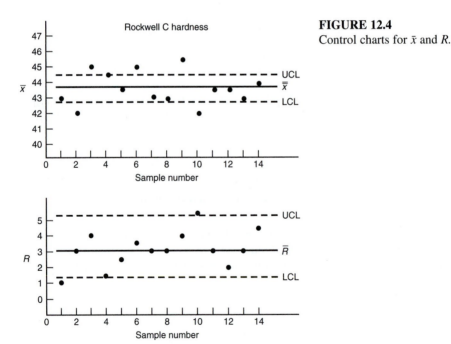

FIGURE 12.4
Control charts for $\bar{x}$ and R.

1. E. L. Grant and R. S. Leavenworth, "Statistical Quality Control," 6th ed., McGraw-Hill, New York, 1988; W. S. Messina, "Statistical Quality Control for Manufacturing Managers," Wiley, New York, 1987.

The variability of the process usually is determined from the range, $R = x_{max} - x_{min}$. For samples with small n there is little loss in efficiency in estimating σ from the sample range. The mean range $\bar{R}$ is determined from

$$\bar{R} = \frac{1}{k} \sum_{i=1}^{k} R_i \tag{12.2}$$

The upper and lower control limits on the mean are determined by multiplying $\bar{R}$ by a constant A_2, which depends on the sample size n.

$$UCL = \bar{\bar{x}} + A_2\bar{R}$$

$$LCL = \bar{\bar{x}} - A_2\bar{R} \tag{12.3}$$

Also, the limits on $\bar{R}$ are obtained from

$$UCL = D_4\bar{R}$$

$$LCL = D_3\bar{R} \tag{12.4}$$

where typical values of the statistical parameters are given in Table 12.3. Note how increasing the number of observations in the sample reduces the spread between the UCL and LCL on the mean. Thus, a larger sample size permits controlling the process within narrower limits. The same is true for the limits for R, but n must exceed 6 before the lower control limit exceeds zero. Thus, with small samples only the UCL on the range is meaningful.

In interpreting the control chart it is important to start with the R chart because the limits on the $\bar{x}$ chart depend on the magnitude of the chance variation of the process as measured by R. If some points on the R chart are initially out of control the limits on the $\bar{x}$ chart will be inflated.

On examining the R chart we see two points outside of the control limits. We go back into the records to see if we can establish reasonable cause for these events. We discover that the first point was for the initial heat-treat batch on Monday morning when the furnace had been started up and the temperature strip chart record showed that the furnace was not up to the proper temperature when the parts were quenched. No assignable cause can be given to the second point, but with the warning given here the production supervisor decides to give closer scrutiny to the operators on this furnace. Examining the $\bar{x}$ chart

TABLE 12.3
Factors for use in determining control limits
of control charts

No. of observations in sample n	A_2	D_3	D_4
2	1.88	0	3.27
4	0.73	0	2.28
6	0.48	0	2.00
8	0.37	0.14	1.86
10	0.27	0.22	1.78
12	0.22	0.28	1.71

we conclude that the frequent out of control values, even after the $\bar{x}$ values have been recalculated to omit the two points from the R chart, indicate that hardenability variation in the steel is too great to hold such a tight tolerance. This is an example of the use of the control chart to establish whether a process has the capability of being controlled within the specification limits.

The control chart is a model that describes the way the process variability is expected to appear when only chance causes of variability are present. Once this pattern is established it sends a signal when variation occurs that does not fit the model of common cause variation. If a process is operating within the control limits then any improvement of the process by simple machine adjustment is not likely to improve the mean, but it is likely to induce additional variation in the process. The occurrence of points outside the control limits points to the existence of special causes of variation. However, there are other patterns that may also indicate that the process is out of statistical control. These patterns may be trends, cycles, stratification, etc.[1] The control chart can be helpful in tracking down these causes, as in Sec. 12.3. Some powerful statistical techniques to improve the process and product performance are discussed in Sec. 12.5.

The above discussion of control charts was based on a parameter measured on a quantitative scale. Often in inspection it is quicker and cheaper to check the product on a go-no-go basis; the part either passes or fails some predetermined specification. In this type of *attribute testing* different control charts are used. The p chart, based on the binomial distribution, deals with the fraction of defective parts in a sample over a succession of samples. The c chart, based on the Poisson distribution, monitors the number of defects per sample. Other important issues in statistical quality control are the design of sampling plans and the intricacies of sampling parts on the production line.[2]

12.4.2 Process Capability

An in-control manufacturing process cannot be achieved if the characteristics of the part to be made are not within the capability of the process. Statistics provide the insight. We introduce these concepts through examples.

> **EXAMPLE 12.3** A radical design of an automotive engine consists of modular cylinders and cylinder blocks that are joined into an engine block by hot isostatic pressing. Each module is a block 250 mm high by 200 × 200 mm, with a 100-mm-diameter bore. The modules are made in a computer-controlled high-speed milling machine. The critical dimensions are
>
> Width: 200 mm +0.04 mm   Height: 250 mm +0.00 mm
>   −0.00 mm   −0.10 mm

---

1. A. J. Duncan, "Quality Control and Industrial Statistics," 4th ed., pp. 386–393, Irwin, Inc., Homewood, IL, 1974.
2. D. H. Besterfield, "Quality Control," 5th ed., Prentice-Hall, Upper Saddle River, NJ, 1998; A. Mitra, "Fundamentals of Quality Control and Improvement," 2d ed., Prentice-Hall, Upper Saddle River, NJ, 1998; W. J. Kolarik, "Creating Quality," McGraw-Hill, New York, 1995.

The standard deviation for the milling machine was known from previous production runs to be $\sigma = 0.10$ mm.

(a) We assume that height measurements for blocks made on the machine follow a normal distribution. Defect-free parts are made with dimensions in the range 250.00 to 249.90 mm. Further assume that the machine setting is perfectly centered between the upper and lower tolerance limits to give a mean $\mu = 249.95$ mm. Figure 12.5 shows the situation. The probability of making a part with an out-of-tolerance height is

$$P(X < 249.90) + P(X > 250.00) = P\left(\frac{X - \mu}{\sigma} < \frac{249.90 - 249.95}{0.10}\right)$$

$$+ P\left(\frac{X - \mu}{\sigma} > \frac{250.00 - 249.95}{0.10}\right)$$

$$= P(Z < -0.5) + P(Z > 0.5) = 0.3085 + 0.3085 = 0.617 \text{ (from Table 10.2)}$$

It is clear that a 61.7 percent defect rate is entirely unsatisfactory. The tolerances specified are too tight for the capability of the milling machine.

(b) One common practice is to use a $\pm 3\sigma$ limit to set the tolerances. This would mean that the tolerance on height would be $250 \text{ mm}^{+0.00 \text{ mm}}_{-0.60 \text{ mm}}$. Thus, the height of the modules would fall within 250.00 and 249.40 mm. Using Table B-1 in the Appendix, we see that 99.73 percent of the module heights would fall within this limit, assuming that the machine setting is at the midpoint of the tolerance range.

$$P(249.40 < X > 250.00) = P(-3.0 < Z > 3.0) = 0.9973$$

The percent defects will be equal to or less than $1 - P = 1 - 0.9973 = 0.0027 = 0.27$ percent.

(c) The *process capability index*, C_p, is commonly used to establish the relationship between the tolerances specified for the component and the standard deviation for the process that will make it.

$$C_p = \frac{\text{USL} - \text{LSL}}{6\sigma} \tag{12.5}$$

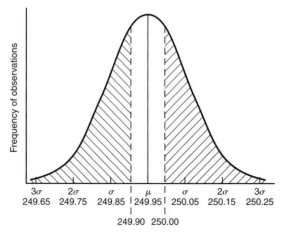

FIGURE 12.5
Normal distribution to estimate the probability of making parts that fall in the tolerance interval 250.00 to 249.90 mm.

where USL is the upper specification limit and LSL is the lower specification limit. Comparing Eq. (12.5) with (b) above, we see that the 6 enters the equation because it represents $2 \times 3\sigma$, used in the $\pm 3\sigma$ rule. Thus, when $C_p = 1$ the expected percent of defects is equal to or less than 0.27 percent, as shown above in (b) when USL $-$ LSL is $\pm 3\sigma$.

The usual expectation is that $C_p \geq 1.30$, especially for mass production where the defect rate is very important. This requires that USL $-$ LSL be greater than 7.8σ. Thus, when $C_p > 1.3$ the expected percent defects is

$$1 - P(-3.9 < Z > 3.9) = 1 - 0.99995 = 0.00005 = 0.005\%$$

$$= 50 \text{ parts per million (ppm)}$$

(d) Motorola set a high hurdle for quality when it announced in the 1980s that its goal would be to achieve "six-sigma quality." Since then, several other major corporations have taken up the same goal. A true six-sigma process is achieved when $\mu \pm 6\sigma$ equals the product specification tolerance interval. Substituting in Eq. (12.5),

$$C_p = \frac{\text{USL} - \text{LSL}}{6\sigma} = \frac{12\sigma}{6\sigma} = 2.0$$

The percentage of defects is given by

$$1 - P(-6 < Z > 6) = 2(1 - 0.999999999) = 0.000000002$$

$$= 0.0000002\% = 2 \text{ ppb}$$

(e) All of the above examples assume that the mean is perfectly centered between the upper and lower tolerance limits, i.e., that the center of the process variation and the tolerance range coincide. As we will see in Sec. 12.5, this is the strategy employed in the Taguchi approach for robust design, but as a process runs for some time the process becomes uncentered due to tool wear and other process variations. The process capability defined in Eq. (12.5) is thus the ideal process capability. The actual process capability is given by C_{pk}

$$C_{pk1} = \frac{\text{USL} - \mu}{3\sigma}$$

$$C_{pk2} = \frac{\mu - \text{LSL}}{3\sigma} \quad \quad (12.6)$$

$$C_{pk} = \min(C_{pk1}, C_{pk2})$$

where μ is the mean of the process. C_{pk} is used by the manufacturing engineers to center the process. In production we seek to make $C_{pk1} = C_{pk2}$ and to keep C_{pk} at a value of 1 or greater.

When the process is not centered in the tolerance range the percentage of defects increases significantly. Consider (b) above, where we set the tolerance range at $\pm 3\sigma$. Now assume that the process mean is not at the tolerance mean, 249.70, but $\mu = 249.90$.

$$C_{pk1} = \frac{250.00 - 249.90}{3(0.10)} = \frac{0.10}{0.30} = 0.33$$

TABLE 12.4
Effect of shift in process mean on defect rate

Tolerance range	Process centered		Process 1.5σ from center	
	Percent good parts	Defective parts ppm	Percent good parts	Defective parts ppm
± One sigma	68.27	317,300	30.23	697,700
± Three sigma	99.73	2,700	93.32	66,810
± Six sigma	99.9999998	0.002	99.99966	3.4

The figure of 3.4 defects per million parts is often taken as the target of a six-sigma quality program.

$$C_{pk2} = \frac{249.90 - 249.40}{3(0.10)} = \frac{0.50}{0.30} = 1.67$$

Clearly the process is not centered in the tolerance interval. $C_{pk} = 0.33$ because it is the lowest of the two values. For $C_{pk} = 1.67$, $Z = 5.01$ and for $C_{pk} = 0.333$, $Z = 1.0$, and

$$P(Z < -5.01) + P(Z > 1.0) = 3 \times 10^{-7} + (1 - 0.8413) = 0.1587 = 16\%$$

Thus, failure to maintain the process centered in the tolerance interval has significantly increased the defect rate. Table 12.4 shows how the defect rate increases significantly when the process shifts from a centered normal distribution to one that is shifted 1.5σ from the center of the tolerance interval.[1]

12.5
TAGUCHI METHOD

A systematized statistical approach to product and process improvement has developed in Japan under the leadership of Dr. Genichi Taguchi.[2] This took a total quality emphasis but developed quite unique approaches and terminology. It emphasizes moving the quality issue upstream to the design stage and focusing on prevention of defects by process improvement. Taguchi has placed great emphasis on the importance of minimizing variation as the primary means of improving quality. Special attention is given to the idea of designing products so that their performance is insensitive to changes in the environment in which the product functions, also called noise. The process of achieving this through the use of statistically designed experiments has been called *robust design* (see Sec. 12.6).

---

1. M. J. Harry, "Quality Progress," May 1998, pp. 60–64; M. J. Harry and J. R. Lawson, "Six Sigma Producibility Analysis and Process Characterization," Addison-Wesley, Reading, MA, 1992.
2. Taguchi methods is a registered trademark of the American Supplier Institute, Dearborn, MI; http://www.amsup.com. Other web sites are http://wally2.rit.edu/internet/subject/tqm.html and http://kernow.curtin.edu.au/www/Taguchi.

12.5.1 Loss Function

Taguchi defines the quality level of a product to be the total loss incurred by society due to the failure of the product to deliver the expected performance and due to harmful side effects of the product, including its operating cost. This may seem a backward definition of quality because the word quality usually connotes desirability, while the word loss conveys the impression of undesirability. In the Taguchi concept some loss is inevitable from the time a product is shipped to the customer so that the smaller the loss the more desirable the product. It is important to be able to quantify this loss so that alternative product designs and manufacturing processes can be compared. This is done with a quadratic loss function (Fig. 12.6a),

$$L(y) = k(y - m)^2 \tag{12.7}$$

where $L(y)$ is the quality loss when the quality characteristic is y, m is the target value for y and k is a constant, the quality loss coefficient.[1]

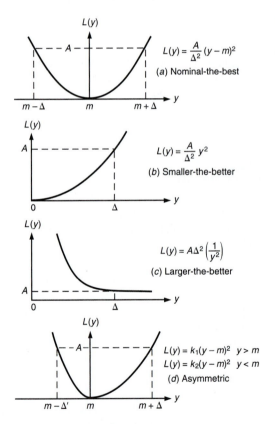

$$L(y) = \frac{A}{\Delta^2}(y - m)^2$$

(a) Nominal-the-best

$$L(y) = \frac{A}{\Delta^2}y^2$$

(b) Smaller-the-better

$$L(y) = A\Delta^2\left(\frac{1}{y^2}\right)$$

(c) Larger-the-better

$$L(y) = k_1(y - m)^2 \quad y > m$$
$$L(y) = k_2(y - m)^2 \quad y < m$$

(d) Asymmetric

FIGURE 12.6
Variations of the quadratic loss function.

1. G. Taguchi, "Introduction to Quality Engineering," Asian Productivity Organization, Tokyo, 1986, available from Kraus Int. Publ., White Plains, NY; G. Taguchi, E. A. Elsayed, and T. Hsiang, "Quality Engineering in Production Systems," McGraw-Hill, New York, 1989; G. Taguchi, "Taguchi on Robust Technology Development," ASME Press, New York, 1993.

We note that when $y = m$ the loss is zero and so is the slope of the loss function. The loss increases slowly when y is near m, but as y deviates further from m the loss increases more rapidly. If $m + \Delta$ and $m - \Delta$ are the customer's tolerance limits, i.e., the product performance is unsatisfactory when y is outside this interval, and if the cost to the customer for repairing or replacing the product is A dollars, then

$$k = A/\Delta^2 \tag{12.8}$$

and
$$L(y) = \frac{A}{\Delta^2}(y - m)^2 \tag{12.9}$$

There is a very important concept of quality engineering inherent in the loss function. In the usual practice of manufacturing quality control the producer specifies a mean (target) value of the performance characteristic and the tolerance interval around that value. *Any value* of the performance characteristic which falls within the interval is defined to be a quality product, even if it is barely inside the -3σ limit. With the loss function as a definition of quality the emphasis is on achieving the target value of the performance characteristic and deviations from that value are penalized. The greater the deviation from the target value the greater the quality loss.

The type of loss function described by Eq. (12.9) is called nominal-the-best. Other situations require a modification of this equation. The smaller-the-better type of function (Fig. 12.6b) describes the case where the ideal value is equal to zero. An example would be if y described pollution from an automobile exhaust. The situation is produced by letting $m = 0$ in Eq. (12.9). Another situation can be called larger-the-better, Fig. 12.6c. This describes the situation where $y = 0$ is the worst case and as y increases the quality loss becomes progressively smaller. A situation where strength is the performance characteristic is a good example. Finally, there is the asymmetric loss function, which has a different k on each side of the target value (Fig. 12.6d).

The average quality loss to customers is obtained by averaging the quadratic loss function. Because of noise factors the performance characteristic y of a product varies from piece to piece and over time during the use of the product. If $y_1, y_2, \ldots, y_n$ are measurements of y taken from n units, then the average quality loss, Q, is given by

$$Q = \frac{1}{n}[L(y_1) + L(y_2) + \cdots + L(y_n)]$$

$$= \frac{k}{n}[(y_1 - m)^2 + (y_2 - m)^2 + \cdots + (y_n - m)^2] \tag{12.10}$$

$$= k\left[(\mu - m)^2 + \frac{n-1}{n}\sigma^2\right]$$

In Eq. (12.10) μ is the mean of y and σ^2 is the variance

$$\mu = \frac{1}{n}\sum_{i=1}^{n} y_i \quad \text{and} \quad \sigma^2 = \frac{1}{n-1}\sum_{i=1}^{n}(y_i - \mu)^2$$

When n is large we can write Eq. (12.10) as

$$Q = k[(\mu - m)^2 + \sigma^2] \tag{12.11}$$

Equation (12.11) shows that the average quality loss consists of two components. The first, $k(\mu - m)^2$, results from the deviation of the average value of y from the target m. The second, $k\sigma^2$, results from the *mean squared error* of y around m. Of the two components of quality loss, it is usually easier to minimize the first by appropriate parameter design. Reducing the second component is more difficult, because it requires reducing the variance. This can be achieved by screening out bad products and searching for and fixing causes of poor process performance. Usually, both of these approaches add cost to the product. The preferred approach under the Taguchi philosophy is to apply the concepts of robust design (Sec. 12.6) to search for design parameters that minimize the product's sensitivity to noise.

EXAMPLE 12.4. A diesel truck engine has a target horsepower of 300 hp. Customer acceptance falls off below that value, and at 260 the agitation with underpower becomes strong enough to require some action. On the high side of 300 hp concern arises because an overpowered engine may cause transmission failure. The average cost to correct the underpower by replacing injectors is $300. What is the loss function?

$$A = \$300 \text{ and } \Delta = 40$$

$$L(y) = k(y - m)^2 = \frac{A}{\Delta^2}(y - m)^2 = \frac{300}{(40)^2}(y - 300)^2 = 0.19(y - 300)^2$$

To inspect and replace injectors in the plant during manufacture costs $200. At what tolerance from the target would it be economical to do this, and in the process increase quality to the customer?

This reduces to the question, when the loss is $200 per unit, what is the value of y?

$$200 = 0.19(y - 300)^2; y = 300 \pm 32.4 \text{ and the rework tolerance is 32 hp.}$$

12.5.2 Tolerance Selection

The proper selection of tolerances is an important quality issue as well as a vital economic issue. Taguchi[1] shows how the quality loss function can deal with this problem.

EXAMPLE 12.5. From Example 12.4, the loss function for the diesel truck horsepower was

$$L(y) = 0.19(y - m)^2$$

The average total quality is given by Eq. (12.11) and if we assume the engine horsepower will be centered around the target value, then $Q = k\sigma^2$. If the present variance of engine horsepower is $\sigma^2 = (30)^2 = 900$, then the average quality loss per engine is $Q = \$0.19(900) = \$171$. One of the variables controlling horsepower is the valve that controls fuel input. A higher-performance control valve would increase cost by $12 and would reduce the standard deviation to 20 hp. Would this be a wise investment?

$$Q = 0.19(20) = \$76$$

Thus, the quality loss is decreased by $95 at a cost of $12. This, obviously, is a wise investment.

1. G. Taguchi, E. A. Elsayed, and T. Hsiang, op. cit., pp. 45–59.

12.5.3 Noise

The input parameters that affect the quality of the product or process may be classi-fied as *design factors* and *disturbance factors*. The former are parameters that can be specified freely by the designer. It is the designer's responsibility to select the opti-mum levels of the design factors. Disturbance factors are the parameters that are either inherently uncontrollable or impractical to control.

The variability of the input and output parameters plays an important role in the Taguchi methodology. These will be classified into four categories.

Output variability
   Variational noise is the short-term unit-to-unit variation due to the manufac-turing process.
   Inner noise is the long-term change in product characteristics over time due to deterioration and wear.
Input variability
   Tolerances (design factor variability) is the normal variability in design factors.
   Outer noise represents the variability of the disturbance factors that contribute to output variability. Examples are temperature, humidity, dust, vibration, and human error.

Taguchi has borrowed the concept of *signal-to-noise ratio* from electrical engi-neering and applied it in a specific way to quality engineering. Since we need to con-trol the performance characteristic with respect to both the mean and the variation around the mean it is important to have a performance measure that combines both of these parameters. Taguchi uses the signal-to-noise ratio, *S/N*, as the objective function to be optimized in many situations.

For the nominal-the-best type problem

$$S/N = 10 \log(\mu^2/\sigma^2) \tag{12.12}$$

where
$$\mu = \frac{1}{n} \sum_{i=1}^{n} y_i \quad \text{and} \quad \sigma^2 = \frac{1}{n-1} \sum_{i=1}^{n} (y_i - \mu)^2$$

and n is the number of outer noise observation combinations used for each design parameter matrix (inner array) combination (see Fig. 12.7).

For a smaller-the-better type of problem

$$S/N = -10 \log\left(\frac{1}{n} \sum y_i^2\right) \tag{12.13}$$

In this case the signal is a constant value aimed at making $y = 0$.

For a larger-the-better type problem the quality performance characteristic is con-tinuous and nonnegative; we would like y to be as large as possible. To find the *S/N* we turn this into a smaller-the-better problem by considering the reciprocal of the per-formance characteristic

$$S/N = -10\log\left(\frac{1}{n} \sum \frac{1}{y_i^2}\right) \tag{12.14}$$

Of the many design factors involved in an experimental design, we look for two particular types.

Control factors affect primarily the *S/N* ratio, but not the mean. These factors are
   first set at the appropriate levels so as to minimize the output variability.
Signal factors affect primarily the mean response of the performance characteristic.

The strategy in setting design parameters is to first use the control factors to minimize output variability and then employ the signal factors to move the mean to the desired target. The design parameters that prove to be neither control nor signal factors are set at their low-cost settings, since they do not affect the performance. Fundamental to the Taguchi method is the approach of economical maximization of the output performance characteristic while minimizing the effect of output variability.

12.6
ROBUST DESIGN

Robust design is the systematic approach to finding optimum values of design factors which lead to economical designs with low variability. Taguchi achieves this goal by first performing *parameter design,* and then, if the conditions still are not optimum, by performing *tolerance design.*

Parameter design is the process of identifying the settings of the design parameters or process variables that reduce the sensitivity of the design to sources of variation. In parameter design an accurate modeling of the mean response is not as important as finding the factor levels that optimize robustness. Thus, once the variance has been reduced the mean response should be easily adjusted by using a suitable design parameter, known as the signal factor. An important tenet of robust design is that a design found optimum in laboratory experiments should also be optimum under manufacturing and service conditions. Also, since product designs are often broken down into subsystems for design purposes, it is vital that the robustness of a subsystem not be affected by changes in other subsystems. Therefore, interactions among control factors are highly undesirable.

12.6.1 Parameter Design

Parameter design makes heavy use of statistically planned experiments. Two-and three-level orthogonal arrays are most often used.[1] All common fractional factorial designs are orthogonal arrays. These arrays have the pairwise balancing property that every setting of a design parameter occurs with every setting of all other design parameters the same number of times. They keep this balancing property while minimizing the number of test runs.

---

1. G. Taguchi, "System of Experimental Design: Engineering Methods to Optimize Quality and Minimize Cost," two vols, Quality Resources, White Plains, NY, 1987; M. S. Phadke, "Quality Engineering Using Robust Design," Prentice-Hall, Englewood Cliffs, NJ, 1989; T. B. Barker, "Engineering Quality by Design: Interpreting the Taguchi Approach," 2d ed., Marcel Dekker, New York, 1994; W. Y. Fowlkes and C. M. Creveling, "Engineering Methods for Robust Product Design," Addison-Wesley, Reading, MA, 1995.

A Taguchi-type parameter design of experiments consists of two parts: (1) a design parameter matrix and (2) a noise matrix (Fig. 12.7). The design parameter matrix specifies the test settings of the design parameters. In Fig. 12.7 there are four parameters $\theta_1, \theta_2, \theta_3, \theta_4$, each tested at three levels for nine test runs in the design parameter matrix. The noise matrix consists of three noise factors, w_1, w_2, w_3, each at two levels. The complete experiment consists of a combination of the design parameter matrix and the noise matrix. Each test run of the design parameter matrix is crossed with all the rows of the noise matrix. Thus, for test run 1, there are four trials, one for each combination of the factors in the noise matrix, like humidity, operator experience, etc. For test run 2, there are another four experiments, etc., so that all told $9 \times 4 = 36$ test conditions will be run. The performance characteristic is evaluated for each of the four trials in the first test run and performance statistics like the mean and the signal-to-noise ratio are determined. This is done for each of the nine trials of the design performance matrix.

The ability to carry out an experimental design of this type will depend on the cost of the experiments and the time required to complete them. If experiments are very expensive to run then it will not be possible to employ a complete outer array and only the noise factor thought to be the most important will be used. In cases where product and process performance can be modeled accurately then an extensive statistical design can be done rather inexpensively. However, experience has found that time and money spent in establishing a robust design or process conditions pays off handsomely in improved quality and reduced costs.

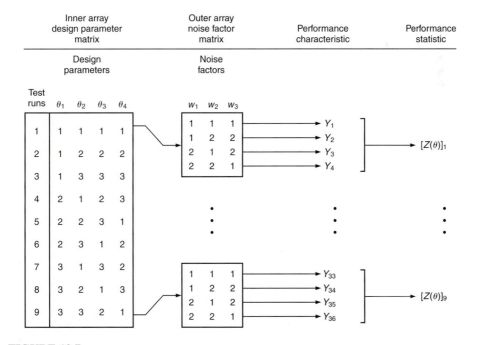

FIGURE 12.7
A typical Taguchi design of experiments for parameter (robust) design. (*After R. N. Kackar, Quality Progress, p. 27, December 1986.*)

EXAMPLE 12.6. The problem is to select the parameters of a compressed-air cooling system (Fig. 12.8) so as to minimize the system cost.[1] The air is cooled first in a precooler, then in a refrigeration unit. Water passes through the condenser of the refrigeration unit, then into the precooler, and finally to the cooling tower, where heat is rejected to the atmosphere. The flow rates of air and water, and critical temperatures are given in Fig. 12.8.

The total cost of the system is the sum of the cost of the refrigeration unit (X_1), the precooler (X_2), and the cooling tower (X_3). Cost equations good for preliminary design have been established.[2]

$$X_1 = 1.20a\,(T_3 - 10)$$

$$X_2 = 1.20b\,\frac{95 - T_3}{T_3 - T_1}$$

$$X_3 = 9.637c\,(T_2 - 24)$$

and the nominal values of the cost parameters are $a = 48$, $b = 50$, and $c = 25$.

Noise factors are factors that cannot be controlled or are too expensive to control. In this problem it was decided that the critical noise factors are:

N_1 = cost parameter for refrigeration unit
N_2 = output temperature of water from cooling tower
N_3 = input temperature of air into precooler

The quality characteristic to be observed is the total cost of the system.

$$C_T = X_1 + X_2 + X_3$$

In optimization terminology, this is the objective function. We wish to find the set of design parameters that minimize C_T subject to the constraints of the mass and energy balances.

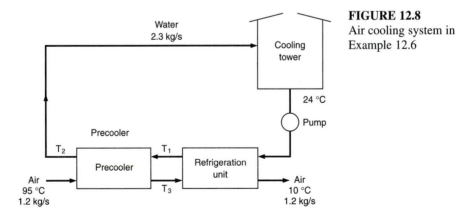

FIGURE 12.8
Air cooling system in Example 12.6

1. R. Umal and E. B. Dean, *Jnl. of Parametrics,* vol. 11, no. 1, 1991. See http://mijuno.larc.nasa.gov for this paper and much more on concurrent engineering methodologies under Design for Competitive Advantage.
2. W. F. Stoecker, "Design of Thermal Systems," 3d ed., McGraw-Hill, New York, 1989, pp. 148–151.

The control factors that can be changed by the designer are the output temperatures T_1, T_2, and T_3. We study these at three levels, low (1), medium (2), and high (3). After a preliminary study it was decided to include three noise factors at two levels. The choice of test conditions for the three control factors and three noise factors is shown below.

	Control factor levels				Noise factor levels		
	1	**2**	**3**			**1**	**2**
T_1	25	28	31	N_1		48	56
T_2	36	39	42	N_2		24	27
T_3	35	38	41	N_3		95	100

Next we need to select the appropriate orthogonal array for the inner array and the outer array. We need to determine the number of degrees of freedom to find the minimum number of experiments that must be performed. One degree of freedom is associated with the overall mean, regardless of the number of control factors. Next we add the degrees of freedom associated with each control factor, i.e., the number of levels minus one. Therefore, the total number of degrees of freedom is $1 + 3(3 - 1) = 7$. We need to conduct at least seven experiments. Taguchi has recorded 18 standard orthogonal arrays.[1] We select the L_9 array (Fig. 12.9a), the smallest three-level orthogonal array. The fact that we have only three control factors and the L_9 array is set up for four factors is not a problem. We just leave one of the columns blank. Orthogonality is not lost by keeping one or more columns of an array empty. For the noise array we select an L_4 array (Fig. 12.9b).

Expt. number	1	2	3	4
1	1	1	1	1
2	1	2	2	2
3	1	3	3	3
4	2	1	2	3
5	2	2	3	1
6	2	3	1	2
7	3	1	3	2
8	3	2	1	3
9	3	3	2	1

Expt. number	1	2	3
1	1	1	1
2	1	2	2
3	2	1	2
4	2	2	1

(a) $L_9(3^4)$ Orthogonal array

(b) $L_4(2^3)$ Orthogonal array

FIGURE 12.9
Two examples of orthogonal arrays for use with Taguchi method. (a) L_9 experimental design for 4 control factors, each at 3 levels; (b) L_4 experimental design for 3 factors, each at 2 levels.

1. G. Taguchi and S. Konishi, "Orthogonal Arrays and Linear Graphs," American Supplier Institute, Dearborn, MI, 1987.

The procedure is as follows. Fig. 12.9a shows the combinations of the three control factors, i.e., whether the factor is set at its low value (1), its nominal value (2), or its high value (3) for each of the 9 experiments. Each of these experiments is conducted four times, with the values of the noise factors set at the values designated by the outer array. Thus, $9 \times 4 = 36$ experiments are required, but this is far less than the $4(3^4) = 324$ experiments that would be needed if a statistically designed experiment had not been used.

We now evaluate the quality characteristic, total cost, using the cost equations. These are the responses listed in Table 12.5. Note that column 3 in the L_9 orthogonal array was left empty since we had only three control parameters in this problem.

In traditional statistical analysis we would utilize the means of the responses. In the Taguchi method we use the signal-to-noise ratio S/N. Using the S/N takes both the mean and variability of the response into account. Because this problem deals with finding the design parameters that minimize cost, we use the smaller-the-better form of S/N [Eq. (12.13)]. The S/N for the four experiments run for row 1 of the control matrix is

$$S/N = -10 \log \left(\frac{1}{n} \sum y_i^2 \right)$$

$$= -10 \log \left\{ \frac{1}{4} \left[(4691)^2 + (3998)^2 + (4961)^2 + (4208)^2 \right] \right\}$$

$$= -10 \log (20,077,067) = -10(7.302700) = -73.03$$

TABLE 12.5
Matrix of experimental responses

	Outer array (noise matrix)		
	N_1	N_2	N_3
1	48	24	95
2	48	27	100
3	56	24	100
4	56	27	95

Inner array (control matrix)

	T_1	T_2	3	T_3	Responses y_{ij}				Mean	σ
1	25	36		35	4691	3998	4961	4208	4468	441
2	25	39		38	5489	4790	5782	5036	5274	445
3	25	42		41	6325	5621	6641	5899	6122	451
4	28	36		41	4926	4226	5247	4501	4725	452
5	28	39		35	5568	4888	5851	5086	5348	440
6	28	42		38	6291	5598	6590	5838	6079	446
7	31	36		38	4993	4312	5304	4539	4787	447
8	31	39		41	5723	5031	6051	5298	5526	452
9	31	42		35	6677	6029	6991	6194	6473	442

The complete results for the calculation of S/N are given below.

Signal-to-noise ratio

	T_1	T_2	3	T_3	S/N average
1	1	1		1	−73.03
2	1	2		2	−74.47
3	1	3		3	−75.76
4	2	1		3	−73.52
5	2	2		1	−74.59
6	2	3		2	−75.70
7	3	1		2	−73.63
8	3	2		3	−74.87
9	3	3		1	−76.24

Control matrix spans the T_1, T_2, 3, T_3 columns.

A standard approach to analyzing these data would be to use the analysis of variance (ANOVA) to determine which factors are statistically significant (see Sec. 10.11). The Taguchi approach uses a simpler graphical technique to determine which factors are significant. Since the L_9 experimental design is orthogonal, it is possible to separate out the effect of each factor. This is done by looking at the control matrix and calculating the average S/N for each factor at each of the three test levels. For example, factor T_3 was at level 2 in experiments 2, 6, and 7. The average S/N for this condition is $(-74.47 -75.70 -73.63)/3 = -223.80/3 = -74.60$. This calculation is made for each of the three control parameters at each of the three test levels, and the results are tabulated in the response table.

Response table

Level	T_1	T_2	T_3
1	−74.42	−73.39	−74.62
2	−74.60	−74.64	−74.60
3	−74.91	−75.90	−74.72

Average S/N spans the T_1, T_2, T_3 columns.

The average S/N ratios are plotted against test level for each of the three control parameters in Fig. 12.10. We note that T_2 has a stronger effect on S/N than the other two control parameters. In fact, T_3 is essentially independent of level of test. Note that the largest value of S/N (least negative) is the preferred value. This is true of all forms of S/N; use them as a guide to move toward the largest value. As a result of the plots in Fig. 12.10 we conclude that the optimum settings for the control parameters are:

Control parameter	Optimum level	Parameter setting, °C
T_1	1	25
T_2	1	36
T_3	2	38

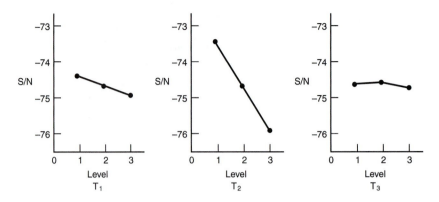

FIGURE 12.10
Linear graphs showing *S/N* for the three control parameters.

Note that none of the experiments carried out in the robust design used exactly this combination of test levels. The initial choice of control parameters was for each to be at level 2. The total cost for this design selection was $5357, with a standard deviation of 445.6 and a *S/N* = −74.60. At the new optimum for control parameters the cost is $4551, about a 15 percent reduction. The standard deviation is a bit less, and *S/N* = −73.19.

In this example we have used a relatively small number of experiments to study a number of design variables, three control parameters, and three noise parameters. The methodology yielded a new set of design parameters that are closer to an optimum than the original "informed guess" and which are robust to the noise factors. This example was one in which a closed form solution for the design model was available. In many cases computer models will have to be used to handle the mathematics. However, in many cases no design model exists, and the robust design is achieved by running physical experiments in which a quality characteristic is measured to determine the effect of various settings of the control parameters and noise factors. This is very often the case when trying to optimize a complex mechanism or a production process.[1]

12.6.2 Tolerance Design

Often, as in the example above, the parameter design results in a design optimized for robustness and with a low variability. However, there are situations when the variability is too large and it becomes necessary to reduce tolerances to decrease variability. Typically, analysis of variance (ANOVA) is used to determine the relative contribution of each control parameter so as to identify those factors that are worth considering for tolerance tightening, substituting an improved material, or some other

1. T. B. Barker, *Quality Progress,* pp. 32–42, December 1986; R. T. Fox and D. Lee, *Int. Jnl. of Powder Metallurgy,* vol. 26, no. 3, pp. 233–243, 1990.

means of improving quality at an increased cost. Tolerance design is beyond the scope of this text. An excellent readable source is available.[1]

Taguchi's methods of quality engineering have generated great interest in the United States as several major manufacturing companies have embraced the approach. While the idea of loss function and robust design is new and important, many of the statistical techniques have been in existence for over 50 years. Statisticians point out[2] that less complicated and more efficient methods exist to do what the Taguchi methods accomplish. However, it is important to understand that before Taguchi systematized and extended these ideas into an engineering context they were largely unused by much of industry. We owe Taguchi a debt of gratitude for demonstrating the way to achieve quality robust designs in the most complex of situations.

12.7
OPTIMIZATION METHODS

The example described in the previous section is a search for the best combination of design parameters using a statistically designed set of experiments. Generally there is more than one solution to a design problem, and the first solution is not necessarily the best. Thus, the need for optimization is inherent in the design process. A mathematical theory of optimization has become highly developed and is being applied to design where design functions can be expressed by mathematical equations or with finite-element computer modeling. These optimization methods require considerable depth of knowledge and mathematical skill to select the appropriate optimization technique and work it through to a solution. The growing acceptance of the Taguchi method comes from its applicability to a wide variety of problems with a methodology that is not highly mathematical.

By the term *optimal design* we mean the best of all feasible designs. Optimization is the process of maximizing a desired quantity or minimizing an undesired one. Optimization theory is the body of mathematics that deals with the properties of maxima and minima and how to find maxima and minima numerically. In the typical design optimization situation the designer has created a general configuration for which the numerical values of the independent variables have not been fixed. An objective function[3] that defines the value of the design in terms of the independent variables is established.

$$U = U(x_1, x_2, \cdots, x_n) \qquad (12.15)$$

Typical objective functions could be cost, weight, reliability, and producibility or a combination of these. Inevitably, the objective function is subject to certain constraints.

---

1. C. M. Creveling, "Tolerance Design: A Handbook for Developing Optimal Specifications," Addison-Wesley Longman, Reading, MA, 1997.
2. R. N. Kackar, *Jnl of Quality Tech.,* vol. 17, no. 4, pp. 176–209, 1985.
3. Also called the criterion function or the payoff function.

Constraints arise from physical laws and limitations or from compatibility conditions on the individual variables. *Functional constraints* ψ, also called equality constraints, specify relations that must exist between the variables.

$$\psi_1 = \psi_1(x_1, x_2, \cdots, x_n) = 0$$

$$\psi_2 = \psi_2(x_1, x_2, \cdots, x_n) = 0$$

$$\vdots$$

$$\psi_n = \psi_n(x_1, x_2, \cdots, x_n) = 0$$

(12.16)

For example, if we were optimizing the volume of a rectangular storage tank, where $x_1 = l_1$, $x_2 = l_2$, and $x_3 = l_3$, then the functional constraint would be that $V = l_1 l_2 l_3$. *Regional constraints* ϕ, also called inequality constraints, are imposed by specific details of the problem,

$$\phi_1 = \phi_1(x_1, x_2, \cdots, x_n) \leq L_1$$

$$\phi_2 = \phi_2(x_1, x_2, \cdots, x_n) \leq L_2$$

$$\vdots$$

$$\phi_p = \phi_p(x_1, x_2, \cdots, x_n) \leq L_p$$

(12.17)

A type of regional constraint that arises naturally in design situations is based on specifications. *Specifications* are points of interaction with other parts of the system. Often a specification results from an arbitrary decision to carry out a suboptimization of the system.

A common problem in design optimization is that there often is more than one design characteristic that is of value to the user. In formulating the optimization problem, one predominant characteristic is chosen as the objective function and the other characteristics are reduced to the status of constraints. Frequently they show up as rather "hard" or severely defined specifications. In reality, such specifications are usually subject to negotiation (soft specifications) and should be considered as target values until the design progresses to such a point that it is possible to determine the penalty that is being paid in trade-offs to achieve the specifications. Siddall[1] has shown how this may be accomplished in design optimization through the use of an interaction curve.

The following example should help to clarify the above definitions. We wish to design a cylindrical tank to store a fixed volume of liquid *V*. The tank will be constructed by forming and welding thin steel plate. Therefore, the cost will depend directly on the area of plate that is used.

---

1. J. N. Siddall and W. K. Michael, *Trans. ASME, J. Mech. Design,* vol. 102, pp. 510–516, 1980.

The design variables are the tank diameter D and its height h. The surface area of the tank is given by

$$A = 2(\pi D^2/4) + \pi Dh$$

If C is the cost per unit area of steel plate, then the objective function can be written

$$U = C[(\pi D^2/2) + \pi Dh] \qquad (12.18)$$

A functional constraint is introduced by the requirement that the tank must hold a specified volume:

$$V = \pi D^2 h/4$$

Regional constraints are introduced by the requirement for the tank to fit in a specified location or to not have unusual dimensions.

$$D_{min} \leq D \leq D_{max}; \qquad h_{min} \leq h \leq h_{max}$$

Siddall,[1] who has reviewed the development of optimal design methods, gives the following insightful description of optimization methods.

1. *Optimization by evolution:* There is a close parallel between technological evolution and biological evolution. Most designs in the past have been optimized by an attempt to improve upon an existing similar design. Survival of the resulting variations depends on the natural selection of user acceptance.
2. *Optimization by intuition:* The *art* of engineering is the ability to make good decisions without being able to formulate a justification. Intuition is knowing what to do without knowing why one does it. The gift of intuition seems to be closely related to the unconscious mind. The history of technology is full of examples of engineers who used intuition to make major advances. Although the knowledge and tools available today are so much more powerful, there is no question that intuition continues to play an important role in technological development.
3. *Optimization by trial-and-error modeling:* This refers to the usual situation in modern engineering design where it is recognized that the first feasible design is not necessarily the best. Therefore, the design model is exercised for a few iterations in the hope of finding an improved design. However, this mode of operation is not true optimization. Some refer to *satisficing,* as opposed to optimizing, to mean a technically acceptable job done rapidly and presumably economically. Such a design should not be called an optimal design.
4. *Optimization by numerical algorithm:* This is the area of current active development in which mathematically based strategies are used to search for an optimum. The chief types of numerical algorithms are listed in the accompanying table.

1. J. N. Sidall, *Trans. ASME, J. Mech. Design,* vol. 101, pp. 674–681, 1979.

Type of algorithm	Example	Reference (see footnotes)
Linear programming	Simplex method	1
Nonlinear programming	Davison-Fletcher-Powell	2
Geometric programming		3
Dynamic programming		4
Variational methods	Ritz	5
Differential calculus	Newton-Raphson	6
Simultaneous mode design	Structural optimization	7
Analytical-graphical methods	Johnson's MOD	8
Monotonicity analysis		9
Genetic algorithms		10
Simulated annealing		11

1. W. W. Garvin, "Introduction to Linear Programming," McGraw-Hill, New York, 1960.
2. M. Avriel, "Nonlinear Programming: Analysis and Methods," Prentice-Hall, Englewood Cliffs, NJ, 1976.
3. C. S. Beightler and D. T. Philips: "Applied Geometric Programming," Wiley, New York, 1976.
4. S. E. Dreyfus and A. M. Law, "The Art and Theory of Dynamic Programming," Academic Press, New York, 1977.
5. M. H. Denn, "Optimization by Variational Methods," McGraw-Hill, New York, 1969.
6. F. B. Hildebrand, "Introduction to Numerical Analysis," McGraw-Hill, 1956.
7. L. A. Schmit (ed.), "Structural Optimization Symposium," ASME, New York, 1974.
8. R. C. Johnson, "Optimum Design of Mechanical Elements," 2d ed., Wiley, New York, 1980.
9. P. Y. Papalambros and D. J. Wilde, "Principles of Optimal Design," Cambridge University Press, New York, 1988.
10. D. E. Goldberg, "Genetic Algorithm," Addison-Wesley, Reading MA, 1989.
11. S. Kirkpatrick, C. D. Gelatt, and M. P. Vecchi, Optimization by Simulated Annealing, *Science,* vol. 220, pp. 671–679, 1983.

There are no "standard techniques" for optimization in engineering design. Linear programming is the most widely applied technique, especially in business and manufacturing situations. However, most design problems in mechanical design are nonlinear. Brief descriptions of various approaches to design optimization are given below. For more depth of understanding about optimization theory consult the various references given within.

12.7.1 Optimization by Differential Calculus

We are all familiar with the use of the calculus to determine the maximum or minimum values of a function. Figure 12.11 illustrates various types of extrema that can occur. A characteristic property of an extremum is that U is momentarily stationary at each point. For example, as E is approached, U increases; but right at E it stops increasing and soon decreases. The familiar condition for a stationary point is

$$\frac{dU}{dx_1} = 0 \tag{12.19}$$

If the curvature is negative, then the stationary point is a maximum. The point is a minimum if the curvature is positive.

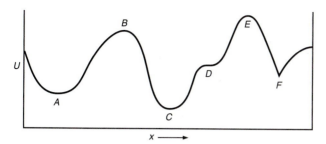

FIGURE 12.11
Different types of extrema in the objective function curve.

$$\frac{d^2U}{dx_1^2} < 0 \quad \text{indicates a local maximum} \tag{12.20}$$

$$\frac{d^2U}{dx_1^2} > 0 \quad \text{indicates a local minimum} \tag{12.21}$$

Both point B and point E are mathematical maxima. Point B, which is the smaller of the two maxima, is called a local maximum. Point E is the global maximum. Point D is a point of inflection. The slope is zero and the curve is horizontal, but the second derivative is zero. When $d^2U/dx^2 = 0$, higher-order derivatives must be used to find a derivative that becomes nonzero. If the derivative is odd, the point is an inflection point, but if the derivative is even it is a local optimum. Point F is not a minimum point because the objective function is *not continuous* at it; the point F is only a cusp in the objective function.

We can apply this simple optimization technique to the tank problem described above. The objective function is given by Eq. (12.18).

$$\frac{dU}{dD} = 0 = C\pi D - \frac{4CV}{D^2} \tag{12.22}$$

$$D = \left(\frac{4V}{\pi}\right)^{1/3} = 1.084 \, V^{1/3} \tag{12.23}$$

The value of diameter established by Eq. (12.23) results in minimum cost, because the second derivative of Eq. (12.22) is positive. In using Eq. (12.23) we apply the regional constraints that were discussed above.

Lagrange multipliers

The Lagrange multipliers provide a powerful method for finding optima in multivariable problems involving functional constraints. We have the objective function $U = U_1(x, y, z)$ subject to the functional constraints $\psi_1 = \psi_1(x, y, z)$ and $\psi_2 = \psi_2(x, y, z)$. We establish a new function, the Lagrange expression (LE)

$$\text{LE} = U_2(x, y, z) + \lambda_1\psi_1(x, y, z) + \lambda_2\psi_2(x, y, z) \tag{12.24}$$

where λ_1 and λ_1 are the Lagrange multipliers. The following conditions must be satisfied at the optimum point.

$$\frac{\partial LE}{\partial x} = 0 \quad \frac{\partial LE}{\partial y} = 0 \quad \frac{\partial LE}{\partial z} = 0 \quad \frac{\partial LE}{\partial \lambda_1} = 0 \quad \frac{\partial LE}{\partial \lambda_2} = 0$$

We will illustrate the determination of the Lagrange multipliers and the optimization method with an example.[1] A total of 300 lineal feet of tubes must be installed in a heat exchanger in order to provide the necessary heat-transfer surface area. The total dollar cost of the installation includes:

1. The cost of the tubes, $700
2. The cost of the shell $= 25D^{2.5}L$
3. The cost of the floor space occupied by the heat exchanger $= 20DL$

The spacing of the tubes is such that 20 tubes will fit in a cross-sectional area of 1 ft^2 inside the shell. The optimization should determine the diameter D and the length L of the heat exchanger to minimize the purchase cost.

The objective function is made up of three costs

$$C = 700 + 25D^{2.5}L + 20DL \tag{12.25}$$

subject to the functional constraint

$$\frac{\pi D^2}{4}L\left(20\,\frac{\text{tubes}}{\text{ft}^2}\right) = 300 \text{ ft}$$

$$5\pi D^2 L = 300 \tag{12.26}$$

$$\psi_1 = L - \frac{300}{5\pi D^2}$$

The Lagrange expression is

$$LE = 700 + 25D^{2.5}L + 20DL + \lambda\left(L - \frac{300}{5\pi D^2}\right) \tag{12.27}$$

$$\frac{\partial LE}{\partial D} = 62.5D^{1.5}L + 20L + 2\lambda\,\frac{60}{\pi D^3} = 0 \tag{12.28}$$

$$\frac{\partial LE}{\partial L} = 25D^{2.5} + 20D + \lambda = 0 \tag{12.29}$$

$$\frac{\partial LE}{\partial \lambda} = L - \frac{300}{5\pi D^2} = 0 \tag{12.30}$$

From Eq. (12.30), $L = 60/\pi D^2$, and from Eq. (12.29), $\lambda = -25D^{2.5} - 20D$. Substituting into Eq. (12.28) yields

$$62.5D^{1.5}\,\frac{60}{\pi D^2} + 20\,\frac{60}{\pi D^2} + 2(-25D^{2.5} - 20D)\,\frac{60}{\pi D^3} = 0$$

$$62.5D^{1.5} + 20 - 50D^{1.5} - 40 = 0$$

1. W. F. Stoecker, "Design of Thermal Systems," 2d ed., McGraw-Hill, New York, 1980.

$$D^{1.5} = 1.6 \quad \text{and} \quad D = 1.37 \text{ ft}$$

Substituting the optimum value of diameter into the functional constraint gives $L = 10.2$ ft. The cost for the optimum heat exchanger is \$1540.

This example is simple and could be solved by direct substitution and finding the derivative, but the method of Lagrange multipliers is perfectly general and could be used for a problem with many variables.

12.7.2 Search Methods

In engineering we often deal with problems that cannot be expressed by analytical functional relations. Therefore, instead of differentiating an analytical expression to find the optimum conditions, we are interested in devising a search strategy for finding the optimum in the fewest possible experiments. We can identify several classes of search problems. A *deterministic search* is assumed to be free from appreciable experimental error, whereas in a *stochastic search* the existence of random errors must be considered. We can have a search involving only a single variable or the more complicated and more realistic situation involving a search over multiple variables. We can have a simultaneous search, in which the conditions for every experiment are specified and all the observations are completed before any judgment regarding the location of the optima is made, or a sequential search, in which future experiments are based on past outcomes. All well-developed search methods are based on the assumption that the behavior is unimodal (a single peak). Some search problems involve *constrained optimization,* in which certain combinations of variables are forbidden. Linear programming and dynamic programming are techniques that deal well with situations of this class. This section considers only single variable search techniques.

Uniform search

In the uniform search method the trial points are spaced equally over the allowable range of values. Each point is evaluated in turn in an exhaustive search. Consider the example of our wanting to optimize the yield of a chemical reaction by varying the concentration x of catalyst. Suppose x lies over the range $x = 0$ to $x = 10$. We have four experiments available, and we distribute them at equidistant spacing over the range $L = 10$ (Fig. 12.12). This divides L into intervals, each of width $L/n + 1$. From inspection of the experimental points we can conclude that the optimum will not lie

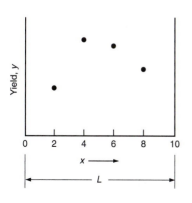

FIGURE 12.12
Example of the uniform search technique.

at $x < 2$ or $x > 6$. Therefore, we know the optimum will lie between $2 < x < 6$. This is known as the *interval of uncertainty*. With four experiments we have narrowed the range of values that require further search to 40 percent of the total range.

Uniform dichotomous search

In the uniform dichotomous search procedure, experiments are performed in pairs to establish whether the function is increasing or decreasing. Since the search procedure is uniform, the experiments are spaced evenly over the entire range of values. Using the same example, we place the first pair of experiments at $x = 3.33$ and the other pair at $x = 6.67$. For the n experiments there will be $n/2$ pairs. The range L is divided into $(n/2) + 1$ intervals each of width $L/[(n/2) + 1]$. The two experiments in each pair are separated in x by an amount slightly greater than the experimental error in x. In this case, $x = 3.30$ and 3.36 and $x = 6.64$ and 6.70. In Fig. 12.13, since $y_a < y_b$, the region $0 < x < 3.33$ is excluded. Also, since $y_c > y_d$, the region $6.67 < x < 10$ is excluded, so the optimum lies in the interval $x = 3.33$ to 6.67. Thus, with four experiments we have narrowed the range of values that require further search to 33 percent of the total range.

Sequential dichotomous search

The two preceding examples were simultaneous search techniques in which all experiments were planned in advance. This example is a sequential search in which the experiments are done in sequence, each taking advantage of the information gained from the preceding one. Using the same example, we first run a pair of experiments near the center of the range, $x = 5.0$. As before, the two experiments are separated just enough in x to be sure that outcomes are distinguishable. Referring to Fig. 12.14, since

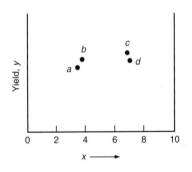

FIGURE 12.13
Example of the uniform dichotomous search technique.

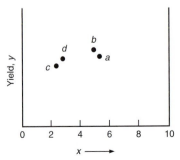

FIGURE 12.14
Example of the sequential dichotomous search technique.

$y_b > y_a$, we eliminate $x > 5.0$ from further consideration. The second pair of experiments are run at the center of the remaining interval, $x = 2.5$. Since $y_d > y_c$, we eliminate the region $0 < x < 2.5$ and the optimum is sought in the interval $2.5 < x < 5.0$. With this search technique four experiments have narrowed the range of values that require further search to 25 percent of the total range.

Fibonacci search

The Fibonacci search is a very efficient sequential technique. It is based on the use of the Fibonacci number series, which is named after a thirteenth-century mathematician. A Fibonacci series is given by $F_n = F_{n-2} + F_{n-1}$, where $F_0 = 1$ and $F_1 = 1$. Thus, the series is

n	0	1	2	3	4	5	6	7	8	9	
F_n	1	1	2	3	5	8	13	21	34	55	...

Note that the nth Fibonacci number is the sum of the preceding two numbers.

The Fibonacci search begins by placing experiments at a distance $d_1 = (F_{n-2}/F_n)L$ from each end of the range of values (Fig. 12.15). For $n = 4$, $d_1 = \frac{2}{5}(10) = 4$. Since $y_4 > y_6$, the interval $6 < x < 10$ is eliminated from further consideration. The value of d_2 is obtained by letting $n_2 = n - 1$, so $d_2 = (F_{n-3}/F_{n-1})L$ and $d_2 = \frac{1}{3}(6) = 2$. Therefore, we run experiments at $x = 2$ and $x = 4$ (where we already have a data point). Since $y_4 > y_2$, we eliminate region $0 < x < 2$ and we are left with the region between $x = 2$ and $x = 6$. Since we have used only three data points, we run the last experiment just to the right of $x = 4$ to determine whether the optimum is $2 < x < 4$ or $4 < x < 6$. This experiment eliminates the former interval and we have narrowed the optimum to the region $4 < x < 6$. In this case, with four experiments we have narrowed the range of values that require further search to 20 percent of the total range. Computer programs for the optimization of a single variable by Fibonacci search are available.[1]

Golden section search

Although the Fibonacci search is very efficient, it has the disadvantage that it requires an advance decision on the number of experiments before we have any

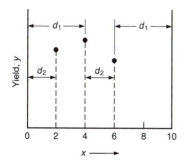

FIGURE 12.15
Example of the Fibonacci search technique.

---

1. J. L. Kuester and J. H. Mize, "Optimization Techniques with Fortran," pp. 285–295, McGraw-Hill, New York, 1973.

information about the behavior of the function near the maximum. If the function turns out to be very steep near the maximum, we should have chosen more trials. With the sequential dichotomous search we would continue placing pairs of trials in the middle of the interval of uncertainty until the change in value of the objective function from one trial to the next was acceptably small.

The golden section search is a good compromise, because although it is slightly less efficient than the Fibonacci search, it does not require an advance decision on the number of trials. This search technique is based on the fact the ratio of two successive Fibonacci numbers, $F_{n-1}/F_n = 0.618$ for all values of $n > 8$. This same ratio was discovered by Euclid, who called it the golden mean. He defined it as a length divided into two unequal segments such that the ratio of the length of the whole to the larger segment is equal to the ratio of the length of the larger segment to the smaller segment. The ancient Greeks felt 0.618 was the most pleasing ratio of width to length of a rectangle, and they used it in the design of many of their buildings.

In using the golden section search the first two trials are located at $0.618L$ from either end of the range. Note that the same value is used, no matter the ultimate number of trials. Based on the value of y_1 and y_2 we eliminate an interval. With the new reduced interval we take additional experiments at $\pm 0.618L_2$. Actually, only one new experiment is needed, since one of the experiments from the preceding trial lies at exactly the proper location for the new series of trials. This procedure is continued until the maximum is located to within as small an interval as desired.

Comparison of methods

A measure of the efficiency of a search technique is the reduction ratio, which is defined as the ratio of the original interval of uncertainty to the interval remaining after n trials. It is shown in Table 12.6.

12.7.3 Multivariable Search Methods

When the objective function is a function of two or more variables, the geometric representation is a *response surface* (Fig. 12.16a). It usually is convenient to deal with contour lines produced by the intersection of planes of constant y with the response surface[1] and projected on the x_1x_2 plane (Fig. 12.16b).

Lattice search

In the lattice search, which is an analog to the single-variable exhaustive search, a two-dimensional grid lattice is superimposed over the contour (Fig. 12.17). In the absence of special knowledge about the location of the maximum, the starting point is selected near the center of the region, at point 1. The objective function is evaluated for points 1 through 9. If point 5 turns out to be the largest value, it becomes the cen-

---

1. A technique for generating contour plots by polynomial curve fitting to the objective function surface has been given by K. A. Afimiwala and R. W. Mayne, *Trans ASME, J. Mech. Design,* vol. 101, pp. 349–354, 1979.

TABLE 12.6
Comparison of univariable search techniques

Search method		For $n = 4$	For $n = 13$
		Reduction ratio	
Uniform	$(n + 1)/2$	2.5	7
Uniform dichotomous	$(n + 2)/2$	3	7.5
Sequential dichotomous	$2^{n/2}$	4	90
Fibonacci	F_n	5	377
Golden search			250

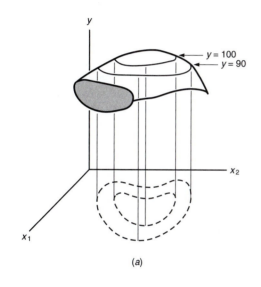

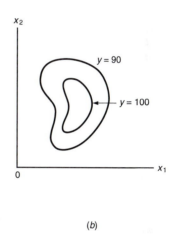

(a)   (b)

FIGURE 12.16
Representation of two variables by contour lines.

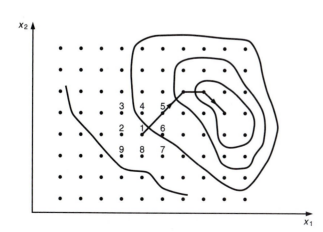

FIGURE 12.17
Procedure for a lattice search. (*From W. F. Stoecker, "Design of Thermal Systems," 2d ed., McGraw-Hill, New York, 1980; used with permission of McGraw-Hill.*)

tral point for the next search. The procedure continues until the location reached is one at which the central point is greater than any of the other eight points. Frequently, a coarse grid is used initially and a finer grid is used after the maximum is approached.

Univariate search

The univariate search is a one-at-a-time method. All of the variables are kept constant except one, and it is varied to obtain an optimum in the objective function. That optimal value is then substituted into the function and the function is optimized with respect to another variable. The objective function is optimized with respect to each variable in sequence, and an optimal value of a variable is substituted into the function for the optimization of the succeeding variables.

Figure 12.18a shows the univariate search procedure. Starting at point 0 we move along $x_2 = $ const to a maximum by using any one of the single variable search techniques. Then we move along $x_1 = $ const to a maximum at point 2 and along $x_2 = $ const to a maximum at 3. We repeat the procedure until two successive moves are less than some specified value. If the response surface contains a ridge, as in Fig. 12.18b, then the univariate search can fail to find an optimum. If the initial value is at point 1, it will reach a maximum at $x_1 = $ const at the ridge, and that will also be a maximum for $x_2 = $ const. A false maximum is obtained.

Steepest ascent

The path of steepest ascent (or descent) up the response surface is the *gradient vector.* Imagine that we are walking at night up a hill. In the dim moonlight we can see far enough ahead to follow the direction of the local steepest slope. Thus, we would tend to climb normal to contour lines for short segments and adjust the direction of climb as the terrain came progressively into view. The gradient method does essentially that with mathematics. We change the direction of the search in the direction of the maximum slope, but we must do so in finite straight segments.

We shall limit our discussion to the situation of two independent variables with understanding that the steepest ascent method is applicable to many variables. The gradient vector is given by

$$\nabla U = \frac{\partial U}{\partial x_1}\mathbf{i}_1 + \frac{\partial U}{\partial x_2}\mathbf{i}_2 \tag{12.31}$$

To move in the direction of the gradient vector, we take the step lengths δx_1 and δx_2 in proportion to the components of the gradient vector (Fig. 12.19).

$$\frac{\delta x_1}{\delta x_2} = \frac{\partial U/\partial x_1}{\partial U/\partial x_2} \tag{12.32}$$

For the general case of n independent variables, $U = U(x_1, x_2, \ldots, x_n)$

$$\frac{\delta x_i}{\delta x_1} = \frac{\partial U/\partial x_i}{\partial U/\partial x_1} \tag{12.33}$$

If the objective function is in analytical form, the partial derivatives can be obtained by calculus. If it is not, a numerical procedure must be used. Starting at the initial

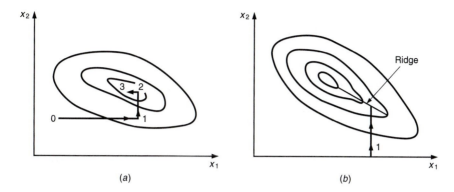

FIGURE 12.18
Univariate search procedure.

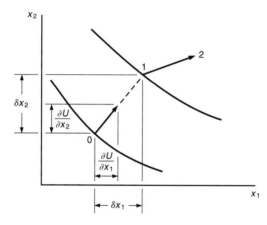

FIGURE 12.19
Definitions in the steepest ascent method.

point, we take a small finite difference Δx_i in each variable and evaluate the function at each $x_i + \Delta x_i$ in turn, holding all other x_i's at their initial value. $\Delta U_i = U_i - U_0$. The partial derivatives are evaluated by $\Delta U_i / \Delta x_i$, so

$$\frac{\delta x_i}{\delta x_1} \approx \frac{\Delta U_i / \Delta x_i}{\Delta U_1 / \Delta x} \tag{12.34}$$

Determining the direction of the gradient vector is standard, but deciding on the length of the step is not as straightforward. Numerous methods for establishing δx_1 have been developed. The most direct is to arbitrarily select δx_1 and compute δx_i from Eq. (12.33) or Eq. (12.34). This method breaks down when one of the partial derivatives becomes zero. The second method is to select the step size such that the objective function improves by a specific amount ΔU.

$$\delta x_1 = \frac{(\partial U / \partial x_1) \Delta U}{(\partial U / \partial x_1)^2 + (\partial U / \partial x^2)^2 + \cdots} \tag{12.35}$$

A third method of establishing the step length is to proceed from the initial point in the direction of the gradient vector until an optimal value is reached. Then another gradient vector is calculated and the experiment moves in that direction until an optimal value is obtained. This procedure is repeated until all partial derivatives become negligibly small.

An important consideration in working with the gradient method is the scaling of variables. Since the units of each variable may be different and arbitrary in magnitude, the relative scale of the variables is arbitrary. Transforming the scale of a variable changes the shape of the contours of the objective function and the magnitude and direction of the gradient vector. When possible, contours that approach circles are to be preferred.

12.7.4 Nonlinear Optimization Methods

The methods discussed above are not really practical techniques for most real problems. However, multivariable optimization of nonlinear problems has been a field of great activity and many computer-based methods are available. Space permits mention of only a few of the more useful methods. These are of two types, those that require information about derivative values in determining the search direction for optimization (indirect methods), and direct methods that rely solely on the evaluation of the objective function. Because an in-depth understanding requires considerable mathematics for which we do not have space, only a brief word description can be given. The interested student is referred to original sources or the text by Arora.[1]

Methods for unconstrained multivariable optimization are discussed first. Newton's method is an indirect technique that employs a second-order approximation of the function.[2] This method has very good convergence properties, but it can be an inefficient method because it requires the calculation of $n(n + 1)/2$ second-order derivatives, where n is the number of design variables. Therefore, methods that require the computation of only first derivatives and use information from previous iterations to speed up convergence have been developed. The DFP (Davidon, Fletcher, and Powell) method is one of the most powerful methods.[3] Updates and improvements on this method were provided by Broyden, Fletcher, Goldfarb, and Shannon in what is called the BFGS method.[4] There is a general feeling among practitioners that the BFGS algorithm is the best one to use.

---

1. J. S. Arora, "Introduction to Optimum Design," McGraw-Hill, New York, 1989.
2. J. S. Arora, op. cit., pp. 319–325.
3. R. Fletcher and M. J. D. Powell, *Computer J.,* vol. 6, pp. 163–180, 1963; Arora, op. cit., pp. 327–330; for a computer program see J. L. Kuester and J. H. Mize, "Optimization Techniques with Fortran," pp. 355–366, McGraw-Hill, New York, 1973 and J. N. Siddall, "Analytical Decision-Making in Engineering Design," Appendix E, Prentice-Hall, Englewood, NJ, 1972.
4. For a computer program see R. W. Pike, "Optimization for Engineering Systems," pp. 285–291, Van Nostrand Reinhold, New York, 1986; Arora, op. cit., pp. 330–332.

One of the earliest direct search methods was the method of Hooke and Jeeves.[1] Because, it does not use derivatives it is fast and is less sensitive to irregular or discontinuous functions. The method does have problems in highly constrained situations where a search can get stalled when the contour lines and the inequality constraint line have a certain orientation. Other direct search techniques are the simplex method[2] (no relation to the technique in linear programming) and Powell's quadratic convergence method.[3]

Optimization of nonlinear problems with constraints is a more difficult area. The more general approach to this class of problems is called the Generalized Reduced Gradient method (GRG). The idea of GRG is to convert the constrained problem into an unconstrained one by direct substitution. However, with nonlinear constraint equations this is not feasible by direct substitution and the procedures of constrained variation and Lagrange multipliers must be used.[4]

Another approach is to successively linearize the constraints and objective function of a nonlinear problem and solve using the technique of linear programming. Sequential linear programming (SLP) is the name of the method.[5] The most recent, and perhaps the best method of solving nonlinear problems is sequential quadratic programming (SQP).[6]

Computer programs with names like GRG2, NLPQL, OPT, and MINOS are available to perform nonlinear optimization on the mainframe, and programs like GINO, NLP Solver, and OPTISOLVE are available for microcomputers.

12.7.5 Other Optimization Methods

Monotonicity analysis

Monotonicity analysis is an optimization technique that may be applied to design problems with monotonic properties, i.e., where the objective function and constraints successively increase or decrease with respect to some design variable. This is a situation that is very common in design problems. Engineering designs tend to be strongly defined by physical constraints. When these specifications and restrictions are monotonic in the design variables then monotonicity analysis can often show the designer which constraints are active at the optimum. An active constraint refers to a design requirement which has a direct impact on the location of the optimum. This information can be used to identify the improvements that could be achieved if the feasible domain were modified, which would point out directions for technological improvement.

1. R. Hooke and T. A. Jeeves, *J. Assoc. Comp. Mach.,* vol. 8, pp. 202–229, 1961; for a computer program see Kuester and Mize, op. cit., pp. 309–319 and Siddall, op. cit., Appendix H.
2. J. A. Nelder and R. Meade, *Computer J.,* vol. 7, pp. 308–318, 1965.
3. M. J. D. Powell, *Computer J.,* vol. 7, pp. 303–307, 1964.
4. J. S. Arora, op. cit., pp. 415–417.
5. J. S. Arora, op. cit., pp. 365–371.
6. J. S. Arora, op. cit., pp. 384–392.

The ideas of monotonicity analysis were first presented by Wilde.[1] Subsequent work by Wilde and Papalambros has applied the method to many engineering problems[2] and to the development of a computer-based method of solution.[3]

Dynamic programming

Dynamic programming is a mathematical technique that is well suited for the optimization of staged processes. The word "dynamic" in the name of this technique has no relationship to the usual use of the word to denote changes with respect to time. Dynamic programming is related to the calculus of variations and is not related to linear and nonlinear programming methods. The method is well suited for allocation problems, as when x units of a resource must be distributed among N activities in integer amounts. It has been broadly applied within chemical engineering to problems like the optimal design of chemical reactors. Dynamic programming converts a large complicated optimization problem into a series of interconnected smaller problems, each containing only a few variables. This results in a series of partial optimizations which requires a reduced effort to find the optimum.

Dynamic programming was developed by Richard Bellman[4] in the 1950s. It is a well-developed optimization method.[5]

Johnson's method

A method of optimum design that is especially suited to the nonlinear problems found in the design of mechanical elements such as gears, roller bearings, and hydrodynamic journal bearings has been developed by R. C. Johnson.[6] The trend in design optimization has been toward methods that require little effort to enter problems into general computer optimization routines. In contrast, Johnson's method often requires significant effort to reduce the system of equations to a form suitable for an optimization study. However, the benefit of the method is that it gives considerable insight into the nature of and possible solutions to the problem.

Genetic algorithms

Genetic algorithms (GA) are based on the laws of evolution. The GA approach, a stochastic optimization method, involves reducing a large number of possible designs to a set of genetic codes, expressed as a string of 1s and 0s. Genetic algorithms follow a survival of the fittest strategy, where members of the design population compete. Different codes are paired by the computer to produce offspring with different design characteristics from both parents. These offspring are evaluated

1. D. J. Wilde, *Trans. ASME, Jnl. of Engr. for Industry,* vol. 94, pp. 1390–1394, 1975.
2. P. Papalambros and D. J. Wilde, "Principles of Optimal Design," Cambridge University Press, New York, 1989.
3. S. Azarm and P. Papalambros, *Trans. ASME, Jnl. of Mechanisms, Transmissions, and Automation in Design,* vol. 106, pp. 82–89, 1984.
4. R. E. Bellman, "Dynamic Programming," Princeton University Press, Princeton, NJ, 1957.
5. G. L. Nernhauser, "Introduction to Dynamic Programming," Wiley, New York, 1960; E. V. Denardo, "Dynamic Programming Models and Applications," Prentice-Hall, Englewood Cliffs, NJ, 1982.
6. R. C. Johnson, "Optimal Design of Mechanical Elements," 2d ed., Wiley-Interscience, New York, 1980; R. C. Johnson, *Trans. ASME, J. Mech. Design,* vol. 101, pp. 667–673, 1979.

against a fitness function, and those designs judged to have the greatest "goodness" are then paired. The process repeats itself, until after several generations of computer simulation, only the fittest designs will have survived.[1] The limitation on the GA method is that it is computationally intensive.

Simulated annealing

Simulated annealing (SA) is a stochastic method that addresses the category of combinatorial optimization problems. The method has been used to automate digital circuit layout[2] in 2-D, and it is being used to study the layout of mechanical and electromechanical components in the product architecture step of design.[3]

SA starts at an initial design state, and the objective function for that state is evaluated. A step is taken to a new state by applying a move from an available move set. The new state is evaluated and if an improvement, it becomes the current design state. However, if the step leads to an inferior state, the state may still be accepted with some probability. The probability is an exponential function of a parameter called *temperature*. It is from this idea that the method gets its name in analogy with the annealing of metals. The temperature starts high and decreases with time. Initially, steps taken through the design space are almost random, but as the probability of accepting inferior steps decreases the algorithm converges to an optimum.

12.8
EVALUATION CONSIDERATIONS IN OPTIMIZATION

We have presented optimization chiefly as a collection of computer-based mathematical techniques. However, of more importance than knowing how to manipulate the optimization tools is knowing where to use them in the design process. In many designs a single design criterion drives the optimization. In consumer products it usually is cost, in aircraft it is weight, and in implantable medical devices it is power consumption. The strategy is to optimize these "bottleneck factors" first. Once the primary requirement has been met as well as possible, there may be time to improve other areas of the design, but if the first is not achieved, the design will fail. In some areas of design there may be no rigid specifications. An engineer who designs a talking, walking teddy bear can make almost any trade-off he or she wants between cost, power consumption, realism, and reliability. The designers and market experts will work together to decide the best combination of characteristics for the product, but in the end the four-year old consumers will decide whether it is an optimal design.

An additional consideration is the evaluation of the results of the optimization. One issue deals with the trade-offs that must be made to arrive at an optimal design. The other concerns finding the sensitivity of the criterion function to variations in each of the design parameters.

1. C. Y. Lin and P. Hajela, *Jnl. of Engr. Optimization,* vol. 19, pp. 309–327, 1992.
2. N. Sherwani, "Algorithms for VLSI Physical Design Automation," Kluwer, Boston, 1988.
3. J. Cagan, D. Degentesh, and S. Yin, *Computer-Aided Design,* vol. 30, no. 10, pp. 781–790, 1998.

12.8.1 Trade-off Analysis

In performing trade-off studies a methodology is needed to put design attributes measured in different units on a common scale. Derringer[1] has combined the desirability function proposed by Harrington[2] with response surface methodology[3] in a technique called desirability optimization methodology (DOM). We shall present this method of dealing with trade-offs with an example.

> **EXAMPLE 12.7.** An advanced supersonic transport plane will use a new steel/titanium laminate composite material in its landing gear. There is a trade-off between the tensile strength and the elongation, as measured with the tensile test. The design team first must decide on the desirability curve, where desirability varies from 0.0 (not at all desirable) to 1.0 (completely desirable). As Fig. 12.20 shows, the shapes of the curves can be very different depending on what is being considered. For tensile strength, we want values balanced around 200 ksi (Fig. 12.20a) while for elongation (ductility) we will not accept a value less than 3 percent and we want values as large as we can get.
>
> For four compositions of laminate composite material we get the values given in Table 12.7. The values of desirability for tensile strength (d_1) and elongation (d_2) are scaled from Fig. 12.20. Also, we decide that the importance weight of tensile strength to elongation is 1:3. We are terribly afraid of brittle fracture of the landing gear, so we give heavy weight to the elongation (ductility) property.

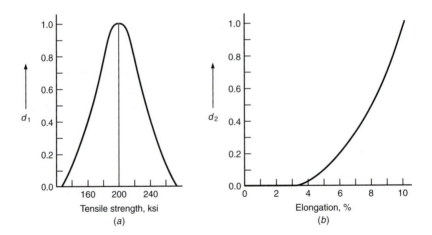

FIGURE 12.20
Desirability curves for two factors considered in Example 12.7. (a) d_1 for tensile strength; (b) d_2 for elongation.

1. G. C. Derringer, *Quality Progress,* June 1994, pp. 51–58.
2. E. C. Harrington, Jr., *Industrial Quality Control,* April 1965, pp. 494–498.
3. G. E. P. Box and N. R. Draper, "Empirical Model Building and Response Surfaces," Wiley, New York, 1987.

TABLE 12.7
Determination of combined desirability

Material	Tensile strength, ksi	d_1	Elongation, %	d_2	Combined desirability D
1	190	0.90	6	0.20	0.29
2	200	1.00	3	0.00	0.00
3	160	0.38	10	1.00	0.78
4	180	0.75	8	0.50	0.55

$$w_1 = 1 \qquad w_2 = 3$$

The combined desirability is given by $D = (d_1^{w_1} d_2^{w_2})^{1/4}$. The general equation for D is

$$D = (d_1^{w_1} d_2^{w_2} \dots d_n^{w_n})^{1/\Sigma w_i} \qquad (12.36)$$

We note that because Eq. (12.36) is in the form of a geometric mean, not an arithmetic mean, any property that scores zero for the desirability of one of the factors disqualifies that option as one of the alternatives. Thus material 2 is disqualified. Material 3 is found to be the best balanced alternative, chiefly because we weighted elongation so highly.

12.8.2 Sensitivity Analysis

With a sensitivity analysis we find which design parameters are most critical to the performance of the design, and what are the critical ranges of those parameters. For some problems the sensitivity analysis can be done analytically, while in others it must be done numerically. To do a sensitivity analysis on an analytically expressed objective function, take the partial derivative with respect to each of the design variables. To put these on a relative basis, each partial derivative should be divided by its actual value.

For example, the cost of producing a chemical in batch runs in given by[1]

$$C^{\text{opt}} = 2\sqrt{K_1 K_2 Q} + K_3 Q$$

where K_1, K_2, and K_3 are cost coefficients and Q is the quantity of material. The sensitivity coefficients (on an absolute basis) are obtained by taking the partial derivative of C^{opt} with respect to each parameter.

$$\frac{\partial C^{\text{opt}}}{\partial K_1} = \sqrt{\frac{K_2 Q}{K_1}} \qquad \frac{\partial C^{\text{opt}}}{\partial K_2} = \sqrt{\frac{K_1 Q}{K_2}}$$

1. T. F. Edgar and D. Himmelblau, "Optimization of Chemical Process," pp. 20–26 McGraw-Hill, New York, 1988.

$$\frac{\partial C^{\text{opt}}}{\partial K_3} = Q \qquad \frac{\partial C^{\text{opt}}}{\partial Q} = \sqrt{\frac{K_1 K_2}{Q}} + K_3$$

If we let $Q = 100{,}000$, $K_1 = 1.0$, $K_2 = 10{,}000$, and $K_3 = 4.0$, then

$$\frac{\partial C^{\text{opt}}}{\partial K_1} = 31{,}620 \qquad \frac{\partial C^{\text{opt}}}{\partial K_2} = 3.162$$

$$\frac{\partial C^{\text{opt}}}{\partial K_3} = 100{,}000 \qquad \frac{\partial C^{\text{opt}}}{\partial Q} = 4.316$$

To put these sensitivities on a more meaningful basis we need to determine the relative sensitivities. The relative sensitivity of C^{opt} to K_1 is

$$S_{K_1}^C = \frac{\partial C^{\text{opt}}/C^{\text{opt}}}{\partial K_1/K_1} = \frac{\partial \ln C^{\text{opt}}}{\partial \ln K_1} = \sqrt{\frac{K_2 Q}{K_1}} \cdot \frac{K_1}{C^{\text{opt}}} = 31{,}620 \cdot \frac{(1.0)}{463{,}240} = 0.0683$$

Numerical values for the other relative sensitivities are:

$$S_{K_2}^C = 0.0683 \quad S_{K_3}^C = 0.863 \quad S_Q^C = 0.932$$

We thus see that changes in the variables Q and K_3 have the greatest influence on C^{opt}.

If the problem could not be obtained with an analytical solution we would determine the sensitivities numerically. First the objective function would be evaluated at the average value of each variable. Then we change each variable, one at a time, by some small amount, e.g., 1 percent, 5 percent, or 10 percent, and determine the change in the objective function. This change is repeated in the opposite direction, i.e., minus 10 percent, to see if the function is symmetrical. This type of calculation is conveniently done with a spreadsheet.

12.9
DESIGN OPTIMIZATION

It has been a natural development to combine computer-aided-engineering (CAE) analysis and simulation tools with computer-based optimization algorithms.[1] Linking optimization with analysis tools creates CAE design tools by replacing traditional trial-and-error approaches with a systematic design-search approach. This extends the designer's capability from being able with FEA to quantify the performance of a particular design to adding information about how to modify the design to better achieve critical performance criteria.

Figure 12.21 shows a general framework for CAE-based optimal design. Starting with an initial design (size and shape parameters) a numerical analysis simulation,

---

1. D. E. Smith, Design Optimization, "ASM Handbook," vol. 20, pp. 209–218, ASM International, Materials Park, OH, 1997.

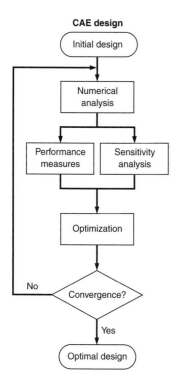

CAE design

FIGURE 12.21
General framework for CAE-based design optimization.
(*From D. E. Smith, "ASM Handbook," vol. 20. p. 211. Used with permission.*)

e.g., FEA, is performed on the design to compute the performance measures, e.g., von Mises stress, and the sensitivity of the performance measures with respect to the design parameters. Then an optimization algorithm computes new design parameters, and the process is continued until an optimum design is achieved. Most FEA packages offer optimization routines that integrate design simulation, optimization, and design-sensitivity analysis into a comprehensive design environment.[1] The user inputs preliminary design data and specifies acceptable variables and required constraints. The optimization algorithm generates successive models, in conjunction with remeshing routines, until it ultimately converges on an optimized design. For example, structural optimization of a turbine wheel design resulted in a 12 percent reduction in mass and a 35 percent reduction in stress.[2]

Structural design optimization can be classified into three groups: (1) size optimization, (2) shape optimization, and (3) topology optimization. Size optimization is the easiest and requires the least computer power. *Size optimization* deals with parameters that do not alter the location of nodal points in the numerical model. Examples are material properties, boundary conditions, and element properties such as plate thickness. *Shape optimization* deals with parameters that describe boundary position in the numerical model. Since these define nodal locations, shape optimization generally requires that the entire numerical model be remeshed. Shape optimization can

---

1. M. Puttre, Putting Optimization Routines in the Loop, *Mechanical Engineering,* July 1993, pp. 76–80.
2. D. E. Smith, op. cit.

lead to important design benefits, but it also can yield shapes that are difficult or impossible to manufacture. To avoid this situation *topology optimization* is being introduced. The goal of topology optimization is to determine where to place material and where to leave it out of the structure. It is best done early in the design process when little has been decided about the design details and when design changes are most easily accommodated.[1] For example, topology optimization could help determine the number and placement of holes or stiffening members. This would provide a good starting point for size optimization and shape optimization. Topology optimization packages being introduced allow the user to specify manufacturing dimensions and other design features as design constraints. They also allow cost to be considered as an optimization parameter.

12.10
SUMMARY

This chapter presents many of the modern views about design. The overarching concept is that quality is built into products during design. Manufacturing cannot compensate for errors in design. Second, we have emphasized that variability during manufacture and in service are the challenges to a quality design. We aim for a robust design that is less sensitive to process variations and to extreme conditions in service.

Quality must be viewed as a total system from the perspective called total quality management (TQM). TQM places the customer at the center and solves problems with a data-driven approach using simple but powerful tools (see Sec. 3.7). It emphasizes continuous improvement where large changes are achieved by many small improvements made over time.

Statistics plays a significant role in achieving quality and robustness. A control chart shows whether the variability of a process is within reasonable bounds. The process capability index C_p tells whether the selected tolerance range is easily achievable by a particular process.

New ways of looking at quality have been introduced by Taguchi. The loss function provides a better way of looking at quality than the traditional upper and lower tolerance limits around a mean value. The signal-to-noise parameter (S/N) provides a powerful metric to search for design situations that minimize variability. Orthogonal experimental designs provide a useful and widely adopted methodology to find the design or process conditions which are most robust.

The search for optimum conditions has been a design goal for many years. A wide selection of optimization methods is described in Sec. 12.7, along with the important topics of trade-off and sensitivity analysis in Sec. 12.8. The coming together of optimization routines with computer modeling methods to provide a powerful design optimization tool ends this chapter.

---

1. M. Chirehdast, H-C. Gea, N. Kikuchi, and P. Y. Papalambros, *Jnl. of Mechanical Design*, vol. 116, pp. 997–1004, 1994.

Bralla[1] lists the following design guidelines that help produce products with a potential for higher levels of quality.

1. Design the product, its major subassemblies, and other components so they can be readily tested.
2. Utilize standard, commercially proven parts whenever possible.
3. Use clear, standardized dimensioning of drawings.
4. Design parts and set tolerances to reduce or eliminate adjustments.
5. Design parts so that critical dimensions can be controlled by tooling, rather than the setup of production equipment or by individual workmanship.
6. Choose the dimensional tolerances with great care.
7. Minimize the number of different but similar part designs. This makes it unlikely that the wrong part is inadvertently assembled into a product.
8. Use modular design. Modularity makes it easier to test and verify quality.
9. Thoroughly analyze quality ramifications of engineering changes.
10. Develop more robust components and assemblies.
11. Design for ease of assembly (see Sec. 9.5).

BIBLIOGRAPHY

Quality

Gevirtz, C. D.: "Developing New Products with TQM," McGraw-Hill, New York, 1994.
Kolarik, W. J.: "Creating Quality," McGraw-Hill, New York, 1995.
Summers, D. C. S.: "Quality," Prentice-Hall, Upper Saddle River, NJ, 1997.
Wesner, J. W., J. M. Hiatt, and D. C. Trimble: "Winning with Quality," Addison-Wesley, Reading MA, 1995.

Optimization

Arora, J. S.: "Introduction to Optimum Design," McGraw-Hill, New York, 1989.
Krottmaier, J.: "Optimizing Engineering Designs," McGraw-Hill, New York, 1993.
Papalambros, P. Y., and D.J. Wilde: "Principles of Optimal Design," Cambridge University Press, New York, 1989.
Vanderplaats, G. N.: "Numerical Optimization Techniques for Engineering Design," McGraw-Hill, New York, 1984.

Robust Design

Ealey, L. A.: "Quality by Design," 2d ed., ASI Press, Dearborn, MI, 1984.
Fowlkes, W. Y., and C. M. Creveling: "Engineering Methods for Robust Product Design," Addison-Wesley, Reading MA, 1995.
Lochner, R. H., and J. E. Matar: "Designing for Quality," Quality Press, ASQ, Milwaukee, WI, 1990.
Ross, P. J.: "Taguchi Techniques for Quality Engineering," 2d ed., McGraw-Hill, New York, 1996.

---

1. J. G. Bralla, "Design for Excellence," Chap. 14, McGraw-Hill, New York, 1996.

PROBLEMS AND EXERCISES

12.1. Discuss as a class how Deming's 14 points could be applied to higher education.

12.2. Divide into teams and use the TQM problem-solving process introduced in Sec. 3.7 to decide how to make several of your courses (one course per team) more robust.

12.3. Discuss the concept of quality circles. What would be involved in implementing a quality circle program in industry? How could the concept be applied to the class-room?

12.4. Use the concept of statistical hypothesis testing to identify and classify the errors that can occur in quality-control inspection.

12.5. Dig deeper into the subject of control charts and find some rules for identifying out-of-control processes.

12.6. For the control chart shown in Fig. 12.4, determine C_p. Note: Hardness is only recorded to the nearest 0.5 RC.

12.7. A product has specification limits of 120 ± 10 MN and a target value of 120 MN. The standard deviation of the products coming off the process line is 3 MN. The mean value of strength is initially 118 MN, but it shifts to 122 MN and then 125 MN without any change in variability. Determine C_p and C_{pk}.

12.8. Construct a model for the cost of poor quality production. Consider a single level process in which q sets of materials each costing M are introduced. They all get processed and tested at a cost of P each. A certain number w fail the quality test and must be reworked, at a cost R each. The probability of success in the process is y, the fractional yield of the process. Assume that binomial statistics applies.

12.9. A grinding machine is grinding the root of gas turbine blades where they attach to the disk. The critical dimension at the root must be 0.450 ± 0.006 in. Thus, a blade falls out of specs in the range 0.444 to 0.456 and has to be scrapped at a cost of $120.
(*a*) What is the Taguchi loss equation for this situation?
(*b*) Samples taken from the grinder had the following dimensions:

  0.451; 0.446; 0.449; 0.456; 0.450; 0.452; 0.449; 0.447; 0.454; 0.453; 0.450; 0.451

What is the average loss function for the parts made on the machine?

12.10. The weather strip that seals the door of an automobile has a specification on width of 20 ± 4 mm. Three suppliers of weather strip produced the results shown below.

Supplier	Mean width $\bar{x}$	Variance s^2	C_{pk}
A	20.0	1.778	1.0
B	18.0	0.444	1.0
C	17.2	0.160	1.0

Field experience showed that when the width of the weather strip is 5 mm below the target the seal begins to leak and about 50 percent of the customers will complain and insist that it be replaced at a cost of $60. When the strip width exceeds 25 mm door closure becomes difficult and the customer will ask to have the weather strip replaced. Historically, the three suppliers had the following number of parts out of spec in deliveries of 250,000 parts:

A: 0.27%; B: 0.135%; C: 0.135%

(a) Compare the three suppliers on the basis of loss function.
(b) Compare the three suppliers on the basis of cost of defective units.

12.11. Part of the pollution control system of an automobile engine consists of a nylon tube inserted in a flexible elastomeric connector. The tubes had been coming loose, so an experimental program was undertaken to improve the robustness of the design. The effectiveness of the design was measured by the pounds of force needed to pull the nylon tube out of the connector. The control factors for this design were:

A—interference between the nylon tube and the elastomer connector
B—wall thickness of the elastomer connector
C—depth of insertion of the tube in the connector
D—the percent, by volume, of adhesive in the connector predip

The environmental noise factors that conceivably could affect the strength of the bond had to do with the conditions of the predip that the end of the connector was immersed in before the tube was inserted. There were three:

X—time the predip was in the pot	24 h	120 h
Y—temperature of the predip	72°F	150°F
Z—relative humidity	25%	75%

(a) Set up the orthogonal arrays for the inner array and the outer array. How many runs will be required to complete the tests?
(b) The S/N ratio for the 9 experimental conditions of the control matrix are, in order:

(1) 24.02; (2) 25.52; (3) 25.33; (4) 25.90; (5) 26.90; (6) 25.32; (7) 25.71; (8) 24.83; (9) 26.15

What type of S/N ratio should be used? What do you learn from these results?

12.12. Conduct a robust design experiment to determine the most robust design of paper airplanes. The control parameters and noise parameters are given in the table below.

Control parameters

Parameter	Level 1	Level 2	Level 3
Weight of paper (A)	One sheet	Two sheets	Three sheets
Configuration (B)	Design 1	Design 2	Design 3
Width of paper (C)	4 in	6 in	8 in
Length of paper (D)	6 in	8 in	10 in

Noise parameter

Parameter	Level 1	Level 2
Launch height (X)	Standing on ground	Standing on chair
Launch angle (Y)	Horizontal to ground	45° above horizontal
Ground surface	Concrete	Polished tile

All planes are launched by the same person in a closed room or hallway with no air currents. When launching a plane, the elbow must be touching the body and only the forearm, wrist, and hand are used to send the plane into flight. Planes are made from ordinary copy paper. The class should decide on the three designs, and once this is decided, the designs will not be varied throughout the experiment. The criterion function to be optimized is the distance the plane flies and glides to a stop on the floor, measured to the nose of the plane.

12.13. We want to design a hot-water pipeline to carry a large quantity of hot water from the heater to the place where it will be used. The total cost is the sum of four items: (1) the cost of pumping the water, (2) the cost of heating the water, (3) the cost of the pipe, (4) the cost of insulating the pipe.

(a) By using basic engineering principles, show that the system cost is

$$C = K_p \frac{1}{D^5} + K_h \frac{1}{\ln[(D + x)/D]} + K_m D + K_i x$$

where x is the thickness of insulation on a pipe of ID = D.

(b) If $K_p = 10.0$, $K_h = 2.0$, $K_m = 3.0$, $K_i = 1.0$, and $D_0 = x_0 = 1.0$, find the value of D and x that will minimize the system cost.

12.14. Determine the value of insulation thickness for a furnace wall that results in minimum cost, given the following data:

Wall temperature 500°F; outside air temperature 70°F
Air film coefficient = 4 Btu/(h)(ft^2)(°F)
Thermal conductivity of insulation = 0.03 Btu/(h)(ft)(°F)
Insulation cost per inch of thickness = $0.75 per ft^2
Value of heat saved = $0.60 per million Btu
Fixed charges = 30 percent per year
Hours of operation = 8700 h/year

Base your calculation on 100 ft^2 of furnace wall.

12.15. Find the maximum value of $y = 12x - x^2$ with the golden mean search method for an original interval of uncertainty of $0 \le x \le 10$. Carry out the search until the difference between the two largest calculated values of y is 0.01 or less.

12.16. A manufacturer of steel cans wants to minimize costs. As a first approximation, the cost of making a can consists of the cost of the metal plus the cost of welding the longitudinal seam and the top and bottom. The can may have any dimension D and L to give a volume V_0. The thickness is δ. By the using the method of Lagrange multipliers, find the dimension of the can that will minimize cost.

12.17. By using the method of steepest ascent, find the minimum of the objective function $U = x^2 + 2y^2 + xy$. Start the search at location $x = 2.0$, $y = 2.0$.

13

ECONOMIC DECISION MAKING

13.1
INTRODUCTION

Throughout this book we have repeatedly emphasized that the engineer is a decision maker and that engineering design basically consists of making a series of decisions over time. We also have emphasized from the beginning that engineering involves the application of science to real problems of society. In this real-world context, one cannot escape the fact that economics (or costs) may play a role as big as, or bigger than, that of technical considerations in the decision making of design. In fact, it sometimes is said, although a bit facetiously, that an engineer is a person who can do for $1.00 what any fool can do for $2.00.

The major engineering infrastructure that built this nation—the railroads, major dams, and waterways—required a methodology for predicting costs and balancing them against alternative courses of action. In an engineering project, costs and revenues will occur at various points of time in the future. The methodology for handling this class of problems is known as engineering economy or engineering economic analysis. Familiarity with the concepts and approach of engineering economy generally is considered to be part of the standard engineering toolkit. Indeed, an examination on the fundamentals of engineering economy is required for professional engineering registration in all disciplines in all states.

The chief concept in engineering economy is that *money has a time value.* Paying out $1.00 today is more costly than paying out $1.00 a year from now. A $1.00 invested today is worth $1.00 plus interest a year from now. Engineering economy recognizes the fact that *use of money* is a valuable asset. Money can be rented in the same way one can rent an apartment, but the charge for using it is called interest rather than rent. This time value of money makes it more profitable to push expenses into the future and bring revenues into the present as much as possible.

Before proceeding into the mathematics of engineering economy, it is important to understand where engineering economy sits with regard to related disciplines like economics and accounting. Economics generally deals with broader and more global issues than engineering economy, such as the forces that control the money supply and trade between nations. Engineering economy uses the interest rate established by the economic forces to solve a more specific and detailed problem. However, it usually is a problem concerning alternative costs in the future. The accountant is more concerned with determining exactly, and often in great detail, what costs have been in the past. One might say that the economist is an oracle, the engineering economist is a fortune teller, and the accountant is an historian.

13.2
MATHEMATICS OF TIME VALUE OF MONEY

If we borrow a present sum of money or principal P at a simple interest rate i, the annual cost of interest is $I = Pi$. If the loan is repaid in a lump sum S at the end of n years, the amount required is

$$S = P + nI = P + nPi = P(1 + ni) \qquad (13.1)$$

where $S =$ future worth (sometimes denoted by F)
$\quad\;\; P =$ present worth
$\quad\;\; I =$ annual cost of interest
$\quad\;\; i =$ annual interest rate
$\quad\;\; n =$ number of years

If we borrow $1000 for 6 years at 10 percent simple interest rate, we must repay at the end of 6 years:

$$S = P(1 + ni) = \$1000\,[1 + 6(0.10)] = \$1600$$

Therefore, we see that $1000 available today is not equivalent to $1000 available in 6 years. Actually, $1000 in hand today is worth $1600 available in only 6 years at 10 percent simple interest.

We can also see that the *present worth* of $1600 available in 6 years and invested at 10 percent is $1000.

$$P = \frac{S}{1 + ni} = \frac{\$1600}{1 + 0.6} = \$1000$$

In making this calculation we have discounted the future sum back to the present time. In engineering economy the terminology *discounted* refers to bringing dollar values *back in time* to the present.

13.2.1 Compound Interest

However, you are aware from your personal banking experiences that financial transactions usually use compound interest. In *compound interest,* the interest due at the end of a period is not paid out but is instead added to the principal. During the next period, interest is paid on the total sum.

First period: $\qquad S_1 = P + Pi = P(1 + i)$

Second period: $\qquad S_2 = P(1 + i) + iP(1 + i) = [P(1 + i)](1 + i) = P(1 + i)^2$

Third period: $\qquad S_3 = P(1 + i)^2 + iP(1 + i)^2 = P[(1 + i)^2](1 + i) = P(1 + i)^3$

nth period: $\qquad S_n = P(1 + i)^n$ $\qquad\qquad\qquad\qquad\qquad\qquad$ (13.2)

Alternatively, we can write this as

$$S = P(F_{PS}) = P(\text{SCAF}) \qquad\qquad (13.3)$$

where F_{PS} is the single-payment compound-amount factor (SCAF) that converts present value P to future worth S.

EXAMPLE 13.1. How long will it take money to double if it is compounded annually at a rate of 10 percent per year?

Solution. $\qquad\qquad\qquad S = PF_{PS,10\%,n} \qquad$ but $S = 2P$
$$\qquad\qquad\qquad 2P = PF_{PS,10\%,n}$$

Therefore, the answer clearly is found in a table of single-payment compound-amount factors at the year n for which $F_{PS} = 2.0$. Examining Table A-2 in Appendix A we see that, for $n = 7$, $F_{PS} = 1.949$ and, for $n = 8$, $F_{PS} = 2.144$. Linear extrapolation gives us $F_{PS} = 2.000$ at $n = 7.2$ years. We can generalize the result to establish the financial rule of thumb that the number of years to double an investment is 72 divided by the interest rate (expressed as an integer).

Usually in engineering economy, n is in years and i is an annual interest rate. However, in banking circles the interest may be compounded at periods other one year. Compounding at the end of shorter periods, such as daily, raises the effective interest rate. If we define r as the nominal annual interest rate and p as the number of interest periods per year, then the interest rate per interest period is $i = r/p$ and the number of interest periods in n years is pn. Using this notation, Eq. (13.2) becomes

$$S = P\left[\left(1 + \frac{r}{p}\right)^p\right]^n \qquad\qquad (13.4)$$

Note that when $p = 1$, the above expression reduces to Eq. (13.2). Standard compound interest tables that are prepared for $p = 1$ can be used for other than annual periods. To do so, use the table for $i = r/p$ and for a number of years equal to $p \times n$. Alternatively, use the interest table corresponding to n years and an effective rate of yearly return equal to $(1 + r/p)^p - 1$.

If the number of interest periods per year p increases without limit, then $i = r/p$ approaches zero.

TABLE 13.1
Influence of compounding period on effective rate of return

Frequency of compounding	No. annual interest periods p	Interest rate for period, %	Effective rate of yearly return, %
Annual	1	12.0	12.0
Semiannual	2	6.0	12.4
Quarterly	4	3.0	12.6
Monthly	12	1.0	12.7
Continuously	∞	0	12.75

$$S = P \lim_{p \to \infty} \left(1 + \frac{r}{p} \right)^{pn} \tag{13.5}$$

From calculus, an important limit is $\lim_{x \to 0} (1 + x)^{1/x} = 2.7178 = e$. If we let $x = r/p$, then

$$pn = \frac{p}{r} rn = \frac{1}{x} rn$$

Since $p = r/x$, as $p \to \infty$, $x \to 0$, so Eq. (13.5) is rewritten as

$$S = P \left[\lim_{x \to 0} (1 + x)^{1/x} \right]^{rn} = Pe^{rn} \tag{13.6}$$

Table 13.1 shows the influence of the number of interest periods per year on the effective rate of return.

Engineering economy was developed to deal with sums of money at different times in the future. These situations can rapidly become quite complex, so that a methodology for setting up problems is needed. A good procedure is to place the dollar amounts on a dollar-time diagram, as shown in Fig. 13.1. Receipts (income) are placed above the line, and disbursements (costs) are placed below the line. The length of the arrow should be proportional to the dollar amount.

13.2.2 Uniform Annual Series

In many situations we are concerned with a uniform series of receipts or disbursements occurring equally at the end of each period. Examples are the payment of a debt

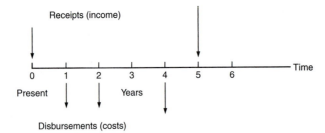

FIGURE 13.1
Dollar-time diagram.

on the installment plan, setting aside a sum S that will be available at a future date for replacement of equipment, and a retirement annuity that consists of a series of equal payments instead of a lump sum payment. We will let R be the equal end-of-the-period payment that makes up the uniform annual series. Other authors may use the symbol A for the same thing.

Figure 13.2 shows that if an annual sum R is invested at the end of each year for 3 years, the total sum S at the end of 3 years will be the sum of the compound amount of the individual investments R

$$S = R(1 + i)^2 + R(1 + i) + R$$

and for the general case of n years.

$$S = R(1 + i)^{n-1} + R(1 + i)^{n-2} + \cdots + R(1 + i)^2 + R(1 + i) + R \quad (13.7)$$

Multiplying by $1 + i$, we get

$$S(1 + i) = R(1 + i)^n + R(1 + i)^{n-1} + \cdots + R(1 + i)^3 + R(1 + i)^2 + R(1 + i) \quad (13.8)$$

Subtracting Eq. (13.7) from Eq. (13.8):

$$(1 + i)S = R[(1 + i)^n + \cancel{(1+i)^{n-1}} + \cdots + \cancel{(1+i)^3} + \cancel{(1+i)^2} + \cancel{(1+i)}]$$

$$S = R[\cancel{(1+i)^{n-1}} + \cancel{(1+i)^{n-2}} + \cdots + \cancel{(1+i)^2} + \cancel{(1+i)} + 1]$$

$$iS = R[(1 + i)^n - 1]$$

$$S = R\frac{(1 + i)^n - 1}{i} \quad (13.9)$$

Equation (13.9) gives the future sum of n uniform payments of R when the interest rate is i. This equation may also be written:

$$S = RF_{RS} = R(\text{USCAF}) \quad (13.10)$$

where F_{RS} is the uniform-series compound amount factor (USCAF) that converts a series R to future worth S.

By solving Eq. (13.9) for R, we have the uniform series of end-of-period payments, that, at compound interest i, provide a future sum S.

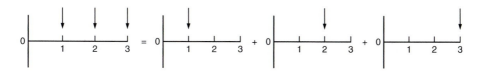

FIGURE 13.2
Equivalence of a uniform annual series.

$$R = S\frac{i}{(1 + i)^n - 1} \tag{13.11}$$

This type of calculation often is used to set aside money in a sinking fund to provide funds for replacing worn-out equipment.

$$R = SF_{SR} = S(\text{SFF}) \tag{13.12}$$

where $F_{SR} = \text{SFF}$ is the sinking fund factor.

By combining Eq. (13.2) with Eq. (13.9), we develop the relation for the present worth of a uniform series of payments R:

$$P = R\frac{(1 + i)^n - 1}{i(1 + i)^n} = RF_{RP} \tag{13.13}$$

Solving Eq. (13.13) for R gives the important relation for capital recovery:

$$R = P\frac{i(1 + i)^n}{(1 + i)^n - 1} = PF_{PR} = P(\text{CRF}) \tag{13.14}$$

where CRF is the capital recovery factor. The R in Eq. (13.14) is the annual payment needed to return the initial capital investment P plus interest on that investment at a rate i over n years.

Capital recovery is an important concept in engineering economy. It is important to understand the difference between capital recovery and sinking fund. Consider the following example:

EXAMPLE 13.2 What annual investment must be made at 10 percent to provide funds for replacing a $10,000 machine in 20 years?

$$R = SF_{SR,10\%,20} = \$10,000(0.01746)$$

$$R = \$174.60 \text{ per year for the sinking fund}$$

What is the annual cost of capital recovery of $10,000 at 10 percent in 20 years?

$$R = PF_{PR,10\%,20} = \$10,000(0.11746)$$

$$R = \$1174.60 \text{ per year for capital recovery}$$

We see that

$$F_{PR} = F_{SR} + i$$

$$0.11746 = 0.01746 + 0.10000$$

Annual cost capital recovery	=	annual cost sinking fund	+	annual interest cost

$$\$1174.60 = \$174.60 + 0.10(\$10,000)$$

With a sinking fund we put away each year a sum of money that, over n years, together with accumulated compound interest, equals the required future amount S. With capital recovery we put away enough money each year to provide for replacement in n years *plus we charge ourselves interest on the invested capital.* The use of capital recovery is a conservative but valid economic strategy. The amount of money invested in capital equipment ($10,000 in the above example) represents an *opportunity cost,* since we are forgoing the revenue that the $10,000 could provide if invested in interest-bearing securities.

In working with a uniform series R it is conventional to assume that R occurs at the end of each period. However, if a beginning-of-period payment R_b is required, it can be determined readily by discounting R one year to the present according to Eq. (13.3), where $S = R$, $P = R_b$, and $n = 1$. Thus,

$$R = R_b(1 + i) \tag{13.15}$$

and this can be substituted into Eq. (13.9), (13.11), or (13.13).

A summary of the compound interest relations among S, P, R, and R_b is given in Table 13.2.

TABLE 13.2
Summary of S, P, R, and R_b relationships

Item	Conversion	Algebraic relation	Relation by factor	Name of factor
1	P to S	$S = P[(1 + i)^n]$	$S = PF_{PS,i,n}$	Compound interest factor
2	S to P	$P = S[(1 + i)^{-n}]$	$P = SF_{SP,i,n}$	Present worth factor
3	R to P	$P = R\dfrac{(1 + i)^n - 1}{i(1 + i)^n}$	$P = RF_{RP,i,n}$	Uniform series present worth factor
4	P to R	$R = P\dfrac{i(1 + i)^n}{(1 + i)^n - 1}$	$R = PF_{PR,i,n}$	Capital recovery factor
5	R to S	$S = R\dfrac{(1 + i)^n - 1}{i}$	$S = RF_{RS,i,n}$ $S = RF_{RP,i,n}F_{PS,i,n}$	Equal payment series future worth factor
6	S to R	$R = S\dfrac{i}{(1 + i)^n - 1}$	$R = SF_{SR,i,n}$ $R = SF_{SP,i,n}F_{PR,i,n}$	Sinking fund factor
7	R_b to P	$P = R_b(1 + i)\dfrac{(1 + i)^n - 1}{i(1 + i)^n}$	$P = R_b(1 + i)F_{RP,i,n}$	
8	P to R_b	$R_b = \dfrac{P}{1 + i}\dfrac{i(1 + i)^n}{(1 + i)^n - 1}$	$R_b = \dfrac{P}{1 + i}F_{PR,i,n}$	
9	R_b to S	$S = R_b(1 + i)\dfrac{(1 + i)^n - 1}{i}$	$S = R_b(1 + i)F_{RS,i,n}$ $S = R_b(1 + i)F_{RP,i,n}F_{PS,i,n}$	
10	S to R_b	$R_b = \dfrac{S}{1 + i}\dfrac{i}{(1 + i)^n - 1}$	$R_b = S\dfrac{1}{1 + i}F_{SR,i,n}$ $R_b = S\dfrac{1}{1 + i}F_{SP,i,n}F_{PR,i,n}$	

13.3
COST COMPARISON

Having covered the usual compound interest relations, we now are in a position to use them to make economic decisions. A typical decision is which of two courses of action is the least expensive when time value of money is considered.

13.3.1 Present Worth Analysis

When the two alternatives have a common time period, a comparison on the basis of present worth is advantageous.

EXAMPLE 13.3. Two machines each have a useful life of 5 years. If money is worth 10 percent, which machine is more economical?

	A	B
Initial cost	$25,000	$15,000
Yearly maintenance cost	2,000	4,000
Rebuilding at end of third year	—	3,500
Salvage value	3,000	
Annual benefit from better quality production	500	

From the cost diagrams given below and on the next page we see that the cash flows definitely are different for the two alternatives. To place them on a common basis for comparison, we discount all costs back to the present time.

$$P_A = 25,000 + (2000 - 500)F_{RP,10\%,5} - 3000F_{SP,10\%,5}$$

$$= 25,000 + 1500(3.791) - 3000(0.621) = \$28,823$$

$$P_B = 15,000 + 4000F_{RP,10\%,5} + 3500F_{SP,10\%,3}$$

$$= 15,000 + 4000(3.791) + 3500(0.751) = \$32,793$$

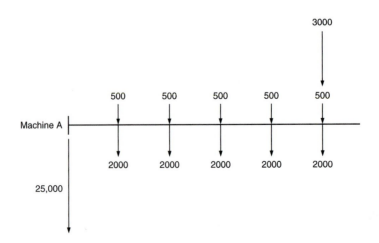

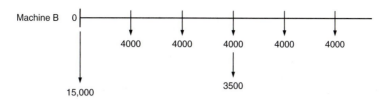

Machine A is the most economical because it has the lowest cost on a present worth basis. In this example we considered both (1) costs plus benefits (savings) due to reduced scrap rate and (2) resale value at the end of the period of useful life. Thus, we really determined the *net present worth* for each alternative. We should also point out that present worth analysis is not limited to the comparison of only two alternatives. We could consider any number of alternatives and select the one with the smallest net present worth of costs.

In Example 13.3, both alternatives had the same life. Thus, the time period was the same and the present worth could be determined without ambiguity. Suppose we want to use present worth analysis for the situation shown below:

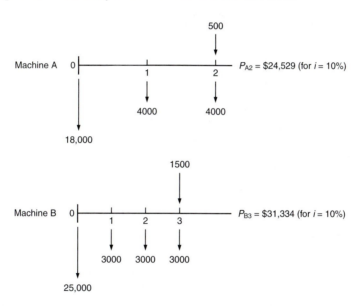

We cannot directly compare P_A and P_B because they are based on different time periods. One way to handle the problem would be to use a common 6-year period, in which we would replace machine A three times and replace machine B twice. This procedure works when a common multiple of the individual periods can be found easily, but a more direct approach is to convert the present worth based on a period n_1 to an equivalent P based on n_2 by[1]

---

1. For a derivation of Eq. (13.16) see F. C. Jelen and J. H. Black, "Cost and Optimization Engineering," 2d ed., p. 28, McGraw-Hill, New York, 1983.

$$P_{n_2} = \frac{P_{n_1} F_{PR,i,n_1}}{F_{PR,i,n_2}} \tag{13.16}$$

For our example, we convert P_B from a 3-year time period to a 2-year period.

$$P_{B_2} = P_{B_3} \frac{F_{PR,10\%,3}}{F_{PR,10\%,2}} = 31,334 \frac{0.402}{0.576} = \$21,867$$

and $P_{A_2} = \$24,529$. Now, when compared on the basis of equal time periods, machine B is the most economical.

13.3.2 Annual Cost Analysis

In the annual cost method, the cash flow over time is converted to an equivalent uniform annual cost or benefit. In this method no special procedures need be used if the time period is different for each alternative, because all comparisons are an annual basis ($n = 1$).

Example	Machine A	Machine B
First cost	$10,000	$18,000
Estimated life	20 years	35 years
Estimated salvage	0	$3000
Annual cost of operation	$4000	$3000

$$R_A = 10,000 F_{PR,10\%,20} + 4000 = 10,000(0.1175) + 4000 = \$5175$$

$$R_B = (18,000 - 3000) F_{PR,10\%,35} + 3000(0.10) + 3000 = \$4855$$

Machine B has the lowest annual cost and is the most economical. Note that in calculating the annual cost of capital recovery for machine B we used the difference between the first cost and the salvage value; for it is only this amount of money that must be recovered. However, although the salvage value is returned to us, we are required to wait until the end of the useful life of the machine to recover it. Therefore, a charge for the annual cost of the interest on the investment tied up in the salvage value is made as part of the annual cost analysis.

Perhaps a more direct way to handle the case of machine B in the above example is to determine the equivalent annual cost based on the cash disbursements minus the annual benefit of the future resale value.

$$R_B = 18,000 F_{PR,10\%,35} + 3000 - 3000 F_{SR,10\%,35}$$

$$= 18,000(0.1037) + 3000 - 3000(0.0037) = \$4855$$

13.3.3 Capitalized Cost Analysis

Capitalized cost is a special case of present worth analysis. The capitalized cost of a project is the present value of providing for that project in perpetuity ($n = \infty$). The concept was originally developed for use with public works, such as dams and waterworks, that have long lives and provide services that must be maintained indefinitely. Capitalized cost subsequently has been used more broadly in economic decision making because it provides a method that is independent of the time period of the various alternatives.

We can develop the mathematics for capitalized cost quite simply from Eq. (13.16). If we let $n_2 = \infty$ and $n_1 = n$, then

$$P_\infty = P_n \frac{F_{PR,i,n}}{F_{PR,i,\infty}}$$

$$F_{PR,i,n} = \frac{i(1 + i)^n}{(1 + i)^n - 1} \qquad F_{PR,i,\infty} = \frac{i(1 + i)^\infty}{(1 + i)^\infty - 1} = i$$

Therefore, the capitalized cost K of a present sum P is given by

$$K = P_\infty = P \frac{(1 + i)^n}{(1 + i)^n - 1} = PF_{PK,i,n} \tag{13.17}$$

Since many tables of compound interest factors do not include capitalized cost, we need to note that

$$F_{PK,i,n} = \frac{F_{PR,i,n}}{i} \tag{13.18}$$

In addition, the capitalized cost of an annual payment R is determined as follows:

$$P = RF_{RP}: \qquad K = PF_{PK} = \frac{PF_{PR}}{i}$$

$$\text{Substituting for } P: \qquad K = \frac{RF_{RP}F_{PR}}{i} = \frac{R}{i} \quad \text{since} \quad F_{RP} = \frac{1}{F_{PR}} \tag{13.19}$$

The capitalized cost is the present worth of providing for a capital cost in perpetuity; i.e., we assume there will be an infinite number of renewals of the initial capital investment. Consider a bank of condenser tubes that cost $10,000 and have an average life of 6 years. If $i = 10$ percent, then the capitalized cost is

$$K = PF_{PK,i,n} = P \frac{F_{PR,i,n}}{i} = 10,000 \frac{0.2296}{0.10} = \$22,960$$

We note that the excess over the first cost is $22,960 - 10,000 = \$12,960$. If we invest that amount for the 6-year life of the tubes,

$$S = PF_{PS,10\%,6} = 12,960(1.772) = \$22,960$$

Thus, when the tubes need to be replaced, we have generated $22,960. We take $10,000 to purchase a new set of tubes (we are neglecting inflation) and invest the difference $(22,960 - 10,000 = 12,960)$ at 10 percent for 6 years to generate another $22,960. We can repeat this process indefinitely. The capital cost is provided for in perpetuity.

EXAMPLE 13.4. Compare the continuous process and the batch process on the basis of capitalized cost analysis if $i = 10$ percent.

Solution.

	Continuous process	Batch process
First cost	$20,000	$6000
Useful life	10 years	15 years
Salvage value	0	$500
Annual power costs	$1000	$500
Annual labor costs	$600	$4300

Continuous process:

$$K = 20,000 \frac{F_{PR,10\%,10}}{0.10} + \frac{1000 + 600}{0.10}$$

$$K = 20,000 \frac{0.1627}{0.10} + \frac{1600}{0.10} = \$48,540$$

Batch process:

$$K = 6000 \frac{0.1315}{0.10} - 500 \frac{1}{(1 + 0.10)^{15}} \frac{0.1315}{0.10} + \frac{4800}{0.10}$$

$$= 7890 - 500(0.2394)(1.315) + 48,000 = \$55,733$$

Note that the $500 salvage value is a negative cost occurring in the fifteenth year. We bring this to the present value and then multiply by $F_{PK} = F_{PR}/i$.

Each of the three methods of analysis will give the same result when applied to the same problem. The best method to use depends chiefly on whom you need to convince with your analysis and which technique you feel they will be more comfortable with.

13.4
DEPRECIATION

Capital equipment suffers a loss in value with time. This may occur by wear, deterioration, or obsolescence, which is a loss of economic efficiency because of technological advances. Therefore, a company must lay aside enough money each year to accumulate a fund to replace the obsolete or worn-out equipment. This allowance for loss of value is called depreciation. Another important aspect of depreciation is that the federal government permits depreciation to be deducted from gross profits as a cost of doing business. In a capital-intensive business, depreciation can have a strong influence on the amount of taxes that must be paid.

$$\text{Taxable income} = \text{total income} - \text{allowable expenses} - \text{depreciation}$$

The basic questions to be answered about depreciation are: (1) what is the time period over which depreciation can be taken and (2) how should the total depreciation charge be spread over the life of the asset? Obviously, the depreciation charge in any given year will be greater if the depreciation period is short (a rapid write-off).

The Economic Recovery Act of 1981 introduced the *accelerated cost recovery system* (ACRS) as the prime capital-recovery method in the United States. This was modified in the 1986 Tax Reform Act to MACRS. The statute sets depreciation recovery periods based on the expected useful life. Some examples are:

- Special manufacturing devices; some motor vehicles 3 years
- Computers; trucks; semiconductor manufacturing equipment 5 years
- Office furniture; railroad track; agricultural buildings 7 years
- Durable-goods manufacturing equipment; petroleum refining 10 years
- Sewage treatment plants; telephone systems 15 years,

Residential rental property is recovered in 27.5 years and nonresidential rental property in 31.5 years. Land is a nondepreciable asset, since it is never used up.

We shall consider four methods of spreading the depreciation over the recovery period n: (1) straight-line depreciation, (2) declining balance, (3) sum-of-the-years digits, and (4) the MACRS procedure. Only MACRS and the straight-line method currently are acceptable under the U.S. tax laws but the other methods are useful in classical engineering economic analyses.

13.4.1 Straight-Line Depreciation

In straight-line depreciation an equal amount of money is set aside yearly. The annual depreciation charge D is

$$D = \frac{\text{initial cost} - \text{salvage value}}{n} = \frac{C_i - C_s}{n} \tag{13.20}$$

The *book value* is the initial cost minus the sum of the depreciation charges that have been made. For straight-line depreciation, the book value B at the end of the jth year is

$$B = C_i - \frac{j}{n}(C_i - C_s) \tag{13.21}$$

13.4.2 Declining-Balance Depreciation

The declining-balance method provides an accelerated write-off in the early years. The depreciation charge for the jth year D_j is a fixed fraction F_{DB} of the book value at the beginning of the jth year (or the end of year $j - 1$). For the book value to equal the salvage value after n years,

$$F_{DB} = 1 - \sqrt[n]{\frac{C_s}{C_i}} \qquad (13.22)$$

and the book value at the beginning of the jth year is

$$B_{j-1} = C_i(1 - F_{DB})^{j-1} \qquad (13.23)$$

Therefore, the depreciation in the jth year is

$$D_j = B_{j-1}F_{DB} = C_i(1 - F_{DB})^{j-1}F_{DB} \qquad (13.24)$$

The most rapid write-off occurs for double declining-balance depreciation. In this case $F_{DDB} = 2/n$ and $B_{j-1} = C_i(1 - 2/n)^{j-1}$. Then

$$D_j = C_i\left(1 - \frac{2}{n}\right)^{j-1}\frac{2}{n}$$

Since the DDB depreciation may not reduce the book value to the salvage value at year n, it may be necessary to switch to straight-line depreciation in later years.

13.4.3 Sum-of-Years Digits Depreciation

The sum-of-years digits (SOYD) depreciation is an accelerated method. The annual depreciation charge is computed by adding up all of the integers from 1 to n and then taking a fraction of that each year, $F_{SOYD,j}$.

For example, if $n = 5$, then the sum of the years is $(1 + 2 + 3 + 4 + 5 = 15)$ and $F_{SOYD,2} = 4/15$, while $F_{SOYD,4} = 2/15$. The denominator is the sum of the digits; the numerator is the digit corresponding to the jth year when the digits are arranged in *reverse order.*

13.4.4 Modified Accelerated Cost Recovery System (MACRS)

In MACRS the annual depreciation is computed using the relation

$$D = qC_i \qquad (13.25)$$

where q is the recovery rate obtained from Table 13.3 and C_i is the initial cost. In MACRS the value of the asset is completely depreciated even though there may be a

TABLE 13.3
Recovery rates q used in MACRS method

Year	Recovery rate, q, %				
	$n = 3$	$n = 5$	$n = 7$	$n = 10$	$n = 15$
1	33.3	20.0	14.3	10.0	5.0
2	44.5	32.0	24.5	18.0	9.5
3	14.8	19.2	17.5	14.4	8.6
4	7.4	11.5	12.5	11.5	7.7
5		11.5	8.9	9.2	6.9
6		5.8	8.9	7.4	6.2
7			8.9	6.6	5.9
8			4.5	6.6	5.9
9				6.5	5.9
10				6.5	5.9
11				3.3	5.9
12–15					5.9
16					3.0

n = recovery period, years.

true salvage value. The recovery rates are based on starting out with a declining-balance method and switching to the straight-line method when it offers a faster write-off. MACRS uses a half-year convention which assumes that all property is placed in service at the midpoint of the initial year. Thus, only 50 percent of the first year depreciation applies for tax purposes and requires that a half year of depreciation be taken in year $n + 1$.

Table 13.4 compares the annual depreciation charges for these four methods of calculation.

TABLE 13.4
Comparison of depreciation methods

Year	$C_i = \$6000, C_s = \$1000, n = 5$			
	Straight line	Declining balance	Sum-of-years digits	MACRS
1	1000	1807	1667	1200
2	1000	1263	1333	1920
3	1000	882	1000	1152
4	1000	616	667	690
5	1000	431	333	690
6	—	—	—	348

13.5
TAXES

Taxes are an important factor to be considered in engineering economic decisions. The chief type of taxes that are imposed on a business firm are:

1. *Property taxes:* Based on the value of the property owned by the corporation (land, buildings, equipment, inventory). These taxes do not vary with profits and usually are not too large.
2. *Sales taxes:* Imposed on sales of products. Sales taxes usually are paid by the retail purchaser, so they generally are not relevant to engineering economy studies.
3. *Excise taxes:* Imposed on the manufacture of certain products like tobacco and alcohol. Also usually passed on to the consumer.
4. *Income taxes:* Imposed on corporate profits or personal income. Gains resulting from the sale of capital property also are subject to income tax.

Generally, federal income taxes have the most significant impact on engineering economic decisions. Although we cannot delve into the complexities of tax laws, it is important to incorporate the broad aspects of income taxes into our analysis.

The income tax rates are strongly influenced by politics and economic conditions. Currently the United States has a corporate *graduated tax schedule* as follows:

Taxable income	Tax rate
$1–$50,000	0.15
$50,001–$75,000	0.25
$75,001–$100,000	0.34
$100,001–$335,000	0.39
$335,001–$10 M	0.34
Over $10 M	0.35

Most states and some cities and counties also have an income tax. For simplicity in economic studies a single effective tax rate is often used. This commonly varies from 35 to 50 percent. Since state taxes are deductible from federal taxes, the effective tax rate is given by

$$\text{Effective tax rate} = \text{state rate} + (1 - \text{state rate})(\text{federal rate}) \qquad (13.26)$$

The chief effect of corporate income taxes is to reduce the rate of return on a project or venture.

$$\text{After-tax rate of return} = \text{before tax rate of return} \times (1 - \text{income tax rate})$$

$$r = i(1 - t) \qquad (13.27)$$

Note that this relation is true only when there are no depreciable assets. For the usual case when we have depreciation, capital gains or losses, or investment tax credits, Eq. (13.27) is a rough approximation. The importance of depreciation in reducing taxes is shown in Fig. 13.3. The depreciation charge appreciably reduces the gross profit, and thereby the taxes. However, since depreciation is retained in the corporation, it is available for advancement of the enterprise.

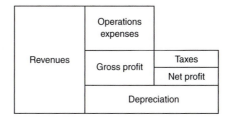

FIGURE 13.3
Distribution of corporate revenues.

EXAMPLE 13.5. High-Tech Pumps has a gross income in 1 year of $15M. Operating expenses (salaries and wages, materials, etc.) are $10M. Depreciation is $2.6M. Also, this year there is a *depreciation recapture* of $800,000 because a specialized CNC machine tool that is no longer needed is sold for more than its book value. (*a*) Compute the company's federal income taxes. (*b*) What is the average federal tax rate? (*c*) If the state tax rate is 11 percent, what is the total income taxes paid?

(*a*) Taxable income (TI) = gross income-operating expenses-depreciation + depreciation recapture

$$TI = 15 - 10 - 2.6 + 0.8 = \$3.2M$$

$$Taxes = (TI \text{ range})(marginal \text{ rate})$$

$$= (50,000)0.15 + (25,000)0.25 + (25,000)0.34$$

$$+ (235,000)0.39 + (3.2M - 0.335M)0.34$$

$$= 7500 + 6250 + 8500 + 91,650 + 974,100 = \$1,088,000$$

(*b*) Average federal tax rate $= \dfrac{1,088,00}{3,200,00} = 0.34$

(*c*) From Eq. (13.26)

$$Effective \text{ tax rate} = 0.11 + (1 - 0.11)(0.34) = 0.11 + 0.3026 = 0.4126$$

$$Total \text{ income taxes} = 3,2000,000(0.4126) = \$1,320,320$$

Consider a depreciable capital investment $C_d = C_i - C_s$. At the end of each year depreciation amounting to $D_f C_d$ is available to reduce the taxes by an amount $D_f C_d t$

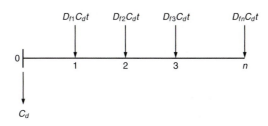

Note that the fractional depreciation charge each year D_f may vary from year to year depending on the method used to establish the depreciation schedule. See, for example, Table 13.4. The present value of this series of costs and benefits is

$$P = C_d - C_d t\left[\frac{D_{f_1}}{1 + r} + \frac{D_{f_2}}{(1 + r)^2} + \frac{D_{f_3}}{(1 + r)^3} + \cdots + \frac{D_{f_n}}{(1 + r)^n}\right] \quad (13.28)$$

The exact evaluation of the term in brackets will depend on the depreciation method selected.

EXAMPLE 13.6. A manufacturing company of modest size is considering an investment in energy-efficient electric motors to reduce its large annual energy cost. The initial cost would be $12,000, and over a 10-year period it is estimated that the firm would save $2200 annually in electricity costs. The salvage value of the motors is estimated at $2000. Determine the after-tax rate of return.

Solution. First we will establish the before-tax rate of return. We need to determine the cash flow for each year. Cash flow, in this context, is the net profit or savings for each year. We shall use straight-line depreciation to determine the depreciation charge. Table 13.5 shows the cash flow results. The before-tax rate of return is the interest rate at which the before-tax cash flow savings just equals the purchase cost of the motors.

$$12,000 = 2200(F_{RP,i,10}) + 2000(F_{SP,i,10})$$

We find the rate of return by trying different values of i in the compound interest tables. For $i = 14$ percent,

$$12,000 = 2200(5.2161) + 2000(0.2697)$$

$$= 11,475 + 539 = 12,014$$

Therefore, the before-tax rate of return is very slightly more than 14 percent. To find the after-tax rate of return, we use the after-tax cash flow in Table 13.5. From Eq. (13.27) we estimate the after-tax rate of return to be 7 percent.

$$12,000 = 1600F_{RP,i,10} + 2000F_{SP,i,10}$$

For $i = 6\%$: $12,000 = 1600(7.3601) + 2000(0.5584)$

$$= 11,776 + 1117 = 12,893 \qquad i \text{ too low}$$

TABLE 13.5
Cash flow calculations for example

Year	Before-tax cash flow	Depreciation	Taxable income	50% income tax	After-tax cash flow
0	−12,000				−12,000
1 to 9	2,200	1000	1200	−600	1,600
10	2,200	1000	1200	−600	1,600
	2,000				2,000

For $i = 8\%$: $12{,}000 = 1600(6.7101) + 2000(0.4632)$

$$= 10{,}736 + 926 = 11{,}662 \qquad i \text{ too high}$$

$$i = 6\% + 2\%\, \frac{12{,}893 - 12{,}000}{12{,}893 - 11662} = 6\% + 2\%(0.72)$$

$$i = 6 + 1.44 = 7.44\%$$

The method of including taxes illustrated above is the most straightforward approach, but it is not the quickest or shortest path to an answer. We can rewrite Eq. (13.28) as

$$P = C_d(1 - t\psi) \tag{13.29}$$

where ψ represents the term in brackets in Eq. (13.28). Jelen[1] presents expressions for annual cost and capitalized cost calculations based on straightline depreciation and sum-of-years digits. The following example illustrates the approach.

EXAMPLE 13.7. (Selection of alternative material based on capitalized cost.) We wish to decide whether carbon-steel or special alloy-steel tubes should be used in a heat exchanger. One material has a higher initial cost but offers longer service life, lower annual maintenance costs, and a saving in product due to less downtime. The comparison is based on capitalized cost using an after-tax rate of return $r = 10$ percent and a tax rate $t = 50$ percent. Depreciation is based on the sum-of-the-years digits method.

Solution.

	Carbon steel	Special alloy steel
Initial cost	$10,000	$95,000
Salvage value	—	20,000
Estimated service life, years	4	10
Tube cleaning inside, yearly	3,000	—
Tube cleaning outside, every third year	4,000	—
Tube repair, every fourth year	1,500	—
Tube maintenance, yearly	—	1,000
Annual savings from less product loss because of less downtime	—	−6,000

Capitalized cost of carbon-steel tubes

Initial cost:

$$K_1 = PF_{PK,r,n} = C_d F_{PK,r,n}(1 - tF_{SDP,r,n'})$$

$$= (10{,}000 - 0)(3.1547)[1 - 0.5(0.83013)] = 18{,}452$$

Note: $F_{SDP,r,n'}$ is ψ evaluated for sum-of-years digits, where n' is IRS depreciation life.

1. F. C. Jelen, op. cit., pp. 58–59.

Yearly cost, starting at the beginning of the year:

$$K_2 = \frac{R_b}{i} F_{PK,r,1}(1 - tF_{SDP,r,1})$$

$$= \frac{3000}{0.10}(1.100)[1 - 0.5(0.90909)] = 17,998$$

Tube cleaning outside, beginning of third year:

$$K_3 = C_{bx} \frac{1 - tF_{SDP,r,1}}{(1 + r)^{x-1}} F_{PK,r,n}$$

$$= \frac{4000}{1.10^{3-1}}[1 - 0.5(0.90909)](3.1547) = 5,688$$

Tube repair, beginning of fourth year:

$$K_4 = C_{bx} \frac{1 - tF_{SDP,r,1}}{(1 + r)^{x-1}} F_{PK,r,n}$$

$$= \frac{1500}{1.10^{4-1}}[1 - 0.5(0.90909)](3.1547) = \frac{1,939}{\$44,077}$$

Capitalized cost of special alloy steel

Initial cost:

$$K_1 = (95,000 - 20,000)(1.6275)[1 - 0.5(0.70099)] = 79,280$$

Salvage value: K_2 20,000

Note annual saving, treated as a beginning-of-year saving:

$$1000 - 6000 = -5000 \text{ net savings}$$

$$K_3 = \frac{R_b}{i} F_{PK,r,1}(1 - tF_{SDP,r,1})$$

$$= \frac{-5000}{0.10}(1.100)[1 - 0.5(0.90909)] \quad \frac{-29,997}{\$69,283}$$

This shows that the higher purchase cost of the special steel does not justify use of the steel with an after-tax return of 10 percent.

The expenditures that a business incurs are divided for tax purposes into two broad categories. Those for facilities and production equipment with lives in excess of one year are called capital expenditures; they are said to be "capitalized" in the accounting records of the business. Other expenses for running the business, such as labor and material costs, direct and indirect costs, and facilities and equipment with a life of one year or less, are ordinary business expenses. Usually they total more than the capital expenses. In the accounting records, they are said to be "expensed." The ordinary expenses are directly subtracted from the gross income to determine the taxable income, but only the annual depreciation charge can be subtracted from the capitalized expenses.

When a capital asset is sold, a capital gain or loss is established by subtracting the book value of the asset from its selling price. Frequently in our modern history capital gains have received special treatment by being taxed at a rate lower than for ordinary income.

Investment in capital is a vital step in the innovation process that leads to increased national wealth. Therefore, the federal government frequently uses the tax system to stimulate capital investment. This most often takes the form of a tax credit, usually 7 percent but varying with time from 4 to 10 percent. This means that 7 percent of the purchase price of qualifying equipment can be deducted from the taxes that the firm owes the U.S. government. Moreover, the depreciation charge for the equipment is based on its full cost.

13.6
PROFITABILITY OF INVESTMENTS

One of the principal uses for engineering economy is to determine the profitability of proposed projects or investments. The decision to invest in a project generally is based on three different sets of criteria.

Profitability: Determined by techniques of engineering economy to be discussed in this section.

Financial analysis: How to obtain the necessary funds and what it will cost. Funds for investment come from three broad sources: (1) retained earnings of the corporation, (2) long-term commercial borrowing from banks, insurance companies, and pension funds, and (3) the equity market through the sale of stock.

Analysis of intangibles: Legal, political, or social consideration or issues of a corporate image often outweigh financial considerations in deciding on which project to pursue. For example, a corporation may decide to invest in the modernization of an old plant because of its responsibility to continue employment for its employees when investment in a new plant 1000 miles away would be economically more attractive.

However, in our free-enterprise system a major goal of a business firm is to maximize profit. It does so by committing its funds to ventures that appear to be profitable. If investors do not receive a sufficiently attractive profit, they will find other uses for their money, and the growth—even the survival—of the firm will be threatened.

Four methods of evaluating profitability are commonly used. Accounting rate of return and payback period are simple techniques that are readily understood, but they do not take time value of money into consideration. Net present value and discounted cash flow are the most common profitability measures in which time value of money is considered. Before discussing them, however, we need to look a bit more closely at the concept of cash flow.

Cash flow measures the flow of funds into or out of a project. Funds flowing in constitute positive cash flow; funds flowing out are negative cash flow. The cash flow for a typical plant construction project is shown in Fig. 13.4. From an accounting point of view, cash flow is defined as

$$\text{Cash flow} = \text{net annual cash income} + \text{depreciation}$$

You might consider cash income as "real dollars" and the depreciation a bookkeeping adjustment to allow for capital expenditures. Table 13.6 shows how cash flow can be determined in a simple situation.

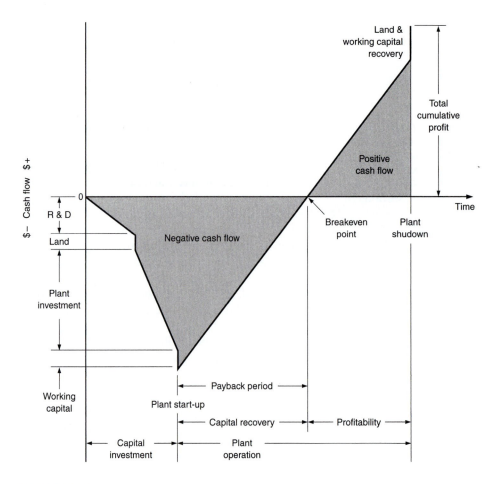

FIGURE 13.4
Typical costs in the cycle of a plant investment.

TABLE 13.6
Calculation of cash flow

(1) Revenue (over 1-year period)	$500,000
(2) Operating costs	360,000
(3) (1) − (2) = gross earnings	140,000
(4) Annual depreciation charge	60,000
(5) (3) − (4) = taxable income	80,000
(6) (5) × 0.35 = income tax	28,000
(7) (5) − (6) = net profit after taxes	52,000
Net cash flow (after taxes)	
(7) + (4) = 52,000 + 60,000	112,000

13.6.1 Rate of Return

The rate of return on the investment (ROI) is the simplest measure of profitability. It is calculated from a strict accounting point of view without consideration of the time value of money. It is a simple ratio of some measure of profit or cash income to the capital investment. There are a number of ways to assess the rate of return on the capital investment. ROI may be based on (1) net annual profit before taxes, (2) net annual profit after taxes, (3) annual cash income before taxes, or (4) annual cash income after taxes. These ratios, usually expressed as percents, can be computed for each year or on the average profit or income over the life of the project. In addition, capital investment sometimes is expressed as the average investment. Thus, although the ROI is a simple concept, it is important in any given situation to understand clearly how it has been determined.

> **EXAMPLE 13.8.** An initial capital investment is $360,000 and has a 6-year life and a $60,000 salvage value. Working capital is $40,000. Total net profit after taxes over 6 years is estimated at $167,000. Find the ROI.

Solution.

$$\text{Annual net profit} = \frac{167,000}{6} = \$28,000$$

$$\text{ROI on initial capital investment} = \frac{28,000}{360,000 + 40,000} = 0.07$$

13.6.2 Payback Period

The payback period is the period of time necessary for the cash flow to fully recover the initial total capital investment (Fig. 13.4). Although the payback method uses cash flow, it does not include a consideration of the time value of money. Emphasis is on rapid recovery of the investment. Also, in using the method, no account is taken of cash flows or profits recovered after the payback period. Consider the following example.

	Cash flow	
Year	Project A	Project B
0	$ −100,000	$ −100,000
1	50,000	0
2	30,000	10,000
3	20,000	20,000
4	10,000	30,000
5	0	40,000
6	0	50,000
7	0	60,000
	$10,000	$110,000
Payback period	3 years	5 years

By the payback period criterion, project A is the most desirable because it recovers the initial capital investment in 3 years. However, project B, which returns a cumulative cash flow of $110,000, obviously is the most profitable overall.

13.6.3 Net Present Worth

In Sec. 13.3, as one of the techniques of cost comparison, we introduced the criterion of net present worth (NPW).

$$\text{Net present worth} = \text{present worth of benefits} - \text{present worth of costs}$$

By this technique the expected cash flows (both + and −) through the life of the project are discounted to time zero at an interest rate representing the minimum acceptable return on capital. The project with the greatest positive value of NPW is preferred. NPW depends upon the project life, so strictly speaking the net present worths of two projects should not be compared if the projects have different service lives.

Obviously, the value of NPW will be dependent upon the interest rate used for the calculation. Low values of interest rate will tend to make NPW more positive, for a given set of cash flows, and large values of interest will push NPW in a negative direction. There will be some value of i for which the sum of the discounted cash flows equals zero; NPW $= 0$. This value of i is called the discounted cash flow rate of return.

13.6.4 Discounted Cash Flow

Discounted cash flow (DCF) is the rate of return for which the net present worth equals zero. We can express this criterion in a variety of ways.

$$\text{PW of benefits} - \text{PW of costs} = 0$$

$$\frac{\text{PW of benefits}}{\text{PW of costs}} = 1$$

$$\text{PW of benefits} = \text{PW of costs}$$

If, for example, the DCF rate of return is 20 percent, it implies that 20 percent per year will be earned on the investment in the project, in addition to which the project will generate sufficient funds to repay the original investment. Depreciation is considered implicitly in NPW and DCF calculations through the definition of cash flow.

Because the decision on profitability is expressed as a percentage rate of return in the DCF method, it is more readily understood and accepted by engineers and businesspeople than the NPW method, which produces a sum of money as an answer. In the NPW method it is necessary to select an interest rate for use in the calculations, and that may be a difficult and controversial thing to do. But by using the DCF method, we compute a rate of return, called the internal rate of return, from the cash flows. One situation in which NPW has an advantage is that individual values of NPW for a series of subprojects may be added to give the NPW for the complete project. That cannot be done with the rate of return developed from DCF analysis.

EXAMPLE 13.9. A machine has a first cost of $10,000 and a salvage value of $2000 after a 5-year life. Annual benefits (savings) from its use are $5000, and the annual cost of operation is $1800. The tax rate is 50 percent. Find the DCF rate of return.

Solution. Using straight-line depreciation, the annual depreciation charge is

$$D = \frac{C_i - C_s}{n} = \frac{10,000 - 2,000}{5} = \$1600$$

The annual cash flow after taxes is the sum of the net receipts and depreciation.

$$A_{CF} = (5000 - 1800)(1 - 0.50) + 1600(0.50)$$

$$= 1600 + 800 = \$2400$$

Year	Cash now
0	−10,000
1	2,400
2	2,400
3	2,400
4	2,400
5	2,400 + 2,000 (C_s)

$$NPW = 0 = -10,000 + 2400F_{RP,i,5} + 2000F_{SP,i,5}$$

If $i = 10$ percent, NPW $= +340$; if $i = 12$ percent, NPW $= -214$. Thus, we have the DCF rate of return bracketed, and

$$i = 10\% + (12\% - 10\%)\frac{340}{340 + 214}$$

$$= 10 + 2\left(\frac{340}{564}\right) = 10 + 1.2 = 11.2\%$$

It is an important rule of engineering economy that *each increment* of investment capital must be justified on the basis of earning the minimum required rate of return.

EXAMPLE 13.10. A company has the option of investing in one of the two machines described in the following table. Which investment is justified?

	Machine A	Machine B
Initial cost C_i	$10,000	$15,000
Useful life	5 years	10 years
Salvage value C_s	$2,000	0
Annual benefits	$5,000	$7,000
Annual costs	$1,800	$4,300

Solution. Assume a 50 percent tax rate and a minimum attractive rate of return of 6 percent. The conditions for machine A are identical with those in Example 13.9, for which $i = 11.2$ percent. Calculation of the DCF rate of return for machine B shows it is slightly in excess of the minimum rate of 6 percent. However, that is not the proper question. Rather, we should ask whether the *increment of investment* ($15,000 − $10,000) is justified. In addition, because machine B has twice the useful life of machine A, we should place them both on the same time basis.

Cash flow

Year	Machine A	Machine B	Difference, B − A
0	−10,000	−15,000	−5,000
1	2,400	2,100	−300
2	2,400	2,100	−300
3	2,400	2,100	−300
4	2,400	2,100	−300
5	2,400 − 10,000 + 2,000	2,100	−300 + 8,000
6	2,400	2,100	−300
7	2,400	2,100	−300
8	2,400	2,100	−300
9	2,400	2,100	−300
10	2,400 + 2,000	2,100	−300 − 2,000

$$\text{NPW} = 0 = -5000 - 300 F_{RP,i,10} + 8000 F_{SP,i,5} - 2000 F_{SP,i,10}$$

But, even at $i = \frac{1}{4}$ percent, NPW $= -2009$, and there is no way that the extra investment in machine B can be justified economically.

When only costs—not income (or savings)—are known, we can still use the DCF method for incremental investments, but not for a single project. We assume that the lowest capital investment is justified without being able to determine the DCF rate of return, and we then determine whether the additional investment is justified.

EXAMPLE 13.11. On the basis of the data in the following table, determine which machine should be purchased.

	Machine A	Machine B
First cost	$3000	$4000
Useful life	6 years	9 years
Salvage value	$500	0
Annual operating cost	$2000	$1600

Solution. This solution will be based on cash flow before taxes. To place the machines on a common time frame, we use a common life of 18 years.

Year	Machine A	Machine B	Difference, B − A
0	−3000	−4000	−1000
1 to 5	−2000	−1600	+400
6	−2000 − 2500	−1600	+400 + 2500
7 to 8	−2000	−1600	+400
9	−2000	−1600 − 4000	+400 − 4000
10, 11	−2000	−1600	+400
12	−2000 − 2500	−1600	+400 + 2500
13 to 17	−2000	−1600	+400
18	−2000 + 500	−1600	+400 − 500

$$\text{NPW} = 0 = -1000 + 400 F_{RP,i,18} + 2500 F_{SP,i,6} + 2500 F_{SP,i,12}$$
$$-400 F_{SP,i,9} - 500 F_{SP,i,18}$$

Trial and error shows that $i \approx 47$ percent, which clearly justifies purchase of machine B.

We have presented information on the four most common techniques for evaluating the profitability of an investment. The rate-of-return method has the advantage of being simple and easy to use. However, it ignores the time value of money and the consideration of cash flow. The payback period also is a simple method, and it is particularly attractive for industries undergoing rapid technological change. Like the rate-of-return method, it ignores time value of money, and it places an undue emphasis on projects that achieve a quick payoff. The net present worth method takes both cash flow and time value of money into account. However, it suffers from the problem of ambiguity in setting the required rate of return, and it may present problems when projects with different service lives are compared. Discounted cash flow rate of return has the advantage of producing an answer that is the real internal rate of return. The method readily permits comparison between alternatives, but it is assumed that all cash flows generated by the project can be reinvested to yield a comparable rate of return.

13.7
OTHER ASPECTS OF PROFITABILITY

Innumerable factors[1] affect the profitability of a project in addition to the mathematical expressions discussed in Sec. 13.6. The purpose of this section is to round out our consideration of the crucial subject of profitability.

We need to realize that profit and profitability are not quite the same concept. Profit is measured by accountants, and its value in any one year can be manipulated in many ways. Profitability is inherently a long-term parameter of economic decision making. As such, it should not be influenced much by short-term variations in profits. In recent years there has been a strong trend toward undue emphasis on quick profits and short payoff periods that work to the detriment of long-term investment in high-technology projects.

Estimation of profitability requires the prediction of future cash flows, which in turn requires reliable estimates of sales volume and sales price by the marketing staff and of material price and availability. The quadrupling of crude oil price in the early 1970s was a dramatic (and traumatic) example of how changes in raw material costs can greatly influence profitability predictions. Similarly, trends in operating costs must be looked at carefully, especially with respect to whether it is more profitable to reduce operating costs through increased investment, as with automation.

The estimated investment in machinery and facilities that is required for the project is usually the most accurate component of the profitability evaluation. (This topic is considered in more detail in Chap. 14.) The depreciation method used influences how the expense is distributed over the years of a project, and that in turn determines what the cash flow will be. However, a more fundamental aspect of depreciation is the effect of writing off a capital investment over a long time period. As a result, costs are underestimated and selling prices are set too low. A long-term write-off combined with inflation results in insufficient cash flow to permit reinvestment. Inflation creates hidden expenses like inadequate allowance for depreciation. When depreciation methods do not allow for inflated replacement costs, those costs must be absorbed on an after-tax basis. Profit and profitability are overstated in an inflationary period.

A number of technical decisions are closely related to the investment policy and profitability. At the design stage it may be possible to ensure a level of product superiority that is more than needed by the current market. Later, when competitors enter the market, the superiority would prove useful, but it is not achieved without an initial cost to profitability. Economics generally favor building as large a production unit as the market can absorb. However, this increased profitability is achieved at some risk to maintaining continuity of production should the unit be down for repairs. Thus, there often is a trade-off between the increased reliability of having a number of small units over which to spread the risk and a single large unit with somewhat higher profitability.

The profitability of a particular product line can be influenced by decisions of cost allocation. Such factors as overhead, utility costs, transfer prices between divi-

---

1. F. C. Jelen, *Hydrocarbon Processing,* pp. 111–115, January 1976.

sions of a large corporation, or scrap value often require arbitrary decisions for allocation between various products. Thus, the situation often favors certain products and discriminates against others because of cost allocation policies. Sometimes corporations take a position of milking an established product line with a limited future (a "cash cow") in order to stimulate the growth of a new but promising product line. Another profit decision is whether to charge a particular item as a current expense or capitalize it to make a future expense. In a period of inflation there is strong pressure to increase present profitability by deferring costs into the future by capitalizing them. It is argued that a fixed dollar amount deferred into the future will have less consequence in terms of future dollars.

The role of the government in influencing profitability is very great. In the broader sense the government creates the general economic climate through its policies on money supply, taxation, and foreign affairs. It provides subsidies to stimulate selected parts of the economy. Its regulating powers have had an increasing influence on profitability in such areas as pollution control, occupational health and safety, consumer protection and product safety, use of federal lands, antitrust, minimum wages, and working hours.

Since profitability analysis deals with future predictions, there is inevitable uncertainty. Incorporation of uncertainty or risk is possible with advanced techniques. Unfortunately, the assignment of the probability to deal with risk is in itself a subjective decision. Thus, although risk analysis (Sec. 13.10) is an important technique, one should always realize its true origins.

13.8
INFLATION

Since engineering economy deals with decisions based on future flows of money, it is important to consider inflation in the total analysis. Surprisingly, however, inflation is considered substantively in few engineering economy texts.[1] From 1967 to 1997 the average inflation has been 5.2 percent. However, there have been years (1974, 1979, and 1980) when the rate of inflation was in double digits.

Inflation exists when prices of goods and services are increasing so that a given amount of money buys less and less as time goes by. Interest rates and inflation are directly related. The basic interest rate is about 2 to 3 percent higher than the inflation rate. Thus, in a period of high inflation, not only does the dollar purchase less each month but the cost of borrowing money also rises.

Price changes may or may not be considered in an economic analysis. For meaningful results, costs and benefits must be computed in comparable units. It would not be sensible to calculate costs in 1990 dollars and benefits in 2000 dollars. However, we could have a situation in which future benefits fluctuate with the inflation of the dollar. In that somewhat rare situation, inflation would have no effect on the before-tax

---

1. B. W. Jones, "Inflation in Engineering Economics Analysis," Wiley, New York, 1982.

economic analysis. However, as will be seen later, adjusting future before-tax benefits for inflation (constant-value money) will not avoid the impact of inflation in an after-tax analysis.

Inflation is measured by the change in the Consumer Price Index (CPI), as determined by the U.S. Department of Labor. The CPI shows the relationship between present and past costs for typical things that consumers must buy, such as food, transportation, energy, and rent. Other industrial cost indices discussed in Sec. 14.4 may be more relevant to design situations.

> **EXAMPLE 13.12.** The CPI in 1967 was 100.0, and in 1976 it was 171.10. Find the inflation rate f.

$$S = PF_{PS,i,10}$$

$$171.1 = 100F_{PS,i,10} \qquad F_{PS,i,10} = 1.71$$

and on interpolation we find $f = 0.055$.

Another way to arrive at f is to use the equation for annualized return.

$$\text{Annualized return} = \left(\frac{\text{current value}}{\text{original value}} \right)^{\frac{1}{n}} - 1$$

$$= \left(\frac{171.1}{100} \right)^{\frac{1}{10}} - 1 = 1.0551 - 1 = 0.055$$

Money in one time period t_1 can be brought to the same value as money in another time period t_2 by the equation

$$\text{Dollars in period } t_1 = \frac{\text{Dollars in period } t_2}{\text{Inflation rate between } t_1 \text{ and } t_2} \tag{13.30}$$

It is useful to define two situations: then-current money and constant-value money. Let the dollars in period t_1 be constant-value dollars. Constant-value represents equal purchasing power at any future time. Current money, in time period t_2, represents ordinary money units which decline in purchasing power with time. For example, if an item cost \$10 in 1998 and inflation was 3 percent during the previous year, in constant 1997 dollars, the cost is equal to \$10/1.03 = \$9.71.

There are three different rates to be considered when dealing with inflation.

> *Real or inflation-free interest rate i:* This is the rate at which interest is earned when effects of changes in the currency have been ignored. This is the interest rate we have used up until now in this chapter.
>
> *Market interest rate i_f:* This is the interest rate that is quoted on the business news every day. It is a combination of the real interest rate i and the inflation rate f. This is also called the *inflated interest rate.*
>
> *Inflation rate f:* This is a measure of the rate of change in the value of the currency.

Consider the equation for the present worth of a future sum S in current dollars. S must be first discounted for the real interest rate and then for the inflation rate.

$$P = \frac{S}{(1 + i)^n} \frac{1}{(1 + f)^n} = S\frac{1}{(1 + i + f + if)^n} = \frac{S}{(1 + i_f)^n} \qquad (13.31)$$

where $i_f = i + f + if$ is the market interest rate, also called the inflated interest rate.

EXAMPLE 13.13. A project requires an investment of \$10,000 and is expected to return, in future, or "then current," dollars, \$2500 at the end of year 1, \$3000 at the end of year 2, and \$7000 at the end of year 3. The monetary interest rate is 10 percent, and the inflation rate is 6 percent per year. Find the net present worth of this investment opportunity.

Solution. The inflated interest rate is $0.10 + 0.06 + (0.10)(0.06) = 0.166$; for simplicity we shall use $i_f = 0.17$.

Current-dollars approach

Year	Cash flow	$F_{SP,17,n}$	Present worth
0	−10,000	1.00	−10,000
1	2,500	0.8547	2,137
2	3,000	0.7305	2,191
3	7,000	0.6244	4,971
		NPW =	−711

Constant-value-dollars approach

Year	Cash flow*	$F_{SP,10,n}$	Present worth
0	−10,000	1.00	−10,000
1	2,358	0.9091	2,144
2	2,670	0.8264	2,206
3	5,877	0.7513	4,415
		NPW =	−1,235

*Adjusted for $f = 0.06$.

The difference in the NPWs found by the two treatments is due to using an approximate combined discount rate instead of the more accurate value of $i_f = 0.166$. However, the approximation is justified in view of the uncertainty in predicting the rate of inflation. It should be noted that, for this example, the NPW is +\$10 if inflation is ignored. That emphasizes the fact that neglecting the influence of inflation overemphasizes the profitability.

When profitability is measured by the DCF rate of return i, the inclusion of the inflation rate f results in an effective rate of return i' based on constant-value money.[1]

$$1 + i = (1 + i')(1 + f)$$

$$i' = i - f - i'f \approx i - f \qquad (13.32)$$

---

1. F. A. Holland and F. A. Watson, *Chem. Eng.*, pp. 87–91, Feb. 14, 1977.

To a first approximation, the DCF rate of return is reduced by an amount equivalent to the average inflation rate.

Interest rates are quoted to investors in current money i, but investors generally expect to cover any inflationary trends and still receive an acceptable return. In other words, investors hope to obtain a constant-value interest rate i'. Therefore, the current interest rate tends to fluctuate with the inflation rate. If the calculation is to be made with constant-value money, discounting should be done with the normal interest rate i. If calculations are in terms of current money, then the discount rate should be $i_f = i + f$.

Note that tax allowance for depreciation has a reduced benefit when constant money is used for profitability evaluations. By law, depreciation is defined in terms of current money. Therefore, under high inflation when constant-money conditions are appropriate, a full tax credit for depreciation is not achieved.

Another effect of inflation is that it increases the cash flow because the prices received for goods and services rise as the value of money falls. Even when constant-value money is used, the yearly cash flows should display the current money situation.

Detailed relations for capitalized cost that include an inflation factor have been developed by Jelen.[1] They can also be used to introduce inflation calculation into annual cost calculations.

13.9
SENSITIVITY AND BREAK-EVEN ANALYSIS

A *sensitivity analysis* determines the influence of each factor in the problem on the final result, and therefore it determines which factors are most critical in the economic decision. Since there is a considerable degree of uncertainty in predicting future events like sales volume, salvage value, and rate of inflation, it is important to see how much the economic analysis depends on the magnitude of the estimates. One factor is varied over a reasonable range and the others are held at their mean (expected) value. The amount of computation involved in a sensitivity analysis of an engineering economy problem can be very considerable, but the use of computers has made sensitivity analysis in this context a much more practical endeavor.[2]

A *break-even analysis* often is used when there is particular uncertainty about one of the factors in an economic study. The break-even point is the value for the factor at which the project is just marginally justified.

> **EXAMPLE 13.14.** Consider a $20,000 investment with a 5-year life. The salvage value is $4000, and the minimal acceptable return is 8 percent. The investment produces annual benefits of $10,000 at an operating cost of $3000. Suppose there is considerable uncertainty as to whether the new machinery will survive 5 years of continuous use. Find the break-even point, in terms of life, at which the project just becomes economically viable.

---

1. F. C. Jelen, op. cit., pp. 122–128; *Chem. Eng.,* pp. 123–128, Jan. 27, 1958.
2. J. C. Agarwal and I. V. Klumpar, *Chem. Eng.,* pp. 66–72, Sept. 29, 1975.

Solution. Using the annual cost method,

$$\$10{,}000 - 3000 - (20{,}000 - 4000)F_{PR,8,n} - 4000(0.08) = 0$$

$$F_{PR,8,n} = \frac{6680}{16{,}000} = 0.417$$

and interpolating in the interest tables gives us $n = 2.8$ years. Thus, if the machine does not last 2.8 years, the investment cannot be justified.

Break-even analysis frequently is used in problems dealing with staged construction. The usual problem is to decide whether to invest more money initially in unused capacity or to add the needed capacity of a later date when needed, but at higher unit costs.

EXAMPLE 13.15. A new plant will cost \$100 million for the first stage and \$120 million for the second stage at n years in the future. If it is built to full capacity now, it will cost \$140 million. All facilities are expected to last 40 years. Salvage value is neglected. Find the preferable course of action.

Solution. The annual cost of operation and maintenance is assumed the same for a two-stage construction and full-capacity construction. We shall use a present worth (PW) calculation with a 10 percent interest rate. For full-capacity construction now, PW = \$140 million (\$140M). For two-stage construction

$$PW = \$100M + \$120M\ F_{SP,10,n}$$

$$n = 5 \text{ years: PW} = 100 + 120(0.6201) = \$174M$$

$$n = 10 \text{ years: PW} = 100 + 120(0.3855) = \$146M$$

$$n = 20 \text{ years: PW} = 100 + 120(0.1486) = \$118M$$

$$n = 30 \text{ years: PW} = 100 + 120(0.0573) = \$107M$$

These results are plotted in Fig. 13.5. The break-even point (12 years) is the point at which the two alternatives have equivalent cost. If the full capacity will be needed before 12 years, then full capacity built now would be the preferred course of action.

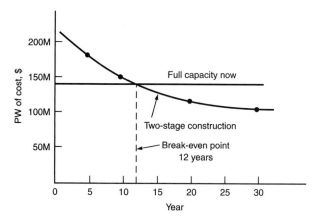

FIGURE 13.5
Break-even plot for Example 13.15.

13.10
UNCERTAINTY IN ECONOMIC ANALYSIS

In the preceding section we discussed the fact that engineering economy deals chiefly with decisions based on future estimates of costs and benefits. Since none of us has a completely clear crystal ball, such estimates are likely to contain considerable uncertainty. In all of the examples presented so far in this chapter we have used a single value that was the implied best estimate of the future.

Now that we are willing to recognize that estimates of the future may not be very precise, there are some ways by which we can guard against the imprecision. The simplest procedure is to supplement your estimated most likely value with an optimistic value and a pessimistic value. The three estimates are combined into a mean value by

$$\text{Mean value} = \frac{\text{optimistic value} + 4(\text{most likely value}) + \text{pessimistic value}}{6} \quad (13.33)$$

In Eq. (13.33) the distribution of values is assumed to be represented by a beta frequency distribution. The mean value determined from the equation is used in the economic analysis.

The next level of advance would be to associate a probability with certain factors in the economic analysis. In a sense, by this approach we are transferring the uncertainty from the value itself to the selection of the probability.

EXAMPLE 13.16. The expected life of a piece of mining equipment is highly uncertain. The machine costs $40,000 and is expected to have $5000 salvage value. The new machinery will save $10,000 per year, but it will cost $3000 annually for operations and maintenance. The service life is estimated to be:

3 years, with probability = 0.3

4 years, with probability = 0.4

5 years, with probability = 0.5

Solution. For 3-year life:

$$\text{Net annual cost} = (10,000 - 3000) - (40,000 - 5000)F_{PR,10,3} - 5000(0.10)$$

$$= 7000 - 35,000(0.4021) - 500 = -8573$$

For 4-year life:

$$\text{Net annual cost} = 7000 - 35,000(0.3155) - 500 = -4542$$

For 5-year life:

$$\text{Net annual cost} = 7000 - 35,000(0.2638) - 500 = -2733$$

$$\text{Expected value of net annual cost} = E(\text{AC}) = \Sigma \text{AC} \times P(\text{AC})$$

$$= -8573(0.3) + [-4542(0.4)] + [-2733(0.3)] = -5207$$

13.11
BENEFIT-COST ANALYSIS

An important class of engineering decisions involves the selection of the preferred system design, material, purchased subsystem, etc., when economic resources are constrained. The methods of making cost comparisons and profitability analysis described in Secs. 13.3 and 13.6 are important decision-making tools in this type of situation.

Frequently, comparisons are based on a *benefit-cost ratio,* which relates the desired benefits to the capital investment required to produce the benefits. This method of selecting alternatives is most commonly used by governmental agencies for determining the desirability of public works projects. A project is considered viable when the net benefits associated with its implementation exceed its associated costs. *Benefits* are advantages to the public (or owner), expressed in terms of dollars. If the project involved disadvantages to the owner, these *disbenefits* must be subtracted from the benefits. The costs to be considered include the expenditures for construction, operation, and maintenance, less salvage. Both benefits, disbenefits, and costs must be expressed in common dollar terms by using the present worth or annual cost concept.

$$\text{Benefit-cost ratio (BCR)} = \frac{\text{benefits} - \text{disbenefits}}{\text{costs}} \tag{13.34}$$

A design or project for which BCR < 1 does not cover the cost of capital to create the design. Generally, only projects for which BCR > 1 are acceptable. The benefits used in the BCR would be factors like improved component performance, increased payload through reduced weight, and increased availability of equipment. Benefits are defined as the advantages minus any disadvantages, i.e., the net benefits. Likewise, the costs are the total costs minus any savings. The costs should represent the initial capital cost as well as costs of operation and maintenance; see Chap. 14.

Very often in problems of choosing between several alternatives the incremental or marginal benefits and costs associated with changes beyond a base level or reference design should be used. The alternatives are ranked with respect to cost, and the lowest-cost situation is taken as the initial reference. This is compared with the next higher cost alternative by computing the incremental benefit and incremental cost. If $\Delta B/\Delta C < 1$, then alternative 2 is rejected because the first alternative is superior. Alternative 1 now is compared with alternative 3. If $\Delta B/\Delta C > 1$, then alternative 1 is rejected and alternative 3 becomes the current best solution. Alternative 3 is compared with number 4, and if $\Delta B/\Delta C < 1$, then alternative 3 is the best choice. We should note that this may not be the alternative with the largest overall benefit-cost ratio.

EXAMPLE 13.17 You are asked to recommend a site for a small dam to generate hydroelectric power. The construction cost at various sites is given below. These vary with topography and soil conditions. Each estimate includes $3M for the turbines and generators. The annual benefits from the sale of electricity vary between sites because of stream velocity.

We require an annual return of 10 percent. The life of the dam is infinite for purposes of calculation. The hydroelectric machinery has a 40-year life.

Site	Construction cost, $M	Cost of machinery, $M	Cost of dam, $M	Annual income, $M
A	9	3	6	1.0
B	8	3	5	0.9
C	12	3	9	1.25
D	6	3	3	0.5

Since the benefit (income) is on an annual basis, we have to convert the cost to an annual basis. Also, we are going to make our decision on an incremental basis. We construct the table below, placing the alternatives in order of increasing cost (from left to right).

For example, the annual cost of capital recovery for $A = P_D i + P_{H-E}F_{PR,10,40}$ where P_D = cost of dam and P_{H-E} = cost of hydroelectric machinery

$$\text{cap. recovery for } A = 6,000,000(0.10) + 3,000,000(0.1023)$$

$$= 600,000 + 306,900 = \$907,000$$

	D	B	A	C
Annual cost of capital recovery ($1000)	607	807	907	1207
Annual benefits ($1000)	500	900	1000	1250
Comparison	D to do nothing	B to do nothing	A to B	C to B
Δ capital recovery	607	807	100	400
Δ annual benefits	500	900	100	350
$\Delta B/\Delta C$	0.82	1.11	1.00	0.87
Selection	Do nothing	B	B	B

We note that when compared to not building a dam, $\Delta B/\Delta C$ for site D is less than 1.0. The next lowest cost dam site is greater than 1.0, so it is selected in comparison to not building a dam. The benefit-cost ratio for sites A and C also is greater than 1.0, but now, having found a low-cost qualifying site (B), we need to determine whether the increment in benefits and costs is better than B. We see that on a $\Delta B/\Delta C$ basis, A and C are not better choices than B. Therefore, we select site B.

When used in a strictly engineering context to aid in the selection of alternative materials, the benefit-cost ratio is a useful decision-making tool. However, it often is used with regard to public projects financed with tax moneys and intended to serve the overall public good. There is a psychological advantage to the BCR concept over the discounted cash flow rate of return in that it avoids the connotation that the government is profiting from public moneys. Here questions that go beyond economic efficiency become part of the decision process. Many of the broader issues are difficult to quantify in monetary terms. Of even greater difficulty is the problem of relating monetary cost to the real values of society.

Consider the case of a hydroelectric facility. The dam produces electricity, but it also will provide flood control and recreational boating. The value of each of the outputs should be included in the benefits. The costs include the expenditures for construction, operation, and maintenance. However, there may be social costs like the loss of virgin timberland or a scenic vista. Great controversy surrounds the assignment of costs to environmental and aesthetic issues.

Although benefit-cost analysis is a widely used methodology, it is not without problems. The assumption is that costs and benefits are relatively independent. Basically, it is a deterministic method that does not deal with uncertainty in a major way. As with most techniques, it is best not to try to push it too far. Although the quantitative ratios provided by Eq. (13.34) should be used to the greatest extent possible, they should not preempt the utilization of common sense and good judgment.

13.12
SUMMARY

Engineering economy is the methodology that allows rational decision making about the allocation of amounts of money occurring at various points in time and in various manners, e.g., a uniform series over time or a single payment in the future. As such, engineering economy accounts for the time value of money.

The basic relationship is the compound interest formula that relates the future sum S to the present sum P over n years at an interest rate i.

$$S = P(1 + i)^n$$

If P is solved for in this equation, we are discounting the future sum S back to the present time. If the money occurs as equal end-of-the-period amounts R, then

$$S = R \frac{(1 + i)^n - 1}{i}$$

If this equation is solved for R, it gives the annual payment to provide a sinking fund to replace worn-out equipment. More important is the annual payment to return the initial capital investment *plus* paying interest on the principal P tied up in the investment, where CRF is the capital recovery factor.

$$R = P \frac{i(1 + i)^n}{(1 + i)^n - 1} = P(CRF)$$

Engineering economy allows rational decisions to be made about alternative courses of action involving money. To do this, each alternative must be placed on an equivalent basis. There are four common ways of doing this.

- *Present-worth analysis:* All costs or receipts are discounted to the present time to calculate the net present worth. This method works best when the alternatives have a common time period.
- *Annual cost analysis:* The cash flow over time is converted to an equivalent annual cost or benefit. This method works well when the alternatives have different time periods.

- *Capitalized cost analysis:* This is a special case of present worth analysis for a project that exists in perpetuity ($n = \infty$).
- *Benefit-cost ratio:* This method analyzes the costs and benefits of a project on one of the above three basis, and then decides to fund the project if the ratio of benefits to costs is greater than 1.0.

Realistic economic analysis requires consideration of *taxes,* chiefly federal income tax. Accurate determination of the taxable income requires allowance for *depreciation,* the reduction in value of owned assets due to wear and tear or obsolescence. Realistic economic analysis also requires allowance for *inflation,* the decrease in the value of currency over time.

An important use of engineering economy is in determining the profitability of proposed projects or investments. This usually starts with estimating the cash flow to be generated by the project.

$$\text{Cash flow} = \text{net annual cash income} + \text{depreciation}$$

Two common methods of estimating profitability are rate of *return on the investment* (ROI) and *payback period.*

$$\text{ROI} = \frac{\text{average annual net profit}}{\text{capital investment} + \text{working capital}}$$

Payback period is the period of time for the cumulative cash flow to fully recover the initial total capital investment. Both of these methods suffer from not considering the time value of money. A better method to measure profitability is *net present worth.*

$$\text{Net present worth} = \text{present worth of benefits} - \text{present worth of costs}$$

With this method the expected cash flows (both $+$ and $-$) through the life of the project are discounted to time zero at an interest rate representing the minimum acceptable return on capital. The *discounted cash flow rate of return* is the interest rate for which the net present worth equals zero.

$$\text{Net PW} = \text{PW(benefits)} - \text{PW(costs)} = 0$$

Since there is considerable uncertainty in estimating future income streams and costs, engineering economic studies often estimate a range of values and utilize a mean value. Another approach is to place probabilities on the values and use an expected value in the analysis.

BIBLIOGRAPHY

Blank, L. T., and A. J. Tarquin: "Engineering Economy," 4th ed., McGraw-Hill, New York, 1998.
Canada, J. R.: "Intermediate Economic Analysis for Management and Engineering," Prentice-Hall, Englewood Cliffs, NJ, 1971.
English, M. J.: "Project Evaluation: A Unified Approach to the Analysis of Capital Investments," Macmillan, New York, 1984.

Holland, F. A., F. A. Watson, and J. K. Wilkinson: "Introduction to Process Economics," Wiley, New York, 1974.

Humphreys, K. K.: "Jelen's Cost and Optimization Engineering," 3d ed., McGraw-Hill, New York, 1990.

Kurtz, M. (ed.): "Handbook of Engineering Economics," McGraw-Hill, New York, 1984.

Newnan, D. G., and J. P. Lavelle: "Engineering Economics Analysis," 7th ed., Engineering Press, San Jose, Calif., 1998.

Park, C. S., "Contemporary Engineering Economics," 2d ed., Addison-Wesley, Reading, MA, 1996.

Riggs, J. L., D. D. Bedworth, and S. U. Randhawa: "Engineering Economics," 4th ed., McGraw-Hill, New York 1996.

Thuesen, G. J., and W. J. Fabrycky: "Engineering Economy," 8th ed., Prentice-Hall, Englewood Cliffs, NJ, 1993.

White, J. A., K. E. Case, D. B. Pratt, and M. H. Agee, "Principles of Engineering Economic Analysis," 4th ed., Wiley, New York, 1997.

PROBLEMS AND EXERCISES

The interest tables in Appendix A are available to help you solve these problems. Also, note that computer spreadsheet software provides most of the financial functions discussed in this chapter. It is recommended that you use a spreadsheet to solve the problems.

13.1. (*a*) Calculate the amount realized at the end of 7 years through annual deposits of $1000 at 10 percent compound interest.
   (*b*) What would the amount be if interest were compounded semiannually?

13.2. A young woman purchases a new car. After down payment and allowances, the amount to be paid is $8000. If money is available at 10 percent, what is the monthly payment to pay off the loan in 4 years? What would it be at 4 percent interest?

13.3. A new machine tool costs $15,000 and has a $5000 salvage value at the end of 5 years. The interest rate is 10 percent. The annual cost of capital recovery is the annual depreciation charge (use straight-line depreciation) plus the equivalent annual interest charge. Work this out on a year-by-year basis and show that it equals the number obtained quickly by using the capital recovery factor.

13.4. A father desires to establish a fund for his new child's college education. He estimates that the current cost of a year of college education is $12,000 and that the cost will escalate at an annual rate of 4 percent.
   (*a*) What amount is needed on the child's eighteenth, nineteenth, twentieth, and twenty-first birthdays to provide for a 4-year college education?
   (*b*) If a rich aunt gives $5000 on the day the child is born, how much must be set aside at 10 percent on each of the first through seventeenth birthdays to build up the college fund?

13.5. A major industrialized nation manages its finances in such a way that it runs an annual trade deficit with other countries of $100B. If the cost of borrowing is 10 percent, how long will it be before the debt (accumulated deficit) is one trillion dollars (1000B)? If nothing is done, how long will it take to accumulate the second $1000B debt?

13.6. Machine A costs $8500 and has annual operating costs of $4500. Machine B costs $7000 and has an annual operating cost of $4800. Each machine has an economic life of 10 years. If the minimum required rate of return is 10 percent, compare the advantages of machine A by (*a*) present worth method, (*b*) annual cost method, and (*c*) rate of return on investment.

13.7. Make a cost comparison between two conveyor systems for transporting raw materials.

	System A	System B
Installed cost	$25,000	$15,000
Annual operating cost	6,000	11,000

The service life of each system is 5 years and the write-off period is 5 years. Use straight-line depreciation and assume no salvage value for either system. At what rate of return *after taxes* would B be more attractive than A?

13.8. A resurfaced floor costs $5000 and will last 2 years. If money is worth 10 percent after taxes, how long must a new floor costing $19,000 last to be economically justified? The tax rate is 52 percent. For tax purposes a new floor can be written off in 1 year. Use sum-of-the-year-digits depreciation. Use the capitalized cost method for your analysis.

13.9. You are concerned with the purchase of a heat-teating furnace for the gas carburizing of steel parts. Furnace A will cost $325,000 and will last 10 years; furnace B will cost $400,000 and will also last 10 years. However, furnace B will provide closer control on case depth, which means that the heat treater can shoot for the low side of the specification range on case depth. That will mean that the production rate for furnace B will be 2740 lb/h compared with 2300 lb/h for furnace A. Total yearly production is required to be 15,400,000 lb. The cycle time for furnace A is 16.5 h, and that for furnace B is 13.8 h. The hourly operating cost is $64.50 per h.

Justify the purchase of furnace B on the basis of (*a*) payout time and (*b*) discounted cash flow rate of return after taxes. Assume money is worth 10 percent and the tax rate is 50 percent.

13.10. The cost of capital has a strong influence on the willingness of management to invest in long-term projects. If the cost of capital in America is 10 percent and in Japan 4 percent, what must the return be after 2 years on a 2-year investment of $1M for each of the situations to provide an acceptable return on the investment? Repeat the analysis for a 20-year period.

13.11. In order to justify investment in a new plating facility, it is necessary to determine the present worth of the costs.

Calculate the present worth given the following information:

Cost of equipment	$350,000
Planning period	5 years
Fixed charges	20 percent of investment each year
Variable charges	40,000 first year, escalating at 6 percent each year with inflation starting at $t = 0$
Rate of return	$i = 10\%$

13.12. Determine the net present worth of the costs for a major construction project under the following set of conditions:
  (*a*) Estimated cost $300 million over 3 years (baseline case).
  (*b*) Project is delayed by 3 years with rate of inflation 10 percent and interest cost 16 percent.
  (*c*) Project is delayed 6 years with rate of inflation 10 percent and interest costs 16 percent.

13.13. Whether a maintenance operation is classified as a repair (expense charged against revenues in current year) or improvement (capitalized expense) can have a big influence on taxes. Determine the net savings for a $10,000 operation using the two different approaches if (*a*) you are in a business that is in the 50 percent tax bracket and (*b*) you are in a small business in the 20 percent tax bracket. Use a 10 percent interest rate and a 10 percent investment tax credit.

13.14. As a new professional employee you need to worry about your retirement many years in the future. Construct a table showing how much you need to have invested, at 4 percent, 8 percent, and 12 percent annual rate of return, to provide each $100 of monthly income. Assume that inflation will increase at 3 percent annually, so the numbers you calculate will be in inflation-adjusted dollars. Calculate the monthly amount needed for a retirement period of 25, 30, 35, and 40 years. Assume that the investments are made in tax-sheltered accounts.

13.15. At what annual mileage is it cheaper to provide your field representatives with cars than to pay them $0.32 per mile for the use of their own cars? The costs of furnishing a car are as follows:

Purchase price	$9000
Life	4 years
Salvage	$1500
Storage	$150 per year
Maintenance	$0.08 per mile

  (*a*) Assume $i = 10$ percent.
  (*b*) Assume $i = 16$ percent.

13.16. To *levelize expenditures* means to create a uniform end-of-year payment that will have the same present worth as a series of irregular end-of-year payments. To illustrate, consider the estimated 5-year maintenance budget for a pilot plant. Develop a levelized cost assuming that $i = 0.10$ and the annual inflation escalation will be 8 percent.

Year	Maintenance budget estimate
1	25,000
2	150,000
3	60,000
4	70,000
5	300,000

13.17. The marketing department made the following estimates about four different product designs. Use benefit-cost analysis to determine which design to pursue.

Design	Unit manufacturing cost	Sales price	Est. annual sales
A	12.50	25.00	250,000
B	22.00	40.00	200,000
C	15.00	25.00	250,000
D	15.00	20.00	300,000

13.18. You buy 100 shares of stock in QBC Corp. at $40 per share. It is a good buy, for 4 years, 3 months later you sell these shares of stock for $114 7/8. What is the annual rate of return on this fortunate investment?

14

COST EVALUATION

14.1
INTRODUCTION

An engineering design is not complete until we have a good idea of the cost required to build the design or manufacture the product. Generally the lowest-cost design will be successful in a free marketplace. The fact that we have placed this chapter on cost evaluation toward the end of the text does not reflect the importance of the subject. For most products and designs, cost is next to performance in importance.

An understanding of the elements that make up cost is vital, because competition between companies and between nations is fiercer than ever before. The world is becoming a single gigantic marketplace in which newly developing countries with very low labor costs are acquiring technology and competing successfully with the well-established industrialized nations. To maintain markets requires a detailed knowledge of costs and an understanding of how new technology can impact to lower costs.

We have seen that the design process makes commitments which determine 70 to 80 percent of the cost of a product. It is in the conceptual and embodiment design stages that a majority of the costs are locked into the product. Thus, in this chapter emphasis is on how cost estimates can be made early in the design process.

Some of the uses to which cost estimates are put are the following:

1. To provide information to be used in establishing the selling price of a product or a quotation for a good or service.
2. To determine the most economical method, process, or material for manufacturing a product.
3. To be used as a basis for a cost-reduction program.
4. To determine standards of production performance that may be used to control costs.
5. To provide input concerning the profitability of a new product.

It can be appreciated that cost evaluation inevitably becomes a very detailed and "nitty-gritty" activity. This type of information rarely is published in the technical literature, partly because it does not make interesting reading but more important, because costs are highly proprietary information. Therefore, the emphasis in this chapter will be on the identification of the elements of costs and on some of the more generally accepted cost evaluation methods. It should be realized that cost estimation within a particular industrial or governmental organization will follow highly specialized and standardized procedures particular to the organization. However, the general concepts of cost evaluation described here will still be valid.

14.2
CATEGORIES OF COSTS

There are two broad categories of costs:

1. *Nonrecurring costs:* These are one-time costs, which we usually call capital costs. They are divided further into capital costs, which include depreciable facilities, such as plant building or manufacturing equipment and tools, and nondepreciated capital costs, such as land.
2. *Recurring costs:* These costs are direct functions of the manufacturing operation and occur over and over again. They usually are called operating costs or manufacturing costs.

Another classification division is into fixed and variable costs. *Fixed costs* are independent of the rate of production of goods; *variable costs* change with the production rate. A cost often is called a *direct cost* when it can be directly assigned to a particular cost center, product line, or part. *Indirect costs* cannot be directly assigned to a product but must be "spread" over an entire factory. The general categories of fixed and variable costs are given below.

14.2.1 Fixed Costs

1. Indirect plant cost
   (*a*) Investment costs
       Depreciation on capital investment
       Interest on capital investment and inventory
       Property taxes
       Insurance
   (*b*) Overhead costs (burden)
       Technical services (engineering)
       Product design and development
       Nontechnical services (office personnel, security, etc.)
       General supplies
       Rental of equipment

2. Management and administrative expenses
   (*a*) Share of corporate executive staff
   (*b*) Legal staff

(c) Share of corporate research and development staff

(d) Marketing staff

3. Selling expenses
   (a) Sales force
   (b) Delivery and warehouse costs
   (c) Technical service staff

14.2.2 Variable Costs

1. Materials

2. Direct labor (including fringe benefits)

3. Direct production supervision

4. Maintenance costs

5. Power and utilities

6. Quality-control staff

7. Royalty payments

8. Packaging and storage costs

9. Scrap losses and spoilage

Fixed costs such as marketing and sales costs, legal expense, security costs, financial expense, and administrative costs are often lumped into an overall category known as general and administrative expenses (G&A expenses). The above list of fixed and variable costs is meant to be illustrative of the chief categories of costs, but it is not exhaustive.[1]

The way the elements of cost build up to establish a selling price is shown in Fig. 14.1 The chief cost elements of direct material and direct labor determine the *prime cost*. To it must be added indirect manufacturing costs such as light, power, maintenance, supplies, and factory indirect labor. The manufacturing cost is made up of the factory cost plus general fixed expenses such as depreciation, engineering, taxes, office staff, and purchasing. The total cost is the manufacturing cost plus the sales expense. Finally, the selling price is established by adding a profit to the total cost.

Another important cost category is *working capital*, the funds that must be provided in addition to fixed capital and land investment to get a project started and provide for subsequent obligations as they come due. It consists of raw material on hand, semifinished product in the process of manufacture, finished product in inventory, accounts receivable, and cash needed for day-to-day operation. The working capital is tied up during the life of the plant, but it is considered to be fully recoverable at the end of the life of the project.

---

1. For an expanded list of fixed and variable costs see R. H. Perry and C. H. Chilton, "Chemical Engineers' Handbook," 5th ed., pp. 25-13 and 25-27, McGraw-Hill, New York, 1973.

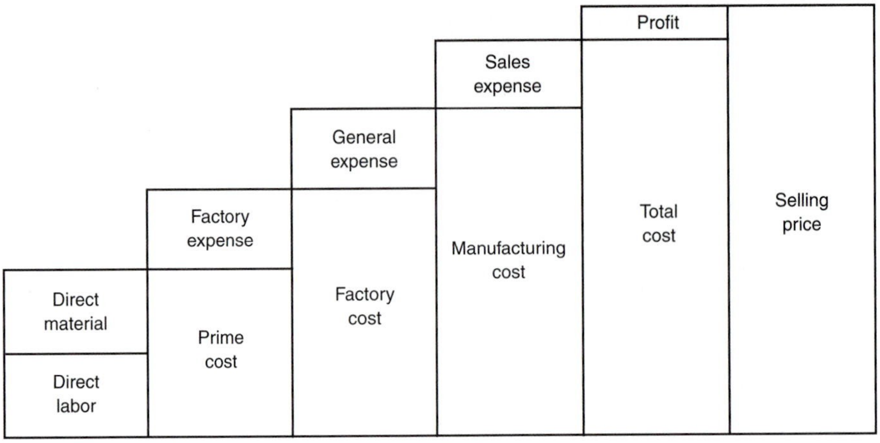

FIGURE 14.1
Elements of cost that establish the selling price.

A concept that provides a rough estimate of the investment cost for a new product is the turnover ratio:

$$\text{Turnover} = \frac{\text{annual sales}}{\text{total investment}} \qquad (14.1)$$

In the chemical industry the turnover ratio for many products is near 1.0; in the steel industry it is around 0.6. Suppose we wanted a quick estimate of the investment for a plant producing 20,000 tons/year of a chemical product that sells for 30¢/lb. Since the total annual sales are

$$0.30 \text{ \$/lb} \times 2000 \text{ lb/ton} \times 2 \times 10^4 \text{ ton/year} = 12 \times 10^6 \text{ \$/year}$$

the total plant investment is of the same magnitude for a turnover ratio of 1.0.

The fact that the variable costs depend on the rate or volume of production and fixed costs do not, leads to the idea of a break-even point (Fig. 14.2). The determination of the production lot size to exceed the break-even point and produce a profit is an important consideration. There are many things to be considered, but a common decision associated with economic lot size is how to allocate production among different machines, plants, or processes of various efficiencies or cost structure to make a product at minimum cost.

14.3
METHODS OF DEVELOPING COST ESTIMATES

The methods used to develop cost evaluations fall into three categories: methods engineering, costs by analogy, and parametric analysis of historical data.

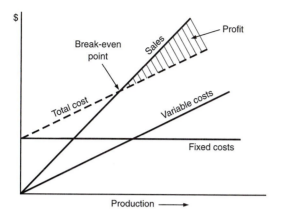

FIGURE 14.2
Break-even curve showing relation between fixed and variable costs and profit before taxes.

14.3.1 Methods Engineering (Industrial Engineering Approach)

In the methods engineering approach[1] the separate elements of work are identified in great detail and summed into the total cost per part. A typical but simplified example is the production of a simple fitting from a steel forging.

Cost per part

Operations	Material	Labor	Overhead	Total
Steel forging	$37.00			$37.00
Set up on milling machine		0.20	0.80	1.00
Mill edges		0.65	2.60	3.25
Set up on drill press		0.35	1.56	1.91
Drill 8 holes		0.90	4.05	4.95
Clean and paint		0.30	0.90	1.20
	$37.00	2.40	9.91	$49.31

Material cost, 75 percent; direct labor, 5 percent; overhead, 20 percent

It illustrates that overhead charges (fixed costs) often are many times more than the direct labor costs. The exact ratio will depend on numerous factors (see Sec. 14.10). Also, in this case material cost is high because the production sequence starts with a finished forging rather than a workpiece cut from bar stock. Using the methods engineering approach on complex systems requires a great deal of effort and computation. However, there are strong trends toward putting material and processing costs into a computer database and using the computer to search out the optimum processing sequence and calculate the costs.

---

1. B. W. Niebel and A. Freivalds, "Methods Standard and Work Design," WCB/McGraw-Hill, New York, 1999.

14.3.2 Analogy

In cost estimation by analogy, the future costs of a project or design are based on past costs of a similar project or design, with due allowance for cost escalation and complexity differences. The method therefore requires a database of experience or published cost data. As will be shown subsequently, this method of cost evaluation commonly is used for feasibility studies of chemical plants and process equipment. When cost evaluation by analogy is used, future costs must be based on the same state of the art. For example, it would be valid to use cost data on a 757 transport aircraft to estimate costs for a larger 777, but it would not be correct to use the same data to predict the cost of a supersonic transport because many discontinuities in technology are involved. One major difference is the change from predominantly aluminum to titanium airframe construction in the SST.

Another concern with determining cost by analogy is to be sure that costs are being evaluated on the same basis. Equipment costs often are quoted FOB (free on board) the manufacturer's plant location, so that delivery cost must be added to the cost estimate. Costs sometimes are given for the equipment not only delivered to the plant site but also installed in place, although it is more usual for costs to be given FOB some shipping point. For example, the purchase cost of a 10-hp electric motor in 1970 was about $1400, but the cost installed in a plant was about $5000.

14.3.3 Parametric Approach

In the parametric or statistical approach to cost estimation, techniques such as regression analysis are used to establish relations between system cost and initial parameters of the system: weight, speed, power, etc. This approach, involving as it does cost estimation at a high level of aggregation, is the direct opposite of the methods engineering approach. For example, the cost of developing a turbofan aircraft engine might be given by

$$C = 0.13937 x_1^{0.7435} x_2^{0.0775}$$

where C is in millions of dollars, x_1 is the maximum engine thrust, in pounds, and x_2 is the number of engines produced. Cost data expressed in this empirical form can be useful in trade-off studies.

14.3.4 Levels of Cost Evaluation

The American Association of Cost Engineers lists five types of cost evaluations for plant or large projects based on the level of detail and accuracy that is involved (see table on next page).

The order-of-magnitude estimate is not based on complete flow sheets or equipment lists. It is developed by analogy using scale-up ratios and cost escalation factors for known costs of similar designs. Depending on the size of the project, an order-of-magnitude estimate takes from a few hours to a few days to complete. The main purpose of such an estimate is to determine whether there is enough likelihood of profit to proceed with further engineering work.

Kind of estimate	Probable error, %
Order of magnitude (ratio)	±40
Study (scope)	±30
Preliminary (budget authorization possible)	±20
Definitive (complete data on project short of detailed drawings and specifications)	±10
Detailed (complete engineering drawings and specifications)	±5

The study estimate proceeds from incomplete but developing detailed engineering design information. The purpose of a study estimate is to provide management with a "ball park figure" so it can decide whether further engineering work is justified.

The preliminary estimate is prepared from well-defined engineering design information such as flow sheets and detailed equipment lists. The accuracy of the cost estimate (±20 percent) is precise enough to allow preliminary budget approval.

The definitive estimate works from complete design information so that the data are sufficiently established to provide numbers for project control.

The detailed estimate adds a complete set of detailed engineering drawings and project specifications to a level of detail suitable to going out for bids or submitting a purchase order.

A number of factors must be given special attention in construction cost estimates:

1. *Allowances for extras:* Every project will incur extra costs depending on its complexity, site location and conditions, and mistakes on and omissions from the construction drawings.
2. *Cost escalation:* In an inflationary economy, if the time for the construction project is long or if the project becomes delayed, the cost of labor and material may have increased considerably before completion. Escalation clauses in labor contracts and purchase orders must be carefully examined.
3. *Design modification:* A complex plant or process often requires changes in the design after start-up so the system can achieve the expected performance. Hopefully, with good engineering design these additions and/or revisions will be minor.
4. *Contingency:* A fund must be set aside to provide for unforeseen cost eventualities. The more detail and work that have gone into a cost estimate the less the contingency fund will be. However, it is not unusual for the contingency fund of a complex project to be between 10 and 20 percent of the total cost of the project.
5. *Expiration date:* Every cost estimate should show a date (3 to 6 months hence) after which the estimate is no longer valid.

14.4
COST INDEXES

Because the purchasing power of money decreases with time, *all published cost data are out of date.* To compensate for this shortcoming, cost indexes are used to convert past costs to present costs.

$$C_{t_2} = C_{t_1} \frac{I_{t_2}}{I_{t_1}} \tag{14.2}$$

In Table 14.1 are listed the most common cost indexes and the source of the data.

You should be aware of some of the pitfalls inherent in using cost indexes. First, you need to be sure that the index you plan to use pertains to the problem you must solve. For example, the cost components in the *Engineering News Record* index would not apply to estimating the cost of an ore beneficiation plant which is more like a chemical plant.[1] Of more basic concern is the fact that the cost indexes reflect the costs of past technology and design procedures. When new radical technology is introduced, it usually changes the cost trends very sharply. For example, through improvements in technology the cost of building a catalytic cracking unit for high-octane gasoline in 1969 was 40 percent less than the cost of a plant of similar capacity in 1946. This cost effect of improved technology, called a learning curve, will be discussed in Sec. 14.13. Finally, it should be noted that published cost indexes reflect national averages and may not accurately reflect local conditions. Information on labor rates and productivity in various sectors of the United States is available from the Department of Labor. Methods of converting cost indexes to another country are available.[2]

Major items of engineering equipment, such as steam turbines, supertankers, and nuclear reactors, require several years to build. With inflation ever-present, it is important to consider that in the cost estimate. A common way is to use a *cost escalation factor* to estimate the cost of equipment at the date it is ready to be shipped to the customer. A typical cost escalation formula is

TABLE 14.1
Engineering cost indexes

Type of cost	Source of cost index
General price indexes	
Consumer (CPI)	Bureau of Labor Statistics
Producer (wholesale)	U.S. Department of Labor
Construction cost indexes	
General construction	*Engineering News Record*
Chemical plants	*Chemical Engineering*
Petroleum refinery	*Oil and Gas Journal*
Industrial equipment	Marshall and Swift index
Plant maintenance	*Factory*

---

1. Details of the composition of the cost indexes are given in the following sources: Chemical Engineering Index: C. H. Chilton, *Chem. Eng.,* vol. 73, no. 9, p. 184, 1966; J. Matley, *Chem. Eng.,* vol. 89, no. 8, p. 153, 1982; Engineering News Record: *Eng. News Record,* 142, no. 11, p. 161, 1949; Marshall and Swift Index: R. W. Stevens, *Chem. Eng.,* p. 124, November 1947; Petroleum Index: W. L. Nelson, *Oil Gas J.,* vol. 65, p. 97, May 15, 1967.
2. A. V. Bridgwater, *Chem. Eng.,* pp. 119–121, Nov. 5, 1979.

$$E = C_0 \times F \times M \times \frac{(\text{BLS})_A}{(\text{BLS})_B} + C_0 \times F \times (1 - M) \times \frac{(\text{BLS})'_A}{(\text{BLS})'_B} \quad (14.3)$$

This formula is based on Bureau of Labor Statistics (BLS) cost indexes. Considered separately are the escalation of material costs through the BLS Producer Price Index $(\text{BLS})_A$ and of wages through the BLS Labor Index $(\text{BLS})'_A$. The use of this formula is shown by an example.[1]

> **EXAMPLE 14.1.** An order is placed for a major equipment item in June 1975 for delivery in December 1976. The quoted cost in June is $500,000. The manufacturer stipulates that 54 percent of the cost is subject to escalation over which he has no control; therefore, $F = 0.54$. These costs typically are purchased components, materials, etc. The materials-to-labor ratio is 70 to 30 percent; therefore, $M = 0.70$. Using the appropriate BLS indexes for material and labor costs:

$(\text{BLS})_A = 199$ at December 1976 $(\text{BLS})'_A = 5.62$ at December 1976

$(\text{BLS})_B = 188.6$ at June 1975 $(\text{BLS})'_B = 5.03$ at June 1975

$$E = 500{,}000(0.54)(0.7)\left(\frac{199}{188.6}\right) + 500{,}000(0.54)(0.30)\left(\frac{5.62}{5.03}\right) = 289{,}920$$

This is the escalated cost in June 1976 for the fraction of the original cost subject to escalation. The fraction not subject to escalation is

$$500{,}000(1 - 0.54) = 230{,}000$$

Therefore, the total cost of the equipment at the time of shipment will be $289,920 + 230,000 = $519,920.

14.5
COST-CAPACITY FACTORS

The cost of most capital equipment is not directly proportional to the size or capacity of the equipment. For example, doubling the horsepower of a motor increases the cost by only about one-half. The economy of scale is an important factor in engineering design. The cost-capacity relation usually is expressed by

$$C_1 = C_0\left(\frac{Q_1}{Q_0}\right)^x \quad (14.4)$$

where C_1 and C_0 are the capital costs associated with the capacity or size Q_1 and Q_0. The exponent x varies from about 0.4 to 0.8, and it is approximately 0.6 for many items of process equipment. For that reason, the relation in Eq. (14.4) often is referred to as the "six-tenths rule." Values of x for different types of equipment are given in Table 14.2.

---

1. A. Pikulik and H. E. Draz, *Chem. Eng.,* pp. 107–21, Oct. 10, 1977.

TABLE 14.2
Typical values of size exponent for equipment

Equipment	Size range	Capacity unit	Exponent x
Blower, single stage	1000–9000	ft^3/min	0.64
Centrifugal pumps, S/S	15–40	hp	0.78
Dust collector, cyclone	2–7000	ft^3/min	0.61
Heat exchanger, shell and tube, S/S	50–100	ft^2	0.51
Motor, 440-V, fan-cooled	1–20	hp	0.59
Pressure vessel, unfired carbon steel	6000–30,000	lb	0.68
Tank, horizontal, carbon-steel	7000–16,000	lb	0.67
Transformer, 3-phase	9–45	kW	0.47

Source: R. H. Perry and C. H. Chilton, "Chemical Engineers' Handbook," 5th ed., p. 25-18, McGraw-Hill, New York, 1973.

The cost-capacity relation also may be used to estimate the effect of plant size on capital cost. In addition, cost indexes and capacity may be combined as shown in the following example.

EXAMPLE 14.2. A 200-ton per day sulfuric acid plant cost $900,000 in 1966. How much would a 400 ton/day plant cost in 1976?

Solution. Using the CE plant cost index for the appropriate years and the six-tenth rule,

$$C_{1976} = \$900,000\left(\frac{400}{200}\right)^{0.6}\left(\frac{192}{107}\right) = \$2,450,000$$

14.6
ESTIMATING PLANT COST

Methods for early estimation of costs have been well developed in building construction[1] and in many areas of chemical plant design and construction.[2] In these areas new designs are mostly variants of existing designs so that historical costs can be used along with cost indexes and cost-capacity factors to estimate the costs of new designs.

Preliminary costing of a chemical plant design is done by making a list of all the needed pieces of equipment and then applying factors to estimate the overall plant cost.

$$C_p = fC_b \tag{14.5}$$

where C_p = installed plant cost
C_b = base cost

---

1. C. Hendrickson and T. Au, "Project Management for Construction," Prentice-Hall, Englewood Cliffs, NJ, 1989; A. Patrusco, "Construction Cost Engineering Handbook," Marcel Dekker, New York, 1988; R. E. Westney (ed.), "The Engineer's Cost Handbook," Marcel Dekker, New York, 1997.
2. M. S. Peters and K. D. Timmerhaus, "Plant Design and Economics for Engineers," 4th ed., McGraw-Hill, New York, 1992; W. D. Baasel, "Preliminary Chemical Engineering Plant Design," 2d ed., Van Nostrand Reinhold, New York, 1990.

$f =$ Lang factor, which depends on the nature of the plant
$f =$ 3.1 for a solids-processing plant
$f =$ 3.6 for a solid-fluid-processing plant
$f =$ 4.7 for a fluid-processing plant

Because it is easy to forget needed items of equipment at an early design stage some contingency allowance, generally between 10 and 50 percent, will be needed.

EXAMPLE 14.3 Estimate the cost of a coal crushing plant in 1995 whose equipment costs based on 1972 prices total $1.6M. The equipment costs are given FOB, so cost must be increased by the freight charge (4 percent) and the sales tax (5 percent).

$$\text{Equipment cost (1972)} = \$1,600,000 \, (1.04)(1.05) = \$1,747,000$$

A coal crushing plant is a solids-processing plant, so the Lang factor is 3.1. We will use a 15 percent contingency factor because the plant is rather simple in design.

$$\$1,747,000 \, (3.1)(1.15) = \$6,228,000 = \$6,200,000 \text{ (1972 costs)}$$

Using the Chemical Engineering Plant Index, the cost of the plant in 1987 would be

$$\$6.2M \times 323.8/137.2 = \$14,600,000$$

Assuming a 3.7 percent increase in prices each year since 1987, the plant cost in 1995 is

$$\$14,600,000 \, (1.037)^8 = 14,600,000 \, (1.3373) = \$19,500,00$$

The factor method has been extended far beyond Lang's simple Eq. (14.5) by including factors for individual types of equipment, different materials of construction, etc.[1]

In making cost estimates for processing and manufacturing plants, there are two different situations to consider. A *grass-roots plant* is a complete plant erected on a new (green field) site. The cost of the plant must include land acquisition, site preparation, and auxiliary facilities, such as rail line, loading dock, and warehouses, in addition to the cost of the processing equipment. A *battery limits plant* is an existing plant to which process equipment has been added. The battery limits circumscribe the new equipment to be added but do not include auxiliary facilities and site preparation, which already exist. When using published cost data, you need to know for which type of situation the costs have been determined.

14.7
DESIGN TO COST

Design to cost is the term that is becoming accepted for the use of costing methods early in the design process to predict the manufacturing costs so that, if they are too high, the design can be changed at an early stage. All too often, past practice has been to wait for a complete cost analysis in the detail design phase. If this proved to be excessive, then the only practical recourse was to try to wring the excess cost out of

---

1. W. D. Baasel, op. cit., pp. 274–316.

the manufacturing process or to substitute a less expensive material, usually at the expense of quality. A common way to make an early cost estimate is to depend on an experienced cost estimator who can compare the part to one made in the past, making due allowance for the differences.[1] But, as more attention has been focused on this important problem a variety of methods for early cost estimation have been developed.

14.7.1 Order of Magnitude Estimates

At the very early stage of product development where the market for a new product is being studied, comparison is usually made with similar products already on the market. This gives bounds on the expected selling price. Often the cost is estimated with a single factor. Weight is most commonly used. For example,[2] products can be divided roughly into three categories:

1. Large functional products—automobile, front-end loader, tractor
2. Mechanical/electrical—small appliances and electrical equipment
3. Precision products—cameras, electronic test equipment

Products in each category cost about the same on a weight basis, but the cost between categories increases by a factor of approximately 10. An automobile is about $5 per pound, a high-end blender is about $50 per pound, and an automatic focusing camera is about $500 per pound.

   A slightly more sophisticated method is to estimate cost on the basis of the percentage of the share of the total cost that is due to materials cost.[3] For example, about 70 percent of the cost of an automobile is material cost, about 50 percent for a diesel engine, about 25 percent for electrical instruments, and about 7 percent for china dinnerware.

> **EXAMPLE 14.4.** What is the total cost of a diesel engine that weighs 300 lb? The engine is made from ductile iron that costs $2/lb. The material cost share for the engine is 0.5.
>
> $$\text{Cost} = (300 \times \$2)/0.5 = \$1200$$

Another rule of thumb is the one-three-nine rule.[4] This states the relative proportions of material cost (1), manufacturing cost (3), and selling price (9). In this rule the material cost is inflated by 20 percent to allow for scrap and tooling costs.

> **EXAMPLE 14.5.** A 2-lb part is made from an aluminum alloy costing $1.50/lb. What is the estimated material cost, part cost, and selling price?

---

1. PC-based software for making cost estimates of machined parts is available from a number of vendors. Two examples are COSTIMATOR from Manufacturing Technologies, W. Springfield, MA, and KAPES, from PS Industry Solutions, Lakeland, FL. While primarily used for making detailed cost estimates when all part details are available, they also can be used early in design to determine relative costs of alternative designs. CostDesigner from AgilTech, Cincinnati, OH, is a feature-based cost estimator that estimates machining costs directly from a solid model.
2. R. C. Creese, M. Adithan, and B. S. Pabla, "Estimating and Costing for the Metal Manufacturing Industries," Marcel Dekker, New York, 1992, p. 101.
3. R. C. Creese et al., op. cit., pp. 102–105.
4. H. F. Rondeau, *Machine Design,* Aug. 21, 1975, pp. 50–53.

$$\text{Material cost} = 1.2 \times 1.50 \ \$/\text{lb} \times 2 \ \text{lb} = \$3.60$$
$$\text{Part cost} = 3 \times \text{material cost} = 3 \times \$3.60 = \$10.80$$
$$\text{Selling price} = 3 \times \text{part cost} = 3 \times \$10.80 = \$32.40 \ \text{or}$$
$$\text{Selling price} = 9 \times \text{material cost} = 9 \times \$3.60 = \$32.40$$

14.7.2 Costing in Conceptual Design

At the conceptual design stage where few details have been decided about the design, costing methods are required that allow for direct comparison between different types of designs that would perform the same functions. An accuracy of ± 20 percent is the goal. Relative costs are often used for comparing the costs of different design configurations, standard components, and materials. The base cost is usually the cost of the lowest cost or most commonly used item. An advantage of relative cost scales is that they change less with time than do absolute costs.

Parametric methods work well where designs tend to be variants of earlier designs. The costing information available at the conceptual design stage usually consists of historical cost for similar products. For example, cost equations for two-engine small airplanes have been developed,[1] and similar types of cost relationships exist for coal-fired power plants and many types of chemical plants. However, for mechanical products, where there is a wide diversity of products, few such relationships have been published. This information undoubtedly exists within most product manufacturing companies.

The methodology[2] for developing a parametric cost estimation equation is given in Fig. 14.3. This methodology was used to determine the cost of producing injection molded ABS plastic parts.

$$C_{0-10\,g} = \left[WC_m(1+s)\right] + (-0.00172W + 0.04217)$$
$$C_{10-60\,g} = \left[WC_m(1+s)\right] + (0.00041W + 0.05358)$$
$$C_{60-500\,g} = \left[WC_m(1+s)\right] + (0.0058W + 0.1214)$$

where W is the part weight (g), C_m is the cost of material (£/g), and s is the scrap percentage.

Note the highly empirical nature of the cost equations, and the fact that a different equation is used in different regimes of total part weight. Another example of the use of parametric relationships to calculate a part cost is given by Hoult and Meador.[3]

An intellectually more satisfying approach to determining costs early in design is *functional costing*.[4] The idea behind this approach is that once the functions to be performed have been determined the minimum cost of the design has been fixed. Since

1. J. Roskam, *J. Aircraft,* vol. 23, pp. 554–560, 1986.
2. A. R. Mileham, G. C. Currie, A. W. Miles, and D. T. Bradford, *Jnl. of Engr. Design,* vol. 4, no. 2, pp. 117–125, 1993.
3. D. P. Hoult and C. L. Meador, Manufacturing Cost Estimating, "ASM Handbook," vol. 20, pp. 716–722, ASM International, Materials Park, OH, 1997.
4. M. J. French, *Jnl. Engr. Design,* vol. 1, no. 1, pp. 47–53, 1990; M. J. French and M. B. Widden, "Design for Manufacturability 1993," DE, vol. 52, pp. 85–90, ASME, New York, 1993.

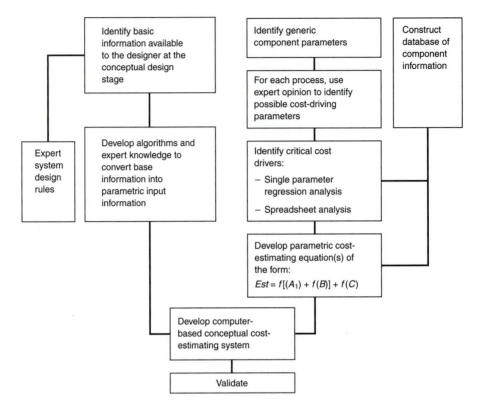

FIGURE 14.3
Procedure for developing a parametric cost-estimating system for the conceptual design stage. (*After A. R. Mileham et al., Jnl. Engr. Design, vol. 4, pp. 117–125, 1993.*)

it is in conceptual design that we identify the needed functions and work with alternative ways of achieving them, linking functions to cost gives us a direct way of designing to cost. A start has been made with standard components like bearings, electric motors, and linear actuators, where the technology is relatively mature and costs have become rather competitive.

14.7.3 Design Guidelines

The following general guidelines for decreasing product costs during conceptual design and embodiment design are given by Hundal,[1] as translated from Ehrlenspiel.[2]

- Ask for fewer demands; only the minimum accuracy, tolerances, and conformance to standards.
- Use concepts which lead to smaller sizes and lighter construction.

1. M. S. Hundal, "Design for Manufacturability 1993," DE, vol. 52, pp. 75–84, ASME, New York, 1993.
2. K. Ehrlenspiel, "Kostenguenstig Konstruieren," Springer-Verlag, Berlin, 1985.

- Use higher speeds for power transmission, thus reducing torque and therefore the amount of material required.
- Use robust physical effects, e.g., mechanical and hydrostatic energy.
- Use concepts with simple construction and fewer parts by the use of function integration.
- Use smaller size parts. Smaller size generally leads to lower costs.
- Use same or similar parts.
- Produce in large quantities, if possible.
- Reduce complexity; use fewer parts and production operations.
- Use safety devices like relief valves so the system does not have to be designed for high loads which occur only occasionally.
- Use higher-strength materials and/or surface treatments to reduce the size and manufacturing costs.
- Reduce scrap generation.
- Use fewer machining operations by using integral design.

14.8
MANUFACTURING COSTS

Manufacturing costs begin to be determined in embodiment design, as design details get firmed up. The methodology developed by Swift and coworkers at the University of Hull, and described in Sec. 9.6, is a good example. This is basically a parametric method that is presented graphically.

Detailed manufacturing costs require considerable specificity in dimensions, tolerances, materials, and process planning. This type of analysis cannot be done before detail design is complete and there is a *bill of materials*. Most commonly this is done in a process planning step that follows detail design. However, with the computerization of the costing process, it becomes easier to move this process to earlier in the sequence. One of the important decisions to be made in costing is the *make/buy decision*. Should the part be made in-house on company production equipment, or should it be subcontracted out to be made by a supplier company?

A detailed estimate of manufacturing cost follows the following steps.

1. Make a detailed analysis of the product and prepare a list of the components that must be made, with an accurate count of the number of parts required.
2. Draw up a manufacturing process plan for each component.
3. Determine the material costs for each component.
4. Determine the manufacturing time (cycle time) for each manufacturing operation listed in step 2.
5. Apply the labor and overhead (burden) rates to each operation. See Sec. 14.10.
6. The manufacturing cost is the sum of steps 3 and 5.

The manufacturing cost per unit C_u is given by

$$C_u = C_M + C_L + OH \qquad (14.6)$$

where C_M is the material cost per unit, C_L is the labor cost per unit, and OH is the overhead. The material cost includes the purchase cost of standard components, like bearings and gears, and the cost of raw material (bar stock, castings, etc.) from which the components are manufactured.

$$C_M = (\eta V_n k_v + OH_M) + (B + OH_b) \qquad (14.7)$$

where V_n is the net volume of the part, as opposed to the gross volume V_g and η is the allowance factor such that $V_g = \eta V_n$. k_v is the cost of the material per unit volume. OH_M is the material overhead to account for the procurement, inspection, storage, interest on this inventory, and material handling costs. B is the purchase cost of components and OH_b is the overhead on B. The labor costs C_L depend on the time t_u to carry out a specific production operation. This is made up of time actually spent on the process, an overhead allowance for time spent changing tools, lubricating, etc., and a nonproductive time allowance for time spent resting, waiting for parts, etc. These times are obtained from cost models of the process (see Sec. 14.14) or, more commonly, from *standard times* obtained from time-and-motion studies of actual processes. The labor cost k_L is the labor rate ($/h) multiplied by the time. The total labor cost is made up of several components.

$$C_L = C_{dl} + C_{su} + OH_L \qquad (14.8)$$

where $C_{dl} = k_L t_u$ is the direct labor cost, C_{su} is the cost of machine setup, and OH_L is the overhead on direct labor. The variable design costs which can be most readily influenced at the design stage are C_M and C_{dl}.

Sometimes costs are allocated by setting the hourly cost of operating a piece of production equipment. This is given by[1]

$$C_{MH} = \frac{C_D + C_I + C_S + C_E + C_R}{t} \qquad (14.9)$$

where C_{MH} is the hourly cost of using a machine, C_D is the yearly depreciation cost, C_I is the annual interest on the purchase of the machine, C_S is the annual cost of the space occupied by the machine, C_E is the annual cost of energy consumed by the machine, C_R is the annual cost for repair and maintenance of the machine, and t is the hours per year that the machine is utilized.

The cost equations given above are for rather disaggregated cost models. Sometimes it is useful to use a more aggregated cost model, such as the model for the total production costs of a product as given by Singh.[2] The total production cost is given by

$$P = M + L + OH \qquad (14.10)$$

where M is the total material cost, L the total labor cost, and OH the overhead. If the overhead is only applied to the labor cost,

1. K. Singh, "Mechanical Principles of Design," Nantel Publications, Melbourne, Australia, 1996.
2. K. Singh, op. cit., p. 354.

$$P = M + (1 + \bar{g})L \tag{14.11}$$

where $\bar{g}$ is the average overhead factor. If we further differentiate between the overhead involved in part production $\bar{g}_p$ and in assembly $\bar{g}_a$ the relationship for total production cost P becomes

$$P = M + (1 + \bar{g}_p)L_p + (1 + \bar{g}_a)L_a \tag{14.12}$$

14.8.1 Complexity Theory

Attempts at replacing the detailed process for the determination of manufacturing cost with a theoretical basis are built around the use of complexity theory, which in turn evolved from Shannon's information theory of communication.[1] The metric used to measure complexity of a product is the information content of the design I.

$$I = \sum_i \log_2 \left(\frac{d_i}{t_i} \right) \tag{14.13}$$

where d_i is the ith dimension of the design and t_i is the tolerance associated with that dimension. This is the same as Eq. (9.2). In Eq. (14.13) i is summed over all dimensions and tolerances in the part or the assembly. The idea of using this metric for expressing design complexity was first proposed by Wilson.[2] Limited experimentation has shown that I is directly proportional to the time to manufacture a part,[3] which is proportional to cost. While this area of costing is just developing, it has the prospect of tying cost estimates directly to the details of the geometry.

14.9
VALUE ANALYSIS IN COSTING

Value analysis[4] was introduced in Sec. 8.11 as an aid in selection of materials. It was shown there that value analysis relates design decisions to the basic functions that the design needs to perform. The value engineering approach is used most frequently in analyzing costs, especially in looking at how a product could be redesigned to reduce cost.

> **EXAMPLE 14.6.** Table 14.3 shows the cost structure for a centrifugal pump.[5] In this table the components of the pump have been classified into three categories, A, B, and C, according to their manufacturing costs. Components in class A comprise 82 percent of the total cost. These "vital few" need to be given the greatest thought and attention.

1. C. E. Shannon and W. Weaver, "Mathematical Theory of Communication," University of Illinois Press, 1949.
2. D. R. Wilson, "An Exploratory Study of Complexity in Axiomatic Design," Doctoral Thesis, MIT, 1980.
3. D. P. Hoult and L. Meador, op. cit.
4. T. C. Fowler, "Value Analysis in Design," Van Nostrand Reinhold, New York, 1990.
5. M. S. Hundal, "Systematic Mechanical Designing," ASME Press, New York, 1997, pp. 175, 193–196.

TABLE 14.3
Cost structure for a centrifugal pump

| Cost category | Part | Manufacturing cost | | Type of cost, % | | |
		$	%	Material	Production	Assembly
A	Housing	5500	45.0	65	25	10
A	Impeller	4500	36.8	55	35	10
B	Shaft	850	7.0	45	45	10
B	Bearings	600	4.9	Purchased	Purchased	Purchased
B	Seals	500	4.1	Purchased	Purchased	Purchased
B	Wear rings	180	1.5	35	45	20
C	Bolts	50	< 1	Purchased	Purchased	Purchased
C	Oiler	20	< 1	Purchased	Purchased	Purchased
C	Key	15	< 1	30	50	20
C	Gasket	10	< 1	Purchased	Purchased	Purchased

From M. S. Hundal, "Systematic Mechanical Design," ASME Press, New York, 1997. Used with permission.

We now focus attention on the functions provided by each component of the pump (Table 14.4).

This table of functions is added to the cost allocation table to create Table 14.5. Note that an estimate has been made of how much each component contributes to each function. For example, the shaft contributes 60 percent to transfer of energy (F2) and 40 percent to supporting the parts (F6). Multiplying the cost of each component by the fraction it serves to provide a given function gives the total cost for each function. For example, the function support parts (F6) is provided partly by the housing, shaft, and bearings.

$$\text{Cost of F6} = 0.5(5500) + 0.4(850) + 1.0(600) = \$3690$$

These calculations are summarized in Table 14.6. This table shows that the expensive functions of the pump are containing the liquid, converting the energy, and supporting the parts. Thus, we know where to focus attention in looking for creative solutions in future designs.

TABLE 14.4
Functions provided by each component of the centrifugal pump

Function	Description	Components
F1	Contain liquid	Housing, seals, gasket
F2	Transfer energy	Impeller, shaft, key
F3	Convert energy	Impeller
F4	Connect parts	Bolts, key
F5	Increase life	Wear rings, oiler
F6	Support parts	Housing, shaft, bearings

From M. S. Hundal, "Systematic Mechanical Design," ASME Press. Used with permission.

TABLE 14.5
Cost structure for centrifugal pump with function cost allocation

Cost class	Part	Manufacturing cost $	Manufacturing cost %	Type of cost, % Material	Type of cost, % Production	Type of cost, % Assembly	Function allocation, %			
A	Housing	5500	45.0	65	25	10	F1	50	F6	50
A	Impeller	4500	36.8	55	35	10	F2	30	F3	70
B	Shaft	850	7.0	45	45	10	F2	60	F6	40
B	Bearings	600	4.9	Purchased	Purchased	Purchased	F6	100		
B	Seals	500	4.1	Purchased	Purchased	Purchased	F1	100		
B	Wear rings	180	1.5	35	45	20	F5	100		
C	Bolts	50	< 1	Purchased	Purchased	Purchased	F4	100		
C	Oiler	20	< 1	Purchased	Purchased	Purchased	F5	100		
C	Key	15	< 1	30	50	20	F2	80	F4	20
C	Gasket	10	< 1	Purchased	Purchased	Purchased	F1	100		

From M. S. Hundal, "Systematic Mechanical Design," ASME Press, New York, 1997. Used with permission.

TABLE 14.6
Calculation of function costs for centrifugal pump

Function	Part	% of part cost for function	Part cost, $	Function cost of individual part, $	Total function cost $	Total function cost %
F1: Contain	Housing	50	5500	2750		
Liquid	Seals	100	500	500		
	Gasket	100	10	10	3260	26.7
F2: Transfer	Impeller	30	4500	1350		
Energy	Shaft	60	850	510		
	Key	80	15	12	1872	15.3
F3: Convert	Impeller	70	4500	3150	3150	25.8
Energy						
F4: Connect	Key	20	15	3		
Parts	Bolts	100	50	50	53	0.4
F5: Increase	Wear rings	100	180	180		
Life	Oiler	100	20	20	200	1.6
F6: Support	Housing	50	5500	2750		
Parts	Shaft	40	850	340		
	Bearings	100	600	600	3690	30.2

From M. S. Hundal, "Systematic Mechanical Design," ASME Press, New York. Used with permission.

14.10
OVERHEAD COSTS

Perhaps no aspect of cost evaluation creates more confusion and frustration in the young engineer than overhead cost. Many engineers consider overhead to be a tax on their creativity and efforts, rather than the necessary and legitimate cost it is.

Overhead can be computed in a variety of ways. Therefore, you should know something about how accountants assign overhead charges.

An overhead cost is any cost not specifically or directly associated with the production of identifiable goods or services. The two main categories of overhead costs are factory or plant overhead and corporate overhead. Factory overhead includes the costs of manufacturing that are not related to direct labor and material. Corporate overhead is based on the costs of running the company that are outside the manufacturing or production activities. Since many manufacturing companies operate more than one plant, it is important to be able to determine factory overhead for each plant and to lump the other overhead costs into corporate overhead. Typical cost contributions to corporate overhead are the salaries and fringe benefits of corporate executives, sales personnel, accounting and finance, legal staff, R & D, corporate engineering and design staff, and the operation of the corporate headquarters building.

EXAMPLE 14.7. A modest-sized corporation operates three plants with direct labor and factory overhead as follows:

Cost	Plant A	Plant B	Plant C	Total
Direct labor	$750,000	400,000	500,000	1,650,000
Factory overhead	900,000	600,000	850,000	2,350,000
Total	1,650,000	1,000,000	1,350,000	4,000,000

In addition, the cost of management, engineering, sales, accounting, etc., is $1,900,000. Find the corporate overhead rate based on direct labor.

Solution.

$$\text{Corporate overhead rate} = \frac{1,900,000}{4,000,000}(100) = 47\%$$

In the next example of overhead costs, we consider the use of factory overhead in determining the cost of performing a manufacturing operation.

EXAMPLE 14.8. A batch of 100 parts requires 0.75 h of direct labor each in the gear-cutting operation. If the cost of direct labor is $14 per h and the factory overhead is 160 percent, determine the total cost of processing a batch.

Solution. The direct labor cost is

$$(100 \text{ parts})(0.75 \text{ h/part})(\$14.00/\text{h}) = \$1050$$

The factory overhead charge is

$$1050\,(1.60) = \$1680$$

The cost of gear cutting for the batch of 100 parts is

$$1050 + 1680 = \$2730$$

The most common basis for allocating factory overhead is direct labor hours. The total overhead charges to a cost center are estimated. The estimated hours of direct labor are divided into the sum to give the overhead rate, in dollars per hour. As production proceeds, the direct labor hours actually used are totaled. The total direct labor hours (DLH) is multiplied by the overhead rate to establish the amount of overhead to be charged to the product.

EXAMPLE 14.9. It is estimated that the factory overhead costs for the powder metallurgy operation of a large manufacturing plant will be $180,000 over the next year. It is further estimated that 30,000 DLH will be used in production in that shop over the same time period. Therefore, the factory overhead for this cost center will be $180,000/30,000 = $6 per DLH. During the past month 900 units of PM gears were started and completed. They required 580 direct labor hours and resulted in the charges for direct labor and direct materials shown below. Find the amount of overhead to be charged to the product.

Solution.

	Total	Per unit
Direct labor	$7,400	$8.22
Direct materials	$9,800	$10.89
Overhead		
580 DLH × $6 per DLH	$3,480	$3.87
Total cost	$20,680	$22.98

We note if the work force were suddenly cut to produce only 15,000 DLH without a corresponding reduction in the overhead costs, the overhead rate would double.

The allocation of overhead on the basis of direct labor hours sometimes can cause confusion as to the real costs when process improvements result in an increase in manufacturing productivity. Consider the following example.

EXAMPLE 14.10. A change from a high-speed steel cutting tool to a new coated WC tool results in a halving of the time for a machining operation. The data for the old tool and the new coated carbide tool are shown below in columns 1 and 2, respectively. Because the cost of overhead is based on the DLH, the cost of overhead apparently is reduced along with the cost of direct labor. The apparent savings per piece is $72.00 − $36.00 = $36.00. However, a little reflection will show that the cost elements that make up the overhead (supervisor, toolroom, maintenance, inspection, etc.) will be constant per piece produced; and if the rate of production doubles, so should the overhead charge. The true costs are given in column 3. Thus, the actual savings per piece is $72.00 − $56.00 = $16.00.

	(1) Old tool	(2) New tool (apparent cost)	(3) New tool (true cost)
Machining time, DLH	4.0	2.0	2.0
Direct labor rate, per hour	$8.00	$8.00	$8.00
Direct labor cost	$32.00	$16.00	$16.00
Overhead rate, per hour	$10.00	$10.00	$20.00
Cost of overhead	$40.00	$20.00	$40.00
Cost of direct labor and overhead	$72.00	$36.00	$56.00

Operating budgets frequently are based on direct labor hours. Thus, in Example 14.10 a manufacturing engineer who reduced DLH could be penalized unless recognition was given to the increased maintenance costs, etc., from the true overhead costs of a more productive situation.

Direct labor hours may not be the best basis for overhead allocation in many manufacturing situations. Consider a plant whose major cost centers are a machine shop, a paint line, and an assembly department. We see that it is reasonable for each cost center to have a different overhead rate in units appropriate to the function that is performed.

Cost center	Est. factory overhead	Est. number of units	Overhead rate
Machine shop	$250,000	40,000 machine hours	$6.25 per machine hour
Paint line	80,000	15,000 gal of paint	$5.33 per gallon of paint
Assembly dept.	60,000	10,000 DLH	$6.00 per DLH

The above examples show that the allocation of overhead on the basis of DLH may not be the best way to do it. This is particularly true of automated production systems where overhead has become the dominant manufacturing cost. In such situations overhead rates are often between 500 and 800 percent of the direct labor cost. In the limit, the overhead rate for an unmanned manufacturing operation would be infinity. There is a trend to base the overhead allocation on the actual machine hours used in production.

There is a danger that improper overhead allocation can lead top management to make the wrong decisions.[1] A company sold one of its product lines because overhead allocations made it appear unprofitable. When the product line was sold profits decreased rather than increased. Moreover, since the overhead from the disposed-of product had to be reallocated to the other existing products several of them now appeared unprofitable.

14.11
ACTIVITY-BASED COSTING

In a traditional cost accounting system indirect costs are assigned to products using direct labor hours or some other unit-based measure to determine overhead cost. We have already seen (Example 14.10) where traditional cost accounting does not accurately represent cost when a large productivity gain has been made. Other types of distortion caused by the cost accounting system are concerned with timing; e.g., R&D costs of future products are charged to products currently being produced, and product mix, e.g., more complex products, will require support costs in greater proportion to their production volume. For these and other reasons a new way of assigning indirect costs called activity-based costing (ABC) has been developed.

---

1. R. Strauss, *Chem. Engr.,* Mar. 16, 1987, pp. 103–106.

Rather than assigning costs to an arbitrary reference like direct labor hours or machine hours, ABC assumes that products incur costs by the *activities* that are required for their design, manufacture, sale, delivery, and service. In turn, these activities create cost by consuming support services such as engineering design, production planning, machine setup, and product packing and shipping. To implement an ABC system you must identify the major activities undertaken by the support departments and select a *cost driver* for each. Typical cost drivers might be hours of engineering design, hours of testing, number of orders shipped, etc.

EXAMPLE 14.11. A company assembles electronic components for specialized test equipment. Two products A75 and B20 require 8 and 10.5 min, respectively, of direct labor, which costs $16 per hour. Product A75 consumes $35.24 of direct materials and product B20 consumes $51.20 of direct materials.

Using a traditional cost accounting system with all overhead costs allocated at the rate of $230 per DLH, the cost of the products would be:

Product A75:   Direct labor + direct material + overhead

$$\$16(8/60) + \$35.24 + \$230(8/60) = 2.13 + 35.24 + 30.59 = \$67.96$$

Product B20: $16(10.5/60) + $51.20 + $230(10.5/60)

$$= 2.80 + 51.20 + 40.25 = \$94.25$$

In an attempt to get a more accurate estimate of costs, the company turns to the ABC approach. Six cost drivers are identified for this manufacturing system.

Activity	Cost driver	Rate
Engineering	Hours of engineering services	$60.00 per h
Production setup	Number of setups	$100.00 per setup
Materials handling	Number of components	$0.15 per component
Automated assembly	Number of components	$0.50 per component
Inspection	Hours of testing	$40.00 per hour
Packing and shipping	Number of orders	$2.00 per order

The activity of the cost drivers must be obtained from cost records.

	Product A75	Product B20
Number of components	36	12
Hours of engineering services	0.10	0.05
Production batch size	50	200
Hours of testing	0.05	0.02
Units per order	2	2.5

In building the cost comparison between products we start with direct labor and direct material costs, as calculated above. Then we turn to ABC in allocating the overhead costs. We apply the activity level of the cost drivers to the cost rate of the driver. For example, for Product A75

Engineering services: 0.10 h/unit $\times$ \$60/h = \$6.00/unit

Production setups: $100 \dfrac{\$}{setup} \dfrac{1}{50} \dfrac{setup}{unit} = 2.00 \dfrac{\$}{unit}$

Materials handling: $36 \dfrac{components}{unit} \cdot 0.15 \dfrac{\$}{component} = 5.40 \dfrac{\$}{unit}$

Packing and shipping: $2.00 \dfrac{\$}{order} \dfrac{1}{2} \dfrac{order}{units} = 1.00 \dfrac{\$}{unit}$

Comparison of the two products on activity-based costing

	A75	B20
Direct labor	2.13	2.80
Direct materials	35.24	51.20
Engineering	6.00	3.00
Production setups	2.00	0.50
Materials handling	5.40	1.80
Assembly	18.00	6.00
Testing	2.00	0.80
Packing and shipping	1.00	0.80
	\$71.77	\$66.90

We see that by using ABC product B20 is the least costly to produce. This shift has come entirely from changing the allocation of overhead costs from DLH to cost drivers based on the main activities in producing the product. B20 incurs lower overhead charges chiefly because it is a less complex product using fewer components and requiring less support for engineering, materials handling, assembly, and testing.

Using ABC leads to improved decisions through more accurate cost data. This is especially important when manufacturing overhead accounts for a large fraction of manufacturing costs. By linking financial costs with activities, ABC provides cost information to complement nonfinancial indicators of performance like quality. By emphasizing the importance of cost drivers like number of components, it provides a link back to better control of costs in design. On the other hand, using only a single cost driver to represent an activity can be too simple. More complex factors can be constructed, but at a considerable cost in complexity of the ABC system.

ABC cost accounting is best used when there is diversity in the product mix of a company in terms of such factors as complexity, different maturity of products, production volume or batch sizes, and need for technical support. Computer-integrated manufacturing is a good example of a place where ABC can be applied to advantage because it has such high needs for technical support and such low direct labor costs.

14.12
PRODUCT PROFIT MODEL

The total cost to produce and market a product can be written

$$C_P = Q(C_M + C_L + OH) + T + S + C_D \qquad (14.14)$$

where Q is the total lifetime volume of production, and the bracketed term is Eq. (14.6). T is the one-time capital costs for equipment and tooling, and S represents corporate overhead costs. C_D is the cost for developing the product up until it enters production, and for providing product updates. The four key objectives associated with developing a new product are the cost of the product, the performance and quality of the product, the quickness with which the product is developed and brought to market, and the cost in developing the product.

A simple product profit model can be used in trade-off studies between these factors.[1]

$$\text{Net sales} = (\text{number of units sold}) \times (\text{sales price}) \qquad (14.15)$$

$$\text{Cost of product sold} = (\text{number of units sold}) \times (\text{unit cost}) \qquad (14.16)$$

$$\text{Gross margin} = (\text{net sales}) - (\text{cost of product sold}) \qquad (14.17)$$

$$\text{Total operating expenses} = \text{development cost} + \text{marketing cost}$$
$$+ \text{ other costs (G\&A)} \qquad (14.18)$$

$$\text{Pretax profit} = (\text{gross margin}) - (\text{total operating expenses}) \qquad (14.19)$$

$$\text{Percentage profit} = [(100 \times (\text{profit/net sales})] \qquad (14.20)$$

Unit cost will be arrived at from Eq. (14.14) and by the methods discussed in Sec. 14.8. The number of units sold will be estimated by the marketing staff. Other costs will be provided by cost accounting or historical records. There obviously will be uncertainties in some of these numbers, as well as delays in developing the product and overruns in budgets. This profit model lends itself to computation in a spreadsheet for trade-off studies, where sales, costs, and profits can be tracked over the expected years to develop the product and the expected life of the product.

> **EXAMPLE 14.12.** Magrab[2] conducted tradeoff studies for a product costing $2M to develop and achieving $81M in sales and $7.5M in cumulative profits over 7 years. The four scenarios studied were:
>
> A: a 50 percent increase in development cost
> B: a 5 percent increase in unit cost of the product
> C: a 10 percent decrease in product sales due to poor performance of the product
> D: a 3-month delay introducing the product to market

1. E. B. Magrab, "Integrated Product and Process Design and Development," pp. 55–63, CRC Press, Boca Raton, FL, 1997.
2. E. B. Magrab, op. cit.

Results of trade-off studies

Scenario	Decrease in cumulative profits	Trade-off rule
A	$1,150,000	$23,000 decrease per 1% increase in development costs
B	$2,682,000	$536,400 decrease per 1% increase in unit product cost
C	$3,169,760	$316,976 decrease per 1% decrease in unit sales
D	$903,050	$301,017 decrease per month delay in introducing the product

(These trade-off rules apply only to the specifics of the particular case.) The trade-off studies can be used in conjunction with benefit-cost analysis. For example: Should we eliminate a secondary balancing operation on the fan?

Benefit

$ saved by eliminating balancing
    operation = $0.50
Average unit cost = $34
Percent saving = 0.50/34 = 0.0147 = 1.47%
1.47% × $536,400= $788,500

Cost

Marketing estimates 5% loss of sales due
    to poor vibration performance

5% × $316,976 = $1,584,000

Benefit/cost = 0.5. Cost clearly outweighs benefit. Keep the secondary balancing operation.

14.12.1 Setting Sales Price

There are two aspects of setting a price for a product. The first is to be sure that all costs are included, not just the obvious costs of labor, materials, and overhead. This calls for understanding your company's cost structure and what the practice is for including costs in the various overhead factors. The second aspect concerns the business strategy of setting price based on the volume-price relation and the estimate of the market potential for the product.

The sales price is the cost of product sold + profit. The two ways of determining profit are return on sales [Eq. (14.20)] and return on investment (Sec. 13.6). Profit should be set to provide a good return to the owners of the company (shareholders) and provide for expansion or modernization. The level of profit will depend on the business strategy. New, innovative products will aim for a high initial profit, but you must be aware that high-profit products attract competition, unless patent protection keeps out competition. While some reduction in the cost of the product can be expected over time because of the learning curve (Sec. 14.13), this may not compensate for the reduction in selling price due to competition. Companies that are trying to enter the market in second or third place may purposely go with a lower profit in order to break into the market as the low-price provider and hope to make up the difference by aggressively building sales volume.

14.12.2 Profit Improvement

The four methods commonly used for profit improvement are: (1) increased prices, (2) increased sales, (3) improved product mix, and (4) reduced cost. Example 14.13 shows the impact of changes in these factors on the profit using the profit model described by Eqs. (14.15) to (14.20).

> **EXAMPLE 14.13.** Case A is the current distribution of cost elements for the product.
>
> Case B shows what would happen if price competition would allow a 5 percent increase in price without loss in units sold. The increased income goes right to the bottom line.
>
> Case C shows what would happen if sales were increased by 5 percent. There would be a 5 percent increase in the four cost elements, while unit cost remains the same. Costs and profits rise to the same degree and percentage profit remains the same.
>
> Case D shows what happens with a 5 percent productivity improvement (5 percent decrease in direct labor) brought about by a process-improvement program. The small increase in overhead results from the new equipment that was installed to increase productivity. Note that the profit per unit has increased by 10 percent.
>
> Case E shows what happens with a 5 percent decrease in the cost of materials or purchased components. About 65 percent of the cost content of this product is materials. This cost reduction could result from a design modification that allows the use of a less expensive material or eliminates a purchased component. In this case, barring a costly development program, all of the cost savings goes to the bottom line and results in a 55 percent increase in the unit profit.

	Case A	Case B	Case C	Case D	Case E
Sales price	$100	$105	$100	$100	$100
Units sold	100	100	105	100	100
Net sales	$10,000	$10,500	$10,500	$10,000	$10,000
Direct labor	$1,500	$1,500	$1,575	$1,425	$1,500
Materials	$5,500	$5,500	$5,775	$5,500	$5,225
Overhead	$1,500	$1,500	$1,575	$1,525	$1,500
Cost of product sold	$8,500	$8,500	$8,925	$8,450	$8,225
Gross margin	$1,500	$2,000	$1,575	$1,550	$1,775
Total operating expenses	$1,000	$1,000	$1,050	$1,000	$1,000
Pretax profit	$500	$1,000	$525	$550	$775
Percentage profit	5%	9.5%	5%	5.5%	7.75%

14.13
LEARNING CURVE

It is a common observation in a manufacturing situation that the workers, as they gain experience in their jobs, can produce more product in a given unit of time. That is due to an increase in the workers' level of skill, to improved production methods that evolve with time, and to better management practices involving scheduling and other aspects of production planning. The extent and rate of improvement also depend on such factors as the nature of the production process, the standardization of the product design, the length of the production run, and the degree of harmony in worker-management relationships. Usually, the processes that are more people-dominated show greater improvement than those that are dominated by machinery, as in a chemical process plant or a central station generating plant.

The improvement phenomenon usually is expressed by a learning curve also called a product improvement curve. Figure 14.4 shows the characteristic features of an 80 percent learning curve. Each time the cumulative production doubles ($x_1 = 1$, $x_2 = 2$, $x_3 = 4$, $x_4 = 8$, etc.) the production time (or production cost) is 80 percent of what it was before the doubling occurred. For a 60 percent learning curve the production time would be 60 percent of the time before the doubling. Thus, there is a constant percentage reduction for every doubled[1] production. Such an obviously exponential curve will become linear when plotted on loglog coordinates (Fig. 14.5).

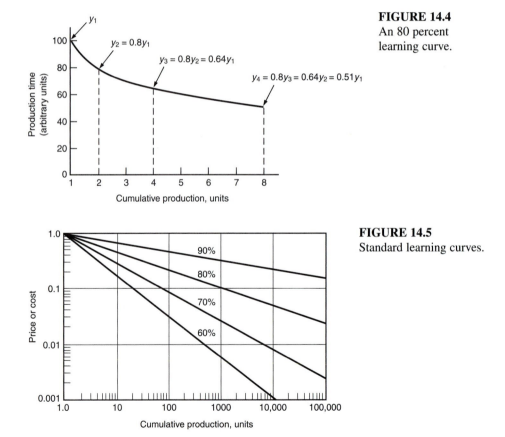

FIGURE 14.4
An 80 percent learning curve.

FIGURE 14.5
Standard learning curves.

---

1. The learning curve could be constructed for a tripling curve of production or any amount, but it is customary to base it on a doubling.

TABLE 14.7
Exponent values for typical learning curve percentages

Learning curve percentage, P	n
65	−0.624
70	−0.515
75	−0.415
80	−0.322
85	−0.234
90	−0.074

The learning curve is expressed by

$$y = kx^n \qquad (14.21)$$

where $y =$ production effort, h/unit or \$/unit
$\quad\quad\ k =$ effort required to manufacture the first unit of production
$\quad\quad\ x =$ cumulative total of units produced
$\quad\quad\ n =$ negative slope

The value of n can be found as follows: For an 80 percent learning curve $y_2 = 0.8y_1$, for $x_2 = 2x_1$; then,

$$\frac{y_2}{y_1} = \left(\frac{x_2}{x_1}\right)^n$$

$$\frac{0.8y_1}{y_1} = \left(\frac{2x_1}{x_1}\right)^n$$

$$n \log 2 = \log 0.8$$

$$n = \frac{-0.0969}{0.3010} = -0.322$$

Other values of n for different percentage learning curves are given in Table 14.7. Note that the learning curve percentage is $P = 2^n$.

EXAMPLE 14.14. The first of a group of 80 machines takes 150 h to build and assemble. If you expect a 75 percent learning curve, how much time would it take to complete the fortieth machine and the last machine?

Solution.

$$y = kx^n$$

For $P = 75\%$, $n = -0.145$, and $k = 150$

$$y = 150(x^{-0.415})$$

For $x = 40$

$$y_{40} = 150(40^{-0.415}) = 32.4 \text{ h}$$

For $x = 80$

$$y_{80} = 150(80^{-0.415}) = 24.3 \text{ h}$$

The learning curve can be expressed with the production time in hours per unit *or* as the cumulative average hours for x units. The distinction between these two ways of expressing the output is shown in the accompanying table. We note that, for a given number of units of output, the cumulative average is greater than the unit values. Note, however, that the learning improvement percentage (80 percent) that applies to the unit values does not apply to the cumulative values. Similarly, if the unit values are derived from cumulative values, the constant percentage does not apply. In constructing learning curves from historical data it is more likely to find records of cumulative total hours than the hours to build each unit.

The total hours required to manufacture a cumulative total of x_e units is given by

$$T = yx_e \tag{14.22}$$

Based on an 80 percent learning curve

x units	y, h/unit	Cumulative total hours	y, cumulative average h/unit
1	100.00	100.00	100.00
2	80.00	180.00	90.00
3	70.22	250.22	83.41
4	64.00	314.22	78.55
5	59.56	373.78	74.76
6	56.16	429.94	71.66
7	53.44	483.38	69.05
8	51.19	534.57	66.82

and since $y = kx^n$

$$T = kx_e^n x_e = k(x_e)^{1+n} \tag{14.23}$$

The total hours to manufacture a group of consecutively produced units when the first unit in the group is x_f and the x_e is given, is[1]

$$
\begin{aligned}
T_1 &= y_e x_e - y_f(x_f - 1) \\
&= kx_e^n x_e - kx_f^n(x_f - 1) \\
&= kx_e^{1+n} - kx_f^n(x_f - 1)
\end{aligned}
\tag{14.24}
$$

The unit time required to manufacture the xth unit is given by

$$U \approx (1 + n)kx^n \qquad \text{if } x > 10 \tag{14.25}$$

---

1. E. M. Malstrom and R. L. Shell, *Manuf. Eng.,* pp. 70–75, May 1979.

14.14
COST MODELS

The importance of modeling in the design process was illustrated in Chap. 7. Modeling can show which elements of a design contribute most to the cost; i.e., it can identify cost drivers. With a cost model it is possible to determine the conditions that minimize cost or maximize production (cost optimization).[1] Another important use for cost models is to aid in the decision process in R&D studies. If competing processes to manufacture a product are being pursued, then cost models may be able to determine which processes have the best chance of achieving the goal at minimum cost.

14.14.1 Machining Cost Model

Extensive work has been done[2] on cost models for metal-removal processes. Broken down into its simplest cost elements, a machining process can be described by Fig. 14.6. The time designated A is the machining plus work-handling costs per piece. If B is the tool cost, including the costs of tool changing and tool grinding, in dollars per tool, then

$$\text{Cost/piece} = \frac{nA + B}{n} = A + \frac{B}{n} \qquad (14.26)$$

where n is the number of pieces produced per tool.

We shall now consider a more detailed cost model for turning a bar on a lathe (Fig. 14.7). The machining time for one cut is

$$t_c = \frac{L}{V_{\text{feed}}} = \frac{L}{fN} = \frac{L}{f}\frac{D}{12v} \qquad (14.27)$$

where $V_{\text{feed}} =$ feed velocity, in/min
$f =$ feed rate, in/rev
$N =$ rotational velocity, rev/min
$D =$ work diameter, in
$v =$ cutting velocity, ft/min

Equation (14.26) holds in detail only for the process of turning a cylindrical bar. For other geometries or other processes such as milling or drilling, different expressions would be used for L or V_{feed}.

---

1. R. Bhatkal and J. Busch, *JOM,* April 1998, pp. 27–28.
2. E. J. A. Armarego and R. H. Brown, "The Machining of Metals," chap. 9, Prentice-Hall, Englewood Cliffs, NJ, 1969; G. Boothroyd, "Fundamentals of Metal Machining and Machine Tools," pp. 161–164, McGraw-Hill, New York, 1975; M. Field, N. Zlatin, R. Williams, and M. Kronenberg, *Trans. ASME,* Ser. B. *J. Eng. Ind.,* vol. 90, pp. 455–466, 585–590, 1968; S. M. Wu and S. S. Ermer, ibid., vol. 88, pp. 435–442, 1966; M. Y. Friedman and V. A. Tipnis, ibid., vol. 98, pp. 481–496, 1976; "Mathematical Modeling of Material Removal Processes," AFML-TR-154, September 1977.

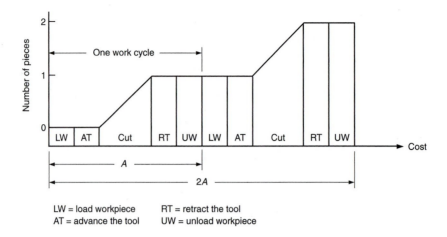

LW = load workpiece   RT = retract the tool
AT = advance the tool   UW = unload workpiece

FIGURE 14.6
Elements of the machining process.

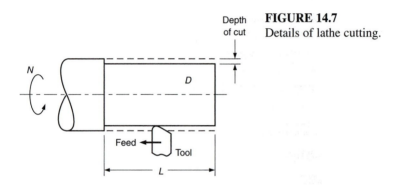

Depth of cut **FIGURE 14.7**
              Details of lathe cutting.

The total cost of a machined part is the sum of the machining cost C_{mc}, the cost of the tooling C_t, and the cost of the material C_m.

$$C_u = C_{mc} + C_t + C_m \qquad (14.28)$$

where C_u is the total unit (per piece) cost. The machining cost usually is the major part of the unit production cost. It depends on the machining time and the costs of the machine, labor, and overhead.

$$C_{mc} = C_1 t_{unit} \qquad (14.29)$$

$$C_1 = \frac{1}{60}\left[\frac{M(1 + OH_m)}{100} + \frac{W(1 + OH_{op})}{100}\right] \qquad (14.30)$$

where C_1 = cost rate, \$/min
        M = machine cost, \$/h
        OH_m = machine overhead rate, %
        W = labor rate for operator, \$/h
        OH_{op} = operator overhead rate, %

The machine cost includes the cost of interest, depreciation, and maintenance. It is found by the methods of Chap. 13 by determining these costs on an annual basis and converting them to per hour costs on the basis of the number of hours the machine is used in the year. The machine overhead cost includes the cost of power and other services and a proportional share of the building, taxes, insurance, etc.

The production time for a unit t_{unit}, is the sum of the machining time t_m and the nonproduction or idle time t_i.

$$t_{unit} = t_m + t_i \tag{14.31}$$

The machining time is the number of cuts times the time for a cut t_c. The idle time is

$$t_i = t_{set} + t_{change} + t_{hand} + t_{down} \tag{14.32}$$

where t_{set} = total time for job setup divided by number of parts in the batch
t_{change} = prorated time for changing the cutting tool

$$= \text{tool change time} \times \frac{t_m}{\text{tool life}}$$

t_{hand} = time the machine operator spends loading and unloading the work on the machine

t_{down} = downtime lost because of machine or tool failure, waiting for material or tools, or maintenance operations. Downtime is prorated per units of production.

The cost of the tooling is the cost of the cutting tools and the prorated cost of any special jigs and fixtures used to hold the workpiece. The cost of the cutting tool per unit depends on the cost of the tool and the life of the tool.

$$C_t = C_{tool} \frac{t_m}{T} \tag{14.33}$$

where C_{tool} = cost of a cutting tool, $
t_m = machining time, min
T = tool life, min

Tool life usually is expressed by the Taylor tool life equation, which relates surface velocity v and feed f to tool life:

$$vT^n f^m = K \tag{14.34}$$

For a cutting tool that is brazed to the toolholder, the tool cost is given by

$$C_{tool} = \frac{K_t + rK_s}{r + 1} \tag{14.35}$$

where K_t = cost of tool, $
r = number of resharpenings
K_s = cost of resharpening, $

For an insert (throwaway) tool

$$C_{tool} = \frac{K_i}{n_i} + \frac{K_h}{n_h} \tag{14.36}$$

where K_i = cost of tool insert, \$
      n_i = number of cutting edges
      K_h = cost of toolholder
      n_h = number of cutting edges in life of toolholder

Substituting for T in Eq. (14.33) results in

$$C_t = C_{tool} \frac{t_m v^{1/n} f^{m/n}}{K^{1/n}} \tag{14.37}$$

Substituting into the original equation [Eq. (14.28)] yields for the total unit cost

$$C_u = C_1 \left[t_m \left(1 + \frac{t_{tool}}{T} \right) + t_0 \right] + C_t \frac{t_m}{T} + C_m \tag{14.38}$$

where t_{tool} = tool change time
      t_0 = the time elements in Eq. (14.32) that are independent of tool life

In Eq. (14.38) both t_m and T depend on cutting velocity and feed, and the former is the more important variable. If we plot unit cost vs. cutting velocity, for a constant feed (Fig. 14.8), there will be an optimum cutting velocity to minimize cost. That is so because machining time decreases with increasing velocity; but as velocity increases, tool wear and tool costs increase also. Thus, there is an optimum cutting velocity. An alternative strategy would be to operate at the cutting speed that results in maximum production rate. Still another alternative is to operate at the speed that maximizes profit. The three criteria do not result in the same operating point.

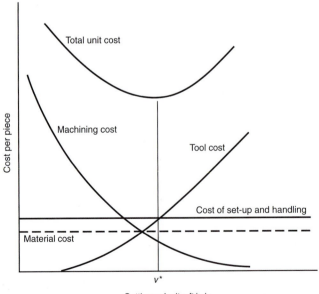

Cutting velocity, ft/min.

FIGURE 14.8
Variation of unit cost with cutting velocity.

14.14.2 Process Cost Model

The comparative cost of competitive processes can be determined directly from the process flowchart.[1] The steps for producing turbine blades from directionally solidified (DS) superalloys are shown in Fig. 14.9 In the directionally solidified structure the mold is slowly withdrawn from the furnace, so that the grains grow in a directional manner parallel to the length of the blade. This produces a structure with practically no grain boundaries perpendicular to the direction of bending stress, so that failure from stress rupture becomes less prevalent.

In this cost model the unit cost is expressed by a modification of Eq. (14.14) in which all overhead costs are lumped into a single factor LOH that is a composite factor of the hourly labor rate plus an overhead rate.

$$C_u = \frac{\text{LOH}}{nm} \Sigma t + C_m \qquad (14.39)$$

where C_u = total unit (per piece) cost
C_m = material cost
n = number of parts per batch

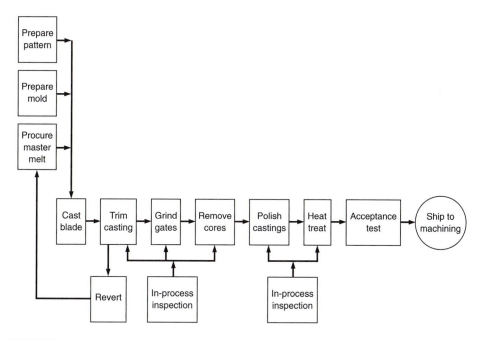

FIGURE 14.9
Process flow chart for directionally solidified turbine blades.

1. C. F. Barth, D. E. Blake, and T. S. Stelson, "Cost Analysis of Advanced Turbine Blade Manufacturing Processes," NASA CR135203, October 1977.

m = number of batch processes operated simultaneously by one operator

t = time required to complete process step for the batch

A detailed cost breakdown is prepared for each process step. The first step in the DS casting process is the preparation of a cluster of wax patterns of turbine blades that will be invested in the next step by coating them with a ceramic slurry. The steps involved in producing the cluster are (1) wax patterns containing internal cores are prepared by injection molding; (2) the parts of the pattern are assembled into a cluster of blades; (3) any defects are smoothed out (dressing the cluster); and (4) the finished cluster is inspected. The cost model for this step in the process sequence is

$$C_{pp} = \frac{1}{n_{cl}} \text{LOH}(t_{amp} + t_{dc} + t_{ci}) + C_p + C_{cl} \tag{14.40}$$

where C_{pp} = cost of pattern preparation

$\quad n_{cl}$ = number of blades per cluster

$\quad t_{amp}$ = time to assemble the patterns in a cluster

$\quad t_{dc}$ = time required to dress the cluster

$\quad t_{ci}$ = time required to inspect the cluster

$\quad C_p$ = cost of a wax pattern

$$C_p = \text{LOH}(t_{dic} + t_{ip} + t_{ifp}) + C_w W_{wp} + C_k \tag{14.41}$$

where t_{dic} = time to dress and inspect cores

$\quad t_{ip}$ = time required to inject the pattern

$\quad t_{ifp}$ = time required to inspect final pattern

$\quad C_w$ = cost of wax, $/lb

$\quad W_{wp}$ = wax weight in pattern

$\quad C_k$ = Cost of ceramic cores

$\quad C_{cl}$ = cost of components in cluster other than the pattern

$$C_{cl} = \text{LOH}\, t_{inp} + C_w W_{wmp} \tag{14.42}$$

where t_{inp} = injection time

$\quad W_{wmp}$ = weight of wax

It is obvious that the costs are established with a high level of detail. An equation similar to Eq. (14.40) is developed for each process step shown in Fig. 14.9. Usually it is just a summation of the time required to carry out an operation in the process step, but occasionally some aspect of the process itself enters into the model.

The process model for the casting (solidification) step is:

$$C_c = \frac{1}{n_c m_f} \text{LOH}\left(t_{ct} + \frac{L}{v}\right) \tag{14.43}$$

where C_c = unit cost for the casting step

$\quad n_c$ = number of blades per casting

$\quad m_f$ = number of furnaces per operator

t_{ct} = constant time needed to cast a batch regardless of the withdrawal rate
L = withdrawal length
v = withdrawal velocity

Thus, since the withdrawal rate of the casting may be very slow, Eq. (14.43) shows that this is a significant cost driver.

An important aspect in process costs, and especially when the technology is new, is the yield obtained in the process. The simple (uncoupled) yield factor is

$$Y_i = \frac{\text{no. of acceptable parts from } i\text{th step}}{\text{no. of parts that enter } i\text{th step}} \qquad (14.44)$$

where Y_i = yield factor for the ith process step. Thus, the cost of producing one acceptable part from the casting step is

$$(C_c)_{ap} = \frac{C_c}{Y_i} \qquad (14.45)$$

The cost of manufacturing DS blades is shown below in terms of four main cost centers. This is important because it shows which aspect of the manufacturing sequence contributes the most to the product cost. Note the importance of including product yield in the cost analysis. Without considering yield the coating of the blade would be the chief cost driver, but since that process has a very high yield the chief contributor to the cost is the casting step.

Percentage contribution of cost

Cost center	Without yield consideration	Considering process yield
Fabrication (casting)	30.2	42.3
Machining and finishing	16.6	24.4
Coating	47.4	25.3
Final acceptance testing	5.8	8.0
	100.0	100.0

14.15
LIFE CYCLE COSTING

Life cycle costing (LCC) is a methodology that attempts to capture all of the costs associated with a product throughout its life cycle.[1] The typical problem, which we have encountered in Chap. 13, is whether it is more economical to spend more money in the initial purchase to obtain a product with lower operating and maintenance costs,

---

1. R. J. Brown and R. R. Yanuck, "Introduction of Life Cycle Costing," Prentice-Hall, Englewood Cliffs, NJ, 1985; W. J. Fabrycky and B. S. Blanchard, "Life-Cycle Cost and Economic Analysis," Prentice-Hall, 1991.

or whether it is less costly to purchase a product with lower first costs but higher operating costs. However, life cycle costing goes into the analysis in much greater detail in an attempt to evaluate all relevant costs, both present and future.

Life cycle costing, also known as "whole life costing," first found strong advocates in the area of military procurement, where it is used for comparison of competing weapons systems.[1] For a typical piece of military hardware, with a service life of 20 years, the operation and maintenance costs can be 60 to 80 percent of the life cycle cost.[2] More recently, life cycle costing has been combined with life cycle assessment (see Sec. 6.8) to consider the costs of energy consumption and pollution during manufacture and service, and the costs of retiring the product when it reaches its useful life. Expansion of the cost models beyond the traditional bounds to include pollution and disposal is an active area of research that will place the design engineer in a better position to make critical trade-offs.

The elements in the life cycle of a product are shown in Fig. 14.10. This figure emphasizes the *overlooked impact on society costs* (OISC) that are rarely quantified and incorporated into a product life cycle analysis.[3] We start with design. The actual costs incurred here are a small part of the LCC but the costs committed here comprise about 75 percent of the avoidable costs within the life cycle of the product. Moreover,

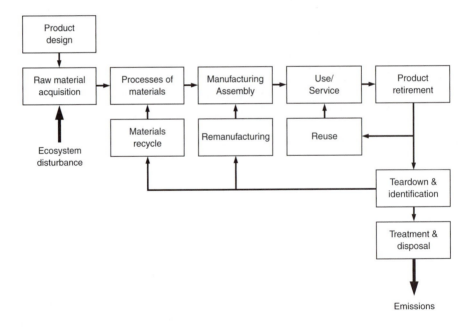

FIGURE 14.10
Total life cycle of a product.

1. MIL-HDBK 259, Life Cycle Costs in Navy Acquisitions.
2. J. J. Griffin, "Engineering Costs and Production Economics," vol. 14, 1988.
3. N. Nasr and E. A. Varel, Total Product Life-Cycle Analysis and Costing, *Proceedings of the 1997 Total Life Cycle Conference,* P-310, pp. 9–15, Society of Automotive Engineers, Warrendale, PA, 1997.

it is about 10 times less costly to make a change or correct an error in design than it is in manufacturing. The costs of acquiring the raw materials, usually by mining or oil extraction, and of processing the materials, can incur large environmental costs. These areas also often have considerable costs of carrying an inventory, and for transportation. We have concentrated in previous sections on the costs in manufacturing and assembly of products.

The cost of ownership of a product is the traditional aspect of LCC. Useful life is commonly measured by cycles of operation, length of operation, or shelf life. In design we attempt to extend life for use and service by using durable and reliable materials and components. Product obsolescence is dealt with through modular products.

Maintenance costs, especially maintenance labor costs, usually dominate other use costs. Most analyses divide maintenance costs into scheduled or preventive maintenance and unscheduled or corrective maintenance. The *mean time between failure* and *the mean time to repair* are important parameters from reliability theory (see Sec. 11.3) that impact LCC. Other costs that must be projected for the operations and support phase are: maintenance of support equipment; maintenance facility costs; pay and fringe benefits for support personnel; warranty costs and service contracts.

Once the product has reached its useful life it enters the retirement stage of the life cycle. We saw in Sec. 6.8 that other options than disposal should be considered. High-value-added products may be candidates for remanufacturing. By *value added* we mean the cost of materials, labor, energy, and manufacturing operations that have gone into creating the product. Products that lend themselves to recycling are those with an attractive *reclamation value,* which is determined by market forces and the ease with which different materials can be separated from the product. Reuse components are subsystems from a product that have not spent their useful life and can be reused in another product. Materials that cannot be reused, remanufactured, or recycled are discarded in an environmentally safe way. This may require labor and tooling for disassembly or treatment before disposal.

14.16
SUMMARY

Cost is a primary factor of design that no engineer can afford to ignore. It is important to understand the basics of cost evaluation so that you can produce high-functioning, low-cost designs. Cost buildup begins in conceptual design and continues through embodiment and detail design.

To be *cost literate* you need to understand the meaning of such terms as nonrecurring costs, recurring costs, fixed costs, variable costs, direct costs, indirect costs, overhead, and activity-based costing.

Cost estimates are developed by three general methods.

1. A detailed breakdown of all the steps required to manufacture a part, or renovate a room, with an association of the cost of each step for materials, labor, and overhead. This method is generally used in the final cost estimate in the detail design stage.
2. Cost estimation by analogy with previous products or projects. This method requires past experience or published cost data. Because this uses historical data,

the estimates must be corrected for price inflation using cost indexes, and for differences of scale using cost-capacity indexes. This method is often used in the embodiment stage of design.

3. The parametric approach uses regression analysis to correlate past costs with critical design parameters like weight, power, speed, etc. This method is useful in conceptual design.

Costs may sometimes be related to the functions performed by the design. This is a situation highly to be desired, because it allows optimization of the design concept with respect to cost.

Manufacturing costs generally decrease with time as more experience is gained in making a product. This is known as a learning curve.

Computer cost models are gaining in popularity as a way to pinpoint the steps in a manufacturing process where cost savings must be achieved. Simple spreadsheet models are useful for determining product profitability and making trade-offs between aspects of the business situation.

Life cycle costing attempts to capture all the costs associated with a product throughout its life cycle from design to retirement from service. Originally LCC focused on the costs incurred in using a product, such as maintenance and repair, but more and more LCC is attempting to capture the costs which impact on society from environmental issues and issues of energy use.

BIBLIOGRAPHY

Creese, R. C., M. Aditan, and B. S. Pabla: "Estimating and Costing for the Metals Manufacturing Industries," Marcel Dekker, New York, 1992.

Hundal, M. S.: "Systematic Mechanical Designing," ASME Press, New York, 1997.

Malstrom, E. M. (ed.): "Manufacturing Cost Engineering Handbook," Marcel Dekker, New York, 1984.

Michaels, J. V., and W. P. Wood: "Design to Cost," Wiley, New York, 1989.

Ostwald, P. F.: "Engineering Cost Estimating," 3d ed., Prentice-Hall, Englewood Cliffs, NJ, 1992.

Sims, E. R., Jr.: "Precision Manufacturing Costing," Marcel Dekker, New York, 1995.

Stewart, R. D.: "Cost Estimating," 2d ed., Wiley, New York, 1990.

Winchell, W. (ed.): "Realistic Cost Estimating for Manufacturing," 2d ed., Society of Manufacturing Engineers, Dearborn, MI, 1989.

PROBLEMS AND EXERCISES

14.1. In an environmental upgrade of a minimill making steel bar it is found that a purchase must be made for a large cyclone dust collector. It is the time of the year for capital budget submissions, so there is no time for quotations from suppliers. The last unit of that type was purchased in 1985. It had a 100 ft^3/min capacity. The new installation in 1998 will require 1000 ft^3/min capacity. The cost escalation for this kind of equipment has been about 5 percent per year. For budget purposes, estimate what it will cost to purchase the dust collector.

14.2. Many consumer items today are designed in the United States and manufactured overseas where labor costs are much lower. A middle range athletic shoe from a name brand manufacturer sells for $70. The shoe company buys the shoe from an off-shore supplier for $20 and sells it to the retailer for $36. The profit margin for each unit in the chain is: supplier—9 percent, shoe company—17 percent; retailer—13 percent. Estimate the major categories of cost breakdown for each unit in the chain. Do this as a team problem and compare the results for the entire class.

14.3. A manufacturer of small hydraulic turbines has the annual cost data given below. Calculate the manufacturing cost and the selling price for a turbine.

Raw material and components cost	$2,150,000
Direct labor	950,000
Direct expenses	60,000
Plant manager and staff	180,000
Utilities for plant	70,000
Taxes and insurance	50,000
Plant and equipment depreciation	120,000
Warehouse expenses	60,000
Office utilities	10,000
Engineering salaries (plant)	90,000
Engineering expenses (plant)	30,000
Administrative staff salaries	120,000
Sales staff, salaries and commissions	100,000

Total annual sales: 60 units

Profit margin: 15%   Plant overhead: 60%

14.4. A jewel case for a compact disk is made from polycarbonate ($2.20 per lb) by a thermoplastic molding process. Each CD case uses 20 grams of plastic. The parts will be made in a 10-cavity mold that makes 1400 parts per hour at an operating cost of $20 per hour. Manufacturing overhead is 40 percent. Since the parts are sold in large lots, the G&A expenses are a low 15 percent. Profit is 10 percent. What is the estimated selling price of each CD case?

14.5. Two competing processes for making high-quality vacuum melted steel are the vacuum arc refining process (VAR) and electroslag remelting (ESR). The estimated costs for operating each of the processes are given below.

Cost component	VAR	ESR
Direct labor, one melter and one helper	$89,000	$89,000
Manufacturing overhead, 140% direct labor	$124,600	$124,600
Melting power	0.3 kWh/lb 1000 lb/h 4.2¢/kWh	0.5 kWh/lb 1250 lb/h 4.2¢/kWh
Cooling water (annual charge)	$5,500	$6,800
Slag	—	$42,000

The capital cost of a VAR system is $1.3M and for an ESR system it is $0.9M. Each melting system has a 10-year useful life. Each uses 1000 ft^2 of factory space, which costs $40 per ft^2. Assume both furnaces operate for 15 8-h shifts per week for 50 weeks in the year. Estimate the cost of melting a pound of high-grade steel for each process.

14.6. The accounting department established the costs given below for producing two products, X and Z, over a given time period. (*a*) Describe the typical costs that would be put in each of the cost categories; (*b*) Determine the overhead and unit cost for each product in terms of direct labor cost; (*c*) Determine the overhead and unit cost for each product on the basis of direct labor hours (DLH); (*d*) Determine the overall overhead rate per DLH and use it to determine the unit cost of product X; (*e*) Determine the overhead and unit cost for each product on the basis of the proportion of direct material costs.

Item	Product X	Product Z
Quantity	3000	5000
Machine hours	70	90
Direct labor hours (DLH)	400	600
Factory floor space	150	50

	Labor rate $/h	Labor amount, h	Material cost $/unit	Material amount, units	Cost $
Product X					
Direct labor	18.00	400			7,200
Direct material			6.50	3000	19,500
Product Z					
Direct labor	14.00	600			8,400
Direct material			7.50	5000	37,500

Cost item	Product X	Product Z	Factory	Admin.	Sales	Total cost, $
1. Direct labor	7,200	8,400				15,600
2. Indirect labor			3,000			3,000
3. Direct material	19,500	37,500				57,000
4. Indirect material			7,000			7,000
5. Direct engineering	900	2,500				3,400
6. Indirect engineering			1,500			1,500
7. Direct expense	1,000	700				1,700
8. Other factory burden			5,500			5,500
9. Admin. expense				11,000		11,000
10. Sales and distribution						
Direct	900	1,100				2,000
Indirect					8,000	8,000
	29,500	50,200	17,000	11,000	8,000	115,700

14.7. Determine the unit cost for making products X and Z in Prob. 14.6 using activity-based costing. Use the cost drivers in Example 14.11, but omit automated assembly. These costs are in the direct labor costs for this example. The cost record, on a per batch basis, is given below.

	Product X	Product Z
Number of components	18	30
Hours of engineering services	15	42
Production batch size	300	500
Hours of testing	3.1	5.2
Units per order	100	200

14.8. Two alternative designs gave three estimates for variable and fixed costs shown below. (*a*) What is the break-even point at which design B becomes more economical than design A? (*b*) What is the break-even point for each design at which the sales begin to generate a profit?

Item	Design A	Design B
Material cost (variable), $/unit	3.50	3.50
Energy cost (variable), $/unit	2.50	1.10
Labor cost (variable), $/unit	4.10	3.85
Maintenance cost (fixed), $	300	800
Depreciation (fixed), $	400	700
Admin. overhead (fixed), $	800	800
Sales overhead (fixed), $	650	650
Selling price	12.00	12.00

14.9. A manufacturer of high-performance pumps has the cost and profit data given below. The company invests $1.2M in an aggressive 2-year design and development program to reduce manufacturing costs by 20 percent. When this is completed, what will be the impact on profit? What business aspects need to be considered that are not covered by this analysis? What questions does it leave unanswered?

	Existing design	Improved design
Sales price	$500	$500
Units sold	200,000	200,000
Revenues	$10M	$10M
Direct labor	1.5M	
Materials	5.0M	
Overhead	2.0M	
Cost of product sold	8.5M	
Gross margin	1.5M	
Total operating expenses	1.0M	
Pretax profit	0.5M	
% Profit	5%	

14.10. A company has received an order for four sophisticated space widgets. The buyer will take delivery of one unit at the end of the first year and one unit at the end of each of the succeeding 3 years. He will pay for a unit immediately upon receipt and not before. However, the manufacturer can make the units ahead of time and store them at no cost for further delivery.

The chief component of cost of the space widget is labor at $15 per h. All units made in the same year can take advantage of an 80 percent learning curve. The first unit requires 100,000 h of labor. Learning occurs only in one year and is not carried over from year to year. If money is worth 16 percent after a 52 percent tax rate, decide whether it would be more economical to build four units the first year and store them or build one unit in each of the 4 years.

14.11. Develop a cost model to compare the cost of drilling 1000 holes in steel plate with a standard high-speed steel drill and a TiN-coated H.S.S. drill. Each hole is 1 in deep. The drill feed is 0.010 per rev. Machining time costs $10 per minute, and the cost of changing a tool is $5.

	Price of a drill	Tool life (no. of holes)	
		500 rpm	900 rpm
Std. H.S.S. drill	$12.00	750	80
TiN-coated H.S.S.	$36.00	1700	750

(*a*) Compare the costs at fixed conditions of 500 rpm.
(*b*) Compare the costs at a constant tool life of 750 holes.

14.12. Determine which system is most economical on a life cycle costing basis.

	System A	System B
Initial cost	$300,000	$240,000
Installation	23,000	20,000
Useful life	12 years	12 years
Operators needed	1	2
Operating hours	2100	2100
Operating wage rate	$8 per h	$8 per h
Parts and supplies cost (% of initial cost)	1%	2%
Power	8 kW at 4¢/kWh	9 kW at 4¢/kWh
Escalation of operating costs	6%	6%
Mean time between failures	600 h	450 h
Mean time to repair	35 h	45 h
Maintenance wage rate	$10 per h	$10 per h
Maintenance escalation rate	6%	6%
Desired rate of return	10%	10%
Tax rate	45%	45%

14.13. Discuss the automobile safety standards and air pollution standards in terms of the concept of life cycle costs.

15

LEGAL AND ETHICAL ISSUES IN ENGINEERING DESIGN

15.1
INTRODUCTION

Engineering is not just applying scientific laws and principles to technical problems. It is basically concerned with improving the lot of society, and as such, it brings engineers into the mainstream of business and industry. Almost all entry-level engineers become involved, at least tangentially, with situations that call for some understanding of the law and situations that call for ethical judgments. Therefore, this chapter presents a brief overview of some legal and ethical issues in engineering. With topics as broad as law and ethics we can only scratch the surface, so we have chosen to focus on those issues that are most pertinent to engineering design.

The following are examples of where a design engineer might be concerned with legal and ethical issues:

- Preparing a contract to secure the services of a specialized software firm.
- Reviewing a contract to determine whether a contractor who built an automated production facility should be paid.
- Deciding whether it is legal and ethical to reverse engineer a product design.
- Managing a design project to avoid the possibility of a product liability suit.
- Protecting the intellectual property created as part of a new product development activity.
- Deciding whether to report a colleague who is taking kickbacks from a subcontractor.

The law is a formalized code of conduct describing what society feels is the proper way to behave. In other words, laws reflect what society values. As society evolves, its attitude toward behavior changes, and the laws change as well. Ethics is the study of human conduct that lays out the moral ground rules. It is intimately related to the values of society. Thus, laws and ethics, while distinct, are not independent. Ethical conduct is the behavior desired by society which is separate from the minimum standards of the law. For example, making a defective product despite taking all due care may subject you to product liability law, but it is not generally considered unethical.

This chapter comes with a warning label. *A little knowledge can be a dangerous thing.* Just as we would not expect a lawyer to practice engineering, so an engineer should not practice law. One objective of this chapter is to give you enough knowledge about the legal aspects of engineering practice to recognize when you need to take certain actions or to know when to seek legal counsel. A second objective is to help you realize that the engineering profession has expectations for your conduct and that a true professional recognizes responsibilities to the employer, the profession, and society, not just those to himself or herself.

15.2
THE ORIGIN OF LAWS

The code of American law has its origin in English *common law.* These laws derived from agreed-upon tradition and custom and were given the authority of law by decisions in the courts. Since each court decision is based on the study of previous court cases, it is often called *case law.*

Statutory law is law that is created by a legislative body, either state or federal. These laws, known as *statutes,* are often codified, as opposed to common law, and prescribe particular actions that apply to specific situations. Statutes may modify, reverse, or abolish common law doctrines. They are subject to both change and extinction by legislative action.

While the two basic types of law are common law and statutory law, many political entities contribute to the body of law. *Constitutional law,* which is based on the Constitution of the United States, defines governmental powers, especially with respect to the states, and secures the rights of the individual citizen. Municipal law is a subdivision of statutory law that is produced by towns and cities. These *municipal ordinances* deal with issues like traffic laws, zoning, and disturbance of the peace. A large body of *administrative law* has been created by rulings and regulations of federal and state agencies, other than the courts. Examples are OSHA and the EPA. Although cases involving administrative law are usually handled within the agency, disputes are finally resolved through the regular court system.

The purpose of the legal system is to protect and make life easier for each member of society. The rule of law is important not only in criminal matters, but it affects other aspects of society like politics, the economy, etc. The slow process of transformation of some former communist countries has demonstrated that an economy cannot grow and flourish without an accepted rule of law.

15.3
CONTRACTS

A contract is a promise by one person to another to do or not to do something. Only promises that the law will enforce are contracts. The three elements of a contract are:

$$Contract = offer + acceptance + consideration$$

An *offer* is an expression made by one person that leads another person to reasonably expect that the promisor wishes to create an agreement. The offer must be clear, definite, and specific, with no room for serious misunderstanding. An *acceptance* of the offer is necessary to make a contract legally binding. Both the offer and the acceptance must be voluntary acts. A contract cannot be forced on anyone. A contract is not enforceable by law unless it contains an agreement to exchange promises with value, the consideration. For example, if A and B enter into a contract in which A promises to pay B $1000 for modifying a CAD software package, both the money and the service are considerations.

15.3.1 Types of Contracts

Contracts can take many forms. They may be classified as express or implied, bilateral or unilateral. Also, a contract may be either written or verbal. Examples of contracts are purchase contracts, leases, a contract to perform a service, or an employment contract.

- An *express contract* is a contract in which all of the terms are agreed upon and expressed in words, either written or oral. An oral contract, once made, can be just as legal as a written contract, but it is much more difficult to prove and enforce. However, many states have statutes of frauds that require writing for certain contracts to be enforceable.
- An *implied contract* is a contract in which the agreement between parties is inferred by the legal system wholly or in part by their actions. For example, Jim goes to the local convenience store, where he has an account. He picks up a Sunday *New York Times* and holds it up so the clerk sees him take it and the clerk nods in return as he leaves the store with the paper. Jim has made an implied contract to pay $2 for the newspaper.
- A *bilateral contract* is a contract in which two parties have both made a promise to each other. A promise is made in return for a promise. Each party is both a promisor and a promisee.
- A *unilateral contract* is one in which the promisor does not receive a promise as consideration for her promise but instead agrees to pay if she receives an act or service. For example, Mrs. Jones says to Johnny Smith, "I promise to pay you $100 tomorrow if you will clean out my basement and garage today." Johnny immediately goes to work. This constitutes acceptance of the offer and creates a unilateral contract.

If more than one promisor or promisee is involved, the contract can take different forms (Table 15.1). The chief implication is with respect to the liabilities incurred by the different parties.

An engineer will have to deal with contracts in a number of different situations. Contracts for the purchase or sale of property are common. On taking a job you may be asked to sign a contract stating that all technical ideas that you develop belong to the company, even those conceived while not on the job. These contracts are often

TABLE 15.1
Types of contracts when there is more than one promisor or promisee

Type of contract	Number of parties	Liability
Joint	Two or more persons promise the same performance as a single party	All promisors are liable for the complete fulfillment of the contract
Several	Separate promises made by more than one promisor	Each promisor is liable for his or her individual promise
Joint and several	Two or more parties make a joint contract but also state that they are individually liable for completion of contract	All promisors face cumulative liability

negotiable at the time of employment and are something to consider when you are looking for employment. In technical dealings between companies, one of the parties may be asked to sign a *confidentiality agreement.* This is a contract in which one of the parties agrees to not disclose, make use of, or copy a design or product that the other party is about to disclose.

15.3.2 General Form of a Contract

In general, every business contract should contain the following information:

1. Introduction to the agreement. Include title and date.
2. Name and address of all parties. If one of the parties is a corporation, it should be so stated.
3. Complete details of the agreement. State all promises to be performed. Include such details as specifications, expected outcomes, etc. Give details on promises of payments, including amounts, timing of payments, interest, etc.
4. Include supporting documents such as technical information, drawings, specifications, and statements of any conditions on which the agreement depends.
5. Time and date of the start of the work and of the expected completion.
6. Terms of payment.
7. Damages to be assessed in case of nonperformance. Statement of how disputes are to be arbitrated.
8. Other general provisions of the agreement.
9. Final legal wording. Signatures of parties, witnesses, and notary public.

In addition, it is important to determine whether the contract contains an *integration clause* that establishes that it is an integrated contract and that other unwritten or oral terms are not implied.

15.3.3 Discharge and Breach of Contract

A contract is said to be *discharged* when the agreement has been performed to the satisfaction of both parties. The contracting parties can agree at any time that the con-

Gary Smith is district salesman for Zip-R Engineering Corp., manufacturers of automation equipment. He has submitted a proposal to ABC Mfg. Co. for 20 specialized robots, in response to their request for bids. The next day, Gary was talking with Joe Clark, purchasing agent for ABC. In the course of the phone conversation Joe told Gary, "Congratulations, you are a lucky guy." Gary took this as a signal that he had won the contract, and that a written agreement would be entered into later. Because it was close to the end of the quarter, and he needed this job to make his sales quota, Gary booked the job. Because Zip-R's backlog was low, they started work on the order immediately.

One week later, ABC's VP of Manufacturing decided to buy the robots from another company because of their reputation for requiring low maintenance. Can Zip-R recover damages for breach of contract?

The words "you are a lucky guy" spoken over the phone are too vague to constitute acceptance of an agreement. Without acceptance there is no legal contract. Gary had no justification for interpreting Joe's vague statement as an acceptance.

tract has been discharged. It can be discharged if it becomes impossible to perform due to circumstances outside the control of the contracting parties, e.g., *force majeure.* However, extreme difficulty in executing the contract does not discharge it even if it becomes more costly to carry out than originally anticipated.

A *breach of contract* occurs when one party fails to perform his or her part of the contract. A legal injury is said to have occurred, and the injured party can sue in court for damages.[1] General or compensatory damages are awarded to make up for the damage that occurred. Special damages are awarded for the direct financial loss due to the breach.

Often the terms of the contract contain *liquidated damages* that are agreed to beforehand in case the contract comes into dispute. If during trial these are found to be unreasonable, they will not come into force. The damages awarded at the trial will prevail. Note that the plaintiff has the obligation to act in such a way as to minimize the damages. If the court determines the damages could have been lessened through prudent action, this will be taken into consideration in setting the compensatory damages.

15.4
LIABILITY

Liability means being bound or obligated to pay damages or restitution. Two ways to incur liability are (1) breaking a contract or (2) committing a tort, e.g., fraud or negligence.

---

1. Another way to settle legal disputes is through arbitration. The United States has become a highly litigious society. In 1995 Americans filed more than 14.8M civil lawsuits and paid $121.7B in legal fees (National Center for State Courts, Research Department).

A *breach* of contract refers to violating a contract's promise. Failure to deliver detail drawings of a new machine by the date specified in the contract is a breach of contract. It makes no difference whether this was done intentionally or not.

Fraud is intentional deceitful action aimed at depriving another party of his or her rights or causing injury in some respect. Examples would be double billing or falsely certifying that a component had passed the ASME pressure vessel code.

Negligence is failing to exercise proper care and provide expertise in accordance with the standards of the profession that results in damage to property or injury to persons. This is the most common way for an engineer to incur liability to the public. For example, an engineer fails to include a major source of loading in design calculations of a public project so that the design fails. Note that being honest and well-intentioned does not absolve the engineer from a legal charge of negligence.

To be liable for negligence it must be proved that the defendant did not take reasonable and prudent action. This is determined by a jury. One way to show reasonable care is to show that you acted at the current level of technological development, i.e., the state of the art. A defense allowed in some states is to prove *contributory negligence,* that the plaintiff was negligent or could have prevented the accident had due care been taken. Sometimes it can be shown that the plaintiff willingly took an unnecessary risk that he or she was aware of, as when a person dives into a pool where the depth is clearly marked at 3 ft.

One way to limit business liability is by creating an appropriate business organization. A *corporation* is a legal entity that possesses many of the legal powers of individuals but one that exists independently of the people who own and manage it. A corporation can buy and sell property, enter into agreements, and sue and be sued. The corporation is in marked distinction from the *sole proprietorship,* in which the owner and the business are one and the same. In this case, a distinction is not made between the property of the business and the owner. The same holds for the finances of the business and the owner. A *partnership* is closer to a proprietorship but with many co-owners. All general partners are responsible for the acts and financial dealings of each other.

The corporation will be held liable for the acts of an employee who commits a civil wrong while engaged in corporate business. For example, the corporation can be fined by the EPA for the act of an employee who discharges liquid waste into a stream. Generally speaking, the corporation will incur the penalties of its employees, and the employee, in turn, may face the wrath of the corporation. Thus, working in a corporate structure provides some degree of protection from liability, but it is not absolute protection. Employees of corporations have been sued in the courts for negligence. Moreover, the trend in the courts is toward greater accountability of corporate employees. A corporate form cannot protect a professional from professional negligence.

15.5
TORT LAW

A *tort* is a *civil wrong* that involves damage committed against a person or his or her property, business, or reputation. It is a breach of the rights of an individual to be

Bill Garrison was hired by ABC Mfg. Co. as a consultant in plastic processing, with a chemical engineering degree and 10 years of experience. He was asked to recommend the equipment needed to convert a certain product line from metal to plastic parts. In particular, it was required that the production rate be at least equal to that when made with metal parts. When over $10M of new equipment was installed it was found that because of longer cycle time due to curing the plastic, the plastic line produced only 70 percent of the number of parts as made by the metal line in a given time. Can ABC hold Garrison personally responsible for this development?

Garrison can be held personally responsible to ABC for damages. As a consultant he acted as ABC's agent in designing the production line. By failing to take proper account of the plastic curing time he showed that he had not acted with due care and skill. He could be liable to ABC for negligence in tort, or for breach of contract. In addition, it was found that his 10 years of experience was in the area of polymer formulation, not plastics molding and manufacturing. If it could be shown that he misrepresented his background in order to secure the consulting contract, he could be liable for fraud.

secure in his or her person and property and be free from undue harassment.[1] Tort law is chiefly case law of the state courts, rather than statutory law. A decision in a case based on tort law hinges on three questions:

- Has a person's rights been infringed upon?
- Did the act occur as a result of negligence or actual intent on the part of the defendant?
- Did the plaintiff suffer damages as a result of the act?

Tort law deals with civil cases for which the penalty usually is monetary compensation rather than confinement. The difference between a tort and a *crime* is that a tort is a civil wrong while a crime is a wrong against society that threatens the peace and safety of the community. The victim of a crime may also bring a tort suit against the defendant to recover damages.

Tort suits involving engineers usually are concerned with one of four types of actions: (1) misrepresentation, (2) nuisance, (3) negligence, and (4) product liability. *Misrepresentation* is a false statement by a person of a fact that is known to be false, with the intent to deceive another person. When done under oath, it is called perjury. Misrepresentation is often claimed in a breach of contract suit. *Nuisance* concerns the annoyance or disturbance of a person such that the use of property becomes physically uncomfortable. Nuisances that affect the community, such as a blaring boom box at an open window, become a public nuisance. *Negligence* was defined in Sec. 15.4. *Product liability* is the action whereby an injured party seeks to recover damages for injury to person or property from a manufacturer or seller when the plaintiff alleges that a defective product or design caused the injury. This rapidly growing type of tort suit is discussed in Sec. 15.6.

---

1. "Engineering Law, Design Liability, and Professional Ethics," Professional Publications, Belmont, CA, 1983.

15.6
PRODUCT LIABILITY

Product liability refers to the legal action by which an injured party seeks to recover damages for personal injury or property loss from the producer or seller of a product. Product liability suits are pursued under the laws of tort. In no area of U.S. law has activity increased as dramatically as in personal injury product-liability civil lawsuits,[1] where suits in federal court increased 116 percent from 1995 to 1996, accounting for 14 percent of all federal civil suits filed. These span the gamut from individual suits by a single plaintiff against a single company to industrywide class action suits with thousands of plaintiffs filed against all asbestos manufacturers. Clearly the cost of preventing and defending against product liability has become a major concern for business and industry. An example of the extreme impact of product liability laws is on the manufacture of general aviation aircraft in the United States. In the 1970s the annual production of piston-engine powered light planes was from 10,000 to 15,000 per year, and accounted for more than 100,000 jobs. In the early 1990s production was barely 500 planes per year. This decrease has been blamed by many on the high costs of dealing with product liability litigation in the industry.[2]

15.6.1 Evolution of Product Liability Law

Before the industrial revolution, product liability laws did not exist. The purchaser had the responsibility to buy carefully and to use the product prudently. If the product broke or caused damage, the manufacturer was not required by law to stand behind it, although the better manufacturers gave warranties with their products. Around the mid-1800s the concept of *privity* came into use. Privity means that liability could occur only between those who entered into a contract or a direct transaction. The courts held that the injured party could sue only the party in privity. Thus, if a consumer was blinded by a broken hammer, he or she could sue only the retailer who sold him the tool; the retailer, in turn, could sue only the wholesaler, who in turn could sue the manufacturer.

A significant change occurred in 1916, when a court allowed an automobile owner to sue the manufacturer for negligence. This established the concept that manufacturers are directly liable to consumers. Clearly, from the viewpoint of recovering monetary damages it is an advantage to be able to directly sue the manufacturer, whose resources are likely to be much greater than those of a local retailer. When the Uniform Commercial Code was made law in the 1960s, it stated that there is an *implied warranty* of the fitness of products for their purposes and intended uses.

Also in the early 1960s the case law evolved to what is now called *strict product liability*. Previously manufacturers or sellers were liable only when they could be proved negligent or unreasonably careless in what they made or how they made it. It

---

1. A. S. Weintein, *Machine Design,* May 8, 1997, pp. 95–98.
2. B. E. Peterman, "Product Liability and Innovation," National Academy Press, Washington, DC, 1994, pp. 62–67.

had to be proved that a reasonable manufacturer using prudence would have exercised a higher standard of care. However, today in most states a standard of strict liability is applied. Under this theory of law the plaintiff must prove that: (1) the product was defective and unreasonably dangerous, (2) the defect existed at the time the product left the defendant's control, (3) the defect caused the harm, and (4) the harm is appropriately assignable to the identified defect. Thus, the emphasis on responsibility for product safety has shifted from the consumer to the manufacturer of products.

A related issue is the use for which the product is intended. A product intended to be used by children will be held to a stricter standard than one intended to be operated by a trained professional. Under strict liability a manufacturer may be held liable even if a well-designed and well-manufactured product injured a consumer who misused or outright abused it.

15.6.2 Goals of Product Liability Law

Only 100 years ago it was the practice in American and British law to not respond to accidental losses. It was generally held that the accident victim, not the manufacturer, should bear the economic burdens of injury. Starting in the mid-twentieth century, the law began to assume a more active role. Product liability law evolved to serve four basic societal goals: loss spreading, punishment, deterrence, and symbolic affirmation of social values.[1] Loss spreading seeks to shift the accidental loss from the victim to other parties better able to absorb or distribute it. In a product liability suit the loss is typically shifted to the manufacturer, who theoretically passes this cost on to the consumer in the form of higher prices. Often the manufacturer has liability insurance, so the cost is spread further, but at the price of greatly increased insurance rates.

Another goal of product liability law is to punish persons or organizations responsible for causing needless loss. It is important to recognize that under liability law the designer, not just the company, may be held responsible for a design defect. In extreme cases, the punishment may take the form of criminal penalties, although this is rare. More common is the assessment of punitive damages for malicious or willful acts. A third function is to prevent similar accidents from happening in the future, i.e., deterrence. Substantial damage awards against manufacturers constitute strong incentives to produce safer products. Finally, product liability laws act as a kind of symbolic reaffirmation that society values human safety and quality in products.

15.6.3 Negligence

A high percentage of product litigation alleges engineering negligence. Negligence is the failure to do something that a reasonable person, guided by the considerations that ordinarily regulate human affairs, would do. In product liability law, the seller is liable for negligence in the manufacture or sale of any product that may *reasonably be*

---

1. D. G. Owen, The Bridge, Summer, 1987, pp. 8–12.

expected to be capable of inflicting substantial harm if it is defective. Negligence in design is usually based on one of three factors:

1. That the manufacturer's design has created a concealed danger.
2. That the manufacturer has failed to provide needed safety devices as part of the design of the product.
3. That the design called for materials of inadequate strength or failed to comply with accepted standards.

Another common area of negligence is failure to warn the user of the product concerning possible dangers involved in the product use. This should take the form of warning labels firmly affixed to the product and more detailed warnings of restrictions of use and maintenance procedures in the brochure that comes with the product.

15.6.4 Strict Liability

Under the theory of strict liability, it is not necessary to prove negligence on the part of the manufacturer of the product nor is it necessary to prove breach of warranty or privity of contract. The defect itself regardless of how it got there, is sufficient to create liability under the tort laws. The fact that the injured party acted carelessly or in bad faith is not a defense under strict liability standards. The courts have acted so as to require the manufacturer to design its products in a way as to anticipate foreseeable use and abuse by the user.

Under most court decisions, defects divide into manufacturing defects and design defects.[1] Failure to conform with stated specifications is an obvious manufacturing defect. A manufacturing defect also exists when the product does not satisfy user requirements. Finally, a manufacturing defect exists when a product leaves the assembly line in a substandard condition, differs from the manufacturers intended result, or differs from other, ostensibly identical units of the same product line.

A design defect exists if the product fails to perform as safely as an ordinary consumer would expect. The criteria by which a defective and unreasonably dangerous nature of any product[2] may be tested in litigation are:

1. The usefulness and desirability of the product
2. The availability of other and safer products to meet the same need
3. The likelihood of injury and its probable seriousness
4. The obviousness of the danger
5. Common knowledge and normal public expectation of the danger
6. The avoidability of injury by care in use of the warnings
7. The ability to eliminate the danger without seriously impairing the usefulness of the product or making the product unduly expensive

---

1. C. O. Smith, Product Liability and Design, "ASM Handbook," vol. 20, pp. 146–151.
2. H. R. Piehler, A. D. Twerski, A. S. Weinstein, and W. A. Donaher, *Science,* vol. 186, p. 1093, 1974.

15.6.5 Design Aspect of Product Liability

Court decisions on product liability coupled with consumer safety legislation have placed greater responsibility on the designer for product safety. The following aspects of the design process should be emphasized to minimize potential problems from product liability.

1. Take every precaution that there is strict adherence to industry and government standards. Conformance to standards does not relieve or protect the manufacturer from liability, but it certainly lessens the possibility of product defects.
2. All products should be thoroughly tested before being released for sale. An attempt should be made to identify the possible ways a product can become unsafe (see Sec. 11.5), and tests should be devised to evaluate those aspects of the design. When failure modes are discovered, the design should be modified to remove the potential cause of failure.
3. The finest quality-control techniques available will not absolve the manufacturer of a product liability if, in fact, the product being marketed is defective. However, the strong emphasis on product liability has placed renewed emphasis on quality engineering as a way to limit the incidence of product liability.
4. Make a careful study of the system relations between your product and upstream and downstream components. You are required to know how malfunctions upstream and downstream of your product may cause failure to your product. You should warn users of any hazards of foreseeable misuses based on these system relationships.
5. Documentation of the design, testing, and quality activities can be very important. If there is a product recall, it is necessary to be able to pinpoint products by serial or lot number. If there is a product liability suit, the existence of good, complete records will help establish an atmosphere of competent behavior. Documentation is the single most important factor in winning or losing a product liability lawsuit.
6. The design of warning labels and user instruction manuals should be an integral part of the design process. The appropriate symbols, color, and size and the precise wording of the label must be developed after joint meetings of the engineering, legal, marketing, and manufacturing staffs. Use international warning symbols. (See Sec. 11.9.)
7. Create a means of incorporating legal developments in product liability into the design decision process. It is particularly important to get legal advice from the product liability angle on new innovative and unfamiliar designs.
8. There should be a formal design review before the product is released for production. (See Sec. 16.6.)

15.6.6 Business Procedures to Minimize Risk

In addition to careful consideration of the above design factors, a number of business procedures can minimize product liability risk.

1. There should be an active product liability and safety committee charged with seeing to it that the corporation has an effective product liability loss control and product safety program. This committee should have representatives from the advertising, engineering, insurance, legal, manufacturing, marketing, materials, purchasing, and quality-control departments of the corporation.
2. Insurance protection for product liability suits and product recall expenses should be obtained.
3. Develop a product usage and incident-reporting system just as soon as a new product moves into the marketplace. It will enable the manufacturer to establish whether the product has good customer acceptance and detect early signs of previously unsuspected product hazards or other quality deficiencies.

15.6.7 Problems with Product Liability Law

As product liability has grown so rapidly certain problems have developed in the implementation of the law.[1] There has been a dramatic shift in the doctrine of product liability law from negligence to strict liability but the law has proved incapable of defining the meaning of strict liability in a useful fashion. The rules of law are vague, which gives juries little guidance, and as a result verdicts appear capricious and without any definitive pattern. Another problem concerns the computation of damages once liability is established. There is great uncertainty and diversity in awarding damages for pain and suffering. Our adversarial legal system and the unfamiliarity of juries with even the rudiments of technical knowledge lead to high costs and much frustration.

  The great increases in the number of product liability claims and the dollars awarded by the courts to consumers, other companies, and government have brought a clamor to bring some restraint to the situation before we become a no-fault economy in which producers and sellers will be held for all product-related injuries. Advocates of reform point to product liability insurance costs and damage awards as a significant factor in reducing American competitiveness. National product liability legislation has been introduced in the U.S. Congress to ease the situation. It aims at making tort law on product liability uniform in all the states and on speeding up product liability disputes. It proposes a limit on joint and several liability, a doctrine by which a defendant responsible for only a small portion of harm may be liable for an entire judgment award. It also calls for a limit on a product seller's liability to cases in which the harm was proximately caused by the buyer's own lack of reasonable care or a breach of the seller's warranty.

15.7
PROTECTING INTELLECTUAL PROPERTY

The protection of intellectual property by legal means has become a topic of general interest and international diplomatic negotiations. There are two conflicting motiva-

---

1. D. G. Owen, op. cit.

A small child threw an aerosol can into a blazing fireplace. The can exploded, injured the child, and the child's father sued the manufacturer of the cleaner in the spray can. The manufacturer defended itself by stating that the can contained a label warning the user not to incinerate. The child's father argued that the manufacturer should have anticipated that some cans would accidentally be incinerated and that some sort of fail-safe design should have been provided to prevent explosion.

The manufacturer of the spray can won the case by arguing that the presence of a warning label against incineration should excuse liability for the injury. This is a situation where the present state of technology does not provide for a safe means of preventing an explosion upon rapid rise in temperature. The manufacturer should not be held in liability so long as the users of the product have been clearly warned of potential dangers. In fact, the parents of the child were really negligent for allowing their child to play with an aerosol can near an open fire.

tions for this: (1) creations of the mind are becoming more valuable in the information age, and (2) modern information technology makes it easy to transfer and copy such information. We saw in Sec. 4.7 that intellectual property is protected by patents, copyrights, trademarks, and trade secrets. These entities fall within the area of property law, and as such they can be sold or leased just like other forms of property.[1]

The functional features of a design can be protected with *utility patents.* A utility patent protects not only the specific embodiments of the idea shown in the patent application but functional equivalents as well. A well-written patent is the best protection for a valuable idea. For the criteria of patentability refer to Sec. 4.7. If an idea is worth patenting, it is worth hiring an experienced patent attorney to do the job well.

A different type of patent, the *design patent,* covers the ornamental aspects of a product such as its shape, configuration, or surface decoration. Design patents are easier to obtain than utility patents, and they are easier to enforce in court. If a competitive design has essentially the same overall appearance, then it is in violation of your patent. A design patent can have only one claim, which is a serious disadvantage, because it means that every unique aspect of a product's design requires a separate patent. This can be expensive.

A copyright has only limited usefulness in protecting product designs. This form of intellectual property is primarily intended to protect writing; however, it has become the dominant method of protecting software.

Trademarks are used to protect the names or symbols (logo) of products. A related form of protection is known as *trade dress.* This consists of distinctive features of a product like its color, texture, size, or configuration. Trademark and trade dress are intended to protect the public about the source of a product, i.e., to protect against cheap "knock-offs." Trademark protection is achieved by registration with the U.S. Patent and Trademark Office, or by actual use of the trademark in the marketplace such that it achieves market recognition. Obviously, it is easier to defend against a

---

1. D. A. Gregory, C. W. Saber, and J. D. Grossman, "Introduction to Intellectual Property Law," The Bureau of National Affairs, Inc., Washington, DC, 1994.

competing trademark if it is registered. A registered trademark is issued for 20 years and can be renewed every 20 years as long as the product remains in the marketplace.

An innovation becomes a *trade secret* when a company prefers to forgo legal protection for the intellectual property. The reason for doing this is often a feeling that patents are difficult or costly to defend in the particular area of technology, or an unwillingness to let the public know what the company is doing. If the company takes active steps to protect the trade secret, then the courts will protect it as a form of intellectual property. Process innovations are more often protected by trade secrets than product innovations. Companies sometimes require nondisclosure agreements of their employees and may attempt to legally prevent an employee who leaves their employ with sensitive trade knowledge from working for a competitor in order to protect a trade secret.

15.8
THE LEGAL AND ETHICAL DOMAINS

We move now from considerations of the law to a discussion of ethics, and how ethical issues affect the practice of engineering design. Ethics is the principles of conduct that govern the behavior of an individual or a profession. It provides the framework of the rules of behavior that are moral, fair, and proper for a true professional. Ethical conduct is behavior desired by society and is above and beyond the minimum standards of the law.

The connection between legal and ethical action is illustrated by Fig. 15.1. In this model[1] the solid vertical line presents a clear distinction between what is legal and illegal, as set forth by statute and case law. The location of the dashed horizontal line between ethical and unethical behavior is much less well defined. The actions considered ethical depend on values, some of which are important to society, some to the profession, some to the employer, and some to the individual. The task of the ethical professional is to balance these value responsibilities. These values are clarified for the professional and business world by various codes of ethics (see Sec. 15.9). While

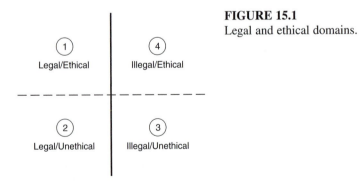

FIGURE 15.1
Legal and ethical domains.

---

1. S. G. Walesh, "Engineering Your Future," Chap. 11, Prentice-Hall, Englewood Cliffs, NJ, 1995; R. H. McCuen and J. M. Wallace, eds., "Social Responsibility in Engineering and Science," Prentice-Hall, Englewood Cliffs, NJ, 1987.

you would find close agreement among engineers as to whether an action is legal or illegal, you would find much greater disagreement as to whether some act is ethical or unethical.

Quadrant 1, legal and ethical behavior, is where you should strive to operate at all times. Most design and manufacturing activities fall within this quadrant. Indeed, a good case can be made that quality is dependent on ethical behavior.[1] "Doing what is right in the first place and doing what is best for all involved, when done at every level of the organization and in every work process, has proven to be the most efficient way of conducting a business."

Quadrant 2, legal and unethical, is the concern of the rest of this chapter. The goal is to explain how to identify unethical behavior and to learn what to do about it when it occurs. There is a feeling that unethical behavior in the workplace is increasing because of increasing workplace pressures and changing societal standards. Most corporations have adopted codes of ethics. Many have established an ethics office and are offering ethics training to their personnel. It is interesting that the prevailing view about ethics instruction has changed substantially. Throughout most of the twentieth century the common view about ethics was that you either learned ethics in the home when you were growing up, or it was too late. This is changing today to a view that ethics is a teachable subject that can be learned by just about everyone.

Quadrant 3, illegal and unethical, is the "go-to-jail" card. In general, most illegal acts also are unethical.

Quadrant 4, illegal and ethical, is a relatively rare event. An example could be an engineer who had signed a secrecy agreement with an employer, but then found that the employer had been engaged in producing a product that was very hazardous to the general public. Unable to get attention focused on the problem within the company, the engineer goes to the press to warn the public. The engineer has breached a contract, but in what is believed to be a highly ethical cause.

15.9
CODES OF ETHICS

We start by making a distinction between *morality* and *professional ethics.* Morality refers to those standards of conduct that apply to all individuals within society rather than only to members of a special group. These are the standards that every rational person wants every other person to follow and include standards such as the following:

- *Respect* the rights of others.
- Show *fairness* in your dealings with others.
- Be *honest* in all actions.
- *Keep promises* and contracts.
- Consider the *welfare* of others.
- Show *compassion* to others.

Note that each of these standards is based on the italicized values.

---

1. L. Bottorff, *Quality Progress,* February 1997, pp. 57–60.

By professional ethics we mean those standards of conduct that every member of a profession expects every other member to follow. These ethical standards apply to members of that group simply because they are members of that professional group. Like morality, standards of ethical conduct are value-based. Some values that are pertinent to professional ethics include[1]:

- *Honesty* and *truth*
- *Honor*—showing respect, *integrity,* and *reputation* for achievement
- *Knowledge*—gained through education and experience
- *Efficiency*—producing effectively with minimum of unnecessary effort
- *Diligence*—persistent effort
- *Loyalty*—allegiance to employer's goals
- *Confidentiality*—dependable in safeguarding information
- Protecting *public safety* and *health*

Note that some of these values are directed toward the employer (e.g., diligence), some toward the customer (e.g., confidentiality), some toward the profession (e.g., honor), and some toward society (e.g., public health and safety). These values reflect the professional's value obligations.

15.9.1 Profession of Engineering

The members of a profession are involved in an intellectual effort that requires special training that benefits society. Collectively, a group of people form a true profession only as long as they command the respect of the public and inspire confidence in their integrity and a belief that they are serving the general welfare.

The nature of professional service varies widely. The physician, lawyer, and clergyman have direct, individual relations with their clients, but an engineer usually is salaried in someone else's employ. About 95 percent of engineers work for either industry or government, and only a small, but important, percentage is in direct contact with the public as consulting engineers. Thus, the service aspect of engineering may be less obvious to the general public than in other professions.

The peculiarities of the engineering profession as compared with the professions of law and medicine carry over into the area of ethics. Because engineering lacks the homogeneous character of such professions as law and medicine, it is not surprising to find that there is no widely accepted code of engineering ethics. Most professional societies have adopted their own codes, and ABET and NSPE have adopted broader-based ethical codes. Again, because engineers who are employees of either business or government are in the great majority, they face ethical problems that self-employed professionals avoid. These arise from the conflict between the engineer's desire to gain a maximum profit for the employer (and thus achieve recognition and promotion) and the desire to adhere to a standard of ethics that places the public welfare ahead of

1. R. H. McCuen, *Ethics Education Program of the Institute for Professional Practice,* Verona, NJ, 1998.

corporate profit. For example, what can an employed engineer[1] do to expose and correct the corrupt practices of an employer? What should an engineer do if employed in a business atmosphere in which kickbacks and bribes are an accepted practice?

15.9.2 Codes of Ethics

Strong parallels exist between problem solving in design and in ethical decision making.[2] In both instances, a uniquely correct solution or response is rarely possible. However, some solutions are better than others, and some solutions are clearly unacceptable.

To provide guidance on how to behave in situations with ethical implications each engineering professional society has published a code of ethics.[3] The Code of Ethics for the American Society of Mechanical Engineers is given in Fig. 15.2. Note that the code is rather brief and quite general in its statements and that it is heavily oriented toward values. It is not a list of do's and don'ts. The three fundamental principles identify goals for the ethical behavior of engineers. Use your knowledge as an engineer for the good of humanity. Do it in an honest and impartial way. Work to increase the competence of the profession of engineering. The seven fundamental canons get a bit more specific, but they still emphasize value statements and leave many things unsaid. This generality is intentional in well-conceived codes of ethics.

The canons[4] present the general duties of an ethical engineer. They start by reinforcing the point that the safety, health, and welfare of the public is the first responsibility of the engineer. Next the canons charge engineers to work only in areas of their competence. The greater our competence, the better we will be able to protect public safety. Recognition of your real competencies is an important attribute. One should work to improve and expand their competence (canon 3), but also know when it is important to bring in other expertise to work on a design. After all, that is why we have design teams. While it is not specifically stated in canon 3, the wording implies that maintaining professional competency applies not just to technical competency but it also applies to competency in knowledge of ethics and values.

Canon 4 charges the engineer to act professionally with respect to the employer or client as faithful agents or trustees. This implies that the engineer places high importance on the values of loyalty, confidentiality, efficiency, and diligence. The second part of this canon talks about avoiding *conflicts of interest,* or the *appearance* of such conflict. A person is in a position of conflict of interest when he or she is in a position to personally benefit from actions under his or her influence, especially when the employer is unaware of this benefit. For example, a design engineer who owns considerable stock in a startup company would be in conflict of interest if he specified that company's product in his new design. Often a conflict of interest is impossible or

---

1. T. S. Perry, *IEEE Spectrum,* pp. 56–61, September 1981.
2. C. Whitbeck, "Ethics in Engineering Practice and Research," Cambridge University Press, New York, 1998, pp. 55–66.
3. A collection of the code of ethics for many professional societies can be found at http://ethics.cwru.edu.
4. A canon is an ecclesiastical or secular rule or law.

AMERICAN SOCIETY of MECHANICAL ENGINEERS
Founded 1880
CODE OF ETHICS OF ENGINEERS

THE FUNDAMENTAL PRINCIPLES

Engineers uphold and advance the integrity, honor, and dignity of the Engineering profession by:

I. using their knowledge and skill for the enhancement of human welfare;

II. being honest and impartial, and serving with fidelity the public, their employers and clients, and

III. striving to increase the competence and prestige of the engineering profession.

THE FUNDAMENTAL CANONS

1. Engineers shall hold paramount the safety, health and welfare of the public in the performance of their professional duties.

2. Engineers shall perform services only in the areas of their competence.

3. Engineers shall continue their professional development throughout their careers and shall provide opportunities for the professional and ethical development of those engineers under their supervision.

4. Engineers shall act in professional matters for each employer or client as faithful agents or trustees, and shall avoid conflicts of interest or the appearance of conflicts of interest.

5. Engineers shall build their professional reputation on the merit of their services and shall not compete unfairly with others.

6. Engineers shall associate only with reputable persons or organizations.

7. Engineers shall issue public statements only in an objective and truthful manner.

BOARD ON PROFESSIONAL PRACTICE AND ETHICS

FIGURE 15.2
The Code of Ethics of ASME International. (*Reprinted with permission.*)

impractical to avoid. In this case, the best practice is to make it known to everyone involved. For example, persons serving on study committees of the National Research Council are asked to disclose any potential conflicts of interest. You might have large stock holdings in your company, yet your expertise is vital to the study that is to be undertaken. Your conflict would be made part of the public record, and you would need to excuse yourself from deliberations if they ever come close to your area of conflict.

Avoiding the appearance of conflict of interest is as important as avoiding the conflict itself. Some areas that could get you in trouble are hiring relatives or close friends, accepting expensive gifts from vendors or customers, accepting a paid trip to a conference, or owning large blocks of stock in a competitor of your company.

In another example of conflict of interest, a law firm is forbidden by the code of ethics for lawyers to represent both parties in a dispute, even if the lawyer is asked to represent the second party in a case unrelated to the first dispute.[1] This part of the code arises from the adversarial nature of the legal system and the role that lawyers play therein. It is interesting to note that none of the engineering ethics codes would prevent a company from designing plastics plants for two directly competing companies as long as confidentiality was upheld. This points out that codes of ethics can differ significantly between different professions since professional practice can be quite different.

The fifth canon deals with how engineers treat each other in professional practice. One should not misrepresent or exaggerate their academic or professional credentials. For example, you should be able to provide documentation for all items on your résumé. Be fair in your dealings with other engineers. A good practice that will win you many friends is to be generous in giving credit for accomplishments when credit is due.

Canon 6 is self-evident. Nothing is more important over the long term than your reputation for following high ethical standards. The old adage, *we are known by the company we keep* applies here.

The last canon has many implications. It charges the engineer to be objective and truthful in professional reports, statements, and public testimony. Engineers may express publicly a professional opinion on technical subjects only when that opinion is founded upon adequate knowledge and competence in the subject matter. When making public statements, any payments you may have received by interested parties to make that statement must be disclosed.

The fundamental principles and canons are by necessity very general. This enables them to be applied to the broad gamut of situations with which engineers are involved. Some codes of ethics provide detailed *rules of practice.* The following are examples from the code of ethics of the National Society of Professional Engineers (NSPE):

- "Engineers shall not solicit or accept financial or other valuable consideration, directly or indirectly from contractors, their agents, or other parties in connection with work for employers or clients for which they are responsible."
- "Engineers shall not solicit or accept a professional contract from a governmental body on which a principal or officer of their organization serves as a member."
- "Engineers shall not reveal facts, data, or information obtained in a professional capacity without the prior consent of the client or employer except as authorized by law or this Code."

Four reasons why engineers should support their profession's code of ethics can be presented.[2] First, supporting the code helps protect engineers from being harmed by what other engineers do. Second, the code helps ensure to each engineer a work environment in which refusing to perform an unethical directive becomes easier to do. One can point to the code in support of your position. Third, supporting the code helps to make engineering a profession about which you need feel no morally justified embarrassment. Finally, supporting the code is the professional thing to do.

---

1. C. Whitbeck, op. cit.
2. M. Davis, "Thinking Like an Engineer," Oxford University Press, New York, 1998, pp. 59–60.

A young engineer on active duty with the Air Force discovers that a component used in three aircraft is overdesigned. A special adapter, costing several hundred dollars, is required for every component. The original purpose of the adapter was to permit the component to be used with a certain aircraft, but that aircraft now has been phased out of service. Thus, the adapter is redundant. The engineer tries to get the specification changed to eliminate the adapter, but she is told that the "system" will not permit this.

The resourceful engineer then submits the proposal through the suggestion system of the maintenance depot where she is assigned. She receives a phone call from a staff member informing her that as a military officer she cannot receive a monetary award for suggestions. The advice is given that she add a civil service engineer to the suggestion, and that they agree to split the award.

The potential for monetary reward through the suggestion system is rather great, but to do as suggested would be unethical. It would be dishonest and not show integrity. She submits the suggestion as originally formulated, forgoing an award that could exceed $100,000. Her reward, as a military officer, was dinner for two at the officer's club, and the start of a career with a reputation for high ethical behavior.

15.9.3 Extremes of Ethical Behavior

Ethical theory considers two extreme types of behavior. *Altruism* is a form of moral behavior in which individuals act for the sake of other people's interests. Ethical altruism is the view that individuals ought to act with each others' interests in mind. This is the viewpoint best summarized by the Golden Rule: Do unto others as you would have others do unto you. *Egoism* is a form of moral behavior in which individuals act for their own advantage. Ethical egoism is the view that individuals ought always to act to satisfy their own interests.[1] Most day-to-day practice of engineering is done in the individual's self-interest and is not in conflict with the codes of ethics. However, the codes of ethics are meant to alert the practicing professional that he or she has altruistic obligations that must be properly balanced with self-interest.

15.10
SOLVING ETHICAL CONFLICTS

It is probably safe to say that every engineer must resolve at least one ethical dilemma over the duration of his or her career. If the engineer mishandles the situation, his or her career can be damaged even in cases where he or she is trying to do the right thing. Therefore, it is important to know how to handle ethical conflicts and to have thought about conflict resolution before being confronted by a problem. A difficult problem

---

1. R. H. McCuen, *Issues in Engineering—Jnl. of Prof. Activities,* ASCE, vol. 107, no. E12, pp. 111–120, April 1981.

for engineers arises from their dual obligation to serve diligently and with loyalty both their employer and society. The vast majority of businesses aim to be honest and responsible corporate citizens, but the conflict between profit and societal good is potentially always present. What should you do when confronted by an ethical conflict where it is obvious that you have competing value responsibilities?

Ethical decision making is not easy. However, the chances for successfully resolving an ethical conflict can be greatly increased by following a systematic procedure. Table 15.2 presents one set of guidelines that will help ensure meeting one's professional responsibilities. Except under the unusual circumstances of imminent danger to the public, it is important that all internal steps should be explored before seeking options outside of the organization. The process of seeking resolution to an ethical conflict within the organization is usually handled through an appeals process within management or by the complaint process through the office of the ombudsman or the ethics officer. Seeking resolution outside of the organization is usually called *whistleblowing*. Table 15.2 gives a step-by-step procedure for resolving an ethical conflict or any conflict for that matter, through an internal appeals process and external to your company.

The steps that the individual should take in preparation for disclosure of unethical behavior are straightforward. Once you have studied and documented the facts and

TABLE 15.2
Procedure for Solving Ethical Conflicts*

I. Internal appeal option
   A. Individual preparation
      1. Maintain a record of the event and details
      2. Examine the company's internal appeals process
      3. Be familiar with the state and federal laws that could protect you
      4. Identify alternative courses of action
      5. Decide on the outcome that you want the appeal to accomplish
   B. Communicate with your immediate supervisor
      1. Initiate informal discussion
      2. Make a formal written appeal
      3. Indicate that you intend to begin the company's internal process of appeal
   C. Initiate appeal through the internal chain of command
      1. Maintain formal contacts as to where the appeal stands
      2. Formally inform the company that you intend to pursue an external solution
II. External appeal option
   A. Individual actions
      1. Engage legal counsel
      2. Contact your professional society
   B. Contact with your client (if applicable)
   C. Contact the media

R. H. McCuen, "Hydrologic Analysis and Design," 2d ed., Prentice-Hall, Englewood Cliffs, NJ, 1998.

formulated a plan for appeal, you should discuss the matter with your immediate supervisor. Failure to fully communicate your concerns to your immediate supervisor or secretly going over his or her head to higher levels is viewed as disloyalty and will be viewed negatively by all involved, even your supervisor's superior. It will also decrease the likelihood of a favorable resolution of the conflict. Often the value difference will be resolved by communicating with the immediate supervisor. However, if after fully discussing the issue with your supervisor, you feel that your supervisor is not willing or able to take appropriate action, then inform your supervisor in writing of your intention to appeal beyond that level.

The process of appealing an ethical conflict within the company is usually similar to the process of interacting with your immediate supervisor on the issue. You should have the facts and a plan of how you would like to see the issue resolved. Formal steps should follow informal discussions, and steps within the appeal chain should not be bypassed. If the internal appeal does not resolve the ethical conflict, then you should notify the company that you intend to continue with an external review of the problem.

Before expressing any public concern legal advice should be obtained. A lawyer can identify courses of action and legal pitfalls in your external appeal. While lawyers understand the legal issues, they may not have the technical background to evaluate the technical adequacy of your arguments. For this reason it might be helpful to involve an engineering professional society as an impartial judge of your arguments. Engineering societies vary widely in their willingness to become involved in these kind of activities.

If your company worked for a client in the issue about which you are concerned, then the client should be approached before going public. The client may pressure your company to resolve the issue internally, or the client may provide the resources to obtain an unbiased review of the issue.

The last resort is public disclosure by contacting the press and news TV. This is often called whistleblowing.

15.10.1 Whistleblowing

Whistleblowing is the act of reporting on unethical conduct within an organization to someone outside of the organization in an effort to discourage the organization from continuing the activity. In the usual case the charges are made by an employee or former employee who has been unable to obtain the attention of the organization's management to the problem. Sometimes whistleblowing is confined to within the organization where the whistleblower's supervision is bypassed in an appeal to higher management. An important issue is to determine the conditions under which engineers are justified in blowing the whistle. DeGeorge[1] suggests that it is morally permissible for engineers to engage in whistleblowing when the following conditions are met.

---

1. R. T. DeGeorge, *Business and Prof. Ethics Jnl.,* vol. 1, pp. 1–14, 1981.

1. The harm that will be done by the product to the public is considerable and serious.
2. Concerns have been made known to their superiors, and getting no satisfaction from their immediate superiors, all channels have been exhausted within the corporation, including the board of directors.
3. The whistleblower must have documented evidence that would convince a reasonable impartial observer that his or her view of the situation is correct and the company position is wrong.
4. There must be strong evidence that releasing the information to the public would prevent the projected serious harm.

Clearly a person engaging in whistleblowing runs considerable risk of being labeled a malcontent or of being charged with disloyalty, and possibly being dismissed. The decision to blow the whistle requires great moral courage. Federal government employees have won protection under the Civil Service Reform Act of 1978, but protection under state laws or active support from the engineering professional societies is still spotty. Some farsighted companies have established the office of ombudsman or an ethics review committee to head off and solve these problems internally before they reach the whistleblowing stage.

15.10.2 Case Studies

Ethics is best taught by looking at real-life situations through case studies. From time to time major incidents occur that catch the public's attention, and these are recorded for posterity in the engineering ethics texts. Prominent examples are the space shuttle *Challenger* tragedy,[1] the Bay Area Rapid Transit (BART)[2] control system failure, and the meltdown of the Chernobyl nuclear reactor.[3] Entire areas of technology and society are the subjects of continuing ethical discussions, for example, genetic engineering. Environmental issues[4] and questions of scientific fraud and integrity in doing research[5] are also prominent areas for discussion.

While these major incidents and cutting-edge activities get most of the attention, the likelihood that the average engineer will be involved heavily in such cases is small. A more typical ethical situation would be:

- Should I authorize the release of production parts that are only marginally out of specification?
- Should I condone the use of pirated design software?

---

1. R. L. B. Pinkus et al., "Engineering Ethics: Lessons Learned from the Space Shuttle," Cambridge University Press, New York, 1997; C. Whitbeck, op. cit., Chap. 4.
2. S. H. Unger, "Controlling Technology: Ethics and the Responsible Engineer," 2d ed., Wiley, New York, 1994, pp. 20–25.
3. S. H. Unger, op. cit., pp. 77–91.
4. P. A. Vesilind and A. S. Gunn, "Engineering, Ethics, and the Environment," Cambridge University Press, New York, 1998.
5. C. Whitbeck, op. cit., Chaps. 6, 7, 9, 10.

A consulting engineer is hired by the county to investigate a bridge collapse. In the course of his investigation he examines a bridge of similar design and finds that it is only marginally safe. He contacts the county engineer to tell him about this discovery. The county official tells him that they know about this condition and that they hope to repair it in the next budget year. However, they must keep the second bridge open because to close it would increase the response time of emergency vehicles by about 30 minutes. What should the consulting engineer do?

He goes back to the marginal bridge and makes a more thorough investigation, taking photographs and measurements. He finds the situation more dangerous than he first thought. Back at the office he makes some calculations and prepares a brief report. He asks for a meeting with the county engineer and lays out the case for closing the bridge. The county engineer is impressed, but points out the political implications of closing the bridge. He suggests a joint meeting with the county supervisor. They meet with the county supervisor, who is impressed with the severity of the situation, and the spirit of civic duty shown by the engineer. They agree to post the bridge to forbid general traffic, but to leave it open to emergency vehicles. The county supervisor schedules a press conference to which he invites the consulting engineer as an honored guest.

- What should I do about the fact that my boss has inflated my credentials on the résumé that went out with the last proposal?

Fortunately, the number and diversity of case studies that deal with just such day-to-day ethical problems is rapidly growing. The NSPE Board of Ethical Review answers ethics questions submitted by members of the society. These have been published in their monthly magazine *Professional Engineer,* and currently in the monthly newspaper *Engineering Times.* Engineering case studies can be found on the worldwide web at a number of locations.[1]

15.11
SUMMARY

Engineers mostly work in the business world, and they are therefore required to perform their duties within the laws of the nation. But more than this, engineering is a profession that is critical to the advancement of society. How engineers do their jobs determines what kind of world future generations will enjoy. Thus, the practice of engineering without question will involve you in making ethical judgments. Most will be small in nature, involving your relationship with your management and your fellow engineers, but others could be momentous, affecting the safety of a city.

The law is a formalized code of conduct describing what society feels is the proper way to behave. Statutory law is created by a federal or state legislative body. Case law arises from the decisions of the courts. Ethics is the study of human con-

---

1. http://ethics.cwru.edu; http://ethics.tamu.edu/Nsfcases/; http://repont.tcc.virginia.edu/ethics/.

duct that lays out the moral ground rules based on society's values. Ethical conduct is the behavior that is desired by society and is separate from the minimum standards of the law.

Engineers should be familiar with contracts and liability, especially product liability. A contract is an agreement between two parties to do or not to do something. It consists of an offer, an acceptance of the offer, and a consideration, the exchange of something of value. A contract is discharged when the agreement has been performed to the satisfaction of both parties. A breach of contract occurs when one party fails to perform its part of the contract.

Liability means being bound or obligated to pay damages or restitution. Two common ways of incurring liability are breaching a contract or committing a tort. A tort is a civil wrong committed against a person, or the business, property, or reputation that causes damage. Common examples of torts are fraud, misrepresentation, negligence, and product liability.

Product liability is the legal action by which an injured party seeks to recover damages for personal injury or property loss from the producer or seller of a product. The law under which product liability is tried has evolved to a standard of strict product liability. Under this theory of law, the plaintiff must prove that: (1) the product was defective and unreasonably dangerous, (2) the defect existed at the time the product left the defendant's control, (3) the defect caused the harm, and (4) the harm is appropriately assignable to the identified defect. Previously, manufacturers were liable only when they could be proved negligent or unreasonably careless.

To protect against product liability suits, the design procedures described in Secs. 11.4, 11.5, 11.6, and 11.9 must be followed. In addition, documentation of these design methods and of testing and quality activities is vital. There should be an active product safety committee to see that every step is taken to ensure the design and production of safe products. Finally, it may be a wise business decision to obtain insurance protection for product liability suits and product recall expenses.

While the boundary between legal and illegal acts is generally well defined by the law, the distinction between what is ethical and what is unethical is much less well defined. Professional engineering societies provide guidance by means of codes of ethics. Different individuals respond differently depending on their value system. Engineering is a profession, and as such, you are bound by the ethical standards of the profession. It does not matter whether you have individually made that agreement. The profession expects you to behave in a certain ethical way. These rules of conduct are laid down in the code of ethics of each professional engineering society. The existence of a code of ethics is important to you because it gives you an authoritative standard to fall back on if you are engaged in a serious ethical conflict in the workplace.

Table 15.3 suggests typical ethical questions associated with different steps in the product design process.[1]

It is not inconceivable that you will be involved in a serious ethical conflict sometime in your career. This often arises from the competing value responsibilities that engineers have, such as loyalty and diligence to both their employer and the good of

---

1. M. W. Martin and R. Schinzinger, "Ethics in Engineering," 3d ed., McGraw-Hill, New York, 1996.

TABLE 15.3
Typical ethical questions associated with product design

Steps in product design	Possible ethical questions
Market study	Is the study unbiased or has it been embellished to attract investors or management support?
Conceptual design	Will the product be useful or will it be just a gimmick?
Embodiment design	Does the design team have sufficient expertise to properly judge whether computer programs are giving reliable results? Have any patents been violated?
Detail design	Has checking of results been done?
Manufacturing	Is the workplace safe and free of environmental hazards? Is enough time allowed to do quality work?
Product use	Is the product safe to use? Are users informed of possible hazards?
Retirement from service	Has the design allowed for recycling or reuse?

society. If such a conflict does arise, it is important to try to resolve the disagreement internally in your organization. Follow to the letter the prescribed appeal procedure, and document everything. If you must go outside of the organization, obtain competent legal counsel. While it may be necessary to be a whistleblower and release your story to the press, do this only as a last resort.

BIBLIOGRAPHY

Law and the Engineer

Blinn, K. W.: "Legal and Ethical Concepts in Engineering," Prentice-Hall, Englewood Cliffs, NJ, 1989.
Dunham, C. W., R. D. Young, and J. T. Bockrath: "Contracts, Specifications, and Law for Engineers," 3d ed., McGraw-Hill, New York, 1979.
"Engineering Law, Design Liability, and Professional Ethics," Professional Publications, Belmont CA, 1983.

Product Liability

Brown, S., I. LeMay, J. Sweet, and A. Weinstein, eds.: "Product Liability Handbook: Prevention, Risk, Consequence, and Forensics of Product Failure," Van Nostrand Reinhold, New York, 1990.
Hunziker, J. R., and T. O. Jones: "Product Liability and Innovation," National Academy Press, Washington, DC, 1994.
Smith, C. O.: "Products Liability: Are You Vulnerable?" Prentice-Hall, Englewood Cliffs, NJ, 1981.

Engineering Ethics

Davis, M.: "Thinking Like an Engineer: Studies in the Ethics of a Profession," Oxford University Press, Oxford, 1998.

Harris, C. E., M. S. Pritchard, and M. Rabins: "Engineering Ethics: Concepts and Cases," Wadsworth Publishing Co., Belmont, CA, 1995.

Martin, M. W., and R. Schinzinger: "Ethics in Engineering," 3d ed., McGraw-Hill, New York, 1996.

Unger, S. H.: "Controlling Technology. Ethics and the Responsible Engineer," 2d ed., Wiley, New York, 1994.

Whitbeck, C.: "Ethics in Engineering Practice and Research," Cambridge University Press, New York, 1998.

PROBLEMS AND EXERCISES

15.1. John Williams, a professional engineer, agrees to testify as an expert witness for the firm of Jones & Black in a court case. In return, the firm promises to pay Williams $1500 plus expenses for his services. (*a*) Is this a lawful contract? State the reasons for your decision. (*b*) Suppose Williams agrees to accept $2500 if Jones wins, but only expenses if Jones loses. Is this a lawful contract? State your reasons.

15.2. ABC Electric agreed by fax on Monday to buy 100 fractional-horsepower motors for $3000 from Amalgamated Electric. On Wednesday the purchasing agent from ABC calls and says he is canceling the order. Amalgamated says the motors have already been shipped and they want their money. (*a*) What is the legal responsibility of ABC Electric in this transaction? (*b*) Would it have been any different if the motors had not already been shipped?

15.3. A car designer specified steel bolts of the highest quality and strength when designing a connection for the front-end steering rods. The manufacturer of the bolts used an inadequate sampling plan for inspecting the bolts, and several defective bolts caused failure of the steering mechanism. Several deaths resulted and there was a major product recall. Discuss the liability of the designer, the auto company, and the bolt manufacturer.

15.4. Read the story of the failure of the General Electric refrigerator with the revolutionary rotary compressor (*Wall Street Journal,* May 7, 1990, p. A1, A5). What lessons does this teach us about product design? What implications does it have for product liability?

15.5. Aristotle put forth the precept that humanity should follow four virtues: (1) prudence, (2) justice, (3) fortitude, and (4) temperance. Define each virtue broadly and give examples of ethical behavior for each virtue.

15.6. Make a list of business practices that signal whether an organization is an ethical corporation. What role does the CEO of the corporation play in this?

15.7. We are in a period where the desire for steady increase in corporate earnings, driven to a large degree by the stock market, sometimes causes management to require layoffs even when profits are good. Discuss the ethics of this from the viewpoint of both corporate management and the individual engineer.

15.8. Imagine what it would be like if there were no codes of ethics for engineers. What would be the consequences?

15.9. A trend in sports equipment has been to improve the players' performance by introducing new products. Examples are the graphite-composite shaft and titanium head in golf drivers, lighter-weight composite tennis rackets with a larger "sweet" spot, and an aluminum baseball bat with built-in damping. Discuss the ethics of compensating for personal inadequacies in performance with technology in competitive sports.

15.10. Discuss the ethics of the following situation. You are a design engineer for the Ajax Manufacturing Co., a large multiplant producer of plastic parts. As part of your employment, you were required to sign a secrecy agreement that prohibits divulging information that the company considers proprietary.

Ajax has modified a standard piece of equipment that greatly increases the efficiency in cooling viscous plastic slurries. The company decides not to patent the development but instead to keep it as a trade secret. As part of your regular job assignment, you work with this proprietary equipment and become thoroughly familiar with its enhanced capabilities.

Five years later you leave Ajax and go to work for a candy manufacturer as chief of production. Your new employer is not in any way in competition with Ajax. You quickly realize that Ajax's trade secret can be applied with great profit to a completely different machine used for cooling fudge. You order the change to be made. Discuss the ethics.

15.11. Discuss the ethics in the following situation. You have been on the job for nine months as an assistant research engineer working with a world-famous authority on heat transfer. It is an ideal job, because you are learning a great deal under his sympathetic tutelage while you pursue an advanced degree part-time.

You are asked to evaluate two new flame-retardant paints A and B. Because of late delivery of some constituents of paint A, the test has been delayed and your boss has been forced to make a tentative recommendation of paint A to the design group. You are asked to make the after-the-fact tests "for the record." Much to your surprise, the tests show that your boss was wrong and that formulation B shows better flame resistance. However, a large quantity of paint A already has been purchased. Your boss asks you to "fudge the data" in favor of his original decision, and since there is reasonable possibility that your data were in error, you reluctantly change them to favor his decision. Discuss the ethics.

16

DETAIL DESIGN

16.1
INTRODUCTION

We have finally come to the last of the three stages into which we have divided the design process. As mentioned in the introduction to Chap. 6, the boundary between embodiment design and detail design has become blurred by the emphasis on reducing the product development cycle time by the use of concurrent engineering methods (Design for X), aided and abetted by computer-aided engineering (CAE). In many engineering organizations it is no longer correct to say that detail design is the phase of design where all of the dimensions, tolerances, and details are finalized. However, detail design, as the name implies, is the phase where all of the details are brought together, all decisions are finalized, and a decision is made by management to release the design for production. Poor detail design can ruin a brilliant design concept and lead to manufacturing defects, high costs, and poor reliability in service. The reverse is not true. A brilliant detail design will not rescue a poor conceptual design.

Figure 16.1 shows the stages of design by which we have organized this book. The numbers of the Chaps. 7 through 14 have been superimposed in order to show you where in the process this knowledge is generally applied. Detail design is the lowest level in the design abstraction hierarchy. Many decisions have been made to get to this point. Most of these decisions are very basic to the design and to change them now would be costly in time and effort. Thus, as the name implies, detail design is mainly concerned with filling in the details to ensure that a proven and tested design can be manufactured.

16.2
DETAIL DESIGN

We have defined the end of the design process to be detail design. Figure 16.2 shows the principal tasks to be completed in detail design.

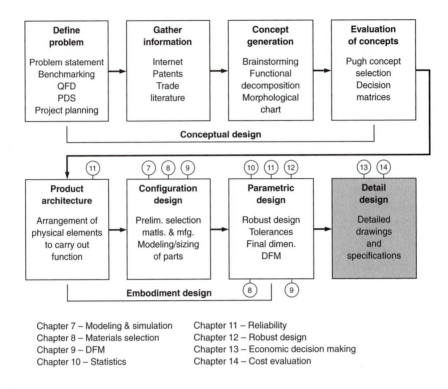

Chapter 7 – Modeling & simulation   Chapter 11 – Reliability
Chapter 8 – Materials selection   Chapter 12 – Robust design
Chapter 9 – DFM   Chapter 13 – Economic decision making
Chapter 10 – Statistics   Chapter 14 – Cost evaluation

FIGURE 16.1
Steps in the design process, showing where Chaps. 7 through 14 are chiefly applied.

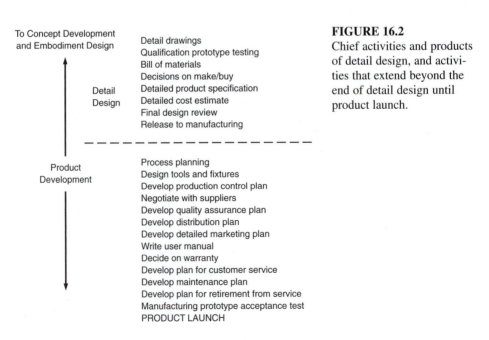

FIGURE 16.2
Chief activities and products of detail design, and activities that extend beyond the end of detail design until product launch.

The first task of detail design is to complete the detail drawings. Much of the technical analysis will have been done in embodiment design, but there will still be calculations to perform, questions to answer, and decisions to make. One important decision is to decide whether to make a component in-house or to buy it from an outside vendor. There will be many tests to run to assure that the components and assemblies meet the requirements laid down in the product design specification.

As each component, subassembly, and assembly is completed it is documented completely with detail drawings (see Sec. 16.3) and specifications. For individual parts the drawing and the specification are often the same document. The specification contains information on the technical performance of the part, its dimensions, test requirements, materials requirements, reliability requirement, design life, packaging requirement, and marking for shipment. The specification should be sufficiently detailed to avoid confusion as what is expected from the supplier.

If the product design is at all complex it most likely will be necessary to impose a *design freeze* at some point prior to completion. This means that beyond a certain point in time no changes to the design will be permitted unless they go through a formal review by a design control board. This is necessary to prevent the human tendency of always wanting to make a slight improvement, which unless controlled by some external means results in the job's never actually being completed. With a design freeze, only those last-minute changes that truly affect performance, safety, or cost get made.

Once the design is finalized, a final prototype is built and qualification tested to ensure that the design functions as required and is safe and reliable. Depending on the complexity of the product the qualification testing may simply be to run the product during an expected duty cycle and under overload conditions, or it may be a more elaborate series of staged tests. The detail drawings allow the calculation of detailed cost estimates. To make these calculations a bill of materials (see Sec. 16.4) is drawn up that lists all of the parts needed for making the product.

Once these steps have been successfully completed it is time to review the complete product design in a final design review (see Sec. 16.6). This involves top corporate management, the chief design personnel, the responsible manufacturing personnel, and the customer (if the product has a defined customer). The successful outcome of the design review is the release of the design to the manufacturing department.

While design release to manufacturing formally ends the detail design phase, as Fig. 16.2 shows, it does not end the product development process. There are many additional tasks, some carried out by engineers, some by other professionals, that must be done to achieve product launch. Most of these tasks are self-explanatory and do not require further elaboration.

Process planning and production control are related tasks. Process planning is the task of selecting the processes, and the individual steps that must be performed by each process in a prescribed sequence, to make each component. Ultimately, this information is needed to definitively calculate manufacturing cost. Process planning also includes laying out the production flow line. Production control is the task of scheduling the flow of work into production and providing the materials, supplies, and technical data needed for carrying out the manufacturing operation. One popular way of doing this today is *just-in-time* (JIT) *manufacturing.* With JIT a company minimizes inventory by receiving parts and subassemblies in small lots just as they are needed on the production floor. With this method of manufacturing the supplier is an extension of the production line. JIT manufacturing obviously requires close and harmonious relations with the supplier companies. The supplier must be reliable, ethical, and capable of delivering quality parts.

Just as successful testing of a qualification prototype ends the design phase of product development, the successful testing of the pilot runs from manufacturing ends the product development process. The proven ability to manufacture the product to specification and within cost budget makes possible the product launch in which the product is released to the general public or shipped to the customer.

16.3
DETAIL DRAWINGS

The historical goal of detail design has been to produce drawings that contain the information needed to manufacture the product. These drawing should be so complete that they leave no room for misinterpretation. The information on a detail drawing includes:

- Standard views of orthogonal projection—top, front, side views
- Auxiliary views such as sections, enlarged views, or isometric views that aid in visualizing the component and clarifying the details
- Dimensions—presented according to the GD&T standard ANSI Y14.5M
- Tolerances
- Material specification, and any special processing instructions
- Manufacturing details, such as parting line location, draft angle, surface finish

Sometimes a specification sheet replaces the notes on the drawing and accompanies it. Figure 16.3 is an example of a detail drawing. Note the level of detail contained in it. If the design is developed digitally in a CAD system, then the digital model becomes the authority for the component definition. In this case it is not necessary to include quite so much detail in the drawings since it resides in the computer and can be pulled up when needed.

Assembly drawings are part of the detail design. These are of two kinds. *Design layouts* show the spatial relationships of all components in the assembled product (the system). These have evolved from the rough sketches of conceptual design into often elaborate three-dimensional drawings. The solid model in CAD is a design layout. The design layout serves to visualize the functioning of the product and to ensure that there is physical space for all of the components. Design layout starts with the first glimmer of the concept and is developed more fully in the product architecture step of embodiment design. Design layouts serve as an input in the creation of detail drawings for each component.

Assembly drawings are created in detail design as tools for passing design intent to the production department, as well as the user. They show how the part is related in space and connected to other parts of the assembly. Dimensional information is limited to that necessary for the assembly. Reference is made to the detail drawing number for each part for this information. Also, the detail drawing will give the drawing

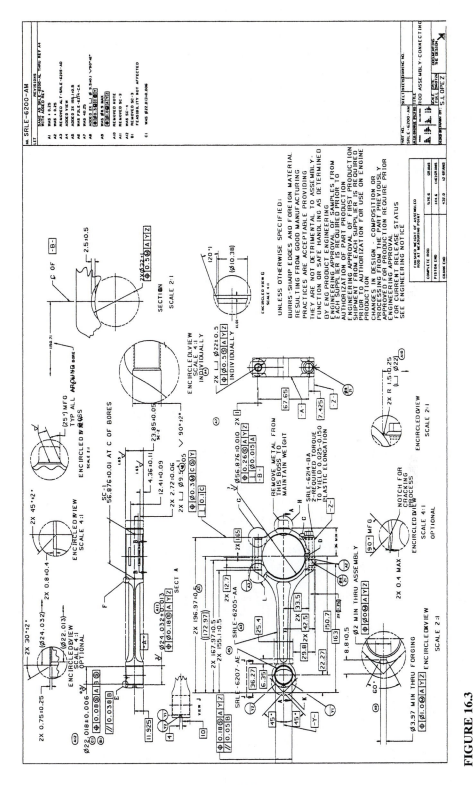

FIGURE 16.3

An example of a detail drawing. ("*ASM Handbook,*" *vol. 20, p. 227, ASM International, Materials Park, OH, 1997. Used with permission.*)

number of the assembly drawing that shows how the part fits into the next higher assembly. Often an assembly drawing utilizes exploded views (Fig. 16.4) to show how the components are assembled together, including the assembly sequence.

When a detail drawing is finished, it must be checked to ensure that the drawing correctly portrays the function and fit of the design.[1] Checking should be performed by someone not initially involved with the project who can bring a fresh but experienced perspective. Since design is an evolutionary process, it is important to record the history of the project and the changes that are made along the way. This should be done in the title block and revision column of the drawing. A formalized drawing release process must be in place so that everyone who needs to know is informed about design changes. An advantage of using a digital model is that if changes are only made there, then everyone who can access the model has up-to-date information.

An important issue in detail design is managing the volume of information. Ensuring that design information can be retrieved is very important. For example, we may have made a design change that was expected to ease the assembly process, but it did not work as expected. Now, we want to go back to the previous design. This should be readily retrievable. The process of storing and retrieving design information is particularly complicated today. Much is still on paper, but more and more is moving to digital form. The problem comes from the fact that digital data may be stored in different formats, which may be difficult to be communicated across different CAD platforms.

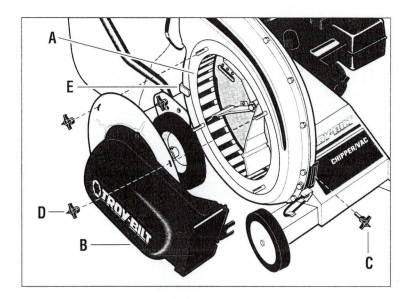

FIGURE 16.4
An example of an exploded assembly drawing showing how to replace the shredder screen (A) in a garden implement. (*Drawing courtesy of Garden Way Inc., 1 Garden Way, Troy, NY, 12180.*)

---

1. G. Vrsek, Documenting and Communicating the Design, "ASM Handbook," vol. 20, ASM International, Materials Park, OH, pp. 222–230, 1998.

Unfortunately, there is not a strong tradition of recording the decisions made during design so that the knowledge is not lost with the designer, and so that novices can learn from it. The place where this information is captured is the design notebook (see Sec. 17.2).

16.4
BILL OF MATERIALS

The bill of materials (BOM) or the parts list is a list of each individual component in the product. As Fig. 16.5 shows, it lists the part description, quantity needed for a complete assembly, part, number, the source of the part, and purchase order number if outsourced to a suppler. This version of the bill of materials also lists the name of the engineer responsible for the detail design of each part, and the name of the project engineer who is responsible for tracking the parts through manufacture and assembly.

The bill of materials has many uses. It is essential for determining the cost of the product. A bill of materials will be started early in the embodiment design phase, when the product architecture has been established, as a way of checking whether the product costs are in line with that called for in the PDS.[1] The bill of materials will be finalized in the detail design phase and will be used in the detailed cost analysis.

The bill of materials is vital for tracking the parts during manufacture and assembly. It is an important archival document for the design that needs to be preserved and be available for retrieval.

Qty / Engine	ENGINE PROGRAM PARTS LIST DOCUMENTING THE DESIGN — PART DESCRIPTION	PART NUMBER					Delivery	RESPONSIBILITY	
		Prefix	Base	End	P.O. #	Source	Date	Design	Engineer
	PISTON								
6	**PISTON (CAST/MACH)**	SRLE	6110	24093	RN0694	Ace	11/17/95	S. LOPEZ	M. Mahoney
6	**PISTON RING - UP COMPRESSION**	SRLE	6150	AC	RN0694	Ace	rec'd FRL	S. LOPEZ	M. Mahoney
6	**PISTON RING - LOWER COMPRESSION**	SRLE	6152	AC	RN0694	Ace	rec'd FRL	S. LOPEZ	M. Mahoney
12	**PISTON RING - SEGMENT OIL CONTROL**	SRLE	6159	AC	RN0694	Ace	rec'd FRL	S. LOPEZ	M. Mahoney
6	**PISTON RING - SPACER OIL CONTROL**	SRLE	6161	AB	RN0694	Ace	rec'd FRL	S. LOPEZ	M. Mahoney
6	**PIN - PISTON**	SRLE	6135	AA		BN Inc.		S. LOPEZ	M. Mahoney
6	**PISTON & CONNECTING ROD ASSY**	SRLE	6100	AG				S. LOPEZ	M. Mahoney
6	**CONNECTING ROD - FORGING**	SRLE	6205	AA		Formall		S. LOPEZ	M. Mahoney
6	**CONNECTING ROD ASSY**	SRLE	6200	CI		MMR Inc.		S. LOPEZ	M. Mahoney
12	**BUSHINGS - CONNECTING ROD**	SRLE	6207	AE		Bear Inc.		S. LOPEZ	M. Mahoney
12	**RETAINER - PISTON PIN**	SRLE	6140	AC		Spring Co.		S. LOPEZ	M. Mahoney

FIGURE 16.5
An example of a bill of materials. (*"ASM Handbook," vol. 20, p. 228, ASM International. Used with permission.*)

---

1. K. T. Ulrich and S. D. Eppinger, "Product Design and Development," pp. 69–70, McGraw-Hill, New York, 1995.

16.6.1 Input Documents

The input for the review consists of documents such as the PDS, the QFD analysis, key technical analyses like FEA and CFD, FMEAs, the quality plan, including robustness analysis, the results of the qualification tests, the detail and assembly drawings, and the product specifications, and cost projections. This documentation can be voluminous and it is not all covered in the final review. Important elements will have been reviewed previously and they will be certified at the final review. Another important input to the meeting is the selection of the people who will attend the review. They must be authorized to make decisions about the design and have the ability and responsibility to take corrective action.

Everyone attending the design review must receive a package of information well before the meeting. An ideal way to conduct a review is to hold a briefing session at least 10 days before the formal review. Members of the design team will make presentations to review the PDS and design review checklist to ensure that the review team has a common understanding of the design requirements. Then an overview of the design is given, describing how the contents of the design review information package relate to the design. Finally, members of the design review team will be assigned questions from the design checklist for special concentration. This is an informational meeting. Criticism of the design is reserved for later.

16.6.2 Review Meeting Process

The design review meeting should be formally structured with a well-planned agenda. The final design review is more of an audit in contrast to the earlier reviews, which are more multifunctional problem-solving sessions. The meeting is structured so it results in a documented assessment of the design. The review uses a checklist of items that need to be considered. Each item is discussed and it is decided whether it passes the review. The drawings, simulations, test results, FMEAs, etc., are used to support the evaluation. Sometimes a 1–5 scale is used to rate each requirement, but in a final review an "up or down" decision needs to be made. Any items that do not pass the review are tagged as action items and with the name of the responsible individual. Table 16.1 shows an abbreviated checklist for a final design review. A new checklist should be developed for each new product.

While the checklist in Table 16.1 is not exhaustive, it is illustrative of the many details that need to be considered in the final design review.

16.6.3 Output from Review

The output from the design review is a decision as to whether the product is ready to release to the manufacturing department. Sometimes the decision to proceed is tentative, with several open issues that need to be resolved, but in the judgment of management the fixes can be made before product launch.

TABLE 16.1
Typical items in a design checklist

1. Overall requirements—does it meet:
   Customer requirements
   Product design specification
   Applicable industry and governmental standards
2. Functional requirements—does it meet:
   Mechanical, electrical, thermal loads
   Size and weight
   Mechanical strength
   Projected life
3. Environmental requirements—does it meet:
   Temperature extremes, in operation and storage
   Extremes of humidity
   Extremes of vibration
   Shock
   Foreign material contamination
   Corrosion
   Outdoor exposure extremes (ultraviolet radiation, rain, hail, wind, sand)
4. Manufacturing requirements—does it meet:
   Use of standard components and subassemblies
   Tolerances consistent with processes and equipment
   Materials well defined and consistent with performance requirements
   Materials minimize material inventory
   Have critical control parameters been identified
   Manufacturing processes use existing equipment
5. Operational requirements
   Is it easy to install in the field?
   Are items requiring frequent maintenance easily accessible?
   Has serviceperson safety been considered?
   Have human factors been adequately considered in design?
   Are servicing instructions clear? Are they derived from FMEA or FTA?
6. Reliability requirements
   Have hazards been adequately investigated?
   Have failure modes been investigated and documented?
   Has a thorough safety analysis been conducted?
   Have life integrity tests been completed successfully?
   Has derating been employed in critical components?
7. Cost requirements:
   Does the product meet the cost target?
   Have cost comparisons been made with competitive products?
   Have service warranty costs been quantified and minimized?
   Has value engineering analysis been made for possible cost reduction?
8. Other requirements:
   Have critical components been optimized for robustness?
   Has a search been conducted to avoid patent infringement?
   Has prompt action been taken to apply for possible patent protection?
   Does the product appearance represent the technical quality and cost of the product?
   Has the product development process been adequately documented for defense in possible product liability action?
   Does the product comply with applicable laws and agency requirements?

The design review builds a paper trail of meeting minutes, the decisions or ratings for each design requirement, and a clear action plan of what will be done by whom and by when to fix any deficiencies in the design. This is important documentation to be used in any future product liability or patent litigation.

16.7
SUMMARY

Detail design is the phase of the design process where all of the details are brought together, decisions finalized, and a decision is made by management whether to release the design for production. It is the design phase where the greatest design cost is involved because the work to be done requires a heavy personnel contribution.

The first task of detail design is to complete the detail drawings and the assembly drawings. These documents, together with the design specifications, should contain the information to unambiguously manufacture the product. In order to finish the detail drawings, calculations and decisions not completed in the embodiment design phase need to be made. Often, in order to complete all these myriad details it is necessary to impose a *design freeze.* Once a freeze has been imposed, no changes can be made to the design unless they have been approved by a formal design control board.

The detail design phase also involves qualification testing of a prototype, the generation of a bill of materials (BOM) from the assembly drawings, a detailed cost estimate, and decisions on whether to make each part in-house or to obtain it from an outside supplier.

Detail design ends when the design is reviewed and accepted by a formal design review process. The review consists of comparing the design documentation (drawings, analyses, simulations, QFD, FAMAs, etc.) against a checklist of design requirements.

While detail design is the end of the *design process,* it is not the end of the *product development process.* Some of the tasks that must be completed before *product launch* are process planning, design of tooling, negotiate with suppliers, develop a quality assurance plan, marketing plan, distribution plan, customer service plan, maintenance plan, and a plan for retirement of the product from service. Product launch depends on the first batch of product from the production line passing a manufacturing prototype acceptance test.

BIBLIOGRAPHY

AT&T: "Moving a Design into Production," McGraw-Hill, New York, 1993.
"Detail Design," The Institution of Mechanical Engineers, London, 1975.
Hales, C.: "Managing Engineering Design," Wiley, New York, 1993.
Sekine, K., and K. Arai: "Design Team Revolution," Productivity Press, Portland, OR, 1994.
Vrsek, G.: Documenting and Communicating the Design "ASM Handbook," vol. 20: "Materials Selection and Design," pp. 222–230, ASM International, Materials Park, OH, 1997.

PROBLEMS AND EXERCISES

16.1. Examine the detail drawings for a product designed by a nearby manufacturing company. Be sure you can identify the actual shape, dimensions, and tolerances. What other information is contained in the drawing?

16.2. Look at an automotive mechanics manual. Identify a subassembly like a fuel-injection system or a front suspension. From the assembly drawings, write up a bill of materials.

16.3. Visualize the impact of CAE in a world that is even more electronic connected than it is today. How might the practice of detail design change?

16.4. Create a numbering system for use in drawing retrieval.

16.5. Prepare a final design review for your design project.

17

COMMUNICATING THE DESIGN

17.1
THE NATURE OF COMMUNICATION

You've completed your design, and you've come up with some innovative concepts and a cost estimate that predicts a nice healthy profit. What next? Now you must be able to communicate your findings to the people who matter. There is an old adage that a tree falling in a forest doesn't make a sound unless there is someone to listen. Similarly, the best technical design in the world might never be implemented unless you can communicate it to the proper people in the right way. To be successful with a design project, you must be able to communicate with your peers, your subordinates, and your superiors.

Communication can be simply described as *the flow of information from one mind to another.* Communication occurs through a common system of symbols, signs, and behavior that utilize one or more of the five human senses. We communicate by actual physical touch, as in a handshake or a pat on the back. We also communicate by visual movements of the body (body language), as with the wink of an eye or a smile. Sometimes we even communicate via taste and smell. In most technical communication, however, symbols or signs that are either heard with the ear or seen with the eyes are used.

The basic elements of all communication activities are shown in Fig. 17.1. The source of the communication arises somewhere in the organization: someone has information to communicate to someone else. An encoding process occurs when the information is translated into a systematic set of symbols (language) that expresses what the source wishes to transmit. The product of encoding is the message. The form of the message depends on the nature of the communication channel. The channel is the medium through which the message will be carried from source to receiver. The message can be a written report, a face-to-face communication, a telephone call, or a transmission via a computer network. It may or may not undergo a transformation in the decoder. When it reaches the receiver, it is interpreted in light of the person's pre-

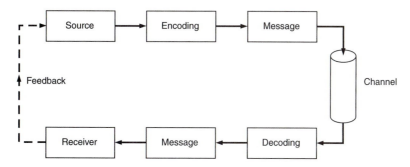

FIGURE 17.1
Basic elements of communications system.

vious experience or frame of reference. Feedback from the receiver to the source aids in determining how faithfully the message has been transmitted. Feedback gives the source an opportunity to determine whether the message has been received and whether it has produced the intended purpose.

Any communications system is less than perfect. Strong electric currents or atmospheric disturbances will produce some error in the transmission. These disturbances are called noise. In the same way, any communication between two persons will be subject to semantic noise and psychological noise. *Semantic noise* has to do with the meaning of words and ideas. In technical fields common words take on specialized meanings that can be completely unknown to the layman. *Psychological noise* arises from the particulars of the situation. A person who feels threatened by a new technology may have difficulty learning about it. Some people are intimidated by authority, and thus they have difficulty communicating with people above them in the organization. For communication to occur, the sender must have the authority of knowledge and a purpose for generating and transmitting the message. In addition, the recipient must be willing to receive the message and be capable of understanding it.

Studies of communication in business organizations have shown that geographic proximity is a major factor in enhancing communication. There usually is more information than the channel has capacity to handle. Sometimes selective filtration of information occurs. At any rate, the managements of many organizations make decisions on the basis of inadequate information. Communication in an organization is a multidimensional process. Many managers attempt to communicate only with the people under them and forget to communicate laterally with those who are assisting them. Lateral communication between project teams is particularly important in engineering organizations.

This chapter offers help and hints for communication of technical information. We deal first with written communication. Suggestions are presented for keeping a design notebook and organizing and writing a technical report, a technical paper, and a proposal for support. For oral communication, suggestions are made for giving successful presentations at business meetings and for presenting technical talks. Special emphasis is given to visual aids and electronic mail as enablers in effective communication.

17.2
THE DESIGN NOTEBOOK

We digress briefly from discussing communication of results to consider how ideas and experimental results should be recorded during the progress of the design project. If the recording task is treated with care and thoroughness, communicating the results of a design project will be made easier.

The chief tool for recording results is the design notebook. It should be an $8\frac{1}{2}$- by 11-in bound notebook (not spiral bound), preferably with a hard cover. It should be the repository for all of your planning (including plans that were not carried out), all analytical calculations, all records of experimental data, all references to sources of information, and all significant thinking about your project.

You should not use your notebook as a diary; but at the same time, you and your notebook should become an intimate communication system. Information should be entered directly into the notebook, not recopied from rough drafts. However, you should organize the information you enter in the notebook. Use main headings and subheadings; label key facts and ideas; liberally cross-reference your material; and keep an index at the front of the book to aid in your organization. About once a week, review what you have done and write a summary of your progress that emphasizes the high points. Whenever you do anything that may seem the least bit patentable, have your notebook read and witnessed by a knowledgeable colleague.

The following are good rules[1] for keeping a design notebook.

1. Keep an index at the front of the book.
2. Make your entries at the time you do the work. Include favorable and unfavorable results and things not fully understood at the time. If you make errors, just cross them out. Do not erase, and never tear a page out of the notebook.
3. All data must be in their original primary form (strip charts, oscilloscope pictures, photomicrographs, etc.), not after recalculation or transformation.
4. Rough graphs should be drawn directly in the notebook, but more carefully prepared plots on graph paper also should be made and entered in the book.
5. Give complete references to books, journals, reports, patents, and any other sources of information.
6. Entries should be made in ink and, of course, must be legible. Do not be obsessed with neatness at the expense of faithfully recording everything as it happens. Do not crowd your material on the pages. Paper is very much less expensive than engineering time.

A good engineering design notebook is one from which, several years after the project is completed, the project can be reconstructed. Critical decisions will be apparent, and the reasons for the actions taken will be backed up by facts. It should be possible to show where every figure, statement and conclusion of the published report of the project can be substantiated by original entries in the design notebook.

---

1. Adapted from T. T. Woodson, "Engineering Design," app. F, McGraw-Hill, New York, 1966.

17.3
WRITING THE TECHNICAL REPORT

In no other area of professional activity will you be judged so critically as your first technical report. The quality of a report generally provides an image in the reader's mind that, in large measure, determines the reader's impression of the quality of the work. Of course, an excellent job of report writing cannot disguise a sloppy investigation, but many excellent design studies have not received proper attention and credit because the work was reported in a careless manner. You should be aware that written reports carry a message farther than the spoken word and have greater permanence. Therefore, technical workers often are known more widely for their writings than for their talks.

17.3.1 Organization of Reports

Written communications take the form of letters, brief memorandum reports, formal technical reports, technical papers, and proposals. In terms of the communications model shown in Fig. 17.1, the source is the mind of the writer. The process of encoding consists of translating the idea from the mind to words on a paper. The channel is the pile of manuscript papers. Decoding the message depends on the reader's ability to understand the language and familiarity with the ideas presented in the message. The final receiver is the mind of the reader. Noise is present in the form of poor writing mechanics, incomplete diagrams, incorrect references, etc. Since there is no direct feedback, the writer must anticipate the needs of the receiver and attempt to minimize the noise.

The first principle of written communication is to know your audience so that you can anticipate and fulfill its needs. The purpose of engineering writing is to present information, not to entertain. Therefore, the information should be easy to find. Always when writing your report, keep in mind the busy reader who has only a limited amount of time for your report and may, in addition, not be familiar with your subject.

Memorandum reports

The memorandum report usually is written to a specific person or group of persons concerning a specific topic with which both the writer and recipient(s) are familiar. For example, it may be a report of a trip you took to observe a competitor's new product, or it could be a disclosure of a new idea you have for an improved product. It is written in memorandum form.

- Date
- To:
- From:
- Subject:
- Introduction (brief and indicating why the study was carried out)
- Discussion (includes data and its analysis)
- Conclusions (includes what was concluded from the study and recommendations made on the conclusions)

Memorandum reports are short (one to three pages). Sometimes a report that is more than one page has a summary section before the introduction so the reader does not have to read the entire memorandum to get its message. The purpose in writing a memorandum report is to get a concise report to interested parties as quickly as possible. The main emphasis is on results, discussion, and conclusions with a minimum of writing about experimental details unless, of course, those details are critical to the analysis of the data. Very often a more detailed report follows the memorandum report.

Formal technical reports

A formal technical report usually is written at the end of a project. Generally, it is a complete, stand-alone document aimed at persons having widely diverse backgrounds. Therefore, much more detail is required. The outline of a typical formal report[1] might be:

- Covering letter (letter of transmittal)
- Title page
- Summary (containing conclusions)
- Table of contents, including list of figures and tables
- Introduction (containing background to the work to acquaint reader with the problem and the purpose for carrying on the work)
- Experimental procedure
- Experimental results
- Discussion (of results)
- Conclusions
- References
- Appendixes
- Tables
- Figures

The *covering letter* is provided so that persons who might receive the report without prior notification will have some introduction to it. The *title page* provides a concise title and the names, affiliations, and addresses of the authors. The *summary* is provided early in the report to enable the busy reader to determine if it is worth the effort to read the entire report or send it to someone else who may be interested in the topic. It is generally less than a page in length and contains three paragraphs. The first briefly describes the objective of the study and the problems studied. Paragraph two describes your solution to the problem. The last paragraph addresses its importance to the business in terms of cost savings, improved quality, or new business opportunities.

The *introduction* should contain the pertinent technical facts that might be unknown to the reader but will be used in the report. It sets the stage in the same way as the summary, but in greater detail. The *experimental procedure section* is usually included to indicate how the data were obtained and to describe any nonstandard types of apparatus or techniques that were employed. The *experimental results sec-*

---

1. The contribution of Professor Richard W. Heckel for much of the material in this section is acknowledged.

tion describe's the results of the study. Data in the form of tables or figures are usually placed at the end of the report and are referred to by number in the results section. This section should also indicate any uncertainties in the data and possible errors in their sources. The *discussion section* is normally concerned with analyzing the data to make a specific point, develop the data into some more meaningful form, or relate the data to theory described in the introduction. All arguments based on the data are developed here. The section may also be used to discuss the effects of experimental error on the analysis of the data. The *conclusion section* states in as concise a form as possible the conclusions *that can be drawn from the study. No new information should be introduced here.* In general, this section is the culmination of the work and the report. The conclusions are directly related to the purpose in undertaking the study. *Appendixes* are used for mathematical developments, sample calculations, etc., that are not directly associated with the subject of the report and that, if placed in the main body of the report, would seriously impede the logical flow of thought. Final equations developed in the appendixes are then placed in the body of the report with reference to the particular appendix in which they were developed. The same procedure applies to mathematical calculations; the results are used in the report and the proper appendix is referenced.

Technical papers. These usually have an outline similar to this:

- Abstract
- Introduction
- Experimental procedure   } These may be combined in a single section
- Experimental results
- Discussion
- Summary and/or conclusions
- Acknowledgments
- Appendixes
- Tables
- Figures

The content of the above sections is the same as for a formal technical report. The *abstract section* is similar to the formal report summary except that it doesn't contain reference to orientation of the work to particular company problems. A technical paper often contains a *summary* at its end that allows the writer to "assemble" the paper prior to making *conclusions.* Some papers omit the summary section if it is not felt to be necessary and if it would be essentially the same as the abstract.

Proposals. A proposal is a report written to a sponsor in solicitation of financial support. The object of a proposal is to convince the sponsor of the value of your idea and to convince him that your organization has the capability (laborpower and facilities) to deliver the expected results. A typical proposal to a federal agency might be organized as follows:

- Introduction
- Purpose and objectives

- Technical background
- Program approach (your ideas and approach)
- Statement of work
- Program schedule
- Program organization
- Personnel qualifications
- Facilities and equipment
- Summary
- References
- Budget (often submitted as a separate document)
- Appendixes

The secret to writing winning proposals is to align your interests with those of the sponsor. Be sure the proposal carefully indicates the magnitude of the expected gain from doing the proposed work. It usually is helpful to know who will evaluate the proposal and what their standards for performance and excellence will be.

17.3.2 Steps in Writing a Report

The five operations involved in the writing of a high-quality report are best remembered with the acronym POWER.

P	Plan the writing
O	Outline the report
W	Write
E	Edit
R	Rewrite

The planning stage of a report is concerned with assembling the data, analyzing the data, drawing conclusions from the data analysis, and organizing the report into various logical sections. The planning of a report is usually carried out by considering the various facets of the work and providing a logical blend of the material. The initial planning of a report should begin *before* the work is carried out. In that way the planning of the work and planning of the report are woven together, which facilitates the actual writing operation.

Outlining the report consists of actually formulating a series of headings, subheadings, sub-subheadings, etc., which encompass the various sections of the report. The outline can then be used as a guide to the writing. A complete outline can be detailed to the point at which each line consists of a single thought or point to be made and will then represent one paragraph in the report. The main headings and subheadings of the outline are usually placed in the report to guide the reader.

The writing operation should be carried out in the form of a rough draft using the maximum technical and compositional skill at the command of the writer. However, do not worry about perfection at this stage. Once you get going, don't break stride to check out fine details of punctuation or sentence structure.

Editing is the process of reading the rough draft and employing self-criticism. It consists of strengthening the rough draft by analyzing paragraph and sentence structure, economizing on words, checking spelling and punctuation, checking the line of logical thought, and, in general, asking oneself the question "Why?" Editing can be the secret of good writing. It is better for writers to ask themselves embarrassing questions than to hear them from their technical readers, a supervisor, or an instructor. In connection with editing, it has often been said that the superior writer makes good use of both ends of the pencil.

It is generally good practice to allow at least a day to elapse after writing the rough draft before editing it. That allows the writer to forget the logical pattern used in writing the report and appear more in the role of an unbiased reader when editing. Many mistakes or weak lines of thought that would normally escape unnoticed are thereby uncovered. The rewriting operation consists of retyping or rewriting the edited rough draft to put it in a form suitable for the reader.

17.3.3 Word Processing Software

Word processing software is a very useful tool for the engineer. The ability to move sections of a report by "cut and paste," to insert or change words, to check spelling and grammar, and to increase the readability by changing the format or size of type have made writing easier. Most engineers in industry function today without major secretarial support because they have mastered simple typing and editing skills.

However, there are two pitfalls to watch out for.[1] *Nice-output syndrome* occurs when engineers are so enthralled by the beautiful printed pages coming out of the laser printer that they fail to do a proper job of editing and rewriting. The *hurry-up-and-wait disease* occurs when writers are composing at the keyboard and are such good touch typists that they can easily outrun the speed at which they can think. This interrupts the natural flow of ideas in writing. It is far better to match the flow at which the ideas are put to paper with the speed at which they are generated by the mind. Some writers purposely write the first draft by hand to achieve this synchronization.

17.3.4 Mechanics of Writing

The following suggestions are presented as a guide to writing and an aid in avoiding some of the most common mistakes. You also should avail yourself of one of the popular guides to English grammar and style.[2]

Title
The title should be a meaningful description of what you have written.

Basic ideas
State basic ideas early; give the reader an overview of your study at the beginning. Enough background must be given to allow the least-informed reader to understand

---

1. D. E. Goldberg, "Life Skills and Leadership for Engineers," chap. 2, McGraw-Hill, New York, 1995.
2. W. Strunk and E. B. White, "The Elements of Style," 3d ed., Macmillan, New York, 1978; S. W. Baker, "The Practical Stylist," 8th ed., Addison-Wesley, Reading, MA, 1997.

why the design was undertaken. The reader must be able to fit whatever is being reported into a context that is meaningful.

Whole vs. part

Describe the whole before the part; be sure you present the reader with the whole picture before you lead into the details. You should describe the essence, the function, and the purpose of a device, process, or product before you go into details about the parts.

Important information

Emphasize important information; beware of the common error of burying it under a mass of details. Put the important ideas early in your writing; use appropriate capitalization and underlining; and relegate information of secondary importance to the appendix.

Headings

Use headings liberally; headings and subheadings are signposts that help the reader understand the organization of your ideas. If you have prepared a good outline, the headings will be self-evident.

Fact vs. opinion

Separate fact from opinion. It is important for the reader to know what your contributions are, what ideas you obtained from others (the references should indicate that), and which are opinions not substantiated by fact.

Paragraph structure

Each paragraph should begin with a topic sentence that provides an overall understanding of the paragraph. Since each paragraph should have a single theme or conclusion, the topic sentence states that theme or conclusion. Any elaboration is deferred to subsequent sentences in the paragraph. Don't force the reader to wade through many sentences of disconnected verbiage to arrive at the conclusion in the last sentence of the paragraph. The reader should be able to get an understanding of the report by reading the first sentence of each paragraph.

Sentence length

Sentences should be kept as short as possible so that their structure is simple and readable. Generally a sentence should not exceed about 35 words. Long sentences require complex construction, provide an abundance of opportunity for grammatical errors, take considerable writing time, and slow the reader down. Long sentences are often the result of putting together two independent thoughts that could be stated better in separate sentences.

Pronouns

There is no room for any degree of ambiguity between a pronoun and the noun for which it is used. Novices commonly use "it," "this," "that," etc., where it would be better to use one of several nouns. It may be clear to the writer, but it is often ambiguous to the reader.

In general, personal pronouns (I, you, he, she, we, my, mine, our, us) are not used in technical reports. The only exception is when the writer *must* be involved personally in the report, as when writing a report to be used as a basis for a patent. In that instance, the writer must state definitely that he or she was the inventor.

Tense

The choice of the tense of verbs is often confusing to student writers. The following simple rules are usually employed by experienced writers:

Past tense: Use to describe work done in the laboratory or in general to past events. "Hardness readings *were* taken on all specimens."

Present tense: Use in reference to items and ideas in the report itself. "It *is* clear from the data in Figure 4 that strain energy *is* the driving force for recovery" or "The group recommends that the experiment be repeated" (present opinion).

Future tense: Use in making prediction from the data that will be applicable in the future. "The market data given in Table II indicate that the sales *will continue* to increase in the next ten years."

Spelling and punctuation

Errors in these basic elements of writing in the final draft of the report are inexcusable.

Appendixes, tables, and figures

Appendixes, tables, and figures are placed at the end of a report to speed the reading of the text and to allow separate preparation of tables and figures without having to bother with leaving the required space in the text. The following numbering conventions are often used:

Appendix A, B, C, D (capital letters)
Table I, II, III, IV (Roman numerals)
Figure 1, 2, 3, 4 (numbers)

Each appendix, table, or figure should have a title and should be self-explanatory. For example, the reader should not have to refer to the text for an understanding of the variables plotted on a graph. Graphs should be drawn on suitable graph paper (linear, semilog, loglog, etc.), and the axes should be properly labeled (including units). If data points are labeled by code numbers, the code should appear on the graph, not in the text. The independent and dependent variables are usually placed on the abscissa (x axis) and ordinate (y axis), respectively. Generally, only one table or figure is placed on a page. However, several small photographs may make up one figure.

Reference to data in tables and figures is often a cause of difficulty. The statement "It is obvious from the data that . . ." is used much too often when nothing is obvious before the data have been explained properly. The statement given above should be used *only* when an obvious conclusion can be made, not when it is the hope of the writer that the reader thinks something is obvious.

The *initial* reference to a set of data in a table or figure as: "The data are given in Table III" exerts an extreme hardship on readers and in many instances forces them to grope unguided through the data. In such instances the reader must digest the data in a few minutes, whereas the digestion may have taken the writer days or weeks. (The writer also runs the risk of the reader's reaching an "improper" conclusion before being biased by the text.) It is better form to first describe what a set of results shows *prior* to telling the reader where the data can be found. This saves the reader's time by indicating what to look for when exposed to the data.

References

References are usually placed at the end of the written text. Those to the technical literature (described as readily available on subscription and included in most library collections) are made by author and journal reference (often with the title of article omitted) as shown by the following example. There is no single universally accepted format for references, but many journals follow the style given here.

Journal article
   R. M. Horn and Robert O. Ritchie: *Metall. Trans. A,* 1978 vol. 9A, pp. 1039–1053.
Book
   Thomas T. Woodson: "Introduction to Engineering Design," pp. 321–346, McGraw-Hill, New York, 1966.
A private communication
   J. J. Doe, XYZ Company, Altoona, PA, unpublished research, 1981.
Internal reports
   J. J. Doe: Report No. 642, XYZ Company, Altoona, PA, February 1980.

Be scrupulous about references as you acquire them. A few minutes spent in recording the full reference (including the inclusive page numbers) can prevent hours of time being spent in the library when you are finishing the final report.

17.4
MEETINGS AND PRESENTATIONS

The business world is full of meetings which are held to communicate on a variety of levels and subjects. Most of these involve some kind of prepared oral presentation.

At the lowest level of this hierarchy is the design team meeting. Those present are focused on a common goal and have a generally common background. The purpose of the meeting is to share the progress that has been made, identify problems, and hopefully, find help and support in solving the problems. This is a group discussion, with an agenda and probably some visual aids, but the presentation is informal and not rehearsed. Detailed tips for effectively holding this type of meeting were given in Sec. 3.5.

Next up in the meeting hierarchy would be a design briefing or design review. The size and diversity of the audience would depend on the importance of the project. It could vary from 10 to 50 people, and include company managers and executives. A design briefing for high-level management must be brief and to the point. Such people are very busy and not at all interested in the technical details that engineers love to talk about. Usually you will have only 5 to 10 min to get your point across to the top executive. If you are speaking to technical managers, they will be more interested in the important technical details, but don't forget also to cover information on schedule and costs. Generally, they will give you 15 to 30 min to get your points across. A presentation similar to the latter type of design briefing is a technical talk before a professional or technical society. Here you will generally have 15 to 20 min to make your presentation before an audience of 30 to 100 people.

At the top of the pyramid is a formal speech. A speech is carefully scripted and discussion often is not allowed. It is one-way communication from the speaker to the audience, as when the President of the United States gives the State of the Union address. As a young engineer you probably will not be asked to give speeches, but as you advance in your career such opportunities may appear, e.g., on giving the major lecture at your professional society annual meeting or a speech via satellite video as corporate VP to all members of your division on winning a major contract.

17.5
ORAL PRESENTATIONS

Impressions and reputations (favorable or unfavorable) are made most quickly by audience reaction to an oral presentation. There are a number of situations in which you will be called upon to give a talk. Progress reports, whether to your boss in a one-on-one situation or in a more formal setting to your customer, are common situations in which oral communication is used. Selling an idea or a proposal to your management budget committee or a sponsor is another common situation. In the more technical arena, you may be asked to present a talk to a local technical society chapter or present a paper at a national technical meeting.

Oral communication has several special characteristics: quick feedback by questions and dialogue; impact of personal enthusiasm; impact of visual aids; and the important influence of tone, emphasis, and gesture. A skilled speaker in close contact with an audience can communicate far more effectively than the cold, distant, easily evaded written word. On the other hand, the organization and logic of presentation must be of a higher order for oral than for written communication. The listener to an oral communication has no opportunity to reread a page to clarify a point. Many opportunities for noise exist in oral communication. The preparation and delivery of the speaker, the environment of the meeting room, and the quality of the visual aids all contribute to the efficiency of the oral communication process.

17.5.1 The Business-Oriented Technical Talk

The purpose of your talk may be to present the results of the past 3 months of work by a 10-person design team, or it may be to present some new ideas on computer-aided design to an audience of upper management who are skeptical that their large investment in CAD will pay off. Whatever the reason, you should know the purpose of your talk and have a good idea of who will be attending your presentation. This information is vital if you are to prepare an effective talk.

The most appropriate type of delivery for most business-oriented talks is an *extemporaneous-prepared talk.* All the points in the talk are thought out and planned in detail. However, the delivery is based on a written outline, or alternatively, the text of the talk is completely written but the talk is delivered from an outline prepared from the text. This type of presentation establishes a more natural, closer contact with the audience that is much more believable than if the talk is read by the speaker.

Develop the material in your talk in terms of the interest of the audience. Organize it on a thought-by-thought rather than a word-by-word basis. Write your conclusions first. That will make it easier to sort through all the material you have and select only the pieces of information that support the conclusions. If your talk is aimed at selling an idea, list all of your idea's strengths and weaknesses. That will help you counter arguments against adopting your idea.

The opening few minutes of your talk are vital in establishing whether you will get the audience's attention. You need to "bring them up to speed" by explaining the reason for your presentation. Include enough background that they can follow the main body of your presentation, which should be carefully planned. Stay well within the time allotted for the talk so there is an opportunity for questions. Include humorous stories and jokes in your talk only if you are very good at telling them. If you are not, it is best to play it straight. Also, avoid specialized technical jargon in your talk. Before ending your presentation, summarize your main points and conclusions. The audience should have no confusion as to the message you wanted to deliver.

Visual aids are an important part of any technical presentation; good ones can increase the audience retention of your ideas by 50 percent. The type of visual aid to use depends upon the nature of the talk and the audience. For a small informal meeting of up to 10 or 12 people, handouts of an outline, data, and charts usually are effective. Transparencies used with an overhead projector are good for groups from 10 to 200 people. Slides are the preferred visual aids for large audiences. The important subject of visual aids is discussed at greater length in Sec. 17.6.

The usual reason a technical talk is poor is lack of preparation. It is a rare individual who is able to give an outstanding talk without practicing it. Once you have prepared the talk, the first stage is individual practice. Give the talk out loud in an empty room to fix the thoughts in your mind and check the timing. You may want to memorize the introductory and concluding remarks. If at all possible, videotape your individual practice. The dry run is a dress rehearsal before a small audience. If possible, hold the dry run in the same room where you will give the talk. Use the same visual aids that you will use in your talk. The purpose of the dry run is to help you work out any problems in delivery, organization, or timing. There should be a critique following the dry run, and the talk should be reworked and repeated as many times as are necessary to do it right.

When delivering the talk, if you are not formally introduced, you should give your name and the names of any other team members. You should speak loudly enough to be easily heard. For a large group that will require the use of a microphone. Work hard to project a calm, confident delivery, but don't come on in an overly aggressive style that will arouse adversarial tendencies in your audience. Avoid annoying mannerisms like rattling the change in your pocket and pacing up and down the platform. Whenever possible, avoid talking in the dark. The audience might well go to sleep or, at worst, sneak out. Maintaining eye contact with the audience is an important part of the feedback in the communication loop.

The questions that follow a talk are an important part of the oral communication process; they show that the audience is interested and has been listening. If at all possible, do not allow interruptions to your talk for questions. If the "big boss" interrupts with a question, compliment him for his perceptiveness and explain that the point will be covered in a few moments. Never apologize for the inadequacy of your results. Let a questioner complete the questions before breaking in with an answer. Avoid being argumentative or letting the questioner see that you think the question is stupid. Do not prolong the question period unnecessarily. When the questions slack off, adjourn the meeting.

17.5.2 The Technical Society Talk

It is an honor to be asked to present the results of your work at a national meeting of a professional society, but you may be surprised to find that you have only 15 or 20 minutes in which to present your work. That calls for good planning and organization. Frequently an abstract of your talk will be available in the meeting announcement or printed book of abstracts that is available to registrants of the meeting. However, that does not absolve you from the responsibility for starting your talk with a good introduction, which should set forth the scope and objectives of the talk. It may deal with the history of events leading up to the work to be discussed. Any coauthors of the work described should be mentioned, and acknowledgments should be made to sponsors and those who may have given special help.

The main body of the technical talk covers the following items:

- Experimental or analytical procedure
- Results and observations
- Discussions and conclusions
- Ongoing research and/or future studies

The talk should end with a summary that repeats the main information given in the body of the talk. A good scheme for the organization of a technical talk is to:

1. Tell them what you are going to tell them.
2. Tell them.
3. Tell them what you told them!

No matter what kind of talk you give there are two factors that will control your success. The first is to be in complete command of your facts. Know the material cold. The quickest way to lose an audience is if they think you are faking it. Second, show enthusiasm. We don't mean to "go ape," but neither should you be a deadpan "cold fish." For your talk to succeed you must command the attention of the audience. You do this by being enthusiastic and knowledgeable.

17.6
VISUAL AIDS

Except in a small meeting, visual aids make the difference between effective and ineffective oral communication. The selection of the visual aid medium will depend upon the size and importance of the audience and the number of times the talk will be given. Not only are good visual aids important in transmitting the message, so that the dual senses of hearing and seeing are employed, they also assist greatly in reducing the nervous tension (psychological noise) of inexperienced speakers.

The chief purpose of a visual aid is to improve and simplify communication. There should be no intention to have a slide or viewgraph stand alone, unsupported by oral communication. The spoken word should complement the limited information that can be contained on the visual. On the other hand, it is a waste of time for the speaker to read the words on the slide to the audience.

The three most commonly used visual aids are handouts, overhead transparencies, and 35-mm slides. There are advantages and disadvantages to each. A photocopied handout is inexpensive, does not require much lead time for preparation, and provides a take-away from the presentation. However, it has the disadvantage that it can serve as a distraction from the presentation as people page through the handout at a pace that differs from the pace of the presentation. Handouts are most commonly used at small, informal presentations. The other types of visual aids are discussed below.

17.6.1 Graphics Presentation Software

The preparation of transparencies and slides has been greatly facilitated by the development of graphics presentation programs (GPP) for use with the personal computer. Perhaps the best-known GPP is Microsoft PowerPoint. Not only does this software provide great versatility in available fonts, styling, and colors, but it provides advice in graphic design through on-line help screens and tutorials. GPPs can be used to produce professional-level black-and-white transparencies, 35-mm slides, and full-screen projected electronic color slides. The outlining feature built into the software assists in preparing the presentation, and the ability to print out reduced-scale copies of the slides gives a take-away note-taking capability.

17.6.2 Overhead Transparencies

Transparencies used with an overhead projector provide a flexible visual aid system for groups up to about 100 people. This is the most common way of presenting visual material. Because transparencies can be used in a semilighted room, they permit the speaker to face the audience and maintain eye contact. They can be made quickly and inexpensively with a standard office copier. Color transparencies can be made with a color laser printer or color copier, but at greater expense. Paper copies of the transparencies can be used as handouts for small groups. The chief disadvantage of transparencies is that they are not as good as slides for reproducing photographs.

17.6.3 35-mm Slides

Slides prepared from 35-mm transparency film (2- by 2-in size) are the standard visual aid for presentations given to large audiences since they can be projected with a large distance from the projector to the screen. Photographs are reproduced sharp and clear. The preparation of slides generally requires several days of lead time. Proper viewing of slides requires a darkened room.

17.6.4 Computer Projected Slides

A modern alternative to 35-mm slides is computer projected slides. The text material is prepared with a GPP, like PowerPoint, and copied to a floppy disk. The disk is loaded into the computer attached to the projection system, and a mouse is used to click through the slides, line by line. This system not only provides vivid color and graphics, but it has the feature of advanced visual effects. For example, you can build each slide line by line, as the words fly in from the side, or you can transition from slide to slide with eye-catching dissolves.

17.6.5 Other Visual Aids

There are several other kinds of visual aids that may be appropriate. A *flip chart,* a large pad of paper mounted on an easel, can be effective in a small meeting of up to 12 people. Each page is flipped over to reveal the neatly lettered message.

Video tapes are often used to display group dynamics or high-speed motion. These work well in small groups but are difficult to manage with large groups because of the small size of most TV monitors.

It is common to bring small pieces of hardware or test specimens for the audience to view and touch and feel. It will be distracting to your talk if these are passed around during the talk. It is better to have them on display for viewing after the talk.

17.6.6 Tips on Talks

1. Limit slides (or transparencies) to *no more than* one per minute.
2. Each slide should contain one idea.
3. Slides that present more than three curves on a single graph or 20 words or numbers on a slide are too complicated.
4. The first slide in the presentation should show the title of your talk and the names and affiliations of the authors.
5. The second slide should give a brief outline of your talk.
6. The last slide should summarize the message you delivered.
7. If you need to show a slide more than once, use a second copy so the projectionist will not have to back up, and in the process distract the audience.
8. Avoid leaving on the screen a slide that you have finished discussing while you go on to something else. Put in a blank slide, which will allow the audience to turn their attention to your words.

9. Prior to the presentation, make sure the slides are in order and oriented properly.
10. Place a blank sheet of paper between each pair of transparencies to avoid sticking. The sheet of paper also enables you to read the transparency prior to placing on the overhead projector.
11. If a transparency includes a number of unrelated items, such as a list of conclusions, use a piece of paper to conceal those you have not yet discussed. Move the paper down the transparency to reveal the items as they are discussed.
12. When making your presentation, do not stand in a place that blocks the audience's view of the screen. A good place to stand is to the right of the screen, facing the audience.
13. Do not try to use notecards when working with transparencies. They should contain sufficient clues for you to make a smooth talk—with practice.
14. Stay with one type of visual aid if possible. If you must switch between 35-mm slides and transparencies, arrange the talk to do this only once.

It is important to remember that the projected area of an overhead projector is square. When preparing the transparencies, position the letters in a square format, not in long lines across the page. Make the letters as large as possible, at least 18 point. Use no more than seven lines on a transparency. A combination of upper- and lowercase letters is easier to read than all uppercase letters. Also, fonts with serifs are easier to read than sans-serif fonts. Give each transparency a title at the top. Use a horizontal format with a height-to-width ratio of 4:5 and a maximum area of 7.5×9.5 in. If a vertical format must be used, keep the text in the upper two-thirds of the transparency. Otherwise it will be out of focus.

17.7
USING ELECTRONIC MAIL

No form of communication has grown so rapidly as electronic mail (e-mail). It is estimated that Americans will send four trillion e-mail messages in 1998 and that this will grow to seven trillion by 2000. Electronic mail has proved to be invaluable for scheduling meetings, communicating between engineers at different plant sites, communicating with the office while on a trip, or just keeping up with the activities of the relatives.

Because e-mail is so instant and personal there is a tendency to treat it differently from other written communication. People feel free to write and send things they would never put in a business letter. E-mail seems to free people from their normal inhibitions. It is so easy to hit the reply key without thinking about the consequences. There are many documented instances of two business friends "having fun" in their e-mail exchange, only to discover that the message inadvertently was given mass circulation.

Electronic mail also can easily lead to information overload. Top management, in their desire to communicate with those below them, can easily dilute the message by overuse. It is all too easy to invite unwelcome advertisements into your e-mail address. And, there are those coworkers who just cannot resist the temptation to share the latest joke with you.

17.7.1 E-mail Etiquette

There are easy things you can do to help your colleagues cope with the mass of messages coming over their computer screens. Following these suggestions will also give a professional tone to your business messages.

- For formal business correspondence, write as you would in a business letter.
- For communications to close friends or colleagues, write as you would speak.
- Use both upper- and lowercase letters, as you would in a business letter. If you type all capitals it means you are angry and shouting.
- Don't use tabs. Use spaces for indenting.
- Be very careful in using the "redirect" key. Be sure you know who will be on the receiving end.
- If you think formatting will be a problem, send the message first to yourself.
- Always treat the person you are writing to with respect. But remember, that person might forward your message to someone else.
- Keep your messages short.
- Realize that deleting a message by sending it to the trash bin probably means that it is still retrievable, since most companies keep backup tapes on e-mail.
- If you write a message in anger or frustration, wait 24 hours before sending it. This gives you time to cool down and possibly rethink your action. This also is good practice in ordinary nonelectronic communication.

17.7.2 Business Issues

Many technology companies view their open e-mail environment as conducive to creativity. They are reluctant to regulate its use by their employees in any way. However, more companies are beginning to realize that e-mail communication can be a time bomb. Electronic messages are not secure. For example, e-mail messages have become important evidence in sexual harassment suits and antitrust suits. Many companies are developing strict e-mail policies so they can limit their exposure to lawsuits. These typically involve company monitoring of e-mail and prohibition against using company e-mail to share jokes, photographs, and nonbusiness information of any kind.

17.7.3 Groupware

The popularity of e-mail as a communication medium has led to its widespread use managing group projects. However, most e-mail systems are not well suited for interactive discussion because messages can only be seen in the order in which they were written. Thus, a response to a question may be separated from the original query by a large number of unrelated messages. Generally, the responder will reply to the messages in serial order, losing the continuity of the discussion. Also, there is difficulty with the transfer of documents in arbitrary formats, using many e-mail systems.

APPENDIX B

STATISTICAL TABLES

Cumulative distribution function for the standard normal distribution (SND)

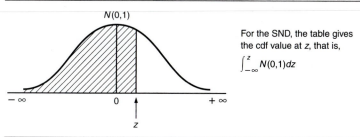

$N(0,1)$

For the SND, the table gives the cdf value at z, that is,

$$\int_{-\infty}^{z} N(0,1)dz$$

z	0.00	0.01	0.02	0.03	0.04	0.05	0.06	0.07	0.08	0.09
−3.5	0.00023	0.00022	0.00022	0.00021	0.00020	0.00019	0.00019	0.00018	0.00017	0.00017
−3.4	0.00034	0.00033	0.00031	0.00030	0.00029	0.00028	0.00027	0.00026	0.00025	0.00024
−3.3	0.00048	0.00047	0.00045	0.00043	0.00042	0.00040	0.00039	0.00038	0.00036	0.00035
−3.2	0.00069	0.00066	0.00064	0.00062	0.00060	0.00058	0.00056	0.00054	0.00052	0.00050
−3.1	0.00097	0.00094	0.00090	0.00087	0.00085	0.00082	0.00079	0.00076	0.00074	0.00071
−3.0	0.00135	0.00131	0.00126	0.00122	0.00118	0.00114	0.00111	0.00107	0.00104	0.00100
−2.9	0.0019	0.0018	0.0017	0.0017	0.0016	0.0016	0.0015	0.0015	0.0014	0.0014
−2.8	0.0026	0.0025	0.0024	0.0023	0.0023	0.0022	0.0021	0.0021	0.0020	0.0019
−2.7	0.0035	0.0034	0.0033	0.0032	0.0031	0.0030	0.0029	0.0028	0.0027	0.0026
−2.6	0.0047	0.0045	0.0044	0.0043	0.0041	0.0040	0.0039	0.0038	0.0037	0.0036
−2.5	0.0062	0.0060	0.0059	0.0057	0.0055	0.0054	0.0052	0.0051	0.0049	0.0048
−2.4	0.0082	0.0080	0.0078	0.0075	0.0073	0.0071	0.0069	0.0068	0.0066	0.0064
−2.3	0.0107	0.0104	0.0102	0.0099	0.0096	0.0094	0.0091	0.0089	0.0087	0.0084
−2.2	0.0139	0.0136	0.0132	0.0129	0.0125	0.0122	0.0119	0.0116	0.0113	0.0110
−2.1	0.0179	0.0174	0.0170	0.0166	0.0162	0.0158	0.0154	0.0150	0.0146	0.0143
−2.0	0.0228	0.0222	0.0217	0.0212	0.0207	0.0202	0.0197	0.0192	0.0188	0.0183
−1.9	0.0287	0.0281	0.0274	0.0268	0.0262	0.0256	0.0250	0.0244	0.0239	0.0233
−1.8	0.0359	0.0351	0.0344	0.0336	0.0329	0.0322	0.0314	0.0307	0.0301	0.0294
−1.7	0.0446	0.0436	0.0427	0.0418	0.0409	0.0401	0.0392	0.0384	0.0375	0.0367
−1.6	0.0548	0.0537	0.0526	0.0516	0.0505	0.0495	0.0485	0.0475	0.0465	0.0455
−1.5	0.0668	0.0655	0.0643	0.0630	0.0618	0.0606	0.0594	0.0582	0.0571	0.0559
−1.4	0.0808	0.0793	0.0778	0.0764	0.0749	0.0735	0.0721	0.0708	0.0694	0.0581
−1.3	0.0968	0.0951	0.0934	0.0918	0.0901	0.0885	0.0869	0.0853	0.0838	0.0823
−1.2	0.1151	0.1131	0.1112	0.1093	0.1075	0.1057	0.1038	0.1020	0.1003	0.0985
−1.1	0.1357	0.1335	0.1314	0.1292	0.1271	0.1251	0.1230	0.1210	0.1190	0.1170
−1.0	0.1587	0.1562	0.1539	0.1515	0.1492	0.1469	0.1446	0.1423	0.1401	0.1379
−0.9	0.1841	0.1814	0.1788	0.1762	0.1736	0.1711	0.1685	0.1660	0.1635	0.1611
−0.8	0.2119	0.2090	0.2061	0.2033	0.2005	0.1977	0.1949	0.1922	0.1894	0.1867
−0.7	0.2420	0.2389	0.2358	0.2327	0.2297	0.2266	0.2236	0.2207	0.2177	0.2148
−0.6	0.2743	0.2709	0.2676	0.2643	0.2611	0.2578	0.2546	0.2514	0.2483	0.2451
−0.5	0.3085	0.3050	0.3015	0.2981	0.2946	0.2912	0.2877	0.2843	0.2810	0.2776
−0.4	0.3446	0.3409	0.3372	0.3336	0.3300	0.3264	0.3228	0.3192	0.3156	0.3121
−0.3	0.3821	0.3783	0.3745	0.3707	0.3669	0.3632	0.3594	0.3557	0.3520	0.3483
−0.2	0.4207	0.4168	0.4129	0.4090	0.4052	0.4013	0.3974	0.3936	0.3897	0.3859
−0.1	0.4602	0.4562	0.4522	0.4483	0.4443	0.4404	0.4364	0.4325	0.4286	0.4247
−0.0	0.5000	0.4960	0.4920	0.4880	0.4840	0.4801	0.4761	0.4721	0.4681	0.4641

Reprinted, with permission, from E. L. Grant and R. S. Leavenworth, "Statistical Quality Control," McGraw-Hill, New York, 1972.

z	0.00	0.01	0.02	0.03	0.04	0.05	0.06	0.07	0.08	0.09
+0.0	0.5000	0.5040	0.5080	0.5120	0.5160	0.5199	0.5239	0.5279	0.5319	0.5359
+0.1	0.5398	0.5438	0.5478	0.5517	0.5557	0.5596	0.5636	0.5675	0.5714	0.5753
+0.2	0.5793	0.5832	0.5871	0.5910	0.5948	0.5987	0.6026	0.6064	0.6103	0.6141
+0.3	0.6179	0.6217	0.6255	0.6293	0.6331	0.6368	0.6406	0.6443	0.6480	0.6517
+0.4	0.6554	0.6591	0.6628	0.6664	0.6700	0.6736	0.6772	0.6808	0.6844	0.6870
+0.5	0.6915	0.6950	0.6985	0.7019	0.7054	0.7088	0.7123	0.7157	0.7190	0.7224
+0.6	0.7257	0.7291	0.7324	0.7357	0.7389	0.7422	0.7454	0.7486	0.7517	0.7549
+0.7	0.7580	0.7611	0.7642	0.7673	0.7704	0.7734	0.7764	0.7794	0.7823	0.7852
+0.8	0.7881	0.7910	0.7939	0.7967	0.7995	0.8023	0.8051	0.8079	0.8106	0.8133
+0.9	0.8159	0.8186	0.8212	0.8238	0.8264	0.8289	0.8315	0.8340	0.8365	0.8389
+1.0	0.8413	0.8438	0.8461	0.8485	0.8508	0.8531	0.8554	0.8577	0.8599	0.8621
+1.1	0.8643	0.8665	0.8686	0.8708	0.8729	0.8749	0.8770	0.8790	0.8810	0.8830
+1.2	0.8849	0.8869	0.8888	0.8907	0.8925	0.8944	0.8962	0.8980	0.8997	0.9015
+1.3	0.9032	0.9049	0.9066	0.9082	0.9099	0.9115	0.9131	0.9147	0.9162	0.9177
+1.4	0.9192	0.9207	0.9222	0.9236	0.9251	0.9265	0.9279	0.9292	0.9306	0.9319
+1.5	0.9332	0.9345	0.9357	0.9370	0.9382	0.9394	0.9406	0.9418	0.9429	0.9441
+1.6	0.9452	0.9463	0.9474	0.9484	0.9495	0.9505	0.9515	0.9525	0.9535	0.9545
+1.7	0.9554	0.9564	0.9573	0.9582	0.9591	0.9599	0.9608	0.9616	0.9625	0.9633
+1.8	0.9641	0.9649	0.9656	0.9664	0.9671	0.9678	0.9686	0.9693	0.9699	0.9706
+1.9	0.9713	0.9719	0.9726	0.9732	0.9738	0.9744	0.9750	0.9756	0.9761	0.9767
+2.0	0.9773	0.9778	0.9783	0.9788	0.9793	0.9798	0.9803	0.9808	0.9812	0.9817
+2.1	0.9821	0.9826	0.9830	0.9834	0.9838	0.9842	0.9846	0.9850	0.9854	0.9857
+2.2	0.9861	0.9864	0.9868	0.9871	0.9875	0.9878	0.9881	0.9884	0.9887	0.9890
+2.3	0.9893	0.9896	0.9898	0.9901	0.9904	0.9906	0.9909	0.9911	0.9913	0.9916
+2.4	0.9918	0.9920	0.9922	0.9925	0.9927	0.9929	0.9931	0.9932	0.9934	0.9936
+2.5	0.9938	0.9940	0.9941	0.9943	0.9945	0.9946	0.9948	0.9949	0.9951	0.9952
+2.6	0.9953	0.9955	0.9956	0.9957	0.9959	0.9960	0.9961	0.9962	0.9963	0.9964
+2.7	0.9965	0.9966	0.9967	0.9968	0.9969	0.9970	0.9971	0.9972	0.9973	0.9974
+2.8	0.9974	0.9975	0.9976	0.9977	0.9977	0.9978	0.9979	0.9979	0.9980	0.9981
+2.9	0.9981	0.9982	0.9983	0.9983	0.9984	0.9984	0.9985	0.9985	0.9986	0.9986
+3.0	0.99865	0.99869	0.99874	0.99878	0.99982	0.99886	0.99889	0.99893	0.99896	0.99900
+3.1	0.99903	0.99906	0.99910	0.99913	0.99915	0.99918	0.99921	0.99924	0.99926	0.99929
+3.2	0.99931	0.99934	0.99936	0.99938	0.99940	0.99942	0.99944	0.99946	0.99948	0.99950
+3.3	0.99952	0.99953	0.99955	0.99957	0.99958	0.99960	0.99961	0.99962	0.99964	0.99965
+3.4	0.99966	0.99967	0.99969	0.99970	0.99971	0.99972	0.99973	0.99974	0.99975	0.99976
+3.5	0.99977	0.99978	0.99978	0.99979	0.99980	0.99981	0.99981	0.99982	0.99983	0.99983

TABLE B-2
The t distribution

Given v, the table gives (a) the one-tail t_0 value with α of the area above it, that is, $P(t \geq t_0) = \alpha$, or (b) the two-tail $+t_0$ and $-t_0$ values with $\alpha/2$ in each tail, that is, $P(t \leq -t_0) + P(t \geq +t_0) = \alpha$

(a) One-tail α

(b) Two-tail α

	One-tail α							One-tail α					
	0.10	0.05	0.025	0.01	0.005	0.001		0.10	0.05	0.025	0.01	0.005	0.001
	Two-tail α							Two-tail α					
v	0.20	0.10	0.05	0.02	0.01	0.002	v	0.20	0.10	0.05	0.02	0.01	0.002
1	3.078	6.314	12.706	31.821	63.657	318.300	19	1.328	1.729	2.093	2.539	2.861	3.579
2	1.886	2.920	4.303	6.965	9.925	22.327	20	1.325	1.725	2.086	2.528	2.845	3.552
3	1.638	2.353	3.182	4.541	5.841	10.214	21	1.323	1.721	2.080	2.518	2.831	3.527
4	1.533	2.132	2.776	3.747	4.604	7.173	22	1.321	1.717	2.074	2.508	2.819	3.505
5	1.476	2.015	2.571	3.305	4.032	5.893	23	1.319	1.714	2.069	2.500	2.807	3.485
6	1.440	1.943	2.447	3.143	3.707	5.208	24	1.318	1.711	2.064	2.492	2.797	3.467
7	1.415	1.895	2.365	2.998	3.499	4.785	25	1.316	1.708	2.060	2.485	2.787	3.450
8	1.397	1.860	2.306	2.896	3.355	4.501	26	1.315	1.706	2.056	2.479	2.779	3.435
9	1.383	1.833	2.262	2.821	3.250	4.297	27	1.314	1.703	2.052	2.473	2.771	3.421
10	1.372	1.812	2.228	2.764	3.169	4.144	28	1.313	1.701	2.048	2.467	2.763	3.408
11	1.363	1.796	2.201	2.718	3.106	4.025	29	1.311	1.699	2.045	2.462	2.756	3.396
12	1.356	1.782	2.179	2.681	3.055	3.930	30	1.310	1.697	2.042	2.457	2.750	3.385
13	1.350	1.771	2.160	2.650	3.012	3.852	40	1.303	1.684	2.021	2.423	2.704	3.307
14	1.345	1.761	2.145	2.624	2.977	3.787	60	1.296	1.671	2.000	2.390	2.660	3.232
15	1.341	1.753	2.131	2.602	2.947	3.733	80	1.292	1.664	1.990	2.374	2.639	3.195
16	1.337	1.746	2.120	2.583	2.921	3.686	100	1.290	1.660	1.984	2.365	2.626	3.174
17	1.333	1.740	2.110	2.567	2.898	3.646	∞	1.282	1.645	1.960	2.326	2.576	3.090
18	1.330	1.734	2.101	2.552	2.878	3.611							

Reprinted, with permission, from L. Blank, "Statistical Procedures for Engineering, Management and Science," McGraw-Hill, New York, 1980.

TABLE B-3
The F distribution ($\alpha = 0.10$, 0.05, and 0.01)

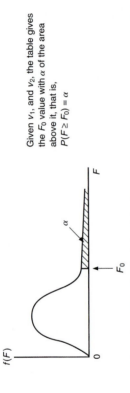

Given v_1, and v_2, the table gives the F_0 value with α of the area above it, that is,

$$P(F \geq F_0) = \alpha$$

v_1 (numerator)

v_2	α	1	2	3	4	5	6	7	8	9	10	11	12	14	15	19	20	24	30	50	100	500	∞
1	.10	39.9	49.5	53.6	55.8	57.2	58.2	58.9	59.4	59.9	60.2	60.5	60.7	61.1	61.2	61.6	61.7	62.0	62.3	62.7	63.0	63.3	63.3
	.05	161	200	216	225	230	234	237	239	241	242	243	244	245	246	248	248	249	250	252	253	254	254
2	.10	8.53	9.00	9.16	9.24	9.29	9.33	9.35	9.37	9.38	9.39	9.40	9.41	9.42	9.42	9.44	9.44	9.45	9.46	9.47	9.48	9.49	9.49
	.05	18.5	19.0	19.2	19.2	19.3	19.3	19.4	19.4	19.4	19.4	19.4	19.4	19.4	19.4	19.4	19.4	19.5	19.5	19.5	19.5	19.5	19.5
	.01	98.5	99.0	99.2	99.2	99.3	99.3	99.4	99.4	99.4	99.4	99.4	99.4	99.4	99.4	99.4	99.4	99.5	99.5	99.5	99.5	99.5	99.5
3	.10	5.54	5.46	5.39	5.34	5.31	5.28	5.27	5.25	5.24	5.23	5.22	5.22	5.20	5.20	5.18	5.18	5.18	5.17	5.15	5.14	5.14	5.13
	.05	10.1	9.55	9.28	9.12	9.10	8.94	8.89	8.85	8.81	8.79	8.76	8.74	8.71	8.70	8.67	8.66	8.64	8.62	8.58	8.55	8.53	8.53
	.01	34.1	30.8	29.5	28.7	28.2	27.9	27.7	27.5	27.3	27.2	27.1	27.1	26.9	26.9	26.7	26.7	26.6	26.5	26.4	26.2	26.1	26.1
4	.10	4.54	4.32	4.19	4.11	4.05	4.01	3.98	3.95	3.94	3.92	3.91	3.90	3.88	3.87	3.84	3.84	3.83	3.82	3.80	3.78	3.76	3.76
	.05	7.71	6.94	6.59	6.39	6.26	6.16	6.09	6.04	6.00	5.96	5.94	5.91	5.87	5.86	5.81	5.80	5.77	5.75	5.70	5.66	5.64	5.63
	.01	21.2	18.0	16.7	16.0	15.5	15.2	15.0	14.8	14.7	14.5	14.4	14.4	14.2	14.2	14.0	14.0	13.9	13.8	13.7	13.6	13.5	13.5
5	.10	4.06	3.78	3.62	3.52	3.45	3.40	3.37	3.34	3.32	3.30	3.28	3.27	3.25	3.24	3.21	3.21	3.19	3.17	3.15	3.13	3.11	3.10
	.05	6.61	5.79	5.41	5.19	5.05	4.95	4.88	4.82	4.77	4.74	4.71	4.68	4.64	4.62	4.57	4.56	4.53	4.50	4.44	4.41	4.37	4.36
	.01	16.26	13.27	12.06	11.39	10.97	10.67	10.46	10.29	10.16	10.05	9.96	9.89	9.77	9.72	9.58	9.55	9.47	9.38	9.24	9.13	9.04	9.02
6	.10	3.78	3.46	3.29	3.18	3.11	3.05	3.01	2.98	2.96	2.94	2.92	2.90	2.88	2.87	2.84	2.84	2.82	2.80	2.77	2.75	2.73	2.72
	.05	5.99	5.14	4.76	4.53	4.39	4.28	4.21	4.15	4.10	4.06	4.03	4.00	3.96	3.94	3.88	3.87	3.84	3.81	3.75	3.71	3.68	3.67
	.01	13.74	10.92	9.78	9.15	8.75	8.47	8.26	8.10	7.98	7.87	7.79	7.72	7.60	7.56	7.42	7.40	7.31	7.23	7.09	6.99	6.90	6.88

The degrees of freedom are v_1 for the numerator and v_2 for the denominator.

TABLE B-3
(continued)

v_1 (numerator)

v_2	α	1	2	3	4	5	6	7	8	9	10	11	12	14	15	19	20	24	30	50	100	500	∞
7	.10	3.59	3.26	3.07	2.96	2.88	2.83	2.78	2.75	2.72	2.70	2.68	2.67	2.64	2.63	2.60	2.59	2.58	2.56	2.52	2.50	2.48	2.47
	.05	5.59	4.74	4.35	4.12	3.97	3.87	3.79	3.73	3.68	3.64	3.60	3.57	3.53	3.51	3.46	3.44	3.41	3.38	3.32	3.27	3.24	3.23
	.01	12.25	9.55	8.45	7.85	7.46	7.19	6.99	6.84	6.72	6.62	6.54	6.47	6.36	6.31	6.18	6.16	6.07	5.99	5.86	5.75	5.67	5.65
8	.10	3.46	3.11	2.92	2.81	2.73	2.67	2.62	2.59	2.56	2.54	2.52	2.50	2.47	2.46	2.43	2.42	2.40	2.38	2.35	2.32	2.30	2.29
	.05	5.32	4.46	4.07	3.84	3.69	3.58	3.50	3.44	3.39	3.35	3.31	3.28	3.24	3.22	3.16	3.15	3.12	3.08	3.02	2.97	2.94	2.93
	.01	11.26	8.65	7.59	7.01	6.63	6.37	6.18	6.03	5.91	5.81	5.73	5.67	5.56	5.52	5.38	5.36	5.28	5.20	5.07	4.96	4.88	4.86
9	.10	3.36	3.01	2.81	2.69	2.61	2.55	2.51	2.47	2.44	2.42	2.40	2.38	2.35	2.34	2.31	2.30	2.28	2.25	2.22	2.19	2.17	2.16
	.05	5.12	4.26	3.86	3.63	3.48	3.37	3.29	3.23	3.18	3.14	3.10	3.07	3.03	3.01	2.95	2.94	2.90	2.86	2.80	2.76	2.72	2.71
	.01	10.56	8.02	6.99	6.42	6.06	5.80	5.61	5.47	5.35	5.26	5.18	5.11	5.00	4.96	4.83	4.81	4.73	4.65	4.52	4.42	4.33	4.31
10	.10	3.28	2.92	2.73	2.61	2.52	2.46	2.41	2.38	2.35	2.32	2.30	2.28	2.25	2.24	2.21	2.20	2.18	2.16	2.12	2.09	2.06	2.06
	.05	4.96	4.10	3.71	3.48	3.33	3.22	3.14	3.07	3.02	2.98	2.94	2.91	2.86	2.85	2.78	2.77	2.74	2.70	2.64	2.59	2.55	2.54
	.01	10.04	7.56	6.55	5.99	5.64	5.39	5.20	5.06	4.94	4.85	4.77	4.71	4.60	4.56	4.43	4.41	4.33	4.25	4.12	4.01	3.93	3.91
11	.10	3.23	2.86	2.66	2.54	2.45	2.39	2.34	2.30	2.27	2.25	2.23	2.21	2.18	2.17	2.13	2.12	2.10	2.08	2.04	2.00	1.98	1.97
	.05	4.84	3.98	3.59	3.36	3.20	3.09	3.01	2.95	2.90	2.85	2.82	2.79	2.74	2.72	2.66	2.65	2.61	2.57	2.51	2.46	2.42	2.40
	.01	9.65	7.21	6.22	5.67	5.32	5.07	4.89	4.74	4.63	4.54	4.46	4.40	4.29	4.25	4.12	4.10	4.02	3.94	3.81	3.71	3.62	3.60
12	.10	3.18	2.81	2.61	2.48	2.39	2.33	2.28	2.24	2.21	2.19	2.17	2.15	2.11	2.10	2.07	2.06	2.04	2.01	1.97	1.94	1.91	1.90
	.05	4.75	3.89	3.49	3.26	3.11	3.00	2.91	2.85	2.80	2.75	2.72	2.69	2.64	2.62	2.56	2.54	2.51	2.47	2.40	2.35	2.31	2.30
	.01	9.33	6.93	5.95	5.41	5.06	4.82	4.64	4.50	4.39	4.30	4.22	4.16	4.05	4.01	3.88	3.86	3.78	3.70	3.57	3.47	3.38	3.36
14	.10	3.10	2.73	2.52	2.39	2.31	2.24	2.19	2.15	2.12	2.10	2.08	2.05	2.02	2.01	1.97	1.96	1.94	1.91	1.87	1.83	1.80	1.80
	.05	4.60	3.74	3.34	3.11	2.96	2.85	2.76	2.70	2.65	2.60	2.57	2.53	2.48	2.46	2.40	2.39	2.35	2.31	2.24	2.19	2.14	2.13
	.01	8.86	6.51	5.56	5.04	4.69	4.46	4.28	4.14	4.03	3.94	3.86	3.80	3.70	3.66	3.53	3.51	3.43	3.35	3.22	3.11	3.03	3.00
15	.10	3.07	2.70	2.49	2.36	2.27	2.21	2.16	2.12	2.09	2.06	2.04	2.02	1.98	1.97	1.93	1.92	1.90	1.87	1.83	1.79	1.76	1.76
	.05	4.54	3.68	3.29	3.06	2.90	2.79	2.71	2.64	2.59	2.54	2.51	2.48	2.42	2.40	2.34	2.33	2.29	2.25	2.18	2.12	2.08	2.07
	.01	8.68	6.36	5.42	4.89	4.56	4.32	4.14	4.00	3.89	3.80	3.73	3.67	3.56	3.52	3.40	3.37	3.29	3.21	3.08	2.98	2.89	2.87
16	.10	3.05	2.67	2.46	2.33	2.24	2.18	2.13	2.09	2.06	2.03	2.01	1.99	1.95	1.95	1.90	1.89	1.87	1.84	1.79	1.76	1.73	1.72
	.05	4.49	3.63	3.24	3.01	2.85	2.74	2.66	2.59	2.54	2.49	2.46	2.42	2.37	2.35	2.29	2.28	2.24	2.19	2.12	2.07	2.02	2.01
	.01	8.53	6.23	5.29	4.77	4.44	4.20	4.03	3.89	3.78	3.69	3.62	3.55	3.45	3.41	3.28	3.26	3.18	3.10	2.97	2.86	2.78	2.75

TABLE B-3
(continued)

v_2	α	\multicolumn{22}{c}{v_1 (numerator)}
		1	2	3	4	5	6	7	8	9	10	11	12	14	15	19	20	24	30	50	100	500	∞
18	.10	3.01	2.62	2.42	2.29	2.20	2.13	2.08	2.04	2.00	1.98	1.96	1.93	1.90	1.89	1.85	1.84	1.81	1.78	1.74	1.70	1.67	1.66
	.05	4.41	3.55	3.16	2.93	2.77	2.66	2.58	2.51	2.46	2.41	2.37	2.34	2.29	2.27	2.20	2.19	2.15	2.11	2.04	1.98	1.93	1.92
	.01	8.29	6.01	5.09	4.58	4.25	4.01	3.84	3.71	3.60	3.51	3.43	3.37	3.27	3.23	3.10	3.08	3.00	2.92	2.78	2.68	2.59	2.57
19	.10	2.99	2.61	2.40	2.27	2.18	2.11	2.06	2.02	1.98	1.96	1.94	1.91	1.87	1.86	1.82	1.81	1.79	1.76	1.71	1.67	1.64	1.63
	.05	4.38	3.52	3.13	2.90	2.74	2.63	2.54	2.48	2.42	2.38	2.34	2.31	2.26	2.23	2.17	2.16	2.11	2.07	2.00	1.94	1.89	1.88
	.01	8.18	5.93	5.01	4.50	4.17	3.94	3.77	3.63	3.52	3.43	3.36	3.30	3.19	3.15	3.03	3.00	2.92	2.84	2.71	2.60	2.51	2.49
20	.10	2.97	2.59	2.38	2.25	2.16	2.09	2.04	2.00	1.96	1.94	1.92	1.89	1.85	1.84	1.80	1.79	1.77	1.74	1.69	1.65	1.62	1.61
	.05	4.35	3.49	3.10	2.87	2.71	2.60	2.51	2.45	2.39	2.35	2.31	2.28	2.22	2.20	2.14	2.12	2.08	2.04	1.97	1.91	1.86	1.84
	.01	8.10	5.85	4.94	4.43	4.10	3.87	3.70	3.56	3.46	3.37	3.29	3.23	3.13	3.09	2.96	2.94	2.86	2.78	2.64	2.54	2.44	2.42
24	.10	2.93	2.54	2.33	2.19	2.10	2.04	1.98	1.94	1.91	1.88	1.85	1.83	1.79	1.78	1.74	1.73	1.70	1.67	1.62	1.58	1.54	1.53
	.05	4.26	3.40	3.01	2.78	2.62	2.51	2.42	2.36	2.30	2.25	2.21	2.18	2.13	2.11	2.04	2.03	1.98	1.94	1.86	1.80	1.75	1.73
	.01	7.82	5.61	4.72	4.22	3.90	3.67	3.50	3.36	3.26	3.17	3.09	3.03	2.93	2.89	2.76	2.74	2.66	2.58	2.44	2.33	2.24	2.21
30	.10	2.88	2.49	2.28	2.14	2.05	1.98	1.93	1.88	1.85	1.82	1.79	1.77	1.73	1.72	1.68	1.67	1.64	1.61	1.55	1.51	1.47	1.46
	.05	4.17	3.32	2.92	2.69	2.53	2.42	2.33	2.27	2.21	2.16	2.13	2.09	2.04	2.01	1.95	1.93	1.89	1.84	1.76	1.70	1.64	1.62
	.01	7.56	5.39	4.51	4.02	3.70	3.47	3.30	3.17	3.07	2.98	2.91	2.84	2.74	2.70	2.57	2.55	2.47	2.39	2.25	2.13	2.03	2.01
50	.10	2.81	2.41	2.20	2.06	1.97	1.90	1.84	1.80	1.76	1.73	1.70	1.68	1.64	1.63	1.58	1.57	1.54	1.50	1.44	1.39	1.34	1.33
	.05	4.03	3.18	2.79	2.56	2.40	2.29	2.20	2.13	2.07	2.03	1.99	1.95	1.89	1.87	1.80	1.78	1.74	1.69	1.60	1.52	1.46	1.44
	.01	7.17	5.06	4.20	3.72	3.41	3.19	3.02	2.89	2.79	2.70	2.63	2.56	2.46	2.42	2.29	2.27	2.18	2.10	1.95	1.82	1.71	1.68
100	.10	2.76	2.36	2.14	2.00	1.91	1.83	1.78	1.73	1.70	1.66	1.63	1.61	1.57	1.56	1.50	1.49	1.46	1.42	1.35	1.29	1.23	1.21
	.05	3.94	3.09	2.70	2.46	2.31	2.19	2.10	2.03	1.97	1.93	1.89	1.85	1.79	1.77	1.69	1.68	1.63	1.57	1.48	1.39	1.31	1.28
	.01	6.90	4.82	3.98	3.51	3.21	2.99	2.82	2.69	2.59	2.50	2.43	2.37	2.26	2.22	2.09	2.07	1.98	1.89	1.73	1.60	1.47	1.43
500	.10	2.72	2.31	2.10	1.96	1.86	1.79	1.73	1.68	1.64	1.61	1.58	1.56	1.52	1.50	1.45	1.44	1.41	1.36	1.28	1.21	1.12	1.09
	.05	3.86	3.01	2.62	2.39	2.23	2.12	2.03	1.96	1.90	1.85	1.81	1.77	1.71	1.69	1.61	1.59	1.54	1.48	1.38	1.28	1.16	1.11
	.01	6.69	4.65	3.82	3.36	3.05	2.84	2.68	2.55	2.44	2.36	2.28	2.22	2.12	2.07	1.94	1.92	1.83	1.74	1.56	1.41	1.23	1.16
∞	.10	2.71	2.30	2.08	1.94	1.85	1.77	1.72	1.67	1.63	1.60	1.57	1.55	1.51	1.49	1.43	1.42	1.38	1.34	1.26	1.18	1.08	1.00
	.05	3.84	3.00	2.60	2.37	2.21	2.10	2.01	1.94	1.88	1.83	1.79	1.75	1.69	1.67	1.59	1.57	1.52	1.46	1.35	1.24	1.11	1.00
	.01	6.63	4.61	3.78	3.32	3.02	2.80	2.64	2.51	2.41	2.32	2.25	2.18	2.08	2.04	1.90	1.88	1.79	1.70	1.52	1.36	1.15	1.00

Reprinted with permission, from L. Blank, "Statistical Procedures for Engineering, Management and Science," McGraw-Hill, New York, 1980.

APPENDIX C

USEFUL OPERATING
CHARACTERISTIC CURVES

The abscissa scale on the operating characteristic curves is interpreted differently depending on the form of H_1:

Two-sided	$H_1: \mu \neq \mu_0$	$d = |\mu - \mu_0|/\sigma$
One-sided	$H_1: \mu < \mu_0$	$d = \mu_0 - \mu/\sigma$
One-sided	$H_1: \mu > \mu_0$	$d = \mu - \mu_0/\sigma$
Two-sided, paired test	$H_1: \mu_1 - \mu_2 \neq \delta$	$d = \dfrac{\mu_1 - \mu_2}{(\sigma_1^2 + \sigma_2^2)^{1/2}}$

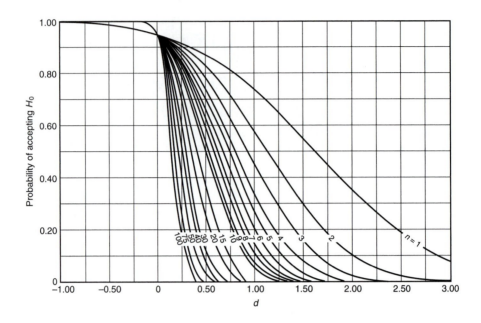

Operating characteristic curves for *one-sided normal* test with $\alpha = 0.05$. (A. H. Bowker and G. J. Lieberman, "Engineering Statistics," 2d ed., 1972, p. 190. Reproduced by permission of Prentice-Hall, Inc., Englewood Cliffs, NJ.)

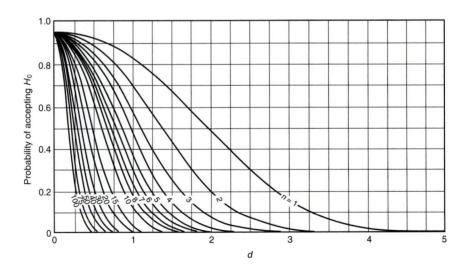

Operating characteristic curves for a *two-sided normal* test with $\alpha = 0.05$. (Reproduced with permission from "OC of the Common Statistical Tests of Significance." C. D. Ferris, F. E. Grubbs, and C. L. Weaver, *The Annals of Mathematical Statistics,* vol. 17, no. 2, June 1946.)

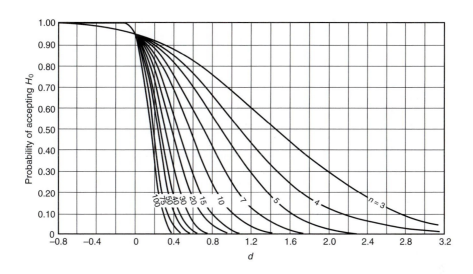

Operating characteristic curves for *one-sided normal* test with $\alpha = 0.05$. (A. H. Bowker and G. J. Lieberman, "Engineering Statistics," 2d ed., 1972, p. 190. Reproduced by permission of Prentice-Hall, Englewood Cliffs, NJ.)

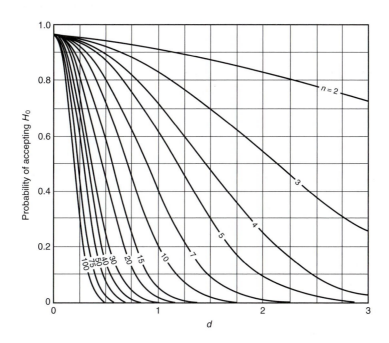

Operating characteristic curves for a *two-sided t* test with $\alpha = 0.05$. (Reproduced with permission from "OC of the Common Statistical Tests of Significance," C. D. Ferris, F. E. Grubbs, and C. L. Weaver, *The Annals of Mathematical Statistics,* vol. 17, no. 2, June 1946.)

AUTHOR INDEX

SUBJECT INDEX